AF183622

T. J. Mabry, K. R. Markham
and M. B. Thomas

The Systematic Identification of Flavonoids

With 325 Figures

Springer-Verlag
Berlin · Heidelberg · New York 1970

Tom J. Mabry
Professor of Botany

K. R. Markham

M. B. Thomas

The Cell Research Institute and
Department of Botany
The University of Texas at Austin

ISBN 978-3-642-88460-3 ISBN 978-3-642-88458-0 (eBook)
DOI 10.1007/978-3-642-88458-0

This work is subject to copyright. All rights are reserved, whether the whole or part of the material is concerned specifically those of translation, reprinting, re-use of illustrations, broadcasting, reproduction by photocopying machine or similar means, and storage in data banks.

Under § 54 of the German Copyright Law where copies are made for other than private use, a fee is payable to the publisher, the amount of the fee to be determined by agreement with the publisher.

© by Springer-Verlag New York Inc. 1970. Library of Congress Catalog Card Number 72-95565
Softcover reprint of the hardcover 1st edition 1970

The use of general descriptive names, trade names, trade marks etc. in this publication, even if the former are not especially identified, is not to be taken as a sign that such names, as understood by the Trade Marks and Merchandise Marks Act, may accordingly be used freely by anyone. Title No. 1622

The authors wish to acknowledge contributions
and comments by Drs. Heinz Rösler and Jacques Kagan

Preface

About 1958, the late Professor R. E. ALSTON and Professor B. L. TURNER, both of the Department of Botany, The University of Texas at Austin, initiated a general systematic investigation of the legume genus *Baptisia*. They found that flavonoid patterns, as revealed by two-dimensional paper chromatography, were valid criteria for the recognition of the *Baptisia* species and for the documentation of their numerous natural hybrids. Later, they showed that the flavonoid chemistry could be used for the analysis of gene flow among populations. At that time no attempt was made to even partially identify the flavonoids which were detected chromatographically. Nevertheless, it soon became apparent that the full value of the chemical data for systematic purposes required knowledge of the structures of the flavonoids.

In 1962, one of us (T. J. M.) in collaboration with Drs. ALSTON and TURNER began the chemical analysis of the more than 60 flavonoids which had been chromatographically detected in the 16 *Baptisia* species. In the intervening years, a number of chemists and botanists, including Drs. K. BAETCKE, B. BREHM, M. CRANMER, D. HORNE, J. KAGAN, B. KROSCHEWSKY, J. MCCLURE, H. RÖSLER, and J. WALLACE, participated in the development of techniques and procedures for the rapid identification of known flavonoids and in the structure determination of new flavonoids. In addition, the flavonoid chemistry of many plants other than *Baptisia* was investigated.

Two of us (K. R. M. and M. B. T.) joined the group in 1965 and were recipients of Post-doctoral Fellowships from the University of Texas at Austin during the period (1965 – 1967) when most of the information presented in this volume was assembled. This volume presents, for the most part, procedures which were most useful in our flavonoid studies, together with our collection of ultraviolet and nuclear magnetic resonance spectra of flavonoids. Thus, no attempt has been made to describe all the information available in the literature regarding the isolation and identification of flavonoids (J. B. HARBORNE's "Comparative Biochemistry of the Flavonoids" provides an excellent summary of the literature up to 1966). Moreover, a number of classes of flavonoids are either not treated at all (anthocyanins) or are only covered briefly (for example, chalcones and aurones). The quantity of data presented for each of the various classes of flavonoids corresponds roughly to the frequency with which we have encountered them.

The book is divided into three parts (I, II and III); the first deals mostly with the isolation and purification of flavonoids while the second and third comprise a spectra section in which flavonoid UV and NMR spectra are discussed.

Before an analysis of the flavonoids in a given plant is initiated, we place in the University of Texas at Austin Herbarium a voucher specimen representing the plant population under investigation. The importance of properly vouchering the plant material before beginning the chemical studies cannot be over emphasized for only in this way can later investigators ascertain with certainty the plant for which the chemical results are reported.

Our first step in a typical investigation of the flavonoids in a plant is to extract the flavonoids from a few dried leaves with methanol or aqueous methanol; the extract is then used to determine the two-dimensional paper chromatographic flavonoid pattern.

Chapter I describes in detail the two-dimensional paper chromatographic analysis of flavonoids. Although sufficient pure material can usually be eluted from the paper chromatograms to obtain the ultraviolet spectra, in some instances the purification of the flavonoids can only be achieved by other techniques such as thin-layer or column chromatography. In Chapter II the column and thin-layer chromatographic procedures which are commonly employed in our laboratory for the separation of flavonoids are described. Chapter III, the last chapter in Part I, presents gas and paper chromatographic procedures for the identification of the sugar moieties in flavonoid glycosides together with comments on the various methods available for determining the structures of flavonoid aglycones.

Part II presents data and procedures for the ultraviolet spectral analysis of flavonoids. Once a pure flavonoid is obtained, its ultraviolet spectra in methanol alone and methanol with each of five diagnostic reagents are always recorded. Chapter IV outlines the steps for obtaining the UV data while Chapters V, VI and VII present, respectively, the UV spectral curves with interpretations for flavones and flavonols; isoflavones, flavanones, and dihydroflavonols; and chalcones and aurones. For each of the 175 flavonoids examined in the present investigation a set of six UV spectra are presented along with R_f values in the solvents used for two-dimensional paper chromatography and spot colors when viewed on paper under ultraviolet light alone and ultraviolet light in the presence of ammonia vapor.

Finally in Part III we have discussed procedures for obtaining and interpreting NMR spectra of flavonoids (Chapter VIII) as well as presenting 128 NMR spectra (Chapter IX). Most of the NMR spectra were determined for the trimethylsilyl ethers of the flavonoids, all of which are soluble in carbon tetrachloride.

Only with two flavonoids, hymenoxin and scaposin, both of which are highly oxygenated and methoxylated flavones, have we found it necessary to resort to the total synthesis of a flavonoid in order to establish its structure; in almost all other cases the information recorded here was sufficient for the complete structure analysis.

We wish to acknowledge a number of people who, either in correspondence or by providing flavonoid samples, helped us complete this volume: E.M. Bickoff, J. Chopin, J.W. Clark-Lewis, P. Crabbe, E. DeEds, S.E. Drewes, D.L. Dreyer, L. Farkas, T.A. Geissman, J.B. Harborne, M. Hasegawa, J. Herran, W. Herz, W.E. Hillis, L. Hörhammer, R.M. Horowitz, P.R. Jefferies, L. Jurd, N. Kawano, A.R. Kidwai, B.H. Koeppen, M. Komatsu, P. Lebreton, A.C. Neish, R. Neu, A. Nilsson, F.S. Okumura, W.D. Ollis, W. Rahman, D.G. Roux, M. Sainsbury, M.K. Siekel, T.R. Seshadri, E. Sondheimer, H. Suginome, T. Swain, T. Tominaga, E.W. Underhill, H. Wagner, J.E. Watkin, S.H. Wender, E. Wong.

Finally, we are grateful to a number of individuals who helped in the preparation of the manuscript and the running of the UV spectra: Susan Woodland, Linda McMahan, Sidney Morris, Genie Brackenridge, Francis Haynes, Sharon Sutherland, James Mears and Judy Autrey.

January 2, 1970

T.J. Mabry
K.R. Markham
M.B. Thomas

Contents

Part I

The Isolation, Purification and Preliminary Identification of Flavonoids

Chapter I. The Two-Dimensional Paper Chromatographic Analysis of Flavonoids . . . 3

I-1. Reagents and Materials . . . 3

I-2. Experimental Procedures for the Two-Dimensional Paper Chromatographic Analysis of Flavonoid Mixtures . . . 4

I-3. The Determination of R_f Values for Flavonoids . . . 9

I-4. The Effects of Flavonoid Structural Variations on R_f Values . . 10

I-5. Relationships between Spot Color and Flavonoid Structure . . 12

I-6. The Isolation and Purification of Flavonoids by Preparative Two-Dimensional Paper Chromatography . . . 13

I-7. The One-Dimensional Paper Chromatographic Purification of a Partially Purified Flavonoid . . . 14

Chapter II. The Separation of Flavonoids by Column and Thin Layer Chromatography . . . 16

II-1. Preliminary Purification of Flavonoids in a Crude Plant Extract Using Charcoal . . . 16

II-2. The Separation of Flavonoids by Polyamide and Silica Gel Column Chromatography . . . 17

II-3. The Separation of Flavonoids by Silica Gel and Polyamide Thin Layer Chromatography . . . 20

Chapter III. The Aglycone and Sugar Analysis of Flavonoid Glycosides . . . 23

III-1. Procedures for the Acidic and Enzymatic Hydrolysis of Flavonoid Glycosides . . . 24

III-2. The Gas and Paper Chromatographic Procedures for Identifying the Sugars Obtained by Hydrolysis of Flavonoid Glycosides . . 26

III-3. The Identification of the Aglycone and Location of the Sugar in Flavonoid Glycosides . . . 27

III-4. The Identification of the Sugars in C-Glycosylflavonoids . . . 31

Part II

The Structure Analysis of Flavonoids by Ultraviolet Spectroscopy

Chapter IV. Reagents and Procedures for the Ultraviolet Spectral Analysis of Flavonoids . . . 35

IV-1. Preparation of Reagent Stock Solutions and Solids 35

IV-2. Procedures for Determining the Ultraviolet Absorption Spectra of Flavonoids 35

Chapter V. The Ultraviolet Spectra of Flavones and Flavonols 41
V-1. The UV Spectra of Flavones and Flavonols in Methanol . . . 41
V-2. The UV Spectra of Flavones and Flavonols in the Presence of NaOMe 45
V-3. The UV Spectra of Flavones and Flavonols in the Presence of NaOAc 48
V-4. The Detection of Ortho-dihydroxyl Groups in Flavones and Flavonols by the Effect of $NaOAc/H_3BO_3$ on the UV Spectrum . 50
V-5. The UV Spectra of Flavones and Flavonols in the Presence of $AlCl_3$ and $AlCl_3/HCl$ 51
V-6. Index of Ultraviolet Absorption Spectra of Flavones and Flavonols 57

Chapter VI. The Ultraviolet Spectra of Isoflavones, Flavanones, and Dihydroflavonols 165
VI-1. The UV Spectra of Isoflavones, Flavanones and Dihydroflavonols in Methanol 165
VI-2. The UV Spectra of Isoflavones, Flavanones and Dihydroflavonols in the Presence of NaOMe 167
VI-3. The UV Spectra of Isoflavones, Flavanones and Dihydroflavonols in the Presence of NaOAc 169
VI-4. The Detection of A-Ring Ortho-dihydroxyl Groups in Isoflavones, Flavanones and Dihydroflavonols by the Effect of $NaOAc/H_3BO_3$ on the UV Spectrum 170
VI-5. The UV Spectra of Isoflavones, Flavanones and Dihydroflavonols in the Presence of $AlCl_3$ and $AlCl_3/HCl$ 171
VI-6. Index of Ultraviolet Absorption Spectra of Isoflavones, Flavanones and Dihydroflavonols 172

Chapter VII. The Ultraviolet Spectra of Chalcones and Aurones 227
VII-1. The UV Spectra of Chalcones and Aurones in MeOH 227
VII-2. The UV Spectra of Chalcones and Aurones in the Presence of NaOMe 228
VII-3. The UV Spectra of Chalcones and Aurones in the Presence of NaOAc 228
VII-4. The Detection of Ortho-dihydroxyl Groups in Chalcones and Aurones by the Effect of $NaOAc/H_3BO_3$ on the UV Spectrum . . 228
VII-5. The UV Spectra of Chalcones and Aurones in the Presence of $AlCl_3$ and $AlCl_3/HCl$ 229
VII-6. Index of Ultraviolet Absorption Spectra of Chalcones and Aurones 230

Part III

The Structure Analysis of Flavonoids by Proton Nuclear Magnetic Resonance Spectroscopy

Chapter VIII. The Determination and Interpretation of NMR Spectra of Flavonoids 253
VIII-1. Introduction 254

VIII-2. The Use of DMSO-d_6 as Solvent for Flavonoid NMR Spectroscopy 254
VIII-3. Preparation of Trimethylsilyl Ether Derivatives of Flavonoids . . 255
VIII-4. Interpretation of the NMR Spectra of Fully and Partially Trimethylsilylated Flavonoids 260

Chapter IX. The NMR Spectra of Flavonoids 274

Subject-Index 345

Skeletons and Numbering Schemes for the Classes of Flavonoids Discussed in this Volume

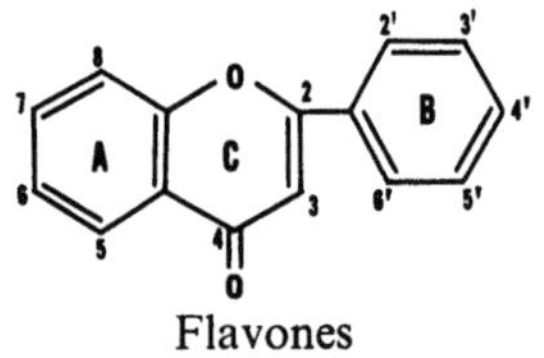

Flavones

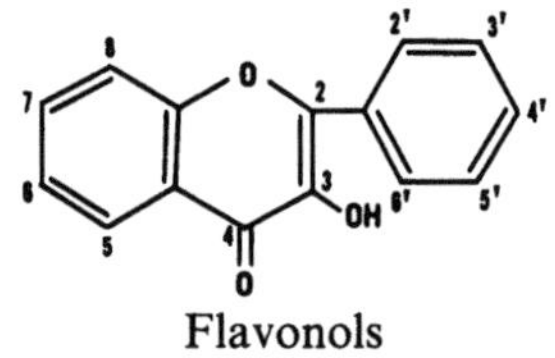

Flavonols

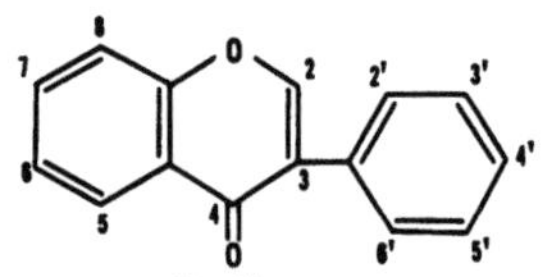

Isoflavones

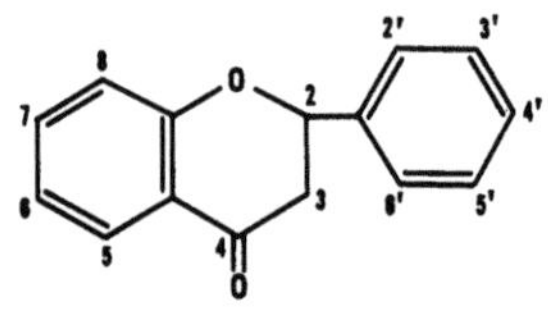

Flavanones

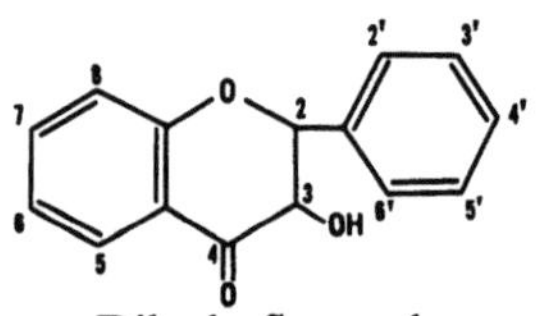

Dihydroflavonols

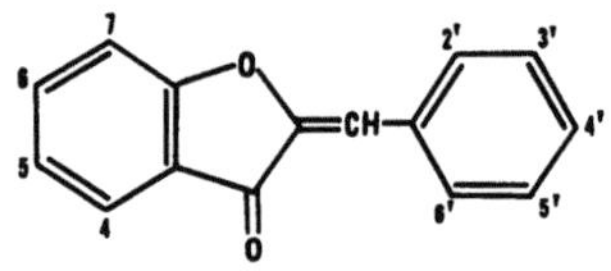

Aurones

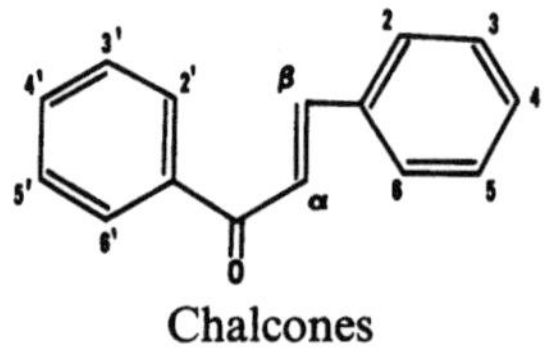

Chalcones

Part I

The Isolation, Purification and Preliminary Identification of Flavonoids

Chapter I

The Two-Dimensional Paper Chromatographic Analysis of Flavonoids

I-1. Reagents and Materials 3
Paper 3
The TBA and HOAc Solvent Systems 4
Chromatographic Cabinet (Chromatocab) 4
Ultraviolet Viewing Lamp 4
Drying Rack 4
I-2. Experimental Procedures for the Two-Dimensional Paper Chromatographic Analysis of Flavonoid Mixtures 4
(A) The Two-Dimensional Paper Chromatographic Analysis of *Baptisia lecontei* Flavonoids 4
(B) The Two-Dimensional Paper Chromatographic Analysis of *Hymenoxys scaposa* Flavonoids 9
I-3. The Determination of R_f Values for Flavonoids 9
I-4. The Effects of Flavonoid Structural Variations on R_f Values 10
I-5. Relationships between Spot Color and Flavonoid Structure 12
I-6. The Isolation and Purification of Flavonoids by Preparative Two-Dimensional Paper Chromatography 13
I-7. The One-Dimensional Paper Chromatographic Purification of a Partially Purified Flavonoid 14
References 15

Two-dimensional paper chromatography represents one of the best methods for the rapid separation of mixtures of flavonoids from crude methanol or methanol-water extracts of dried plant material. Furthermore, sufficient quantities of the separated compound for both hydrolytic and ultraviolet spectral analyses often can be isolated from about 30 chromatograms. The present discussion will emphasize the chromatographic techniques routinely employed in this laboratory for the analysis of flavonoids; however, a number of other procedures and variations have been published elsewhere [*1*, *2*].

I-1. Reagents and Materials

Paper

Whatman 3MM chromatographic paper (46 × 57 cm) has proved to be satisfactory for both qualitative and quantitative analysis of crude plant extracts containing complex mixtures of flavonoids.

The TBA and HOAc Solvent Systems

a) *TBA* 3:1:1 Solution of reagent-grade tertiary butanol:reagent-grade glacial acetic acid:water.

b) *HOAc* 15 Ml of reagent grade glacial acetic acid mixed with 85 ml of H_2O.

The TBA and HOAc solvent systems were satisfactory for the two-dimensional paper chromatographic analysis of most flavonoid extracts encountered in our laboratory. The TBA solvent is unstable when stored for long periods, and it is recommended that it be prepared fresh each month and stored in the dark.

Chromatographic Cabinet (Chromatocab)

Chromatocabs[1] may be constructed according to the plans presented in Fig. 1. A chromatocab is required for each solvent system. The plans call for (i) a glass window in one end of the chromatocab for viewing chromatograms, (ii) a foam neoprene gasket around the top of the cabinet to ensure air-tight sealing of the lid, and (iii) a completely waxed interior to protect the cabinet and to assist in solvent equilibration. Five of each of the following items[2], with the exception of the anti-siphon rods, are required for each cabinet (thus 10 chromatograms may be developed simultaneously in each cab): 24″ (62 cm) glass solvent troughs, $26\frac{3}{4}$″ (68 cm) glass anti-siphon rods (10 rods are required for each chromatocab), $23\frac{1}{2}$″ (59.7 cm) glass anchor rods and $26\frac{1}{2}$″ (67.5 cm) stainless steel trough holders.

Ultraviolet Viewing Lamp

A long wavelength (3,660 angstroms) ultraviolet lamp[3] equipped with two 15-watt Blak-Ray tubes and covered with a glass plate is satisfactory for viewing the developed chromatograms. It is recommended that protective glasses be worn when working with the UV viewing lamp.

Drying Rack

A wooden frame [24″ (61 cm) × 24″ (61 cm) × 18″ (45.6 cm)], open on all sides and fitted with 10 strings placed at $2\frac{1}{2}$″ (6.35 cm) intervals along the top, is suitable for drying chromatograms in a fume hood. The wet chromatograms are suspended with clothespins.

I-2. Experimental Procedures for the Two-Dimensional Paper Chromatographic Analysis of Flavonoid Mixtures

Experimental details are presented below for the two-dimensional paper chromatographic analysis of the flavonoids present in *Baptisia lecontei* (Leguminosae) and *Hymenoxys scaposa* (Compositae) air-dried plant material[4].

(A) The Two-Dimensional Paper Chromatographic Analysis of *Baptisia lecontei* Flavonoids [*3*]. Dried stem and leaf material (24 g), which had been finely ground in a Waring Blendor, was extracted at room temperature for 3 days with cold 25% aqueous methanol (180 ml). After the plant material was removed by filtration, the extract, on evaporation under water pump vacuum, yielded a sticky green residue (5.3 g). (Although the methanol present in the extract is readily removed under water-pump vacuum with the aid of a rotary evaporator, the residual water is best removed under oilpump high vacuum.)

---

1 Suitable cabs can be purchased from Kensington Scientific Corp., Oakland, Calif.
2 The items may be obtained from E.H. Sargent and Co., Dallas, Texas.
3 We used model XX-15 lamps marketed by W.H. Curtin and Co., Houston, Texas.
4 Results similar to those described here were also obtained with fresh plant material.

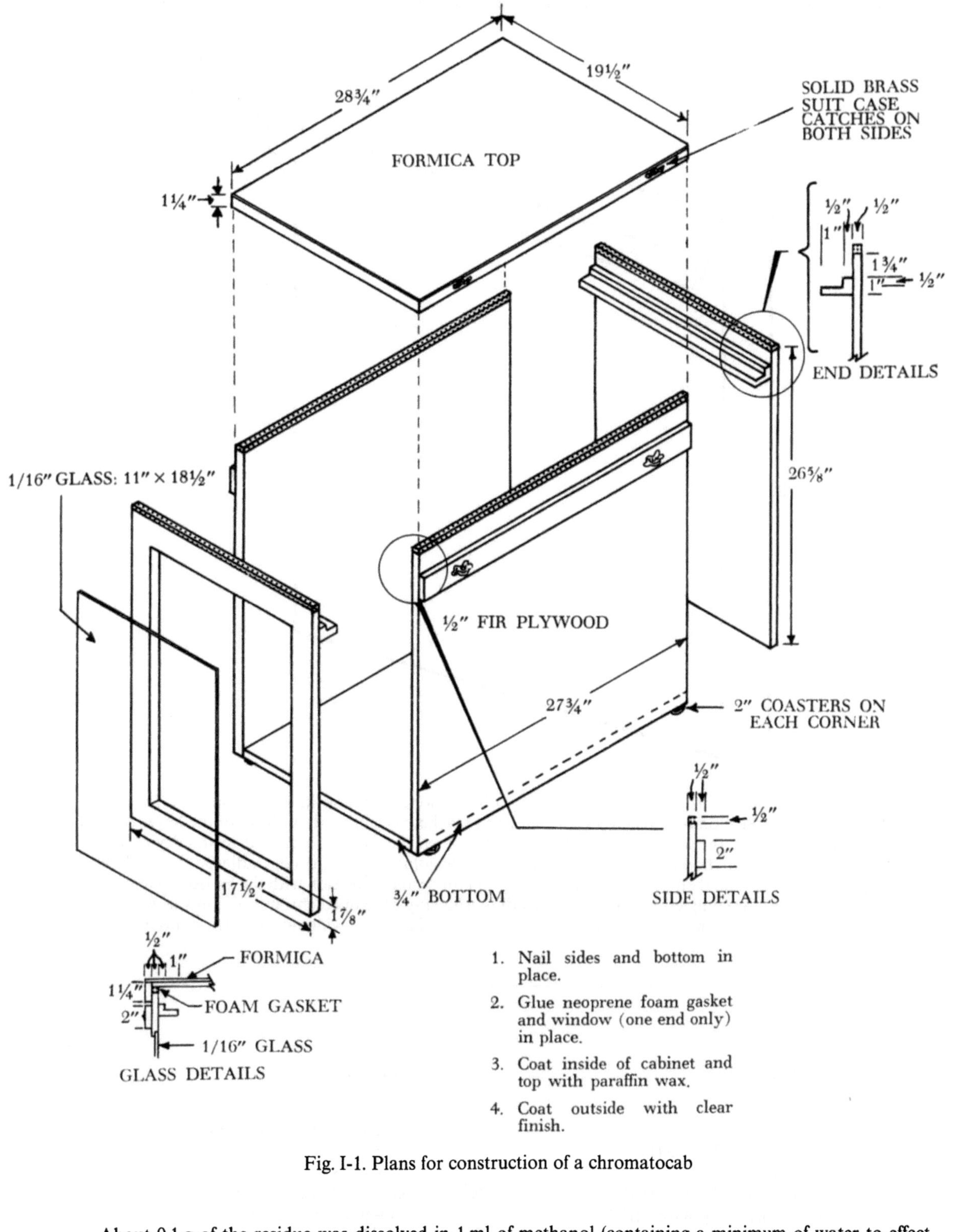

Fig. I-1. Plans for construction of a chromatocab

About 0.1 g of the residue was dissolved in 1 ml of methanol (containing a minimum of water to effect solution). This solution was then spotted (using an ungraduated pipette) on the lower right-hand corner of a sheet of Whatman 3MM chromatographic paper. A hair drier was used for solvent evaporation between repeated applications of the solution to the paper. The final spot, which appeared deep purple when viewed under a 3,660 angstrom UV lamp, was about 4 cm in diameter and 10 cm from each edge of the paper. The chromatogram was folded along the 46 cm edge adjacent to the spot containing the flavonoids in the manner shown in the following sketch:

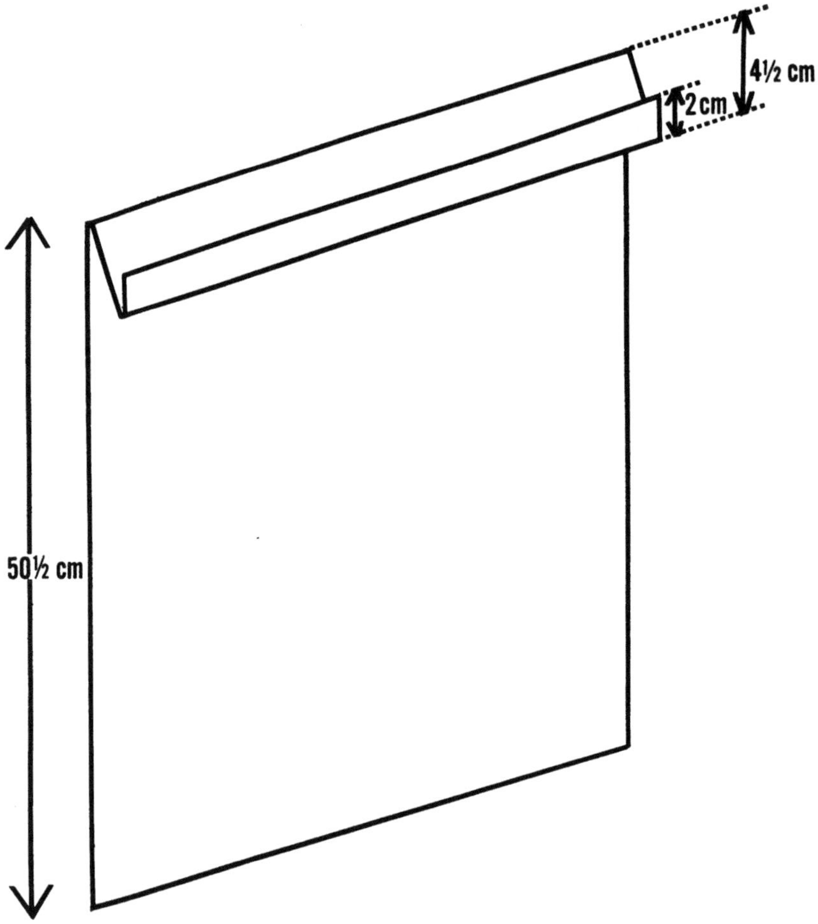

The chromatogram was developed descendingly in the long direction in a chromatocab (Fig. I-1) using TBA as solvent (see Section I-1). When the solvent front reached to within about 3 cm of the lower edge of the paper (after 22–26 hr), the chromatogram was removed from the cabinet, attached to the drying rack, and allowed to dry in a fume hood. The dry chromatogram was folded along the edge adjacent to the band containing the flavonoids and then developed descendingly in the second direction with the HOAc solvent. This run required about 4 hr for completion. The dried two-dimensionally developed chromatogram was viewed in UV light alone and in the presence of ammonia fumes (the mouth of a 100 ml widemouth bottle containing concen-

---

Caption to Fig. I-2a on p. 7

Spot No.	Compound	Spot No.	Compound
1.	Apigenin (II)	5d.	Probably liquiritigenin (4′,7-Dihydroxy-dihydroflavone)
2.	4′,7-Dihydroxyflavone (IV)	5e.	Orobol (X)
3.	Luteolin (I)	6.	3′,4′,7-Trihydroxyflavone 7-O-glucoside (IIIa)
3a.	4′,7-Dihydroxyflavonol (VI)	7.	4′,7-Dihydroxyflavone 7-O-glucoside (IVa)
4.	3′,4′,7-Trihydroxyflavone (III)	8.	Luteolin 7-O-rutinoside (Ib)
4a.	Fisetin (V)	8a.	Luteolin 7-O-glucoside (Ia)
5.	Pseudobaptigenin (XI)	8b.	Fisetin 7-O-rhamnoglucoside (Vb)
5a.	Calycosin (XIII)		
5b.	Daidzein (XII)		
5c.	Genistein (IX)		

trated ammonium hydroxide was held in contact with each spot for about 5 sec). All spots which were detected by this procedure were circled with a lead pencil (Fig. I-2a). The isolation of these compounds by column chromatography and the methods used for identifying them are described in Chapter II [3].

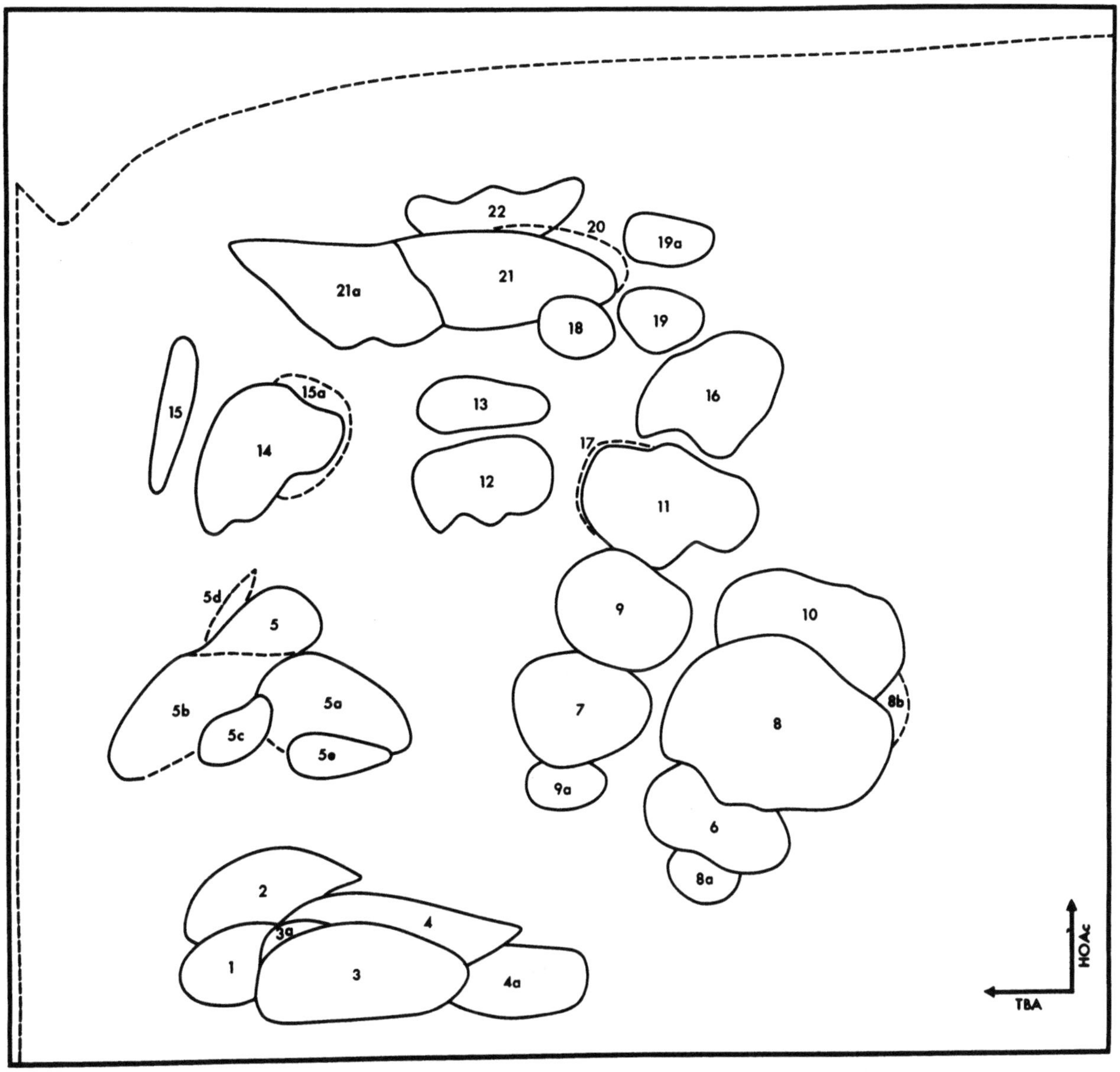

Fig. I-2a. The two-dimensional paper chromatographic pattern of flavonoids obtained from *Baptisia lecontei* plant material [3]

Spot No.	Compound
9.	Apigenin 7-O-rhamnoglucoside (IIb)
10.	3′,4′,7-Trihydroxyflavone 7-O-rhamnoglucoside (IIIb)
11.	4′,7-Dihydroxyflavone 7-O-rhamnoglucoside (IVb)
12.	Calycosin 7-O-glucoside (XIIIa)
13.	Daidzein 7-O-glucoside (XIIa)
14.	Scopoletin (XIV)
15.	(+)-4′,7-Dihydroxy-dihydroflavonol (VII)
15a.	(+)-Fustin (VIII)
16.	Orobol 7-O-rutinoside (Xa)
17.	3′,4′,7-Trihydroxyflavonol 3-O-glucoside (Va)
18.	Sphaerobioside (IXa)
19.	Calycosin 7-O-rhamnoglucoside (XIIIb)
19a.	Daidzein 7-O-rhamnoglucoside (XIIb)
20.	(+)-Fustin 3-O-glucoside (VIIIa)
21.	Pseudobaptisin (XIa)
21a.	Lecontin (VIIa)
22.	Scopoletin 7-O-glucoside (XIVa)

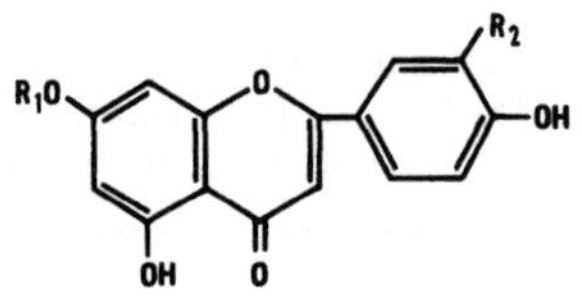

	R_1	R_2	
I	H	OH	luteolin
Ia	glu	OH	luteolin 7-O-glucoside
Ib	rh-glu	OH	luteolin 7-O-rutinoside
II	H	H	apigenin
IIa	glu	H	apigenin 7-O-glucoside
IIb	rh-glu	H	apigenin 7-O-rhamnoglucoside

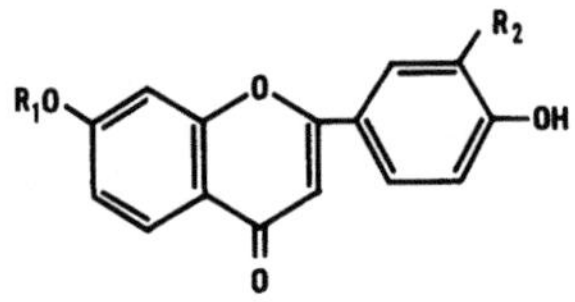

	R_1	R_2	
III	H	OH	3′,4′,7-trihydroxyflavone
IIIa	glu	OH	3′,4′,7-trihydroxyflavone 7-O-glucoside
IIIb	rh-glu	OH	3′,4′,7-trihydroxyflavone 7-O-rhamnoglucoside
IV	H	H	4′,7-dihydroxyflavone
IVa	glu	H	4′,7-dihydroxyflavone 7-O-glucoside
IVb	rh-glu	H	4′,7-dihydroxyflavone 7-O-rhamnoglucoside

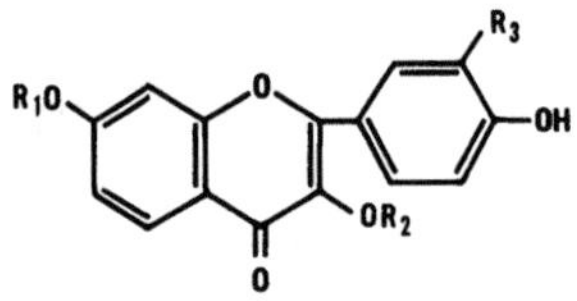

	R_1	R_2	R_3	
V	H	H	OH	fisetin
Va	H	glu	OH	3′,4′,7-trihydroxyflavonol 3-O-glucoside
Vb	rh-glu	H	OH	fisetin 7-O-rhamnoglucoside
VI	H	H	H	4′,7-dihydroxyflavonol

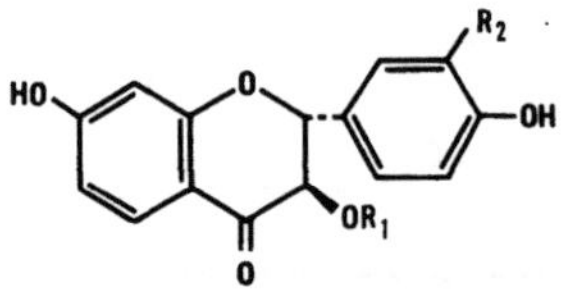

	R_1	R_2	
VII	H	H	(+)-4′,7-dihydroxy-dihydroflavonol
VIIa	glu	H	(lecontin)
VIII	H	OH	(+)-fustin
VIIIa	glu	OH	(+)-fustin 3-O-glucoside

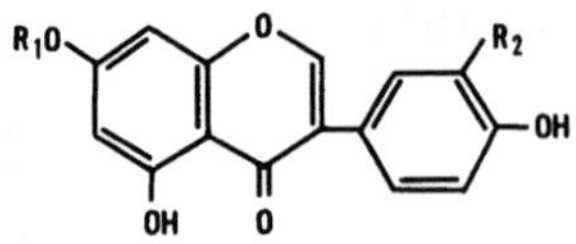

	R_1	R_2	
IX	H	H	genistein
IXa	rh-glu	H	sphaerobioside
X	H	OH	orobol
Xa	rh-glu	OH	orobol 7-O-rutinoside

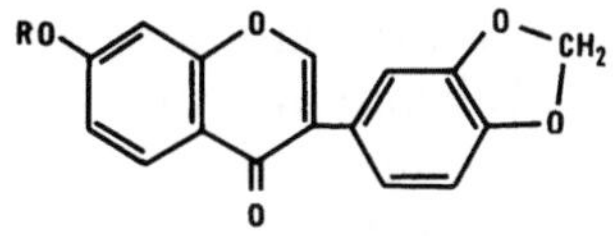

XI	R=H	pseudobaptigenin
XIa	R=rh-glu	pseudobaptisin

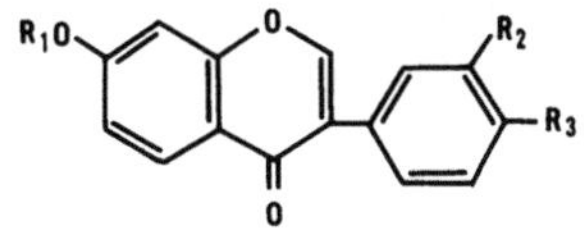

	R_1	R_2	R_3	
XII	H	H	OH	daidzein
XIIa	glu	H	OH	daidzein 7-O-glucoside
XIIb	rh-glu	H	OH	daidzein 7-O-rhamnoglucoside
XIII	H	OH	OCH_3	calycosin
XIIIa	glu	OH	OCH_3	calycosin 7-O-glucoside
XIIIb	rh-glu	OH	OCH_3	calycosin 7-O-rhamnoglucoside

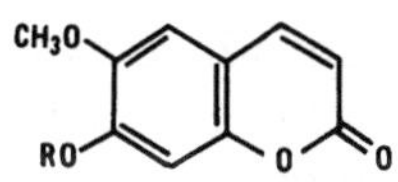

XIV	R=H	scopoletin
XIVa	R=glu	scopoletin 7-O-glucoside

Fig. I-2b. The structures of flavonoids and coumarins detected by the two-dimensional paper chromatographic analysis of an extract of *Baptisia lecontei* plant material (see Fig. I-2a); the compounds were subsequently isolated by polyamide column chromatography (see Chapter II, Section II-2a and Table II-1) [3]

(B) The Two-Dimensional Paper Chromatographic Analysis of *Hymenoxys scaposa* Flavonoids [4]. Dried, ground leaves of *Hymenoxys scaposa* (196 g) were extracted successively in the cold with petroleum ether, b.p. range 35–60° (2 × 1.5 l; 48 hr each), methylene chloride (2 × 2 l; 48 hr each), methanol (2 × 1 l; 48 hr each) and 50% aqueous methanol (2 × 1 l; 48 hr each). All solvents were reagent grade. The leaves were air-dried between extractions which involved different solvents. The two-dimensional paper chromatographic analysis of each extract, using the procedures described for the *Baptisia lecontei* extracts, showed that the petroleum ether had not extracted any flavonoids, the methylene chloride had removed a few flavonoid aglycones, and the methanol and aqueous methanol extracts were rich in flavonoid glycosides and contained aglycones as well.

The petroleum ether pre-extraction procedure described for the *Hymenoxys scaposa* plant material removes a number of non-flavonoid constituents such as fats and chlorophyll. In some instances the petroleum ether pre-extraction is an essential step in order to obtain a workable extract of flavonoids with aqueous methanol.

I-3. The Determination of R_f Values for Flavonoids

The R_f value of a flavonoid as used here is defined as follows:

$$R_f = \frac{\text{distance between origin and the center of concentration of the flavonoid spot}}{\text{distance between origin and solvent front}}.$$

By the following procedure we have determined R_f values (in TBA and HOAc) for all the flavonoids listed in the spectral section of this book; the values for each flavonoid are recorded in the upper left-hand corner of the page which presents the UV spectra for that flavonoid.

An aqueous methanol solution (1 ml) containing a few mg of rutin was applied to the lower right-hand corner of a sheet of chromatographic paper (as described in procedure A, Section I-2). The center of the spot was marked with a lead pencil, and the chromatogram was developed in TBA in the long dimension for about 22 hrs. After drying, the chromatogram was viewed in UV light alone and in the presence of ammonia vapor and the center of concentration of the spot and the solvent front were marked. The TBA R_f value for rutin (0.44) was calculated by dividing the distance the flavonoid spot had moved (using the center of concentration of the spot for the measurement) by the distance from the origin to the solvent front. By a similar procedure the HOAc R_f value of rutin (0.56) was determined.

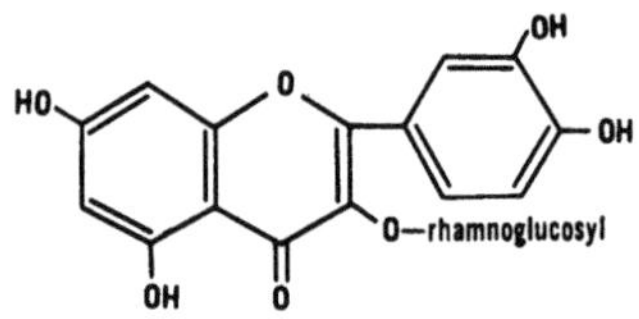

XV. Rutin

R_f values are often difficult to reproduce exactly and the values published in this volume are only accurate within about $\pm 5\%$. In all examples reported here, the R_f values were determined from the center of concentration of the spot (the point which appears to be the center of the material present in the spot).

When R_f values (in two or more solvent systems) of an unknown flavonoid are identical with those obtained for a known compound, the two compounds are often but not always identical; the paper chromatographic evidence should be confirmed by other information such as UV and NMR spectral data.

I-4. The Effects of Flavonoid Structural Variations on R_f Values

Different types of flavonoids usually occur in characteristic areas on a two-dimensional paper chromatogram. A schematic outline of the relationships of R_f values (in TBA and HOAc) of the more commonly encountered types of flavonoids (flavones, flavonols, isoflavones, flavanones, chalcones, dihydroflavonols, aurones, and their O-glycosides) is presented in Fig. I-3. Although there is considerable overlap of the areas assigned to each type of flavonoid, the outline does serve as a useful guide for the preliminary interpretation of the flavonoid pattern.

In the paper chromatogram of a crude plant extract, such as that of *B. lecontei* (Fig. I-2a), the geometrical relationship of one spot to another is often indicative of certain structural differences between the compounds concerned [*3*]. The 5-deoxy flavonoids, for example, run consistently slower in TBA and a little faster in HOAc relative to their 5-hydroxylated equivalents (cf. the pairs of spots: 1, 2; 3, 4; 5e, 5a; 8, 10; 8a, 6; 9a, 7; 9, 11; 18, 19a). Another such geometrical relationship can be ascribed to B-ring oxidation patterns. For instance, 4′-monohydroxy flavonoids run faster in both TBA and HOAc than do their 3′,4′-dihydroxy equivalents (based on R_f values). This is illustrated in Fig. I-2a for flavones (spots 1, 3), 5-deoxyflavones (spots 2, 4) and their glycosides (spots 9, 8; 11, 10), isoflavones (spots 5c, 5e) and isoflavone glycosides (spots 18, 16), 5-deoxyflavonols (spots 3a, 4a), 5-deoxydihydroflavonols (spots 15, 15a) and 5-deoxydihydroflavonol glycosides (spots 21a, 20). Spots of monoglycosides and diglycosides of the same aglycone also show a consistent geometrical relationship to one another, and this is exemplified by the apigenin (spots 9a and 9) and luteolin (spots 8a and 8) glucosides and rhamnoglucosides (Fig. I-2a). The diglycoside always runs slower in TBA and faster in HOAc than do the equivalent monoglycoside, a relationship which also holds for other types of flavonoids (cf. spots 12, 19; 13, 19a; 6, 10 and 7, 11).

Fig. I-4 illustrates the chromatographic behavior of a number of the more common flavone C-glycosides. In addition to the relationships of structure to R_f value discussed above (which are still valid with C-glycosylflavones), 6- and 8-C-glycosyl isomers of the same aglycone exhibit still another chromatographic relationship. The 6-C-glycosides consistently run faster in both TBA and HOAc than do the 8-C-glycoside isomers (cf. spot pairs 1, 15; 2, 23). Partial isomerization of one isomer to the other can usually be achieved by heating the flavonoid in 2N hydrochloric acid for about one hour at 100°; the chromatographic pattern exhibited by the mixture of isomers so produced is diagnostic for 6- and 8-C-glycosylflavonoids [5].

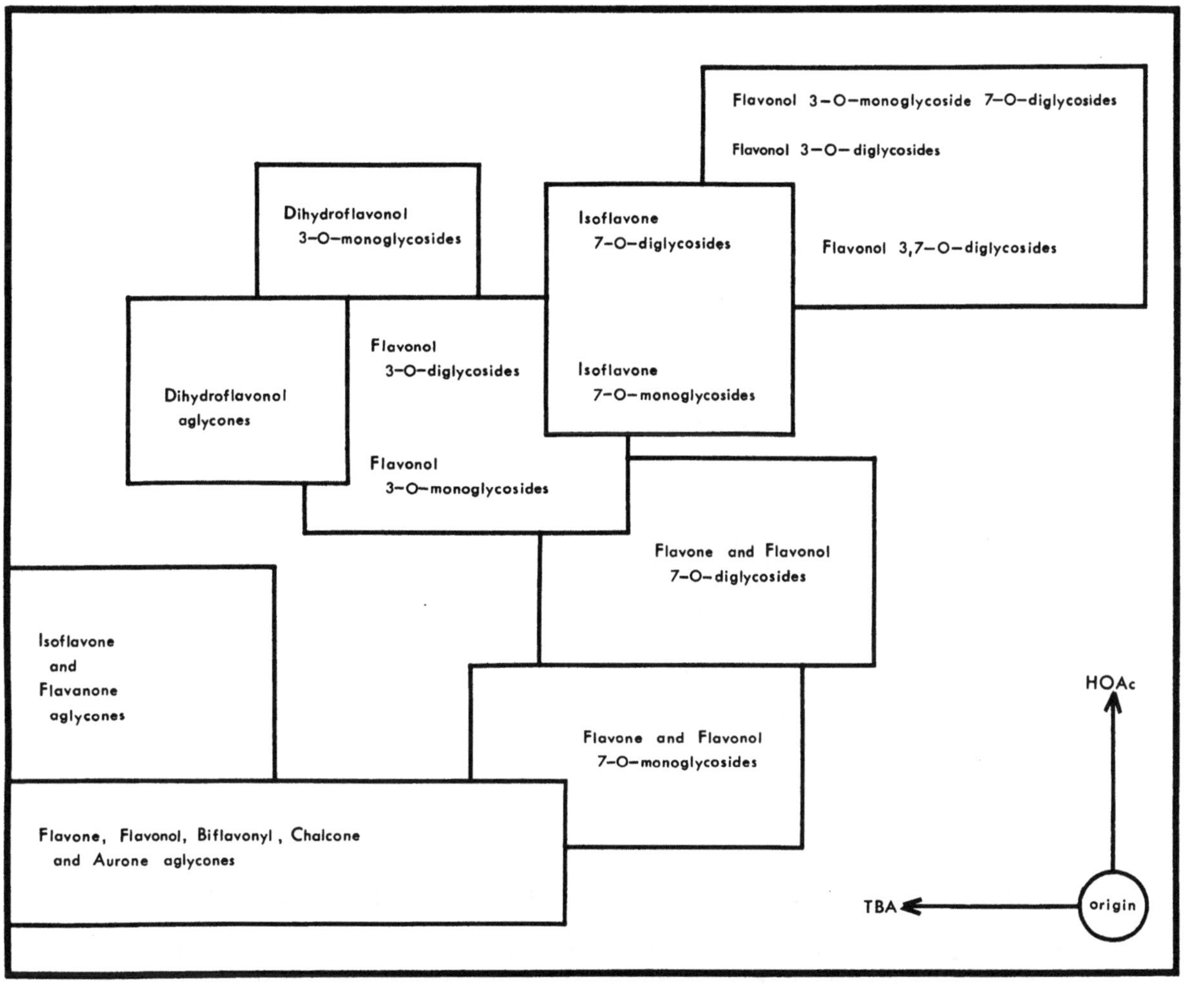

Fig. I-3. The distribution of flavonoids[5] on a TBA/HOAc, two-dimensional paper chromatogram

[5] The chromatographic information presented is for the more commonly encountered flavonoids which possess a 5,7-dihydroxylated A-ring and either a 4′-mono- or a 3′,4′-di-hydroxylated B-ring. The glycosides are the 3- and 7-mono- and 3,7-di-glycosides unless otherwise stated. Flavonoids with other oxygenation and/or glycosylation patterns may exhibit somewhat different chromatographic properties.

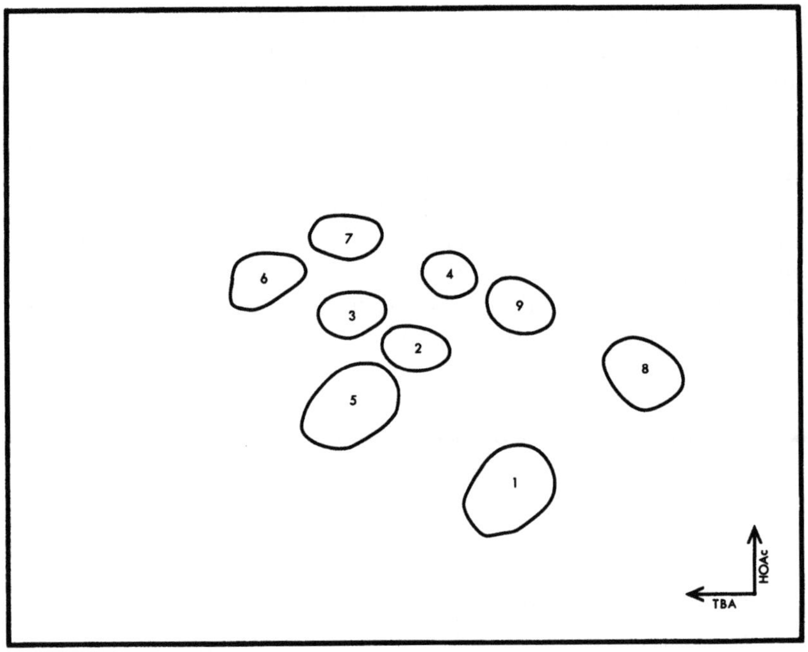

Fig. I-4. The distribution of *Lemna minor* C-glycosylflavones on a TBA/HOAc two-dimensional paper chromatogram [5]

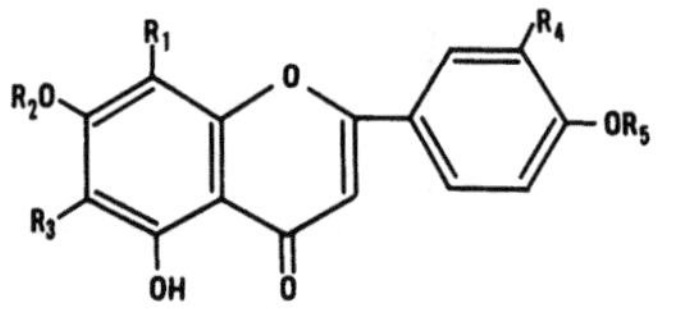

Spot No.	C-Glycosylflavone
1	Orientin, R_1=C-glucosyl; R_2, R_3, R_5=H; R_4=OH
2	Isoorientin, R_3=C-glucosyl; R_1, R_2, R_5=H; R_4=OH
3	Isoorientin 3′-methyl ether
4	Isoorientin 4′-O-glucoside (Lutonarin)
5	Vitexin, R_1=C-glucosyl; R_2, R_3, R_4, R_5=H
6	Isovitexin, R_3=C-glucosyl; R_1, R_2, R_4, R_5=H
7	Isovitexin glycoside
8	Lucenin−1[6], R_1, R_3=C-glycosyl; R_2; R_5=H; R_4=OH
9	Vicenin−1[6], R_1, R_3=C-glycosyl; R_2, R_4, R_5=H

I-5. Relationships between Spot Color and Flavonoid Structure

The spot appearance in UV light alone and in the presence of ammonia for each flavonoid is recorded in the upper left-hand corner of the page presenting the UV spectra of the compound (see Chapters V, VI and VII). Dry chromatograms were viewed over a long wavelength ultraviolet lamp (see Section I-1) for these determinations. A summary of the spot appearance data for different types of flavonoids is presented in Table I-1. As with R_f values, the spot appearance information can be a useful guide in the preliminary identification of the flavonoids detected by paper chromatography.

[6] A number of lucenins and vicenins occur which differ in their C-glycosyl moieties and therefore exhibit different chromatographic properties from those encountered in *Lemna minor*. For a recent review of these and other C-glycosylflavonoids see reference 6.

Table I-1. *Relationships between spot color and flavonoid structure*

Flavonoid spot color		Flavonoid type
UV light	UV/NH_3	
Deep purple	yellow, yellow-green or brown	(a) Usually flavones with 5-OH and 4′-OH or 3-OH substituted flavonols with 5-OH and 4′-OH (b) Some 5-OH flavanones and 4′-OH chalcones lacking B-ring hydroxyl groups
	little or no color change	(a) Flavones or flavonols with 5-OH but with the 4′-OH absent or substituted (b) Isoflavones, dihydroflavonols and some flavanones with 5-OH (c) Chalcones with 2′- or 6′-OH but without a free 2- or 4-OH
	light blue	Some 5-OH flavanones
	red or orange	Chalcones with a free 2- and/or 4-OH
Fluorescent light blue	fluorescent yellow-green or fluorescent blue-green	(a) Flavones and flavanones lacking a free 5-OH (b) Flavonols lacking a free 5-OH but with the 3-OH substituted
	little or no color change	Isoflavones lacking a free 5-OH
	bright fluorescent light blue	Isoflavones lacking a free 5-OH
Invisible	fluorescent light blue	Isoflavones lacking a free 5-OH
Dull yellow and yellow or orange fluorescence	little or no color change	Flavonols with a free 3-OH and with or without a free 5-OH
Fluorescent yellow, yellow-green, blue-green or green	orange or red	Aurones with a free 4′-OH and some 2- or 4-OH chalcones
	little or no color change	(a) Aurones lacking a free 4′-OH and flavanones lacking a free 5-OH (b) Flavonols with a free 3-OH and with or without a free 5-OH
Pale yellow	light yellow-purple	Dihydroflavonols lacking a free 5-OH

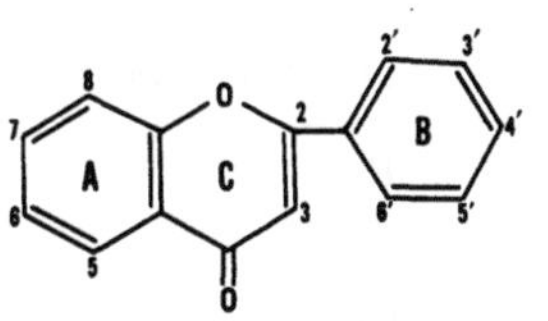

Numbering scheme for flavonoids other than chalcones and aurones

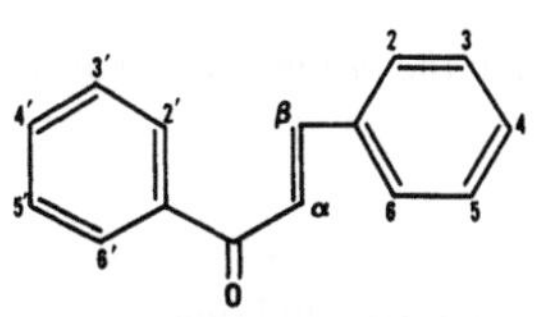

Numbering scheme for chalcones

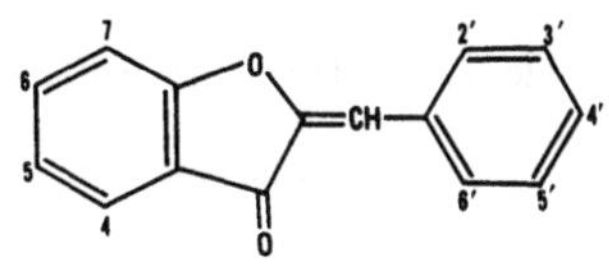

Numbering scheme for aurones

I-6. The Isolation and Purification of Flavonoids by Preparative Two-Dimensional Paper Chromatography

The isolation and purification of a flavonoid can often be achieved by preparative two-dimensional paper chromatography. If a flavonoid is a component of a complex plant extract, it can usually be separated from the mixture by cutting out the area in which it occurs on a two-dimensional chromatogram. Usually 20–50 chromatograms are required to obtain sufficient material for identification of the flavonoid. The compound can be isolated by extraction of the pieces of paper so obtained with reagent grade methanol or 20% aqueous reagent grade methanol. It is convenient to cut the pieces of paper into small sections which may then be mixed with excess solvent in a 125 ml Erlenmeyer flask. Constant mechanical agitation for several hours facilitates the extraction. Filtration and water pump vacuum evaporation of the extract yields the required flavonoid. If the original spot which corresponded to the flavonoid overlapped other flavonoid spots, an additional two-dimensional chromatographic purification is required.

It is important to minimize contamination of a flavonoid to be used for UV spectral studies; therefore, in the purification of the standard flavonoid samples used in the spectroscopic investigations presented in this book, we found it best to extract the flavonoid from the chromatographic paper with spectroscopic methanol for only a few minutes. For most of the flavonoids used in the present study, purification was achieved by one dimensional paper chromatography (see following section).

I-7. The One-Dimensional Paper Chromatographic Purification of a Partially Purified Flavonoid

The solvent to be used in the one-dimensional paper chromatographic purification of a flavonoid was selected after a preliminary two-dimensional run by observing which solvent system (TBA or HOAc) most effectively separated the mixture. The flavonoid mixture (about 4 mg) in methanol (1 – 2 ml) was applied as a band, 10 cm from the top of the chromatographic paper, extending across the width of the paper. A descending run in the selected solvent gave UV-detectable bands, the major one being subsequently cut out and eluted with methanol. When the sample was required for UV spectroscopy, it was eluted with spectroscopic methanol for only a few minutes (i.e. 10 min or less)[7].

Either a single one-dimensional or 30 two-dimensional runs will yield sufficient pure flavonoid for most spectral analyses.

The isolation of flavonoid aglycones, particularly isoflavone aglycones, requires a different chromatographic procedure. Isoflavone aglycones run as a large complex of overlapping spots in the lower left corner of the TBA/HOAc chromatogram (see spots 5, 5a, 5b, 5c and 5d in Fig. I-2a), and generally cannot be separated from one another using the TBA/HOAc solvent systems. However a solvent system consisting of the organic layer of a mixture of benzene:acetic acid:water, 6:7:3, has been used success-

Table I-2. R_f *Values for isoflavone aglycones in the organic phase of a benzene:acetic acid:water, 6:7:3, mixture*

Isoflavone	R_f Value
Afrormosin (7-hydroxy-4',6-dimethoxyisoflavone)	0.85
Biochanin A (5,7-dihydroxy-4'-methoxyisoflavone)	0.70
Formononetin (7-hydroxy-4'-methoxyisoflavone)	0.58
Pseudobaptigenin (7-hydroxy-3',4'-methylenedioxyisoflavone)	0.57
Tectorigenin (4',5,7-trihydroxy-6-methoxyisoflavone)	0.45
Texasin (6,7-dihydroxy-4'-methoxyisoflavone)	0.37
Calycosin (3',7-dihydroxy-4'-methoxyisoflavone)	0.20
Genistein (4',5,7-trihydroxyisoflavone)	0.15
6-Hydroxygenistein (4',5,6,7-tetrahydroxyisoflavone)	0.05
Orobol (3',4',5,7-tetrahydroxyisoflavone)	0.03
3',4',7-Trihydroxyisoflavone	0.01

[7] For further comments on the paper chromatographic purification of a flavonoid for UV spectral studies, see Chapter IV, Section 2.

fully for the separation of isoflavone aglycones by one-dimensional paper chromatography. Non-polar isoflavones run faster in this system than those which are polyhydroxylated (Table I-2). For isoflavone aglycone identifications, it is recommended that known standards be run along side the unknowns.

Further purification of paper chromatographically isolated isoflavone aglycones can often be achieved by sublimation at about 180° and 0.03 mm [7].

References

1. Seikel, M. K., in: The Chemistry of Flavonoid Compounds (edited by T. A. Geissman), p. 34–69. Oxford: Pergamon Press 1962.
2. a) Harborne, J. B.: J. Chromatography **1**, 473 (1958). b) Harborne, J. B.: J. Chromatography **2**, 581 (1959).
3. Markham, K. R., and T. J. Mabry: Phytochemistry **7**, 791 (1968).
4. a) Thomas, M. B., and T. J. Mabry: Phytochemistry **7**, 787 (1968). b) Seeligman, P., and R. E. Alston: Brittonia **19**, 205 (1967).
5. Wallace, J. W., Jr.: Ph. D. Dissertation, University of Texas at Austin, 1967, "Investigations of Flavone Biosynthesis in the Lemnaceae".
6. Alston, R. E., in: Recent Advances in Phytochemistry (edited by T. J. Mabry, R. E. Alston, and V. C. Runeckles), p. 305–327. New York: Appleton-Century-Crofts 1968.
7. Markham, K. R., T. J. Mabry, and W. T. Swift III: Phytochemistry **7**, 803 (1968).

Chapter II

The Separation of Flavonoids by Column and Thin Layer Chromatography

II-1. Preliminary Purification of Flavonoids in a Crude Plant Extract Using Charcoal 16
(A) Adsorption of Flavonoids onto Charcoal from a Crude Plant Extract . 17
(B) Recovery of Flavonoids from the Charcoal 17
II-2. The Separation of Flavonoids by Polyamide and Silica Gel Column Chromatography 17
2a. Polyamide Column Chromatography of Flavonoids 17
(A) Polyamide Column Chromatographic Separation of 27 Flavonoids in a *Baptisia lecontei* Extract 17
(B) Preparation of Polyamide for Flavonoid Column Chromatography . . 18
2b. Silica Gel Column Chromatography of Non-Polar Type Flavonoids 19
Silica Gel Column Chromatographic Separation of *Hymenoxys scaposa* Methoxylated Flavones 19
2c. The Separation of Flavonoids by Sephadex Column Chromatography . . . 20
II-3. The Separation of Flavonoids by Silica Gel and Polyamide Thin Layer Chromatography 20
3a. The Separation of Flavonoids by Silica Gel TLC 21
3b. The Separation of Flavonoids by Polyamide TLC 21
(A) Preparation of Polyamide for Polyamide TLC Plates 21
(B) Preparation of Polyamide TLC Plates 22
References 22

II-1. Preliminary Purification of Flavonoids in a Crude Plant Extract Using Charcoal

Charcoal is useful for the preliminary purification of a mixture of flavonoids, particularly flavonoid glycosides, which are usually present in a crude aqueous or aqueous-methanolic extract of plant material [*1*]. The charcoal procedure described below separates flavonoids from most non-aromatic plant constituents such as the common carbohydrates. The method is especially useful for flavonoid glycosides which are readily recovered almost quantitatively from the charcoal with water containing 7% phenol (i.e. a saturated aqueous solution at room temperature). Although many aglycones can also be recovered, at least in part, from the charcoal, the procedure is not recommended for their purification. Aglycones can often be extracted directly from a crude syrup obtained from a plant extract with a solvent such as ethyl acetate, while their complete recovery from charcoal many require the use of pyridine as eluent. A typical charcoal procedure is presented below for the preliminary purification of the flavonoids in a crude extract obtained from *Baptisia lecontei* plant material.

(A) Adsorption of Flavonoids onto Charcoal from a Crude Plant Extract. About 200 g of dried ground *Baptisia lecontei* leaf and stem material was extracted with excess cold 20% aqueous methanol for 3 days; on evaporation of the solvent the extract yielded about 36 g of a sticky syrup, which was subsequently dissolved in 125 ml of hot methanol. This solution was mixed with 5 g of celite and filtered through a Buchner funnel. The celite-residue material was suspended in 50 ml of hot methanol and filtered again. The two filtrates (about 300 ml including all washings) were combined, left standing overnight, and then refiltered.

About 250 ml of the clear filtrate was mixed with activated charcoal (common commercial type) using a mechanical stirrer. Charcoal was added in portions until the supernatant liquid showed no flavonoids as determined by polyamide TLC. A total of 80 g of charcoal was added, two 20 g and four 10 g portions.

The charcoal-flavonoid material was filtered onto a small Buchner funnel, and the residue was washed with 2 l of boiling methanol. The methanol filtrate yielded, on concentration, 16.3 g of a flavonoid-free residue. The charcoal-flavonoid material was next washed with 1 l of boiling water; the water yielded another 2.7 g of flavonoid-free residue.

(B) Recovery of Flavonoids from the Charcoal. The charcoal-flavonoid material from procedure (A), which had been collected on a Buchner funnel, was now washed (in a fume hood) with 1 l of a boiling solution of phenol:water (7:93). After the phenol-water solution had been concentrated to a small volume on a rotary evaporator (at about 80° and 12 mm pressure), the remaining traces of phenol were removed by ether extraction. Concentration of the phenol-free solution gave a flavonoid-rich residue (4.4 g).

II-2. The Separation of Flavonoids by Polyamide and Silica Gel Column Chromatography

Column chromatography does not, in most cases, separate complex mixtures of flavonoids which may be present in crude plant extracts as well as paper chromatography; nevertheless when larger quantities of the flavonoids are required, column chromatography may be the method of choice.

2a. Polyamide Column Chromatography of Flavonoids

Although a number of different adsorbents have been used for column chromatography of flavonoids (e.g. silica gel, magnesol, cellulose powder, polyamide, charcoal and starch) [*2*], the best adsorbent for the chromatographic separation of all types of flavonoids appears to be polyamide. A polyamide-type adsorbent used in conjunction with various mixtures of water and methanol as eluents has been used successfully for the separation of complex mixtures of glycosides and aglycones of isoflavones, flavones, flavonols, dihydroflavonols and flavanones (see procedure A below).

(A) Polyamide Column Chromatographic Separation of 27 Flavonoids in a *Baptisia lecontei* Extract [*3*]. The polyamide adsorbent, polyvinyl-pyrrolidone (Polyclar AT, General Aniline and Film Corp.)[1], was passed through a No. 120 (U.S. Standard) sieve to remove particles smaller than 0.002 cm. The sieved polyamide, sufficient to half fill a 5 × 50 cm column, was made into a slurry with water and poured into the column which had been plugged with a small amount of glass wool. After the adsorbent had settled, it was drained of excess water. Five grams of a residue obtained from the extract of *Baptisia lecontei* leaves and stems (which had been extracted with cold 20% aqueous methanol for three days) was dissolved in a minimum of aqueous methanol, and the solution was carefully applied to the column. Elution was initiated with 100% water; however, 20% methanol in water was used as soon as uneluted bands (observed in visible and UV light) failed to move down the column at an acceptable rate. This process was continued through a number of steps, involving 30, 40, 50 and 75% aqueous methanol, finishing with 100% methanol. The amount of each solvent mixture used was determined by the rate of movement of the visible and UV-detectable bands. Following 100% methanol, a series of dilute aqueous hydrochloric acid solutions, 0.3, 1.1 and 4.5 N, were required to complete the elution of the UV-detectable bands. Each fraction (about 150 ml) was subsequently analyzed by two dimensional paper chromatography (Table II-1).

Fractions produced from a large polyamide column (Table II-1) often yield pure flavonoids or simple mixtures which may be further separated by additional column

---

[1] We have also successfully used other commercial polyamide powders, for example Polypenco 66D from the Polymer Corp., Reading, Pa. With this material excellent separation of flavonoid mixtures was obtained using as eluent Egger's solvent: chloroform:methanol:methyl ethyl ketone; 12:2:1.

Table II-1. *Approximate flavonoid composition of fractions obtained*

Solvent	Fraction No.[b]	Flavonoids[a]											
		XIVa	XIa	XIIIb	XIIb	XIV	IXa	IVb	XIIIa	XIIa	IIIb	Xa	VIIa
H_2O	1	+											
20% MeOH	2	+	+										
20% MeOH	3	+	+	+	+	+							
20% MeOH	4	+	+			+							
20% MeOH	5					+	+	+	+				
20% MeOH	6						+	+	+	+	+	+	+
30% MeOH	7						+	+	+	+	+	+	+
30% MeOH	8								+		+	+	+
40% MeOH	9										+	+	+
50% MeOH	10											+	+
50% MeOH	11												+
75% MeOH	12												
100% MeOH	13												
0.3 N HCl	14												
1.1 N HCl	15												
4.5 N HCl	16												

[a] For the structures of these flavonoids (I–XIII) and coumarins (XIV and XIVa) see Chapter I, Fig. I-2b.
[b] The volume of each fraction was approximately 150 ml; however, this amount was varied to permit

or paper chromatography. Polyamide columns, in contrast to cellulose columns, use solvent systems which are different from those used in paper chromatography. Thus, many compounds which are inseparable by paper chromatography often can be separated on a polyamide column. For example, compare the *B. lecontei* paper chromatogram results (Fig. I-2a) with the separation observed for the same mixture on a polyamide column (Table II-1). The compound pairs: IVb, Va; XIV, VIII; XIa, VIIa and XIa, IXa (see Fig. I-2b for structures), which were inseparable by paper chromatography in TBA/HOAc, were clearly separated by polyamide column chromatography.

Two problems are often associated with polyamide columns, namely, slow elution rates and the elution with the methanolic solvents of a mixture of flavonoids and low molecular weight polymer material. Because of slow elution rates large polyamide columns may require 3 weeks or longer to run unless steps are taken to alleviate this problem. Methods commonly employed to increase the flow rate in a polyamide column include seiving the adsorbent to remove fine particles (as described above), packing the column with a 1:2 mixture of polyamide and celite, and applying pressure or vacuum to the running column. The problem of low molecular weight polyamide material being eluted during the chromatographic run can be minimized by a thorough prewashing of the adsorbent with 50% aqueous methanol. However, it is possible to eliminate both of the above mentioned difficulties by dissolving and reprecipitating commercially available polyamide (of the highly polymerized nylon or polycaprolactam type) under strictly controlled conditions as described below.

(B) Preparation of Polyamide for Flavonoid Column Chromatography [*4*]. A 3 l 3-necked round-bottom flask containing 2 l of reagent grade conc. HCl was equipped with a powerful stirrer and placed in a fume hood. Polycaprolactam pellets (600 g of Durethan BK 40F, Bayer Co., Leverkusen, West Germany) were added gradually to the continuously-stirred HCl solution via a wide-necked funnel. About 5 hrs of continuous stirring were required to completely dissolve the polycaprolactam. The highly viscous solution was washed into a 20 l battery jar with 5 × 100 ml of methanol. An additional 5 l of methanol were then added. At this point, 200 g of celite were added. Methanol-water (first 1:1, 3 l and then 3:7, 7 l) was added slowly with vigorous mechanical stirring to produce a voluminous precipitate. This precipitate was removed by filtration onto a Buchner funnel and then washed with cold water until the washings were neutral. After final washings with 4 l of hot water and 4 l of distilled water, the polyamide was ready for column chromatographic use. The polyamide, covered with water, was stored in a stoppered jar.

by polyamide column chromatography of a Baptisia lecontei extract

IIb	III/IV	IVa	Ib	Ia	VIIIa	VII	Vb	Va	XI	XII	XIII	IIIa	VIII	VI	V
+															
+	+	+	+												
+	+	+	+	+	+	+									
+		+	+		+	+	+	+	+	+	+				
+			+						+	+	+	+			
			+						+	+		+	+	+	
			+									+		+	
			+											+	
														+	+
														+	+

collection of each band (detected in UV light) in a separate fraction.

The polyamide material prepared by the above method has a number of desirable properties:

(1) It contains almost no water/methanol soluble monomers and oligomers.
(2) It has a uniform grain size (unlike commercial material prepared by grinding).
(3) It forms a column with a satisfactory flow rate.
(4) It has a high adsorption capacity.

2b. Silica Gel Column Chromatography of Non-polar Type Flavonoids

Silica gel may be used for the separation of relatively non-polar flavonoid aglycones such as isoflavones and methoxylated flavones and flavonols.

Silica Gel Column Chromatographic Separation of *Hymonoxys scaposa* Methoxylated Flavones [5]. The residue (13 g) obtained from the methylene chloride extract of ground *Hymenoxys scaposa* leaf material (196 g) was dissolved in a minimum of chloroform, and applied to the top of a silica gel (Baker Analyzed Reagent) column (4.5 x 40 cm), which had previously been packed in the same solvent. The column was initially eluted with chloroform (500 ml fractions were collected) and was observed periodically under UV light. A green chlorophyll band (orange in UV) was followed by a band which appeared dark under UV light. After fraction 6 had been collected, the polarity of the solvent was increased by the addition of methanol (0.5%). A total of fourteen 500 ml fractions were taken from the column with the latter solvent; the elution of the flavonoids was monitored by thin-layer chromatography on silica gel G using $CHCl_3$:MeOH (15:1) as the developing solvent. Evaporation of fraction 8 gave hymenoxin (I, 76 mg), which was purified by recrystallization from chloroform. Fractions 10–13 yielded scaposin (II, 400 mg), and fraction 15 gave demethoxysudachitin (III, 13 mg), both of which were recrystallized from chloroform-benzene.

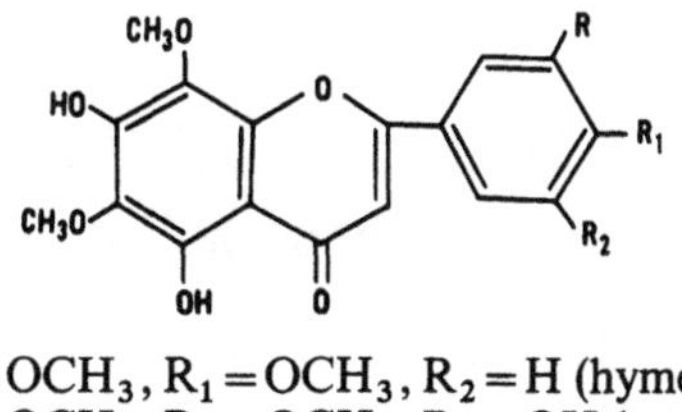

I. $R=OCH_3$, $R_1=OCH_3$, $R_2=H$ (hymenoxin);
II. $R=OCH_3$, $R_1=OCH_3$, $R_2=OH$ (scaposin);
III. $R=H$, $R_1=OH$, $R_2=H$ (demethoxysudachitin)

Silica gel column chromatography is not suitable for the separation of polar flavonoids such as polyhydroxyflavonols or glycosides but does provide a convenient method for the purification of many flavonoid aglycones obtained by the hydrolysis of glycosides. An increase in the methanol content of the eluting solvent will allow the removal of most flavonoid aglycones from silica gel. Isoflavone aglycones can be separated on silica gel by using as eluent chloroform which is gradually increased in polarity by the addition of ether or ethyl acetate. This system separated the isoflavones formononetin (IV), afrormosin (V) and texasin (VI), which were isolated from *Baptisia australis* [6].

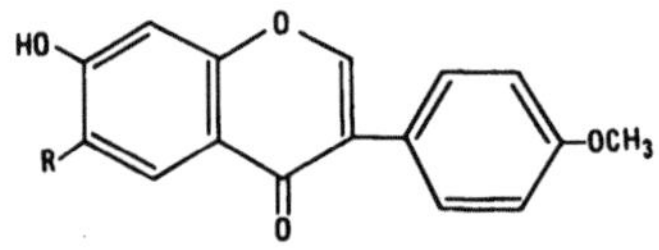

IV. R=H (formononetin);
V. R=OCH_3 (afrormosin);
VI. R=OH (texasin)

2c. The Separation of Flavonoids by Sephadex Column Chromatography

Johnston, Stern and Waiss [7] have described a procedure for the separation of flavonoids, both aglycones and glycosides, on Sephadex LH-20 (available from Pharmacia Inc.) columns using methanol as eluent. Generally the flavonoids were dissolved in methanol and then added to the column; however, in a few instances, a 1:1 dioxane-methanol solution was used to dissolve the flavonoids. To illustrate the effectiveness of the procedure, the separation of a mixture of 166 mg of rutin (VIIa) and 75 mg of quercetin (VIIb) was described. The flavonoids were dissolved in 22 ml of methanol and then added to 40 g column (2.5 x 33 cm) of Sephadex LH-20 (previously packed in methanol). With methanol as eluant and with a flow rate of 4 ml/min, rutin was recovered in the 190–250 ml fraction and quercetin in the 390–460 ml fraction.

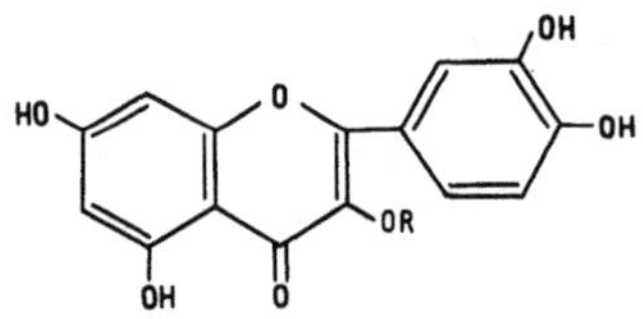

VIIa. R=rutinosyl (rutin);
VIIb. R=H (quercetin)

The authors [7] suggested that the degree of adsorption of flavonoid aglycones onto Sephadex depends generally on the number of hydroxyl groups but not on their acidity while with flavonoid glycosides, with much larger molecular weights, both gel sieving and adsorption are important. Sephadex appears to be an efficient, high capacity medium for both analytical and preparative flavonoid work.

II-3. The Separation of Flavonoids by Silica Gel and Polyamide Thin Layer Chromatography

Thin layer chromatography (TLC) is more commonly used for the analysis of mixtures than for the isolation of pure flavonoids. Polyamide is probably the best TLC adsorbent for all types of flavonoids; however, a number of others [8] (e.g. silica gel G, microcrystalline cellulose and talc) may also be used. The detection of flavonoid spots on thin layer plates may be achieved, as in paper chromatography, by viewing the plate under UV light, with and without the aid of ammonia fumes. A number of adsorbents are now available which contain UV-fluorescent phosphors and these provide a highly sensitive method for the detection of flavonoids. These phosphors are available commercially (Kensington Scientific Corp., California) and may be added to any thin layer adsorbent.

3a. The Separation of Flavonoids by Silica Gel TLC

Silica gel thin layer plates (prepared by standard procedures or purchased commercially as chromatostrips) may be used for the separation of most flavonoid aglycones. For example, the highly methoxylated flavones hymenoxin (I), scaposin (II) and demethoxysudachitin (III) were separated on TLC silica gel G plates with a solvent mixture of chloroform:methanol (15:1) [*5*], and the isoflavones daidzein (VIII), formononetin (IV), genistein (IX) and biochanin A (genistein 4′-methyl ether) were separated on silicic acid chromatostrips using such solvents as ethyl acetate:petroleum ether (3:1 and 1:1) and ethanol:chloroform (1:3 and 1:1) [*9*]. Silica gel TLC of flavonoid glycosides requires a polar solvent such as ethyl acetate:methyl ethyl ketone:formic acid:water (5:3:1:1) [*10*] or benzene:pyridine:formic acid (36:9:5) [*11*].

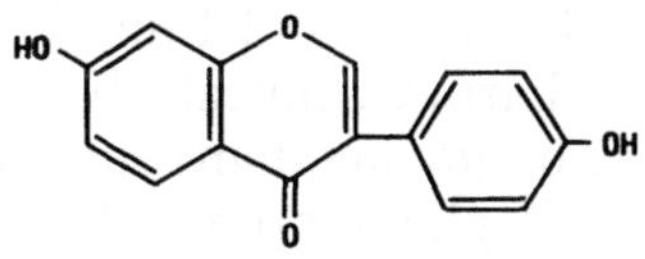

VIII. Daidzein

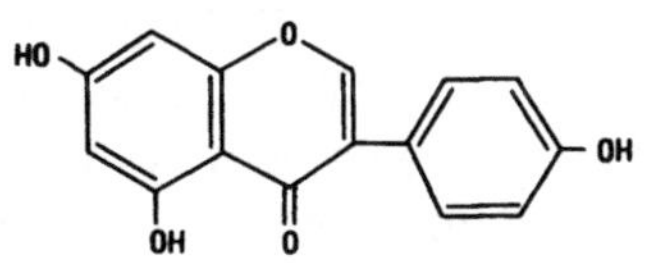

IX. Genistein

3b. The Separation of Flavonoids by Polyamide TLC

By far the most successful adsorbent for the TLC separation of flavonoid glycosides and aglycones is polyamide. A wide variety of commercial polyamides are available differing in chemical composition and in the extent of polymerization. Many of those which are marketed as TLC adsorbents vary widely in their chromatographic properties. Some, due to their chemical composition, are water repellent and are therefore not suitable for use with aqueous solvents and spray-reagents; others do not adhere well to glass plates. Among the many commercially available polyamides, both the Merck and the Macherey and Nagel & Co. polyamides were satisfactory for most flavonoid TLC work. However, we found that an excellent polyamide for the TLC analysis of flavonoids could be prepared by the following procedure, which differs slightly from the procedure previously outlined (Section II-2a) for the preparation of polyamide for use in column chromatography.

(A) Preparation [*4*] of Polyamide for Polyamide TLC Plates[2]. A 3 l, 3-necked round bottom flask containing 1.5 l of about 25% HCl (1 l of conc. HCl plus 500 ml of H_2O) was equipped with a reflux condenser and a powerful stirrer. The solution was refluxed (using a heating mantle) under a fume hood until HCl fumes were no longer lost through the condenser (after about 30 min). The top of the condenser was connected by a glass joint and rubber tubing to a H_2O aspirator. Polycaprolactam pellets (450 g of Durethan BK 40F, Bayer Co., Leverkusen, West Germany) were added to the refluxing and vigorously stirred solution through a wide-mouthed funnel placed in the third neck. By applying suction with the H_2O aspirator, the pellets could be added within 1 – 2 min. The solution was gently refluxed with stirring until all of the pellets dissolved (about 20 min[3]), then the hot solution was quickly transferred to a 20 l Pyrex battery jar and rapidly cooled to room temperature by the addition of small pieces of dry ice; 2.4 l of methanol were then added to the solution. The solution was next vigorously stirred with a mechanical stirrer while tap water was added rapidly until the jar was full. After the

[2] The procedure yields sufficient polyamide to coat about 400 5 × 20 cm plates each with 1 g of polyamide. For fewer plates, the quantities may be proportionally reduced.

[3] An additional heating period equivalent to half the dissolving time should be used with Durethan BK 40 F.

fine white precipitate of polyamide had settled overnight, the supernatant liquid was removed by siphoning. The polyamide was washed to neutrality by repeatedly refilling the jar with tap water. The suspension was filtered onto a Buchner funnel, and the polyamide, which was finally washed with 4 l of distilled water[4], was then ready for use. The material, covered with water, was stored in a stoppered jar.

(B) Preparation of Polyamide TLC Plates. The polyamide (prepared as described above) was slowly filtered onto a Buchner funnel and washed successively with distilled water, methanol and finally thoroughly with ethyl acetate[4] (all traces of water and methanol must be removed). The washed polyamide was shaken vigorously with ethyl acetate to produce a dilute slurry which was poured onto a glass TLC plate (5 x 20 or 10 x 20 cm). The plate, which was about half covered with the slurry, was gently tilted until the material was evenly distributed over the surface. After air drying, the plate was ready for use.

The polyamide prepared by the above procedure gave TLC plates which provided good resolution of most flavonoids (both glycosides and aglycones). One solvent system that is used extensively in our laboratory for polyamide TLC of flavonoids is methanol: acetic acid: water (90:5:5). This solvent has been used successfully with glycosides and aglycones of aurones, chalcones, flavanones, flavones, flavonols and isoflavones. Also, we have found that Egger's solvent [*12*] (chloroform: methanol: butan-2-one; 12:2:1) gives excellent separation of most flavonoids on polyamide TLC plates. Other solvent systems such as methanol, methanol: water (4:1), acetone: water (1:1) and isopropanol: water (3:2), have also been used for the polyamide TLC analysis of certain flavonoids [*8*].

Wender and co-workers [*13*] separated a number of flavanone glycosides by both column chromatography on Polyclar AT (General Aniline and Film Corp., Grasselli, N.J.) and polyamide TLC (Woelm polyamide, Alupharm Chemicals, New Orleans, La.). For the latter, they used a solvent system consisting of nitromethane-methanol (5:2, v/v). For the same flavanones they also employed TLC plates prepared from Avicel SF Technical Grade microcrystalline cellulose (FMC Corporation, American Viscose Division, Marcus Hook, Pa.) with the following developing solvents: benzene-ethyl acetate-formic acid-water (9:21:6:5, v/v/v/v) and n-butanol-acetic acid-water (6:1:2, v/v/v).

References

1. Rösler, H., T.J. Mabry, and J. Kagan: Chem. Ber. **98**, 2193 (1965).
2. Seikel, M.K., in: The chemistry of flavonoid compounds (edited by T.A. Geissman), p. 34–69. Oxford: Pergamon Press 1962.
3. Markham, K.R., and T.J. Mabry: Phytochemistry **7**, 791 (1968).
4. Rösler, H.: Ph. D. Dissertation, University of Munich, Germany (1960). See also Rösler's procedure in H. Wyler, H. Rösler, M. Mercier, and A.S. Dreiding: Helv. chim. Acta **50**, 545 (1967).
5. a) Thomas, M.B., and T.J. Mabry: J. Org. Chem. **32**, 3254 (1967). b) Thomas, M.B., and T.J. Mabry: Tetrahedron **24**, 3675 (1968).
6. a) Lebreton, P., K.R. Markham, W.T. Swift III, Oung-Boran, and T.J. Mabry: Phytochemistry **6**, 1675 (1967). b) Markham, K.R., W.T. Swift III, and T.J. Mabry: J. Org. Chem. **33**, 462 (1968).
7. Johnston, K.M., D.J. Stern, and A.C. Waiss Jr.: J. Chromatog. **33**, 539 (1968).
8. Kirchner, J.G.: Thin Layer Chromatography in: Techniques of organic chemistry series, vol. XII (edited by A. Weissberger), p. 558. New York: Interscience Publishers 1967.
9. Guggolz, J., A.L. Livingston, and E.M. Bickoff: J. Agr. Food Chem. **9**, 135 (1961).
10. Stahl, E., and P.J. Schorn: Hoppe-Seylers Z. Physiol. Chem. **325**, 263 (1961).
11. Hörhammer, L., H. Wagner, and K. Heini: J. Chromatog. **13**, 235 (1964).
12. Egger, K., and M. Keil: Z. Anal. Chem. **210**, 201 (1965).
13. a) Mizelle, J.W., W.J. Dunlap, R.E. Hagen, S.H. Wender, B.J. Lime, R.F. Albach, and F.P. Griffiths: Anal. Biochem. **12**, 316 (1965). b) Hagen, R.E., W.J. Dunlap, J.W. Mizelle, S.H. Wender, B.J. Lime, R.F. Albach, and F.P. Griffiths: Anal. Biochem. **12**, 472 (1965).

[4] The polyamide should not be allowed to dry during these washings!

Chapter III

The Aglycone and Sugar Analysis of Flavonoid Glycosides

III-1. Procedures for the Acidic and Enzymatic Hydrolysis of Flavonoid Glycosides . . . 24

1a. Acidic Hydrolysis of Flavonoid Glycosides . . . 24

(A) Acidic Hydrolysis of Luteolin 7-O-Rhamnoglucoside . . . 24

(B) Acidic Hydrolysis of the O-Xylosyl Linkage in Xylosylisovitexin without Significant Formation of Vitexin . . . 25

1b. Enzymatic Hydrolysis of Flavonoid Glycosides . . . 25

β-Glucosidase Hydrolysis of Luteolin 7-O-β-D-Glucoside . . . 25

III-2. The Gas and Paper Chromatographic Procedures for Identifying the Sugars Obtained by Hydrolysis of Flavonoid Glycosides . . . 26

2a. The Gas Chromatographic Analysis of Sugars Obtained from Flavonoid Glycosides . . . 26

(A) Hydrolysis of the Sugars in Hesperidin and their Trimethylsilylation 26

(B) Gas Chromatography of the Trimethylsilyl Ethers of the Sugars Obtained from Hesperidin . . . 26

2b. The Paper Chromatographic Analysis of the Sugars Obtained from Flavonoid Glycosides . . . 27

Quantitative Paper Chromatographic Analysis of the Sugars in Rutin . . 27

(A) Identification of the Sugars . . . 27

(B) Quantitative Analysis of the Sugars . . . 27

III-3. The Identification of the Aglycone and Location of the Sugar in Flavonoid Glycosides . . . 27

3a. The Identification of the Aglycone in Flavonoid Glycosides . . . 28

(A) The Degradation and Synthesis of the Flavonoid Aglycone, Scaposin 28

(B) The Stereochemistry of Dihydroflavonols . . . 30

3b. The Location of the Sugar in a Flavonoid Glycoside . . . 30

(A) Methylation of Patulitrin and Hydrolysis of the Methyl Derivative . 30

(B) Methylation of 6-Hydroxygenistein 7-O-Rhamnoglucoside and Hydrolysis of the Methyl Derivatives . . . 30

III-4. The Identification of the Sugars in C-Glycosylflavonoids . . . 31

References . . . 32

The procedures discussed in Chapters I and II were concerned with the isolation and purification of flavonoids. If the pure flavonoid is suspected to be a glycoside (e.g. from solubility and paper chromatographic properties), the standard procedure is to hydrolyze it and proceed with the identification of the aglycone and sugar (or sugars) so produced.

III-1. Procedures for the Acidic and Enzymatic Hydrolysis of Flavonoid Glycosides

Two hydrolytic procedures, enzymatic and acidic, are commonly used for the analysis of flavonoid glycosides.

1a. Acidic Hydrolysis of Flavonoid Glycosides

(A) Acidic Hydrolysis of Luteolin 7-O-Rhamnoglucoside (I). Luteolin 7-O-rhamnoglucoside (2 mg) was mixed with 6% aqueous hydrochloric acid (5 ml) using a minimum of methanol to effect complete solution. The solution was heated on a steam bath for 45 min and then cooled and extracted thoroughly by shaking with ether. Evaporation of the aqueous layer yielded rhamnose and glucose (see Section III-2 for procedures used for the identification of the sugars). The ether layer, after drying over sodium sulfate, yielded luteolin on evaporation.

If information is required only about the aglycone, the hydrolytic solution may be taken directly to dryness under high vacuum, and the residue subsequently analyzed by two-dimensional paper chromatography and then by ultraviolet spectroscopy.

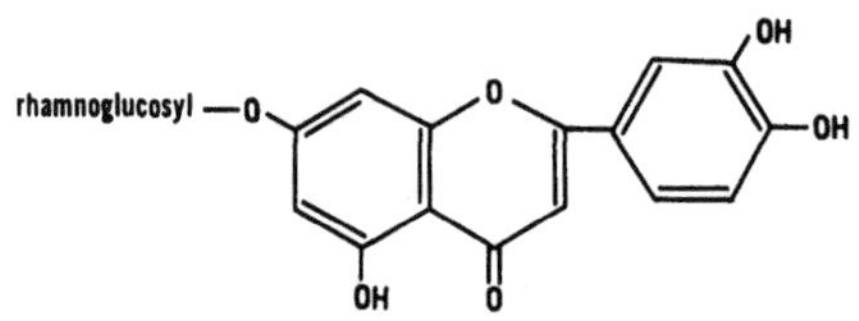

I. Luteolin 7-O-rhamnoglucoside

The above procedure may be used for the hydrolysis of all types of flavonoid O-glycosides. With large samples the aglycone often precipitates from the cooled hydrolysis mixture and may be removed by filtration. Ether extraction of the filtrate will yield the last traces of the aglycone. If the amount of glycoside hydrolyzed is small, a suitable work-up procedure is to evaporate the hydrolysis solution to dryness under high vacuum, and chromatograph the residue on a small polyamide column [see Procedure (A), Section III-2a below]. Elution of the column with a few milliliters of water removes the sugars, and the aglycone can then be eluted with methanol. Paper chromatography can also be used to separate the aglycone and sugars obtained from a small scale acidic hydrolysis. For example, with a typical sugar solvent such as ethyl acetate:pyridine:water, 12:5:4, the flavonoid aglycone will have an R_f value of almost zero on a paper chromatogram while the sugars will have relatively high R_f values.

Since all of the commonly encountered flavonoid O-glycosides will hydrolyze under the acidic hydrolysis conditions described above, those flavonoid glycosides that do not hydrolyze are almost certainly of the C-glycosyl type. (See Section III-4 for a procedure which cleaves C-glycosyl groups from the flavonoid nucleus.) A number of less drastic acid hydrolysis procedures, involving the use of more dilute mineral acids [*1*], formic acid in cyclohexane, or 10% acetic acid, often at lower temperatures, are also available for the partial hydrolysis of di- and triglycosides [*2*]. The relative rates of hydrolysis of sugars attached at different positions on the flavonoid nucleus are significantly different, and this can be useful in the analysis of di- and triglycosides [*1*]. Flavonols having glucuronic acid or glucose attached to the 7-hydroxyl group are readily distinguished from those having a 7-O-rhamnosyl moiety by their resistance to hydrolysis with 1 N HCl; the times required for the complete hydrolysis were 180, 25 and 5 min, respectively [*1*].

Side reactions occur with a few flavonoids during acid hydrolysis. For example, when chalcone glycosides hydrolyze, the aglycone produced is usually the equivalent flavanone (or a mixture of flavanone and chalcone) [*3*]. Significantly, both 6- and 8-C-glycosyl flavonoids interconvert [*4, 5*] (i.e. Wessely-Moser rearrangement) to a mixture of the

6- and 8-isomers under the standard acid hydrolysis conditions. This interconversion can be avoided by using mild hydrolysis conditions such as those outlined below for the hydrolysis of xylosylisovitexin (II) [5a].

(B) Acidic Hydrolysis of the O-Xylosyl Linkage in Xylosylisovitexin (II) without Significant Formation of Vitexin (III) [5a]. Xylosylisovitexin (2 mg) was dissolved in 10% acetic acid (20 ml) and allowed to stand at room temperature for 18 hr[1]. Removal of the solvent under high vacuum at 40–50° gave a residue which was subsequently paper chromatographed. The single spot detected on the two-dimensional chromatogram was chromatographically and spectrally indistinguishable from isovitexin.

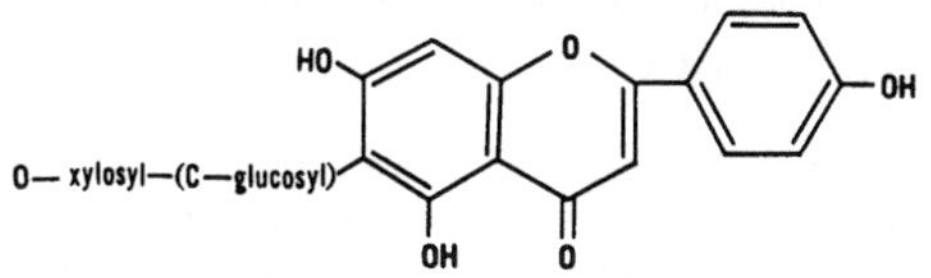

II. Xylosylisovitexin

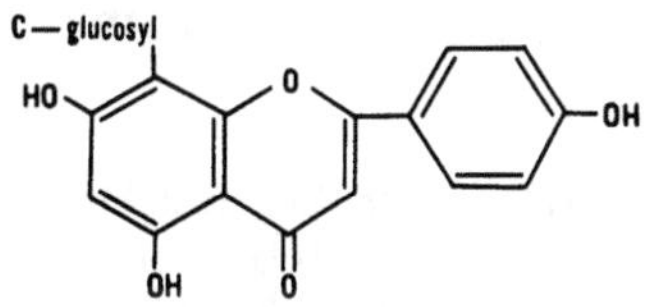

III. Vitexin

Flavonoids containing isoprenoid side-chains or other uncommon substituents may also be altered by acid treatment, and thus, special care may be necessary when dealing with these unusual flavonoids.

1b. Enzymatic Hydrolysis of Flavonoid Glycosides

Enzymatic hydrolysis is limited by the availability of purified enzymes. β-Glucosidase is the most commonly employed enzyme and is widely used since about half of the naturally occurring flavonoid glycosides are β-D-glucosides. β-Glucosidase will only hydrolyze a terminal glucose unit and not an in-the-chain glucose. A readily available semipure form of β-glucosidase is emulsin and although this is usually satisfactory, in some instances it has been found to hydrolyze glycosides other than glucosides (notably galactosides). Other enzymes which are available for hydrolytic studies include β-glucuronidase and anthocyanase [1].

β-Glucosidase Hydrolysis of Luteolin 7-O-β-D-Glucoside (IV). The luteolin glucoside (1 mg) was dissolved in 2 ml of pH 5 buffer (an aqueous 0.5 M NaOAc solution adjusted to pH 5 with HOAc) and about 1 mg (usually a small amount of the powdered enzyme on the tip of a spatula is added) of β-glucosidase (Sigma Corp., St. Louis, Mo.) was added. The mixture was allowed to stand overnight at 37°. When the solution was concentrated under high vacuum and paper chromatographed, only luteolin was detected.

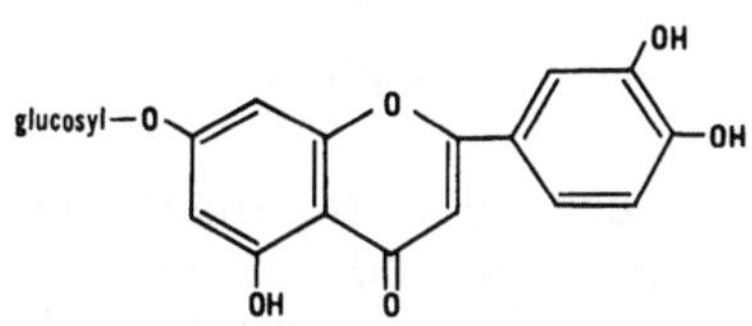

IV. Luteolin 7-O-glucoside

The enzymatic hydrolysis procedure can also be used in conjunction with the mild acid hydrolysis conditions for the analysis of di- and triglycosides. For example, once the rhamnose has been removed from rutin (quercetin 3-O-rhamnoglucoside) with formic acid in boiling cyclohexanol [6], the linkage of the glucose to the aglycone can then be determined by hydrolysis with β-glucosidase. We and others [1] have observed that although most flavonoid 3-,4′- and 7-O-glucosides are readily hydrolyzed by β-glucosidase (sometimes only a few minutes are required), some flavonol 3-O-β-D-glucosides (e.g. sophorosides) require more than 24 hrs at 37°.

[1] These conditions are not sufficiently vigorous to hydrolyze all O-glycoside linkages. Indeed, in one instance, we found that the hydrolysis of an O-rhamnosyl group attached to the C-glycosyl portion of a 6-C-glycosylflavone did not occur even with refluxing 10% acetic acid.

III-2. The Gas and Paper Chromatographic Procedures for Identifying the Sugars Obtained by Hydrolysis of Flavonoid Glycosides

In addition to the enzymatic procedures described above, the other two methods of sugar analysis which we have employed are paper and gas-liquid chromatography.

2a. The Gas Chromatographic Analysis of Sugars Obtained from Flavonoid Glycosides

(A) Hydrolysis of the Sugars in Hesperidin and their Trimethylsilylation [7]. In a typical experiment, 0.5 mg of hesperidin (V) was hydrolyzed in 5 ml of 2 N HCl[2]. The solution was heated on a steam bath for 30 min. (Occasionally, the complete hydrolysis of other compounds required a longer hydrolysis time.)

The reaction mixture was passed through a 2 × 1 cm column of polyamide powder[3] (Polypenco 66D from the Polymer Corp., Reading, Pa.), packed in water. Flavonoids and other phenolic compounds are strongly adsorbed by hydrogen bonding to this support. The column was washed with water until the eluate was neutral. This solution contained the sugars rhamnose and glucose. Subsequent elution of the column with methanol afforded the aglycone, hesperetin.

The acidic aqueous sugar solution was taken to dryness at room temperature under high vacuum. The residue was dissolved in 0.5 ml of dry pyridine and the trimethylsilyl ethers were prepared by the successive addition of 0.2 ml of hexamethyldisilazane and 0.1 ml of trimethylchlorosilane. After the excess solvents and reagents were removed under high vacuum, a few drops of dry heptane were added. The white, insoluble material was filtered off and the clear solution was concentrated to about 0.1 ml for gas chromatography. Upon injection of this concentrated clear solution into the gas chromatograph, a straight base line was displayed almost immediately after the appearance of the solvent peaks.

(B) Gas Chromatography of the Trimethylsilyl Ethers of the Sugars Obtained from Hesperidin [7]. A Research Specialties model 600 gas chromatograph with an argon ionization detector was equipped with a 6-foot × 0.25 inch U-shaped stainless steel column packed with acid washed silanized chromosorb W coated with 3% SE-52 (Applied Science Labs, State College, Pa.). Retention times of 4 and $5\frac{1}{2}$ min for α- and β-rhamnose, and 12 and 18 min for α- and β-glucose, were obtained using a column temperature of 180° and an inlet pressure of 15 p.s.i. These identifications were confirmed by cochromatography with authentic rhamnose and glucose trimethylsilyl ethers.

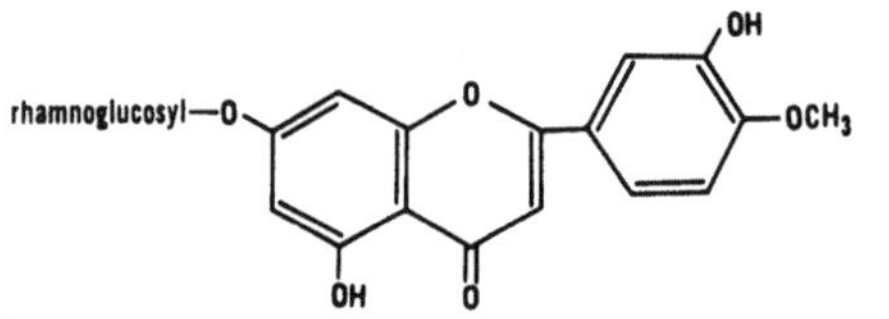

V. Hesperidin

The gas chromatographic procedure for the sugar analysis is often less convenient to use than the paper chromatographic method, but offers greater sensitivity. For example, Kagan and Mabry report [7] the limit of sensitivity of the gas chromatographic procedure at somewhat less than 0.5 μg of sugar, which can be compared with a limit of about 5 μg reported by Pridham [8] for his paper chromatographic method. Gas chromatography also provides more reliable identifications because the gas chromatographic identity of each sugar depends upon positions of two peaks (the α- and β-anomeric forms of the sugar) rather than the single spot identity used in paper chromatography.

Kits containing trimethylsilyl ethers of the common sugars required for comparison in the gas chromatographic analyses are now commercially available (Pierce Chemical Company, Rockford, Illinois). Since the trimethylsilyl ether derivatives of the sugars are rapidly hydrolyzed, contact with water must be avoided.

[2] In the original procedure [7], we often used a small amount of methanol to effect complete solution of the flavonoid; however we have since found that the gas chromatogram displays only the peaks for the sugars when the glycoside is hydrolyzed without added methanol.

[3] Other polyamide powders, e.g. Polyclar AT, General Aniline and Film Corp., can also be used.

2b. The Paper Chromatographic Analysis of the Sugars Obtained from Flavonoid Glycosides

The paper chromatographic method described here for the identification of the sugars obtained on hydrolysis of a flavonoid glycoside is essentially that of Pridham [*8*]. The chief advantage of the method over gas chromatography is that it requires inexpensive equipment and can readily be made quantitative.

Quantitative Paper Chromatographic Analysis of the Sugars in Rutin

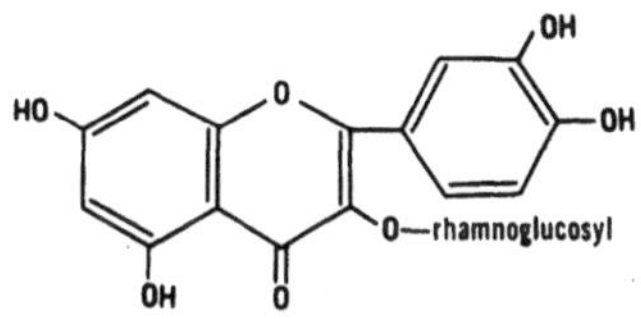

VI. Rutin

(A) Identification of the Sugars. After the acidic hydrolysis (see Section III-1a, procedure A) of about 0.5 mg of rutin (VI) the resultant sugar/aglycone mixture was paper chromatographed ascendingly in a solvent of ethyl acetate:pyridine:water, 12:5:4, alongside some of the more common sugars such as glucose and rhamnose. The chromatogram was dried and sprayed uniformly with a solution of p-anisidine hydrochloride (1 g) and sodium hydrosulfite (0.1 g) in methanol (10 ml) diluted to 100 ml with n-butanol. The sprayed chromatogram was then heated at 130° in an oven for 10 min during which time the sugar spots turned brown; the R_f values for the colored spots of the unknown sugars were compared directly with those of the known sugars. (The R_f value for the sugar in a sugar/aglycone mixture may be slightly lower than that of an authentic sample.)

(B) Quantitative Analysis of the Sugars. After the sugars had been determined qualitatively as rhamnose and glucose, the following procedure gave the aglycone/sugar ratio. A known weight of rutin was hydrolyzed with acid (since only 0.1 – 0.2 mg of the glycoside is required, the weight used is best determined by UV spectroscopy). The resultant sugar/aglycone mixture was paper chromatographed (as described in procedure A above) alongside three different quantitative amounts of both rhamnose and glucose. After the chromatogram had been sprayed and heated each colored sugar spot was cut out, the size of the paper pieces being kept constant for each sugar series. A blank of equal size was also cut out from the same chromatogram. Each spot, including the blank, was eluted by shaking the paper section for 5 min with a 3 ml aliquot of a solution of 1% stannous chloride in 5% aqueous methanol. The optical density of each of these extracts was measured on a spectrophotometer at 510 nm (for aldopentoses) and 400 nm (for aldohexoses). In the present experiment, all extracts were measured at 400 nm.

Next, a graph of optical density versus weight of glucose was prepared for the three spots for which the weight of sugar was known; these values produced a straight line curve which was used to read off directly the weight of glucose in the sample being analyzed. This weight when compared with the weight of rutin originally hydrolyzed allowed the calculation of the aglycone/glucose ratio. In a similar manner the aglycone/rhamnose ratio was determined.

Pridham's method, in our experience, gave sugar values which were about 10% lower than theoretical. Such an error, however, still permits the distinction between mono-, di- and triglycosides. It is important to prepare an optical density/sugar weight graph for every chromatographic run because of the difficulty of achieving comparable results in different determinations.

Considerable information about the sugar moiety in a flavonoid glycoside can also be obtained directly (without hydrolyzing the glycoside) by NMR spectroscopy (see Section VIII-4d).

III-3. The Identification of the Aglycone and Location of the Sugar in Flavonoid Glycosides

Once the sugar moiety has been identified in a flavonoid glycoside, it is necessary to identify the aglycone and to determine the position of attachment of the sugar or sugars to the flavonoid nucleus.

3a. The Identification of the Aglycone in Flavonoid Glycosides

Identification of the aglycone obtained on hydrolysis of a flavonoid glycoside involves such data as R_f values, spot appearance on a paper chromatogram, UV spectra in a number of diagnostic reagents, and the NMR spectrum. Much of the required UV and NMR spectral data will be found in Chapters IV–IX.

Only rarely have we found it necessary to degrade or synthesize a flavonoid in order to determine its structure; therefore this aspect of flavonoid structure elucidation is not emphasized here. Nevertheless, synthesis and/or degradation are important procedures for the structure analysis of certain flavonoid aglycones as well as some glycosides. One example from our investigations, the degradation and synthesis of the flavonoid aglycone scaposin [*9*], is discussed below in part A.

We used ORD (optical rotatory dispersion) and CD (circular dichroism) studies to determine the absolute configuration at C-2 and C-3 in two dihydroflavonol glycosides [*14*]. Since the results concern the absolute structure of the aglycone moiety, we briefly discuss the data in part B.

(A) The Degradation and Synthesis of the Flavonoid Aglycone Scaposin (VII)

On the basis of NMR and UV spectral data, scaposin was assigned a trihydroxy-tetramethoxyflavone structure which contained a hydroxyl group at C-5 and the other substituents (two hydroxyl and four methoxyl groups) at the 3′-,4′-,5′-,6-,7- and 8-positions. However, it was not possible to determine the location of the other two hydroxyl groups from the UV and NMR data and for this reason the structure proof of scaposin required degradation and/or synthesis.

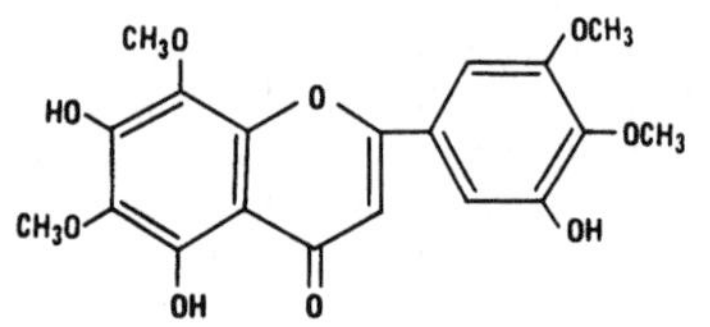

VII. Scaposin

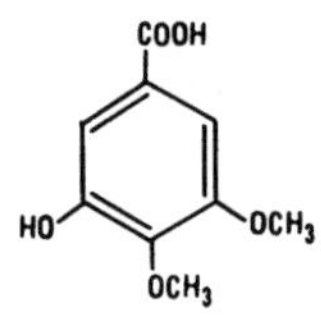

VIII. 3,4-di-O-methylgallic acid

Alkaline Degradation of Scaposin (VII). The B-ring substitution pattern in scaposin was shown to be 5′-hydroxy-3′,4′-dimethoxy by alkaline degradation of scaposin to 3,4-di-O-methylgallic acid (VIII). Scaposin (126 mg) was refluxed with 50% aqueous potassium hydroxide (10 ml) containing a few drops of methanol for 16.5 hr under nitrogen. The solution was cooled, acidified with 3 N HCl and extracted with ether (2 × 15 ml). The ethereal solution was extracted with 8% aqueous sodium bicarbonate (25 ml) and the alkaline extract was acidified and extracted with ether (2 × 15 ml). After removal of the solvent, the residue (about 20 mg) was crystallized from hot water; white needles were obtained which had NMR and IR spectra and m.p. and m.m.p. identical with a sample of 3,4-di-O-methylgallic acid (VIII) prepared by the method of Späth and Röder [*10*]. At this stage of the structure determination of scaposin only the A-ring substitution pattern remained in doubt. The suspected structure VII was shown to be correct by a total unambiguous synthesis of scaposin [*9*].

Synthesis of Scaposin

The final proof of structure of scaposin was established by the following synthesis [*9*]. 4-Benzyloxy-2,5-dihydroxy-3,6-dimethoxyacetophenone [*11*] (IX) was converted to XI on treatment with 5-benzyloxy-3,4-dimethoxybenzoyl chloride [*10*, *12*] (X) in pyridine. Rearrangement to the dibenzoylmethane derivative XII (Baker-Venkataraman transformation [*13*]) followed by ring closure with ethanolic sulfuric acid afforded the flavone XIII. All attempts to crystallize XIII failed but its NMR spectrum (on material obtained from preparative TLC) indicated that ring closure to the flavone had been achieved [singlet at 6.78 ppm (H-3)]. Saponification of XIII followed by methylation gave 5′,7-dibenzyloxy-3′,4′,5,6,8-pentamethoxyflavone (XIV). Debenzylation of XIV and concurrent selective demethylation of the 5-methoxyl group using refluxing acetic acid/HCl yielded 5,5′,7-trihydroxy-3′,4′,6,8-tetramethoxyflavone, which was identical in all respects with scaposin.

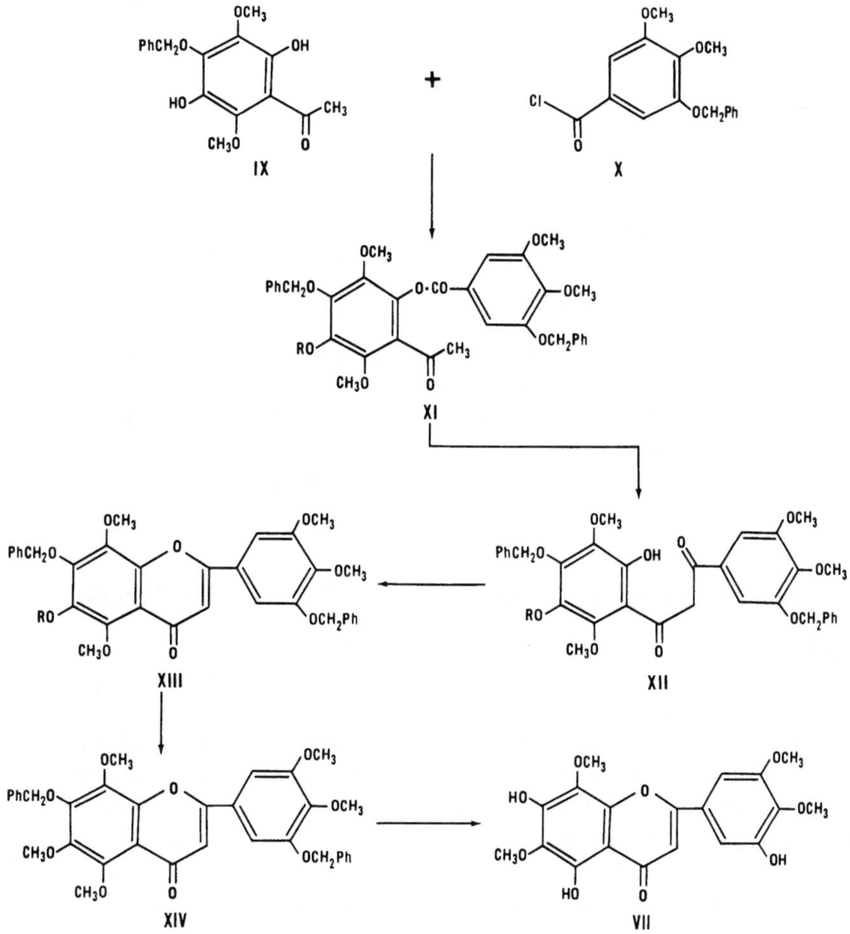

R = 5-benzyloxy-3,4-dimethoxybenzoyl in XI, XII and XIII

4-Benzyloxy-3,6-Dimethoxy-2,5-Di(5-Benzyloxy-3,4-Dimethoxybenzoyloxy)Acetophenone (XI). 5-Benzyloxy-3,4-dimethoxybenzoyl chloride (X) (2.7 g) and IX [*11*] (0.8 g) were heated on a steam bath in pyridine (3 ml) for 30 min. The mixture was poured into 5% HCl (20 ml), and the oil which separated was extracted with chloroform (2 × 10 ml). The residue obtained on evaporation of the chloroform solution crystallized from chloroform-methanol as colorless prisms, yield 1.2 g, m.p. 146–147°.

5′,7-Dibenzyloxy-6-(5-Benzyloxy-3,4-Dimethoxybenzoyloxy)-3′,4′,5,8-Tetramethoxyflavone (XIII). A mixture of XI (500 mg), powdered potassium hydroxide (ca. 60 mg) and pyridine (3 ml) was heated at 60° for 3 hr with stirring. The mixture was poured into 2.5% HCl (15 ml) and the oil which separated was extracted with chloroform. The residue obtained on evaporation of the chloroform was refluxed with 2.5% ethanolic sulfuric acid (8 ml) for 1 hr. Purification was achieved by preparative TLC.

5′,7-Dibenzyloxy-3′,4′,5,6,8-Pentamethoxyflavone (XIV). A mixture of the crude material (XIII) from the previous reaction (about 450 mg) and 1 N sodium methoxide (6 ml) was refluxed for 30 min, acidified with acetic acid and evaporated to dryness under high vacuum. The residue was refluxed in acetone (10 ml) containing dimethyl sulfate (0.8 ml) and potassium carbonate (2 g) for 1 hr. The inorganic salts were then filtered off and the solvent removed from the filtrate. The residue crystallized from methanol as colorless needles, m.p. 134° (yield 72 mg).

5,5′,7-Trihydroxy-3′,4′,6,8-Tetramethoxyflavone (Scaposin, VII). Compound XIV (30 mg) was refluxed with acetic acid (1.5 ml) and HCl (1.5 ml) for 2 hr. The product was steam distilled and the residue extracted with chloroform. The flavone obtained from the chloroform extract crystallized from chloroform-benzene (1:1) as fine yellow needles, yield 7 mg (35%) m.p. and m.m.p. with natural scaposin 210–212°. The IR and UV spectra and TLC behavior were identical with those observed for scaposin (VII) isolated from *Hymenoxys scaposa* [9].

(B) The Stereochemistry of Dihydroflavonols

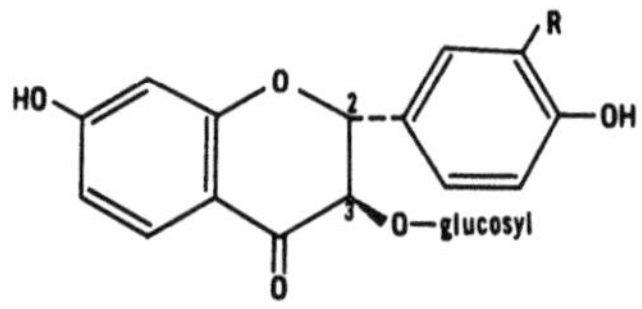

XVa. R=H (lecontin);
XVb. R=OH [(+)-fustin 3-O-β-D-glucoside]

The absolute configuration at C-2 and C-3 in both lecontin (XVa) and (+)-fustin 3-O-glucoside (XVb) were shown to be 2R:3R by ORD (the spectra arè presented elsewhere [14]) and CD studies [14]. Furthermore, it was observed that the dihydroflavonol aglycone from XVa gave an ORD curve with the same general form as XVa; thus the C-3 glucosyl group appears to have no effect on the sign and shape of the ORD curve of the dihydroflavonols. Finally, additional ORD and CD studies suggested that all simple polyoxygenated dihydroflavonols (and their 3-O-glycosides) that possess the trans 2R:3R absolute configuration will give CD and ORD curves which show four Cotton effects in the order (+), (−), (+), (+) from 400 to 200 nm [14].

3b. The Location of the Sugar in a Flavonoid Glycoside

Two of the simplest methods for obtaining information about the position of attachment of a sugar to a flavonoid glycoside are (i) by comparing the UV spectra (in various diagnostic reagents) of a glycoside with those of its aglycone (see Chapters IV – VII) and (ii) by examining the NMR spectrum of the trimethylsilyl ether of the glycoside (see Chapters VIII and IX). A more reliable procedure involves complete methylation of the free phenolic hydroxyl groups before hydrolysis of the glycoside. Any hydroxyl groups free (as determined by UV spectroscopy) after hydrolysis of the methylated flavonoid will represent sites of glycosylation in the original flavonoid. Complete methylation of free phenolic hydroxyl groups in flavonoid glycosides is usually carried out by methods such as those described below.

(A) Dimethyl Sulfate Methylation of Patulitrin and Hydrolysis of the Methyl Derivative [15]. A solution of patulitrin (XVI, 45 mg) in dry acetone (25 ml) was mixed with dimethylsulphate (0.5 ml) and oven-dried potassium carbonate (1.5 g). The mixture was refluxed for 24 hrs with a slow stream of dry nitrogen passing through the apparatus during the reflux. The cooled solution was filtered and evaporated under high vacuum to give a yellow oil, which was then refluxed with 7% sulfuric acid (25 ml) for 1 hr. The hydrolysate yielded an aglycone on repeated ether extraction. Recrystallization of the material obtained from the ether gave 7-hydroxypentamethoxyquercetagetin (XVII), m.p. 234 – 235° (20 mg).

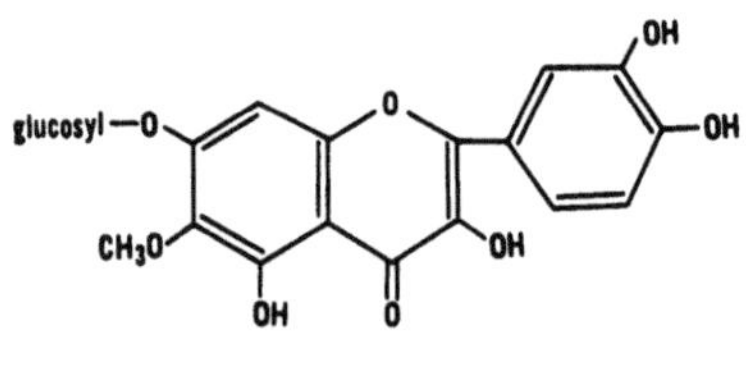

XVI. Patulitrin

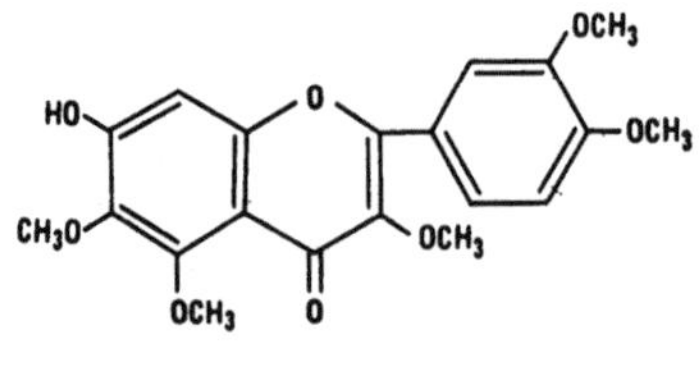

XVII. 7-Hydroxypentamethoxyquercetagetin

(B) Diazomethane Methylation of 6-Hydroxygenistein 7-O-Rhamnoglucoside and Hydrolysis of the Methyl Derivatives [16]. Diazomethane is a good reagent for methylating hydroxyl groups in flavonoids with the exception of the C-5 hydrogen-bonded hydroxyl group. 6-Hydroxygenistein 7-O-rhamnoglucoside (XVIII, 4 mg) dissolved in reagent-grade methanol (3 ml) was treated in excess with a solution of diazomethane in diethyl ether (prepared by the standard procedure [17] from diazald, p-toluene-N-dimethylsulfonamide). The solution was stored overnight at room temperature in a loosely stoppered flask. More of the diazomethane solution was added the next morning and still more, 8 hrs later. The solution was again left overnight. Vacuum evaporation of the product gave an oil which was then acid hydrolyzed. The resultant aglycone material was isolated as described in procedure A above. Thin layer chromatographic analysis on polyamide (see Section II-3b) of the products showed the presence of two compounds as a 1:1 mixture. The products were subsequently shown to be irisolidone (XIX) and 4',5,6-trimethoxy-7-hydroxyisoflavone (XX).

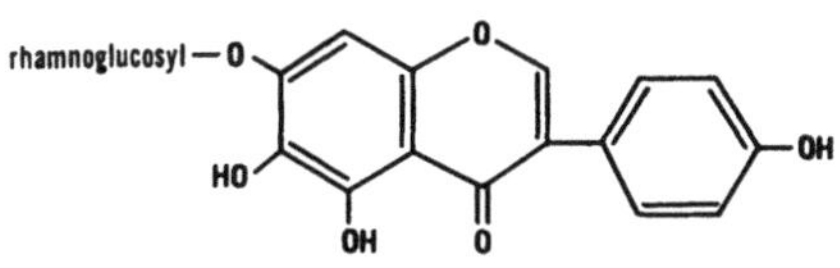

XVIII. 6-Hydroxygenistein 7-rhamnoglucoside

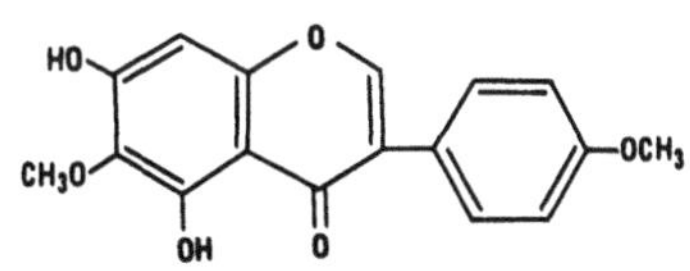

XIX. Irisolidone

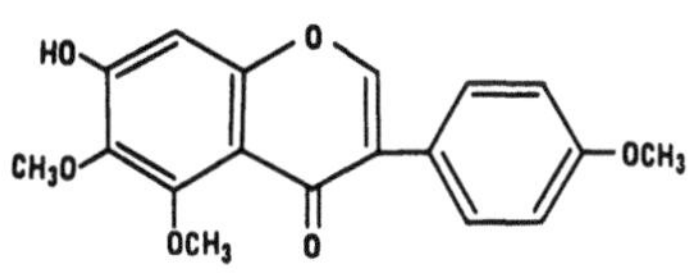

XX. 4',5,6-Trimethoxy-7-hydroxyisoflavone

The rate of methylation of flavonoid hydroxyl groups varies considerably. For example, the acidic 3,7, and 4'-hydroxyl groups methylate readily under conditions which do not methylate the strongly hydrogen-bonded 5-hydroxyl group. The progress of the methylation reaction can conveniently be monitored by UV spectroscopy or by thin layer chromatography. Once the methylated aglycone is obtained pure from the reaction mixture, the free hydroxyl group (or groups) can be located by comparison with authentic material and/or by UV spectral analysis. The UV spectral analysis is particularly valuable since it can be used to detect free hydroxyl groups at the 3,5,7 and 4'-positions on the flavonoid nucleus, and also at the 3',6 and 8-positions if these form part of an ortho-dihydroxyl group (see Chapters V, VI and VII for further details).

NMR spectroscopy may also be used for identifying the methylated aglycone. One procedure involves comparison of the NMR spectrum of the methylated aglycone with that of the product obtained by first acetylating and then methylating the aglycone. Acetylation of the free hydroxyl group in the aglycone obtained from the methylated glycoside causes downfield shifts of about 0.3, 0.15 and 0.5 ppm, respectively, in the signals of aromatic protons which are ortho, meta and para positions to the acetylated hydroxyl group [*18*].

III-4. The Identification of the Sugars in C-Glycosylflavonoids

A large number of C-glycosylflavonoids are now known [*5b*] and several techniques including paper chromatography and NMR and UV spectroscopy have proved of value in identifying them. If, however, the chromatographic, NMR and UV data (as discussed in other sections of this book) do not provide the identity of the sugar moiety in a C-glycosylflavonoid, then it may be necessary to cleave the sugar from the flavone nucleus and then analyze the sugar fragment by standard procedures. Koeppen and Roux [*19*] were able to obtain glucose mixed with arabinose by the ferric chloride oxidation of either orientin (XXI), isoorientin (XXII) or their 3',4',5,7-tetra-O-methyl ethers. Their procedure (which was a modification of one previously employed by Hay and Haynes [*20*]) is outlined below.

The Ferric Chloride Oxidation of Orientin (XXI). About 25 mg of orientin and 0.2 g of $FeCl_3$ in 0.8 ml of water were refluxed for 6 hr. The mixture was cooled, the pH adjusted to 8.0 with an aqueous solution of NaOH, and a precipitate was removed by centrifugation. The pH of the supernatant was adjusted to 7.0 with aqueous

HCl. The solution was then desalted in Shandon electrolytic desalting apparatus (Mark II). The salt-free solution was concentrated to a small volume (about 0.5 ml) and examined by paper chromatography. Approximately a 1:1 mixture of glucose and arabinose was detected by paper chromatography in benzene-butan-1-ol-pyridine-water (1:5:3:3; by vol.), in butan-1-ol-acetic acid-water (20:5:11, by vol.) and in aqueous 75% (w/w) phenol.

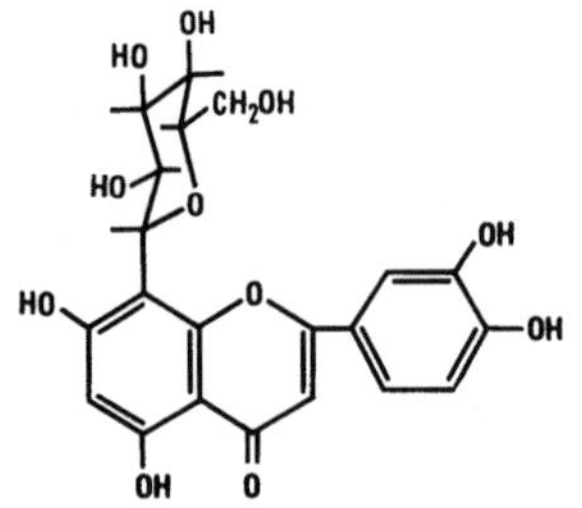

XXI. Orientin

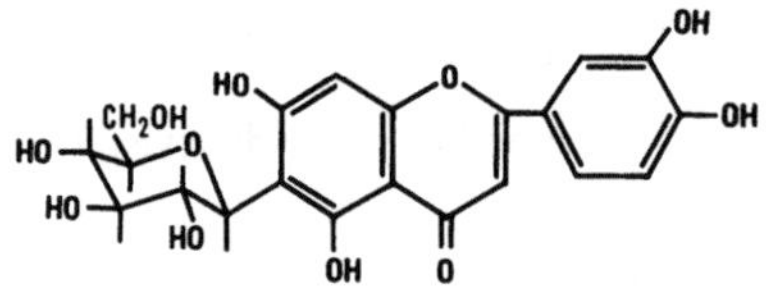

XXII. Isoorientin

References

1. Harborne, J. B.: Phytochemistry **4**, 107 (1965).
2. Venkataraman, K., in: The chemistry of flavonoid compounds (edited by T. A. Geissman), p. 99. Oxford: Pergamon Press 1962.
3. Seshadri, T. R., in: The chemistry of flavonoid compounds (edited by T. A. Geissman), p. 159. Oxford: Pergamon Press 1962.
4. Seikel, M. K., and T. A. Geissman: Arch. Biochem. Biophys. **71**, 17 (1957).
5. a) Krochewsky, B., in: Investigations of flavonoids in the Genus *Tragopogon*. Ph. D. Dissertation, University of Texas (1967). b) Alston, R. E., in: Recent advances in phytochemistry (edited by T. J. Mabry, R. E. Alston, and V. C. Runeckles), p. 305–327. New York: Appleton-Century-Crofts 1968.
6. Fox, D. W., W. L. Savage, and S. H. Wender: J. Am. Chem. Soc. **75**, 2504 (1953).
7. Kagan, J., and T. J. Mabry: Anal. Chem. **37**, 288 (1965).
8. Pridham, J. B.: Anal. Chem. **28**, 1967 (1956).
9. Thomas, M. B., and T. J. Mabry: Tetrahedron **24**, 3675 (1968).
10. Späth, E., and H. Röder: Monatsh. **43**, 93 (1922).
11. Farkas, L., M. Nogradi, V. Sudarsanam, and W. Herz: J. Org. Chem. **31**, 3228 (1966).
12. Farkas, L., M. Nogradi and J. Strelisky: Chem. Ber. **99**, 3218 (1966).
13. Baker, W.: J. Chem. Soc. 1381 (1933). – Mahal, H. S., and K. Ventakaraman: J. Chem. Soc. 1767 (1934).
14. Markham, K. R., and T. J. Mabry: Tetrahedron **24**, 823 (1968).
15. Thomas, M. B., and T. J. Mabry: Phytochemistry **7**, 787 (1968).
16. Markham, K. R., T. J. Mabry, and W. T. Swift, III: Phytochemistry **7**, 803 (1968).
17. Boer, Th. J. de, and H. J. Backer: Org. Syn. **36**, 16 (1956); Col. Vol. IV, p. 225, 1963.
18. Highet, R. J., and P. F. Highet: J. Org. Chem. **30**, 902 (1965).
19. Koeppen, B. H., and D. G. Roux: Biochem. J. **97**, 444 (1965).
20. Hay, J. E., and L. J. Haynes: J. Chem. Soc. 3141 (1956).

Part II

The Structure Analysis of Flavonoids by Ultraviolet Spectroscopy

Each of the 175 flavonoids examined in the present study is represented by six ultraviolet spectra, and these are presented at the end of the appropriate chapter. Each set of six spectra contains one in methanol and five obtained by adding diagnostic reagents to the flavonoid in methanol. Our data complement previous investigations and reviews of the UV spectral analysis of flavonoids such as that of L. Jurd [in *The Chemistry of Flavonoid Compounds* (edited by T. A. Geissman) pp. 107 – 155, Pergamon Press, Oxford (1962)], who reviewed the literature prior to 1962 and assembled the UV spectral data available at that time.

Chapter IV

Reagents and Procedures for the Ultraviolet Spectral Analysis of Flavonoids

IV-1. Preparation of Reagent Stock Solutions and Solids 35
IV-2. Procedures for Determining the Ultraviolet Absorption Spectra of Flavonoids 35
Steps in the UV Spectral Analyses 36

IV-1. Preparation of Reagent Stock Solutions and Solids

Sodium methoxide (NaOMe). Freshly cut metallic sodium (2.5 g) was added cautiously in small portions to dry spectroscopic methanol (100 ml). The solution was stored in a glass container with a tightly fitting plastic stopper.

Aluminum chloride ($AlCl_3$). Five grams of fresh anhydrous reagent grade $AlCl_3$ (which appeared yellow-green and reacted violently when mixed with water) were added cautiously to spectroscopic methanol (100 ml). formed initially, dissolved after about 24 hrs.)

Hydrochloric acid (HCl). Concentrated reagent grade HCl (50 ml) was mixed with distilled water (100 ml); the solution was stored in a glass stoppered bottle.

Sodium acetate (NaOAc). Anhydrous powdered reagent grade NaOAc was used[1].

Boric acid (H_3BO_3). For Procedure I: Anhydrous powdered reagent grade H_3BO_3 was used.

For Procedure II: Spectroscopic methanol (100 ml) was saturated with anhydrous reagent grade H_3BO_3.

When the stock solutions were prepared and stored as described above, they had a shelf life of about 6 months. For convenience as well as to avoid excessive exposure of the stock solutions to the atmosphere, four 30 ml dropping bottles, each containing about 15 ml of one of the stock solutions, were kept near the spectrophotometer. The solutions in the dropping bottles were used for the spectral analyses and were always replaced monthly.

IV-2. Procedures for Determining the Ultraviolet Absorption Spectra of Flavonoids

The methods used for all the UV spectra presented in this volume are outlined below. All spectra were measured on a Beckman DB-G spectrophotometer equipped with a Sargent model SRL recorder. The wavelength calibration of the spectrophotometer was carried out with a Holmium Oxide Filter (supplied by Beckman Instruments), which has λ_{max}'s at 279.3, 287.6, 333.8, 360.8, 418.5, 536.4 and 637.5 nm. For convenience, spectroscopic grade methanol without added reagent was used as reference. It is useful (but not necessary) to have available four matched standard silica cuvettes of 1 cm path length in addition to the reference cuvette.

---

[1] The presence or absence of H_2O was not critical to the spectrum; however, the traces of HOAc which are often present in anhydrous reagent grade NaOAc can markedly affect Band I (see Fig. IV-3) but usually have little or no effect on Band II. Since we were only interested in the Band II data from the NaOAc spectrum, we have not used HOAc-free (fused) NaOAc in this compilation.

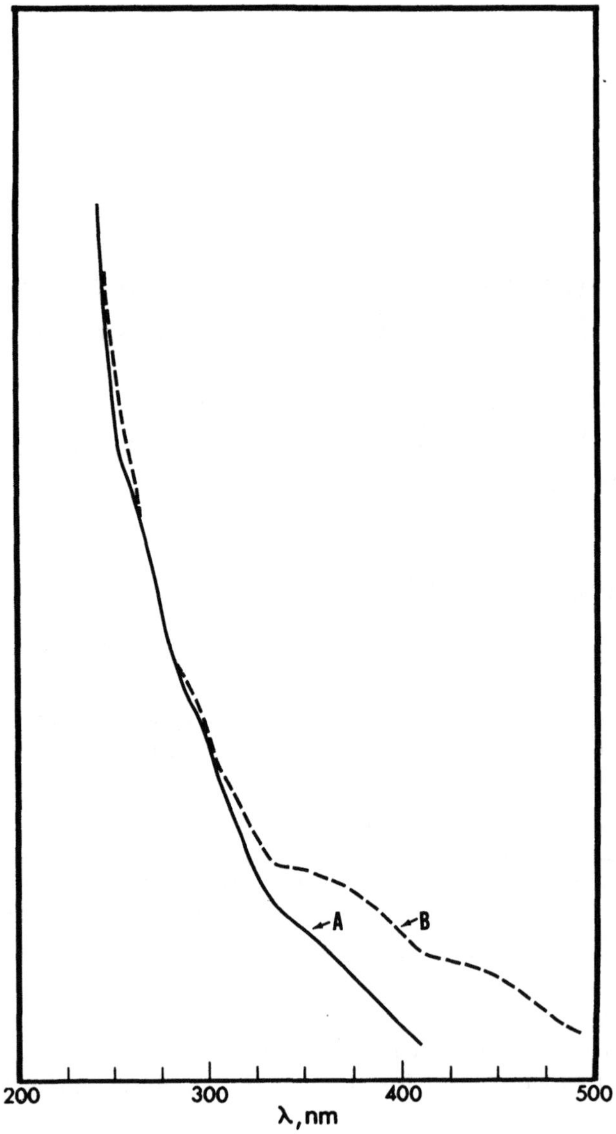

Fig. IV-1. A 150 cm^2 piece of Whatmann 3 mm chromatographic paper was eluted for 10 min with 100 ml of spectroscopic grade methanol. The solution was taken to dryness and redissolved in 10 ml of spectral quality methanol. The UV spectra of the latter solution alone (spectrum A) and after the addition of three drops of the NaOMe stock solution (spectrum B) are shown here; the spectra with all of the other diagnostic reagents were similar to those observed with NaOMe. The reference solution was spectral methanol

Steps in the UV Spectral Analyses.

(1) *A stock solution of the flavonoid* was prepared by dissolving a small amount of the compound (about 0.1 mg) in about 10 ml of spectroscopic methanol. The concentration was then adjusted so that the optical density of the major absorption peak between 250 and 400 nm gave an optical density (OD) reading in the region 0.6 to 0.8.

When the flavonoid was purified by paper chromatography (Whatman 3 MM paper), the following procedure was employed. The zone (usually about 150 cm^2) of the one-dimensional chromatogram (developed in either TBA or HOAc, usually the latter) was cut into small pieces which were then shaken for 10 min or less with 50 to 100 ml of the highest quality spectral grade methanol in a 250 ml Erlenmeyer flask. (Reagent grades of methanol contain traces of non-volatile substances which absorb in the 280–220 nm range.) The solution was filtered and then taken to dryness on a rotary evaporator; the residue thus obtained was redissolved in 10 ml of spectral grade methanol. The latter solution was used directly (or further diluted as necessary) for the spectral analyses. Because methanol elutes some UV-absorbing compounds from the paper itself (Fig. IV-1), we found that the best spectral results were obtained when a reference solution was prepared by extracting a

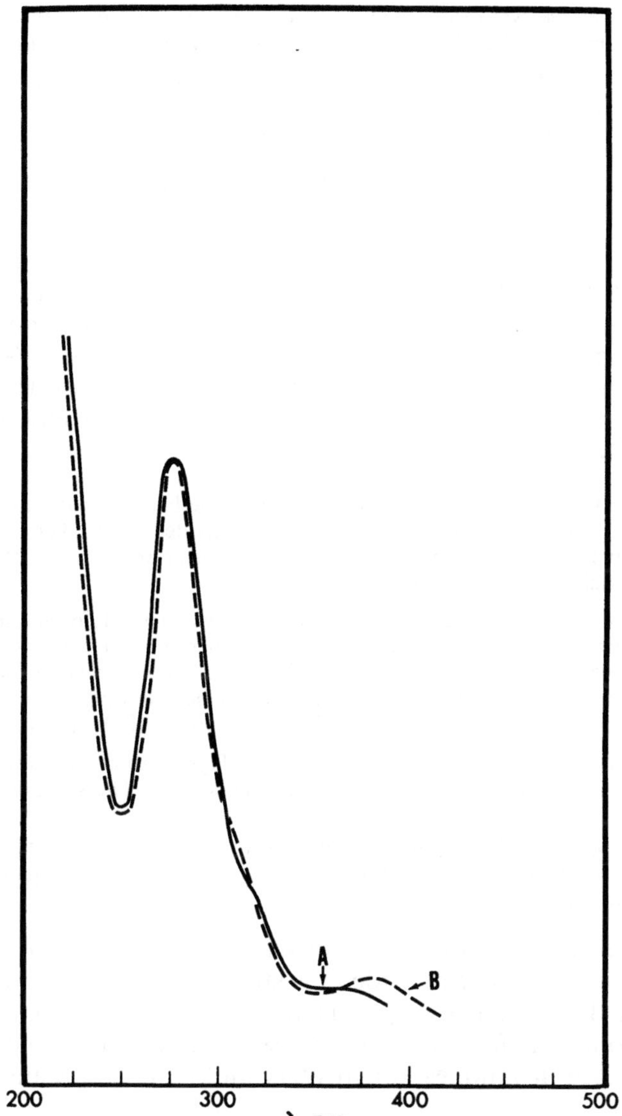

Fig. IV-2. A few isoflavones and dihydroflavonols required about a minute for the $AlCl_3$ to produce its maximum effect, particularly upon the long wavelength band. The spectra illustrate the effect of $AlCl_3$ in MeOH upon iridin (3′,5,7-trihydroxy-4′,5′,6-trimethoxyisoflavone 7-O-glucoside): A. immediately; B. after 1 min

piece of blank chromatographic paper from the same chromatogram (equal in size to the piece which contained the flavonoid) using the same procedure.

(2) *The methanol spectrum* was measured at normal scan speed (about 50 nm per min) using 2–3 ml of the stock solution of the flavonoid.

(3) *The methanol spectrum* was rerun at slow scan speed (about 10 nm per min) in the regions of the peak maxima in order to determine the wavelength (λ) of each maximum more accurately.

(4) *The NaOMe spectrum* was measured immediately after the addition of three drops of the NaOMe stock solution to the solution used for steps 2 and 3. After 5 min, the spectrum was rerun to check for flavonoid decomposition (only the initial spectrum is presented in this compilation). The solution was then discarded.

(5) *The $AlCl_3$ spectrum*[2] was measured immediately after the addition of six drops of the $AlCl_3$ stock solution to 2–3 ml of fresh stock solution of the flavonoid. For a few isoflavones and dihydroflavonols it required about a minute for the $AlCl_3$ to produce its maximum effect on the UV spectrum (Fig. IV-2).

[2] If the flavonoid used for this spectrum had been eluted from a paper chromatogram that had a distinct odor of acetic acid, the spectrum observed with the anhydrous $AlCl_3$ was similar to that obtained with $AlCl_3$/HCl. Therefore, for this spectrum we found it useful to routinely dry the material obtained from a paper chromatogram under high vacuum (oil pump) for about 10 min in order to remove the last traces of acid and water.

(6) *The $AlCl_3/HCl$ spectrum* was recorded immediately after the addition of three drops of the stock HCl solution to the cuvette containing the $AlCl_3$ (from step 5). The solution was then discarded.

(7) *The NaOAc spectrum* of the flavonoid was determined as follows. Excess coarsely powdered, anhydrous[3] reagent grade NaOAc was added with shaking to a cuvette containing 2–3 ml of fresh stock solution of the flavonoid. About a 2 mm layer of NaOAc remained on the bottom of the cuvette. All the NaOAc spectra presented in this volume were recorded within 2 min after the addition of the NaOAc to the solution (with decomposing compounds the time factor is critical). A second spectrum was run after 5–10 min to check for decomposition (again, only the initial spectrum is presented here).

Commercial anhydrous reagent-grade NaOAc contains varying amounts of HOAc, which notably alters the Band I region of the NaOAc spectrum (Fig. IV-3) (the HOAc mainly retards ionization of hydroxyl groups at positions other than C-7). Since we use the NaOAc spectrum to determine the presence or absence of a 7-hydroxyl group by observing the shift in Band II, we have not routinely used fused (HOAc-free) NaOAc. However, if fused NaOAc is employed, the spectrum will more closely resemble the one obtained with NaOMe.

Fused NaOAc. The HOAc present in anhydrous, reagent-grade NaOAc may be removed by the following procedure: The commercial anhydrous reagent-grade NaOAc is melted in a beaker using a Bunsen burner and allowed to remain molten for about 10 min. It is poured into a mortar and stirred with a pestle during solidification. Finally, the material is powdered and stored in a vial.

(8) *The $NaOAc/H_3BO_3$ spectrum* was determined as follows. Two methods for obtaining the $NaOAc/H_3BO_3$ spectra were used in the present investigation depending on whether or not decomposition of the flavonoid was observed during the recording of the NaOMe spectrum. If no decomposition was observed when the NaOMe spectrum was rerun after 5 min, procedure I was employed. When decomposition of the flavonoid in the presence of NaOMe did occur procedure II was used for the $NaOAc/H_3BO_3$ spectrum.

Procedure I. Sufficient powdered anhydrous reagent grade H_3BO_3 to give a saturated solution was added with shaking to the cuvette (from step 7) which contained the NaOAc. The solution was discarded after the spectrum was recorded.

Procedure II. Five drops of the H_3BO_3 stock solution were added to 2–3 ml of fresh stock solution of the flavonoid. The solution was then quickly saturated with coarsely powdered reagent grade NaOAc and the $NaOAc/H_3BO_3$ spectrum was recorded.

The above procedures produced a set of six spectra for each of the 175 compounds examined and these spectra have provided the statistical data on shifts that are discussed in detail in the next chapters. The emphasis in the present investigation has been placed on flavones, isoflavones and flavonols since these are commonly encountered flavonoids. However a few selected flavanones, dihydroflavonols, chalcones and aurones were also studied.

We have found that the methanol and sodium methoxide spectra are generally reproduceable. In contrast, the other four reagents have sometimes produced slightly different spectra when the spectra were redetermined. Therefore, we caution other investigators not to be discouraged to find that their $AlCl_3/HCl$, NaOAc and $NaOAc/H_3BO_3$ spectra have λ_{max} values differing by a few nm from those reported here. The shapes of the spectral curves appear to be more reliable guides for identification purposes when these reagents are used. However, the Band I region of the NaOAc spectrum is quite variable depending upon the amount of HOAc present.

It is customary when reporting a new flavonoid to record not only the absorption peak positions but also their intensities. The intensity of a peak is expressed either in terms of ε (the molecular extinction coefficient) or $\log_{10}\varepsilon$, and may be calculated from the following relationship (when standard cuvettes with 1 cm path lengths are used):

$$\varepsilon = \frac{OD}{c}$$

[3] Although anhydrous NaOAc was routinely used, it was observed that a drop of water could be added to the cuvette without altering the spectrum.

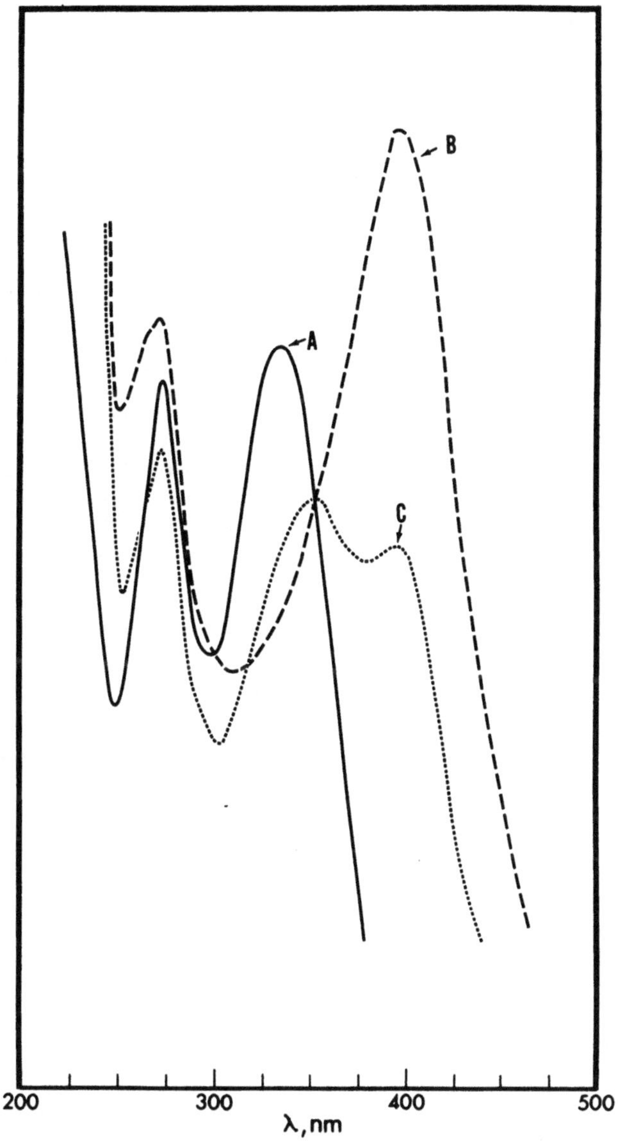

Fig. IV-3. Spectra B and C illustrate the different effects of NaOAc on the spectrum of saponarin (4′,5,7-hydroxy-6-C-glucosylflavone 7-O-glucoside) in MeOH (curve A). Spectrum B was observed when fused (HOAc-free) NaOAc was added to the MeOH solution. Spectrum C was obtained using our standard procedures in which the MeOH solution of saponarin was saturated with anhydrous, reagent-grade NaOAc. Addition of a drop of water to either of the cuvettes used for spectra B and C did not alter the spectra

where *OD* is the optical density of the solution and *c* is the flavonoid concentration in moles per liter. When the extinction coefficient (ε) is known, this expression permits the determination of the concentration of the flavonoid in a solution. We have not determined extinction coefficients in the present study but have, instead, presented all the actual spectral curves which provide directly the relative optical densities (relative intensities) of the peaks observed for a given compound. Extinction coefficients for a number of flavonoids are recorded in the chapter by L. Jurd in *The Chemistry of Flavonoid Compounds* (edited by T.A. Geissman) pp. 107–155, The Macmillan Co., New York, 1962; some of Jurd's data are reproduced in Table IV-1.

Table IV-1. *Molecular extinction coefficients (expressed as* log ε*) for flavonoids*

Flavonoids	EtOH λ_{max}(log ε)
Flavones and Flavonols	
Flavone	297(4.20), 250(4.06)
7-Hydroxyflavone	308(4.50), 250(4.33)
5-Hydroxyflavone	337(3.88), 272(4.35)
3′,4′-Dihydroxyflavone	342.5(4.5), 244(4.46)
4′,7-Di-O-ethylvitexin	326(4.21), 270(4.26)
5,7,4′-Trimethoxyflavone	325(4.33), 265(4.25)
Luteolin	350(4.17), 268[a], 255(4.13)
Diosmetin	354(4.32), 268(4.25), 253(4.28)
3-Methoxyflavone	320[a], 299(4.21), 246(4.25)
3,4′-Dihydroxyflavone	361(4.39)
3,2′-Dihydroxyflavone	353(4.21), 303(3.92), 244(4.28)
3,5,7-Trihydroxyflavone (galangin)	360(4.07), 267.5(4.23)
3,5,7,2′-Tetrahydroxyflavone	360(3.99), 262.5(4.14)
3,7,3′,4′-Tetrahydroxyflavone (fisetin)	370(4.43), 315(4.22), 252.5(4.33)
Amurensin	377(4.23), 270(4.28)
Penduletin	341(4.36), 271(4.28), 212(4.52)
Pendulin	322(4.34), 272(4.36), 212(4.58)
Morin	380(4.15), 263(4.22)
Quercetin 3-L-arabinoside	360(4.24), 260(4.32)
Isoquercitrin	360(4.32), 258(4.41)
3,7,3′-Tri-O-methylquercetin	360(4.31), 268(4.24), 257(4.32)
Myricetin	378(4.29), 255(4.21)
Gossypetin	386(4.15), 341[a], 278(4.23), 262(4.26)
Isoflavones	
Isoflavone	307(3.82)
7,4′-Dihydroxy-5-methoxyisoflavone	256(4.51)
5,7,2′-Trimethoxy-8-methylisoflavone	259(1.48)
	249(1.46)
5,6,7,2′-Tetramethoxyisoflavone	304[a](4.10)
	281(4.29)
	247(4.57)
Flavanones and Dihydroflavonols	
Flavanone	320(3.37), 250(3.86)
5,7-Dihydroxyflavanone (pinocembrin)	314(3.78), 288(4.35)
Naringin	330[a], 284(4.28)
Prunin	330, 283(3.44)
5,7,4′-Trihydroxyflavanone (naringenin)	325[a], 288(4.23)
Hesperetin	330[a], 289(4.27)
Homoeriodictyol	288(4.30), 277.5(4.38)
Astilbin	330[a](3.66), 292(4.21)
Dihydrokaempferol	330[a](3.75), 292(4.29), 252(3.61)
Chalcones and Aurones	
Chalcone	312(4.35), 230(3.91)
2′-Hydroxychalcone	366[a], 316(4.36), 221(4.11)
2′,4′,5′,3,4-Pentahydroxychalcone	393(4.37), 320[a](4.03), 268(4.08)
Aurone	379(4.06), 316.5(4.27), 251(4.10)
4′-Hydroxyaurone	405(4.47), 346(4.07), 260(4.32)
3′-Hydroxyaurone	381(4.29), 316(4.21), 252(4.03)
3′,4′-Dihydroxyaurone	415.5(4.43), 330, 277, 259
3′,4′,4-Trihydroxyaurone	416(4.47), 310(3.92), 276(4.02), 256(3.91)
4′,6,7-Trihydroxyaurone	407(4.39), 355[a](4.22), 241(4.12)

[a] Inflection.

Chapter V

The Ultraviolet Spectra of Flavones and Flavonols

V-1. The UV Spectra of Flavones and Flavonols in Methanol 41
1a. The Effect of Oxidation Patterns on the UV Spectra of Flavones and Flavonols 44
1b. The Effect of Methylation and Glycosylation on the UV Spectra of Flavones and Flavonols 45
1c. The Effect of Acetylation on the UV Spectra of Flavones and Flavonols . . 45
V-2. The UV Spectra of Flavones and Flavonols in the Presence of NaOMe . . 45
2a. The Detection of Either 3- or 4′-Hydroxyl Groups in Flavones and Flavonols by the Effect of NaOMe on the UV Spectrum 46
2b. The Detection of the 3,4′-Dihydroxyl System in Flavonols by the Effect of NaOMe on the UV Spectrum 47
V-3. The UV Spectra of Flavones and Flavonols in the Presence of NaOAc . . 48
3a. The Detection of 7-Hydroxyl Groups in Flavones and Flavonols by the Effect of NaOAc on the UV Spectrum 48
3b. Detection of 4′-Hydroxyl Groups in Flavones and Flavonols by the Effect of NaOAc on the UV Spectrum 49
3c. Degeneration of the UV Spectra of Flavones and Flavonols in the Presence of NaOAc 50
V-4. The Detection of Ortho-dihydroxyl Groups in Flavones and Flavonols by the Effect of $NaOAc/H_3BO_3$ on the UV Spectrum 50
V-5. The UV Spectra of Flavones and Flavonols in the Presence of $AlCl_3$ and $AlCl_3/HCl$ 51
5a. The Structures of the $AlCl_3$ Complexes with Flavones and Flavonols . . 51
5b. The Detection of Ortho-dihydroxyl Groups in Flavones and Flavonols by the Effect of $AlCl_3$ and $AlCl_3/HCl$ on the UV Spectrum 52
5c. The Detection of Either 3- or 5-Hydroxyl Groups in Flavones and Flavonols by the Effect of $AlCl_3/HCl$ on the UV Spectrum 52
V-6. The Index of Ultraviolet Absorption Spectra of Flavones and Flavonols 57
References 61

V-1. The UV Spectra of Flavones and Flavonols in Methanol

The methanol spectra of flavones and flavonols exhibit two major absorption peaks in the region 240 – 400 nm[1]. These two peaks are commonly referred to as Band I (usually 300 – 380 nm, Table V-1 records the λ_{max} values for Band I for all flavones and flavonols examined in the present investigation), and Band II (usually 240 – 280 nm). Band I is considered to be associated with absorption due to the B-ring cinnamoyl system, and Band II with absorption involving the A-ring benzoyl system (see III) [*1*].

[1] Although the UV spectra are reproduced for the range of 220 – 500 nm, only the maxima for those peaks and shoulders occurring at wavelengths longer than 240 nm are tabulated on the pages on which the curves are reproduced.

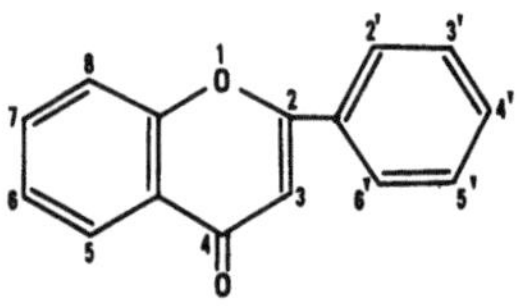

I. Flavone skeleton

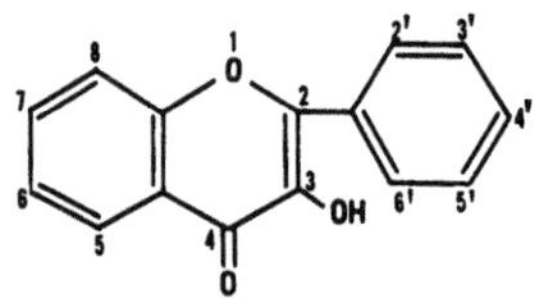

II. Flavonol skeleton

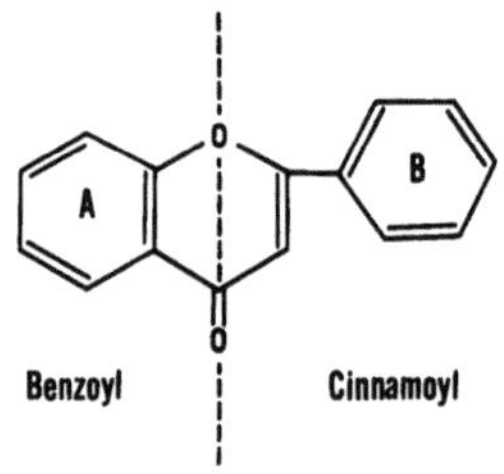

III.

Table V-1. *Band I in the ultraviolet spectra of flavones and flavonols in methanol*

Spectrum No.	Flavonoid	Band I (nm)
100	Gossypitrin	385
98	Gossypetin	385[a]
99	Gossypin	380
64	Herbacetin 8-methyl ether	377
102	Myricetin	374
93	Patulitrin	373
66	Quercetin 7-O-rhamnoside	372
76	Rhamnetin	371
90	Patuletin	371
65	Quercetin	370
77	Isorhamnetin	370
82	Morin[a]	370
80	Tamarixetin 7-O-neohesperidoside	369
58	Kaempferol	367
83	Robinetin	367
61	Kaempferol 4′-methyl ether	367
81	Tamarixetin 7-O-rutinoside	367
56	3,3′,4′-Trihydroxyflavone	366
59	Kaempferol 7-O-neohesperidoside	364
18	5,7,8-Trihydroxyflavone (Norwogonin)	364[c]
62	Fisetin	362
67	Quercetin 3-O-galactoside	362
75	Quercetin 3′,4′,5,7-tetramethyl ether	362
54	Galangin	359
69	Rutin	359
73	Quercetin 3-methyl ether	358
72	Quercetin 3-O-glucoside 7-O-rutinoside	358
71	Quercetin 3-O-glucoside 7-O-rhamnoside	358
78	Isorhamnetin 3-O-galactoside	357
79	Isorhamnetin 3-O-rutinoside	356
53	3,4′,7-Trihydroxyflavone	356
92	Patuletin 3-O-rutinoside	356
57	3-Hydroxy-3′,4′-dimethoxyflavone	355
70	Quercetin 3,7-O-diglucoside	355
52	3-Hydroxy-4′-methoxyflavone	355
91	Patuletin 3-O-glucoside	355
95	Quercetagetin 3′,4′,5,6,7-pentamethyl ether	354
88	Jacein	352
87	Jaceidin	351
101	Gossypetin hexamethyl ether	351
60	Kaempferol 3-O-robinoside 7-O-rhamnoside (Robinin)	350
68	Quercitrin	350
44	Tricin	350
89	Centaurein	350
37	Isoorientin	349
34	Luteolin	349
39	Lucenin-1	349
74	Quercetin 3-methyl ether 4′-O-glucoside 7-O-diglucoside	349

Table V-1 (continued)

Spectrum No.	Flavonoid	Band I (nm)
36	Luteolin 7-O-rutinoside	349
35	Luteolin 7-O-glucoside	348
32	Zapotinin	348[c]
40	Chrysoeriol	347
38	Orientin	346
94	Artemetin	345
41	Scoparin	345
42	Diosmetin	344
51	3-Hydroxyflavone	344
13	3′-4′-7-Trihydroxyflavone	343
19	5,7,8-Trihydroxyflavone 7-O-glucuronide	342[c]
14	3′,4′,7-Trihydroxyflavone 7-O-rhamnoglucoside	341[a]
5	3′,4′-Dihydroxyflavone	340
63	Fisetin 3-O-glucoside	340[a]
55	Galangin 3-methyl ether	340[c]
43	5,7-Dihydroxy-3′,4′-dimethoxyflavone	340
84	Penduletin	340
86	3,5,6,7,8-pentamethoxyflavone	338
96	Quercetagetin 3,3′,4′,5,6-pentamethyl ether	338
48	Hymenoxin	336
23	Isovitexin (Saponaretin)	336
24	Saponarin	336
20	Apigenin	336
26	Rhamnosylvitexin	336
25	Vitexin	336
28	Violanthin	335
49	Amentoflavone	335
103	3′,5,5′-Trihydroxy-3,4′,6,7-tetramethoxyflavone	335
27	2″-O-Xylosylvitexin	335
2	5-Hydroxyflavone	333
6	3′,4′-Dimethoxyflavone	333
21	Apigenin 7-O-glucoside	333
22	Apigenin 7-O-neohesperidoside	333
97	Quercetagetin hexamethyl ether	333
47	Xanthomicrol	332
45	5,7-Dihydroxy-3′,4′,5′-trimethoxyflavone	331
85	Pendulin	330
15	7-Hydroxy-3′,4′-dimethoxyflavone	330
46	Nevadensin	329
9	4′,7-Dihydroxyflavone	328
11	5-Deoxyvitexin (Bayin)	328
29	Acacetin	327
50	Sciadopitysin	326
10	4′,7-Dihydroxyflavone 7-O-rhamnoglucoside	325
31	5,7-Dihydroxy-2′-methoxyflavone	325
33	Zapotin	325
30	Acacetin 7-O-glucoside	324
16	Baicalein (5,6,7-Trihydroxyflavone)	323
12	7-Hydroxy-4′-methoxyflavone (Pratol)	323
4	4′-Methoxyflavone	317
17	Baicalin (5,6,7-Trihydroxyflavone 7-O-glucoronide)	315
7	Chrysin	313
3	7-Hydroxyflavone	307
1	Flavone	307
8	Tectochrysin	303

[a] Band I a.

[b] In MeOH, morin ionizes slightly and gives a band I peak near 385 nm; however, in the presence of even traces of acid, this ionization is suppressed and a peak is observed near 370 nm.

[c] Low intensity band.

These assignments are supported, to some extent, by the UV spectral data of substituted flavonoids. For example, flavones and flavonols oxygenated in the A-ring, but not in the B-ring, tend to give spectra in methanol with a pronounced Band II and a weak Band I (see spectra 2, 7, 8, 16, 17, 18, 19 and 55); but in similar molecules which also possess B-ring oxygenation, Band I is more pronounced and appears at longer wavelengths (see for example spectra 9, 10, 15, 20, 21, 22, 47, 58, 62, and 65).

The methanol spectrum, particularly the position of Band I, provides information about the type of flavonoid as well as its oxidation pattern. It is apparent from the data presented in Table V-2 that the position of Band I distinguishes between flavones and 3-hydroxyflavones (flavonols). Band I of flavones occurs in the range 304 – 350 nm whereas Band I of 3-hydroxyflavones appears at a longer wavelength (352 – 385 nm). However, in flavonols with a substituted 3-hydroxyl group (methylated or glycosylated), Band I (328 – 357 nm) overlaps the region for Band I in flavones, and the general shape of the spectral curves approach those of flavones.

Table V-2. *Band I in the UV spectra of flavones and flavonols*

Flavonoid Type	Number of compounds examined	Range of Band I (nm)
Flavones	50	304 – 350
Flavonols (3-hydroxyl substituted)	26	328 – 357
Flavonols (free 3-hydroxyl)	27	352 – 385

1a. The Effect of Oxidation Patterns on the UV Spectra of Flavones and Flavonols

On increasing the oxygenation of the B-ring in flavones and flavonols, a bathochromic shift in Band I occurs with each additional oxygen function (see Table V-3). On the other hand, while changes in the B-ring oxygenation pattern usually do not produce a shift in Band II, Band II may appear as either one or two peaks (designated IIa and IIb with IIa being the peak at longer wavelength) depending on the B-ring oxidation pattern. For example, the 3′,4′-(and to a lesser extent the 3′,4′,5′-)oxygenated flavones and flavonols usually exhibit two absorption peaks (or one maxima with a shoulder on the long wavelength side of the peak) between 250 – 275 nm, while the 4′-oxygenated equivalents have only one.

Table V-3. *Band I in the UV spectra of flavonols differing in their B-ring Oxidation pattern*

Spectrum No.	Flavonol	Oxidation pattern		Band I (nm)
		A- and C-rings	B-ring	
54	Galangin	3,5,7	–	359
58	Kaempferol	3,5,7	4′	367
65	Quercetin	3,5,7	3′,4′	370
102	Myricetin	3,5,7	3′,4′,5′	374

Increasing hydroxylation of the A-ring in flavones and flavonols produces a notable bathochromic shift in Band II (see Table V-4) and a smaller effect on Band I.

The presence or absence of the hydrogen bonded 5-hydroxyl group has a marked effect on both Band I and II in the UV spectra of flavones. When the 5-hydroxyl group is absent from a flavone or flavonol both bands appear at shorter wavelengths than in the 5-hydroxylated equivalents; 3 – 10 nm in Band I and 6 – 17 nm in Band II (e.g. see spectra pairs: 3, 7; 9, 20; 10, 22; 11, 25; 12, 29; 13, 34; 14, 36; 15, 43; 53, 58; 62, 65 and 83, 102).

Table V-4. *Band II in the UV spectra of flavones having oxidation only in the A-ring*

Spectrum No.	Flavone	A-Ring pattern oxidation	Band II (nm)
1	Flavone	–	250
2	5-Hydroxyflavone	5	268
3	7-Hydroxyflavone	7	252
7	5,7-Dihydroxyflavone	5,7	268
16	Baicalein	5,6,7	274
18	Norwogonin	5,7,8	281

1 b. The Effect of Methylation and Glycosylation on the UV Spectra of Flavones and Flavonols

If a 3-, 5- or 4′-hydroxyl group on the flavone or flavonol nucleus is methylated or glycosylated, hypsochromic shifts (i.e. to shorter wavelengths), especially in Band I, are observed. The shift associated with the substitution of the 3-hydroxyl group is usually of the order of 12 – 17 nm (e.g. see spectra pairs: 54, 55; 58, 60; 65, 69; 65, 73 and 77, 78) but reaches 22 – 25 nm in flavonols which do not contain a free 5-hydroxyl group (see spectra pairs 62, 63 and 95, 97). Methylation of the 5-hydroxyl group results in a 5 – 15 nm hypsochromic shift in both Band I and II [*1*], and methylation or glycosylation of the 4′-hydroxyl group produces a 3 – 10 nm hypsochromic shift in Band I (see spectra pairs: 9, 12; 5, 6; 20, 29; 34, 42; 40, 43; 71, 74; 66, 80; 66, 81; and 84, 85). Substitution of the hydroxyl groups at positions other than 3, 5 and 4′ has little or no effect on the UV spectrum (in methanol).

1 c. The Effect of Acetylation on the UV Spectra of Flavones and Flavonols

Acetylation of a phenolic hydroxyl group is known [*2, 3*] to nullify the effect of that group on the UV spectrum. This principle has been applied by Jurd [*1*] to locate methoxyl groups in polyhydroxyflavones. For example, diosmetin triacetate (IV) was shown to possess a UV spectrum (λ_{max} 257 and 320 nm) similar to that observed for 4′-methoxyflavone (λ_{max} 253 and 317 nm).

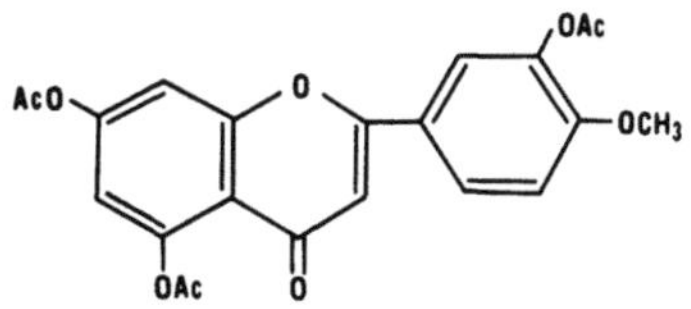

IV. Diosmetin triacetate

V-2. The UV Spectra of Flavones and Flavonols in the Presence of NaOMe

Sodium methoxide is a strong base and ionizes to some extent all hydroxyl groups on the flavonoid nucleus. For this reason, it is difficult to correlate the spectral shifts obtained on the addition of NaOMe with the flavonoid hydroxylation pattern. However, use has been made of the effect of NaOMe on the UV spectra of flavones and flavonols for the detection of free 3- and/or 4′-hydroxyl groups.

2 a. The Detection of Either 3- or 4′-Hydroxyl Groups in Flavones and Flavonols by the Effect of NaOMe on the UV Spectrum

The addition of NaOMe to flavones and flavonols in methanol usually produces bathochromic shifts in all absorption bands. However, a large bathochromic shift of Band I of about 40–65 nm, without a decrease in intensity, is diagnostic for the presence of a free 4′-hydroxyl group. Of the 103 flavone and flavonol spectra obtained in NaOMe, 47 showed 40–65 nm bathochromic shifts in Band I without a decrease in intensity (see Table V-5), and of these all but three (spectra 52, 54 and 57) were spectra of flavonoids containing a free 4′-hydroxyl group. Although flavonols lacking a free 4′-hydroxyl group also give a 50–60 nm bathochromic shift in Band I, there is usually a decrease in intensity of the peak. In these compounds the bathochromic shift results from the presence of a free 3-hydroxyl group (see spectra 51, 61, 75, 80, 81 and 95). (The presence of a 4′-hydroxyl group may also often be detected from the spectrum in NaOAc.)

Table V-5. *Flavones and flavonols in which NaOMe produces Band I shifts of 38–68 nm without a decrease in intensity*

Spectrum No.	Flavonoid	Band I bathochromic shift (nm)
	Flavones	
5	3′,4′-Dihydroxyflavone	64
9	4′,7-Dihydroxyflavone	58
10	4′,7-Dihydroxyflavone 7-O-rhamnoglucoside	60
11	5-Deoxyvitexin (Bayin)	62
13	3′,4′,7-Trihydroxyflavone	52
14	3′,4′,7-Trihydroxyflavone 7-O-rhamnoglucoside	64
20	Apigenin	56
21	Apigenin 7-O-glucoside	53
22	Apigenin 7-O-neohesperidoside	53
23	Isovitexin	62
24	Saponarin	53
25	Vitexin	59
26	Rhamnosylvitexin	60
27	2″-O-xylosylvitexin	60
28	Violanthin	63
34	Luteolin	52
35	Luteolin 7-O-glucoside	46
36	Luteolin 7-O-rutinoside	45
37	Isoorientin	57
38	Orientin	59
39	Lucenin-1	59
40	Chrysoeriol	58
41	Scoparin	61
44	Tricin	66
47	Xanthomicrol	59
49	Amentoflavone	47
	Flavonols	
52	3-Hydroxy-4′-methoxyflavone	54
53	3,4′,7-Trihydroxyflavone	51
54	Galangin	53
57	3-Hydroxy-3′,4′-dimethoxyflavone	57
58	Kaempferol	49
59	Kaempferol 7-O-neohesperidoside	61
60	Kaempferol 3-O-robinoside 7-O-rhamnoside (Robinin)	39
63	Fisetin 3-O-glucoside	68
67	Quercetin 3-O-galactoside	47

Table V-5 (continued)

Spectrum No.	Flavonoid	Band I bathochromic shift (nm)
68	Quercitrin	43
69	Rutin	51
70	Quercetin 3,7-O-diglucoside	41
71	Quercetin 3-O-glucoside 7-O-rhamnoside	38
72	Quercetin 3-O-glucoside 7-O-rutinoside	38
73	Quercetin 3-methylether	49
78	Isorhamnetin 3-O-galactoside	58
79	Isorhamnetin 3-O-rutinoside	58
82	Morin	47
84	Penduletin	48
87	Jaceidin	61
88	Jacein	49
91	Patuletin 3-O-glucoside	55
92	Patuletin 3-O-rutinoside	55

2b. The Detection of the 3,4′-Dihydroxyl System in Flavonols by the Effect of NaOMe on the UV Spectrum

Flavonols which have free hydroxyl groups at both the 3- and 4′-positions are unstable in NaOMe and the absorption peaks in the NaOMe spectrum degenerate in a few minutes (see Table V-6) [*4*, *5*]. Flavonols which contain a 3,3′,4′-trihydroxyl system decompose even faster than those having the 3,4′-dihydroxyl system. Although alkali instability is generally associated with flavonols having the 3,4′-dihydroxyl grouping, other hydroxylation patterns in flavones, notably 5, 6, 7; 5, 7, 8 and 3′,4′,5′ may also cause alkali sensitivity (see spectra 16, 17, 18 and 83).

Table V-6. *The effect of NaOMe and NaOAc on flavonols containing a 3,4′-dihydroxyl system*

Spectrum No.	Flavonol	Alkali[a]	
		NaOMe	NaOAc
53	3,4′,7-Trihydroxyflavone	slow dec	no dec
56	3,3′,4′-Trihydroxyflavone	dec	dec
58	Kaempferol	slow dec	no dec
59	Kaempferol 7-O-neohesperidoside	slow dec	no dec
62	Fisetin	dec	dec
64	Herbacetin 8-methyl ether	dec	dec
65	Quercetin	dec	dec
66	Quercetin 7-O-rhamnoside	dec	dec
76	Rhamnetin	dec	dec
77	Isorhamnetin	dec	dec
82	Morin	slow dec	no dec
83	Robinetin	dec	dec
90	Patuletin	dec	dec
93	Patulitrin	dec	dec
98	Gossypetin	dec	dec
99	Gossypin	dec	dec
100	Gossypitrin	dec	dec
102	Myricetin	dec	dec

[a] dec = spectrum decomposed, as determined by a comparison of the spectrum in alkali measured immediately, with that measured 5–10 min later.

V-3. The UV Spectra of Flavones and Flavonols in the Presence of NaOAc

Sodium acetate is a weaker base than NaOMe, and, as such, ionizes only the more acidic hydroxyl groups in flavones and flavonols, i.e., the 3,7- and 4′-hydroxyl groups. Because ionization of the 7-hydroxyl group mainly affects Band II (whereas ionization of the 3- and/or 4′-hydroxyl groups mainly affects Band I), NaOAc is a particularly useful diagnostic reagent for the specific detection of 7-hydroxyl groups.

We did not employ fused NaOAc since the Band II data is not notably affected by the traces of HOAc present in commercial anhydrous reagent grade NaOAc.

3a. The Detection of 7-Hydroxyl Groups in Flavones and Flavonols by the Effect of NaOAc on the UV Spectrum

The UV spectra of flavones and flavonols containing free 7-hydroxyl groups with few exceptions exhibit a diagnostic 5–20 nm bathochromic shift of Band II in the presence of NaOAc [5] (see Table V-7). However, when 6 and 8 oxygen substituents are present in flavones (but not in flavonols), the bathochromic shift with NaOAc is often small or imperceptable [6, 7] (e.g. spectra 46, 48) presumably because of the reduced acidity of the 7-hydroxyl group. Certain 3′,4′-dioxygenated derivatives are recorded in Table V-7 as having shifts of 20–25 nm (e.g. spectra 37–43), but these results reflect, in part, the fact that the shifts are measured from Band II b (the shorter wavelength peak in those flavonoid spectra which exhibit two Band II peaks). The NaOAc shifts recorded in Table V-7 for the 7-hydroxy-5-deoxyflavones were measured from Band II a because Band II b was usually poorly resolved (see for example spectra 11 and 12). [In the spectra of 5-deoxyflavones which do exhibit well defined Band II b absorption (spectra 9, 13 and 15), bathochromic shifts of approximately 20–35 nm are observed when measurements are made from Band II b.]

Table V-7. *Bathochromic shifts of Band II in the spectra of 7-hydroxyflavones and 7-hydroxyflavonols with added NaOAc*

Spectrum No.	Flavonoid	Bathochromic shift (nm)[a]
	Flavones	
3	7-Hydroxyflavone	14
7	Chrysin	7
9	4′,7-Dihydroxyflavone	8
11	5-Deoxyvitexin (Bayin)	13
12	7-Hydroxy-4′-methoxyflavone (Pratol)	17
16	5,6,7-Trihydroxyflavone (Baicalein)	dec
18	5,7,8-Trihydroxyflavone (Norwogonin)	dec
20	Apigenin	7
23	Isovitexin (Saponaretin)	8
25	Vitexin	10
26	Rhamnosylvitexin	11
27	2″-O-Xylosylvitexin	10
28	Violanthin	7
29	Acacetin	7
31	5,7-Dihydroxy-2′-methoxyflavone	5
34	Luteolin	16
37	Isoorientin	21
38	Orientin	23

Table V-7 (continued)

Spectrum No.	Flavonoid	Bathochromic shift (nm)[a]
39	Lucenin-1	25
40	Chrysoeriol	30
41	Scoparin	28
42	Diosmetin	23
43	5,7-Dihydroxy-3′,4′-dimethoxyflavone	36
44	Tricin	7
45	5,7-Dihydroxy-3′,4′,5′-trimethoxyflavone	7
46	Nevadensin	−1
48	Hymenoxin	4
49	Amentoflavone	5
	Flavonols	
53	3,4′,7-Trihydroxyflavone	10
54	Galangin	8
55	Galangin 3-methyl ether	12
58	Kaempferol	8
61	Kaempferol 4′-methyl ether	7
62	Fisetin	dec
64	Herbacetin 8-methyl ether	dec
65	Quercetin	dec
67	Quercetin 3-O-galactoside	17
68	Quercitrin	16
69	Rutin	12
73	Quercetin 3-methyl ether	16
77	Isorhamnetin	dec
78	Isorhamnetin 3-O-galactoside	19
79	Isorhamnetin 3-O-rutinoside	17
82	Morin	8
83	Robinetin	dec
87	Jaceidin	17
90	Patuletin	dec
91	Patuletin 3-O-glucoside	12
92	Patuletin 3-O-rutinoside	13
96	Quercetagetin 3,3′,4′,5,6-pentamethyl ether	18
98	Gossypetin	dec
99	Gossypin	dec
102	Myricetin	dec

[a] dec = spectrum decomposed as determined by a comparison of the spectrum in NaOAc measured after 2–5 min with that measured after 5–10 min later.

3b. Detection of 4′-Hydroxyl Groups in Flavones and Flavonols by the Effect of NaOAc on the Spectrum

Flavones and flavonols which possess a 4′-hydroxyl group and no free 3- or 7-hydroxyl groups usually show a pronounced shoulder on the long wavelength side of Band I in the presence of NaOAc (not fused) (see spectra 5, 10, 14, 21, 22, 24, 35, 36, 56, 59, 60, 66, 70, 71, 72, 76, 84, 88, 93 and 100). A Band I shift is also observed when the 7-hydroxyl group is free whether or not a 4′-hydroxyl group is present in the flavonoid. If fused (HOAc-free) NaOAc is used with flavones and flavonols containing a 4′-hydroxyl group, Band I appears as a peak similar to that observed with NaOMe.

3c. Degeneration of the UV Spectra of Flavones and Flavonols in the Presence of NaOAc

If the NaOAc spectrum of a flavone or flavonol changes after several minutes then the flavonoid has decomposed due to the presence of an alkali-sensitive grouping (see Tables V-6 and V-7). The most common alkali-sensitive oxygenation patterns in flavones and flavonols are those which contain 5, 6, 7, 5, 7, 8 or 3,3′,4′ hydroxyl groups. In the latter pattern the 3′-function may be a methoxyl group. For this reason it is difficult to determine the presence or absence of free 7-hydroxyl groups in flavonoids possessing these oxygenation patterns unless the NaOAc spectrum is measured immediately after addition of the NaOAc to the cuvette [5].

V-4. The Detection of Ortho-dihydroxyl Groups in Flavones and Flavonols by the Effect of $NaOAc/H_3BO_3$ on the UV Spectrum

In the presence of NaOAc, boric acid will chelate with orthodihydroxyl groups at all locations on the flavonoid nucleus, except at C-5, 6. Such complexes are probably of type V.

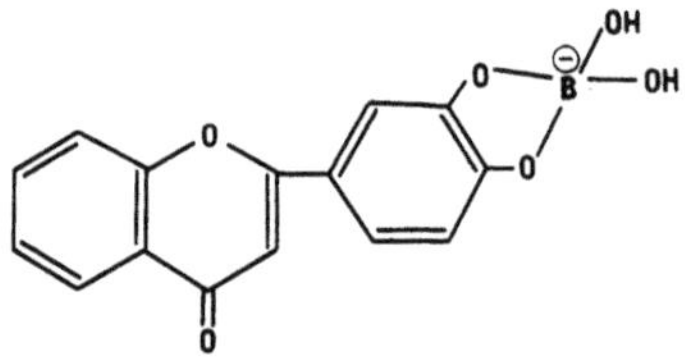

V

Flavones and flavonols containing a B-ring ortho-dihydroxyl group show a consistent 12−30 nm bathochromic shift of Band I in the presence of $NaOAc/H_3BO_3$ (see Table V-8).

Table V-8. *The effect of $NaOAc/H_3BO_3$ and $AlCl_3$ on Band I of the UV spectra of 3′,4′-dihydroxyflavones and 3′,4′-dihydroxyflavonols*

Spectrum No.	Flavonoid	Bathochromic shift with $NaOAc/H_3BO_3$ (nm) relative to MeOH spectrum	Bathochromic shift with $AlCl_3$ (nm) relative to $AlCl_3/HCl$ spectrum
	Flavones		
5	3′,4′-Dihydroxyflavone	25	36
13	3′,4′,7-Trihydroxyflavone	17	31[a]
14	3′,4′,7-Trihydroxyflavone 7-O-rhamnoglucoside	24	39
34	Luteolin	21	41
35	Luteolin 7-O-glucoside	24	45
36	Luteolin 7-O-rutinoside	21	43
37	Isoorientin	28	45
38	Orientin	29	45
39	Lucenin-1	33	46

Table V-8 (continued)

Spectrum No.	Flavonoid	Bathochromic shift with $NaOAc/H_3BO_3$ (nm) relative to MeOH spectrum	Bathochromic shift with $AlCl_3$ (nm) relative to $AlCl_3$/HCl spectrum
	Flavonols		
56	3,3′,4′-Trihydroxyflavone	22	39
62	Fisetin	19	35
63	Fisetin 3-O-glucoside	25	[b]
65	Quercetin	18	30
66	Quercetin 7-O-rhamnoside	14	32
67	Quercetin 3-O-galactoside	15	33
68	Quercitrin	17	29
69	Rutin	28	31
70	Quercetin 3,7-O-diglucoside	25	38
71	Quercetin 3-O-glucoside 7-O-rhamnoside	22	37
72	Quercetin 3-O-glucoside 7-O-rutinoside	22	37
73	Quercetin 3-methyl ether	20	41
76	Rhamnetin	18	28
83	Robinetin	18	21
90	Patuletin	22	32
91	Patuletin 3-O-glucoside	27	31
92	Patuletin 3-O-rutinoside	25	31
93	Patulitrin	21	31
98	Gossypetin	21	45
99	Gossypin	20	11
100	Gossypitrin	14	21
102	Myricetin	18	22

[a] Based on major absorption bands at 340 and 371 nm.
[b] In the presence of $AlCl_3$, fisetin 3-O-glucoside rapidly hydrolyzes to fisetin.

A-Ring ortho-dihydroxyl groups (at C-6, 7 and C-7, 8) in flavonoids are also detectable by the effect of $NaOAc/H_3BO_3$ on the UV spectra. From the limited data available in the present study, it appears that a bathochromic shift of about 5–10 nm in Band I (see the spectra pair 98, 100; and also spectra 16 and 18) may be diagnostic for flavones and flavonols containing either 6,7- or 7,8-dihydroxyl groups.

V-5. The UV Spectra of Flavones and Flavonols in the Presence of $AlCl_3$ and $AlCl_3$/HCl [2]

5a. The Structures of the $AlCl_3$ Complexes with Flavones and Flavonols

With aluminum chloride, flavones and flavonols which contain hydroxyl groups at C-3 or C-5 [*8, 9*] form acid stable complexes; in addition, aluminum chloride forms acid labile complexes with flavonoids which contain ortho-dihydroxyl systems [*10*]. A diagramatic representation of the experimental results is presented in Fig. V-1 together with possible structures VII–XIII for the complexes formed. The complexes formed between $AlCl_3$ and the A- and B-ring ortho-dihydroxyl groups, with few exceptions, decompose in the presence of acid. In contrast, the $AlCl_3$ complex between the C-4 keto function and either the 3- or 5-hydroxyl group is stable in the presence of acid. The $AlCl_3$/HCl shifts given by 3,5-dihydroxyflavones (Table V-11, spectra 58, 65 and 102) are intermediate between those given by their 5-deoxy-3-hydroxy equivalents (Table V-10, spectra 53, 62 and 83) and those observed for their 3-deoxy-5-hydroxy equivalents (Table V-9, spectra 20, 34 and 44).

[2] See note on page 61.

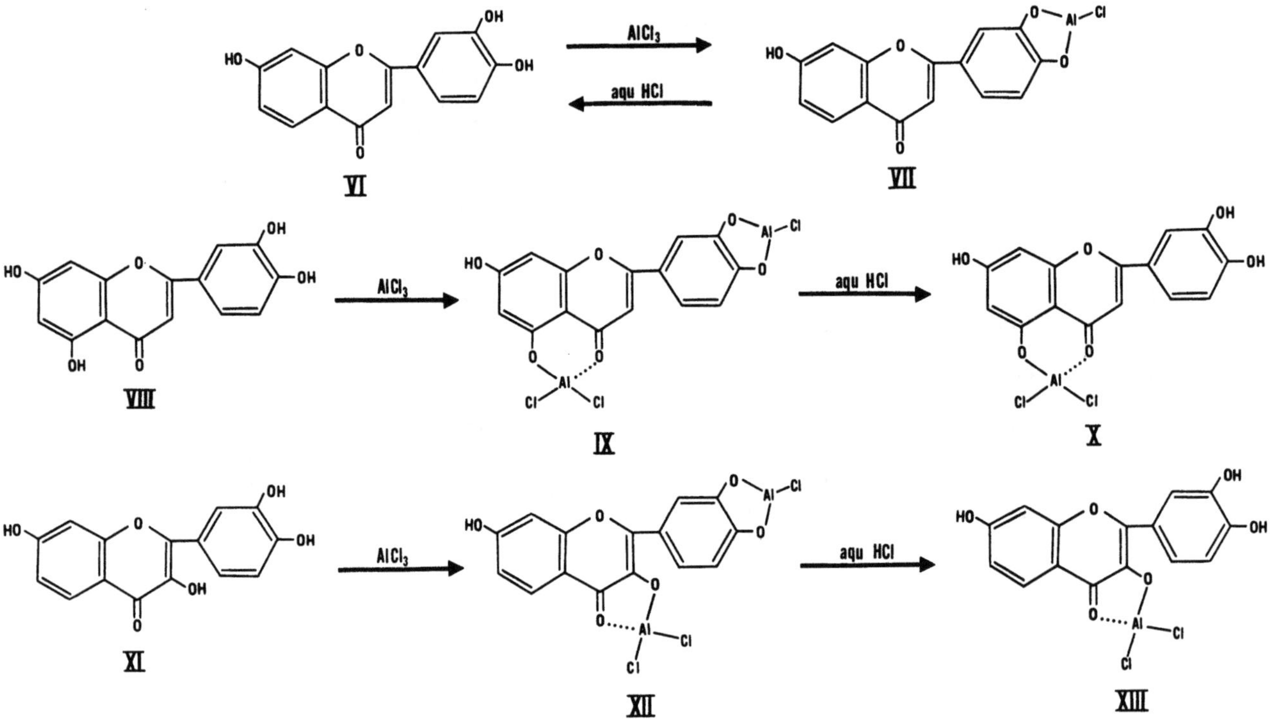

Fig. V-1. Schemes illustrating the types of complexes that $AlCl_3$ could form with certain flavones and flavonols in the presence or absence of acid

5b. The Detection of Ortho-dihydroxyl Groups in Flavones and Flavonols by the Effect of $AlCl_3$ and $AlCl_3$/HCl on the UV Spectrum

For some time $AlCl_3$ has been used as a diagnostic reagent for the detection of ortho-dihydroxyl groups in anthocyanins [*11*]. In 1954 Harborne suggested the use of $AlCl_3$ for recognizing ortho-dihydroxyl groups in other flavonoids [*12*]; however, the method was not fully developed by Harborne or other investigators at that time [*9, 12, 13*]. The presence of an ortho-dihydroxyl group in the B-ring of flavones and flavonols can be detected by a comparison of the spectrum of the flavonoid in the presence of $AlCl_3$ with that obtained in $AlCl_3$/HCl [*10*]. The hypsochromic shift (about 30–40 nm) observed in Band I (or Band Ia if Band I consists of two peaks) of the $AlCl_3$ spectrum on the addition of acid results from the decomposition of the complex of $AlCl_3$ with the ortho-dihydroxyl group (see Table V-8). The presence of three adjacent hydroxyl groups in the B-ring gives only a 20 nm hypsochromic shift on the addition of acid to the $AlCl_3$ solution (spectra 83, 102).

A-Ring ortho-dihydroxyl groups which do not involve the hydrogen-bonded C-5 hydroxyl group (e.g. hydroxylation at C-6, 7 and C-7, 8) can also be detected by this method (see spectra 16, 17 and 18); however, there were insufficient examples available to define the range of the hypsochromic shift for the A-ring systems.

5c. The Detection of Either 3- or 5-Hydroxyl Groups in Flavones and Flavonols by the Effect of $AlCl_3$/HCl on the UV Spectrum

The addition of acid to a methanolic solution of a flavone or flavonol which already contains $AlCl_3$ decomposes complexes between $AlCl_3$ and ortho-dihydroxyl groups; therefore any shift still remaining in Band I or Band II relative to the methanol spectrum

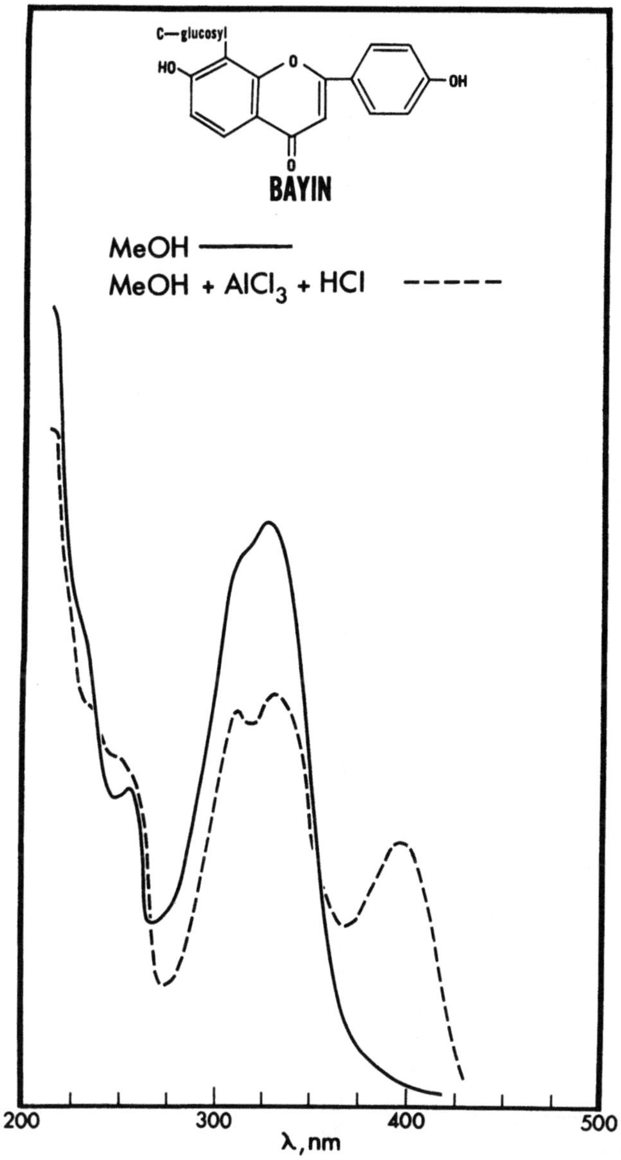

Fig. V-2

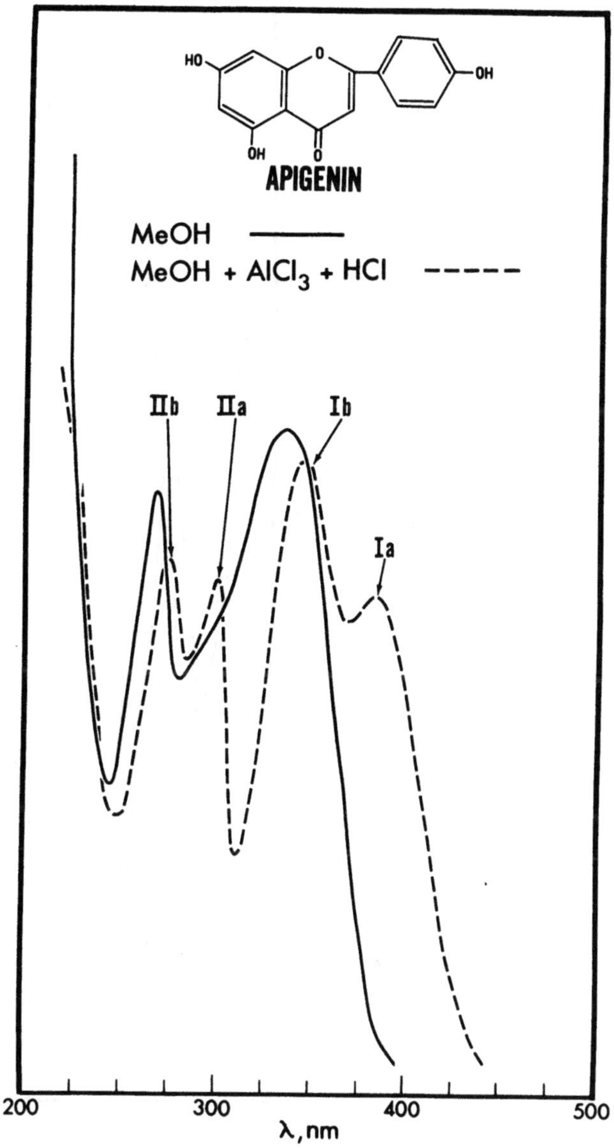

Fig. V-3

Fig. V-2. The effect of $AlCl_3$/HCl on the methanol spectrum of a 5-deoxy-7-hydroxyflavone, bayin. (HCl alone produces the same effect)

Fig. V-3. The effect of $AlCl_3$/HCl on the methanol spectrum of apigenin

will be due to the presence of free 3- and/or 5-hydroxyl groups in the flavonoid. Regeneration of the methanol spectrum on the addition of acid indicates that both the 3- and 5-hydroxyl groups are either absent or substituted. The only difficulty in the interpretation of the $AlCl_3$/HCl spectra was encountered with members of the rare group of 5-deoxy-7-hydroxyflavones. With these compounds the spectrum observed after the addition of HCl to the solution used for the $AlCl_3$ spectrum exhibited all the peaks of the methanol spectrum but, in addition, showed a moderately intense longer wavelength peak about 60 nm from Band I of the methanol spectrum (see Fig. V-2 and spectra 3, 9, 11, 12, 13 and 15). It appears that this moderately intense long wavelength peak is due to protonation of the flavonoid (probably on the C-4 oxygen function) since a methanol/HCl spectrum of 7-hydroxy-5-deoxyflavone was essentially identical with the $AlCl_3$/HCl spectrum obtained for the same compound.

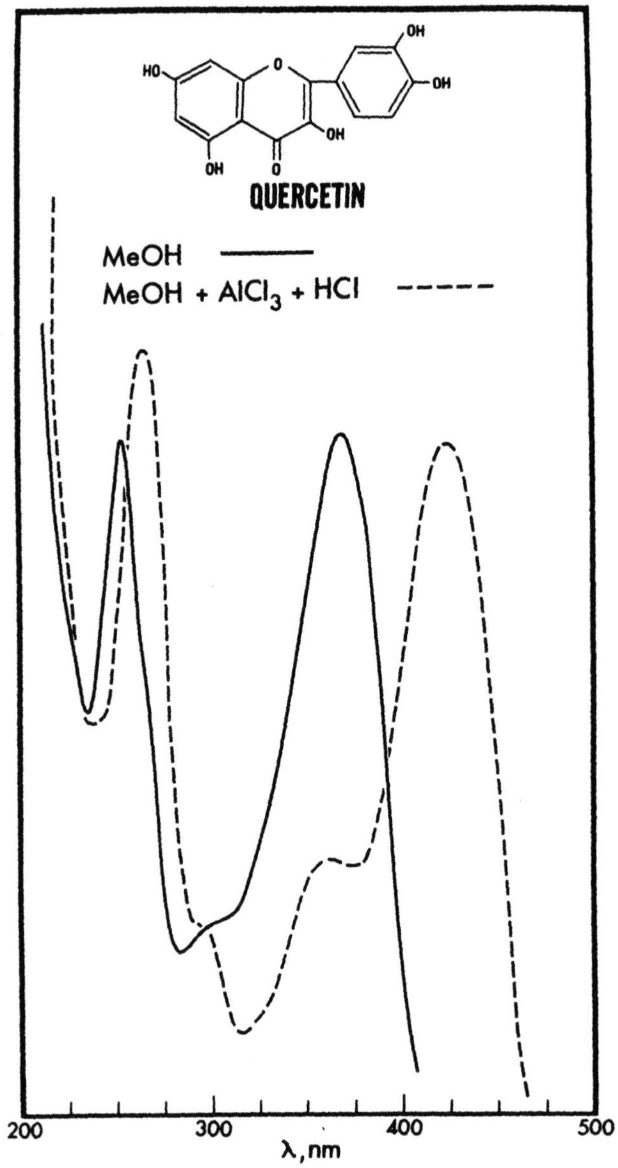

Fig. V-4. The effect of $AlCl_3$/HCl on the methanol spectrum of quercetin

The $AlCl_3$/HCl spectrum of a 5-hydroxyflavone typically consists of four major absorption peaks, Band Ia, Ib, IIa and IIb (see Fig. V-3), which are all shifted bathochromically relative to their Band of origin (presumably Ia and Ib originate from I; and IIa and IIb from II) in the methanol spectrum. It is usual for the general shapes of the $AlCl_3$/HCl spectral curves of both 3-hydroxyflavone and 3,5-dihydroxyflavone to be similar to those of 5-hydroxyflavones (but with Bands Ib and IIa considerably reduced, see Fig. V-4).

There is a clear distinction between the magnitude of the $AlCl_3$/HCl bathochromic shifts associated with 5-hydroxyflavones (and 5-hydroxy-3-substituted flavonols) and those observed for 3-hydroxyflavones. The bathochromic shifts of Band I (in MeOH) to Band Ia (in $AlCl_3$/HCl) in the spectra of 5-hydroxyflavones and 3-substituted flavonols, with few exceptions, are in the range 35–55 nm (Table V-9). In contrast, the shift is consistently around 60 nm for 3-hydroxyflavones (Table V-10). The bathochromic shift of Band I to Ia in 3,5-dihydroxyflavones (range 50–60 nm) (Table V-11) is usually intermediate between those observed for their 3-hydroxy and 5-hydroxy-equivalents.

Table V-9. *The effect of $AlCl_3$/HCl on Band I in the UV spectra of 5-hydroxyflavones and 3-substituted flavonols*

Spectrum No.	Compound	Bathochromic shift (nm) of Band I (in MeOH) to Band Ia (in the presence of $AlCl_3$/HCl)
	5-Hydroxyflavones	
2	5-Hydroxyflavone	60
7	Chrysin	68
8	Tectochrysin[a]	–
16	Baicalein (5,6,7-Trihydroxyflavone)	23
17	Baicalin	28
18	Norwogonin (5,7,8-Trihydroxyflavone)[a]	–
19	5,7,8-Trihydroxyflavone 7-O-glucuronide[a]	–
20	Apigenin	45
21	Apigenin 7-O-glucoside	49
22	Apigenin 7-O-neohesperidoside	47
23	Isovitexin	44
24	Saponarin	42
25	Vitexin	47
26	Rhamnosylvitexin	47
27	2″-O-Xylosylvitexin	47
28	Violanthin	48
29	Acacetin	52
30	Acacetin 7-O-glucoside	57
31	5,7-Dihydroxy-2′-methoxyflavone	53
32	Zapotinin[a]	–
34	Luteolin	36
35	Luteolin 7-O-glucoside	39
36	Luteolin 7-O-rutinoside	40
37	Isoorientin	35
38	Orientin	38
39	Lucenin-1	35
40	Chrysoeriol	39
41	Scoparin	37
42	Diosmetin	39
43	5,7-Dihydroxy-3′,4′-dimethoxyflavone	41
44	Tricin	36
45	5,7-Dihydroxy-3′,4′,5′-trimethoxyflavone	51
46	Nevadensin[a]	–
47	Xanthomicrol[a]	–
48	Hymenoxin[a]	–
49	Amentoflavone	50
50	Sciadopitysin	56
	Flavonols with the 3-hydroxyl substituted	
55	Galangin 3-methyl ether	51
60	Kaempferol 3-O-robinoside 7-O-rhamnoside (Robinin)	48
67	Quercetin 3-O-galactoside	43
68	Quercitrin	51
69	Rutin	43
70	Quercetin 3,7-O-diglucoside	47
71	Quercetin 3-O-glucoside 7-O-rhamnoside	46
72	Quercetin 3-O-glucoside 7-O-rutinoside	46
73	Quercetin 3-methyl ether	44
74	Quercetin 3-methyl ether 4′-O-glucoside 7-O-diglucoside	50
78	Isorhamnetin 3-O-galactoside	46
79	Isorhamnetin 3-O-rutinoside	43
84	Penduletin	62

Table V-9 (continued)

Spectrum No.	Compound	Bathochromic shift (nm) of Band I (in MeOH) to Band I a (in the presence of $AlCl_3$/HCl)
85	Pendulin	72
87	Jaceidin	60
88	Jacein	55
89	Centaurein	54
91	Patuletin 3-O-glucoside	52
92	Patuletin 3-O-rutinoside	48
94	Artemetin	58
103	3′,5,5′-Trihydroxy-3,4′,6,7-tetramethoxyflavone	68

[a] For these compounds the magnitude of the shift was difficult to determine accurately; however, the actual spectra are presented at the end of this chapter.

Table V-10. *The effect of $AlCl_3$/HCl on Band I in the UV spectra of* 3-*hydroxyflavones lacking a* 5-*hydroxyl group*

Spectrum No.	Compound	Bathochromic shift (nm) of Band I (in MeOH) to Band I a (in the presence of $AlCl_3$/HCl
51	3-Hydroxyflavone	56
52	3-Hydroxy-4′-methoxyflavone	62
53	3,4′,7-Trihydroxyflavone	62
56	3,3′,4′-Trihydroxyflavone	61
57	3-Hydroxy-3′,4′-dimethoxyflavone	67
62	Fisetin	61
75	Quercetin 3′,4′,5,7-tetramethyl ether	58
83	Robinetin	59
95	Quercetagetin 3′,4′,5,6,7-pentamethyl ether	67

Table V-11. *The effect of $AlCl_3$/HCl on Band I in the UV spectra of* 3,5-*dihydroxyflavones*

Spectrum No.	Compound	Bathochromic shift (nm) of Band I (in MeOH) to Band I a (in the presence of $AlCl_3$/HCl)
54	Galangin	53
58	Kaempferol	57
59	Kaempferol 7-O-neohesperidoside	58
61	Kaempferol 4′-methyl ether	55
64	Herbacetin 8-methyl ether	57
65	Quercetin	58
66	Quercetin 7-O-rhamnoside	54
76	Rhamnetin	52
77	Isorhamnetin	61
80	Tamarixetin 7-O-neohesperidoside	58
81	Tamarixetin 7-O-rutinoside	56
82	Morin	49
90	Patuletin	56
93	Patulitrin	58
98	Gossypetin	62
99	Gossypin	61
100	Gossypitrin	69
102	Myricetin	54

V-6. Index[a] of Ultraviolet Absorption Spectra of Flavones and Flavonols

Spectrum No.	Flavones	Oxidation pattern 5	6	7	8	2′	3′	4′	5′	6′
1	Flavone									
2	5-Hydroxyflavone	OH								
3	7-Hydroxyflavone			OH						
4	4′-Methoxyflavone							OCH_3		
5	3′,4′-Dihydroxyflavone						OH	OH		
6	3′,4′-Dimethoxyflavone						OCH_3	OCH_3		
7	Chrysin	OH		OH						
8	Tectochrysin	OH		OCH_3						
9	4′,7-Dihydroxyflavone			OH				OH		
10	4′,7-Dihydroxyflavone 7-O-rhamnoglucoside			[O-rh-glu]				OH		
11	5-Deoxyvitexin (Bayin)			OH	C-glu			OH		
12	7-Hydroxy-4′-methoxyflavone (Pratol)			OH				OCH_3		
13	3′,4′,7-Trihydroxyflavone			OH			OH	OH		
14	3′,4′,7-Trihydroxyflavone 7-O-rhamnoglucoside			[O-rh-glu]			OH	OH		
15	7-Hydroxy-3′,4′-dimethoxyflavone			OH			OCH_3	OCH_3		
16	Baicalein 5,6,7-Trihydroxyflavone)	OH	OH	OH						
17	Baicalin (5,6,7-Trihydroxyflavone 7-O-glucuronide)	OH	OH	O-gluc						
18	5,7,8-Trihydroxyflavone (Norwogonin)	OH		OH	OH					
19	5,7,8-Trihydroxyflavone 7-O-glucuronide	OH		O-gluc	OH					
20	Apigenin	OH		OH				OH		
21	Apigenin 7-O-glucoside	OH		O-glu				OH		
22	Apigenin 7-O-neohesperidoside	OH		[O-rh-glu]				OH		
23	Isovitexin (Saponaretin)	OH	C-glu	OH				OH		
24	Saponarin	OH	C-glu	O-glu				OH		
25	Vitexin	OH		OH	C-glu			OH		
26	Rhamnosylvitexin	OH		OH	[C-rh-glu]			OH		
27	2″-O-Xylosylvitexin	OH		OH	[C-xylo-glu]			OH		
28	Violanthin	OH	C-gly	OH	C-gly			OH		
29	Acacetin	OH		OH				OCH_3		

Spectrum No.	Flavones	Oxidation pattern								
		5	6	7	8	2′	3′	4′	5′	6′
30	Acacetin 7-O-glucoside	OH		O-glu				OCH_3		
31	5,7-Dihydroxy-2′-methoxyflavone	OH		OH		OCH_3				
32	Zapotinin	OH	OCH_3			OCH_3				OCH_3
33	Zapotin	OCH_3	OCH_3			OCH_3				OCH_3
34	Luteolin	OH		OH			OH	OH		
35	Luteolin 7-O-glucoside	OH		O-glu			OH	OH		
36	Luteolin 7-O-rutinoside	OH		[O-rh-glu]			OH	OH		
37	Isoorientin	OH	C-glu	OH			OH	OH		
38	Orientin	OH		OH	C-glu		OH	OH		
39	Lucenin-1	OH	C-gly	OH	C-gly		OH	OH		
40	Chrysoeriol	OH		OH			OCH_3	OH		
41	Scoparin	OH		OH	C-glu		OCH_3	OH		
42	Diosmetin	OH		OH			OH	OCH_3		
43	5,7-Dihydroxy-3′,4′-dimethoxyflavone	OH		OH			OCH_3	OCH_3		
44	Tricin	OH		OH			OCH_3	OH	OCH_3	
45	5,7-Dihydroxy-3′,4′,5′-trimethoxyflavone	OH		OH			OCH_3	OCH_3	OCH_3	
46	Nevadensin	OH	OCH_3	OH	OCH_3			OCH_3		
47	Xanthomicrol	OH	OCH_3	OCH_3	OCH_3			OH		
48	Hymenoxin	OH	OCH_3	OH	OCH_3		OCH_3	OCH_3		
49	Amentoflavone	{OH OH		OH OH				OH OH		
50	Sciadopitysin	{OH OH		OCH_3 OH				OCH_3 OCH_3		

	Flavonols	Oxidation pattern								
		3	5	6	7	8	2′	3′	4′	5′
51	3-Hydroxyflavone	OH								
52	3-Hydroxy-4′-methoxyflavone	OH							OCH_3	
53	3,4′,7-Trihydroxyflavone	OH			OH				OH	
54	Galangin	OH	OH		OH					
55	Galangin 3-methyl ether	OCH_3	OH		OH					
56	3,3′,4′-Trihydroxyflavone	OH						OH	OH	
57	3-Hydroxy-3′,4′-dimethoxyflavone	OH						OCH_3	OCH_3	
58	Kaempferol	OH	OH		OH				OH	

Spectrum No.	Flavonols	Oxidation pattern								
		3	5	6	7	8	2′	3′	4′	5′
59	Kaempferol 7-O-neohesperidoside	OH	OH		[O-rh-glu]				OH	
60	Kaempferol 3-O-robinoside 7-O-rhamnoside (Robinin)	[O-rh-gal]	OH		O-rh				OH	
61	Kaempferol 4′-methyl ether	OH	OH		OH				OCH_3	
62	Fisetin	OH			OH			OH	OH	
63	Fisetin 3-O-glucoside	O-glu			OH			OH	OH	
64	Herbacetin 8-methyl ether	OH	OH		OH	OCH_3			OH	
65	Quercetin	OH	OH		OH			OH	OH	
66	Quercetin 7-O-rhamnoside	OH	OH		O-rh			OH	OH	
67	Quercetin 3-O-galactoside	O-gal	OH		OH			OH	OH	
68	Quercitrin	O-rh	OH		OH			OH	OH	
69	Rutin	[O-rh-glu]	OH		OH			OH	OH	
70	Quercetin 3,7-O-diglucoside	O-glu	OH		O-glu			OH	OH	
71	Quercetin 3-O-glucoside 7-O-rhamnoside	O-glu	OH		O-rh			OH	OH	
72	Quercetin 3-O-glucoside 7-O-rutinoside	O-glu	OH		[O-rh-glu]			OH	OH	
73	Quercetin 3-methyl ether	OCH_3	OH		OH			OH	OH	
74	Quercetin 3-methyl ether 4′-O-glucoside 7-O-diglucoside	OCH_3	OH		[O-glu-glu]			OH	O-glu	
75	Quercetin 3′,4′,5,7-tetramethyl ether	OH	OCH_3		OCH_3			OCH_3	OCH_3	
76	Rhamnetin	OH	OH		OCH_3			OH	OH	
77	Isorhamnetin	OH	OH		OH			OCH_3	OH	
78	Isorhamnetin 3-O-galactoside	O-gal	OH		OH			OCH_3	OH	
79	Isorhamnetin 3-O-rutinoside	[O-rh-glu]	OH		OH			OCH_3	OH	
80	Tamarixetin 7-O-neohesperidoside	OH	OH		[O-rh-glu]			OH	OCH_3	
81	Tamarixetin 7-O-rutinoside	OH	OH		[O-rh-glu]			OH	OCH_3	
82	Morin	OH	OH		OH		OH		OH	
83	Robinetin	OH			OH			OH	OH	OH
84	Penduletin	OCH_3	OH	OCH_3	OCH_3				OH	
85	Pendulin	OCH_3	OH	OCH_3	OCH_3				O-glu	
86	3,5,6,7,8-Pentamethoxyflavone	OCH_3	OCH_3	OCH_3	OCH_3	OCH_3				
87	Jaceidin	OCH_3	OH	OCH_3	OH			OCH_3	OH	
88	Jacein	OCH_3	OH	OCH_3	O-glu			OCH_3	OH	

Spectrum No.	Flavonols	Oxidation pattern								
		3	5	6	7	8	2′	3′	4′	5′
89	Centaurein	OCH_3	OH	OCH_3	O-glu			OH	OCH_3	
90	Patuletin	OH	OH	OCH_3	OH			OH	OH	
91	Patuletin 3-O-glucoside	O-glu	OH	OCH_3	OH			OH	OH	
92	Patuletin 3-O-rutinoside	[O-rh-glu]	OH	OCH_3	OH			OH	OH	
93	Patulitrin	OH	OH	OCH_3	O-glu			OH	OH	
94	Artemetin	OCH_3	OH	OCH_3	OCH_3			OCH_3	OCH_3	
95	Quercetagetin 3′,4′,5,6,7-pentamethyl ether	OH	OCH_3	OCH_3	OCH_3			OCH_3	OCH_3	
96	Quercetagetin 3,3′,4′,5,6-pentamethyl ether	OCH_3	OCH_3	OCH_3	OH			OCH_3	OCH_3	
97	Quercetagetin hexamethyl ether	OCH_3	OCH_3	OCH_3	OCH_3			OCH_3	OCH_3	
98	Gossypetin	OH	OH		OH	OH		OH	OH	
99	Gossypin	OH	OH		OH	O-glu		OH	OH	
100	Gossypitrin	OH	OH		O-glu	OH		OH	OH	
101	Gossypetin hexamethyl ether	OCH_3	OCH_3		OCH_3	OCH_3		OCH_3	OCH_3	
102	Myricetin	OH	OH		OH			OH	OH	OH
103	3′,5,5′-Trihydroxy-3,4′,6,7-tetramethoxyflavone	OCH_3	OH	OCH_3	OCH_3			OH	OCH_3	OH

[a] Abbreviations: gal = galactosyl; glu = glucosyl; gluc = glucuronosyl; gly = glycosyl; neohesp = neohesperidosyl; rh = rhamnosyl; rh-glu = rhamnoglucosyl; rut = rutinosyl.

References

1. Jurd, L., in: The chemistry of flavonoid compounds (edited by T. A. Geissman), p. 107–155. Oxford: Pergamon Press 1962.
2. Astill, B. D., and J. C. Roberts: J. Chem. Soc. 3302 (1953).
3. Brockmann, H., E. H. F. Falkenhausen, R. Neeff, A. Dorlars, and G. Budde: Chem. Ber. **84**, 865 (1951).
4. Dechene, E. B.: J. Am. Pharm. Assoc. **40**, 495 (1951).
5. Jurd, L., and R. M. Horowitz: J. Org. Chem. **22**, 1618 (1957).
6. Lee, H. H., and C. H. Tan: J. Chem. Soc. 2743 (1965).
7. Farkas, L., M. Nogradi, V. Sudarsanam, and W. Herz: J. Org. Chem. **31**, 3228 (1966).
8. Hörhammer, L., and R. Hänsel: Arch. Pharm. **285**, 438 (1952).
9. Jurd, L., and T. A. Geissman: J. Org. Chem. **21**, 1395 (1956).
10. Markham, K. R., and T. J. Mabry: Phytochemistry **7**, 1197 (1968).
11. Geissman, T. A., E. C. Jorgenson, and J. B. Harborne: Chem. Ind. (London) 1389 (1953).
12. Harborne, J. B.: Chem. Ind. (London) 1142 (1954).
13. Swain, T.: Chem. Ind. (London) 1480 (1954).

James Mears of our laboratory observed that a bathochromic shift of about 20 nm in Band I (in MeOH) to Band Ia in the presence of $AlCl_3/HCl$ is diagnostic for 5-hydroxyflavones and 3-O-substituted 5-hydroxyflavonols containing an oxygen function (either hydroxyl or methoxyl) at the 6-position. [Dr. Leonard Jurd has informed us that he obtained equivalent results using slightly different conditions; see also L. Jurd, Phytochemistry **8**, 445 (1969).]

1

FLAVONE

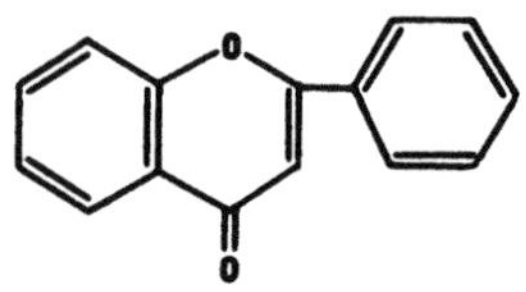

CHROMATOGRAPHIC DATA

Spot Appearance: (UV) blue-purple
(UV/NH_3) blue-purple

R_f Values: 0.92 (TBA), indef. (HOAc)

UV SPECTRAL DATA (λ_{max}, nm)

MeOH	250, 294, 307sh
NaOMe	250, 294, 309sh
$AlCl_3$	250, 293, 306sh
$AlCl_3$/HCl	250, 293, 309sh
NaOAc	248, 292, 307sh
NaOAc/H_3BO_3 (Proc. I)	255sh, 294, 307sh

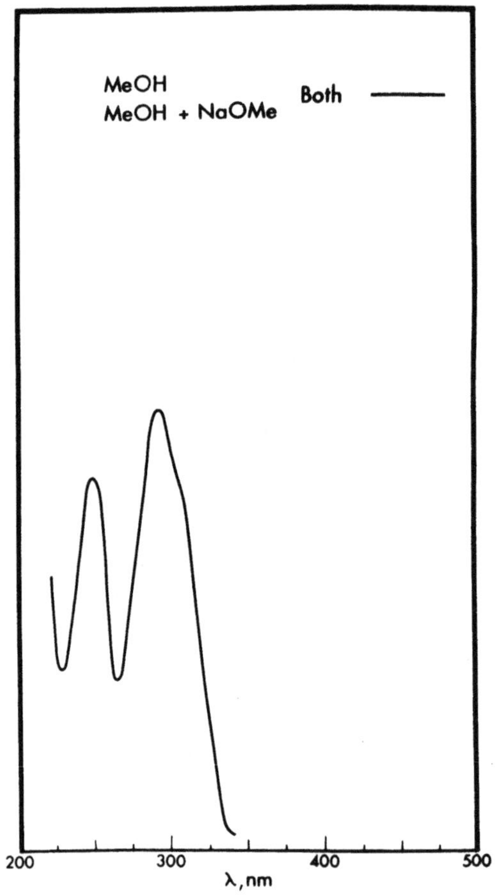

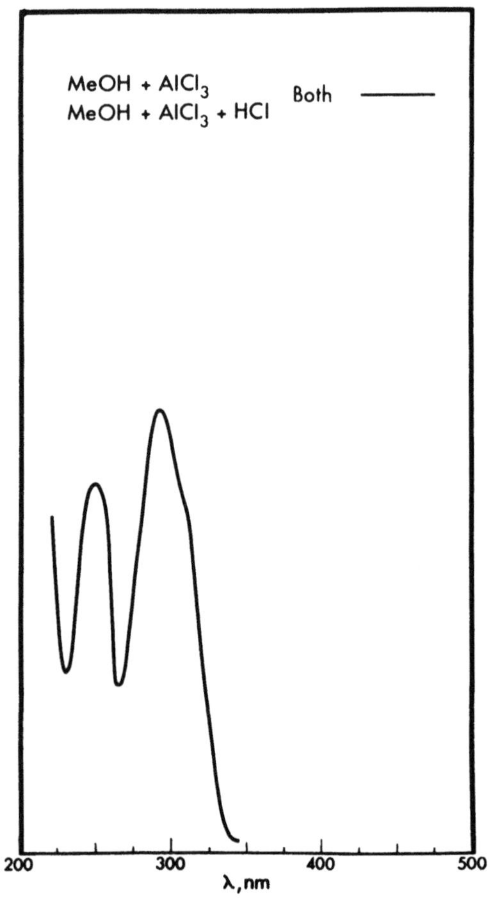

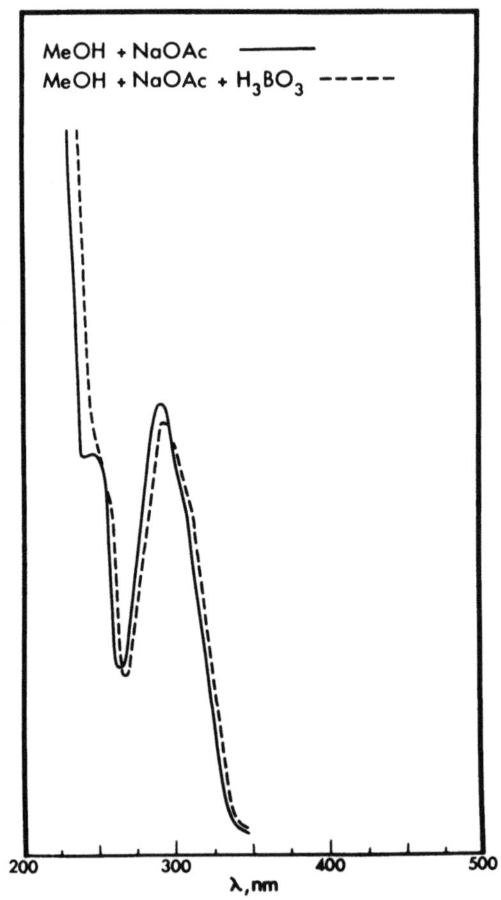

2

5-HYDROXYFLAVONE

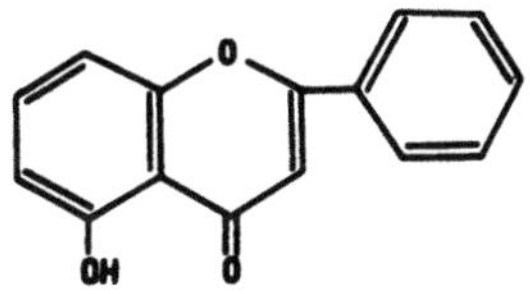

CHROMATOGRAPHIC DATA

Spot Appearance: (UV) deep purple
(UV/NH_3) deep purple

R_f Values: 0.91 (TBA), 0.15 (HOAc)

UV SPECTRAL DATA (λ_{max}, nm)

MeOH	268, 296sh, 333
NaOMe	272, 380
$AlCl_3$	290, 318sh, 394
$AlCl_3$/HCl	291, 319sh, 393
NaOAc	270, 297sh, 335
NaOAc/H_3BO_3	268, 298sh, 334

(Proc. I)

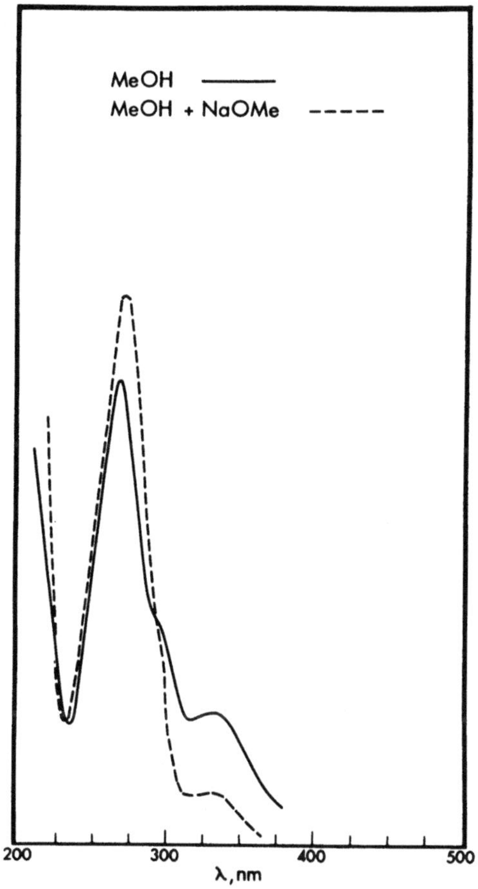

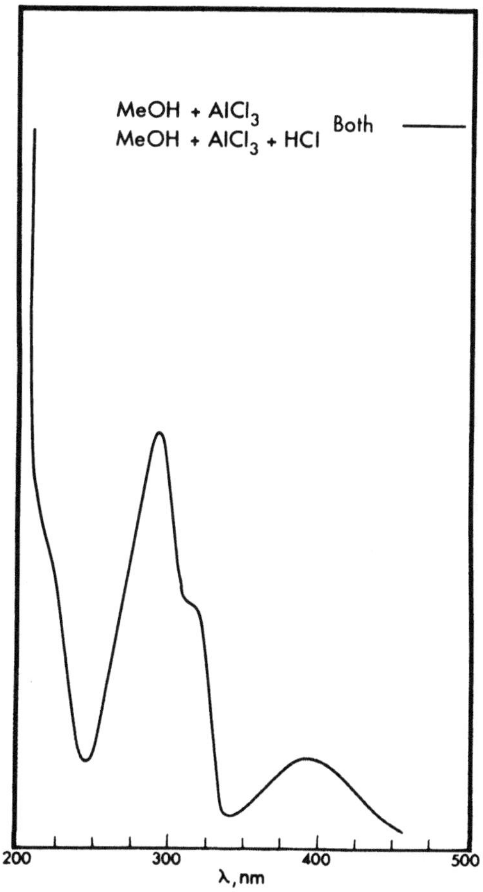

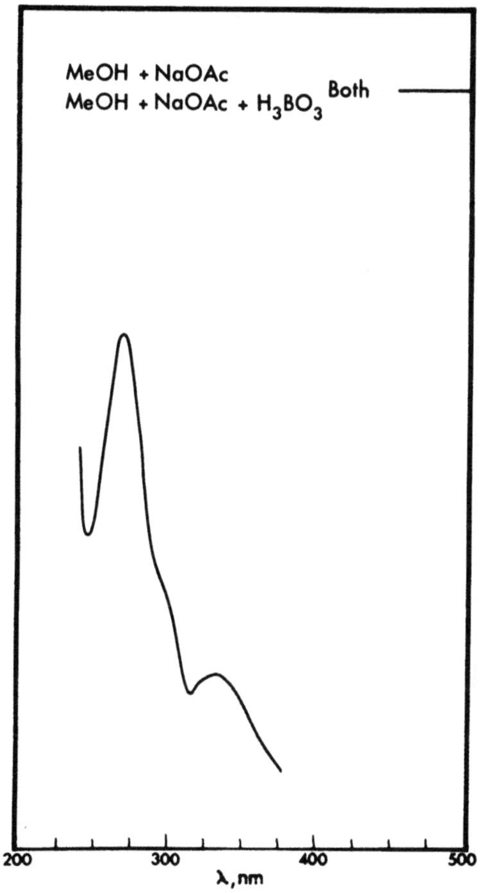

3

7-HYDROXYFLAVONE

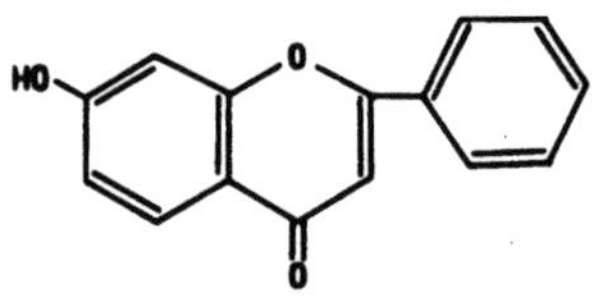

CHROMATOGRAPHIC DATA

Spot Appearance: (UV) fluorescent pale yellow
(UV/NH_3) fluorescent bright yellow

R_f Values: 0.89 (TBA), 0.29 (HOAc)

UV SPECTRAL DATA (λ_{max}, *nm*)

MeOH	252, 268, 307
NaOMe	266, 307, 359
$AlCl_3$	249, 307
$AlCl_3$/HCl	251, 307, 372sh
NaOAc	266, 307, 358
NaOAc/H_3BO_3 (Proc. I)	255sh, 270sh, 309

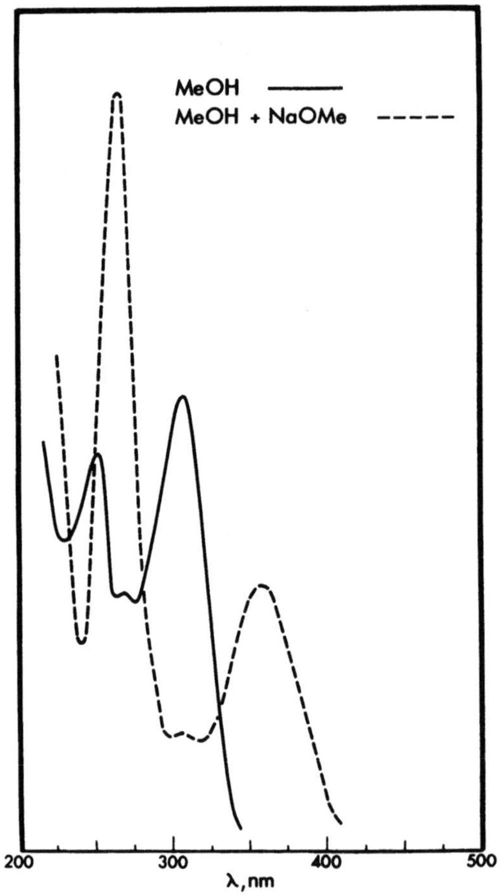

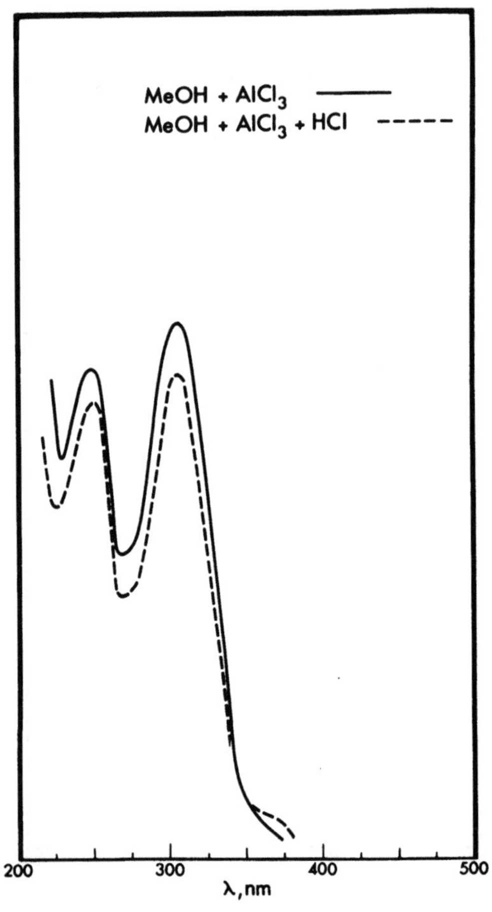

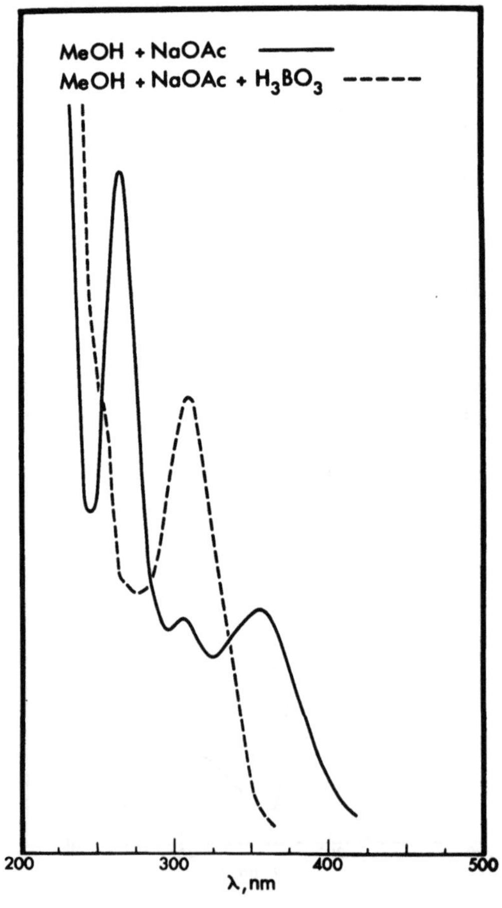

4

4'-METHOXYFLAVONE

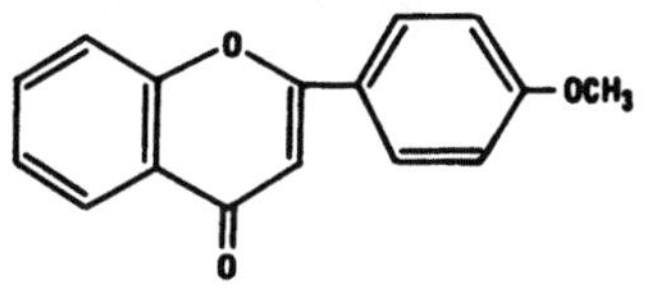

CHROMATOGRAPHIC DATA

Spot Appearance: (UV) fluorescent light blue

(UV/NH_3) fluorescent light blue

R_f Values: 0.88 (TBA), 0.28 (HOAc)

UV SPECTRAL DATA (λ_{max}, *nm*)

MeOH	253, 317
NaOMe	254, 316
$AlCl_3$	253, 317
$AlCl_3$/HCl	253, 319
NaOAc	257sh, 318
NaOAc/H_3BO_3 (Proc. I)	257sh, 319

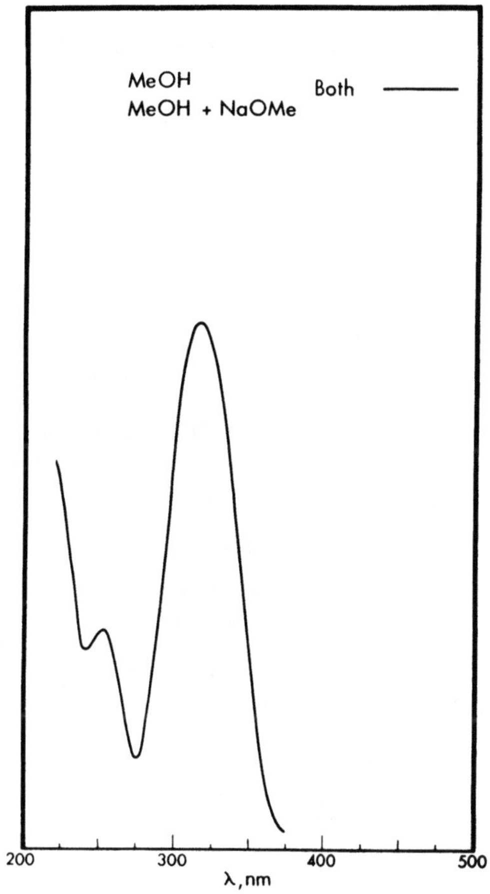

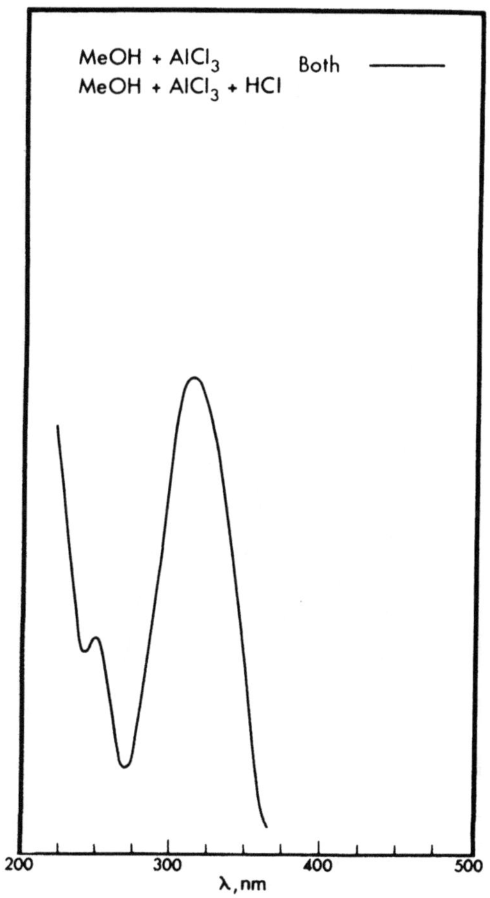

5

3′,4′-DIHYDROXYFLAVONE

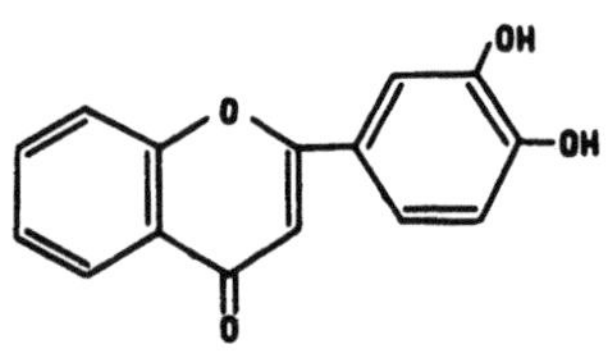

CHROMATOGRAPHIC DATA

Spot Appearance: (UV) fluorescent light blue
(UV/NH_3) fluorescent yellow-green

R_f Values: 0.77 (TBA), 0.18 (HOAc)

UV SPECTRAL DATA (λ_{max}, *nm*)

MeOH	242, 308sh, 340
NaOMe	249sh, 278sh, 302, 404
$AlCl_3$	248sh, 273sh, 304, 378, 468sh
$AlCl_3$/HCl	242, 312sh, 342
NaOAc	305, 348, 400
NaOAc/H_3BO_3 (Proc. I)	306, 365

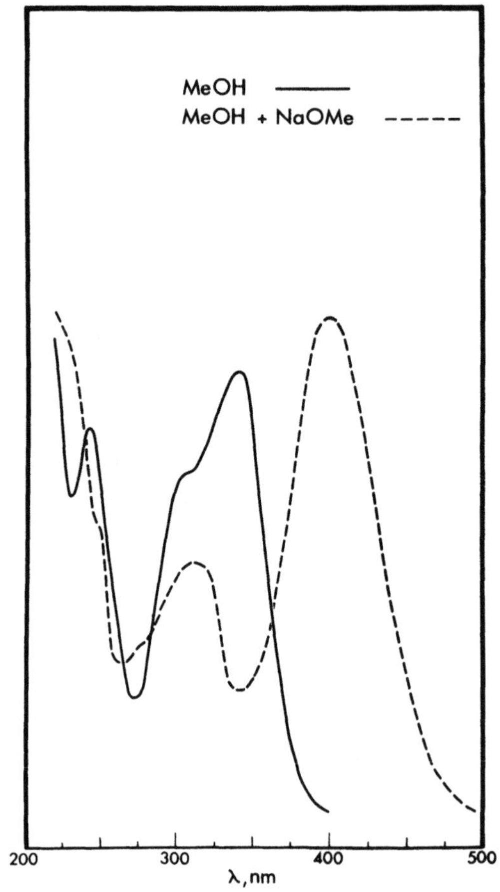

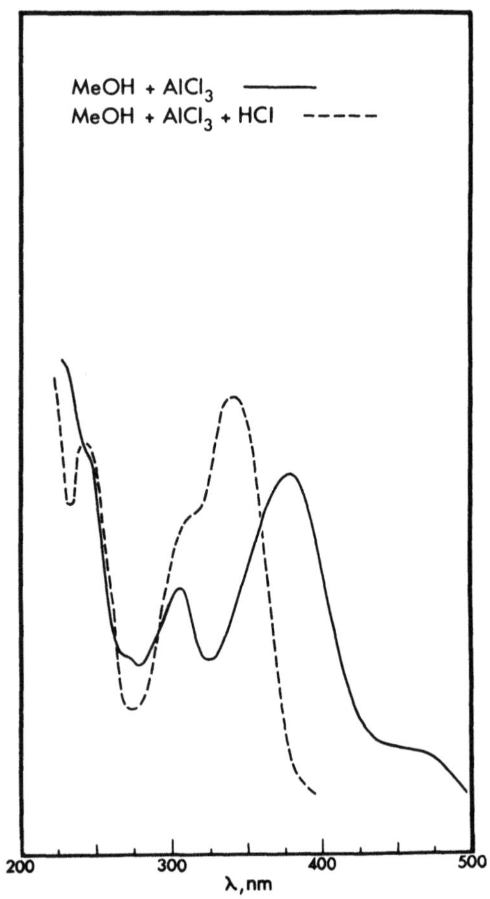

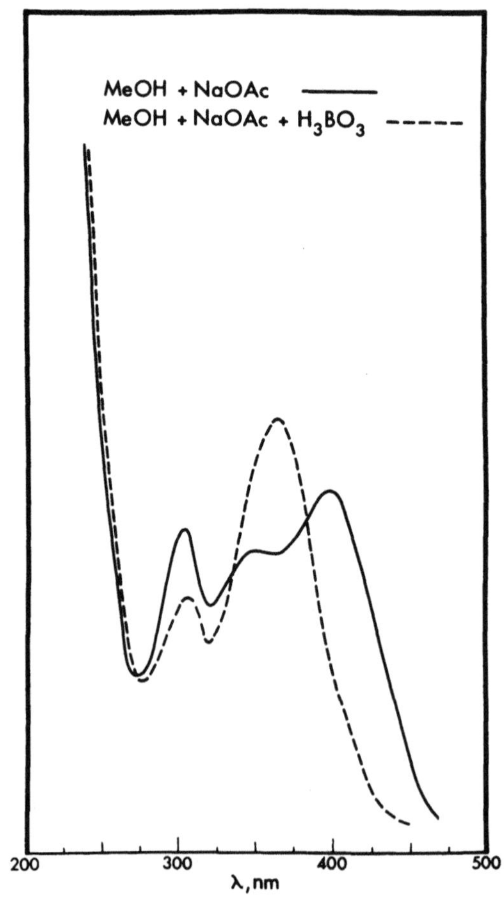

6

3′,4′-DIMETHOXYFLAVONE

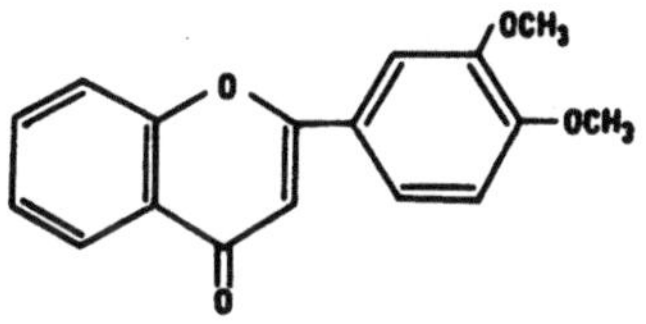

CHROMATOGRAPHIC DATA

Spot Appearance: (UV) fluorescent light blue

(UV/NH_3) fluorescent light blue

R_f Values: 0.84 (TBA), 0.26 (HOAc)

UV SPECTRAL DATA (λ_{max}, *nm*)

MeOH	242, 314sh, 333
NaOMe	241, 314sh, 334
$AlCl_3$	243, 315sh, 333
$AlCl_3$/HCl	242, 315sh, 333
NaOAc	312sh, 334
NaOAc/H_3BO_3 (Proc. I)	314sh, 334

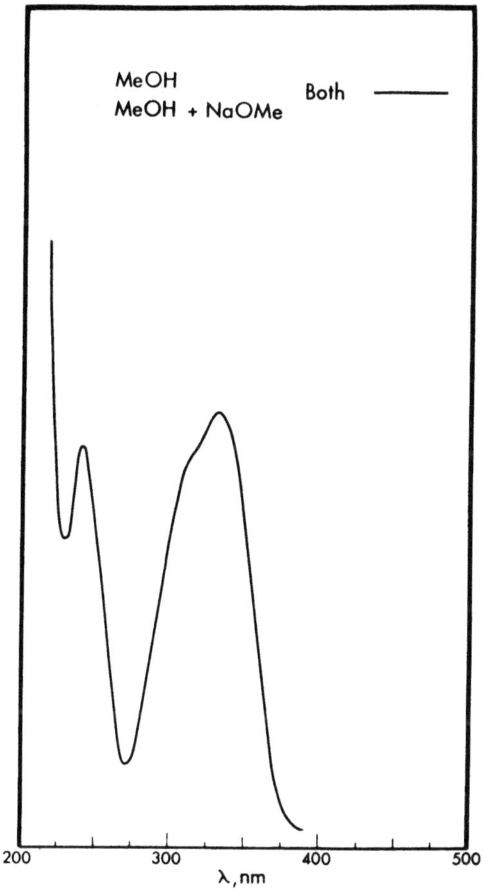

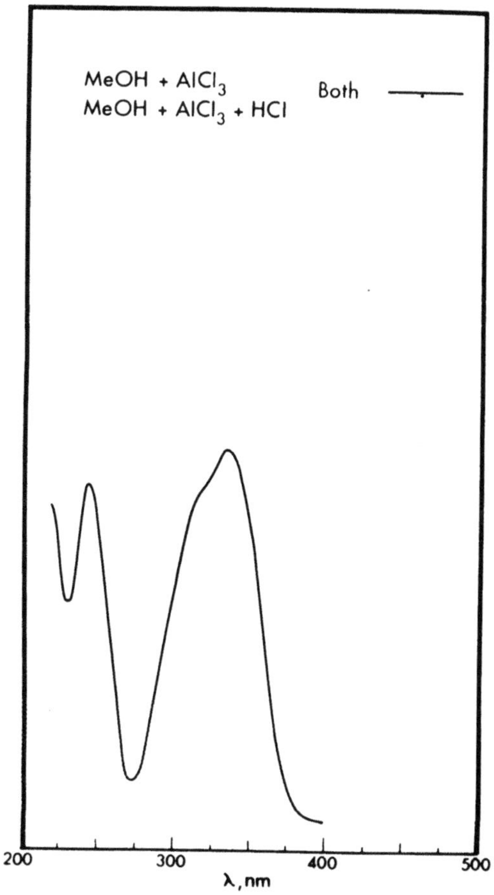

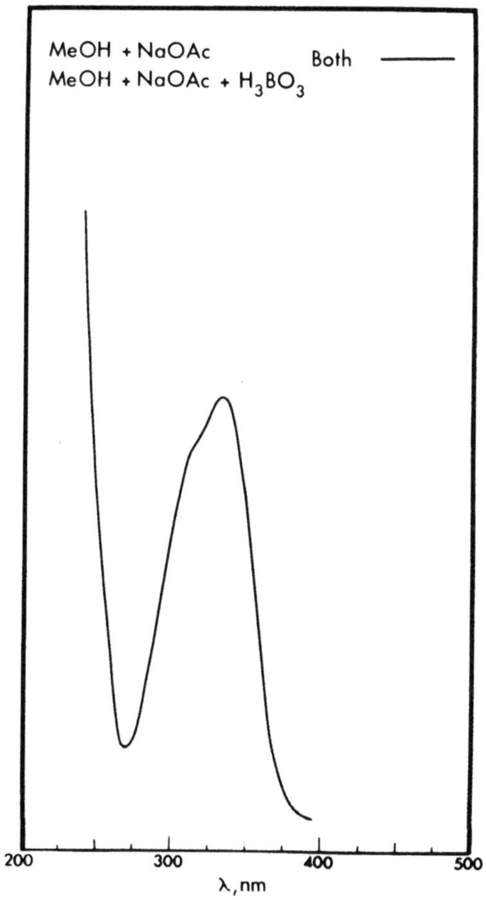

7

CHRYSIN

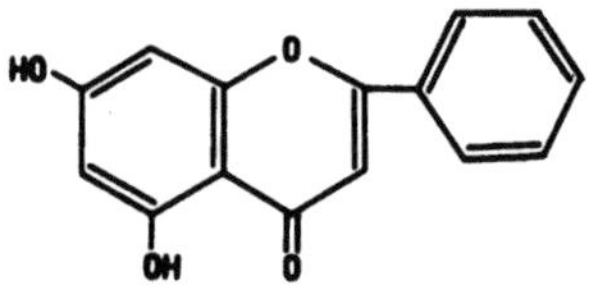

CHROMATOGRAPHIC DATA

Spot Appearance: (UV) deep purple
(UV/NH_3) deep purple

R_f Values: 0.90 (TBA), 0.16 (HOAc)

UV SPECTRAL DATA (λ_{max}, nm)

MeOH	247sh, 268, 313
NaOMe	288, 263sh, 277, 361
$AlCl_3$	252, 279, 330, 380
$AlCl_3$/HCl	251, 280, 326, 381
NaOAc	275, 359
NaOAc/H_3BO_3 (Proc. I)	269, 315

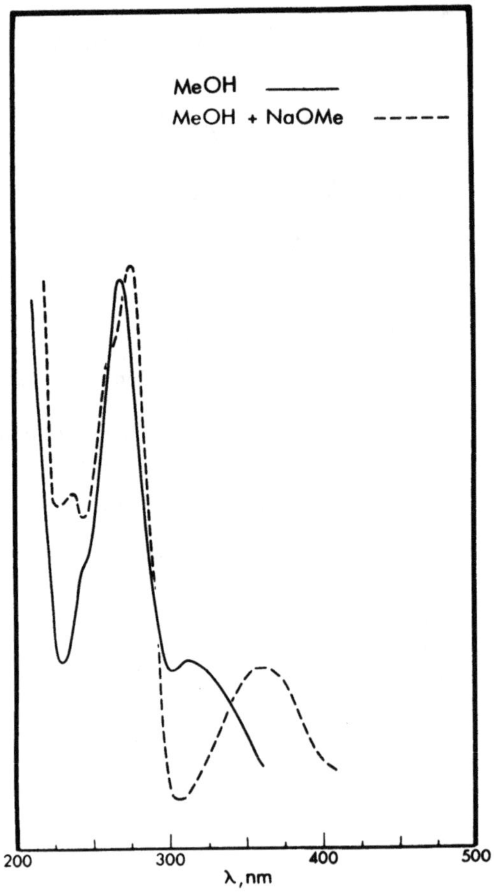

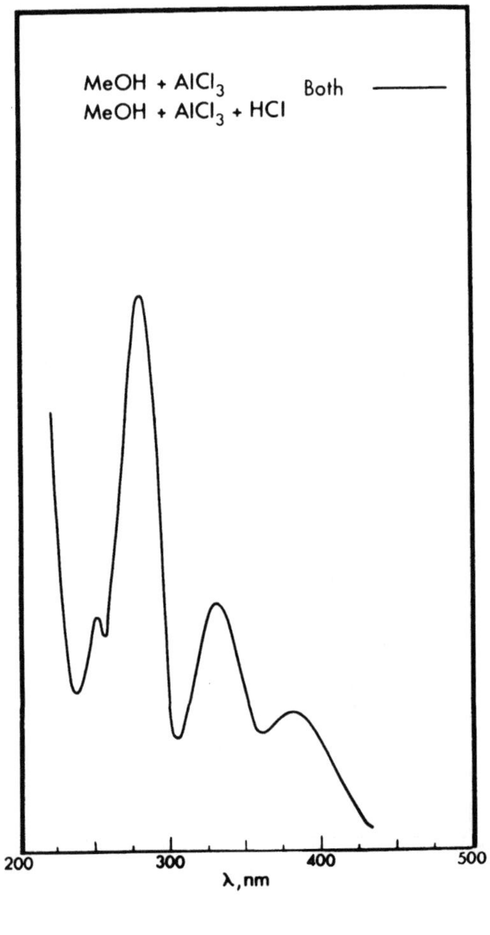

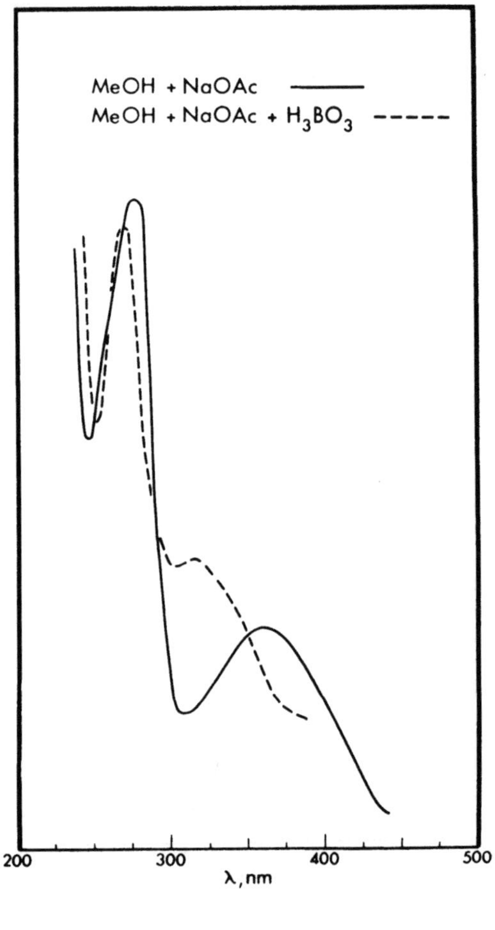

8

TECTOCHRYSIN

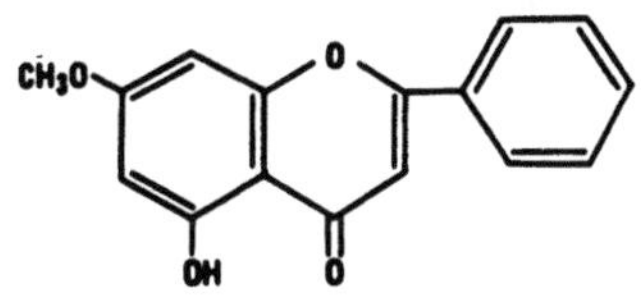

CHROMATOGRAPHIC DATA

Spot Appearance: (UV) deep purple
(UV/NH_3) deep purple

R_f Values: 0.91 (TBA), 0.19 (HOAc)

UV SPECTRAL DATA (λ_{max},*nm*)

MeOH	248sh, 267, 303sh
NaOMe	245, 271
$AlCl_3$	252, 280, 328, 380
$AlCl_3$/HCl	252, 280, 325, 380
NaOAc	268, 308
NaOAc/H_3BO_3 (Proc. I)	268, 309

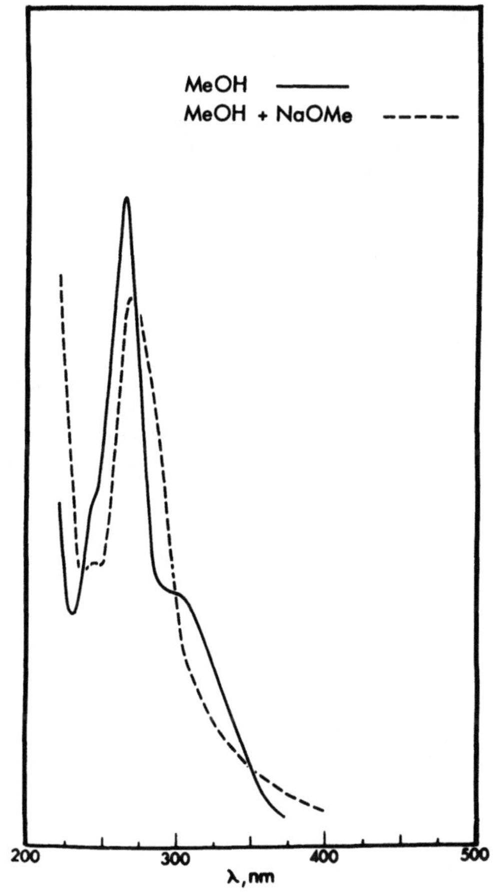

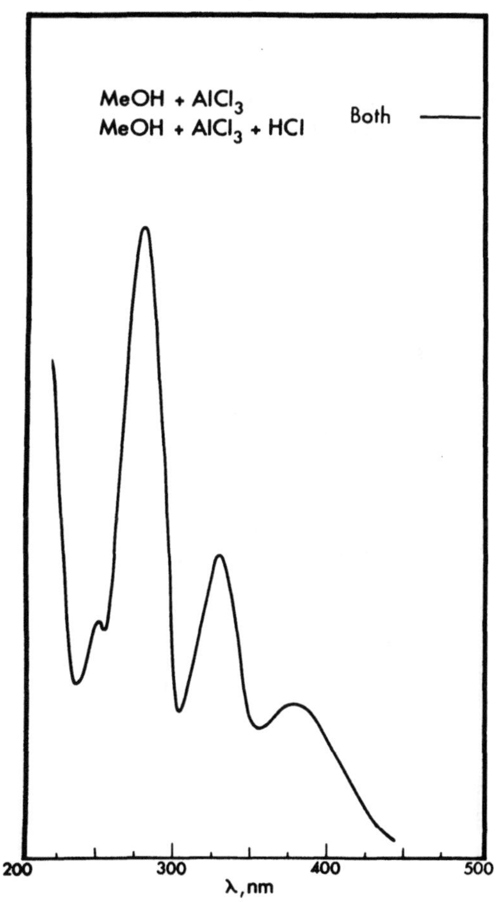

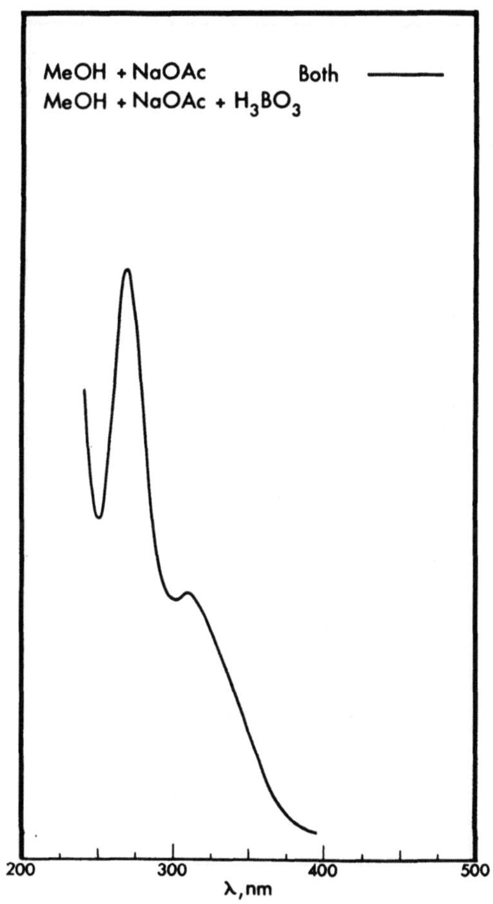

9

4′,7-DIHYDROXYFLAVONE

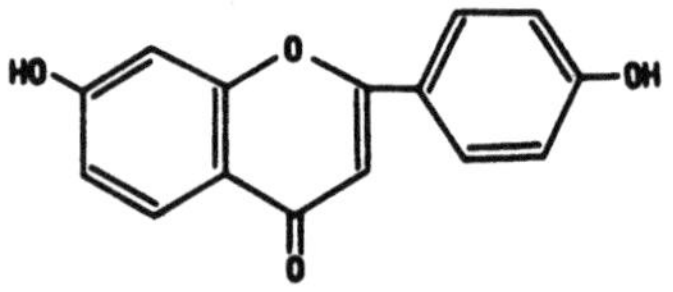

CHROMATOGRAPHIC DATA

Spot Appearance: (UV) fluorescent light blue

(UV/NH_3) fluorescent yellow-green

R_f Values: 0.85 (TBA), 0.11 (HOAc)

UV SPECTRAL DATA (λ_{max}, *nm*)

MeOH	253sh, 312sh, 328
NaOMe	251, 263sh, 329, 386
$AlCl_3$	231sh, 255sh, 313sh, 327, 383sh
$AlCl_3$/HCl	246sh, 255sh, 310sh, 328, 396
NaOAc	261, 309, 320sh, 369
NaOAc/H_3BO_3 (Proc. I)	256sh, 314sh, 329

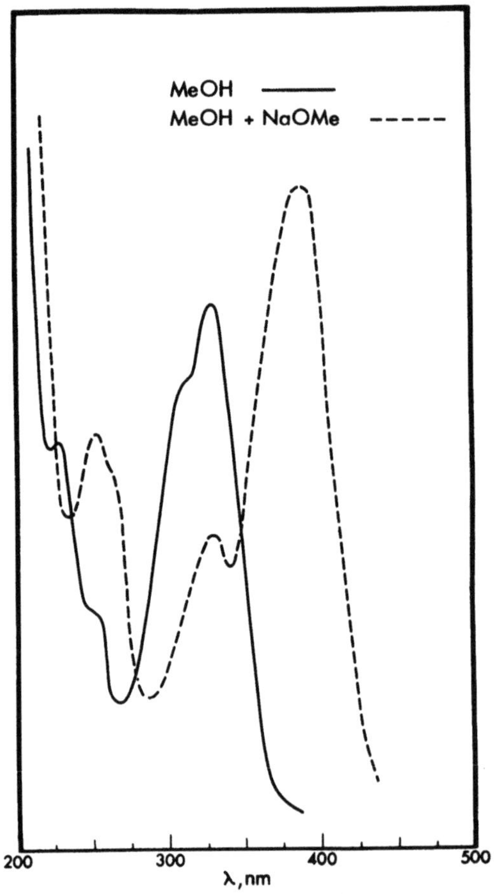

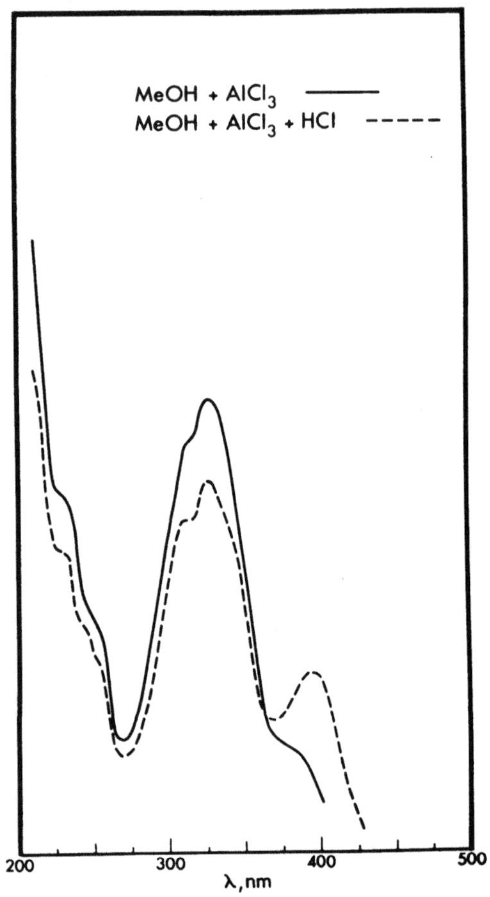

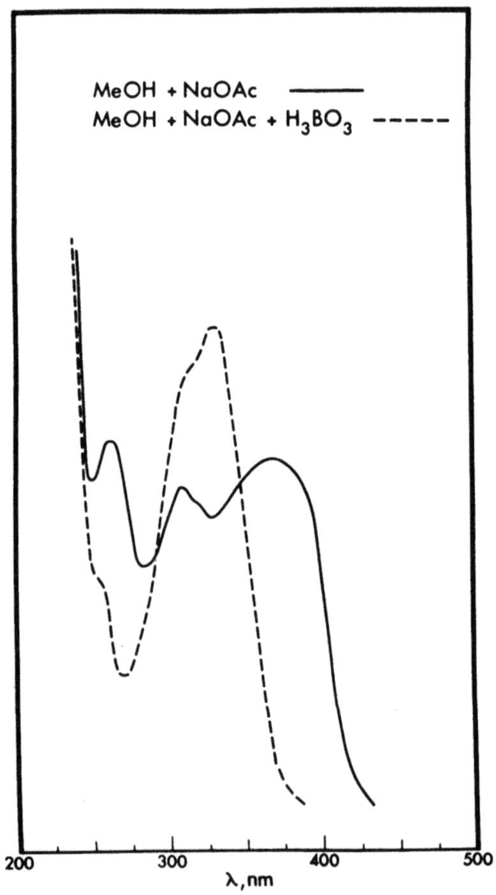

10

4′,7-DIHYDROXYFLAVONE
7-0-RHAMNOGLUCOSIDE

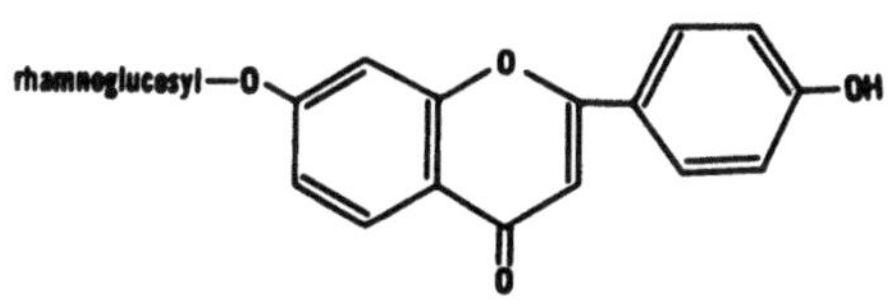

CHROMATOGRAPHIC DATA

Spot Appearance: (UV) fluorescent light blue
(UV/NH_3) yellow-green

R_f Values: 0.44 (TBA), 0.58 (HOAc)

UV SPECTRAL DATA (λ_{max},nm)

MeOH	255sh, 311sh, 325
NaOMe	251sh, 294, 304sh, 385
$AlCl_3$	255sh, 310sh, 327
$AlCl_3$/HCl	253sh, 310sh, 327
NaOAc	257sh, 307, 331, 386sh
NaOAc/H_3BO_3 (Proc. I)	256sh, 312, 328

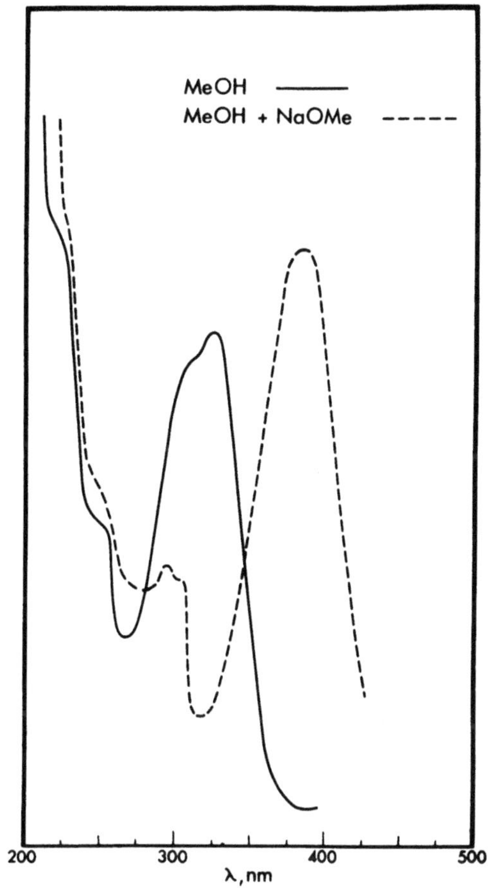

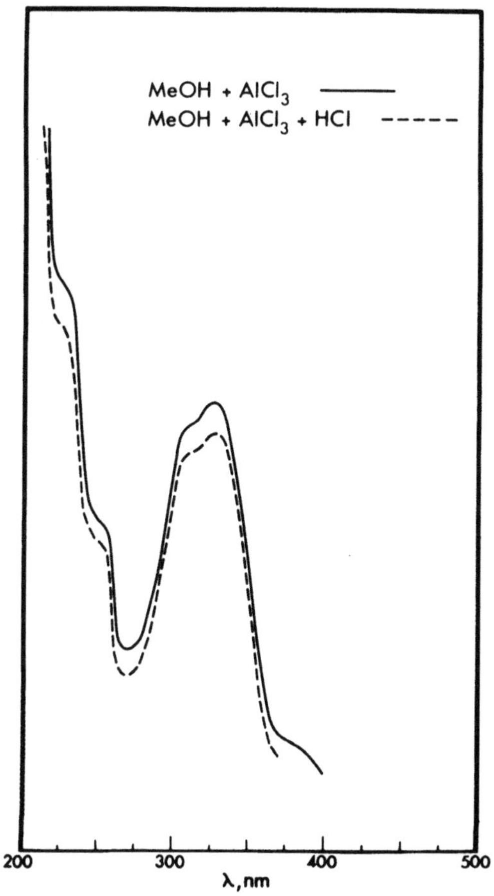

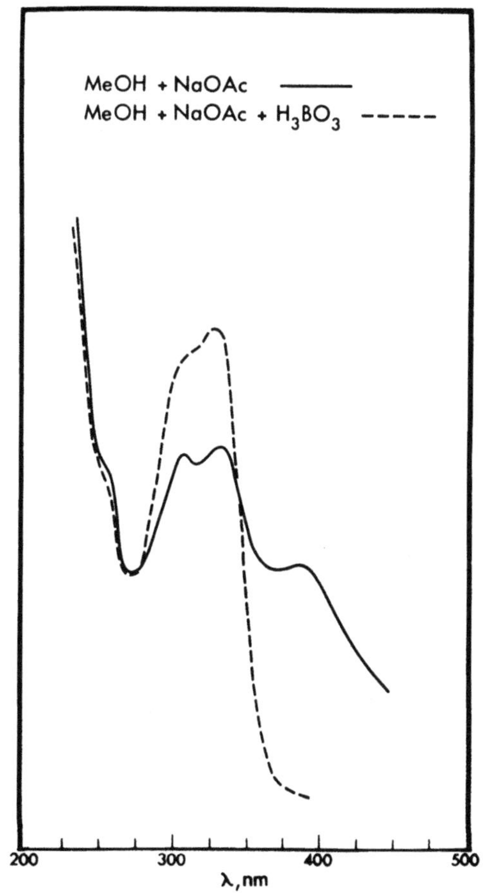

11

5-DEOXYVITEXIN *(BAYIN)*

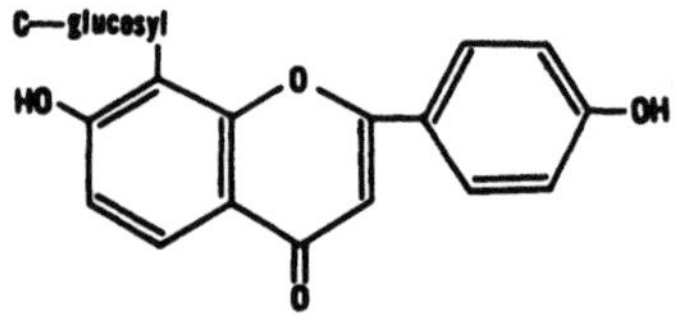

CHROMATOGRAPHIC DATA

Spot Appearance: (UV) fluorescent light blue

(UV/NH_3) fluorescent yellow green

R_f Values: 0.42 (TBA), 0.40 (HOAc)

UV SPECTRAL DATA (λ_{max}, nm)

MeOH	255sh, 312sh, 328
NaOMe	255, 267, 333, 390
$AlCl_3$	254sh, 313, 331, 384
$AlCl_3$/HCl	252sh, 311, 330, 398
NaOAc	268, 310, 320sh, 370
NaOAc/H_3BO_3 (Proc. I)	258, 315sh, 332

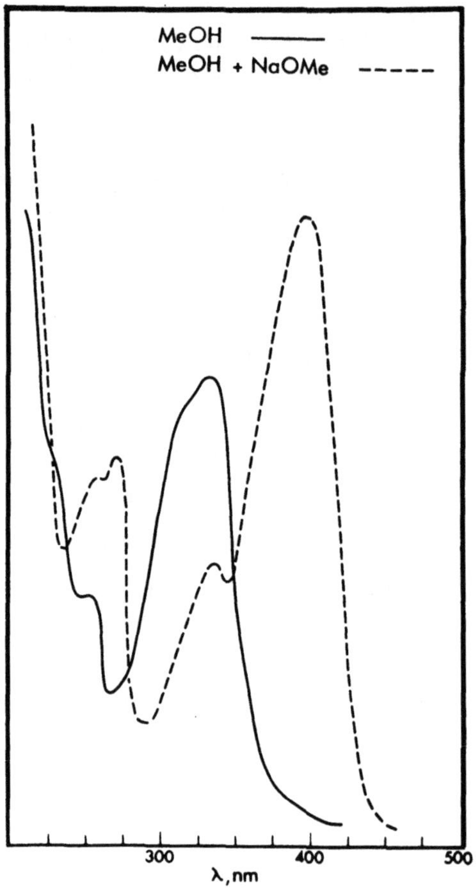

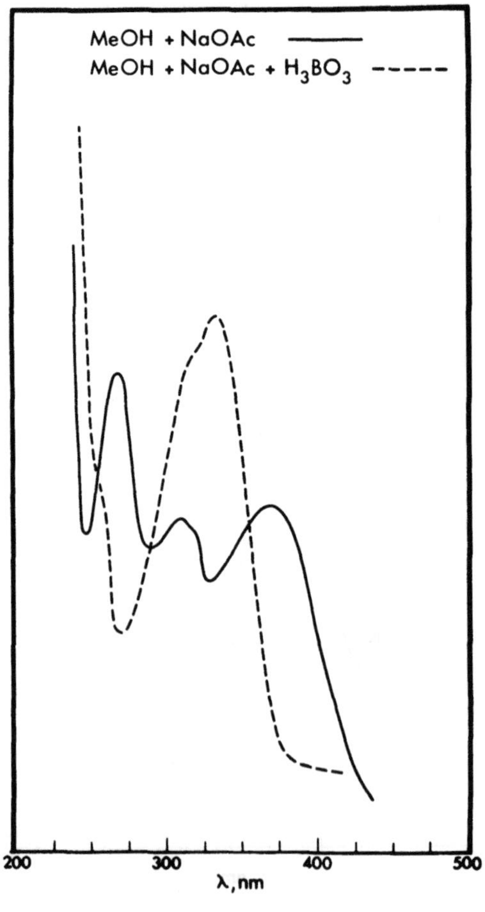

12

7-HYDROXY-4'-METHOXYFLAVONE
(PRATOL)

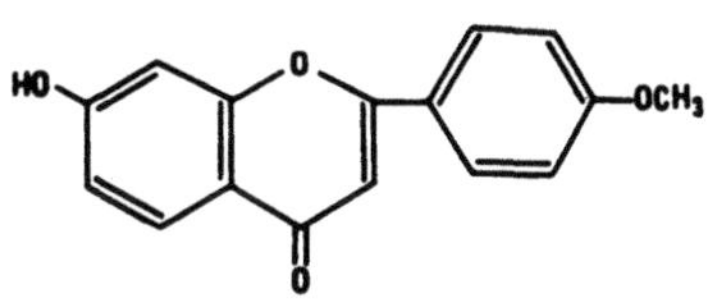

CHROMATOGRAPHIC DATA

Spot Appearance: (UV) fluorescent light blue
(UV/NH_3) fluorescent yellow

R_f Values: 0.88 (TBA), 0.16 (HOAc)

UV SPECTRAL DATA (λ_{max}, *nm*)

MeOH	253, 314sh, 323
NaOMe	266, 301, 319sh, 360
$AlCl_3$	253sh, 314sh, 323, 384sh
$AlCl_3$/HCl	248sh, 255sh, 312, 325, 391
NaOAc	270, 311, 320sh, 344
NaOAc/H_3BO_3 (Proc. I)	257sh, 311sh, 325

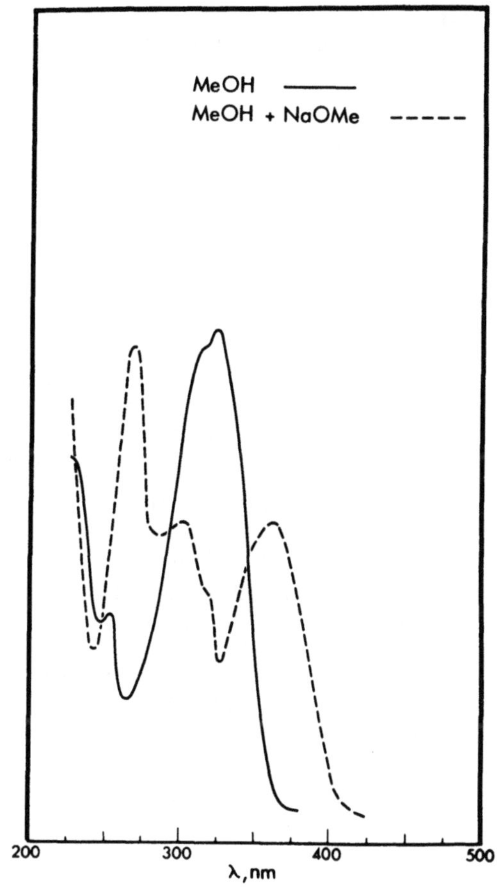

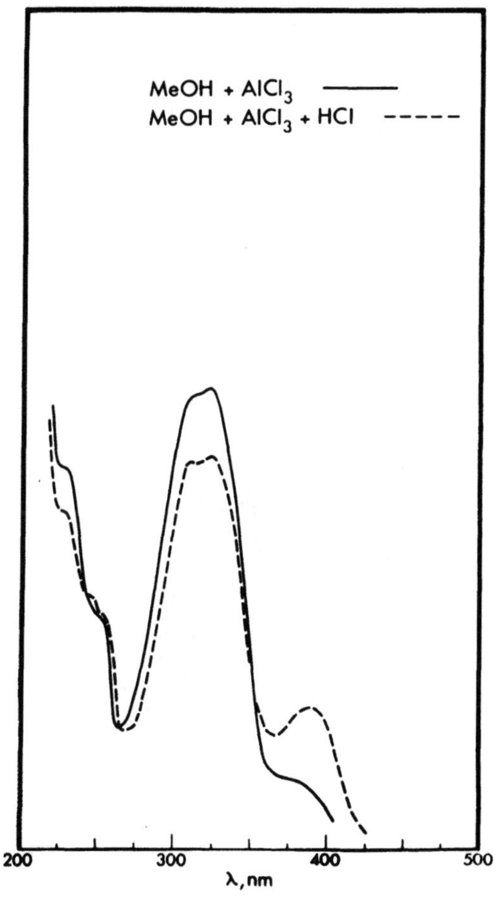

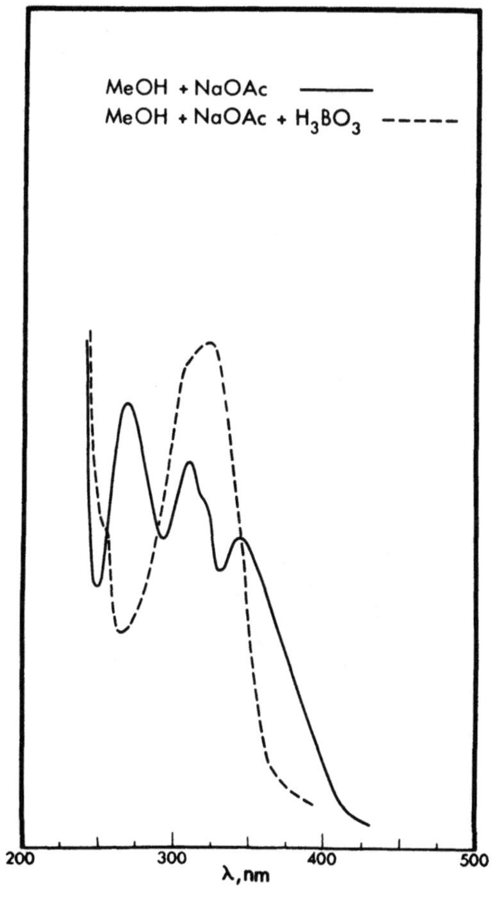

13

3′,4′,7-TRIHYDROXYFLAVONE

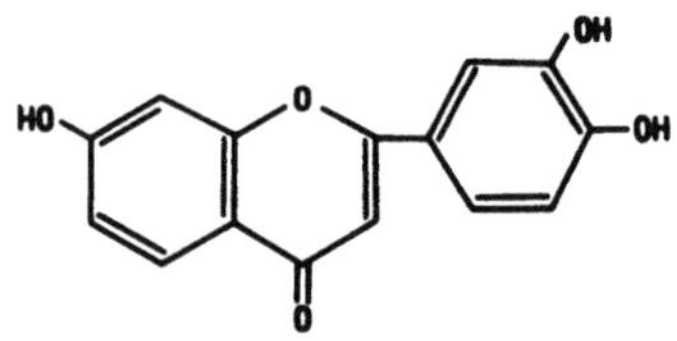

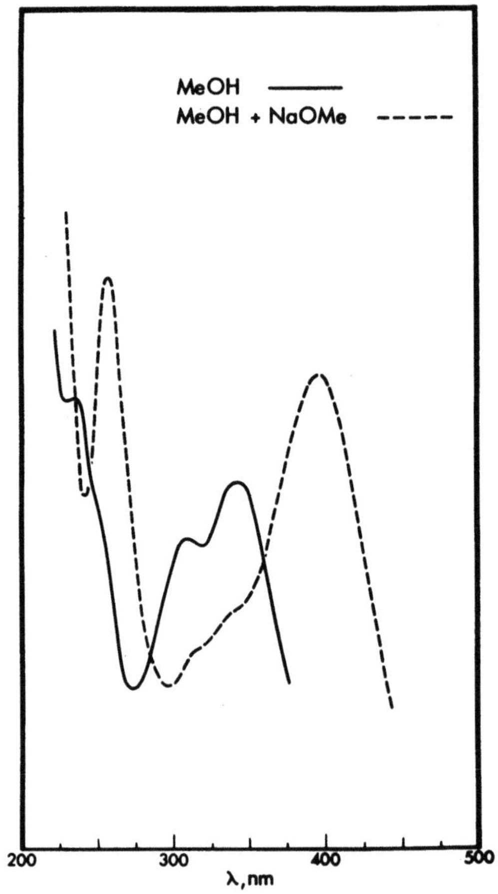

CHROMATOGRAPHIC DATA

Spot Appearance: (UV) fluorescent light blue

(UV/NH_3) fluorescent yellow-green

R_f Values: 0.71 (TBA), 0.07 (HOAc)

UV SPECTRAL DATA (λ_{max},*nm*)

MeOH	235, 250sh, 309, 343
NaOMe	256, 313sh, 338sh, 395
$AlCl_3$	234sh, 305, 371, 458
$AlCl_3$/HCl	235sh, 254sh, 307, 340, 409
NaOAc	255, 310, 373
NaOAc/H_3BO_3 (Proc. I)	258sh, 306, 360

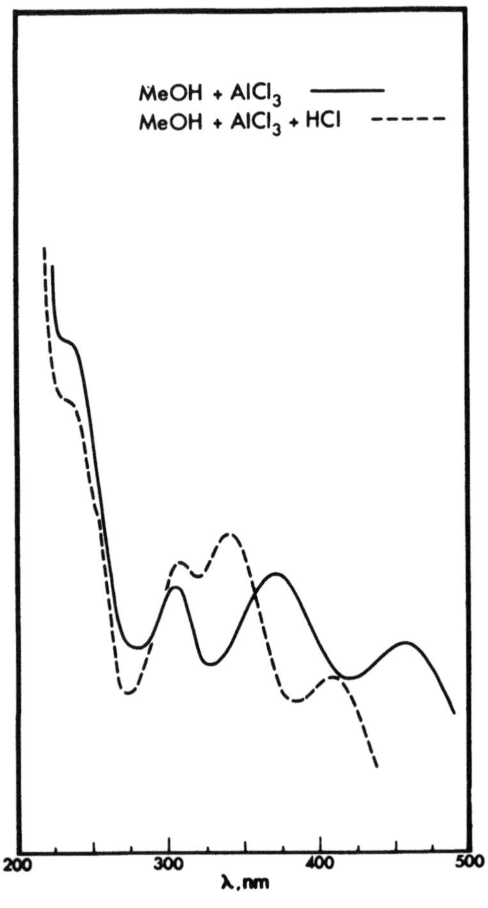

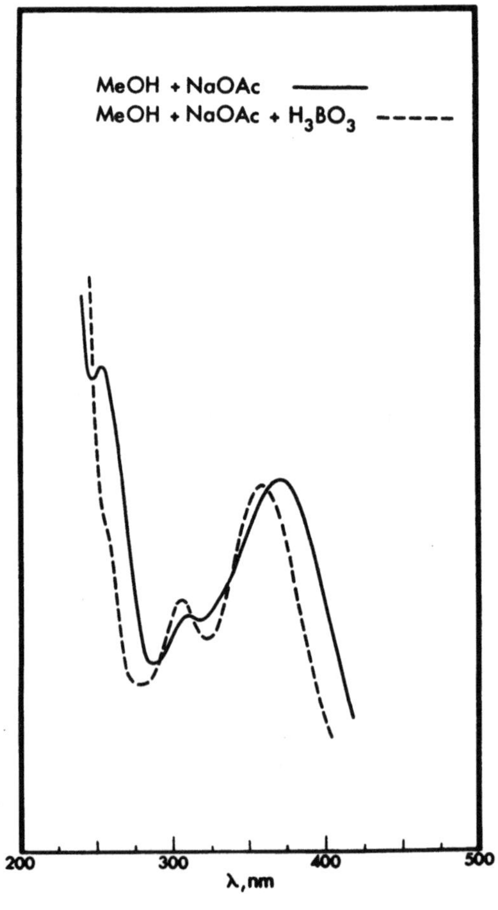

14

3′,4′,7-TRIHYDROXYFLAVONE 7-0-RHAMNOGLUCOSIDE

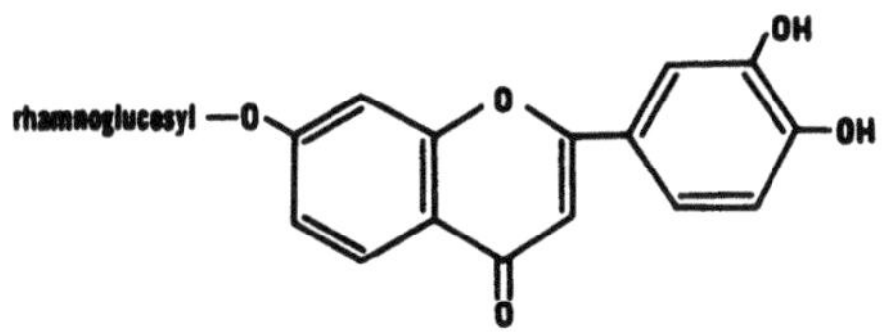

CHROMATOGRAPHIC DATA

Spot Appearance: (UV) fluorescent light blue
(UV/NH_3) fluorescent yellow-green

R_f Values: 0.26 (TBA), 0.38 (HOAc)

UV SPECTRAL DATA (λ_{max}, *nm*)

MeOH	247sh, 255sh, 305, 341
NaOMe	293, 405
$AlCl_3$	244sh, 258sh, 300, 380
$AlCl_3$/HCl	247sh, 257sh, 306, 341
NaOAc	257sh, 299, 350, 401
NaOAc/H_3BO_3 (Proc. I)	257sh, 299, 365

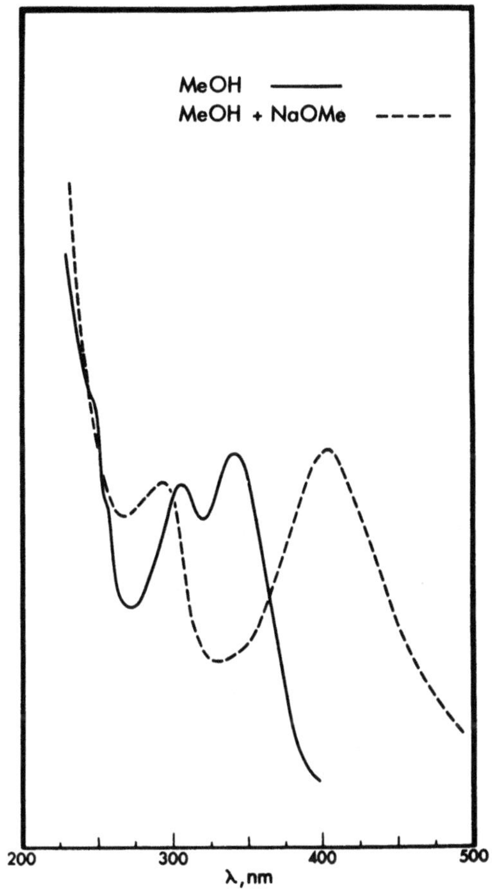

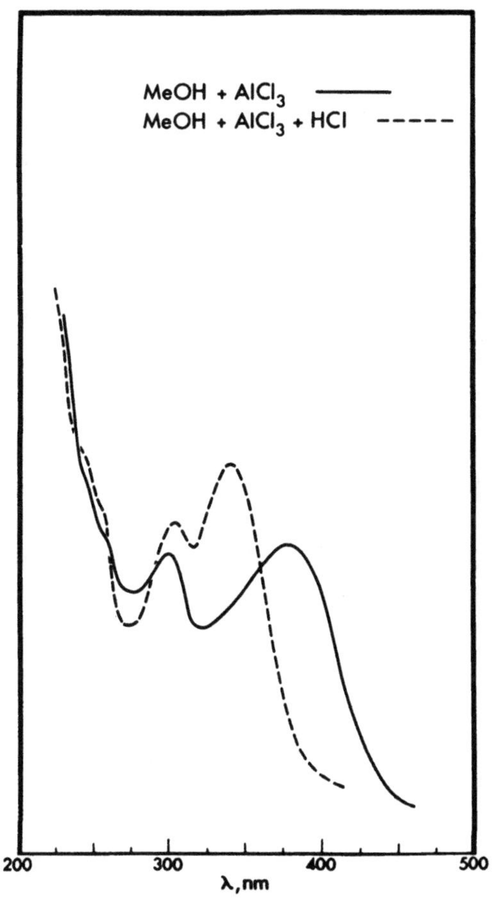

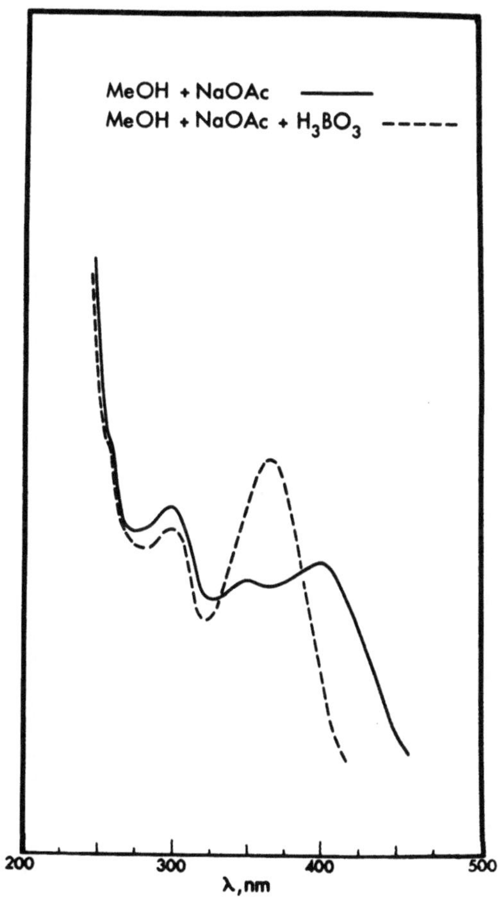

15

7-HYDROXY-3′,4′-DIMETHOXYFLAVONE

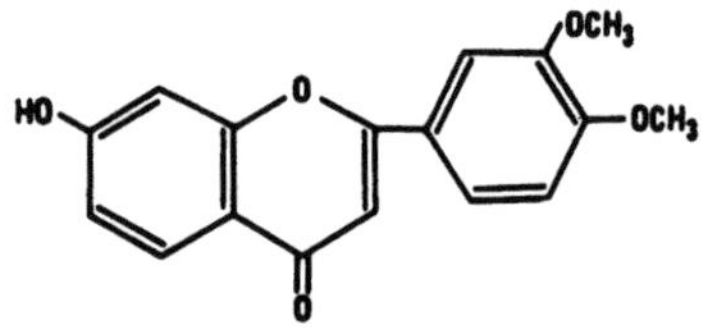

CHROMATOGRAPHIC DATA

Spot Appearance: (UV) fluorescent light blue
(UV/NH_3) fluorescent light blue

R_f Values: 0.86 (TBA), 0.06 (HOAc)

UV SPECTRAL DATA (λ_{max},*nm*)

MeOH	239, 262sh, 330
NaOMe	270, 314, 348
$AlCl_3$	261, 277, 301, 337, 395sh
$AlCl_3$/HCl	259, 277sh, 301, 341, 394
NaOAc	265, 338
NaOAc/H_3BO_3 (Proc. I)	264sh, 331

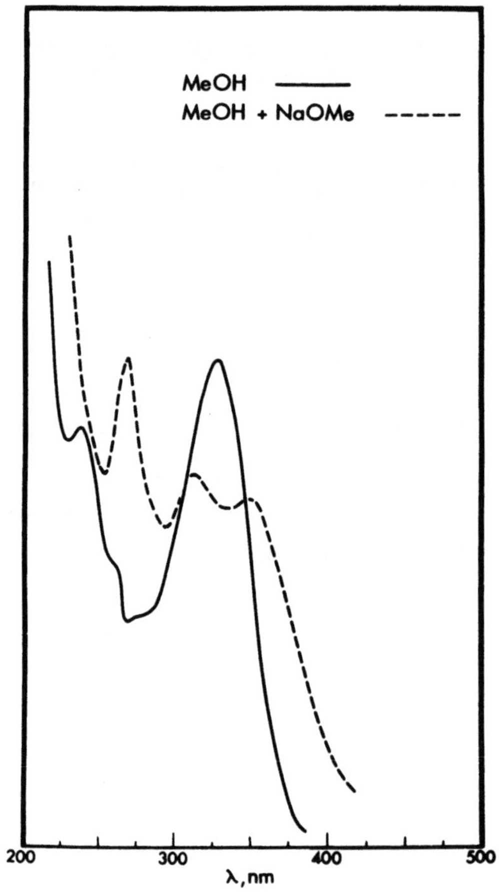

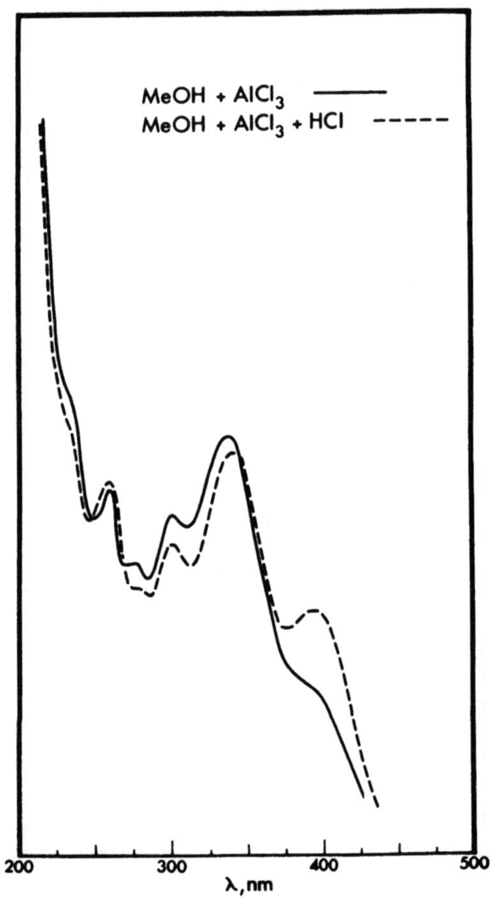

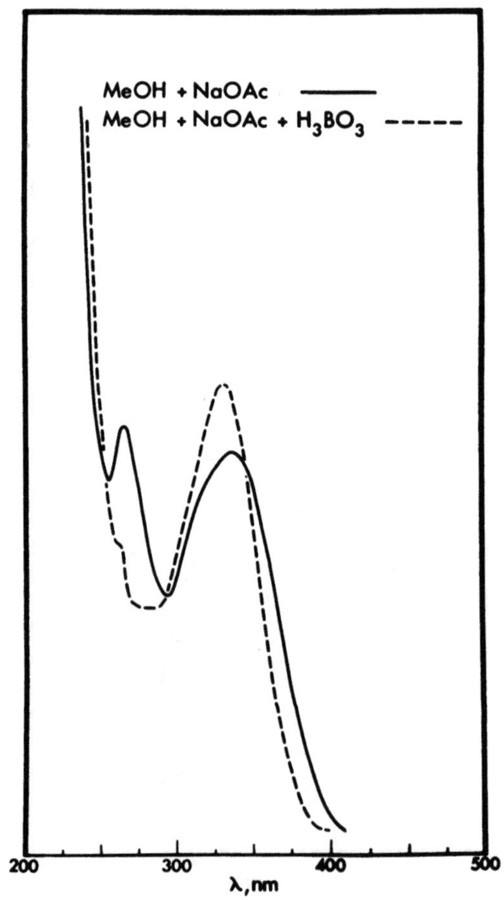

BAICALEIN

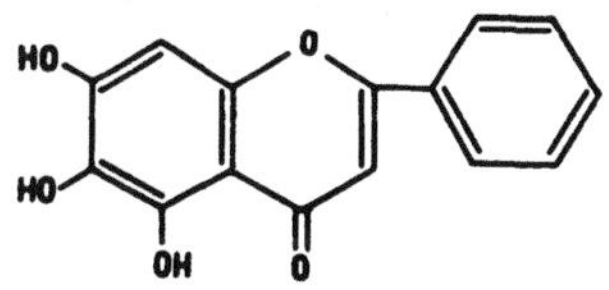

CHROMATOGRAPHIC DATA

Spot Appearance: (UV) deep purple
(UV/NH_3) deep purple

R_f Values: 0.78 (TBA), 0.19 (HOAc)

UV SPECTRAL DATA (λ_{max},nm)

MeOH	247sh, 274, 323
NaOMe	257, 366, 410sh (dec.)
$AlCl_3$	247, 272, 284sh, 375
$AlCl_3$/HCl	255sh, 282, 292sh, 346
NaOAc	257, 360, 405sh (dec.)
NaOAc/H_3BO_3 (Proc. II)	262sh, 277, 333

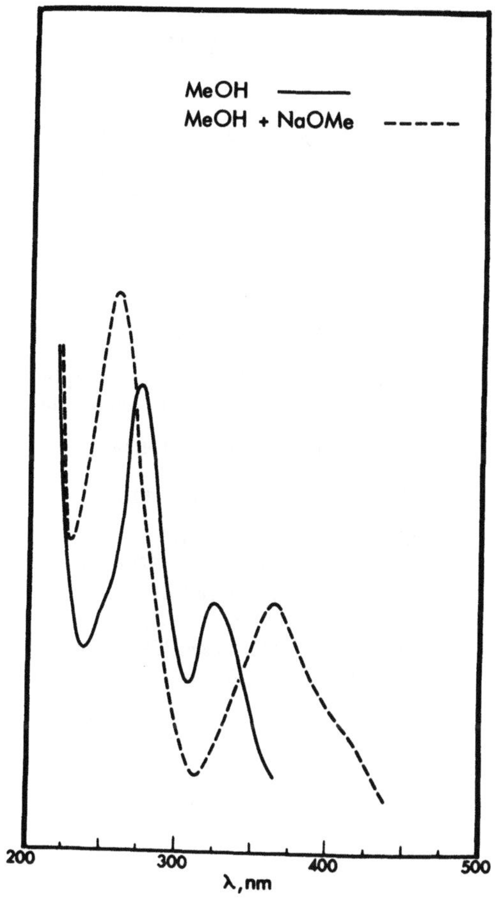

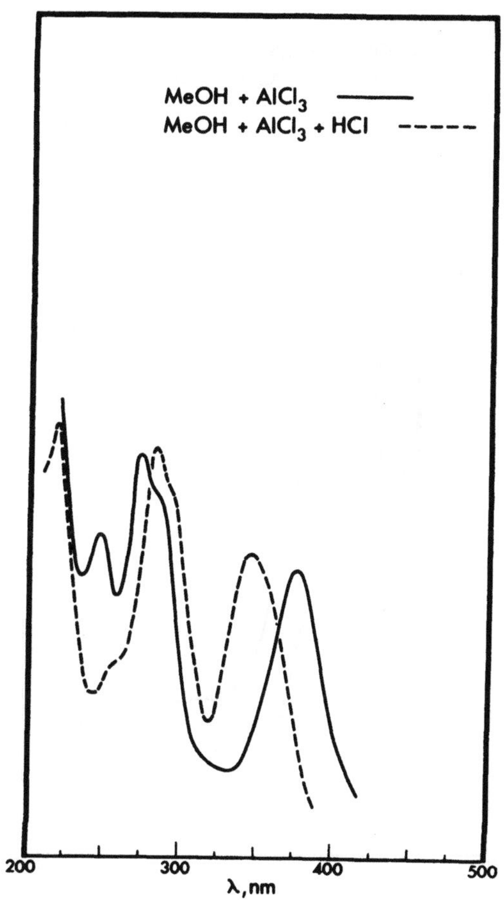

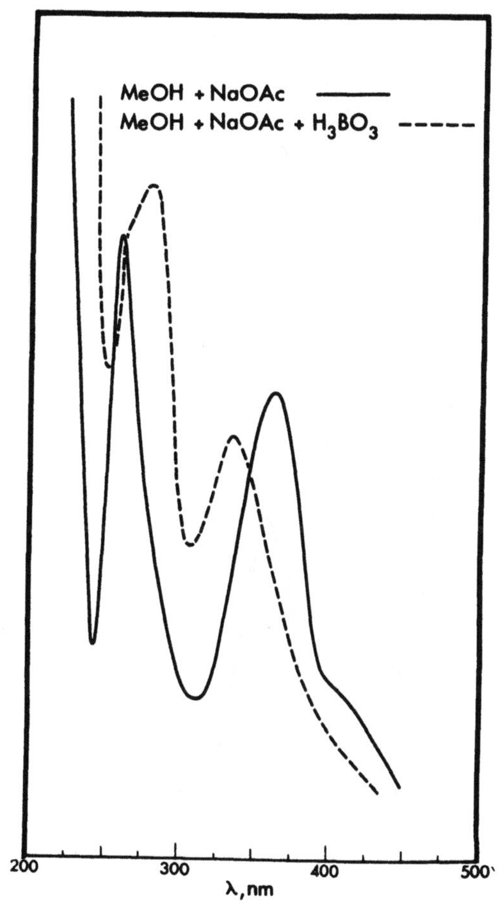

17

BAICALIN

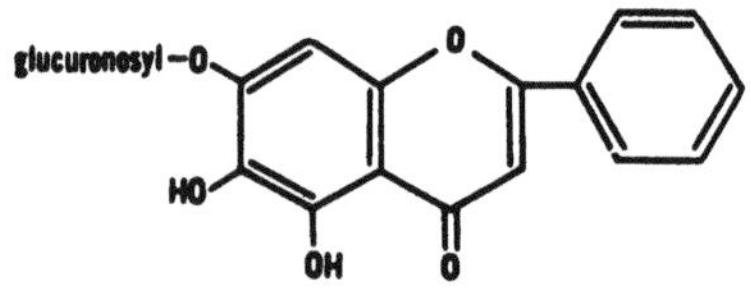

CHROMATOGRAPHIC DATA

Spot Appearance: (UV) deep purple
(UV/NH_3) deep purple

R_f Values: 0.46 (TBA), 0.33 (HOAc)

UV SPECTRAL DATA (λ_{max},*nm*)

MeOH	244, 278, 315
NaOMe	263, 286sh, 357sh (dec.)
$AlCl_3$	249sh, 288, 343
$AlCl_3$/HCl	248sh, 289, 338
NaOAc	277, 305sh, 394sh (dec.)
NaOAc/H_3BO_3 (Proc. II)	283, 318sh

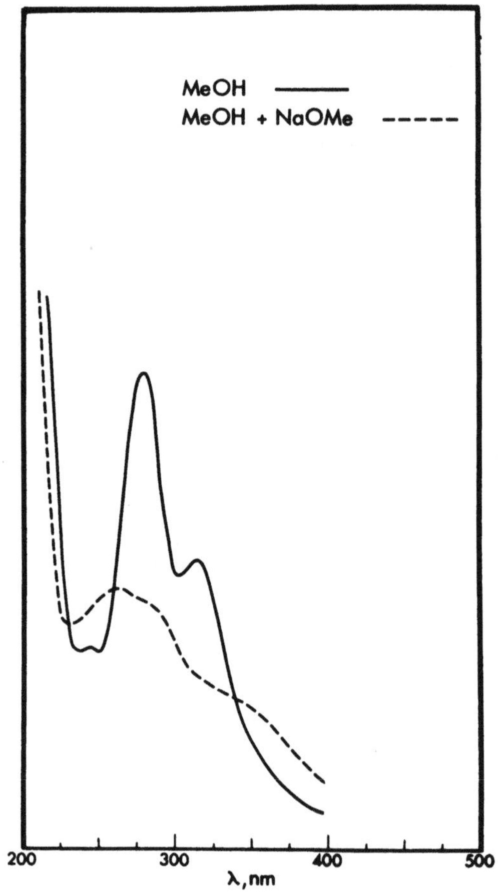

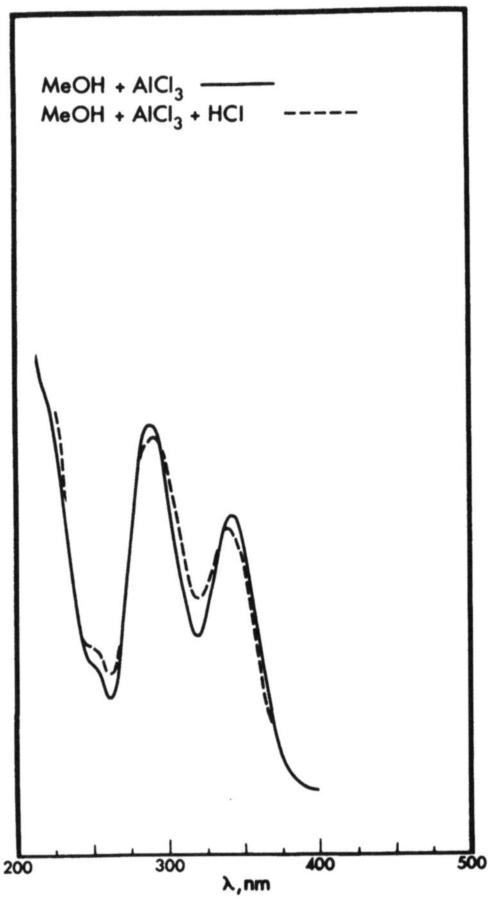

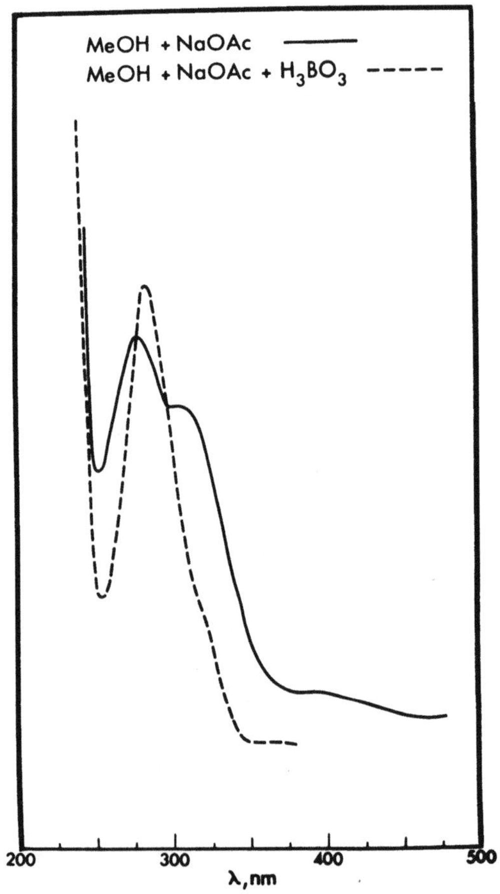

18

5,7,8-TRIHYDROXYFLAVONE (NORWOGONIN)

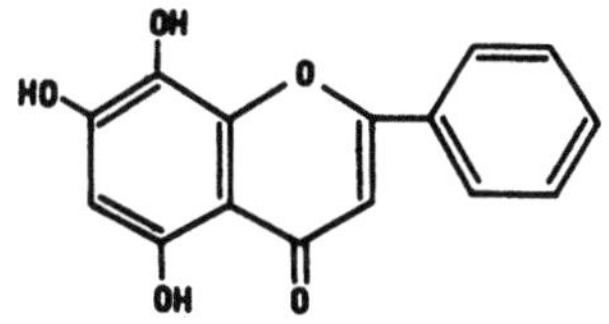

CHROMATOGRAPHIC DATA

Spot Appearance: (UV) deep purple
(UV/NH_3) deep purple

R_f Values: 0.83 (TBA), 0.15 (HOAc)

UV SPECTRAL DATA (λ_{max}, nm)

MeOH	281, 364sh
NaOMe	246, 274 (dec.)
$AlCl_3$	292sh, 315, 366sh
$AlCl_3$/HCl	290sh, 302, 342sh, 395sh
NaOAc	274 (dec.)
NaOAc/H_3BO_3 (Proc. II)	287

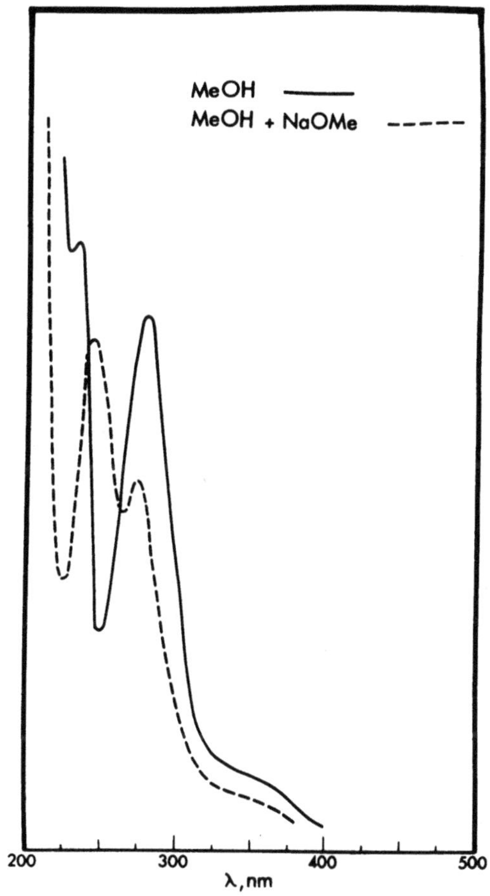

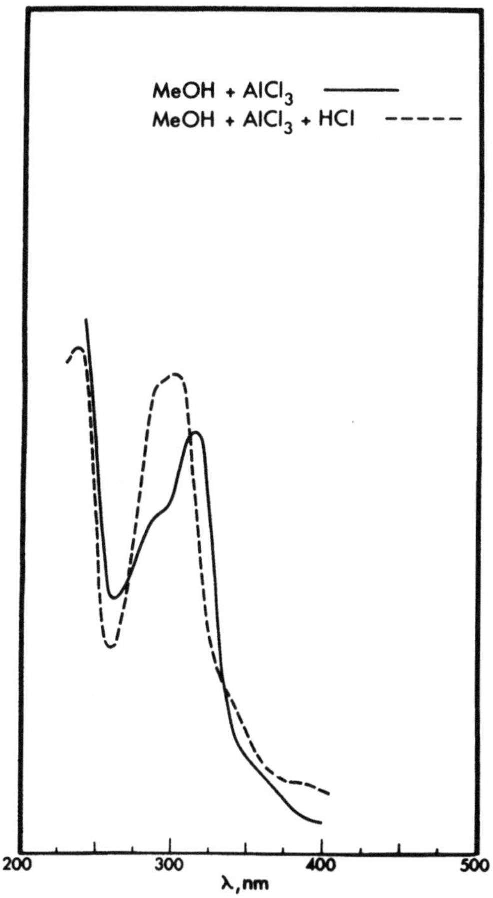

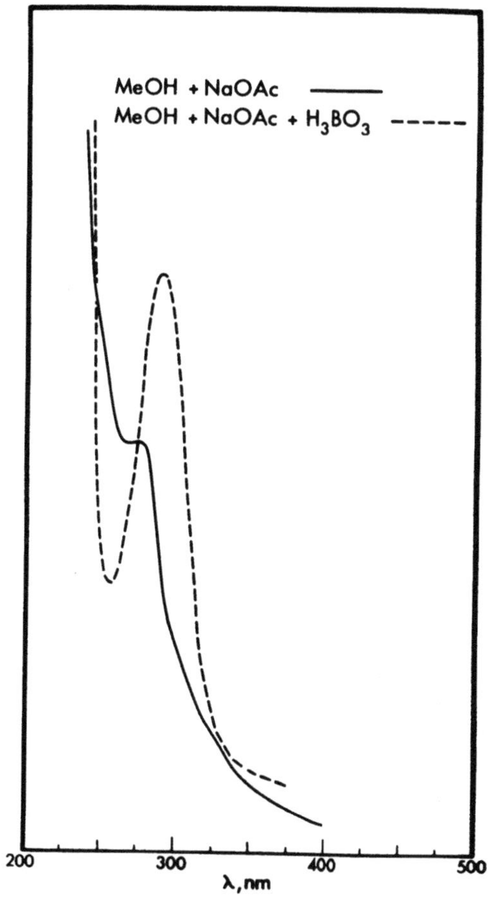

19

5,7,8-TRIHYDROXYFLAVONE 7-0-GLUCURONIDE

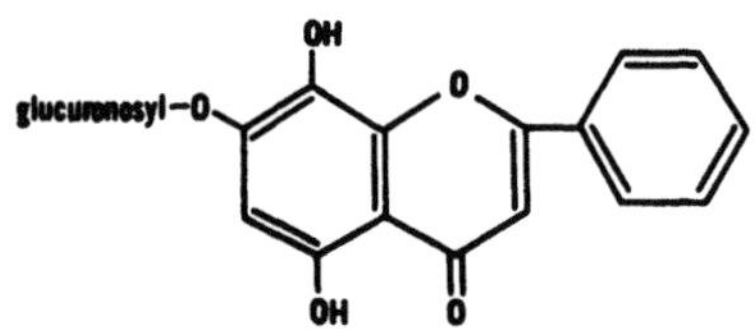

CHROMATOGRAPHIC DATA

Spot Appearance: (UV) deep purple
(UV/NH_3) deep purple

R_f Values: 0.70 (TBA), 0.37 (HOAc)

UV SPECTRAL DATA (λ_{max},*nm*)

MeOH	247, 274, 315sh, 342sh
NaOMe	236sh, 281, 357
$AlCl_3$	252, 286sh, 292, 331, 396
$AlCl_3$/HCl	248, 283sh, 289, 327, 387
NaOAc	264sh, 281, 366
NaOAc/H_3BO_3 (Proc. I)	277, 346

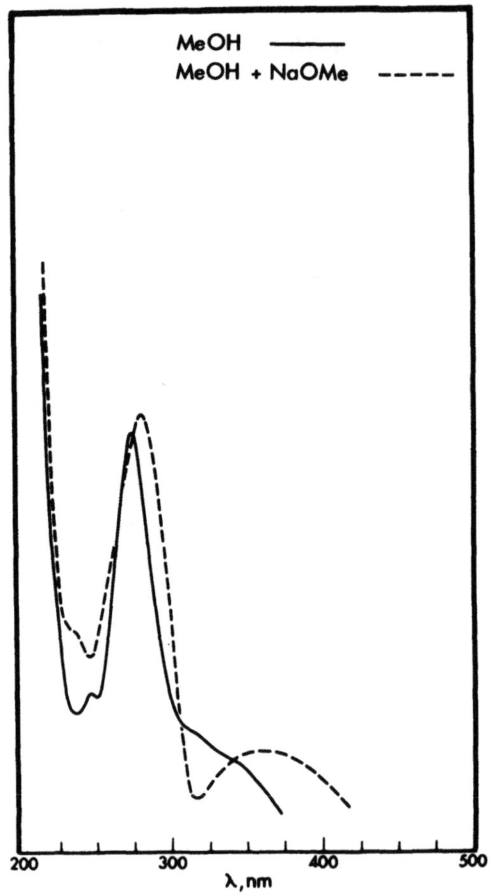

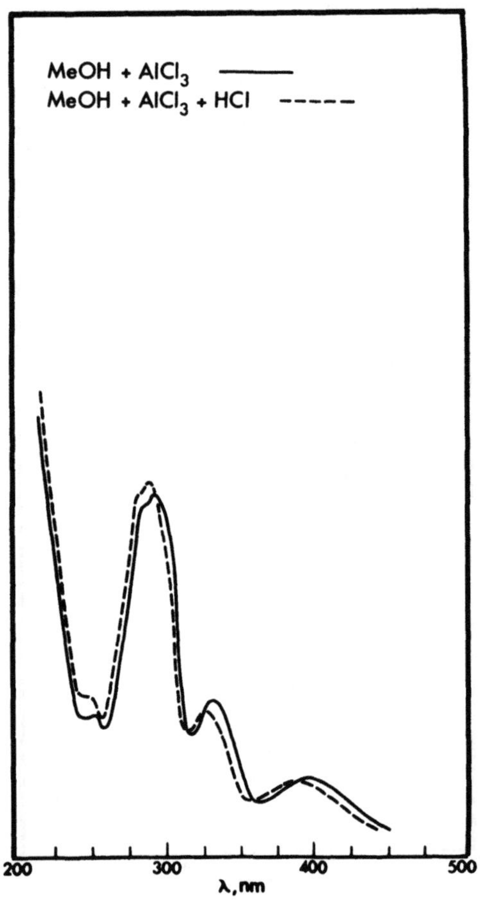

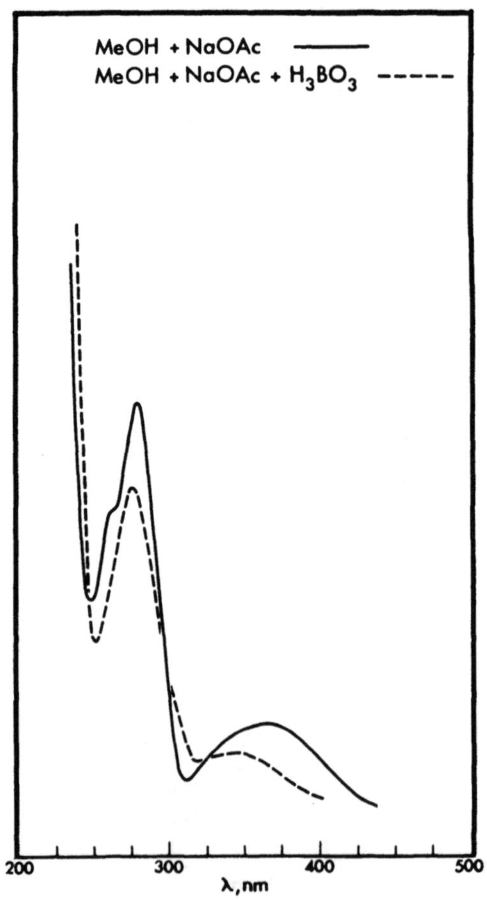

20

APIGENIN

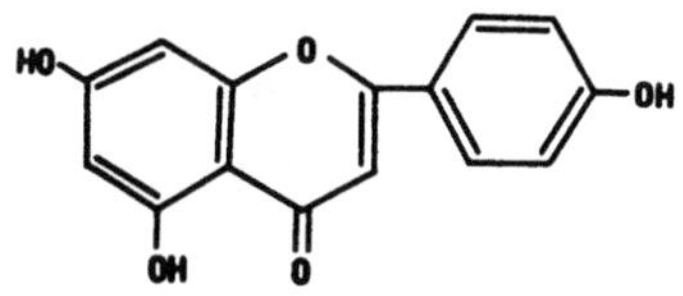

CHROMATOGRAPHIC DATA

Spot Appearance: (UV) deep purple
(UV/NH_3) yellow-green

R_f Values: 0.87 (TBA), 0.11 (HOAc)

UV SPECTRAL DATA (λ_{max}, nm)

MeOH	267, 296sh, 336
NaOMe	275, 324, 392
$AlCl_3$	276, 301, 348, 384
$AlCl_3$/HCl	276, 299, 340, 381
NaOAc	274, 301, 376
NaOAc/H_3BO_3 (Proc. I)	268, 302sh, 338

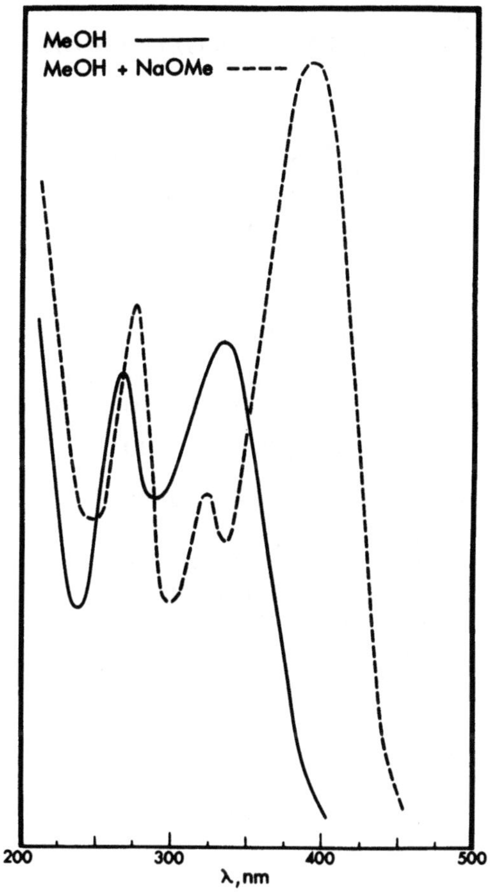

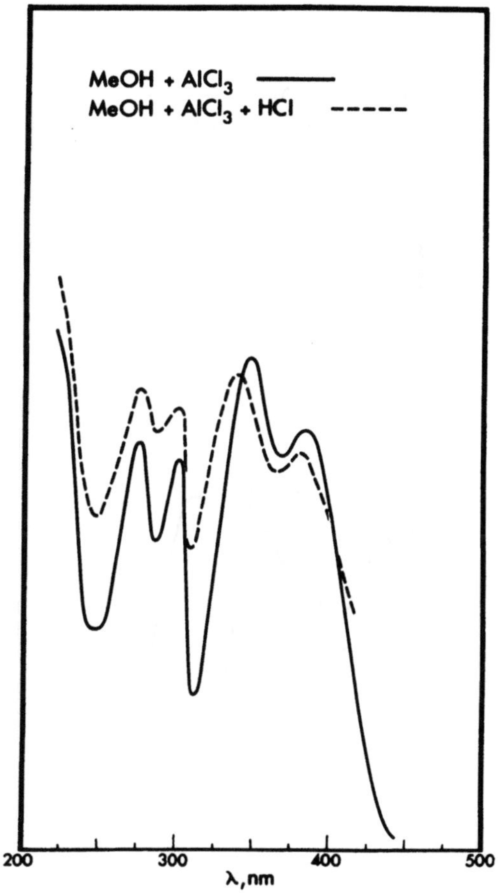

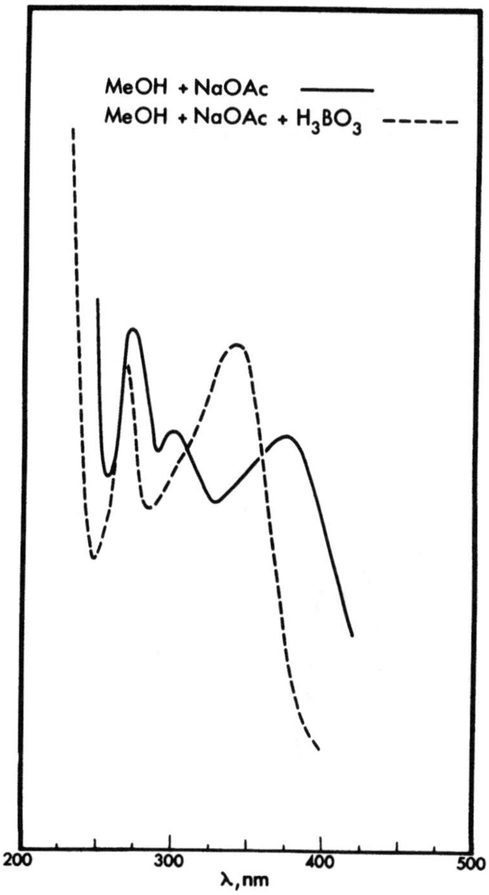

21

APIGENIN 7-0-GLUCOSIDE

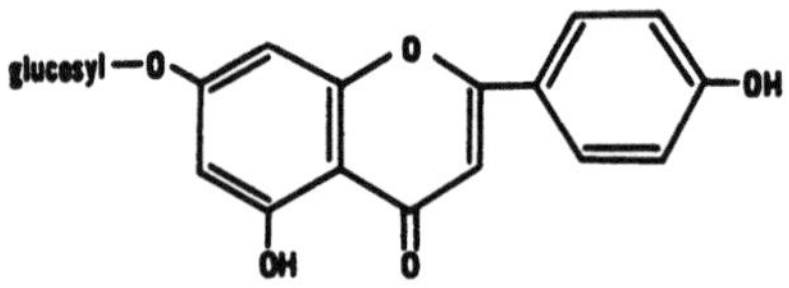

CHROMATOGRAPHIC DATA

Spot Appearance: (UV) deep purple
(UV/NH_3) yellow-green

R_f Values: 0.61 (TBA), 0.23 (HOAc)

UV SPECTRAL DATA (λ_{max}, nm)

MeOH	268, 333
NaOMe	245sh, 269, 301sh, 386
$AlCl_3$	276, 300, 348, 386
$AlCl_3$/HCl	277, 299, 341, 382
NaOAc	256sh, 267, 355, 387
NaOAc/H_3BO_3 (Proc. I)	267, 340

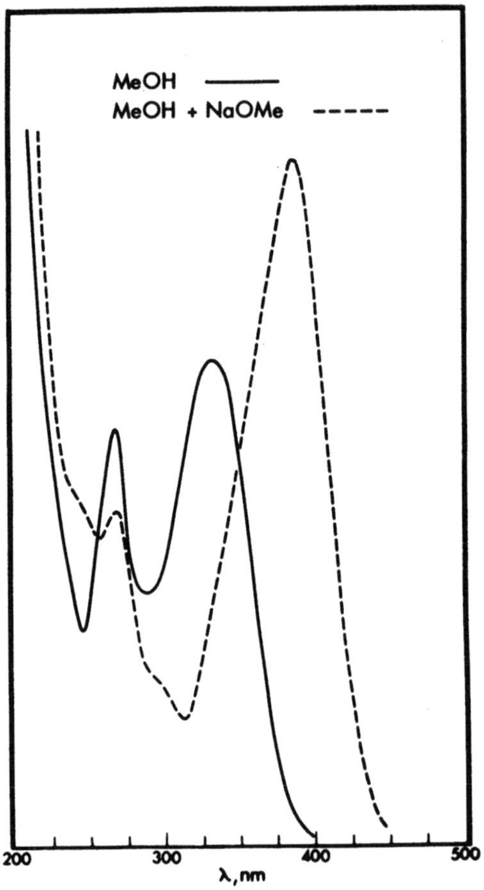

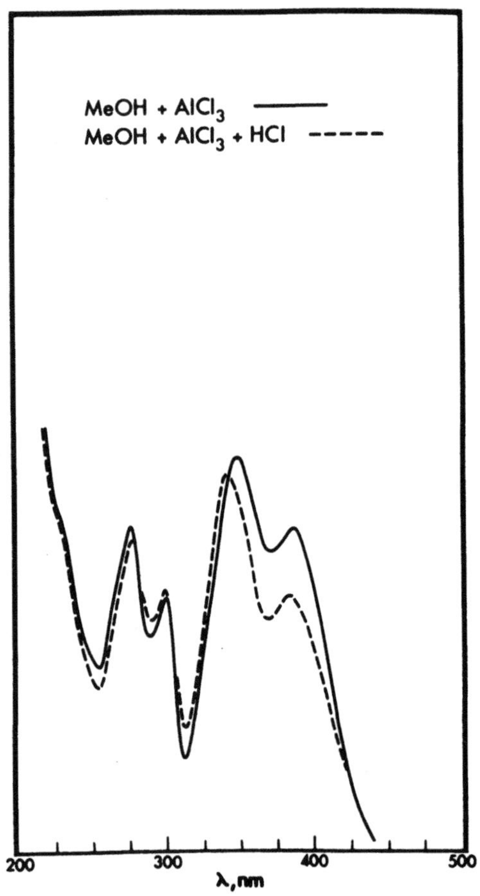

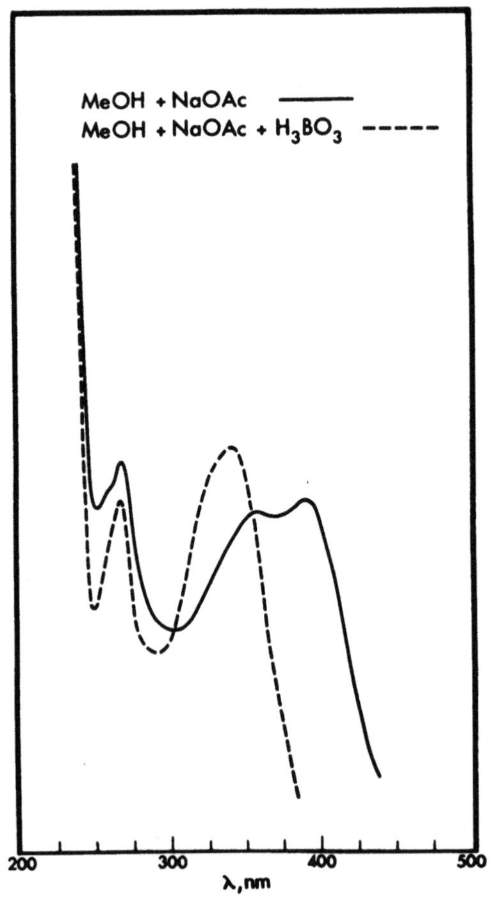

22
APIGENIN 7-0-NEOHESPERIDOSIDE

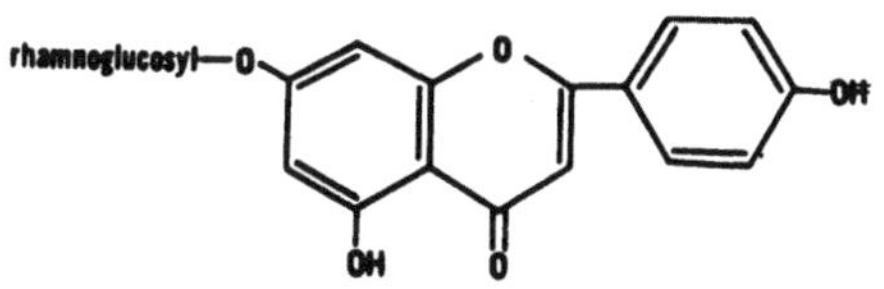

CHROMATOGRAPHIC DATA

Spot Appearance: (UV) deep purple
(UV/NH_3) yellow-green

R_f Values: 0.52 (TBA), 0.49 (HOAc)

UV SPECTRAL DATA (λ_{max},*nm*)

MeOH	268, 333
NaOMe	245sh, 267, 300sh, 386
$AlCl_3$	275, 300, 348, 382
$AlCl_3$/HCl	276, 299, 341, 380
NaOAc	257sh, 267, 354, 387
NaOAc/H_3BO_3 (Proc. I)	267, 341

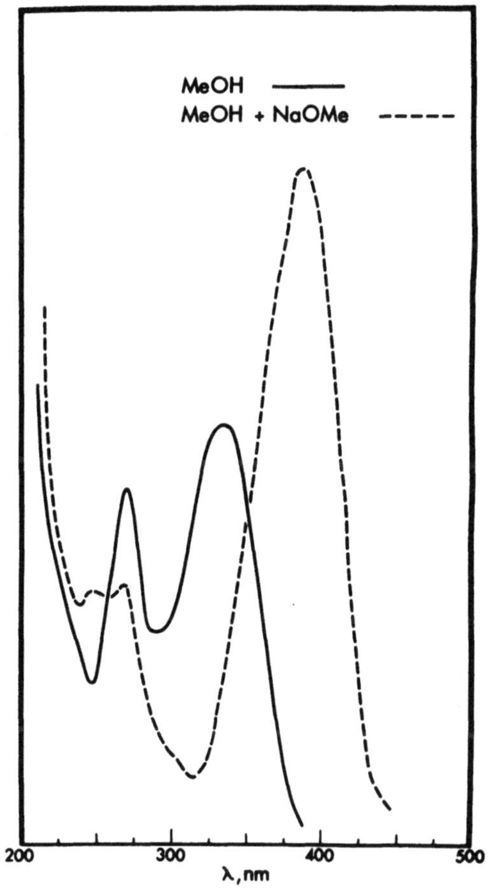

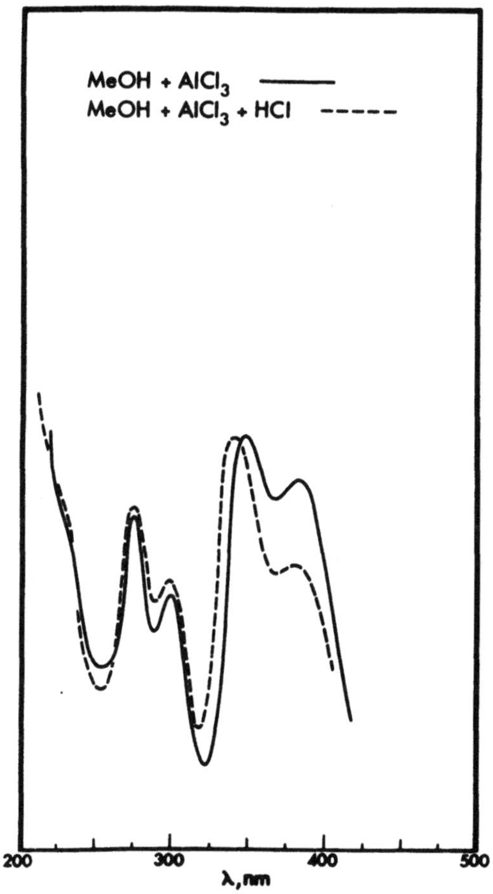

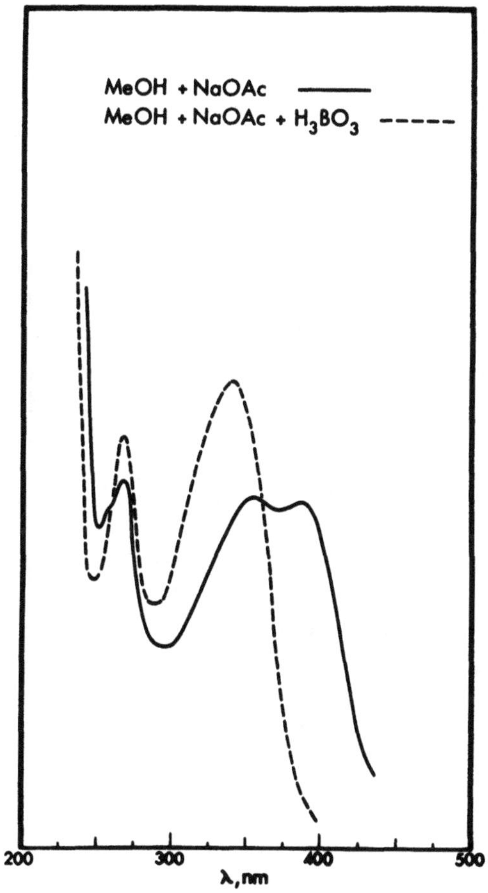

23

ISOVITEXIN
(SAPONARETIN)

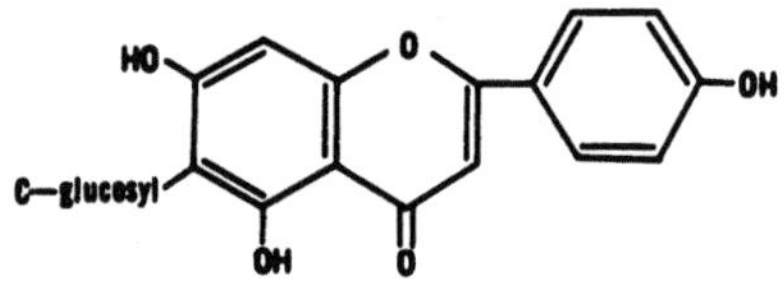

CHROMATOGRAPHIC DATA

Spot Appearance: (UV) deep purple
(UV/NH_3) yellow-green

R_f Values: 0.57 (TBA), 0.55 (HOAc)

UV SPECTRAL DATA (λ_{max},nm)

MeOH	271, 336
NaOMe	278, 329, 398
$AlCl_3$	262sh, 278, 304, 352, 382
$AlCl_3$/HCl	260sh, 280, 302, 344, 380
NaOAc	279, 303, 385
NaOAc/H_3BO_3 (Proc. I)	274, 346, 408sh

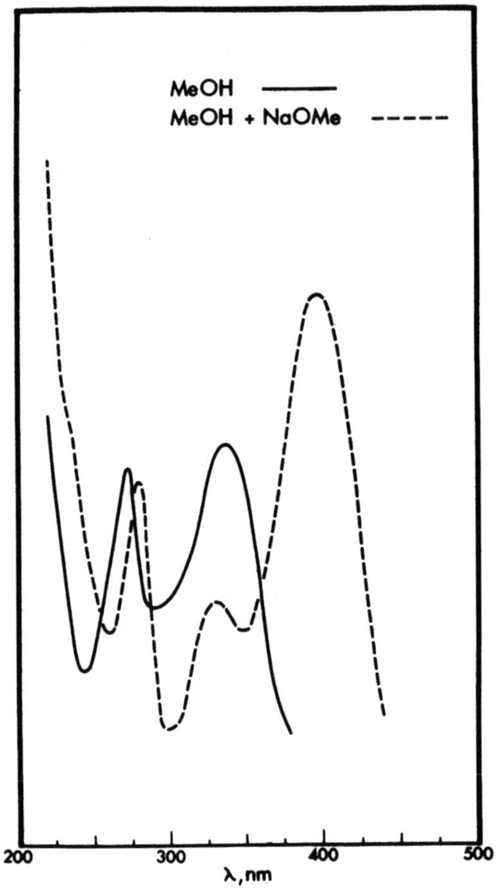

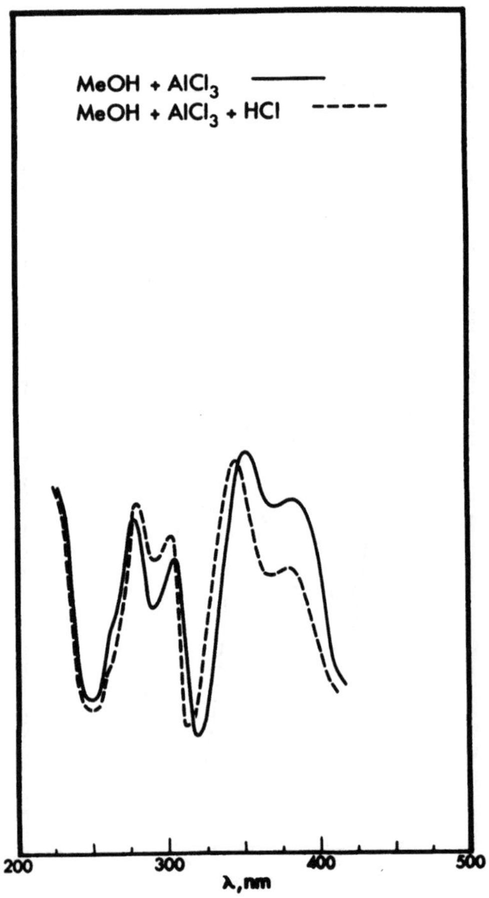

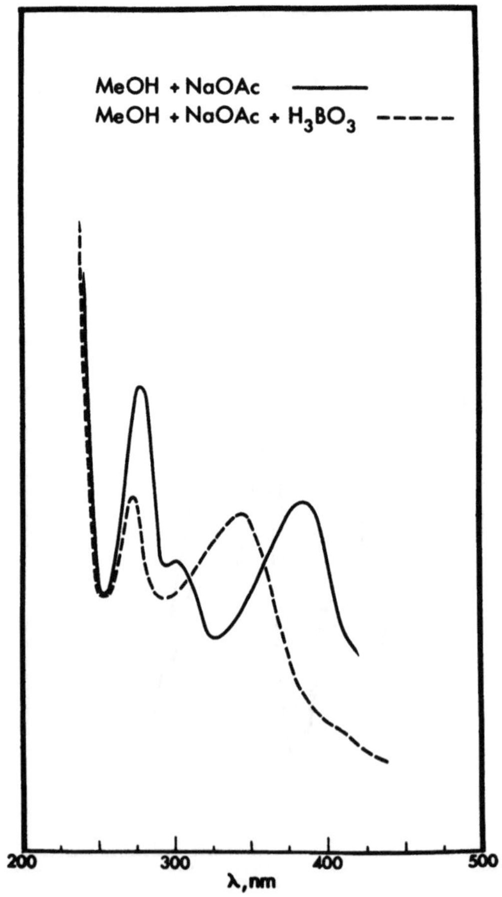

24

SAPONARIN

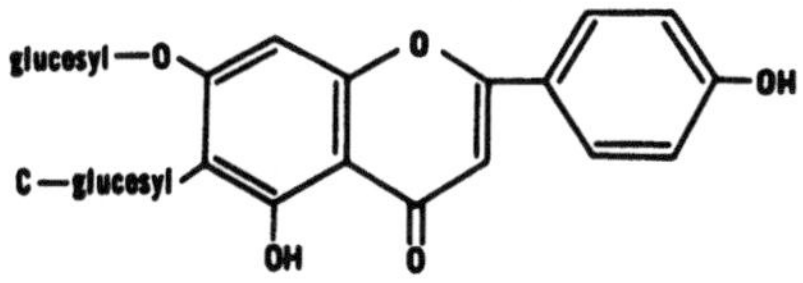

CHROMATOGRAPHIC DATA

Spot Appearance: (UV) deep purple
(UV/NH_3) yellow-green

R_f Values: 0.30 (TBA), 0.68 (HOAc)

UV SPECTRAL DATA (λ_{max}, *nm*)

MeOH	271, 336
NaOMe	249sh, 271, 304sh, 389
$AlCl_3$	268sh, 277, 301, 352, 381
$AlCl_3$/HCl	279, 300, 344, 378
NaOAc	261sh, 271, 350, 392
NaOAc/H_3BO_3 (Proc. I)	269, 341

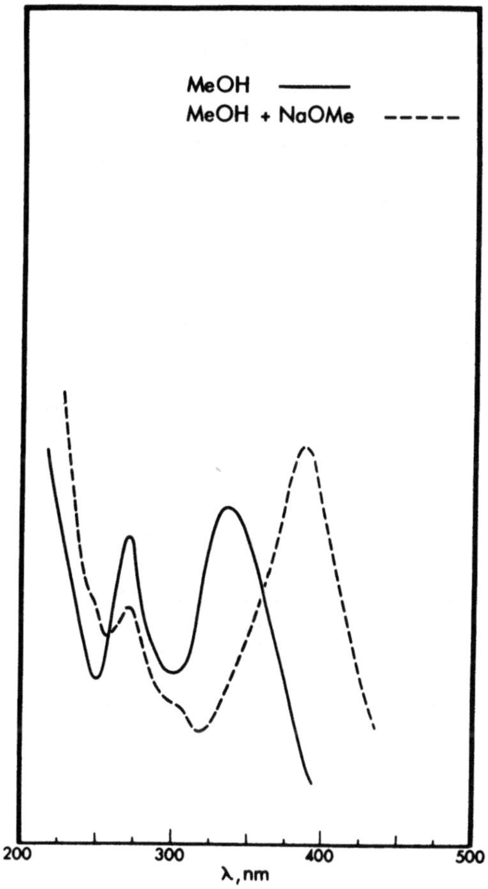

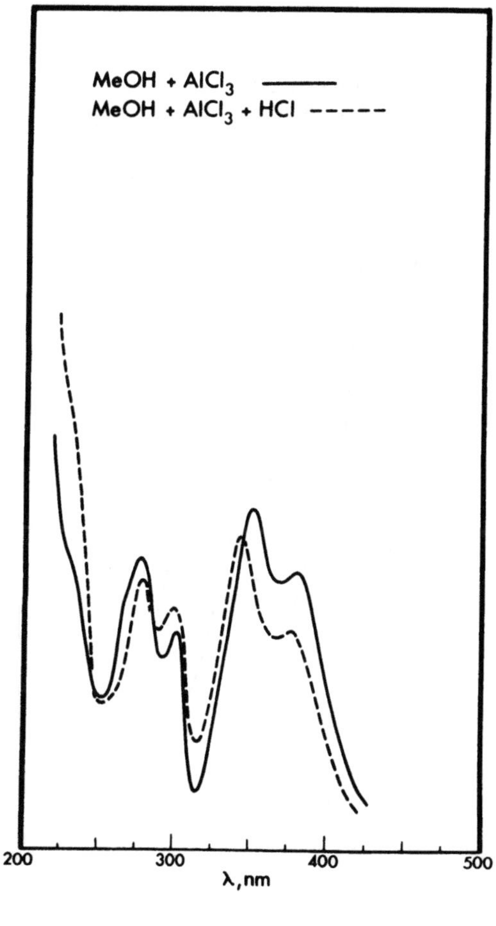

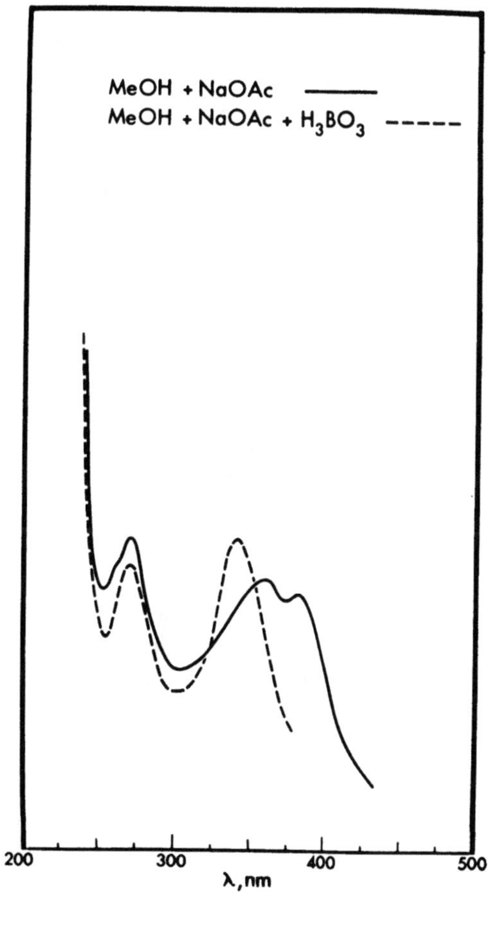

25

VITEXIN

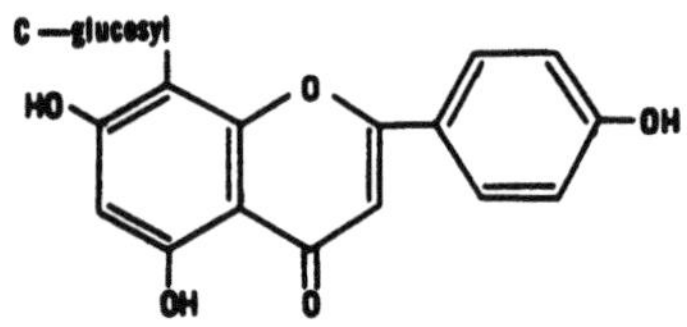

CHROMATOGRAPHIC DATA

Spot Appearance: (UV) deep purple
(UV/NH_3) yellow-green

R_f Values: 0.43 (TBA), 0.29 (HOAc)

UV SPECTRAL DATA (λ_{max},*nm*)

MeOH	270, 302sh, 336
NaOMe	279, 329, 395
$AlCl_3$	277, 305, 350, 386
$AlCl_3$/HCl	278, 303, 343, 383
NaOAc	280, 300, 379
NaOAc/H_3BO_3 (Proc. I)	271, 329sh, 344

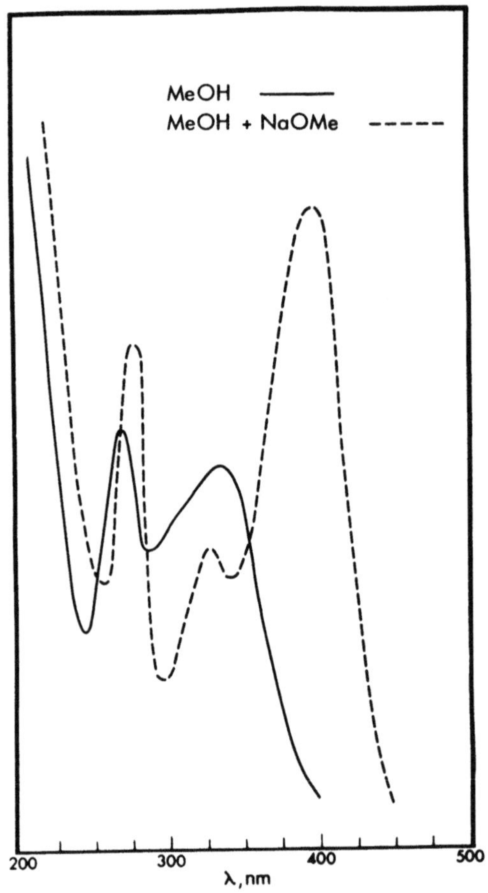

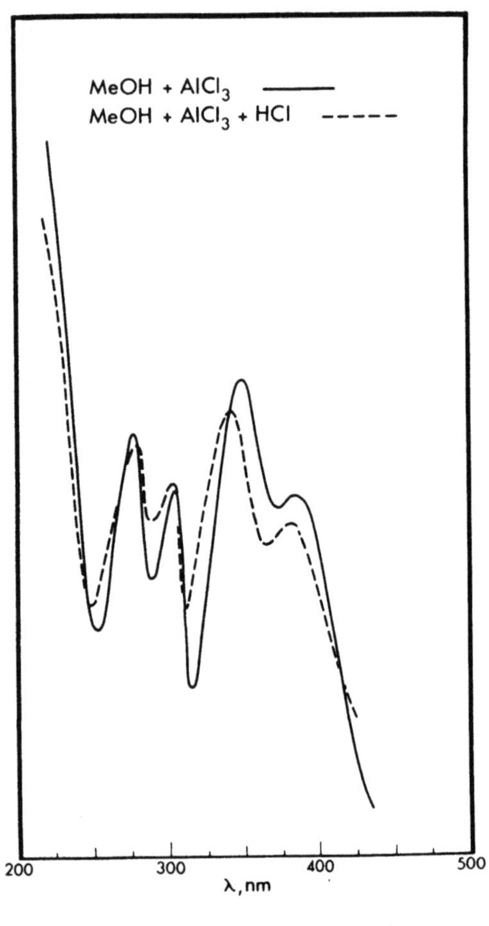

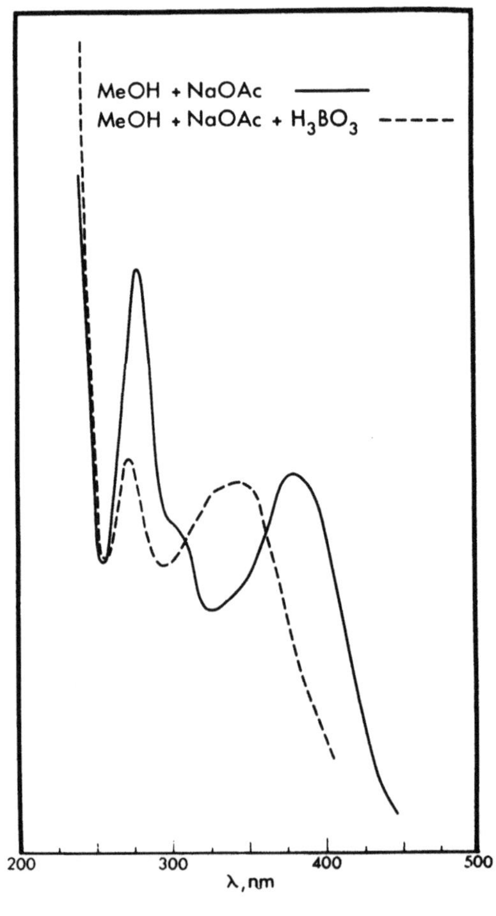

26

RHAMNOSYLVITEXIN

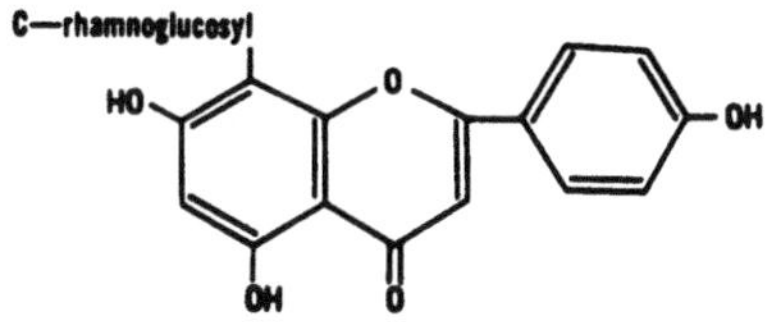

CHROMATOGRAPHIC DATA

Spot Appearance: (UV) deep purple
(UV/NH_3) yellow-green

R_f Values: 0.50 (TBA), 0.72 (HOAc)

UV SPECTRAL DATA (λ_{max},*nm*)

MeOH	270, 303sh, 336
NaOMe	281, 331, 396
$AlCl_3$	277, 305, 349, 386
$AlCl_3$/HCl	278, 303, 343, 383
NaOAc	281, 303sh, 382
NaOAc/H_3BO_3 (Proc. I)	270, 330sh, 344

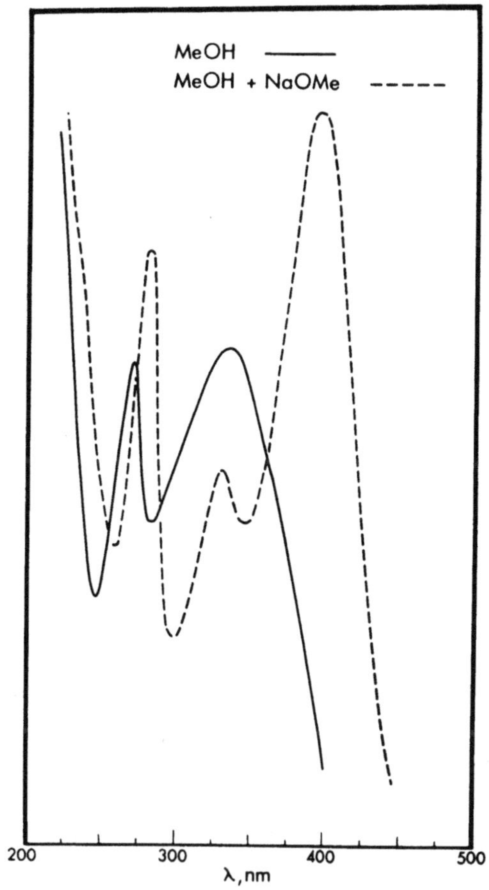

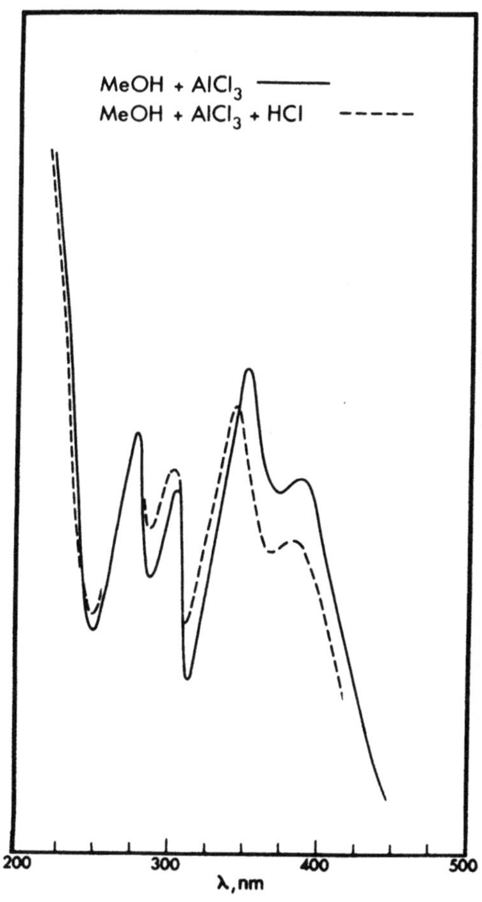

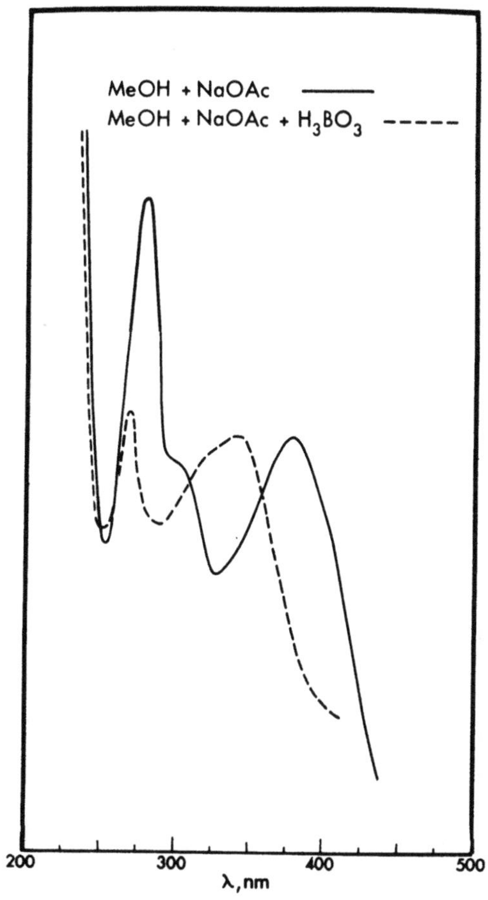

27

2″-0-XYLOSYLVITEXIN

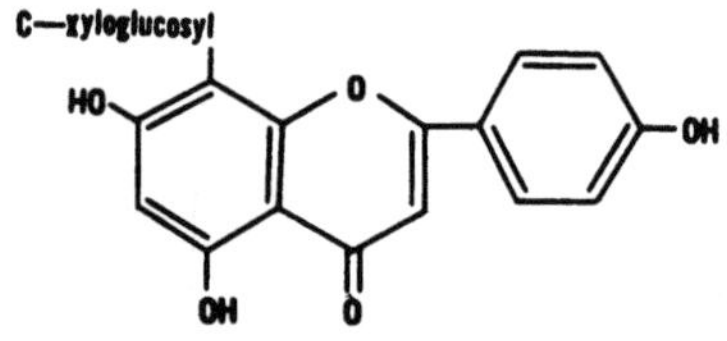

CHROMATOGRAPHIC DATA

Spot Appearance: (UV) deep purple
(UV/NH_3) yellow-green

R_f Values: 0.52 (TBA), 0.71 (HOAc)

UV SPECTRAL DATA (λ_{max}, nm)

MeOH	270, 301sh, 335
NaOMe	280, 329, 395
$AlCl_3$	277, 305, 350, 382
$AlCl_3$/HCl	278, 303, 343, 382
NaOAc	280, 305sh, 381
NaOAc/H_3BO_3 (Proc. I)	272, 284sh, 309sh, 324, 342

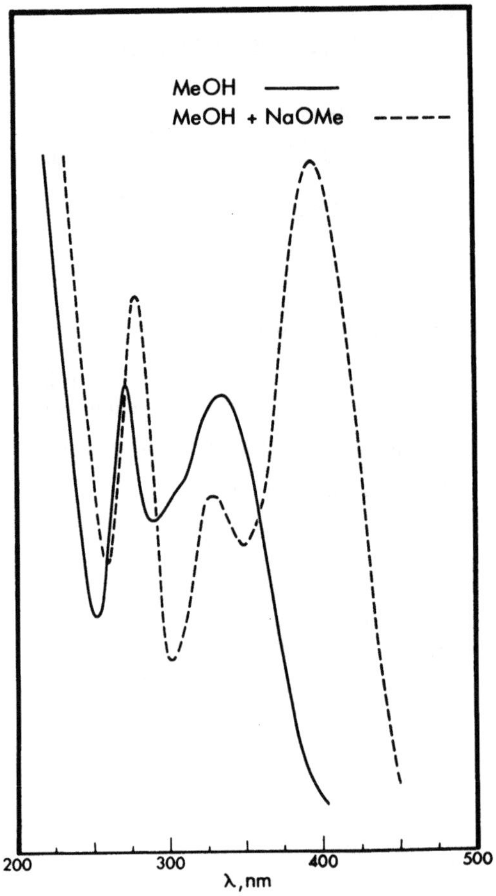

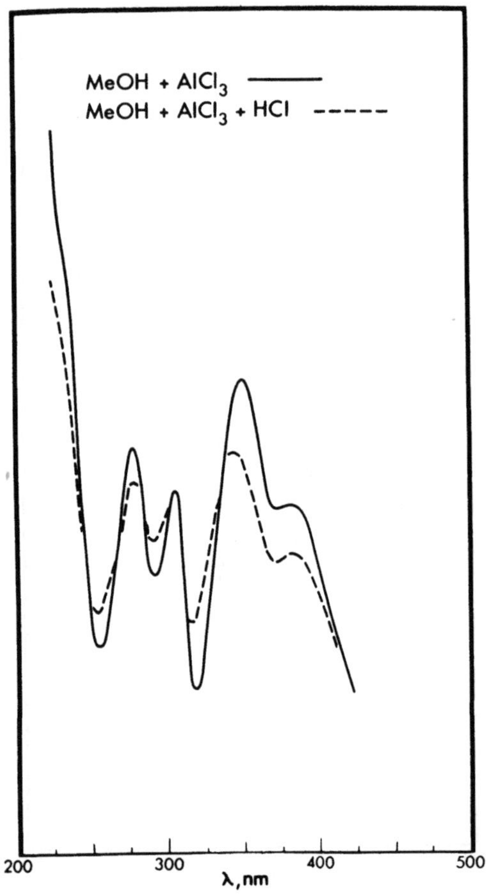

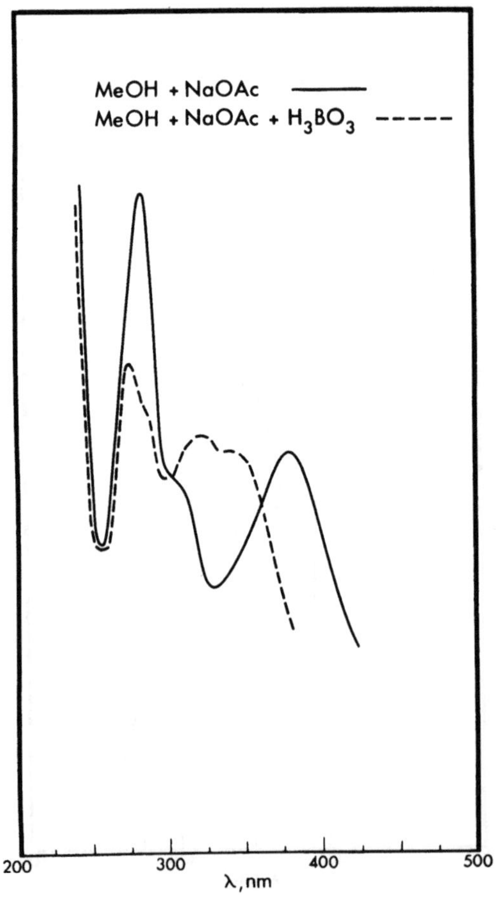

VIOLANTHIN

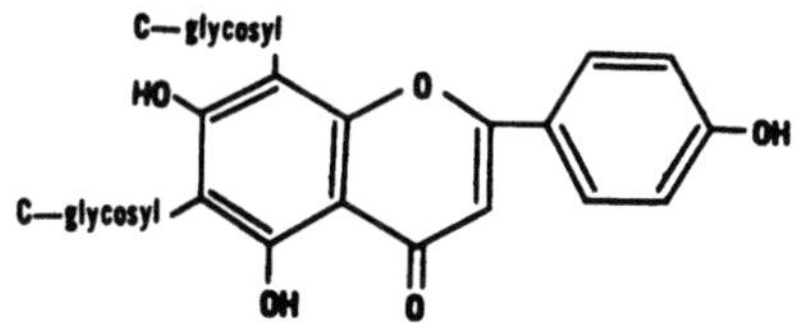

CHROMATOGRAPHIC DATA

Spot Appearance: (UV) deep purple
(UV/NH_3) yellow-green

R_f Values: 0.38 (TBA), 0.72 (HOAc)

UV SPECTRAL DATA (λ_{max},*nm*)

MeOH	274, 311sh, 335
NaOMe	281, 333, 398
$AlCl_3$	265sh, 281, 307, 353, 387
$AlCl_3$/HCl	263sh, 282, 306, 347, 383
NaOAc	281, 304sh, 388
NaOAc/H_3BO_3 (Proc. I)	274, 330sh, 348, 412sh

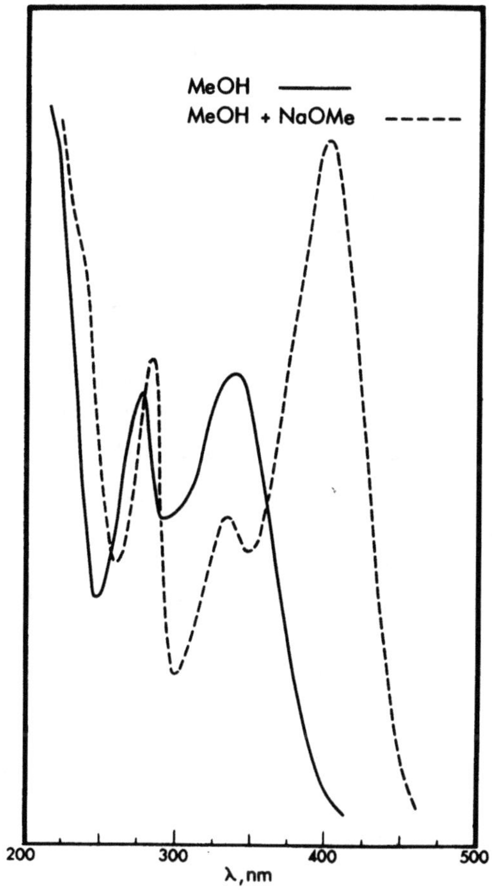

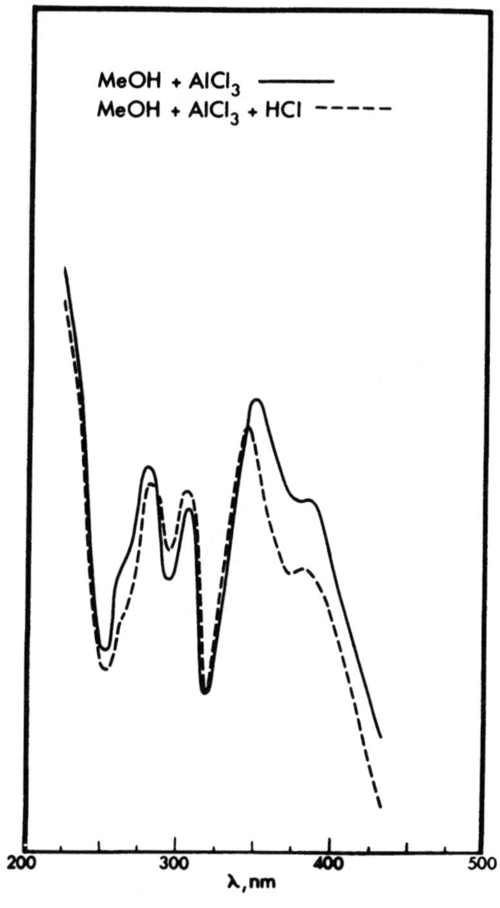

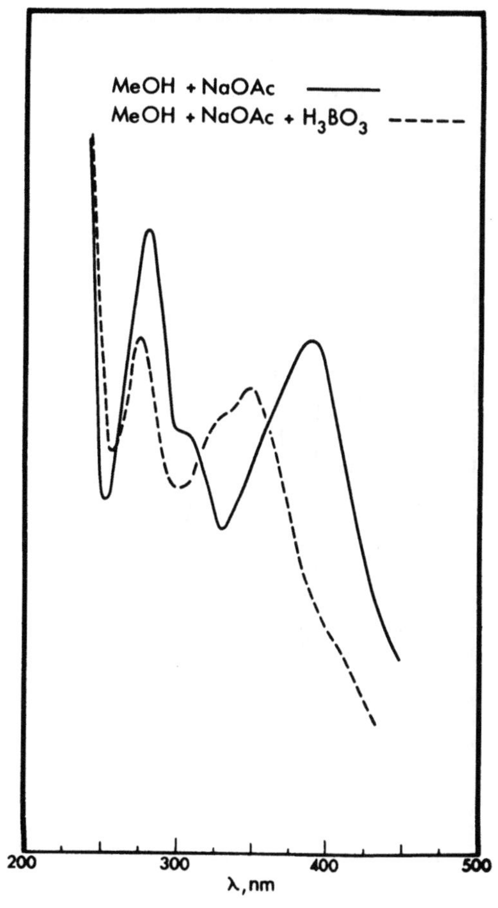

29

ACACETIN

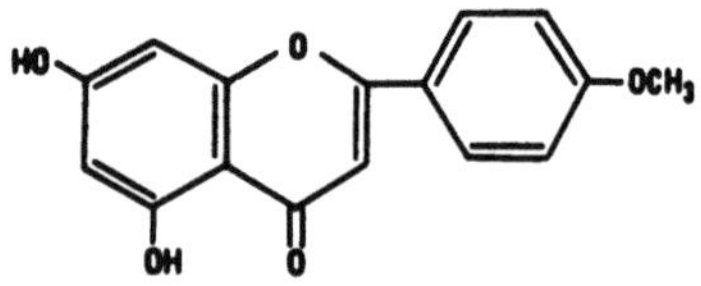

CHROMATOGRAPHIC DATA

Spot Appearance: (UV) deep purple
(UV/NH_3) deep purple

R_f Values: 0.90 (TBA), 0.11 (HOAc)

UV SPECTRAL DATA (λ_{max}, nm)

MeOH	269, 303sh, 327
NaOMe	276, 295sh, 364
$AlCl_3$	259sh, 277, 292sh, 302, 344, 382
$AlCl_3$/HCl	260sh, 279, 294sh, 300, 338, 379
NaOAc	276, 297sh, 358
NaOAc/H_3BO_3 (Proc. I)	269, 309sh, 331

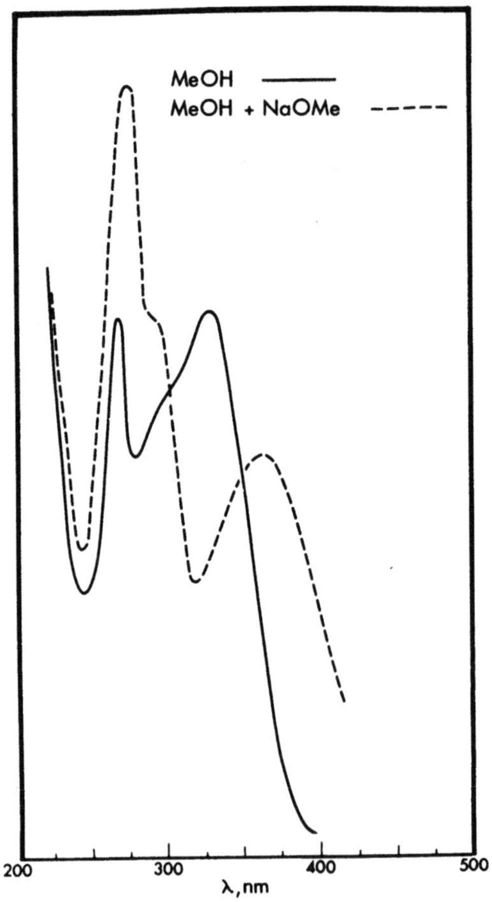

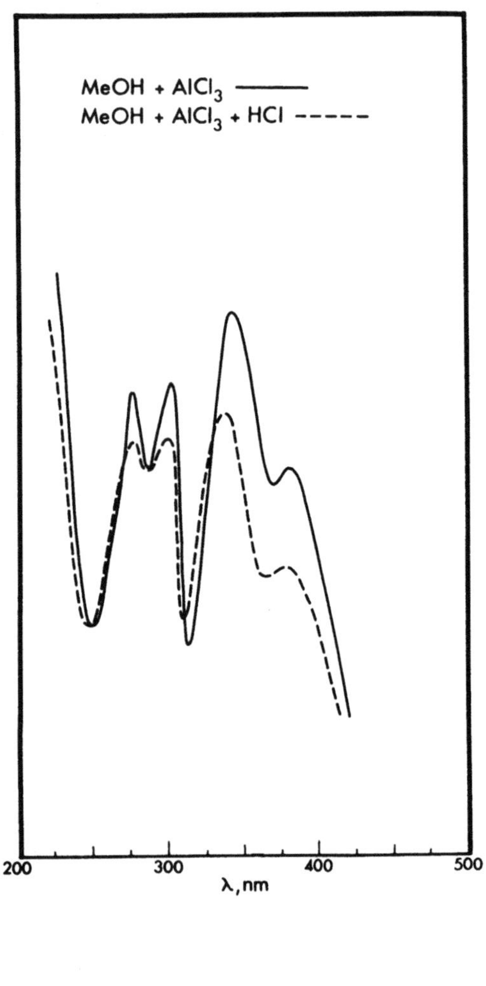

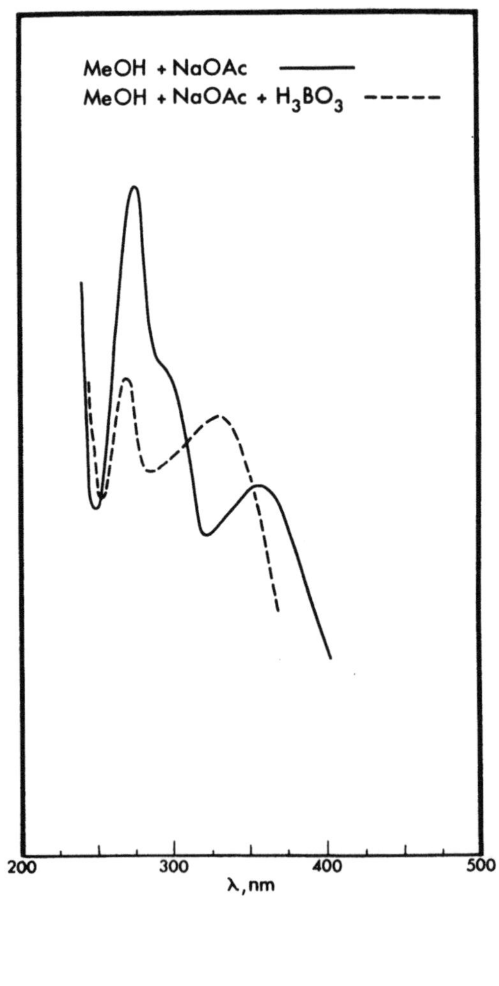

30

ACACETIN 7-0-GLUCOSIDE

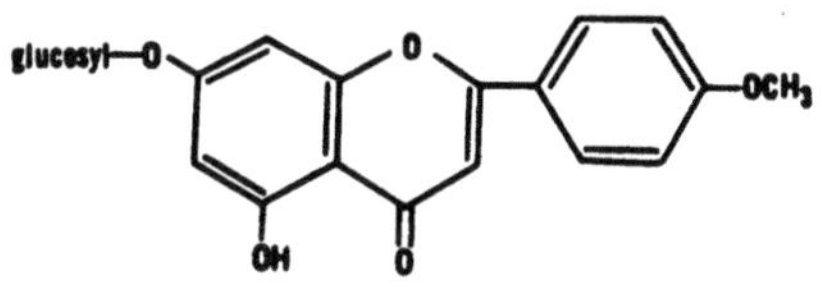

CHROMATOGRAPHIC DATA

Spot Appearance: (UV) deep purple
(UV/NH_3) deep purple

R_f Values: 0.58 (TBA), 0.27 (HOAc)

UV SPECTRAL DATA (λ_{max},*nm*)

MeOH	268, 324
NaOMe	244sh, 287, 357
$AlCl_3$	277, 300, 345, 383
$AlCl_3$/HCl	278, 299, 338, 381
NaOAc	268, 324
NaOAc/H_3BO_3 (Proc. I)	269, 328

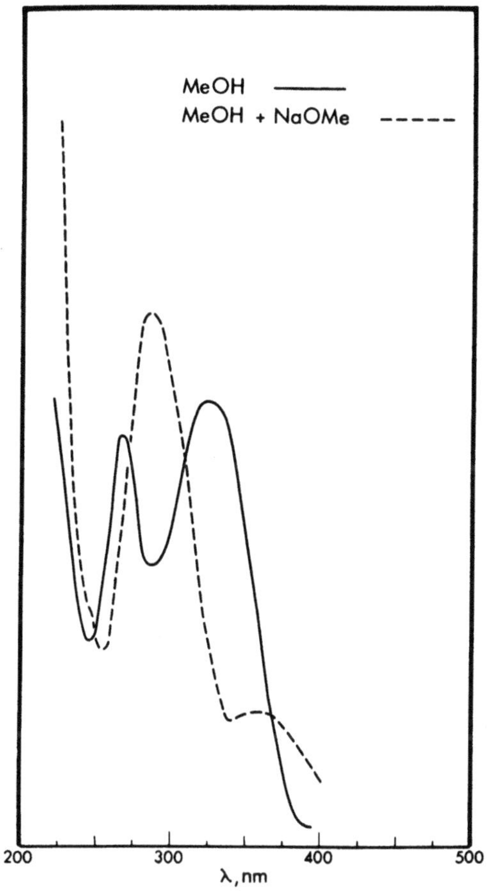

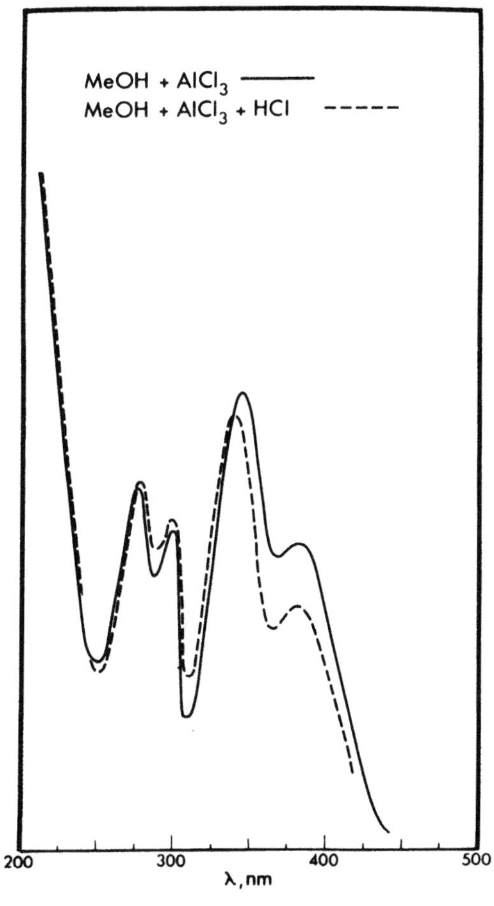

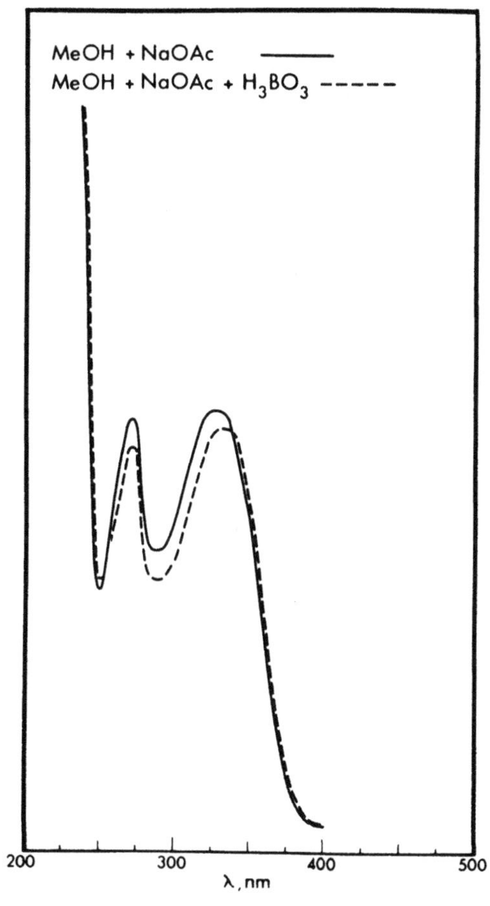

31

5,7-DIHYDROXY-2'-METHOXYFLAVONE

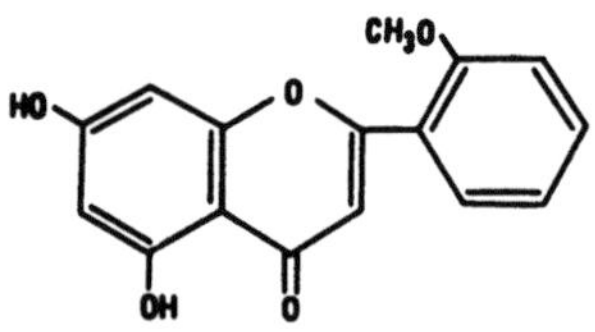

CHROMATOGRAPHIC DATA

Spot Appearance: (UV) deep purple
(UV/NH_3) deep purple

R_f Values: 0.90 (TBA), 0.19 (HOAc)

UV SPECTRAL DATA (λ_{max},nm)

MeOH	266, 325
NaOMe	273, 323sh, 362
$AlCl_3$	252, 276, 344, 375
$AlCl_3$/HCl	252, 277, 337, 378
NaOAc	271, 325sh, 356
NaOAc/H_3BO_3 (Proc. I)	267, 330

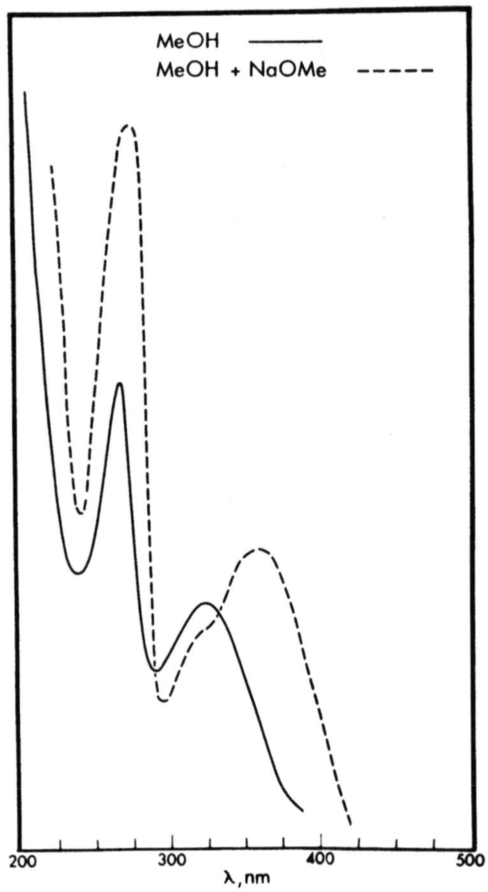

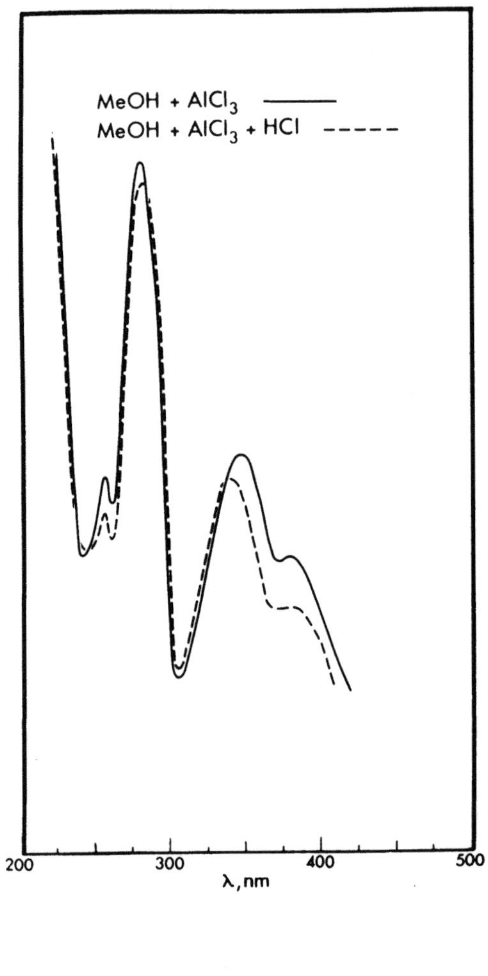

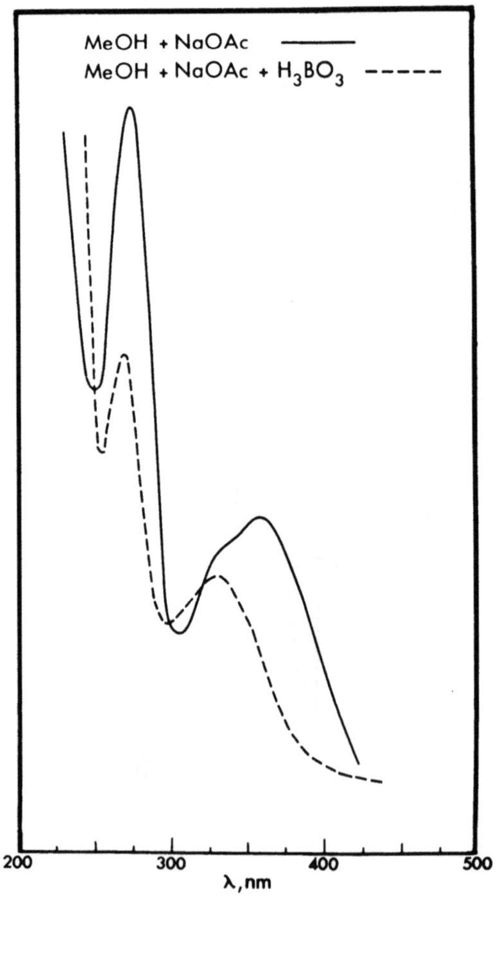

ZAPOTININ

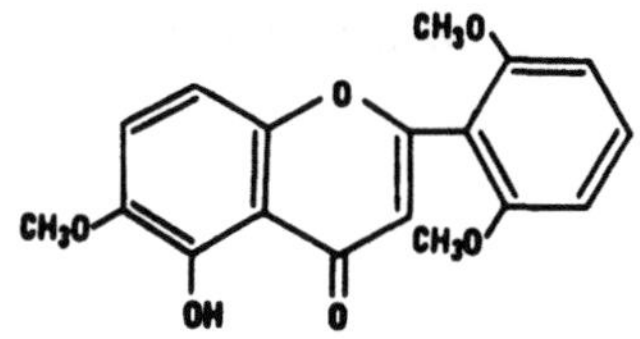

CHROMATOGRAPHIC DATA

Spot Appearance: (UV) deep purple
(UV/NH_3) deep purple

R_f Values: 0.93 (TBA), 0.00 (HOAc)

UV SPECTRAL DATA (λ_{max}, nm)

MeOH	264, 307sh, 348sh
NaOMe	248, 269, 394
$AlCl_3$	236, 255sh, 275, 296sh, 325sh, 411
$AlCl_3$/HCl	236, 274, 293sh, 326sh, 410
NaOAc	263, 349sh
NaOAc/H_3BO_3 (Proc. I)	264, 312sh, 349sh

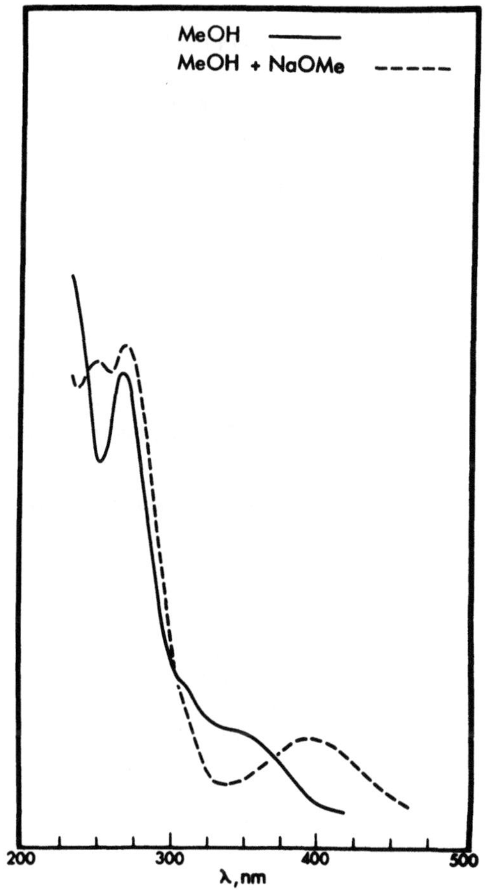

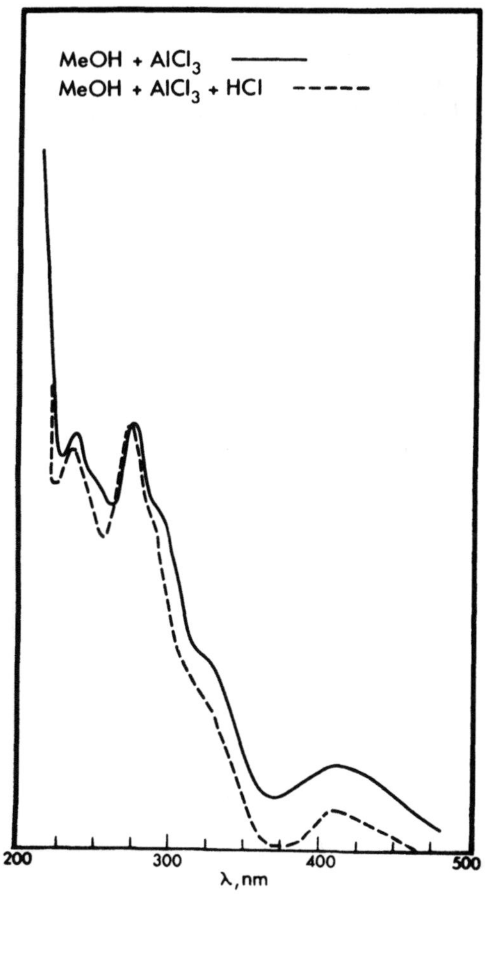

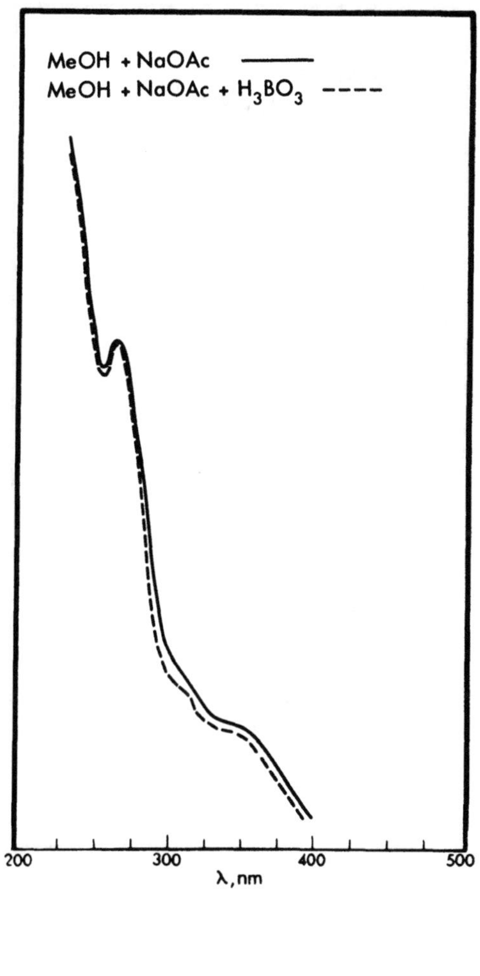

33

ZAPOTIN

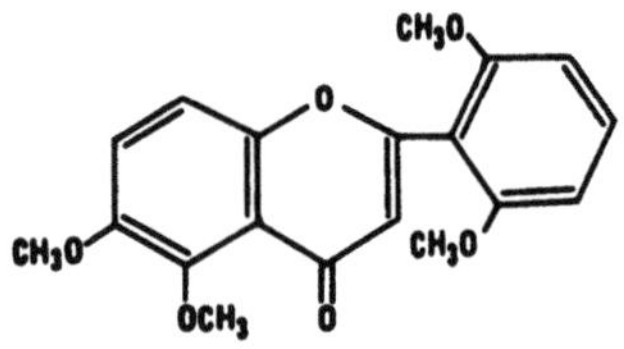

CHROMATOGRAPHIC DATA

Spot Appearance: (UV) fluorescent light blue

(UV/NH_3) fluorescent light blue

R_f Values: 0.91 (TBA), 0.4 (approx.) (HOAc)

UV SPECTRAL DATA (λ_{max},*nm*)

MeOH	255sh, 325
NaOMe	255sh, 295sh, 323
$AlCl_3$	255sh, 325
$AlCl_3/HCl$	255sh, 324
NaOAc	258sh, 324
$NaOAc/H_3BO_3$ (Proc. I)	259sh, 324

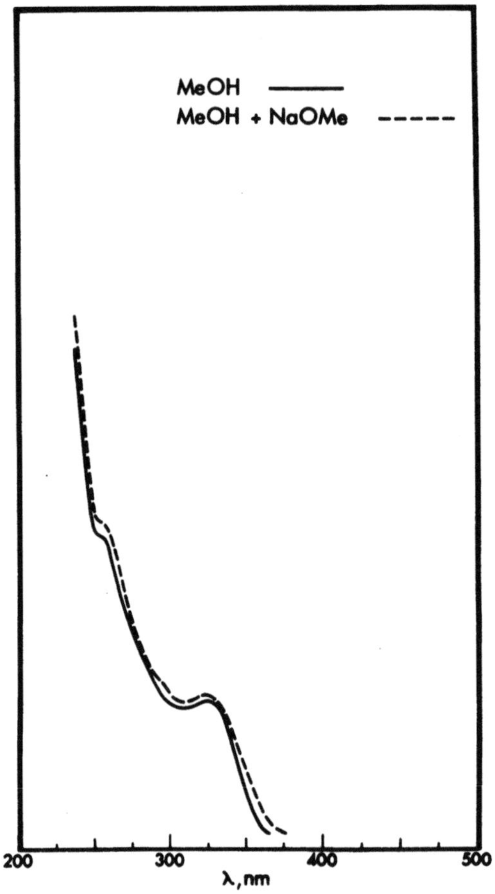

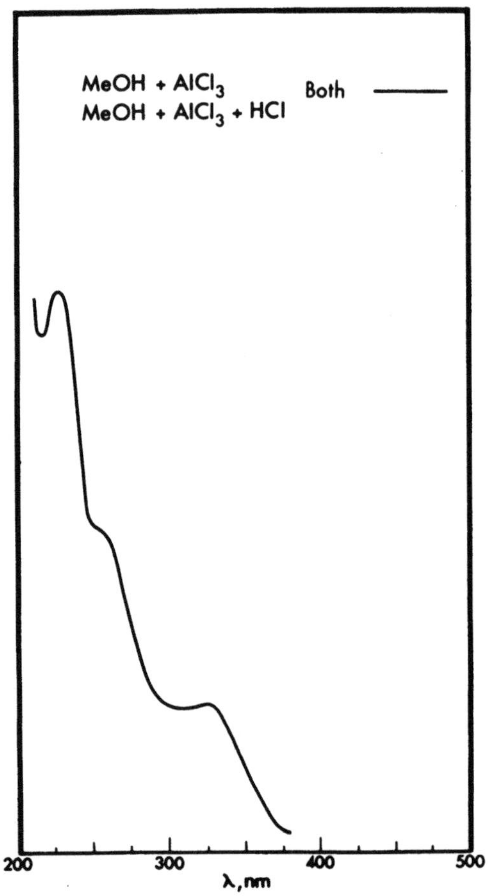

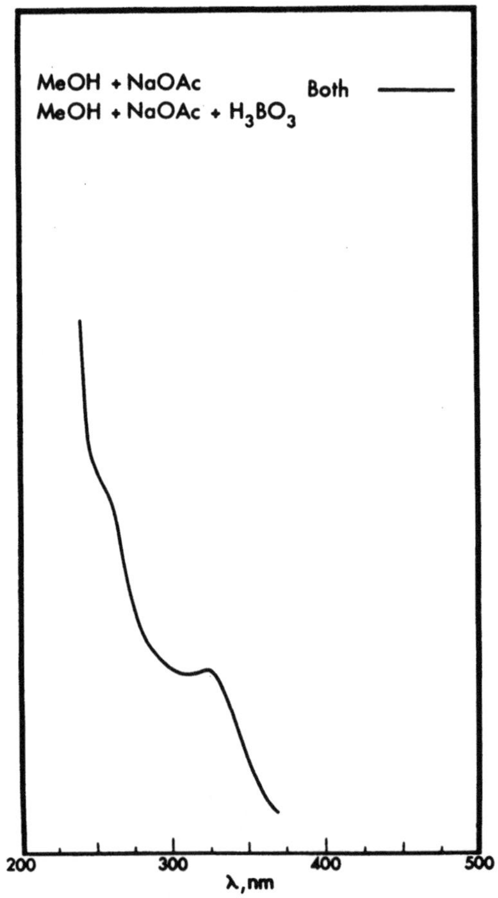

LUTEOLIN

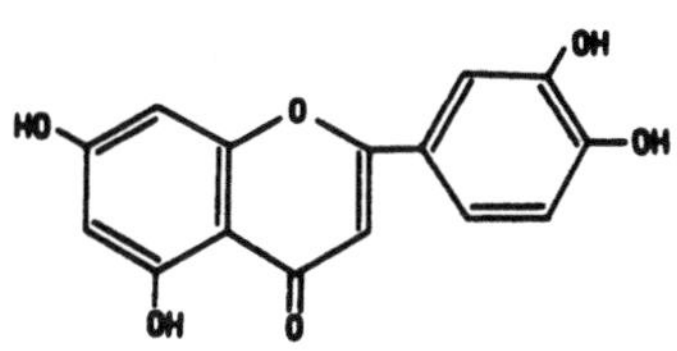

CHROMATOGRAPHIC DATA

Spot Appearance: (UV) deep purple
(UV/NH_3) yellow

R_f Values: 0.77 (TBA), 0.08 (HOAc)

UV SPECTRAL DATA (λ_{max}, nm)

MeOH	242sh, 253, 267, 291sh, 349
NaOMe	266sh, 329sh, 401
$AlCl_3$	274, 300sh, 328, 426
$AlCl_3$/HCl	266sh, 275, 294sh, 355, 385
NaOAc	269, 326sh, 384
NaOAc/H_3BO_3 (Proc. I)	259, 301sh, 370, 430sh

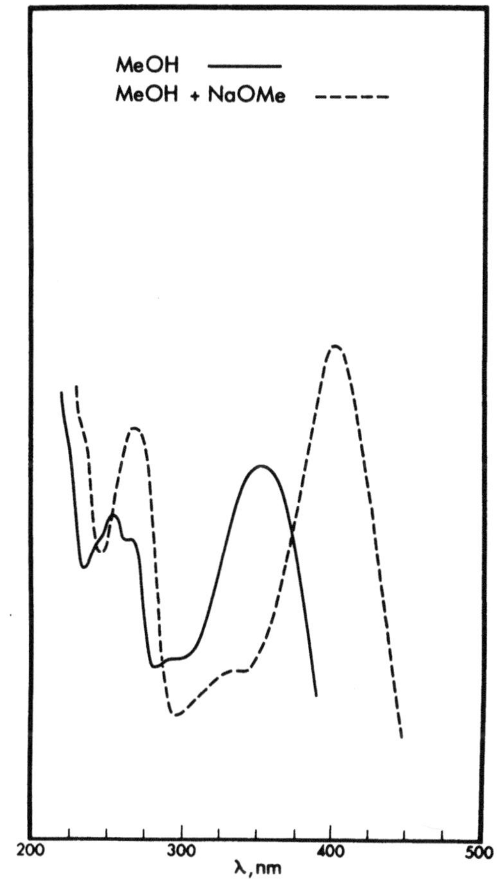

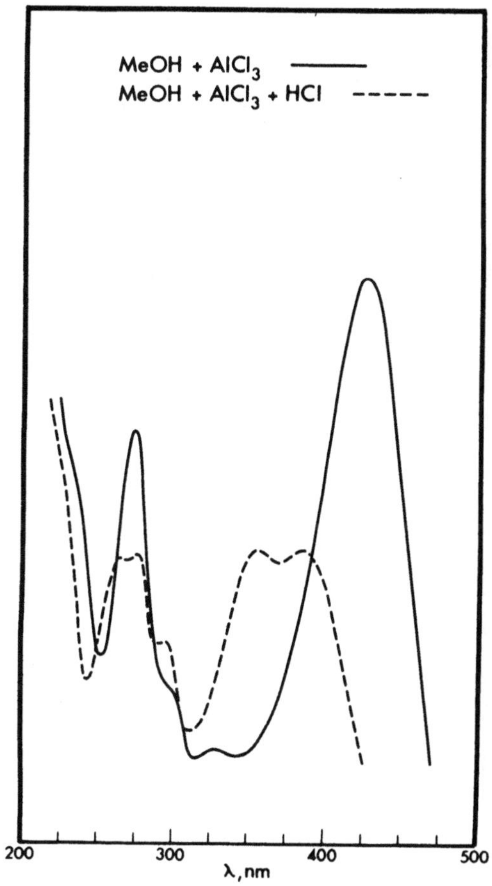

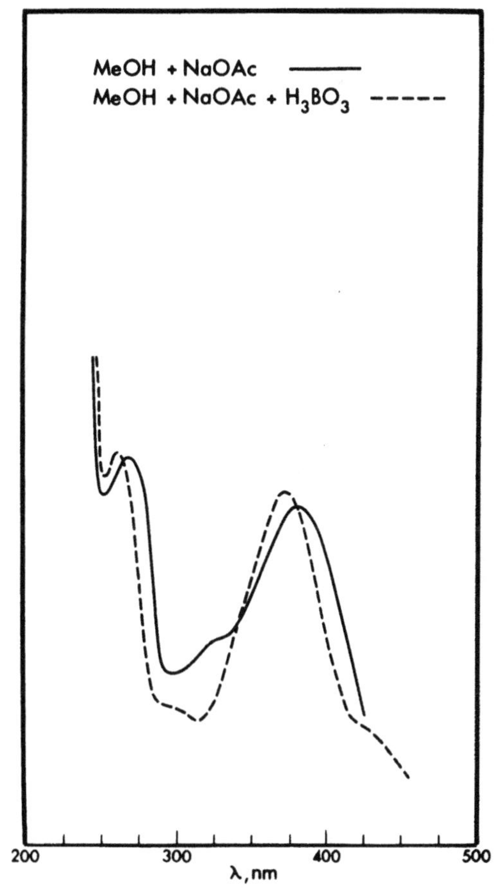

LUTEOLIN 7-0-GLUCOSIDE

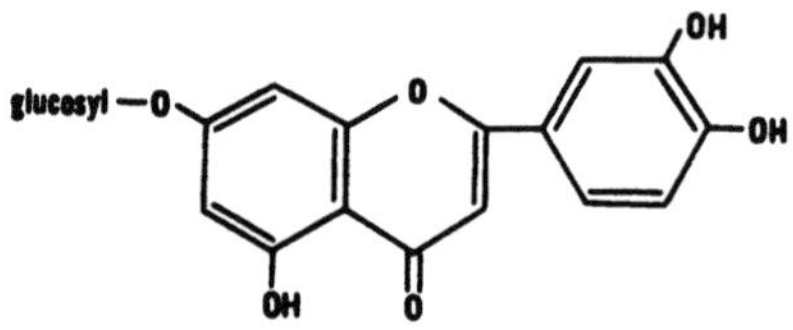

CHROMATOGRAPHIC DATA

Spot Appearance: (UV) deep purple
(UV/NH_3) yellow

R_f Values: 0.43 (TBA), 0.16 (HOAc)

UV SPECTRAL DATA (λ_{max},nm)

MeOH	255, 267sh, 348
NaOMe	263, 300sh, 394
$AlCl_3$	274, 298sh, 329, 432
$AlCl_3$/HCl	273, 294sh, 358, 387
NaOAc	259, 266sh, 365sh, 405
NaOAc/H_3BO_3 (Proc. I)	259, 372

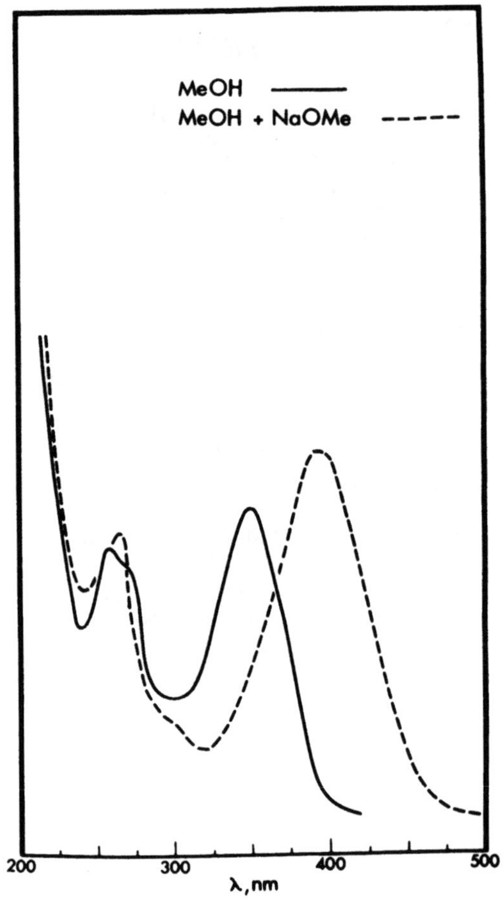

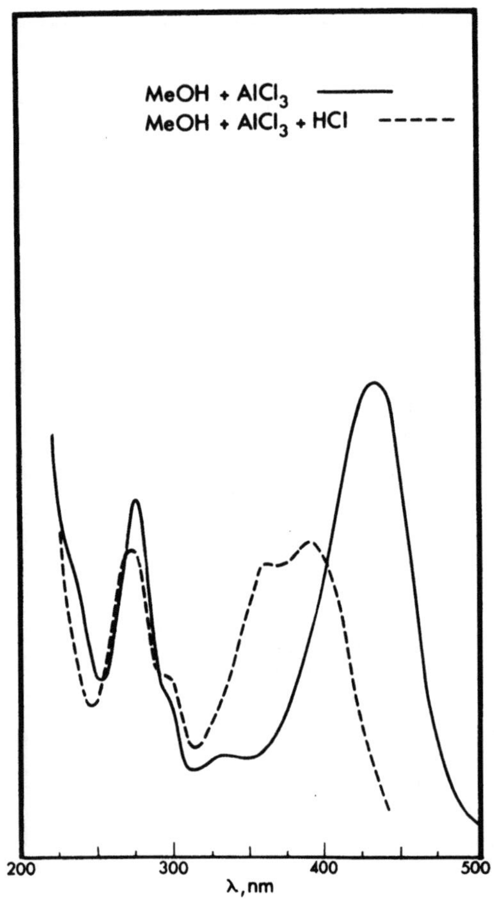

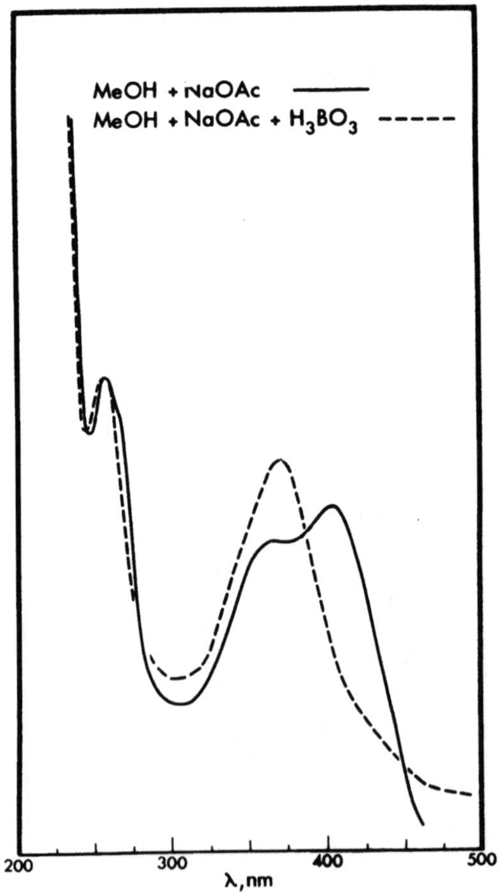

36

LUTEOLIN 7-0-RUTINOSIDE

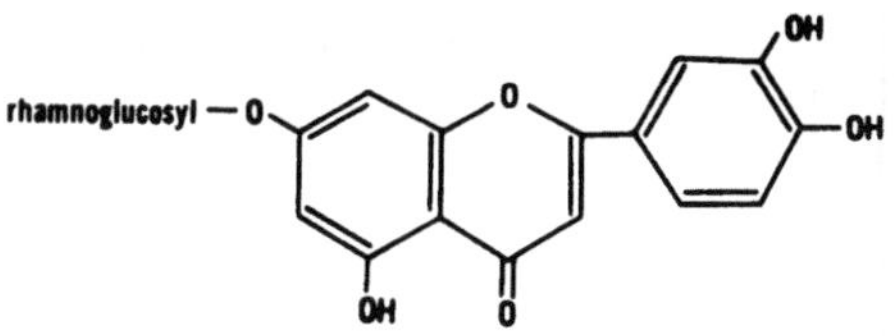

CHROMATOGRAPHIC DATA

Spot Appearance: (UV) deep purple
(UV/NH_3) yellow

R_f Values: 0.26 (TBA), 0.30 (HOAc)

UV SPECTRAL DATA (λ_{max},*nm*)

MeOH	255, 265sh, 349
NaOMe	263, 299sh, 394
$AlCl_3$	272, 296sh, 331, 432
$AlCl_3$/HCl	272, 295, 359, 389
NaOAc	259, 266sh, 366, 403
NaOAc/H_3BO_3 (Proc. I)	258, 370

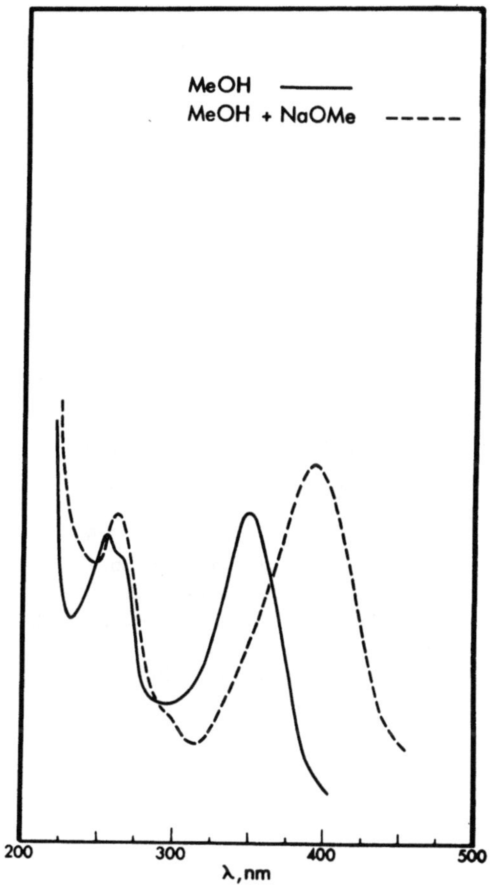

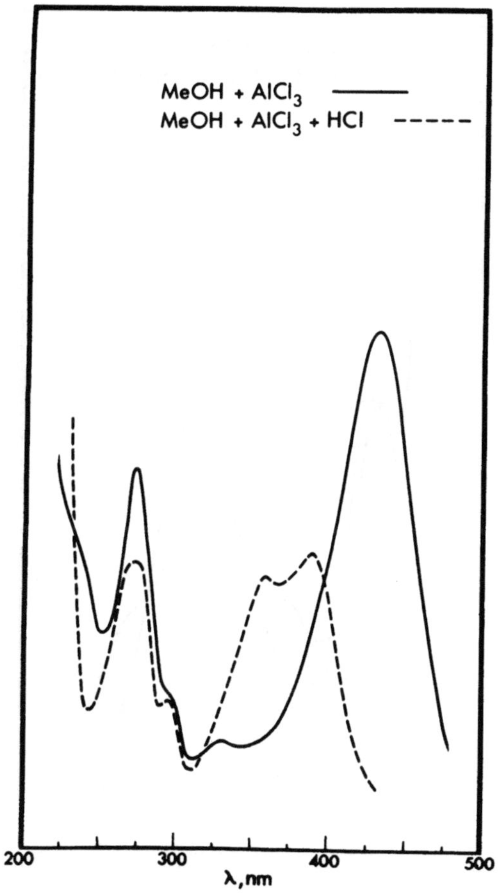

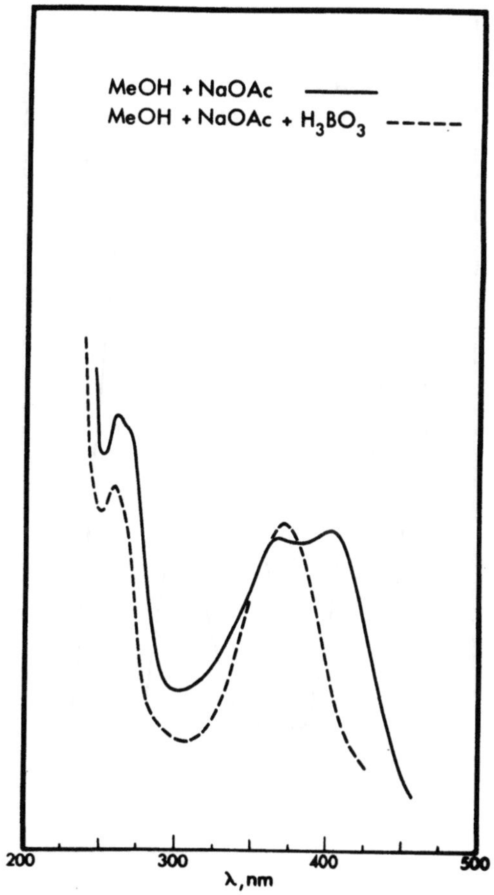

37

ISOORIENTIN

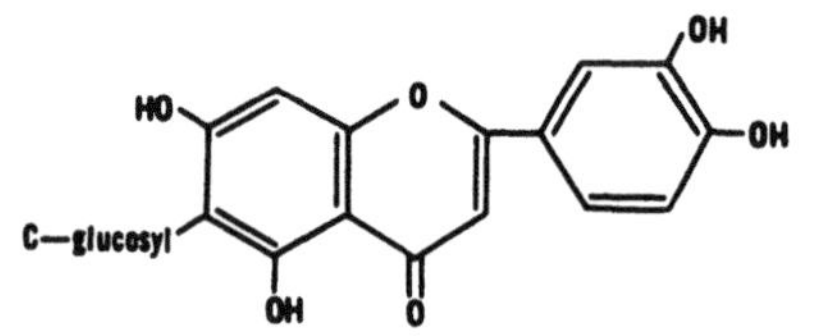

CHROMATOGRAPHIC DATA

Spot Appearance: (UV) deep purple
(UV/NH_3) yellow-green

R_f Values: 0.43 (TBA), 0.39 (HOAc)

UV SPECTRAL DATA (λ_{max}, *nm*)

MeOH	242sh, 255, 271, 349
NaOMe	267, 278sh, 337sh, 406
$AlCl_3$	278, 302sh, 332, 429
$AlCl_3$/HCl	265sh, 279, 296sh, 361, 384
NaOAc	276, 323, 393
NaOAc/H_3BO_3 (Proc. I)	265, 377, 429sh

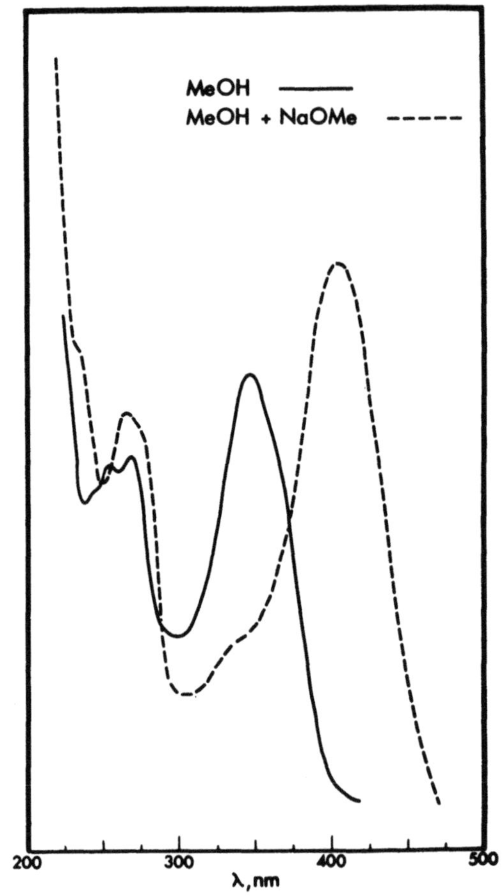

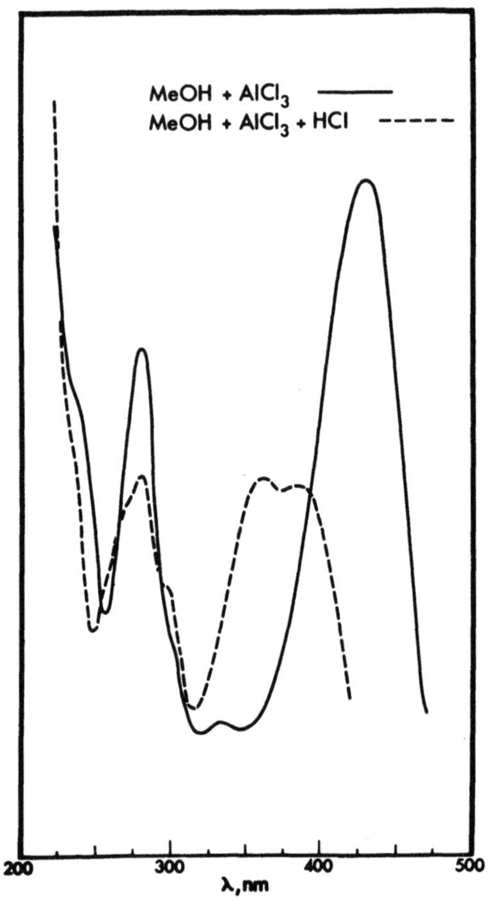

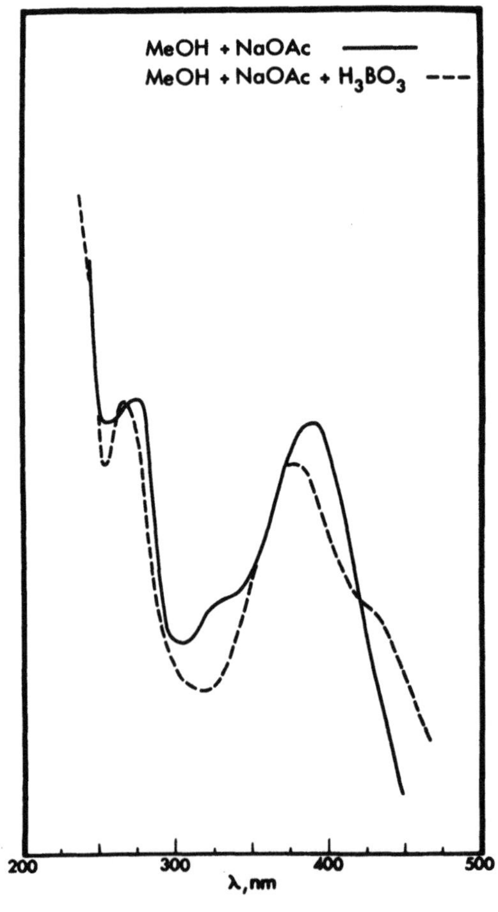

38

ORIENTIN

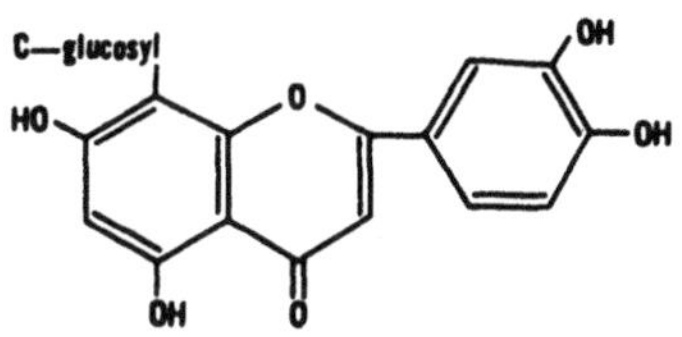

CHROMATOGRAPHIC DATA

Spot Appearance: (UV) deep purple
(UV/NH_3) yellow-green

R_f Values: 0.29 (TBA), 0.21 (HOAc)

UV SPECTRAL DATA (λ_{max}, nm)

MeOH	255, 267, 293sh, 346
NaOMe	268, 278sh, 334sh, 405
$AlCl_3$	276, 302sh, 329, 429
$AlCl_3$/HCl	265sh, 276, 296sh, 357, 384
NaOAc	278, 325, 386
NaOAc/H_3BO_3	264, 375, 430sh
(Proc. I)	

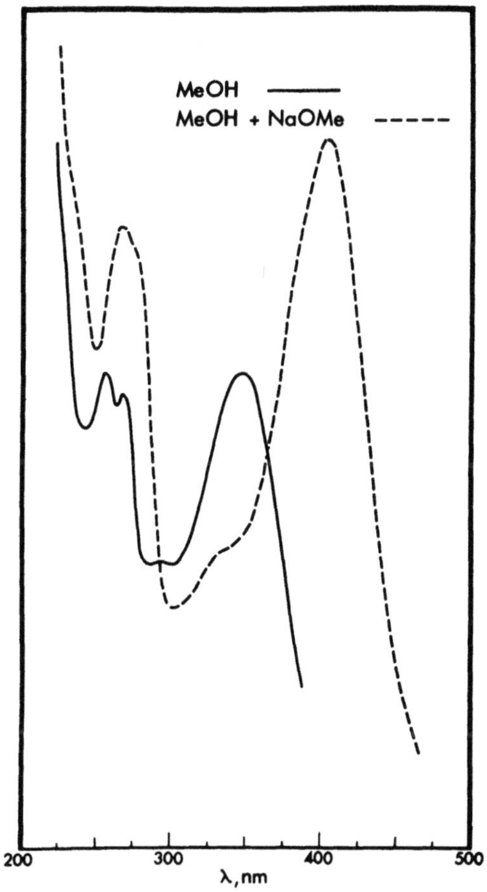

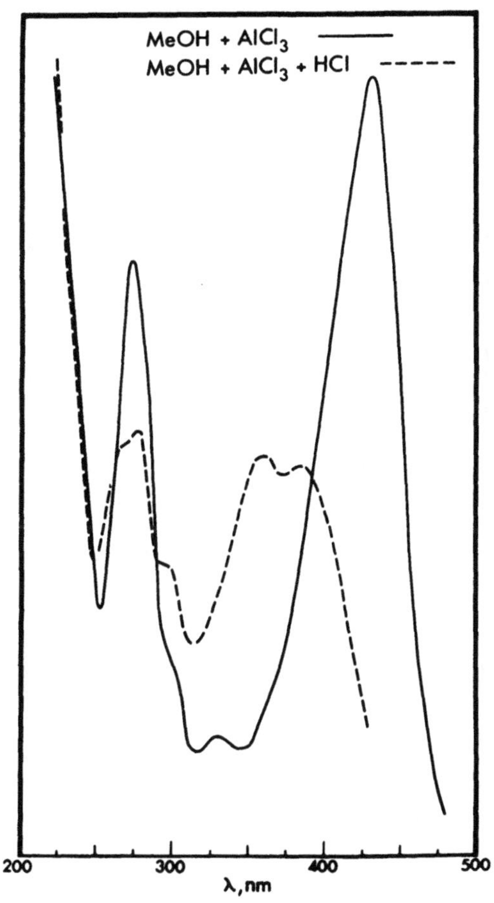

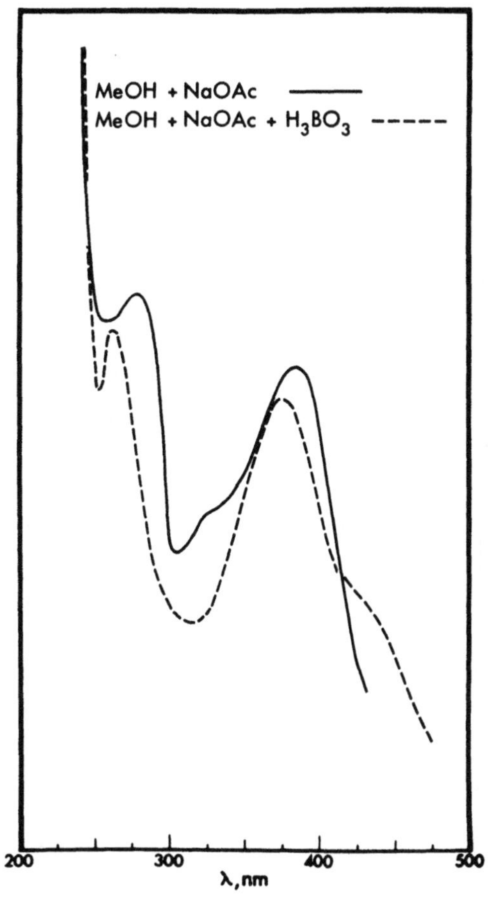

39

LUCENIN-1

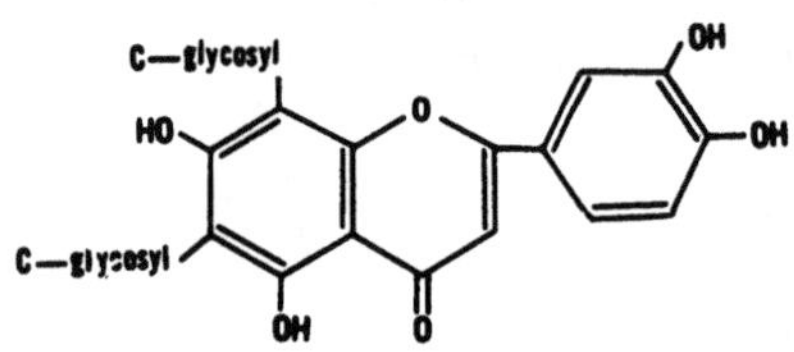

CHROMATOGRAPHIC DATA

Spot Appearance: (UV) deep purple
(UV/NH_3) yellow-green

R_f Values: 0.13 (TBA), 0.38 (HOAc)

UV SPECTRAL DATA (λ_{max},*nm*)

MeOH	257, 272, 349
NaOMe	240sh, 266, 280, 344sh, 408
$AlCl_3$	280, 303sh, 332, 430
$AlCl_3$/HCl	265sh, 278, 297sh, 359, 384sh
NaOAc	271sh, 282, 326, 398
NaOAc/H_3BO_3 (Proc. I)	266, 287sh, 382, 430

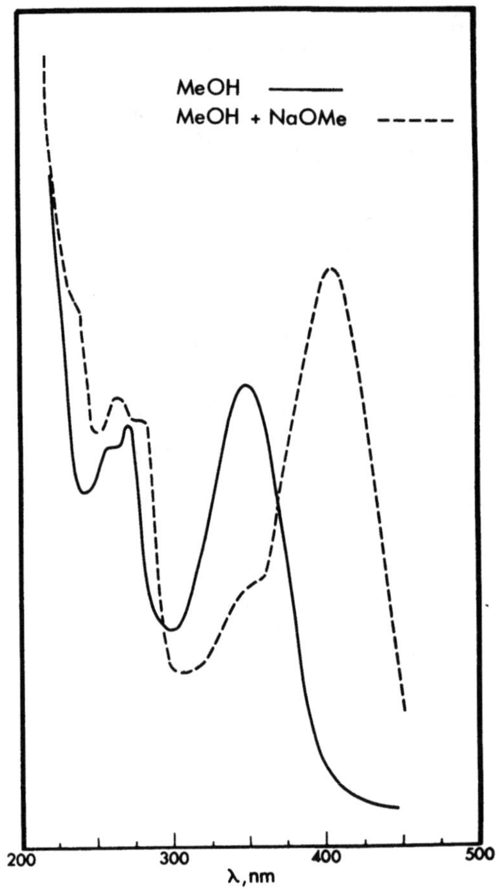

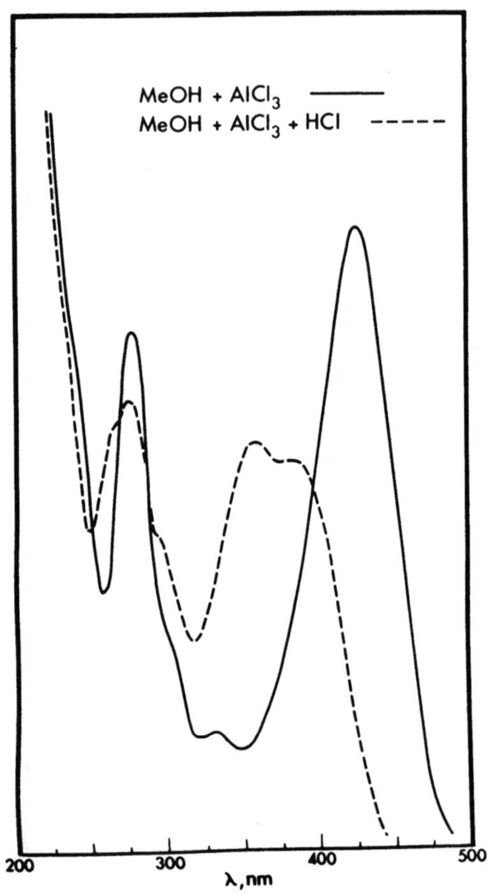

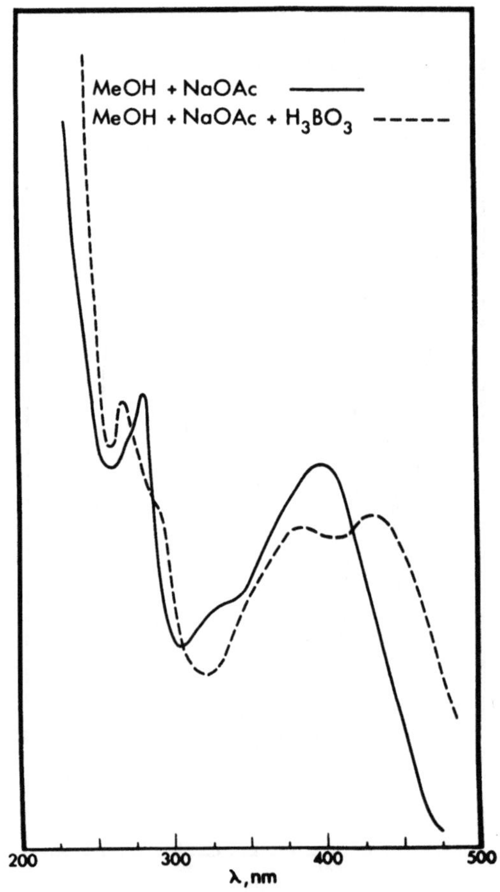

CHRYSOERIOL

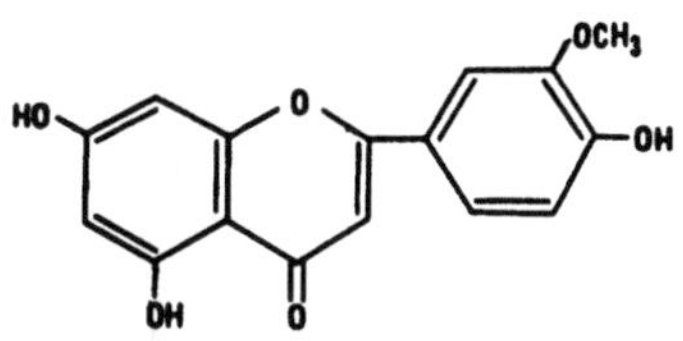

CHROMATOGRAPHIC DATA

Spot Appearance: (UV) deep purple
(UV/NH_3) yellow-green

R_f Values: 0.80 (TBA), 0.05 (HOAc)

UV SPECTRAL DATA (λ_{max}, nm)

MeOH	241, 249sh, 269, 347
NaOMe	264, 275sh, 329sh, 405
$AlCl_3$	262, 274, 296, 366sh, 390
$AlCl_3$/HCl	259, 276, 294, 353, 386
NaOAc	271, 321, 396
NaOAc/H_3BO_3 (Proc. I)	268, 349

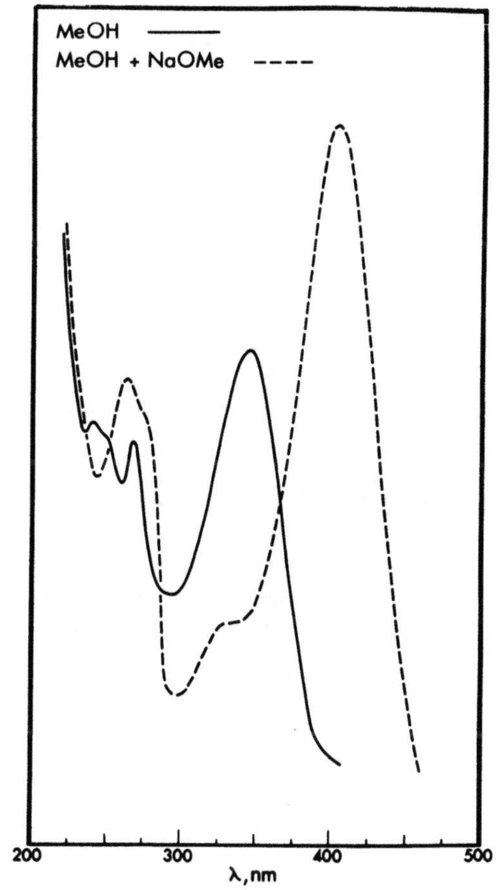

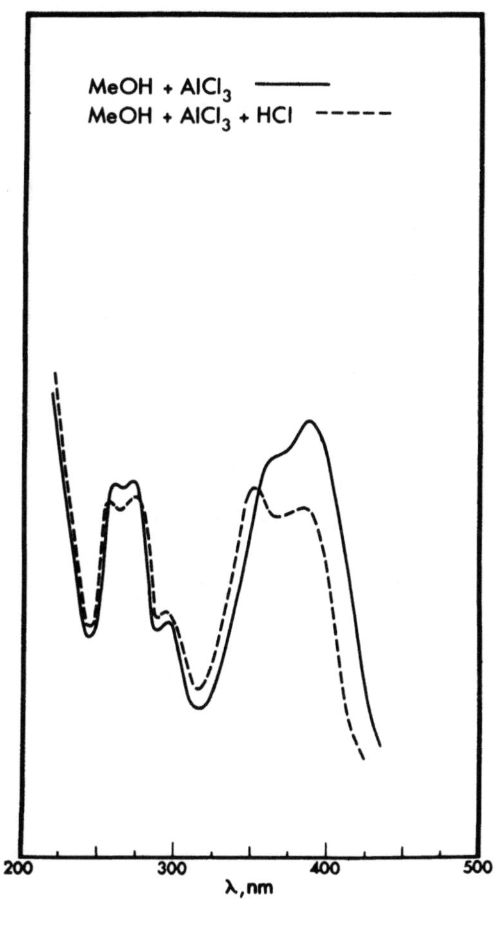

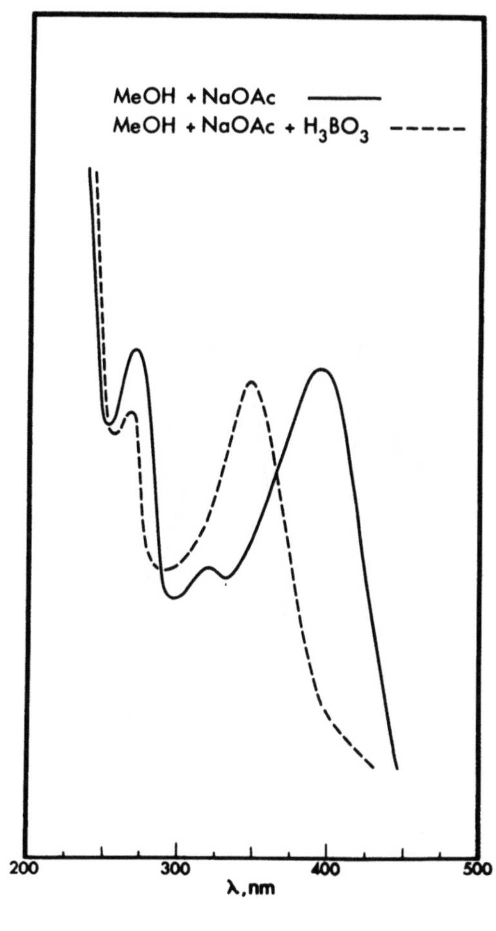

41

SCOPARIN

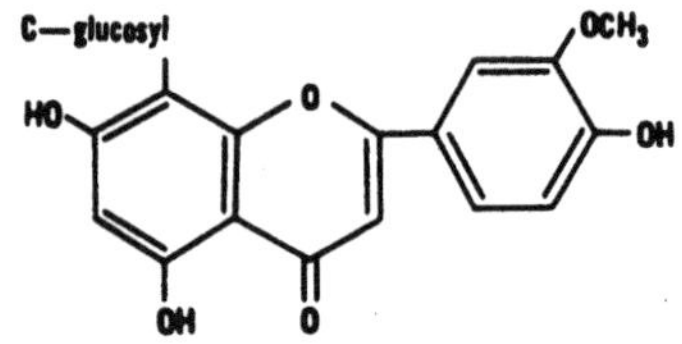

CHROMATOGRAPHIC DATA

Spot Appearance: (UV) purple
(UV/NH_3) yellow-green

R_f Values: 0.29 (TBA), 0.19 (HOAc)

UV SPECTRAL DATA (λ_{max},nm)

MeOH	251, 270, 345
NaOMe	265, 277, 334sh, 406
$AlCl_3$	265sh, 274, 296sh, 364sh, 392
$AlCl_3$/HCl	263sh, 277, 296, 354, 382
NaOAc	271sh, 279, 321, 394
NaOAc/H_3BO_3 (Proc. I)	271, 351

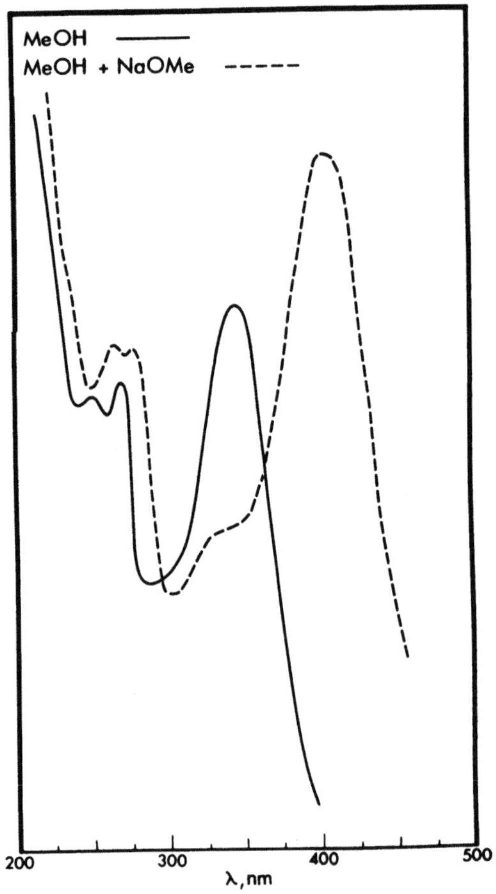

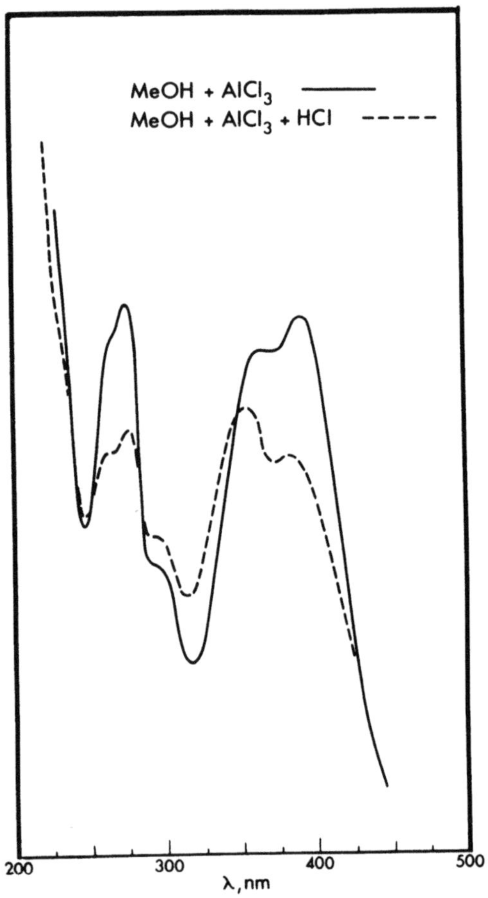

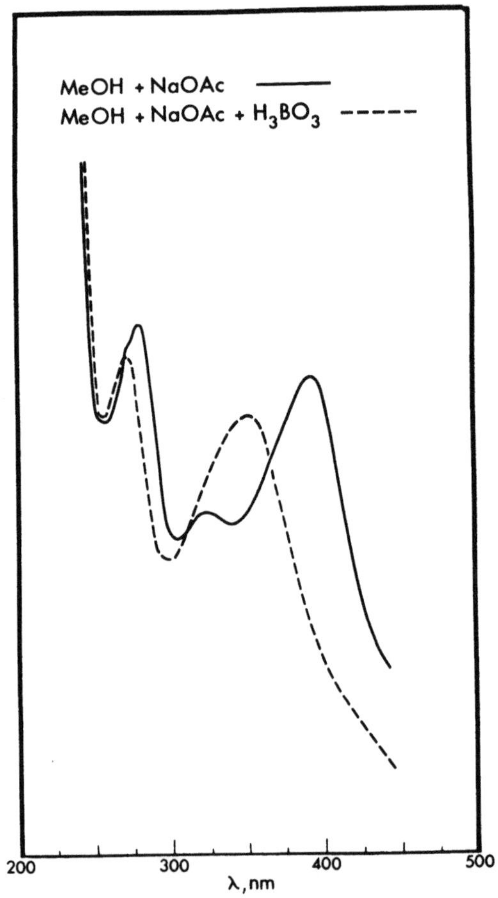

DIOSMETIN

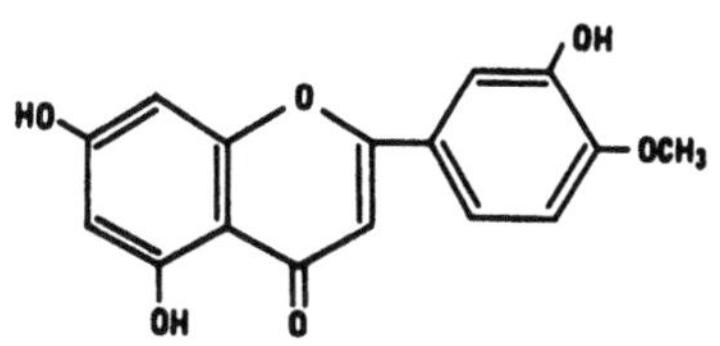

CHROMATOGRAPHIC DATA

Spot Appearance: (UV) deep purple
(UV/NH_3) deep purple

R_f Values: 0.80 (TBA), 0.07 (HOAc)

UV SPECTRAL DATA (λ_{max}, *nm*)

MeOH	240sh, 252, 267, 291sh, 344
NaOMe	270, 303sh, 386
$AlCl_3$	267sh, 273, 296, 362, 390
$AlCl_3$/HCl	264sh, 276, 295, 351, 383
NaOAc	275, 322, 367
NaOAc/H_3BO_3 (Proc. I)	253sh, 268, 348

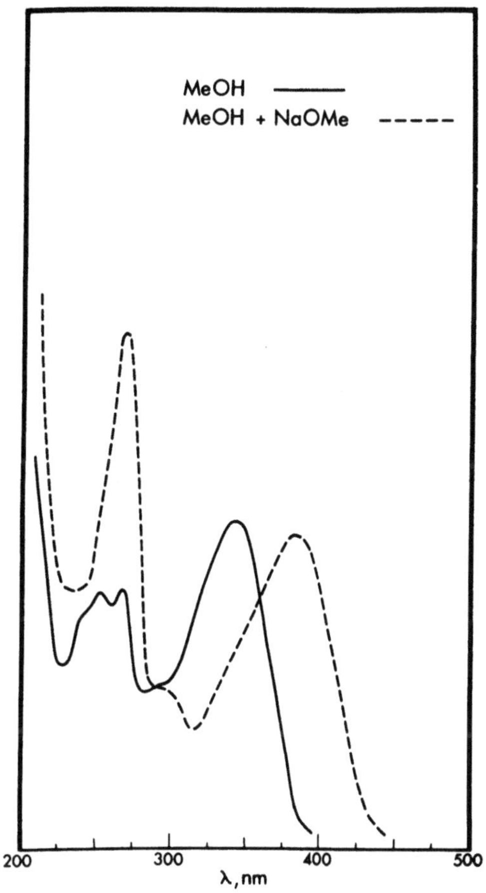

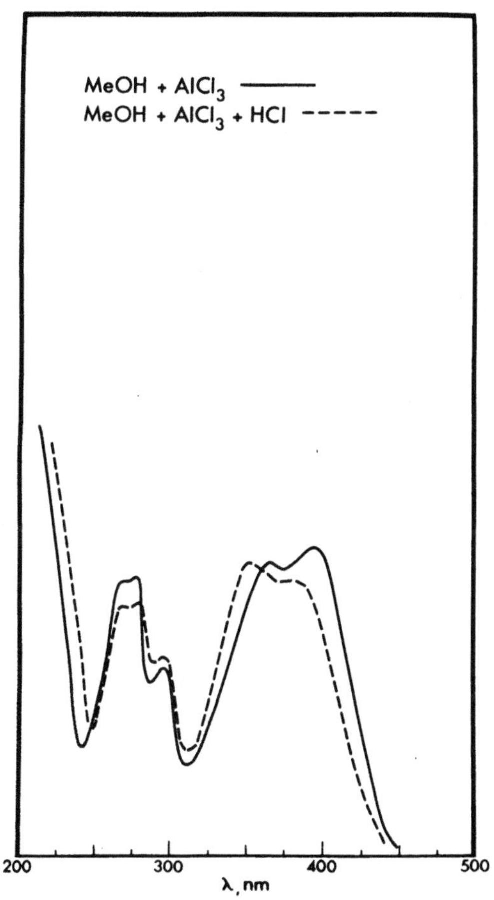

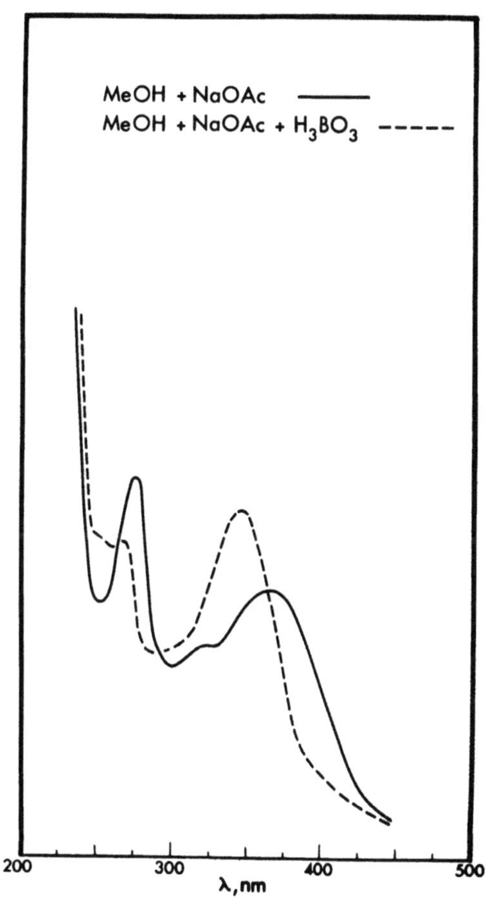

43

5,7-DIHYDROXY-3′, 4′-DIMETHOXYFLAVONE

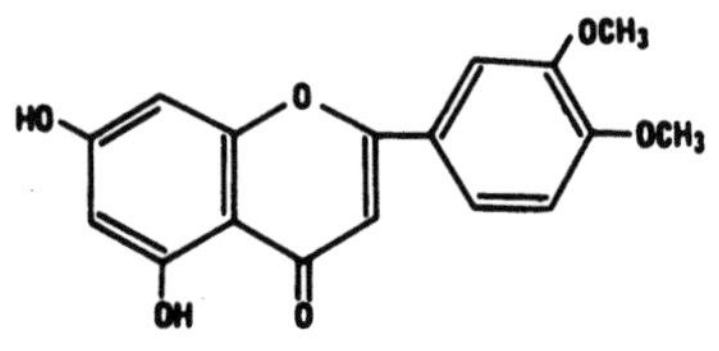

CHROMATOGRAPHIC DATA

Spot Appearance: (UV) deep purple
(UV/NH_3) deep purple

R_f Values: 0.82 (TBA), 0.07 (HOAc)

UV SPECTRAL DATA (λ_{max},nm)

MeOH	240, 248sh, 269, 291sh, 340
NaOMe	277, 312, 369
$AlCl_3$	261, 276, 295, 359, 387
$AlCl_3$/HCl	259, 279, 293sh, 348, 381sh
NaOAc	276, 318, 357
NaOAc/H_3BO_3 (Proc. I)	269, 341

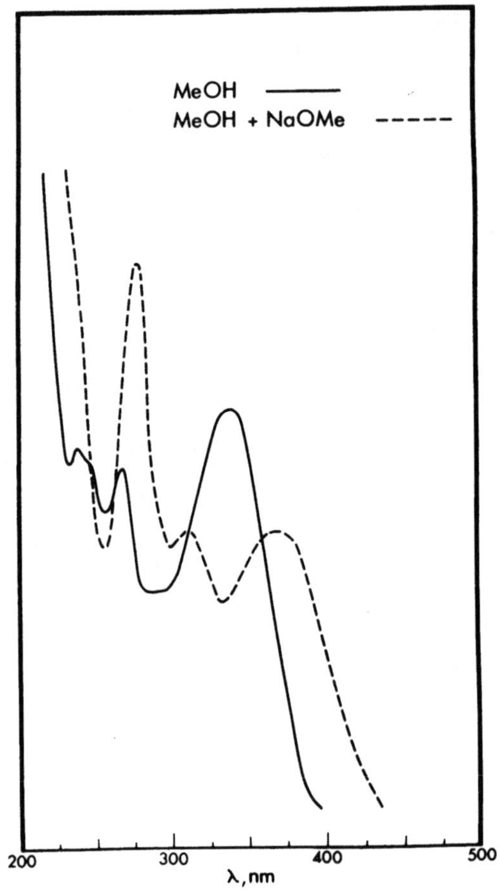

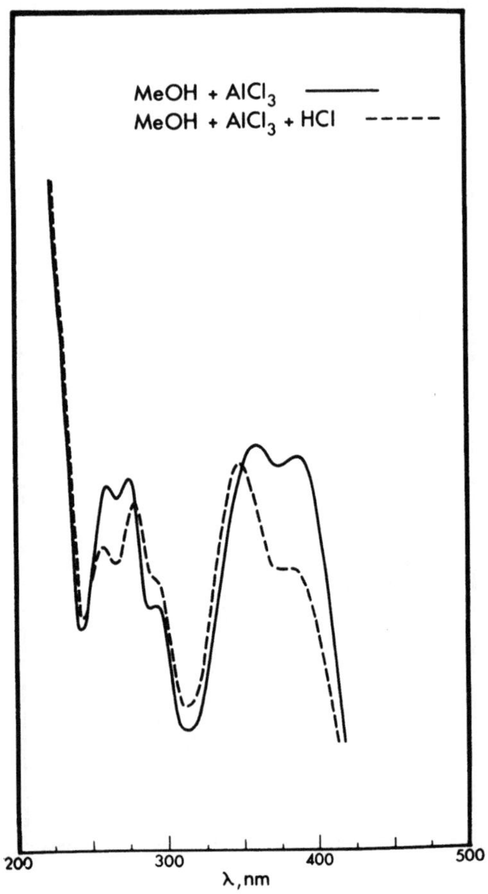

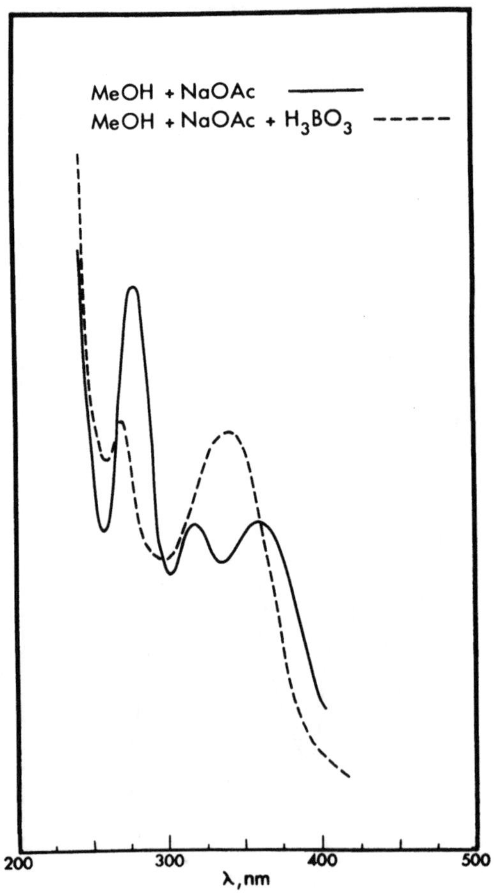

TRICIN

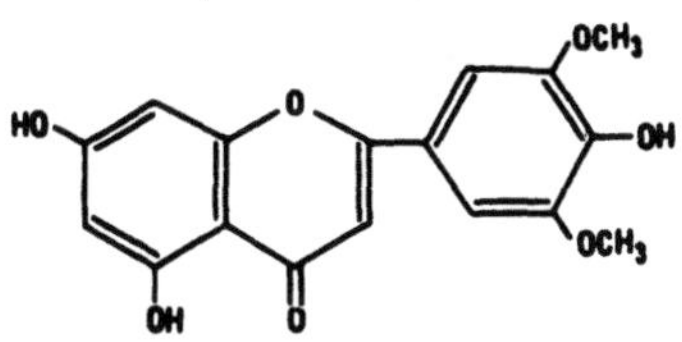

CHROMATOGRAPHIC DATA

Spot Appearance: (UV) deep purple
(UV/NH_3) yellow

R_f Values: 0.68 (TBA), 0.05 (HOAc)

UV SPECTRAL DATA (λ_{max},nm)

MeOH	244, 269, 299sh, 350
NaOMe	263, 275sh, 330, 416
$AlCl_3$	258sh, 277, 303, 366sh, 393
$AlCl_3$/HCl	259sh, 277, 302, 360, 386
NaOAc	264, 276sh, 321, 414
NaOAc/H_3BO_3 (Proc. I)	270, 304sh, 350, 422sh, 482sh

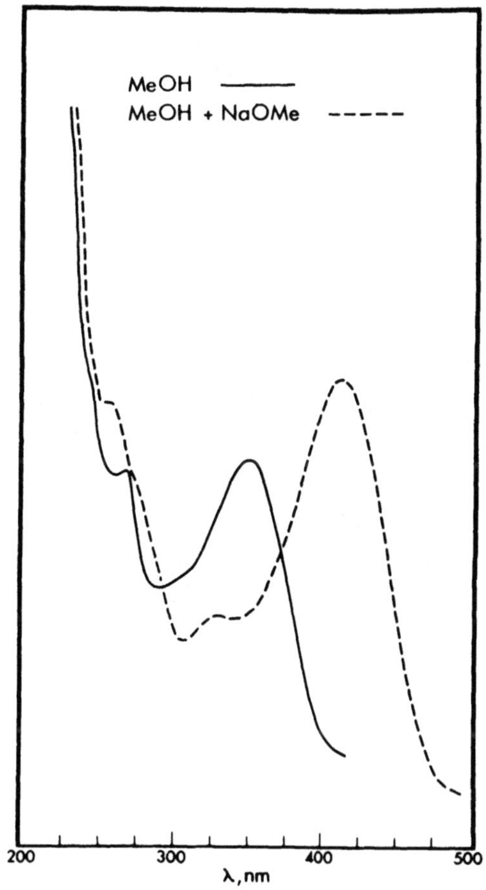

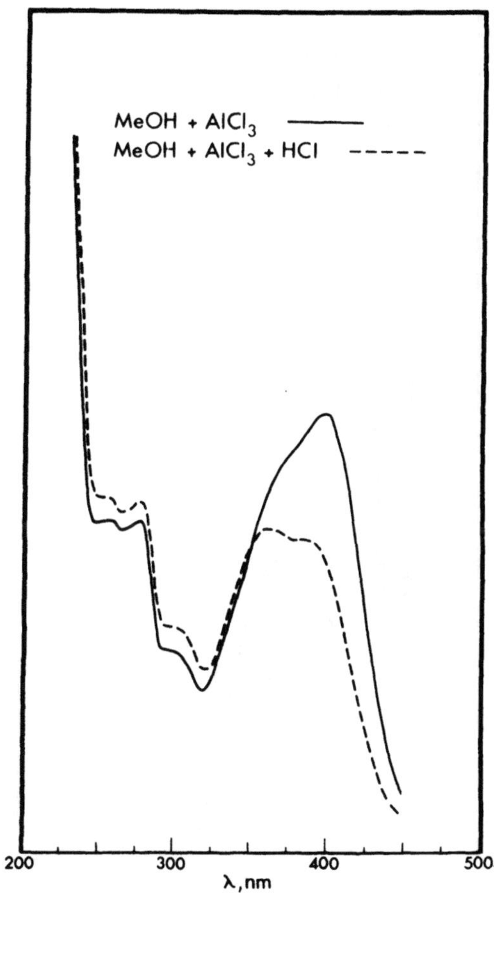

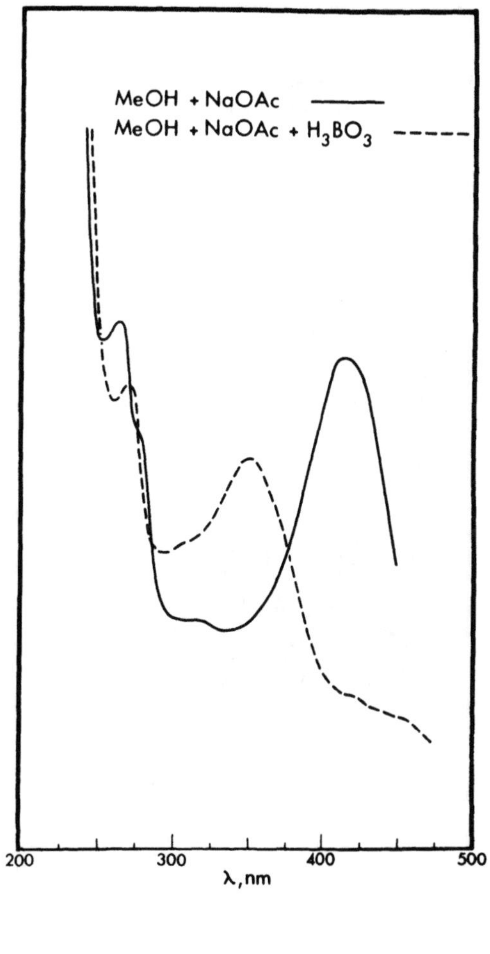

45

5,7-DIHYDROXY-3′,4′,5′-TRIMETHOXYFLAVONE

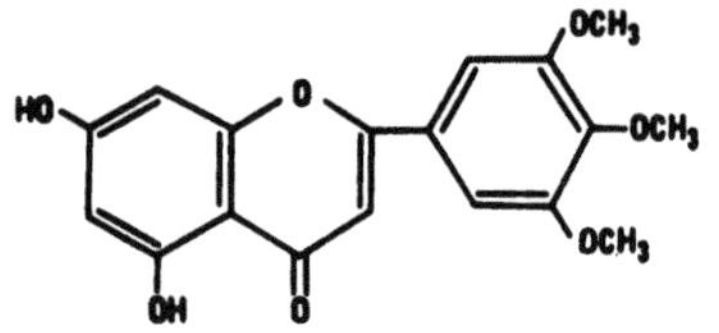

CHROMATOGRAPHIC DATA

Spot Appearance: (UV) deep purple
(UV/NH_3) deep purple

R_f Values: 0.86 (TBA), 0.15 (HOAc)

UV SPECTRAL DATA (λ_{max}, nm)

MeOH	270, 310sh, 331
NaOMe	278, 300sh, 367
$AlCl_3$	253sh, 278, 300, 348, 385sh
$AlCl_3$/HCl	280, 298sh, 340, 382sh
NaOAc	277, 299sh, 359
NaOAc/H_3BO_3 (Proc. I)	272, 313sh, 330

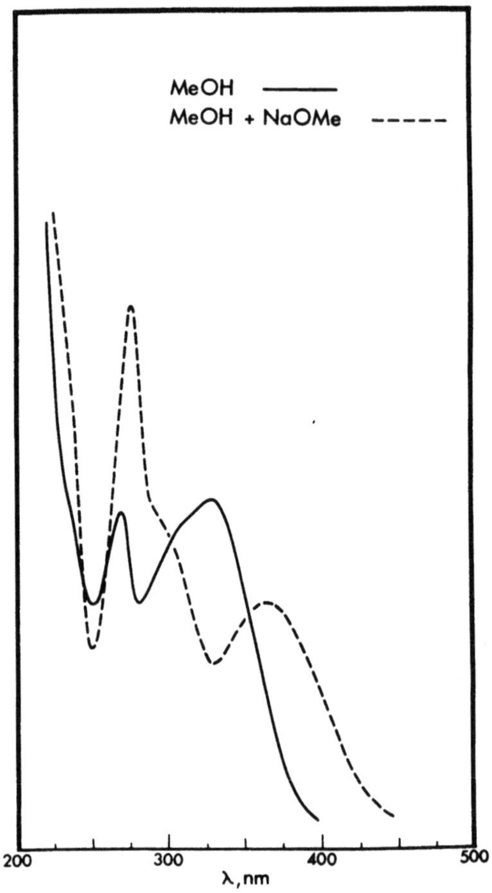

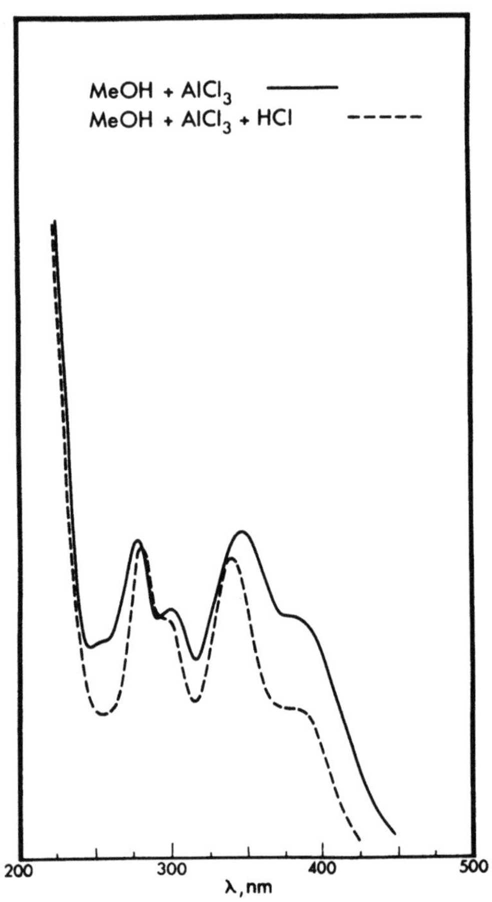

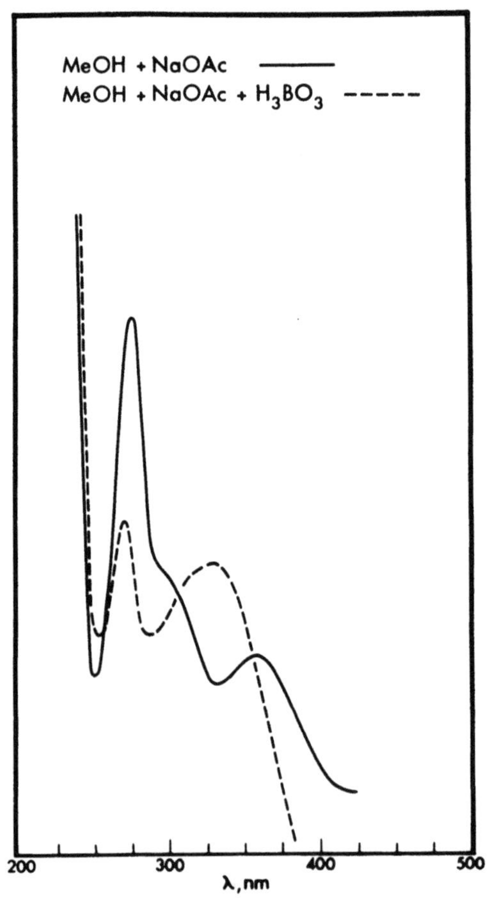

NEVADENSIN

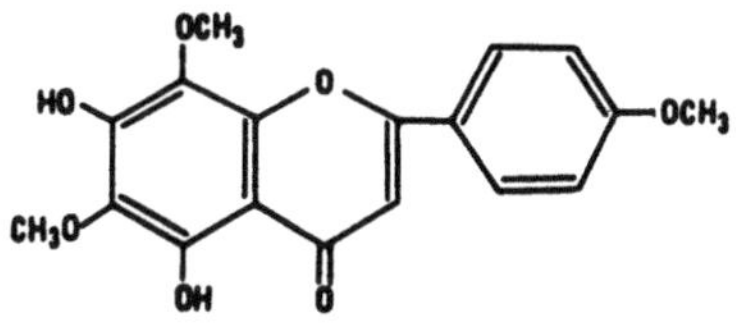

CHROMATOGRAPHIC DATA

Spot Appearance: (UV) deep purple
(UV/NH_3) deep purple

R_f Values: 0.87 (TBA), 0.14 (HOAc)

UV SPECTRAL DATA (λ_{max}, nm)

MeOH	284, 329
NaOMe	283, 300sh, 377
$AlCl_3$	265sh, 290sh, 310, 356, 413sh
$AlCl_3$/HCl	262, 289sh, 309, 351, 404sh
NaOAc	283, 302sh, 376
NaOAc/H_3BO_3 (Proc. I)	286, 322sh, 409sh

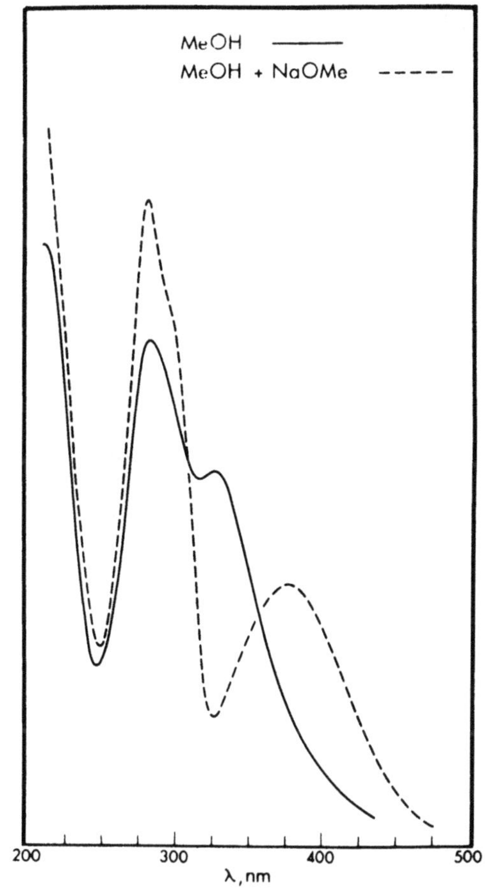

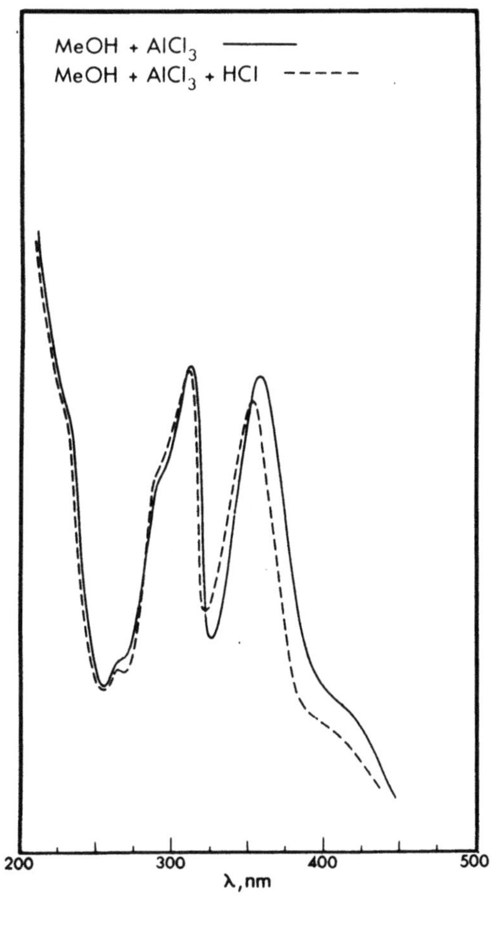

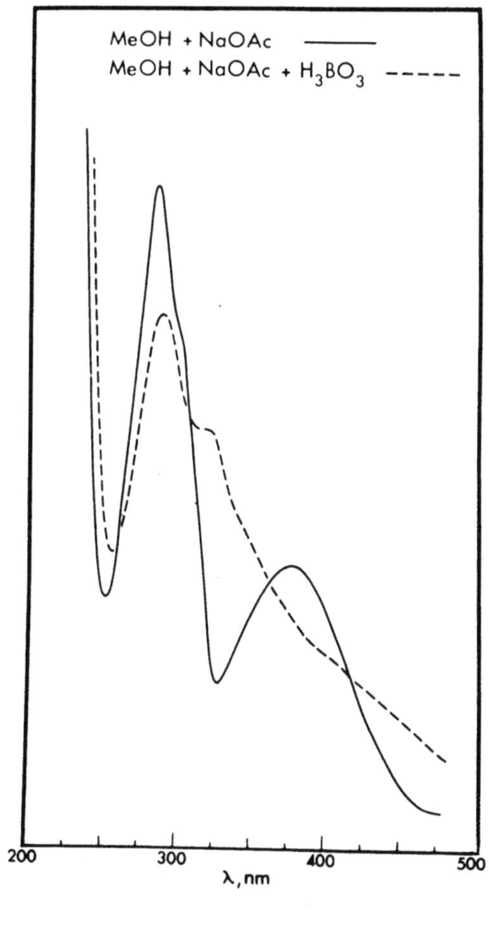

47

XANTHOMICROL

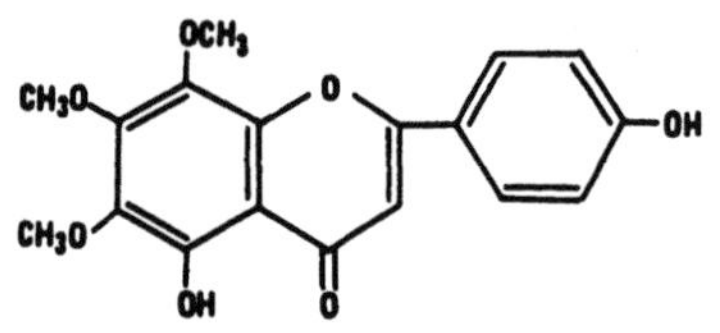

CHROMATOGRAPHIC DATA

Spot Appearance: (UV) deep purple
(UV/NH_3) green

R_f Values: 0.87 (TBA), 0.23 (HOAc)

UV SPECTRAL DATA (λ_{max}, nm)

MeOH	281, 294sh, 332
NaOMe	275, 362sh, 391
$AlCl_3$	267sh, 288, 311, 361, 407sh
$AlCl_3$/HCl	265sh, 290, 311, 354, 408sh
NaOAc	277, 297sh, 339, 390
NaOAc/H_3BO_3 (Proc. I)	279, 296sh, 336

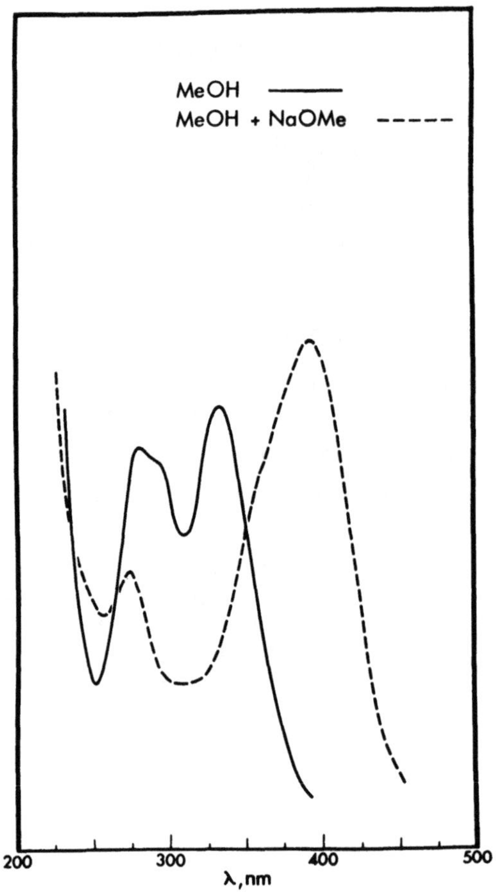

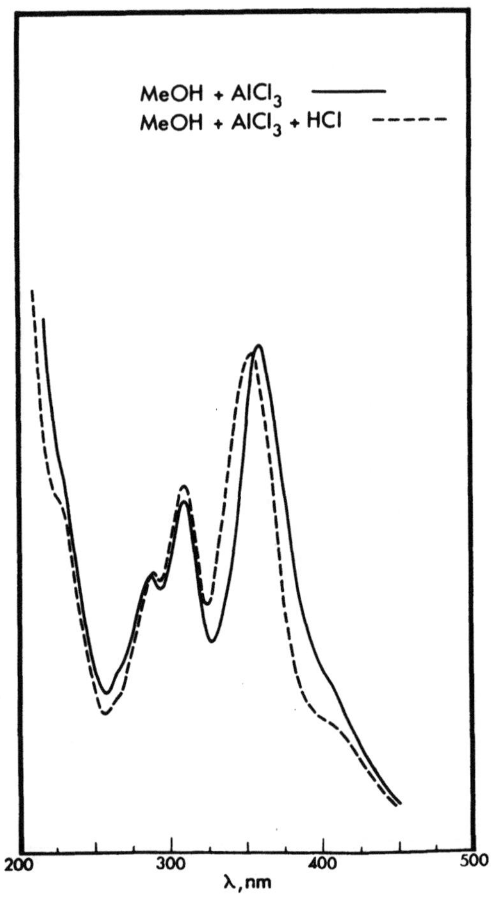

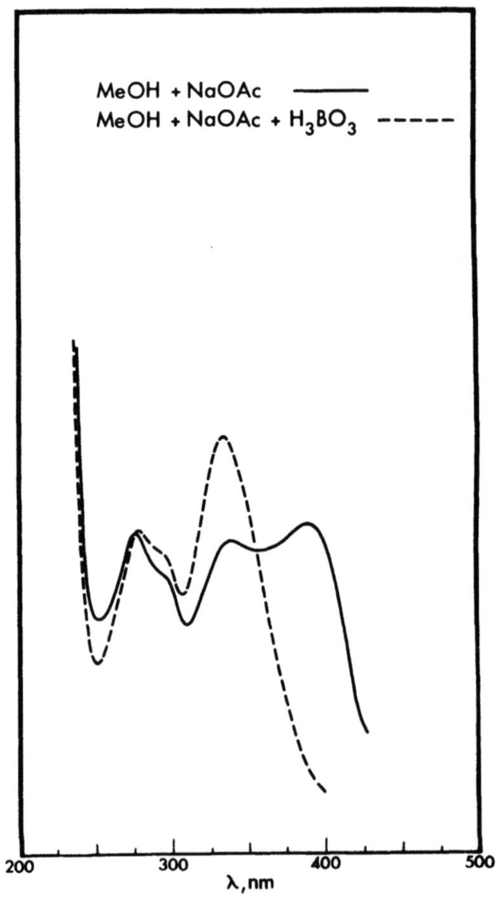

48

HYMENOXIN

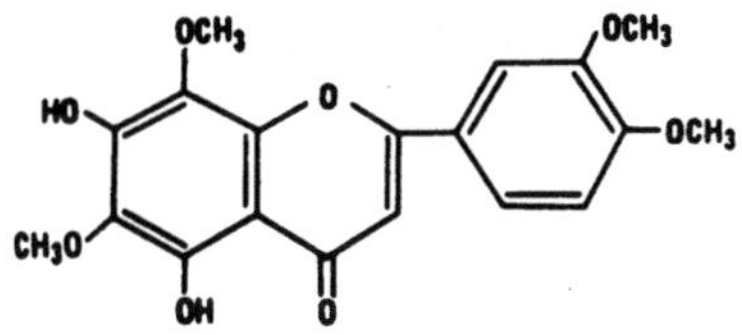

CHROMATOGRAPHIC DATA

Spot Appearance: (UV) deep purple
(UV/NH_3) deep purple

R_f Values: 0.73 (TBA), 0.11 (HOAc)

UV SPECTRAL DATA (λ_{max}, *nm*)

MeOH	250sh, 279, 336
NaOMe	285, 310sh, 363
$AlCl_3$	257sh, 290, 365
$AlCl_3$/HCl	257sh, 293, 357, 412sh
NaOAc	283, 312sh, 378
NaOAc/H_3BO_3 (Proc. I)	281, 329

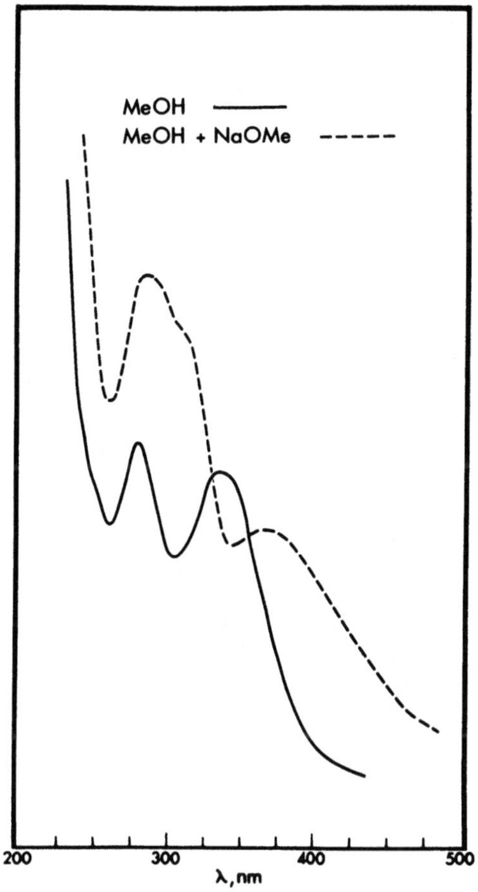

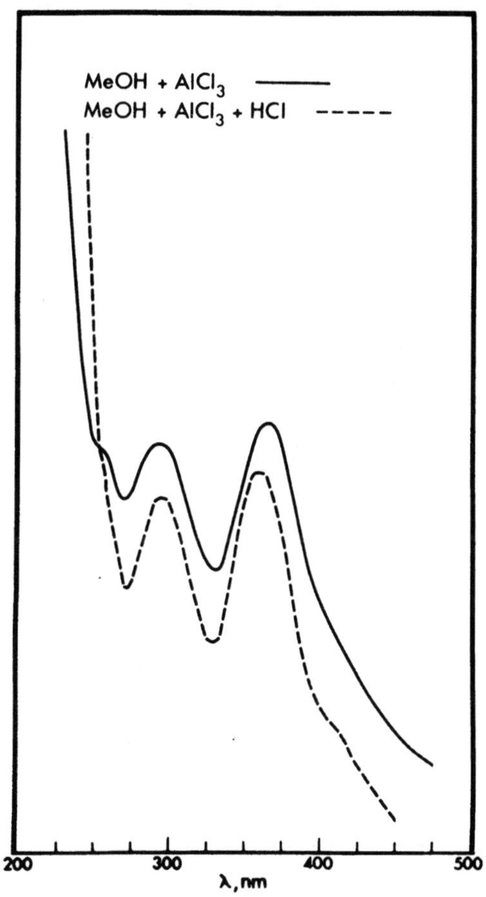

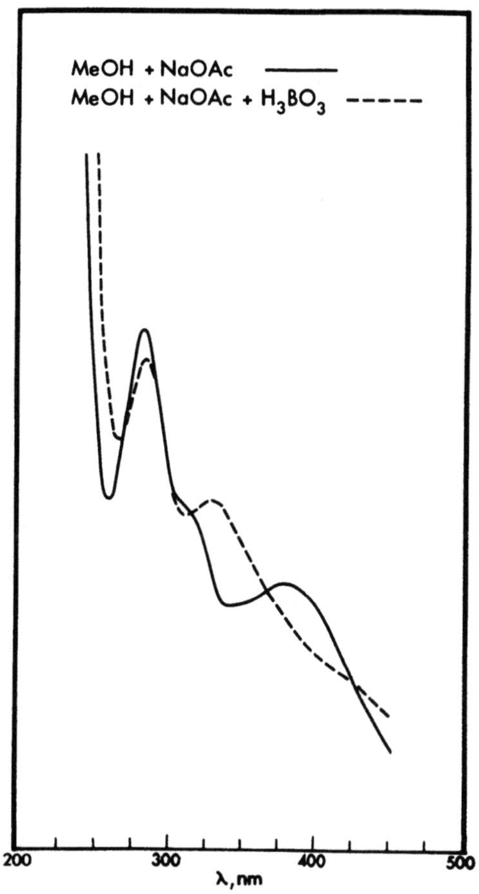

49

AMENTOFLAVONE

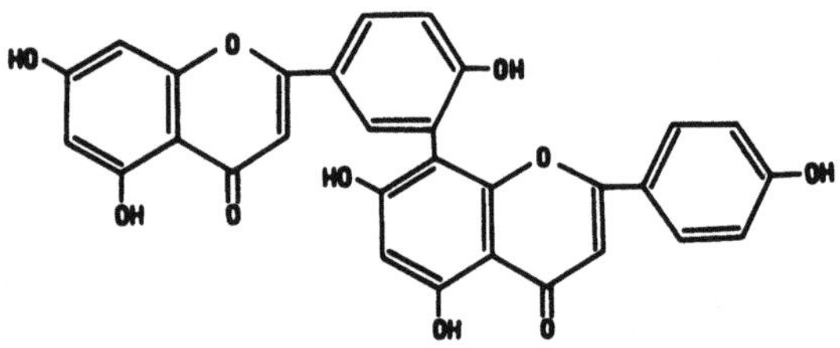

CHROMATOGRAPHIC DATA

Spot Appearance: (UV) deep purple
(UV/NH_3) yellow

R_f Values: 0.91 (TBA), 0.13 (HOAc)

UV SPECTRAL DATA (λ_{max}, *nm*)

MeOH	269, 291sh, 335
NaOMe	275, 295sh, 382
$AlCl_3$	260sh, 277, 299, 350, 386
$AlCl_3$/HCl	262sh, 279, 299, 343, 385
NaOAc	274, 292sh, 369
NaOAc/H_3BO_3 (Proc. I)	271, 332

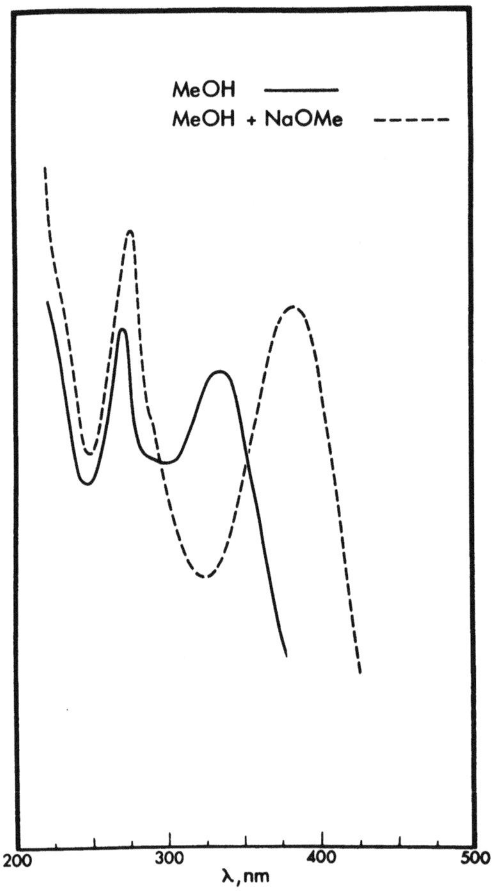

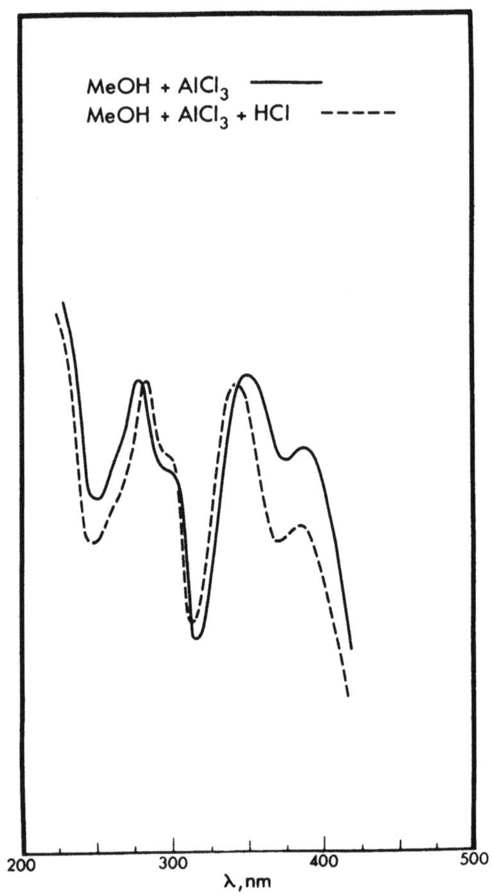

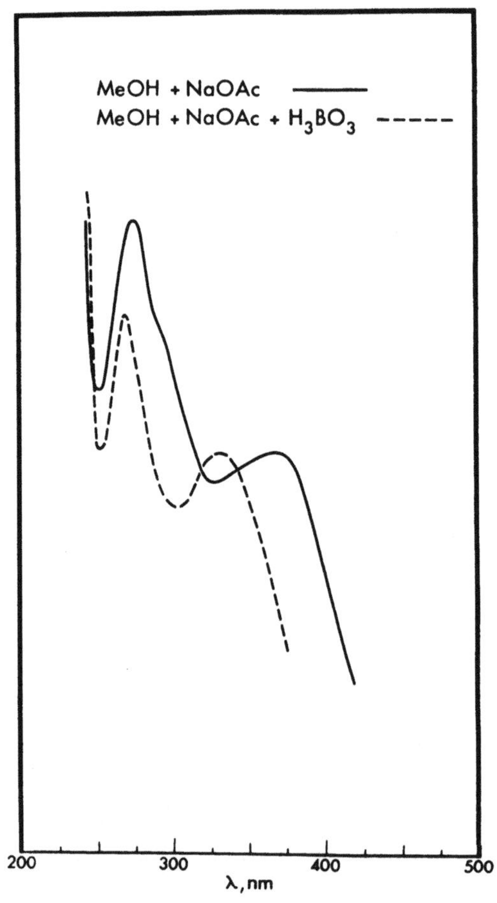

50

SCIADOPITYSIN

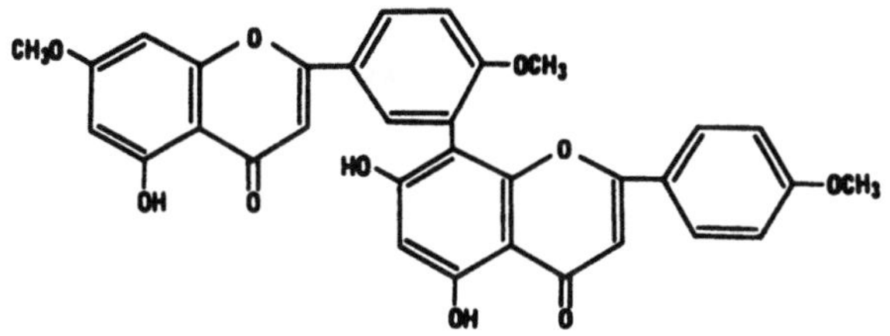

CHROMATOGRAPHIC DATA

Spot Appearance: (UV) deep purple
(UV/NH_3) deep purple

R_f Values: 0.92 (TBA), 0.00 (HOAc)

UV SPECTRAL DATA (*λ_{max},nm*)

MeOH	270, 326
NaOMe	285, 357sh
$AlCl_3$	260sh, 279, 298sh, 345, 383
$AlCl_3$/HCl	259sh, 281, 298sh, 339, 382
NaOAc	271, 282, 316sh, 340
NaOAc/H_3BO_3	271, 327
(Proc. I)	

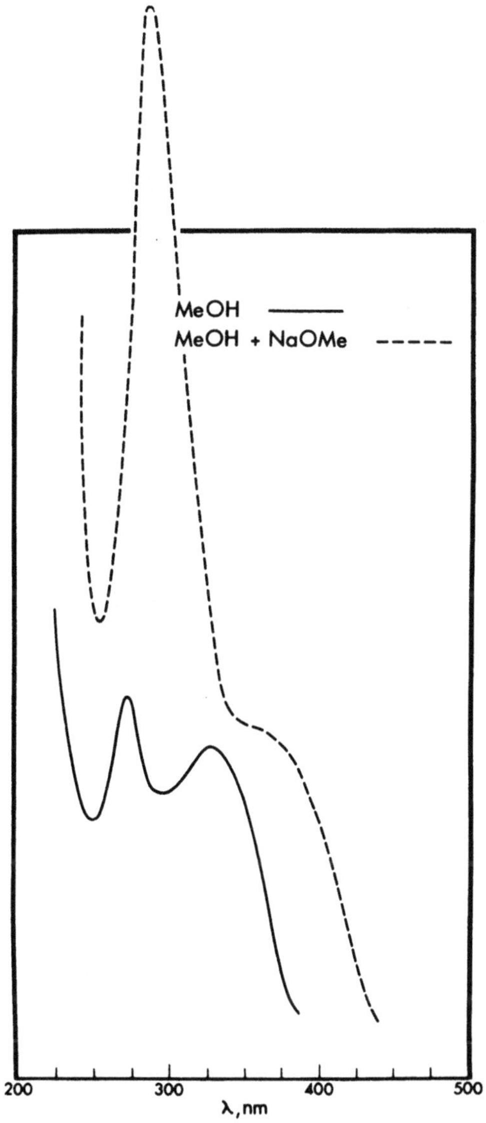

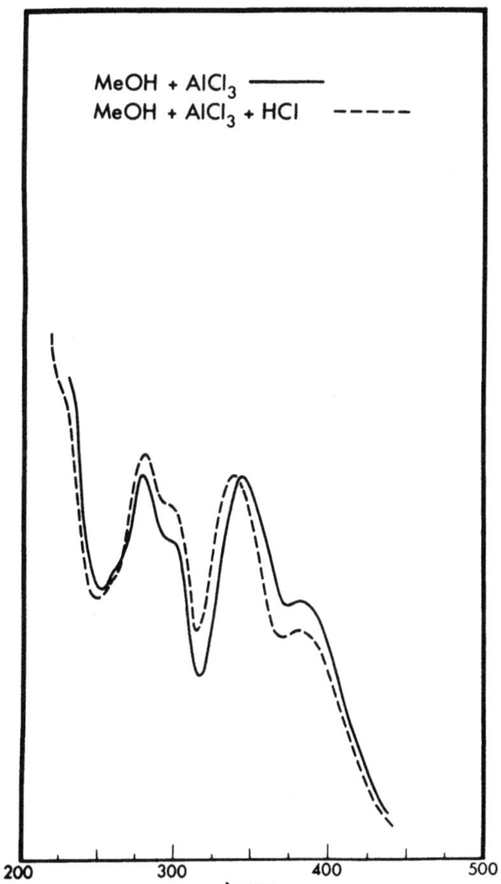

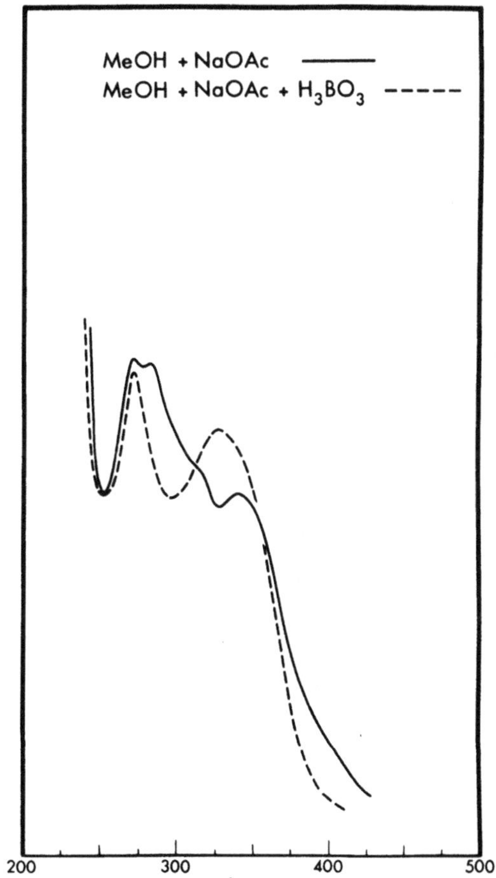

51

3-HYDROXYFLAVONE

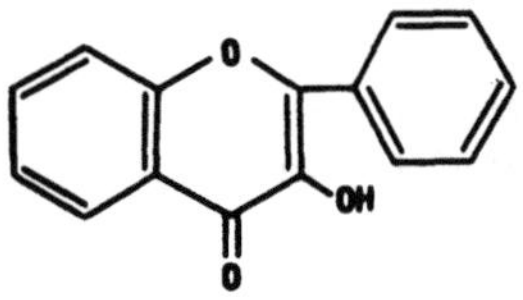

CHROMATOGRAPHIC DATA

Spot Appearance: (UV) fluorescent yellow-green

(UV/NH_3) fluorescent yellow-green

R_f Values: 0.89 (TBA), 0.26 (HOAc)

UV SPECTRAL DATA (λ_{max}, nm)

MeOH	239, 243sh, 306, 344
NaOMe	237, 250sh, 275, 309sh, 405
$AlCl_3$	248, 264sh, 327, 393
$AlCl_3$/HCl	248, 265sh, 325, 400
NaOAc	304, 346, 361sh, 405
NaOAc/H_3BO_3 (Proc. I)	306, 345, 360sh, 407sh

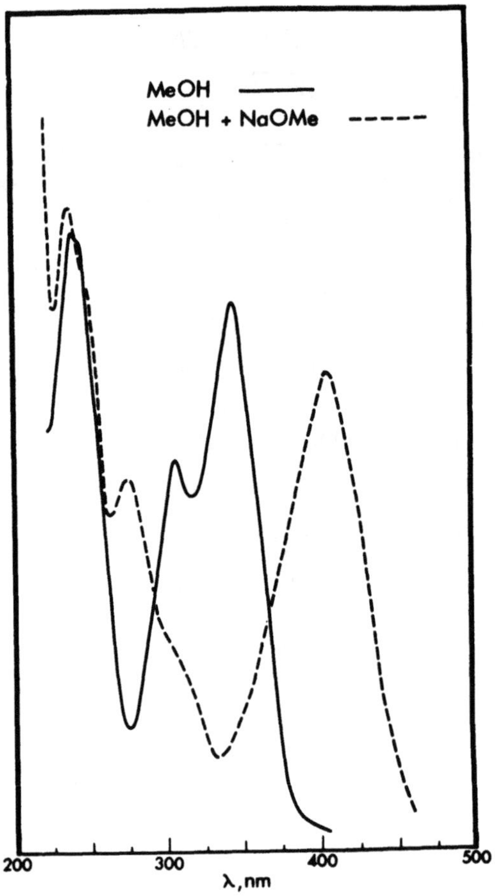

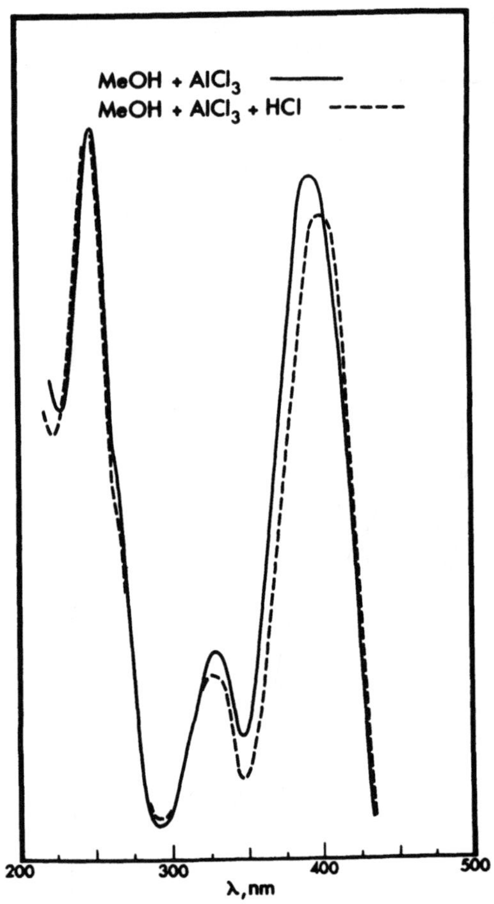

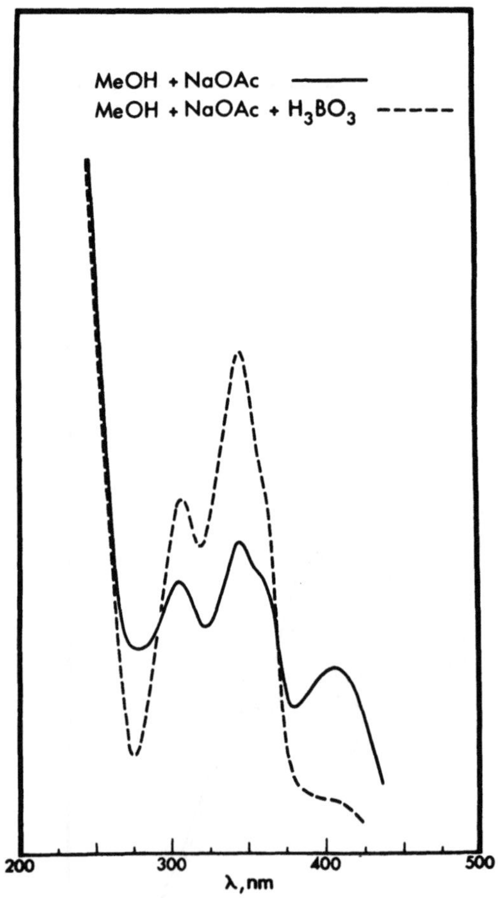

52
3-HYDROXY-4'-METHOXYFLAVONE

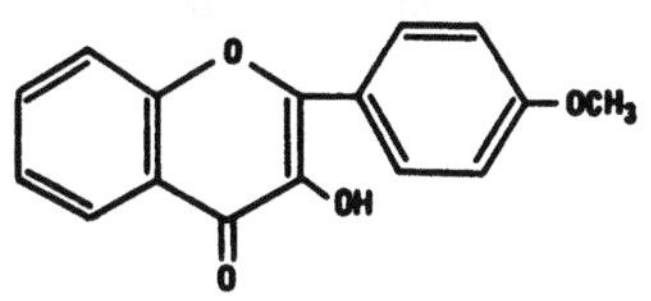

CHROMATOGRAPHIC DATA

Spot Appearance: (UV) fluorescent yellow
(UV/NH_3) fluorescent yellow

R_f Values: 0.85 (TBA), 0.16 (HOAc)

UV SPECTRAL DATA (λ_{max}, nm)

MeOH	232, 252, 318sh, 355
NaOMe	256, 259sh, 277sh, 311sh, 409
$AlCl_3$	232, 251, 263sh, 331, 416
$AlCl_3$/HCl	233, 253, 262sh, 330, 417
NaOAc	254sh, 315, 357, 411sh
NaOAc/H_3BO_3 (Proc. I)	254sh, 319sh, 355

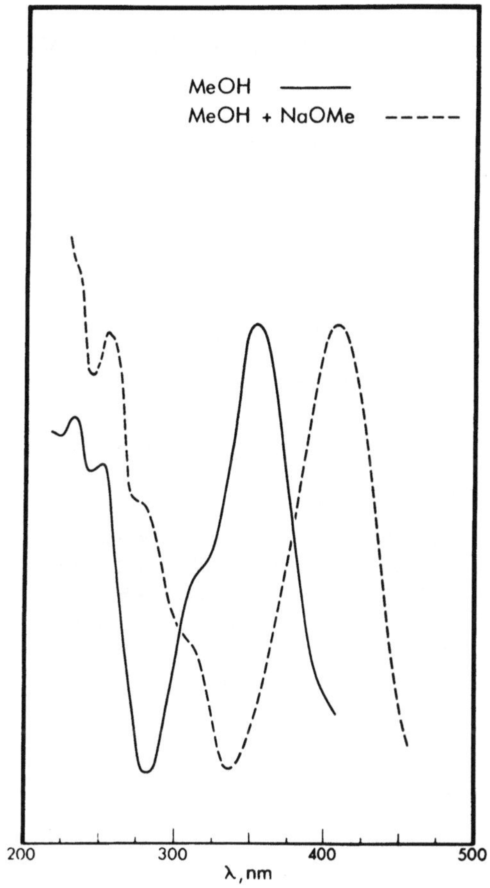

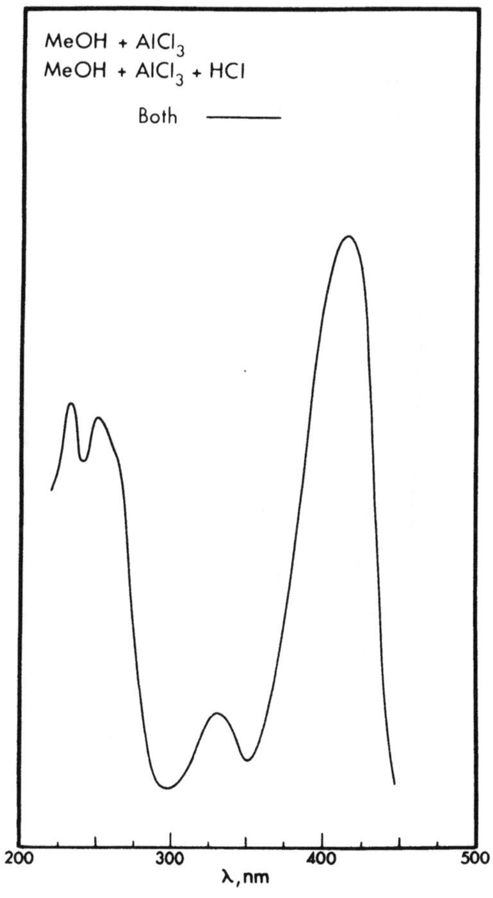

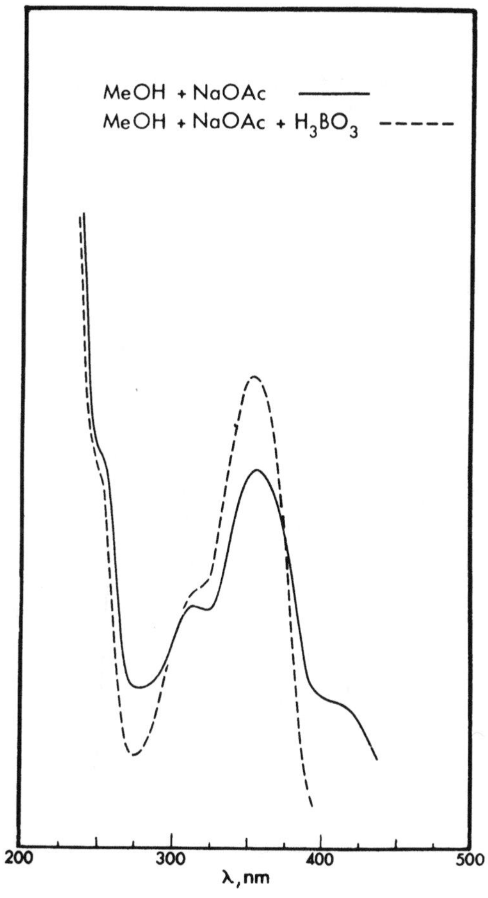

53

3,4',7-TRIHYDROXYFLAVONE

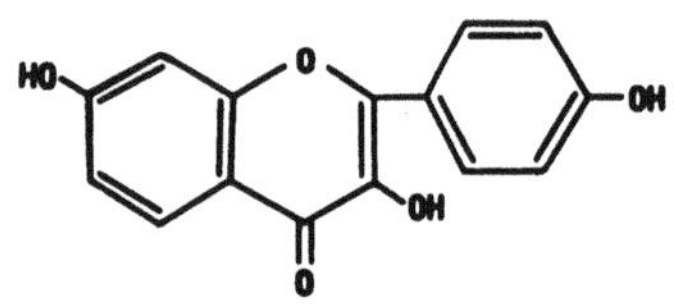

CHROMATOGRAPHIC DATA

Spot Appearance: (UV) fluorescent yellow
(UV/NH_3) fluorescent yellow

R_f Values: 0.80 (TBA), 0.06 (HOAc)

UV SPECTRAL DATA (λ_{max}, *nm*)

MeOH	258, 280sh, 318, 356
NaOMe	275, 289sh, 318, 328, 407 (dec.)
$AlCl_3$	256sh, 271, 306sh, 323, 419
$AlCl_3$/HCl	255sh, 271, 305sh, 323, 418
NaOAc	268, 285sh, 316sh, 327, 378, 430sh
NaOAc/H_3BO_3 (Proc. II)	259, 276sh, 318, 357, 425sh

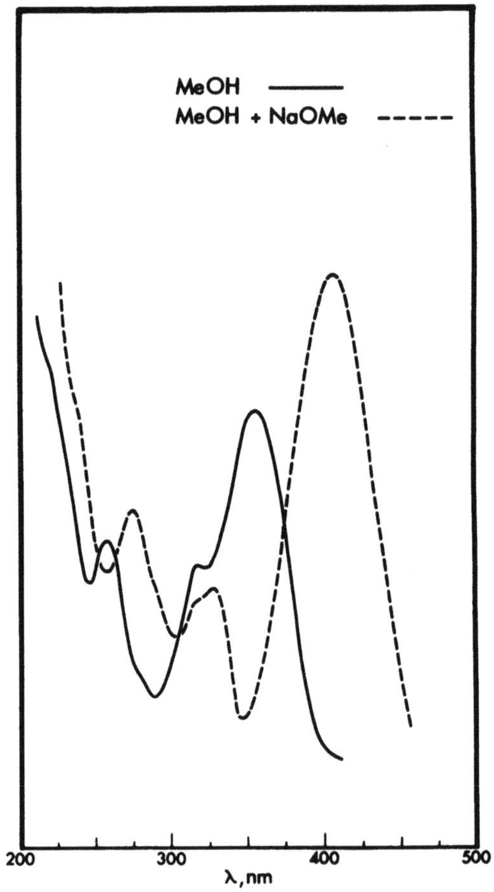

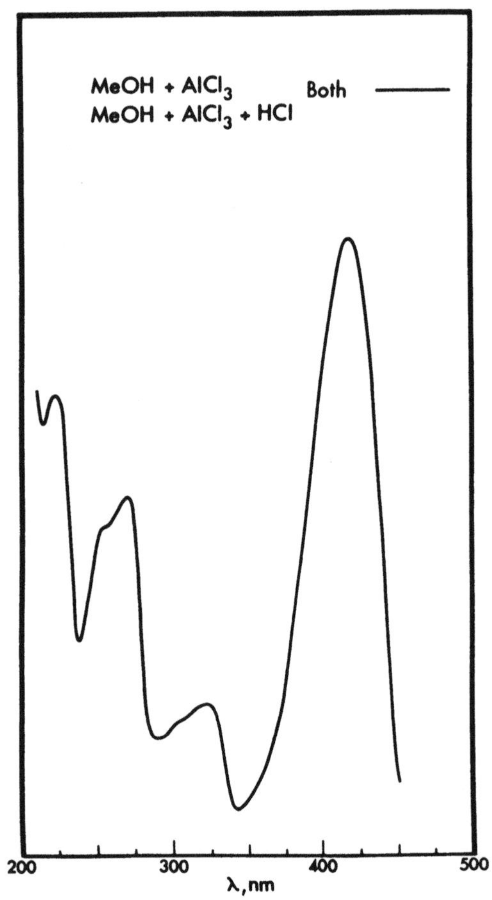

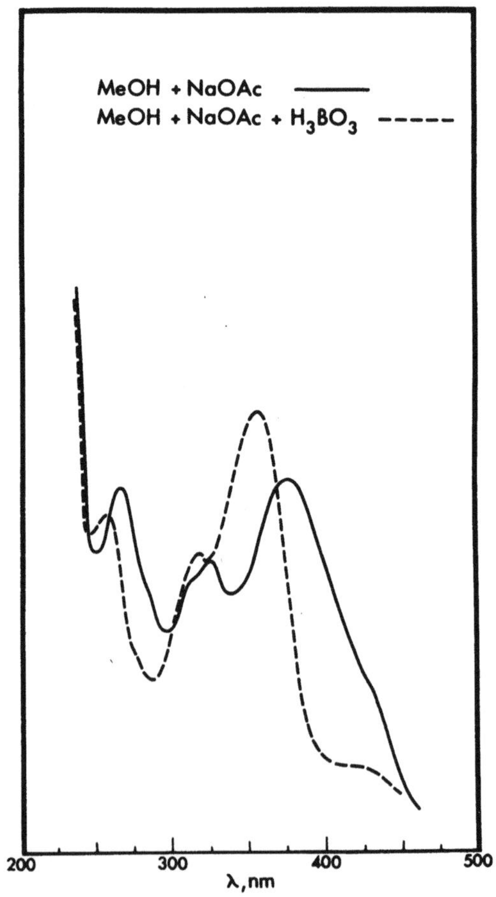

GALANGIN

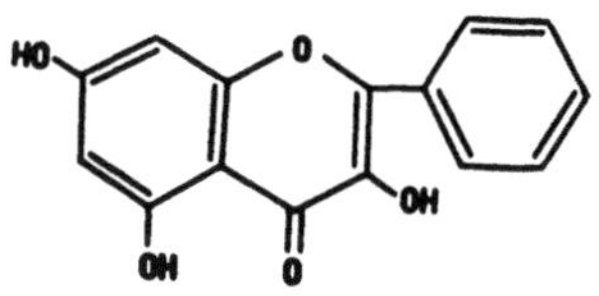

CHROMATOGRAPHIC DATA

Spot Appearance: (UV) purple
(UV/NH_3) yellow-green

R_f Values: 0.88 (TBA), 0.11 (HOAc)

UV SPECTRAL DATA (λ_{max}, nm)

MeOH	267, 305sh, 359
NaOMe	280, 327sh, 412
$AlCl_3$	249, 273, 300sh, 337, 413
$AlCl_3$/HCl	249, 274, 302sh, 334, 412
NaOAc	275, 301sh, 328sh, 388
NaOAc/H_3BO_3 (Proc. I)	267, 300sh, 317sh, 361

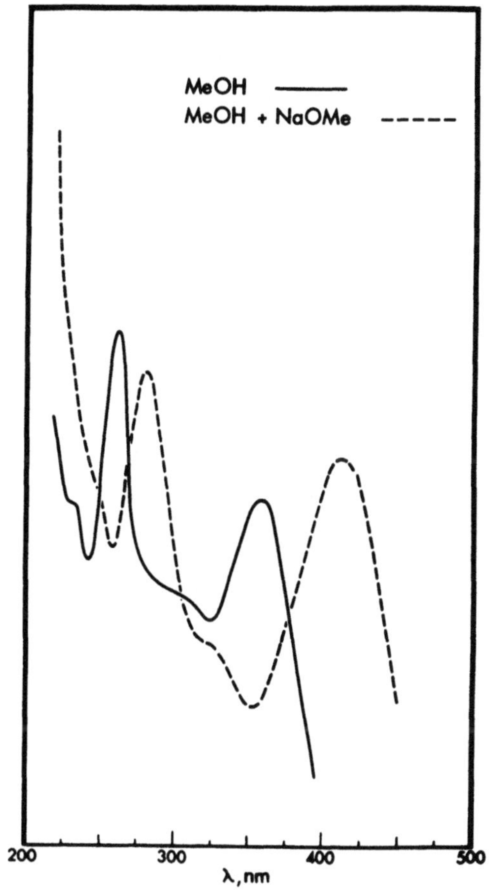

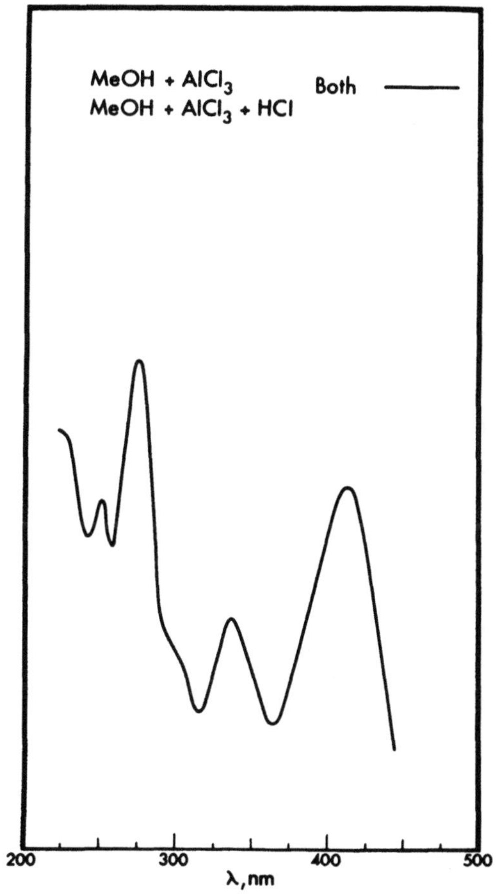

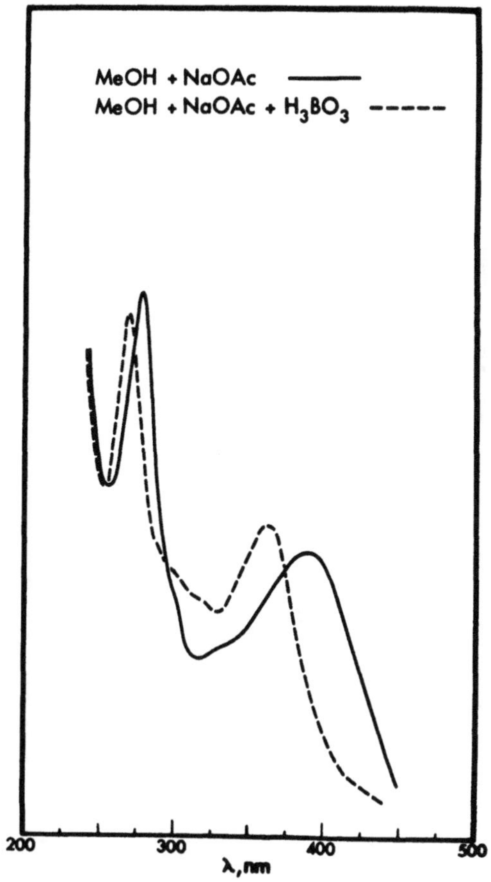

55
GALANGIN 3-METHYL ETHER

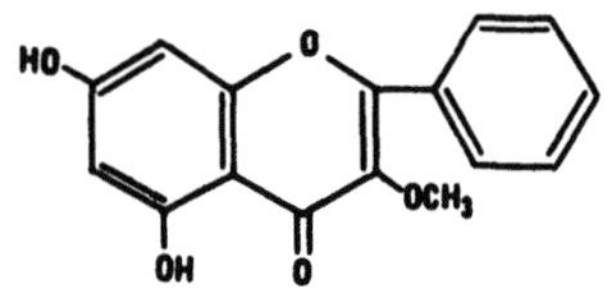

CHROMATOGRAPHIC DATA

Spot Appearance: (UV) deep purple
(UV/NH_3) deep purple

R_f Values: 0.92 (TBA), 0.36 (HOAc)

UV SPECTRAL DATA (λ_{max}, nm)

MeOH	266, 312sh, 340sh
NaOMe	276, 360
$AlCl_3$	278, 333, 393
$AlCl_3$/HCl	278, 329, 391
NaOAc	278, 364
NaOAc/H_3BO_3 (Proc. I)	267, 332sh

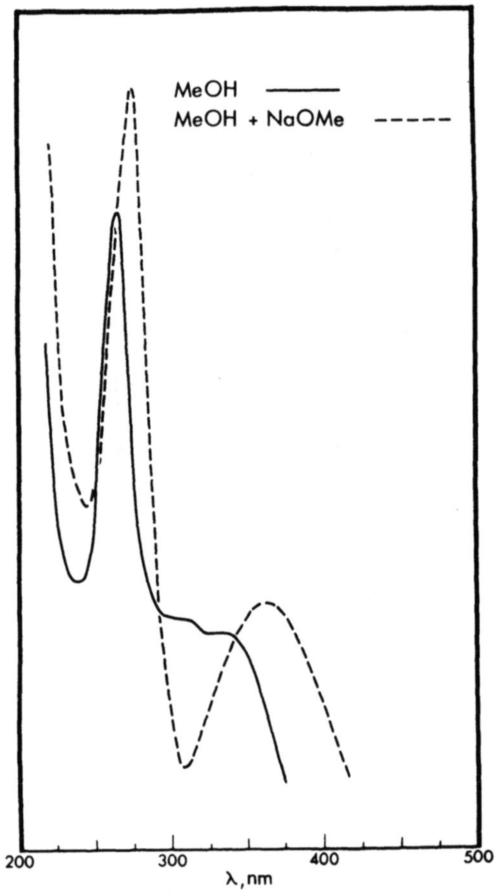

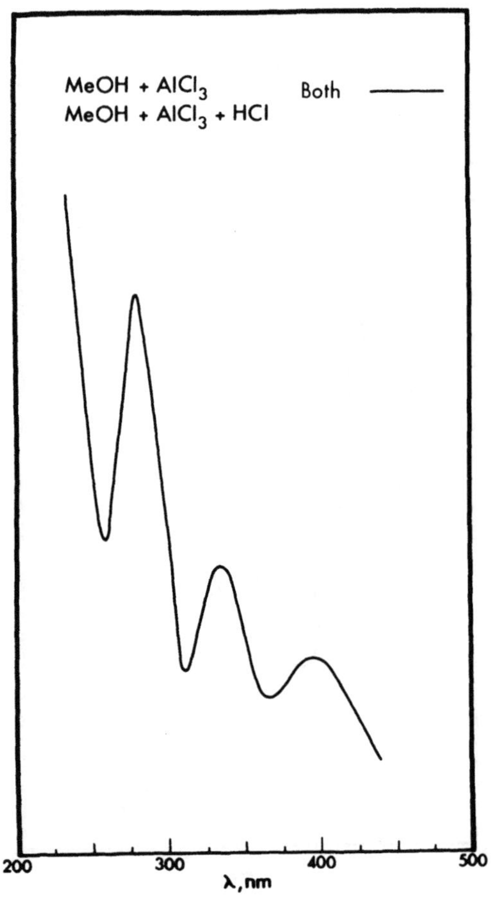

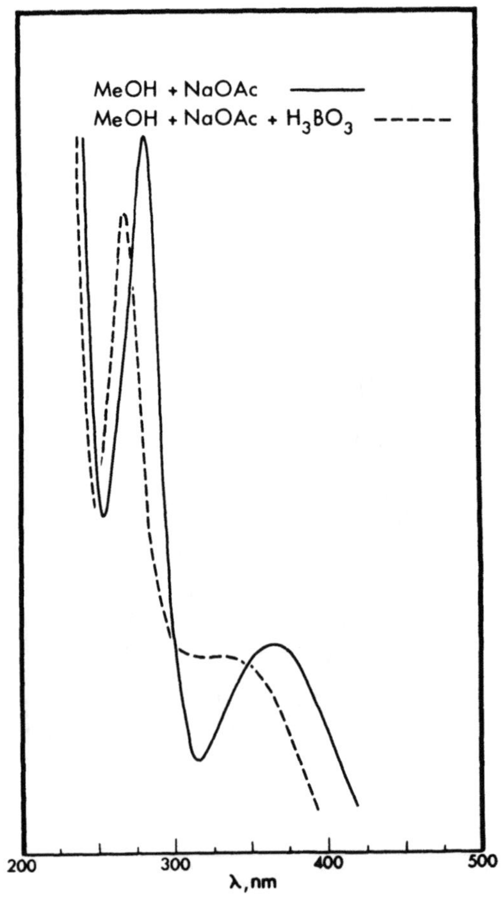

56

3,3′,4′-TRIHYDROXYFLAVONE

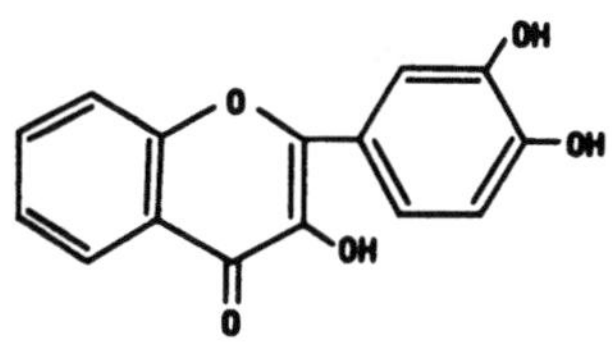

CHROMATOGRAPHIC DATA

Spot Appearance: (UV) fluorescent yellow
(UV/NH_3) fluorescent yellow

R_f Values: 0.70 (TBA), 0.14 (HOAc)

UV SPECTRAL DATA (λ_{max},*nm*)

MeOH	248, 309sh, 366
NaOMe	244, 293, 324sh, 425 (dec.)
$AlCl_3$	235, 270, 319, 371sh, 466
$AlCl_3$/HCl	260, 323, 427
NaOAc	253sh, 322sh, 373, 430 (dec.)
NaOAc/H_3BO_3 (Proc. II)	251sh, 310sh, 326sh, 388

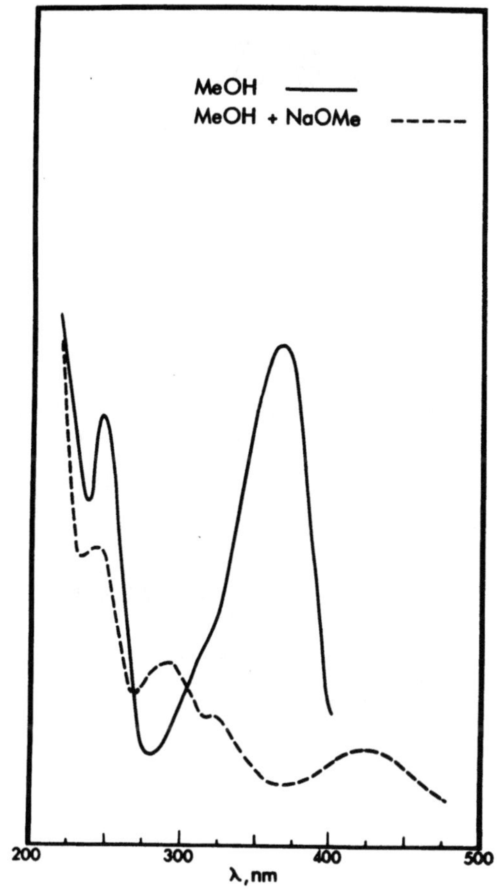

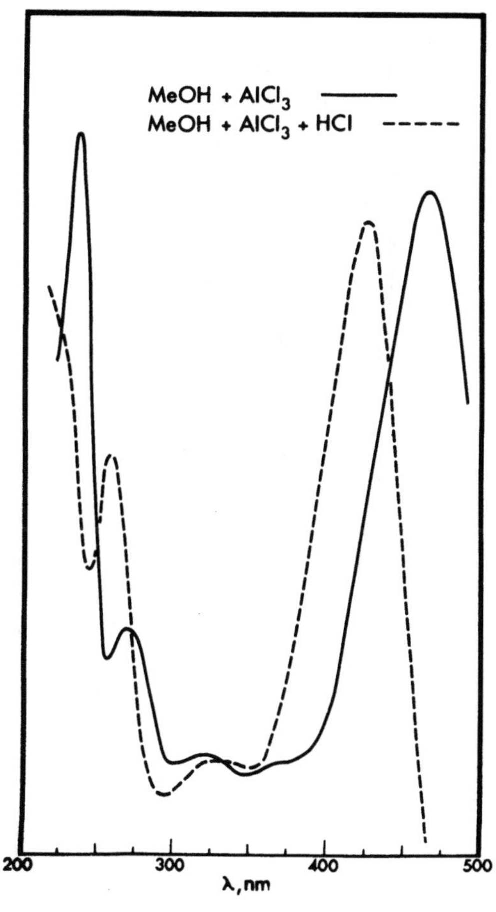

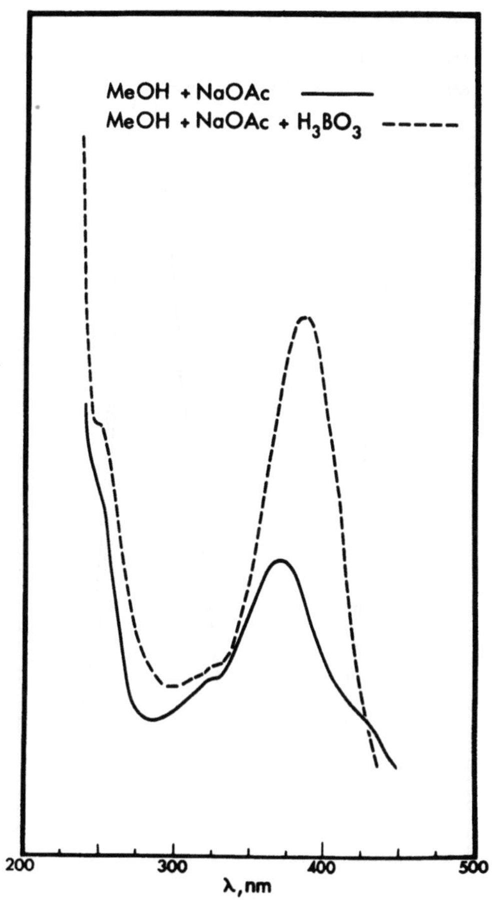

57

3-HYDROXY-3',4'-DIMETHOXYFLAVONE

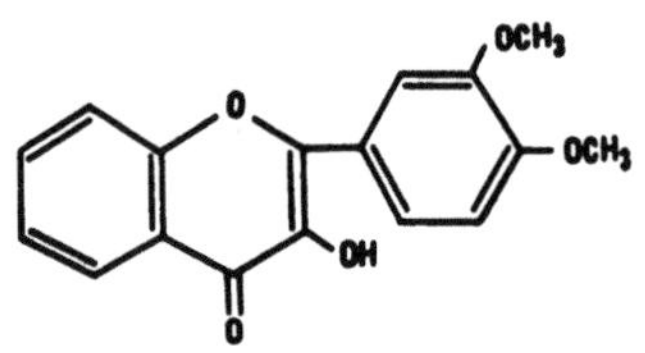

CHROMATOGRAPHIC DATA

Spot Appearance: (UV) fluorescent yellow
(UV/NH_3) fluorescent yellow

R_f Values: 0.81 (TBA), 0.12 (HOAc)

UV SPECTRAL DATA (λ_{max},*nm*)

MeOH	246, 307sh, 320sh, 355
NaOMe	263, 285sh, 317sh, 412
$AlCl_3$	257, 328, 423
$AlCl_3$/HCl	256, 329, 422
NaOAc	320sh, 364, 421sh
NaOAc/H_3BO_3 (Proc. I)	306sh, 323sh, 361

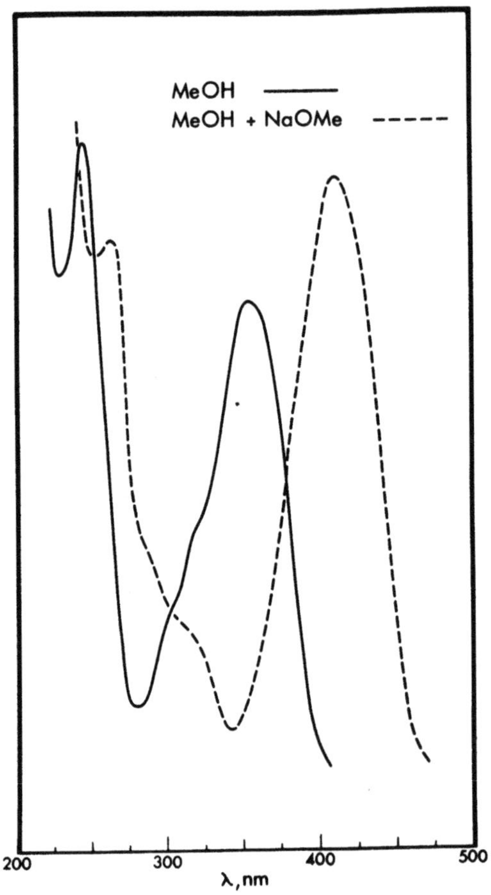

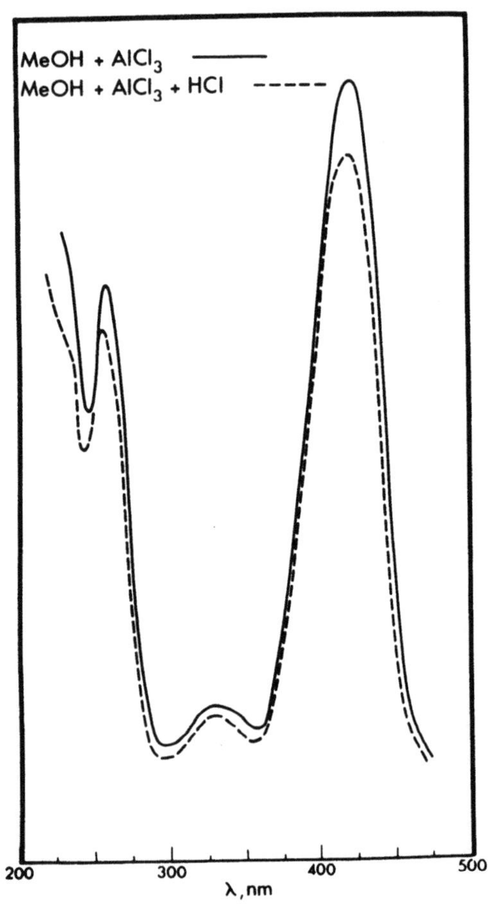

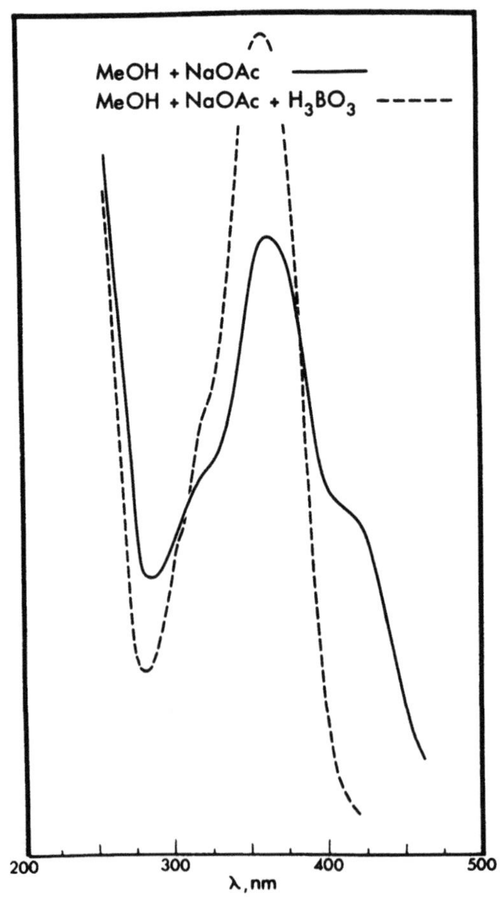

KAEMPFEROL

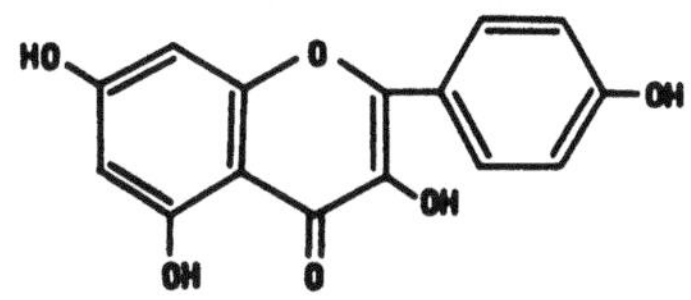

CHROMATOGRAPHIC DATA

Spot Appearance: (UV) dull yellow
(UV/NH_3) dull yellow

R_f Values: 0.79 (TBA), 0.04 (HOAc)

UV SPECTRAL DATA (λ_{max}, *nm*)

MeOH	253sh, 266, 294sh, 322sh, 367
NaOMe	278, 316, 416 (dec.)
$AlCl_3$	260sh, 268, 303sh, 350, 424
$AlCl_3$/HCl	256sh, 269, 303sh, 348, 424
NaOAc	274, 303, 387
NaOAc/H_3BO_3 (Proc. II)	267, 297sh, 320sh, 372

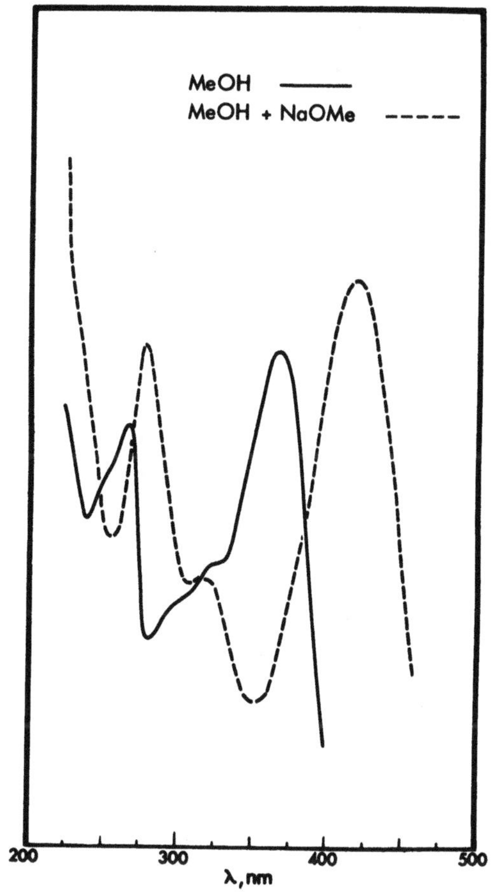

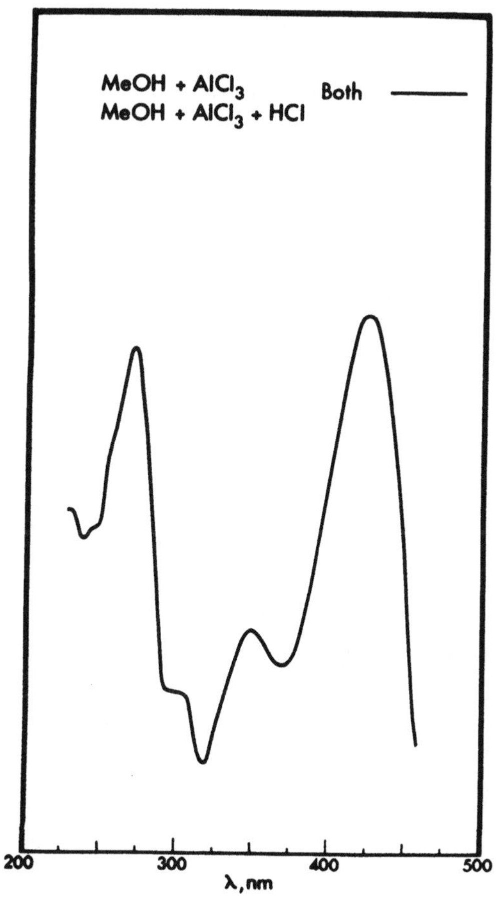

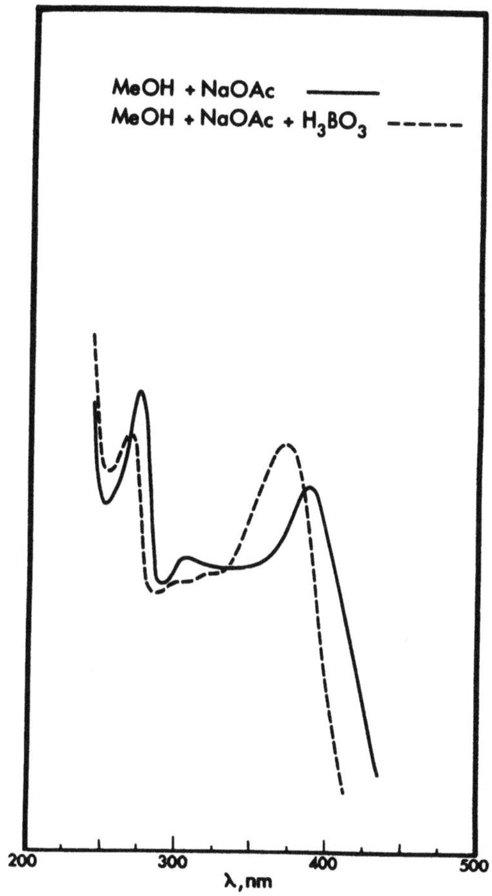

59

KAEMPFEROL 7-0-NEOHESPERIDOSIDE

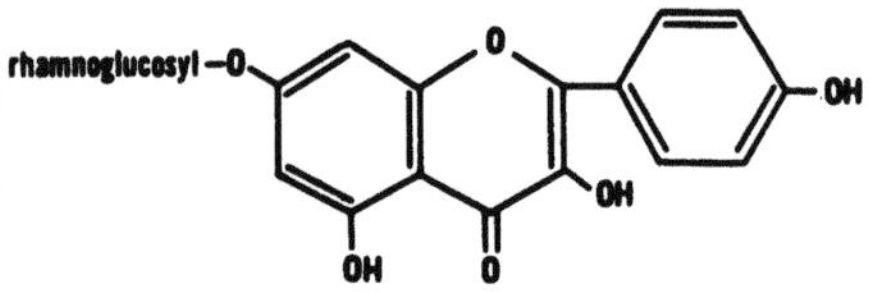

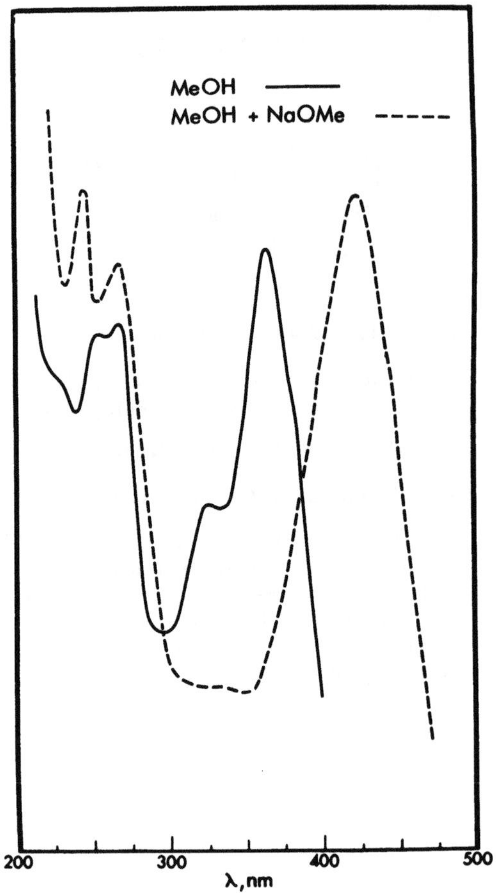

CHROMATOGRAPHIC DATA

Spot Appearance: (UV) dull yellow
(UV/NH_3) dull yellow

R_f Values: 0.30 (TBA), 0.35 (HOAc)

UV SPECTRAL DATA (λ_{max}, *nm*)

MeOH	253, 266, 323, 364
NaOMe	245, 267, 335sh, 425 (dec.)
$AlCl_3$	259sh, 266, 299sh, 353, 424
$AlCl_3/HCl$	244sh, 258sh, 266, 300sh, 350, 422
NaOAc	261, 323, 385, 419sh
$NaOAc/H_3BO_3$ (Proc. II)	265sh, 325sh, 370.

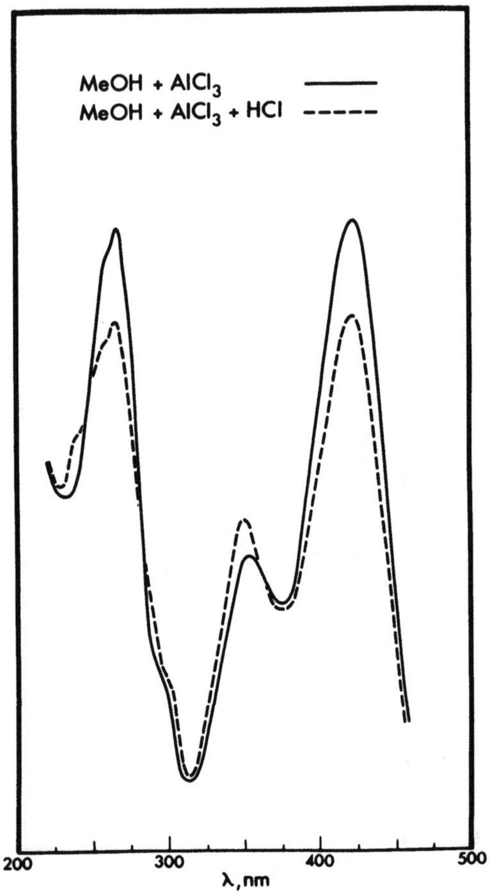

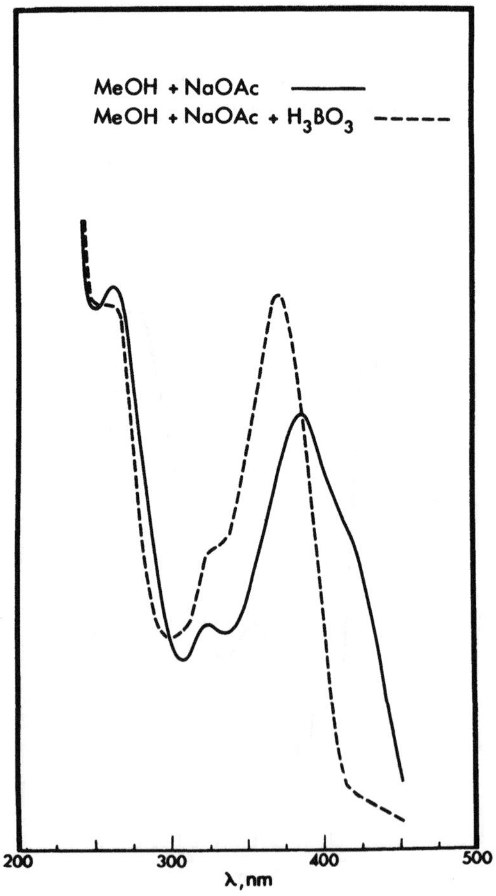

60

KAEMPFEROL 3-O-ROBINOSIDE 7-O-RHAMNOSIDE (ROBININ)

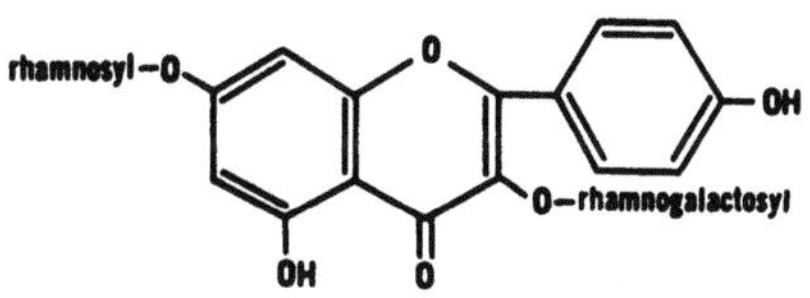

CHROMATOGRAPHIC DATA

Spot Appearance: (UV) deep purple
(UV/NH_3) yellow

R_f Values: 0.40 (TBA), 0.76 (HOAc)

UV SPECTRAL DATA (λ_{max}, nm)

MeOH	244sh, 265, 315sh, 350
NaOMe	246, 269, 301sh, 350sh, 389
$AlCl_3$	255sh, 274, 301, 354, 400
$AlCl_3$/HCl	274, 298sh, 348, 398
NaOAc	265, 318sh, 358, 406sh
NaOAc/H_3BO_3 (Proc. I)	265, 319sh, 352

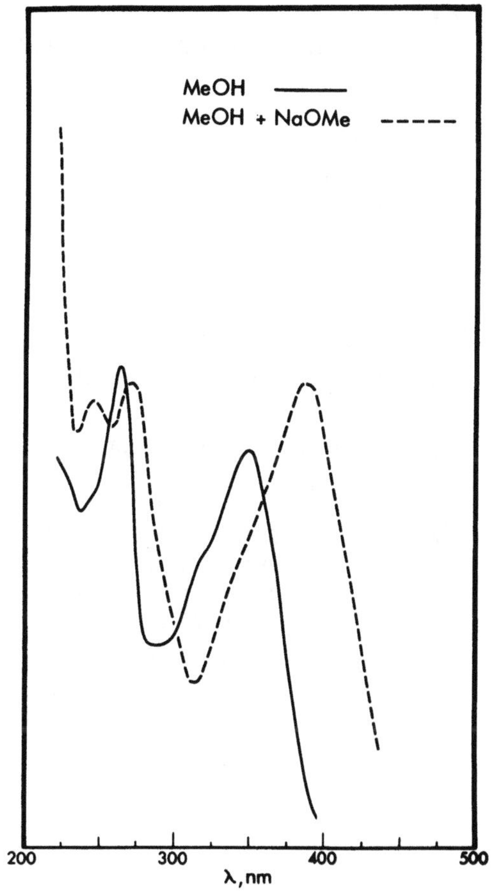

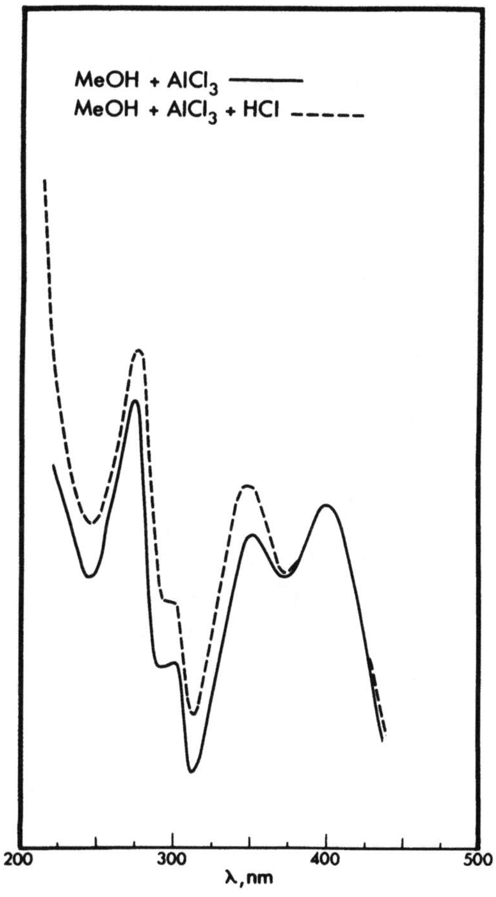

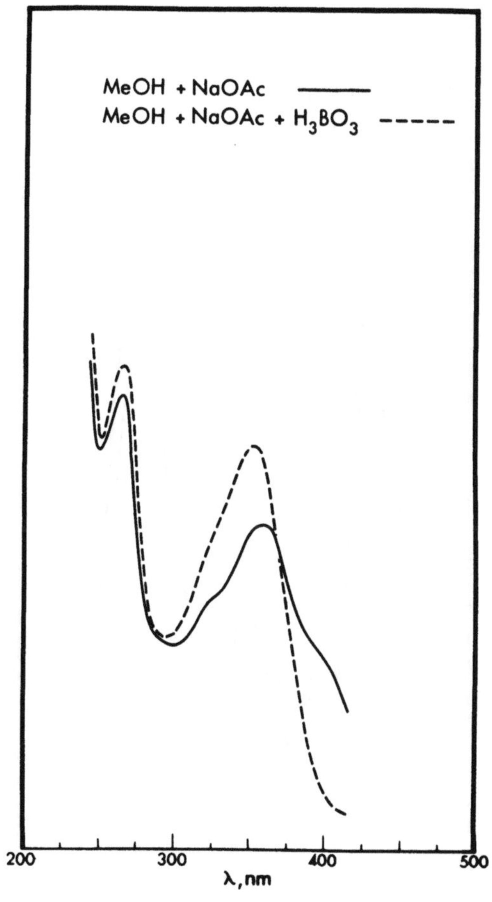

61

KAEMPFEROL 4′-METHYL ETHER

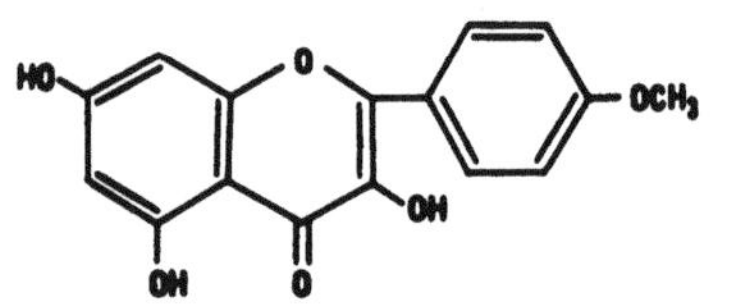

CHROMATOGRAPHIC DATA

Spot Appearance: (UV) dull yellow
(UV/NH_3) dull yellow

R_f Values: 0.86 (TBA), 0.05 (HOAc)

UV SPECTRAL DATA (λ_{max}, *nm*)

MeOH	253sh, 267, 299sh, 320, 367
NaOMe	280, 323sh, 411
$AlCl_3$	254sh, 271, 305, 350, 423
$AlCl_3$/HCl	257sh, 270, 305sh, 347, 422
NaOAc	259sh, 274, 301sh, 384
NaOAc/H_3BO_3 (Proc. I)	268, 299sh, 319, 367

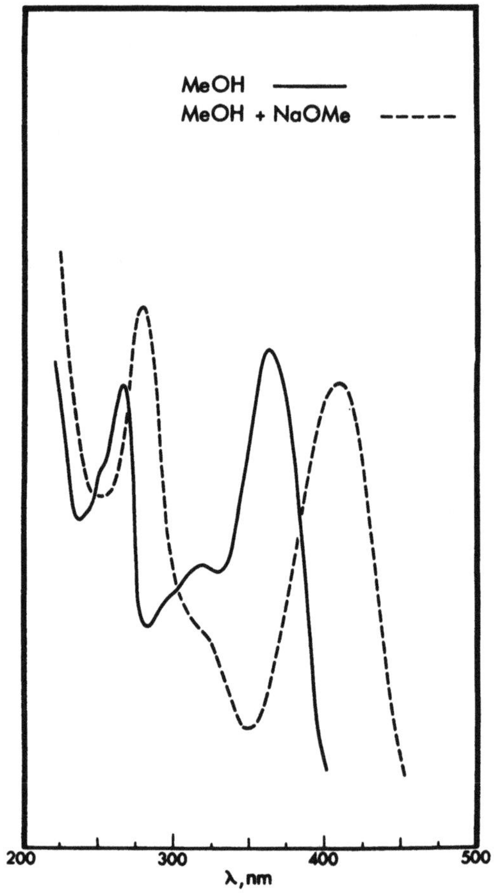

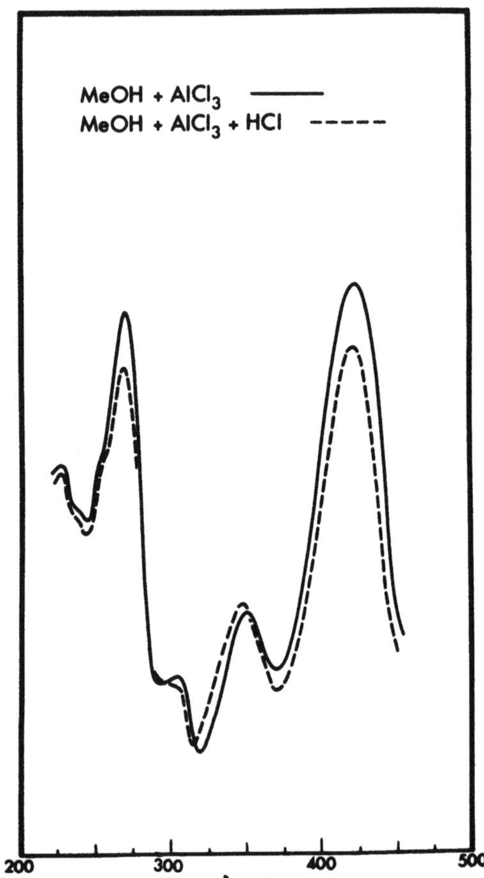

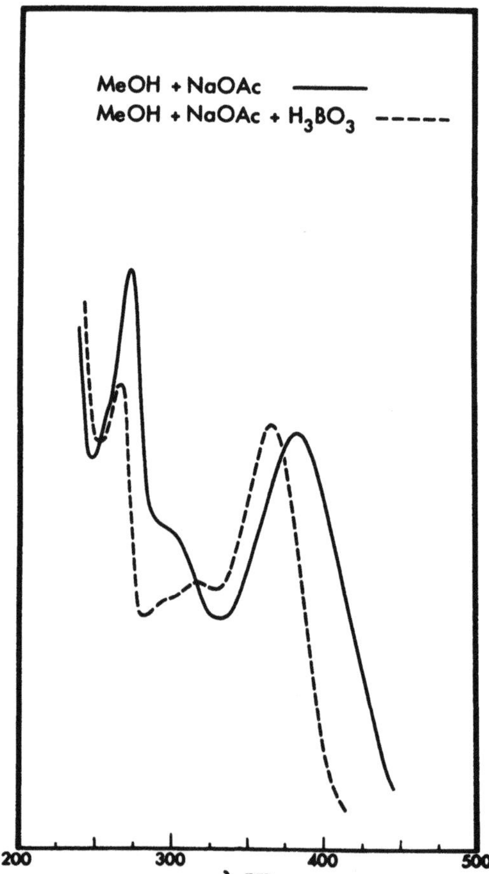

FISETIN

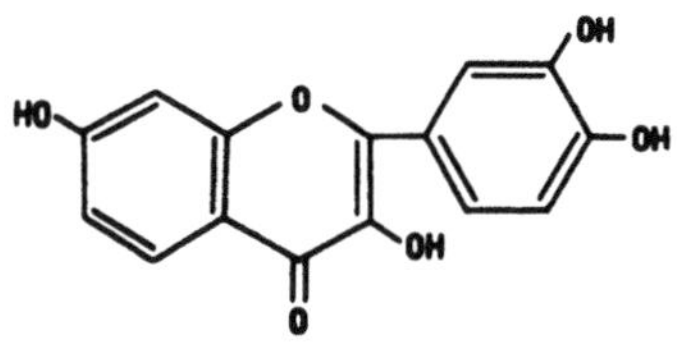

CHROMATOGRAPHIC DATA

Spot Appearance: (UV) fluorescent yellow
(UV/NH_3) fluorescent yellow

R_f Values: 0.56 (TBA), 0.03 (HOAc)

UV SPECTRAL DATA (λ_{max},*nm*)

MeOH	248, 262sh, 307sh, 319, 362
NaOMe	252, 292, 341 (dec.)
$AlCl_3$	268sh, 281, 318sh, 458
$AlCl_3$/HCl	263, 274sh, 322, 423
NaOAc	263sh, 321, 331, 378 (dec.)
NaOAc/H_3BO_3 (Proc. II)	265sh, 315, 381

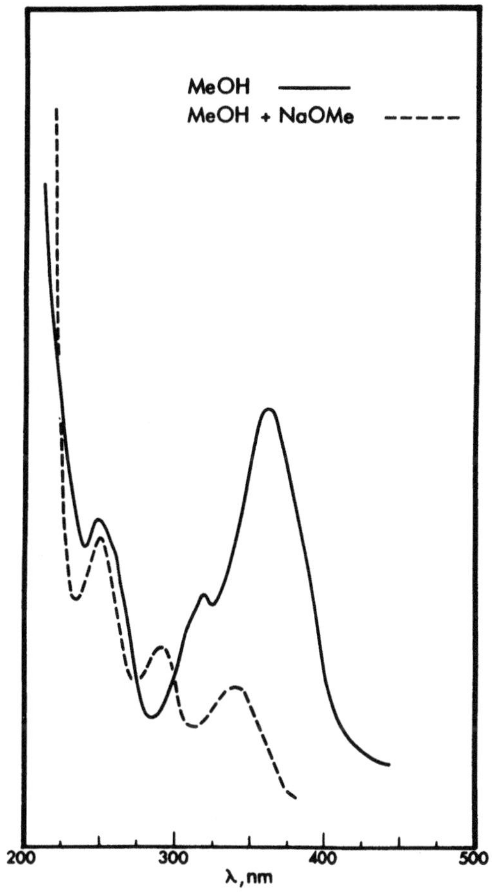

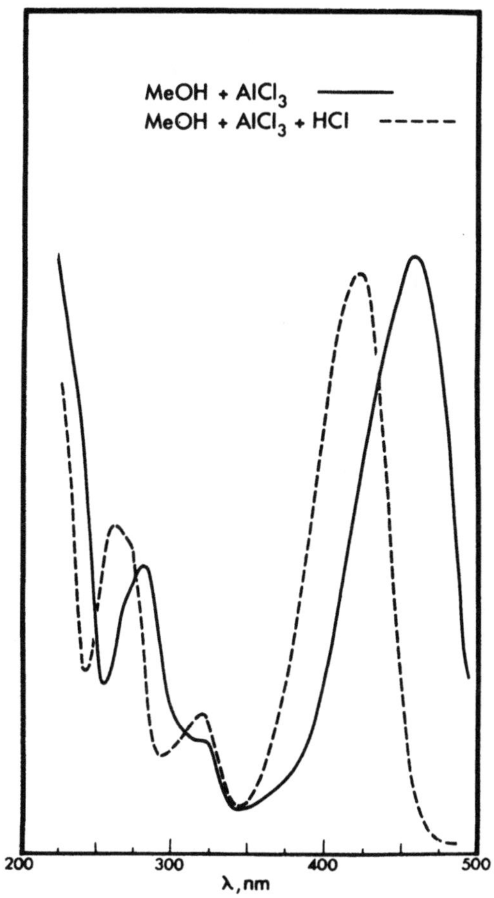

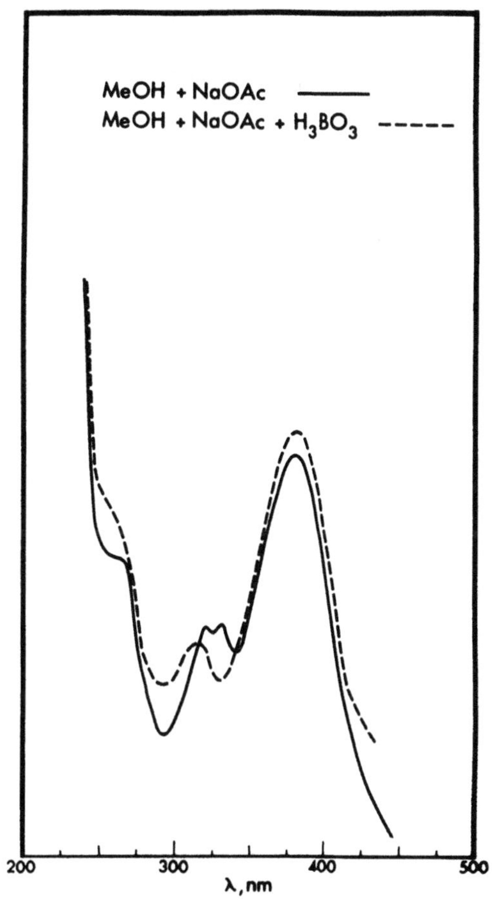

63

FISETIN 3-0-GLUCOSIDE

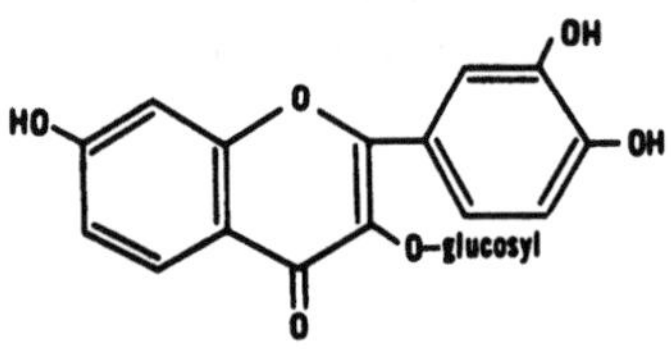

CHROMATOGRAPHIC DATA

Spot Appearance: (UV) fluorescent light blue

(UV/NH_3) fluorescent yellow-green

R_f Values: 0.50 (TBA), 0.55 (HOAc)

UV SPECTRAL DATA (λ_{max}, *nm*)

MeOH	254sh, 310, 340
NaOMe	256, 324, 408
$AlCl_3$	276, 317sh, 381 (hydrolyzes)
$AlCl_3$/HCl	254, 273sh, 307, 352sh, 408sh, 420
NaOAc	256sh, 317, 369
NaOAc/H_3BO_3 (Proc. I)	310, 365

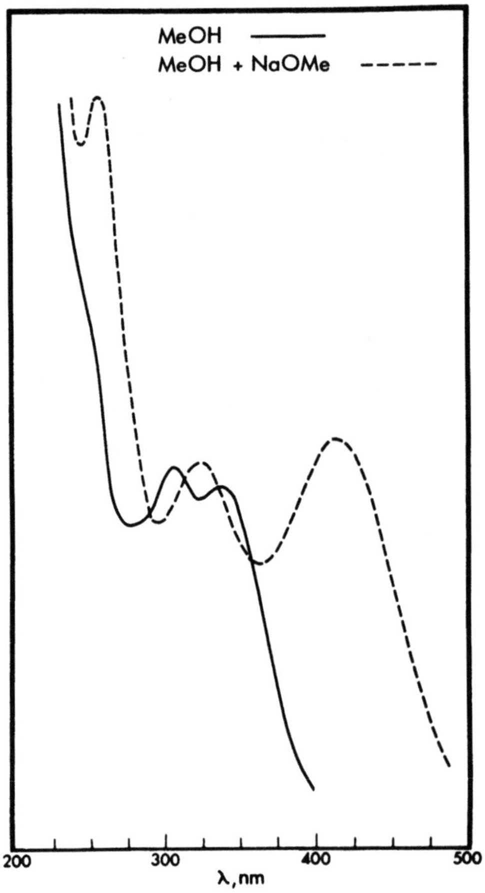

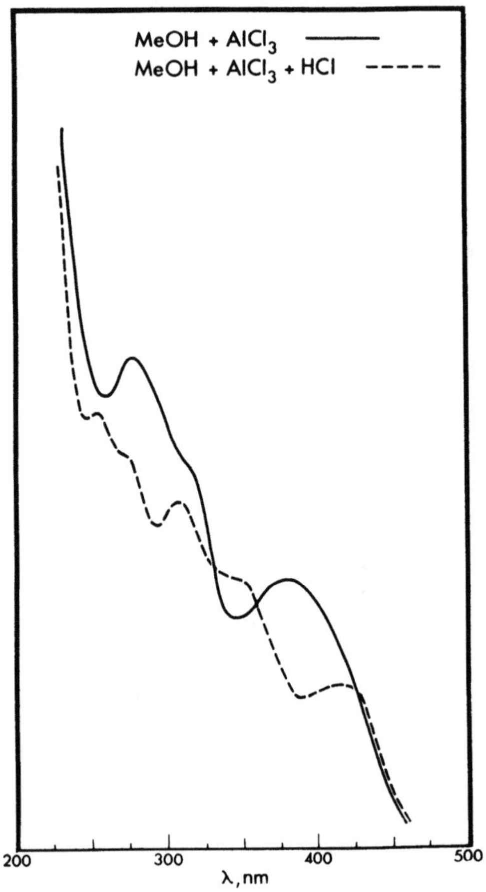

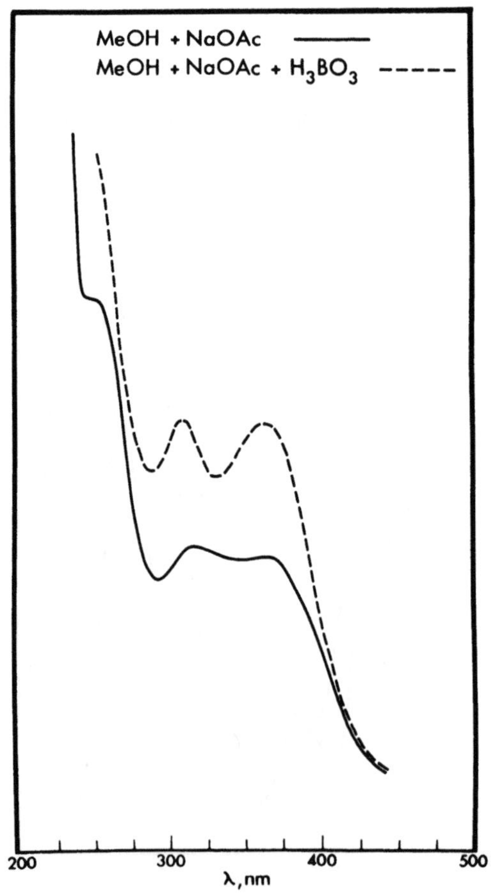

64

HERBACETIN 8-METHYL ETHER

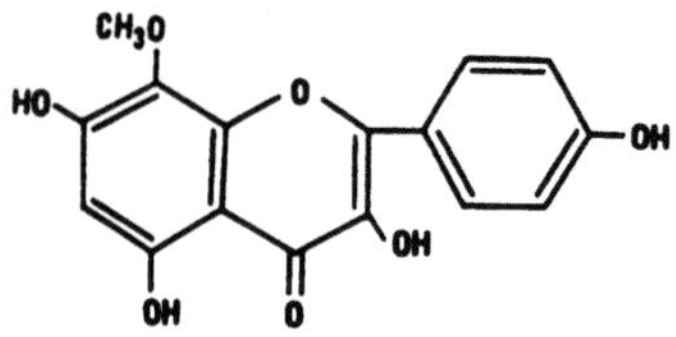

CHROMATOGRAPHIC DATA

Spot Appearance: (UV) brown
(UV/NH_3) brown

R_f Values: 0.78 (TBA), 0.06 (HOAc)

UV SPECTRAL DATA (λ_{max},nm)

MeOH	259sh, 276, 327, 377
NaOMe	289, 338, 430 (dec.)
$AlCl_3$	248sh, 262sh, 276, 310, 359, 435
$AlCl_3$/HCl	247sh, 261sh, 274, 308, 357, 434
NaOAc	257sh, 282, 319, 341sh, 401 (dec.)
NaOAc/H_3BO_3 (Proc. II)	275, 309sh, 322, 382

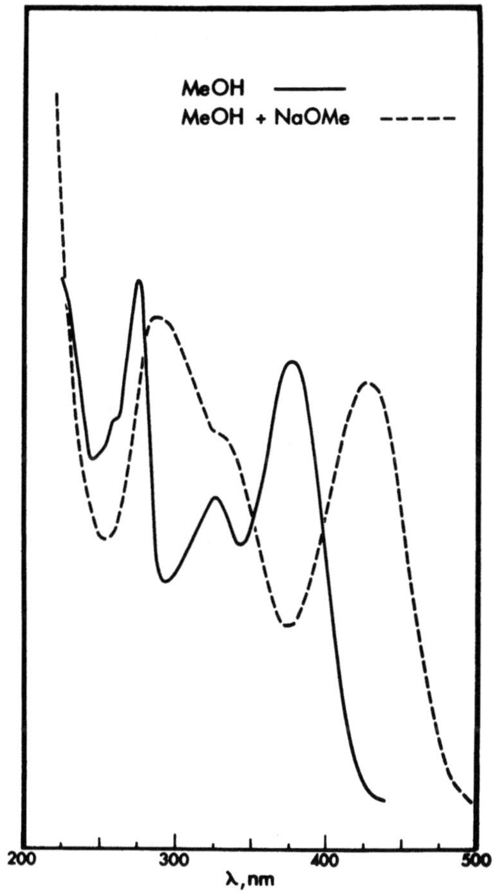

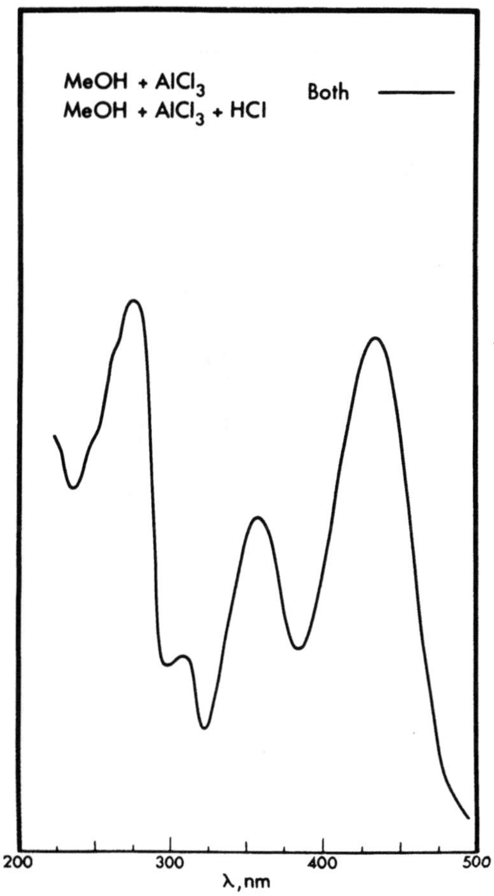

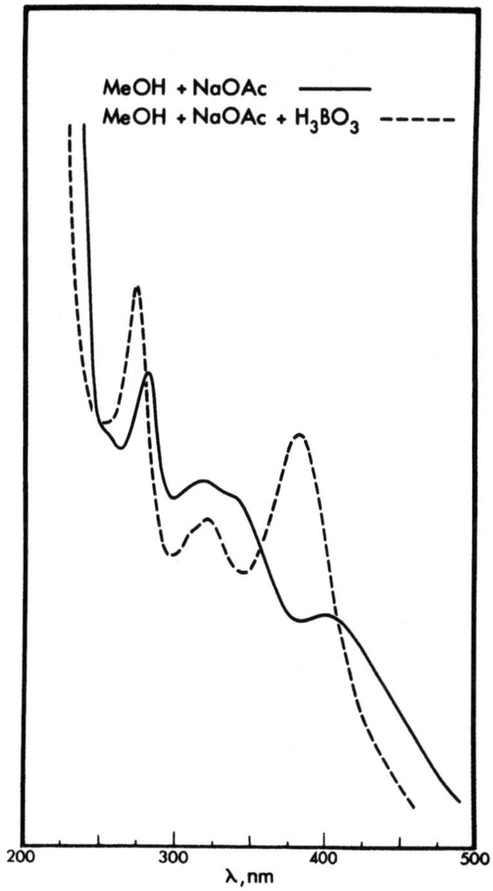

65

QUERCETIN

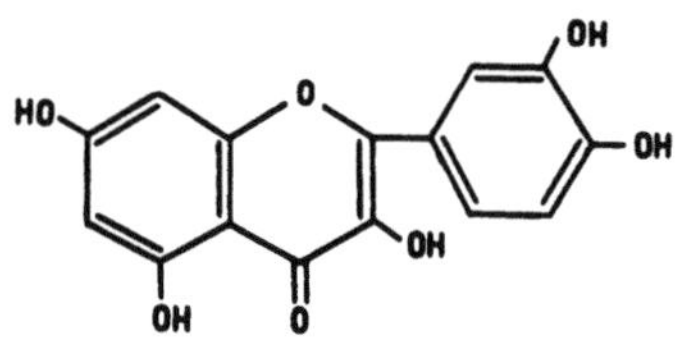

CHROMATOGRAPHIC DATA

Spot Appearance: (UV) yellow
(UV/NH_3) yellow

R_f Values: 0.57 (TBA), 0.03 (HOAc)

UV SPECTRAL DATA (λ_{max},*nm*)

MeOH	255, 269sh, 301sh, 370
NaOMe	247sh, 321 (dec.)
$AlCl_3$	272, 304sh, 333, 458
$AlCl_3$/HCl	265, 301sh, 359, 428
NaOAc	257sh, 274, 329, 390 (dec.)
NaOAc/H_3BO_3 (Proc. II)	261, 303sh, 388

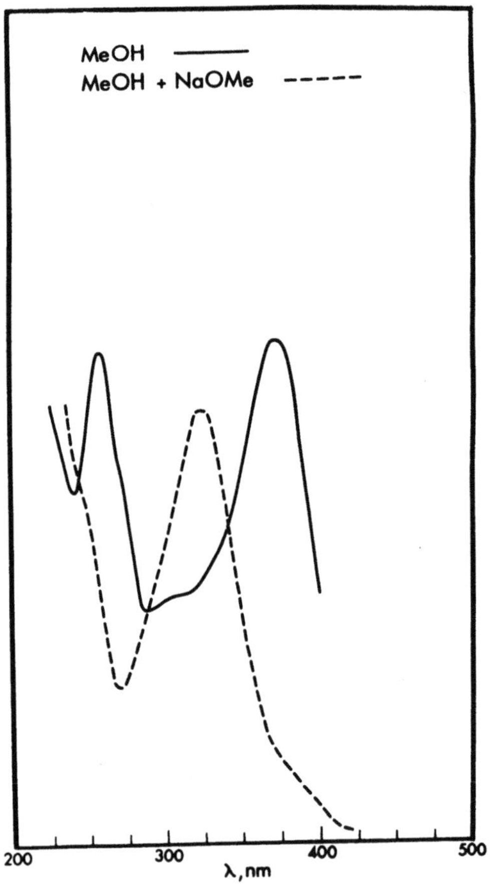

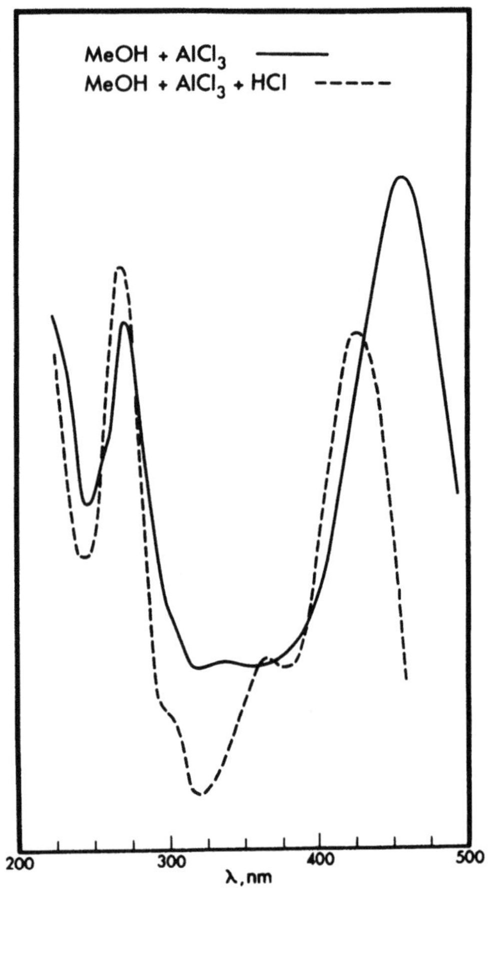

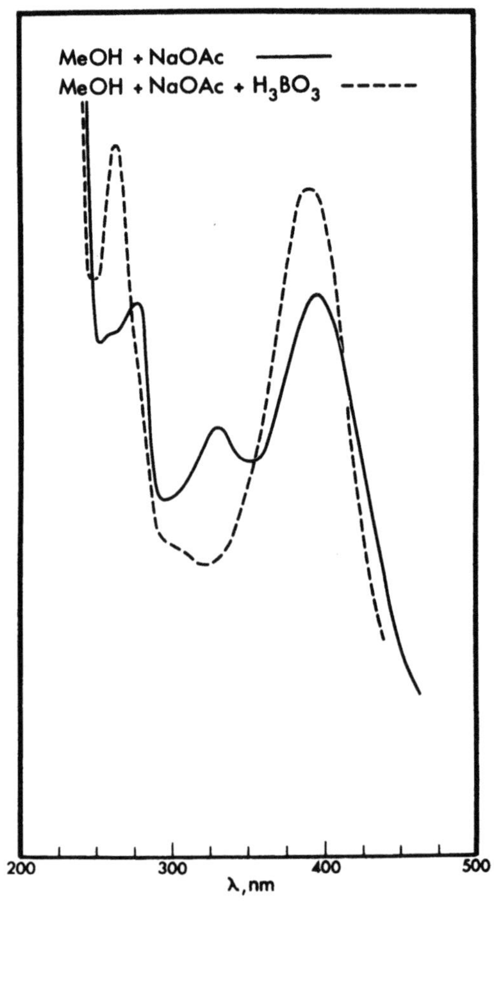

66
QUERCETIN 7-0-RHAMNOSIDE

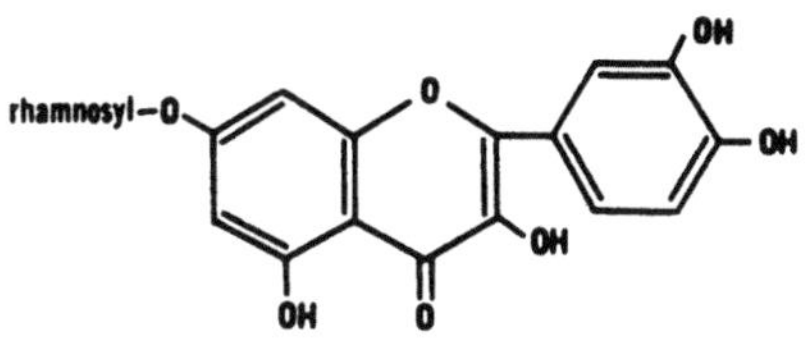

CHROMATOGRAPHIC DATA

Spot Appearance: (UV) yellow
(UV/NH_3) fluorescent yellow

R_f Values: 0.55 (TBA), 0.11 (HOAc)

UV SPECTRAL DATA (λ_{max},*nm*)

MeOH	256, 269sh, 372
NaOMe	241sh, 291, 367, 457 (dec.)
$AlCl_3$	259sh, 273, 339, 458
$AlCl_3$/HCl	268, 303sh, 365, 426
NaOAc	286, 378, 428sh (dec.)
NaOAc/H_3BO_3 (Proc. II)	261, 289sh, 386

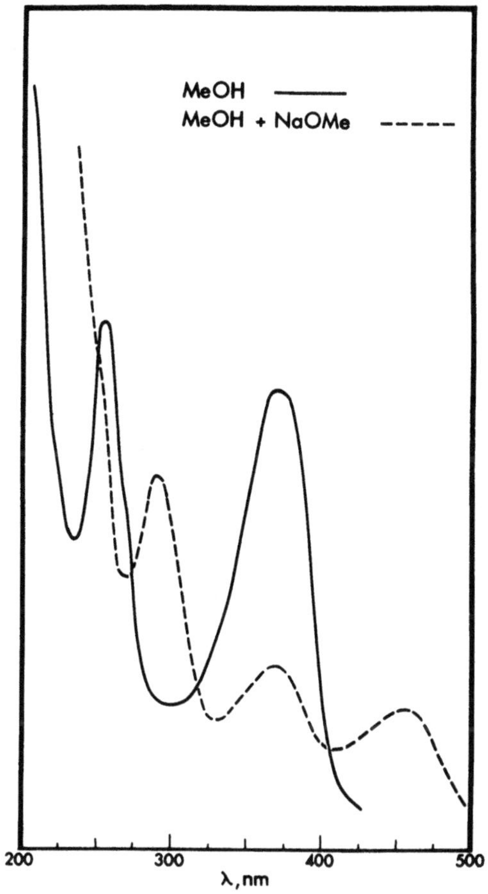

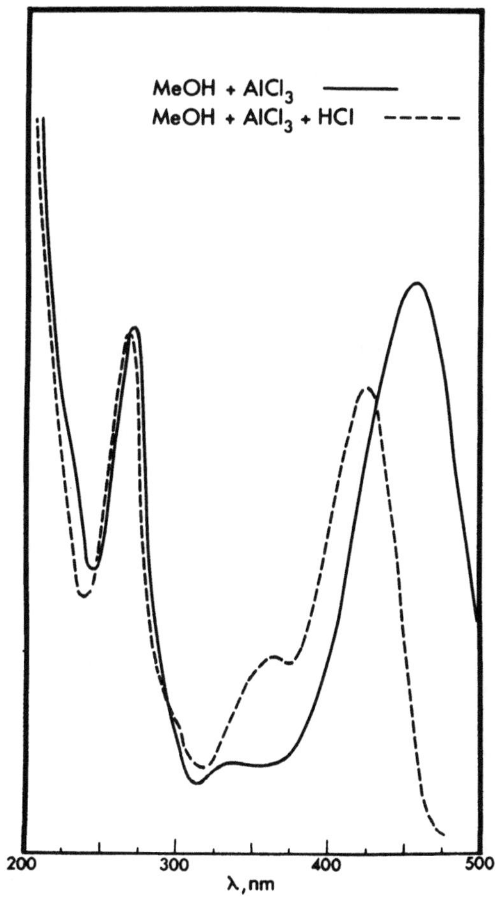

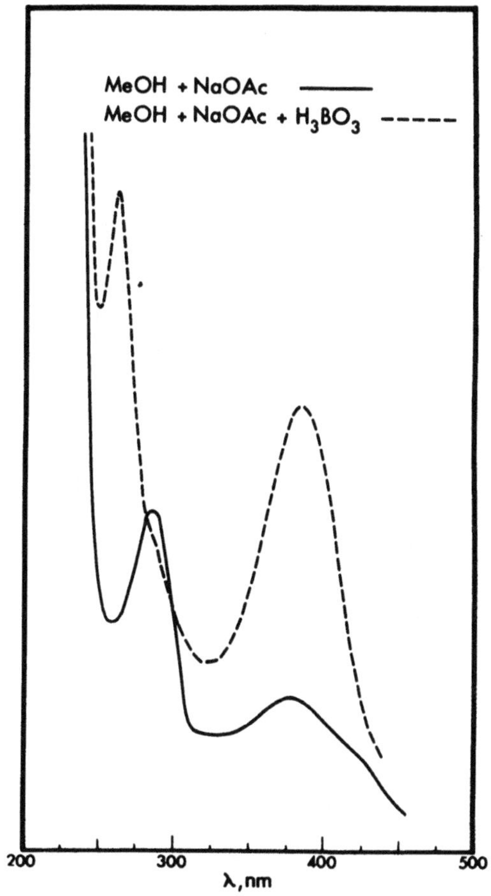

67
QUERCETIN 3-0-GALACTOSIDE

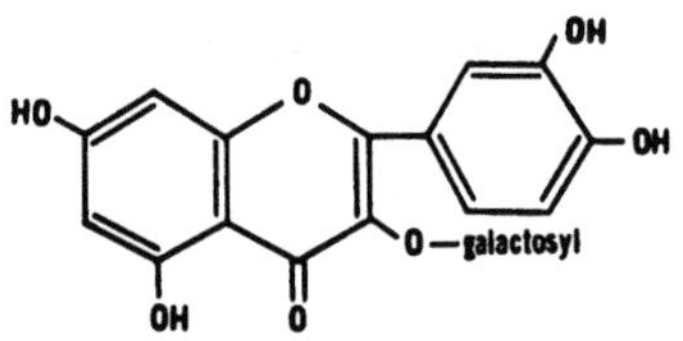

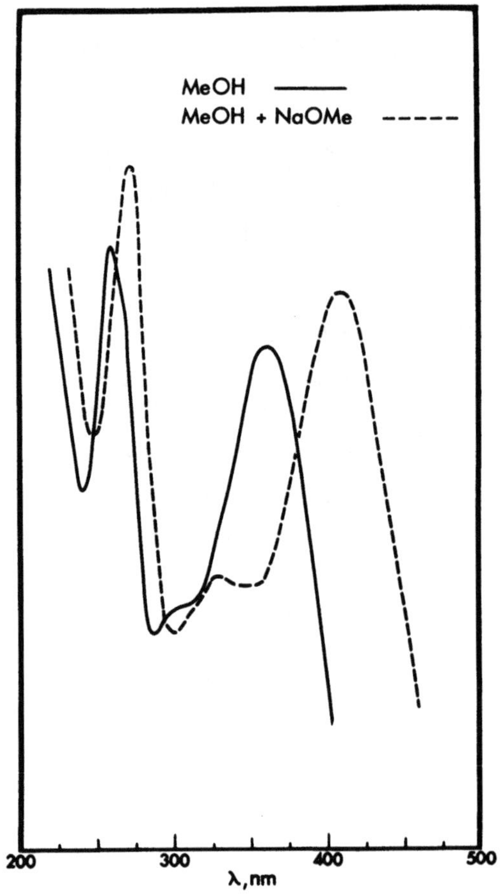

CHROMATOGRAPHIC DATA

Spot Appearance: (UV) deep purple
(UV/NH_3) yellow

R_f Values: 0.48 (TBA), 0.43 (HOAc)

UV SPECTRAL DATA (λ_{max},nm)

MeOH	257, 269sh, 299sh, 362
NaOMe	272, 327, 409
$AlCl_3$	275, 305sh, 331sh, 438
$AlCl_3$/HCl	268, 299sh, 366sh, 405
NaOAc	274, 324, 380
NaOAc/H_3BO_3 (Proc. I)	262, 298sh, 377

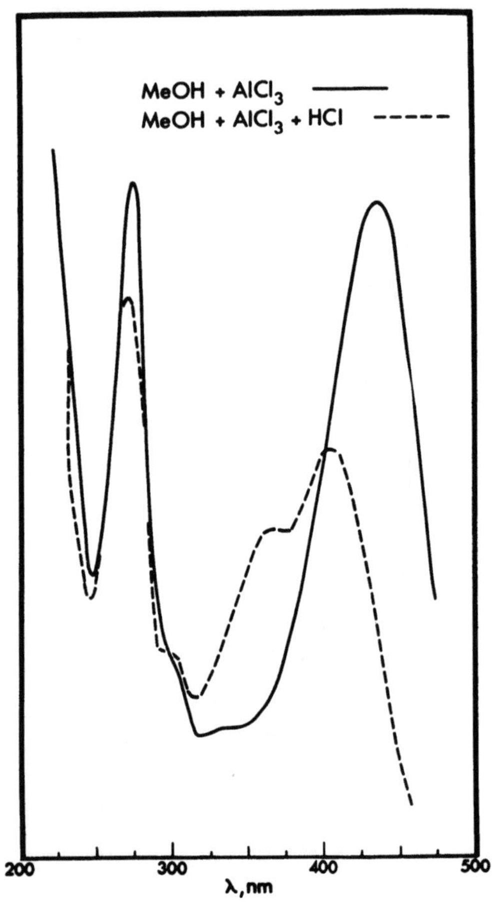

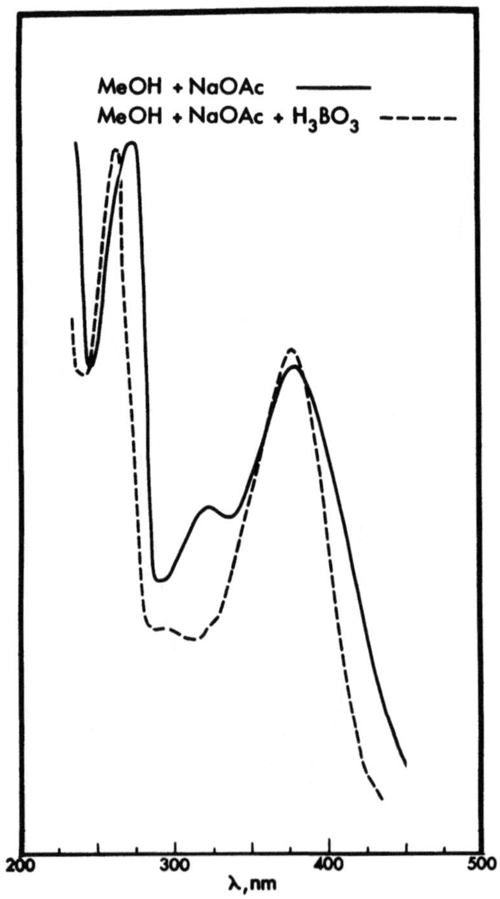

QUERCITRIN

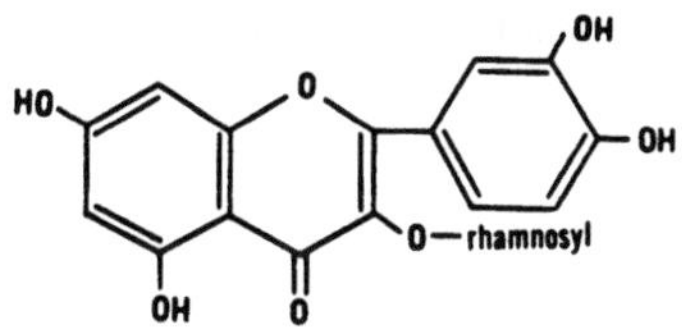

CHROMATOGRAPHIC DATA

Spot Appearance: (UV) deep purple
(UV/NH_3) yellow-green

R_f Values: 0.61 (TBA), 0.58 (HOAc)

UV SPECTRAL DATA (λ_{max},*nm*)

MeOH	256, 265sh, 301sh, 350
NaOMe	270, 326, 393
$AlCl_3$	276, 304sh, 333, 430
$AlCl_3$/HCl	272, 303sh, 353, 401
NaOAc	272, 322sh, 372
NaOAc/H_3BO_3	260, 300sh, 367
(Proc. I)	

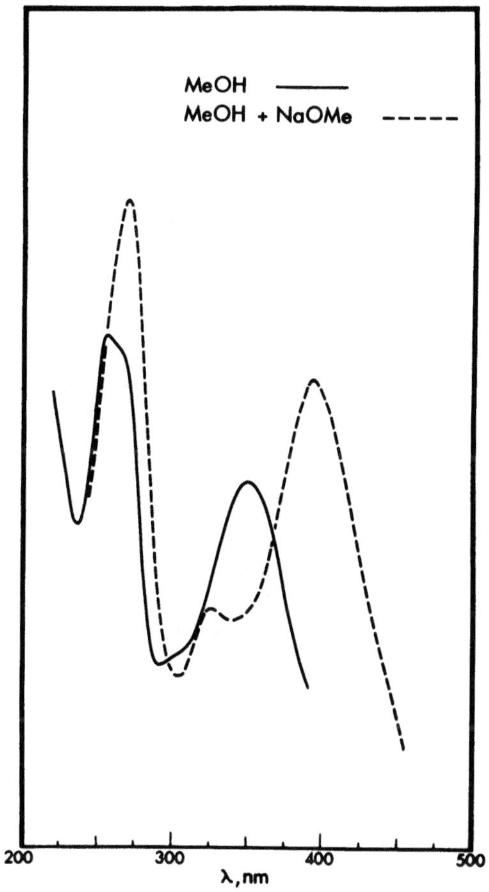

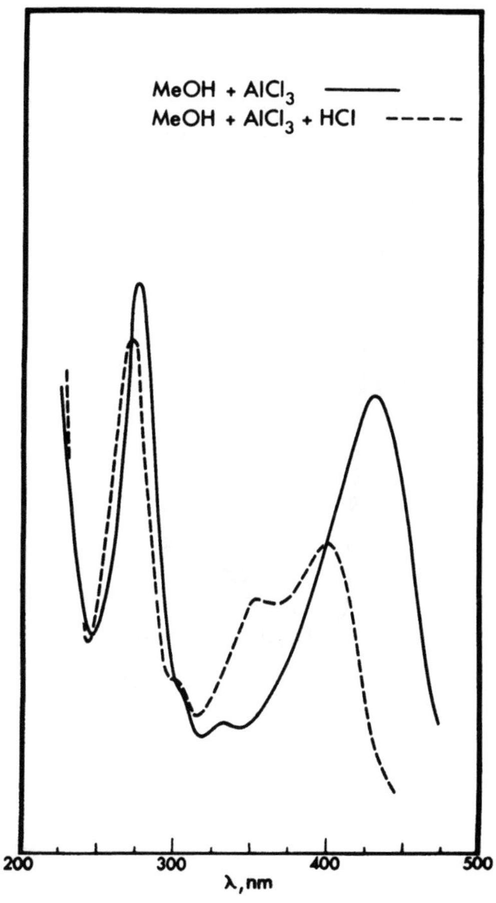

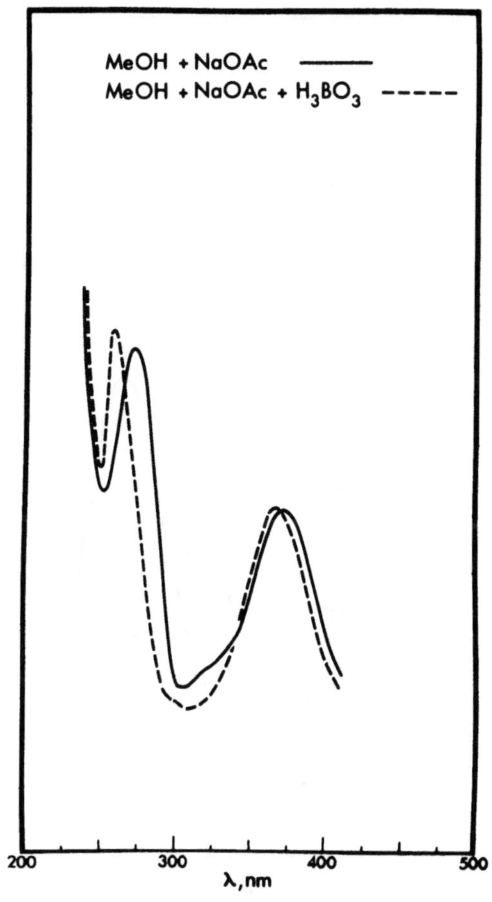

RUTIN

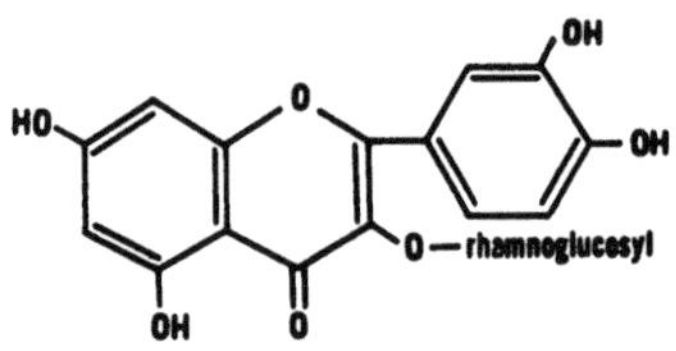

CHROMATOGRAPHIC DATA

Spot Appearance: (UV) deep purple
(UV/NH_3) yellow

R_f Values: 0.44 (TBA), 0.56 (HOAc)

UV SPECTRAL DATA (λ_{max}, *nm*)

MeOH	259, 266sh, 299sh, 359
NaOMe	272, 327, 410
$AlCl_3$	275, 303sh, 433
$AlCl_3$/HCl	271, 300, 364sh, 402
NaOAc	271, 325, 393
NaOAc/H_3BO_3 (Proc. I)	262, 298, 387

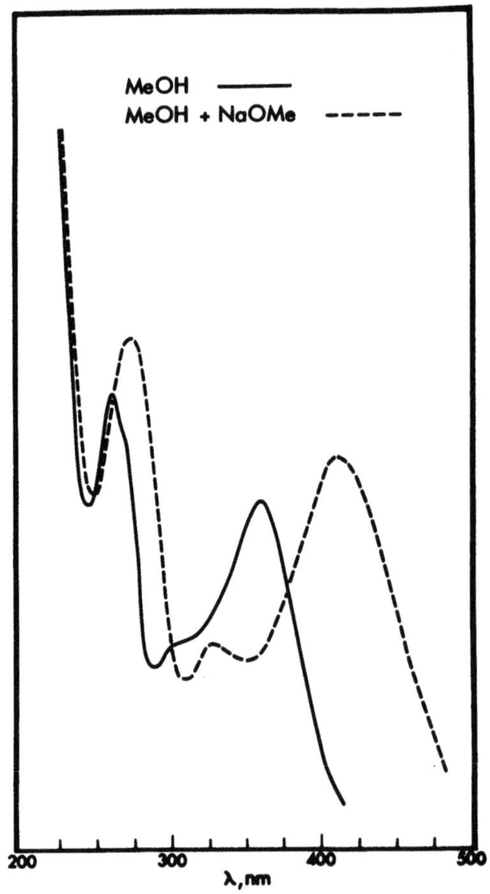

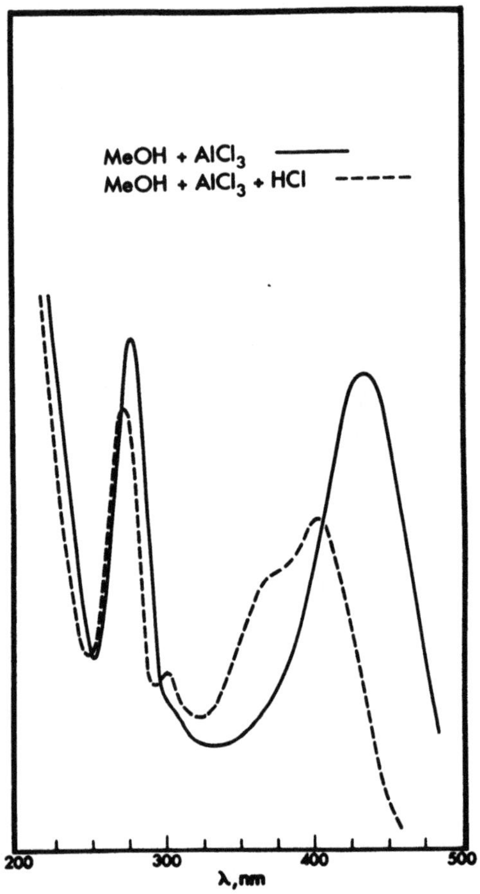

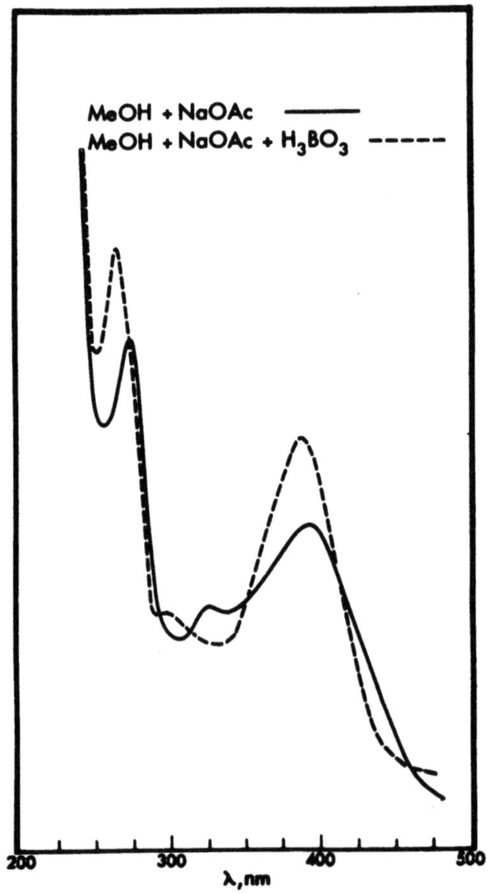

70

QUERCETIN 3,7-0-DIGLUCOSIDE

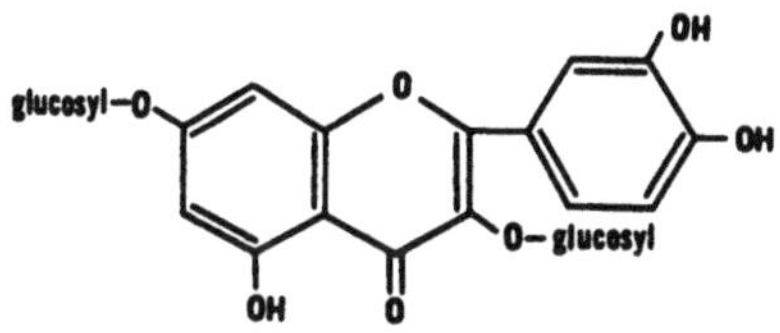

CHROMATOGRAPHIC DATA

Spot Appearance: (UV) deep purple
(UV/NH_3) yellow

R_f Values: 0.13 (TBA), 0.66 (HOAc)

UV SPECTRAL DATA (λ_{max},*nm*)

MeOH	256, 268sh, 355
NaOMe	268, 300sh, 396
$AlCl_3$	275, 298sh, 335, 440
$AlCl_3$/HCl	270, 299sh, 363sh, 402
NaOAc	261, 295sh, 371, 423sh
NaOAc/H_3BO_3 (Proc. I)	261, 380

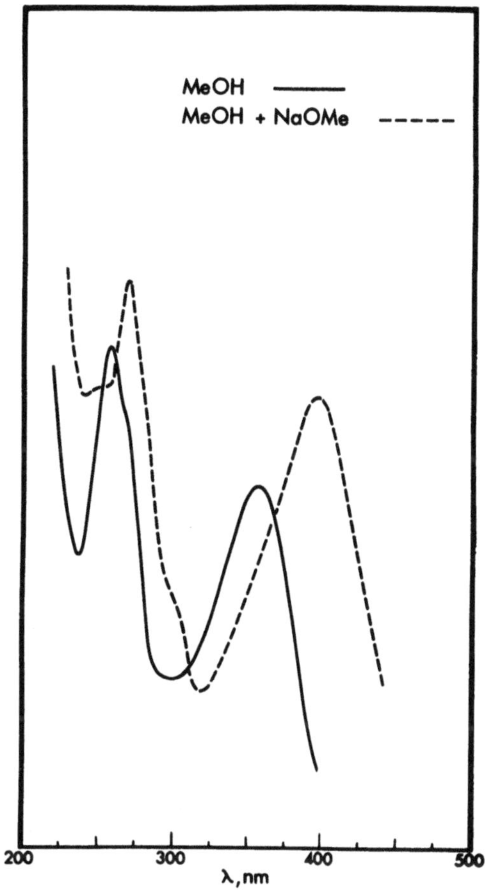

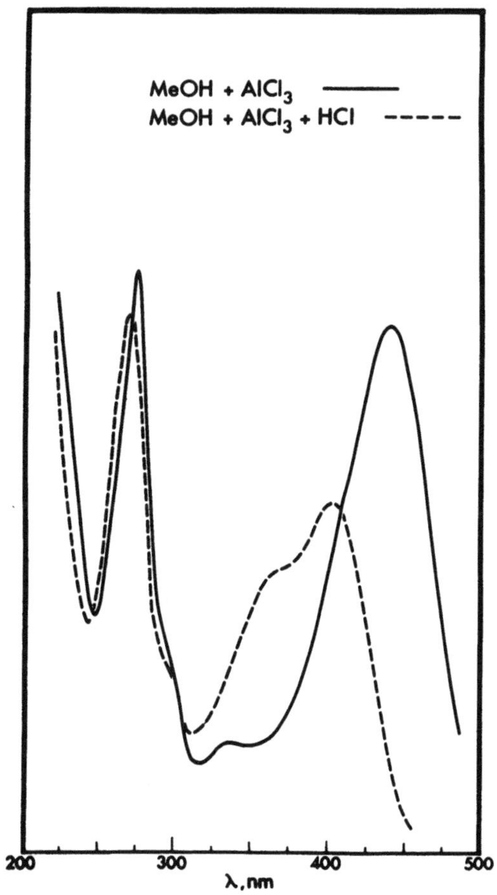

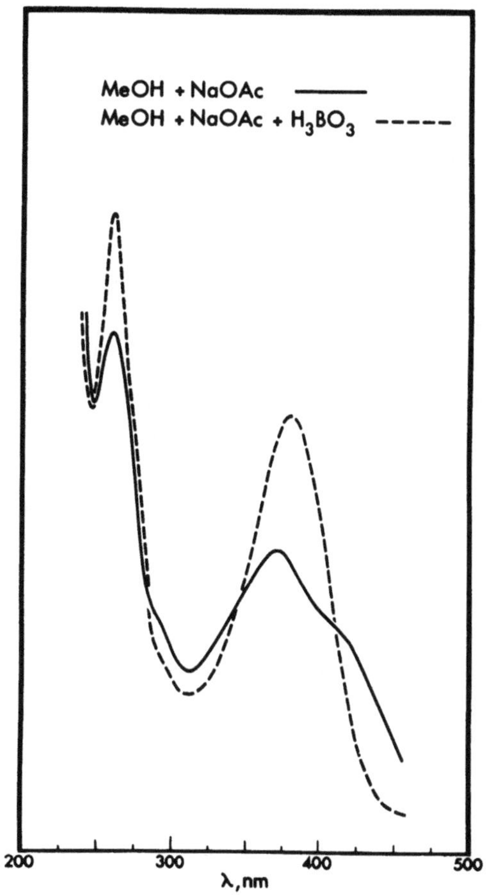

71

QUERCETIN 3-O-GLUCOSIDE 7-O-RHAMNOSIDE

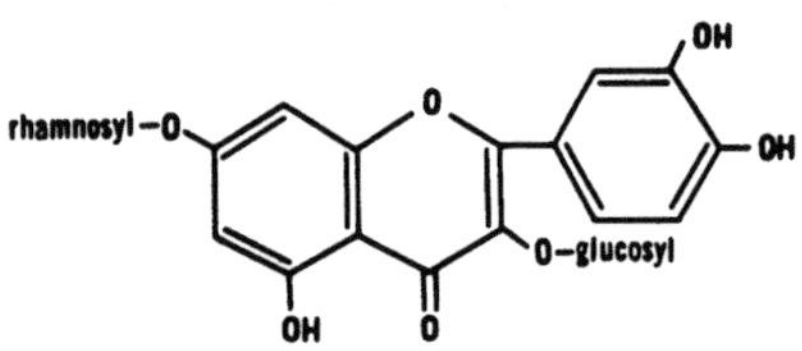

CHROMATOGRAPHIC DATA

Spot Appearance: (UV) deep purple
(UV/NH_3) yellow

R_f Values: 0.38 (TBA), 0.62 (HOAc)

UV SPECTRAL DATA (λ_{max}, nm)

MeOH	257, 269sh, 358
NaOMe	244, 270, 396
$AlCl_3$	276, 300sh, 343sh, 441
$AlCl_3/HCl$	270, 300sh, 366sh, 404
NaOAc	260, 294sh, 370, 416sh
$NaOAc/H_3BO_3$ (Proc. I)	261, 294sh, 380

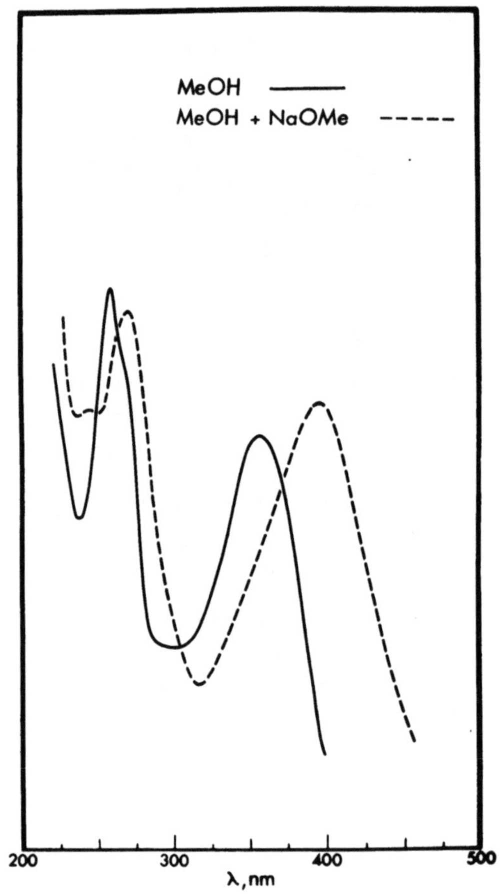

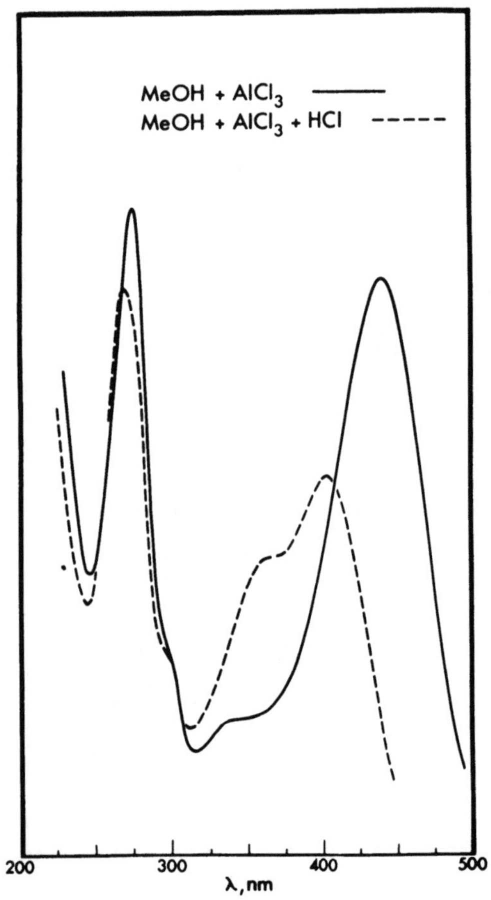

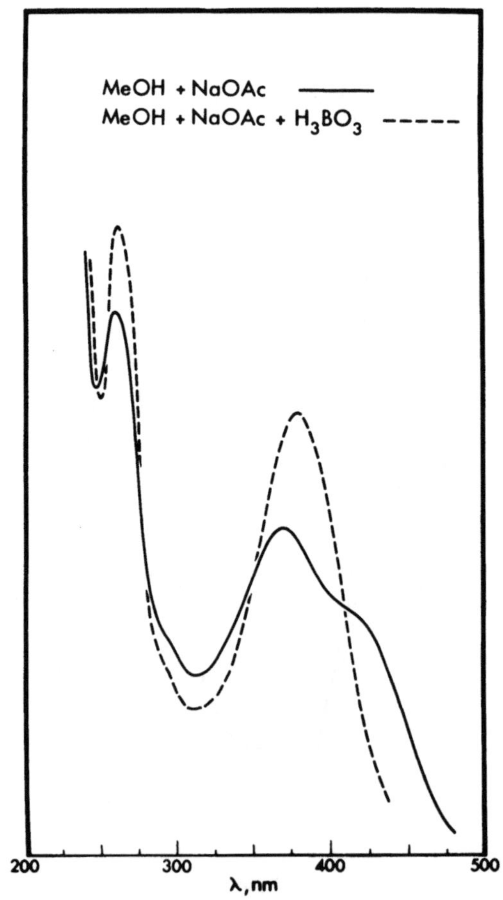

QUERCETIN 3-0-GLUCOSIDE 7-0-RUTINOSIDE

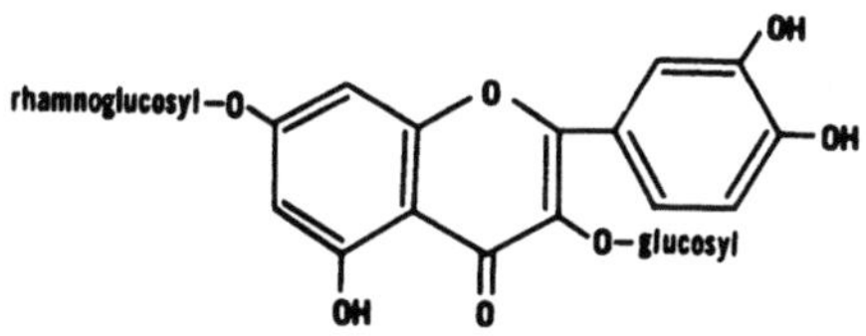

CHROMATOGRAPHIC DATA

Spot Appearance: (UV) deep purple
(UV/NH_3) yellow

R_f Values: 0.09 (TBA), 0.78 (HOAc)

UV SPECTRAL DATA (λ_{max},*nm*)

MeOH	257, 269sh, 358
NaOMe	244, 270, 396
$AlCl_3$	276, 300sh, 343sh, 441
$AlCl_3$/HCl	270, 300sh, 366sh, 404
NaOAc	260, 294sh, 370, 416sh
NaOAc/H_3BO_3 (Proc. I)	261, 294sh, 380

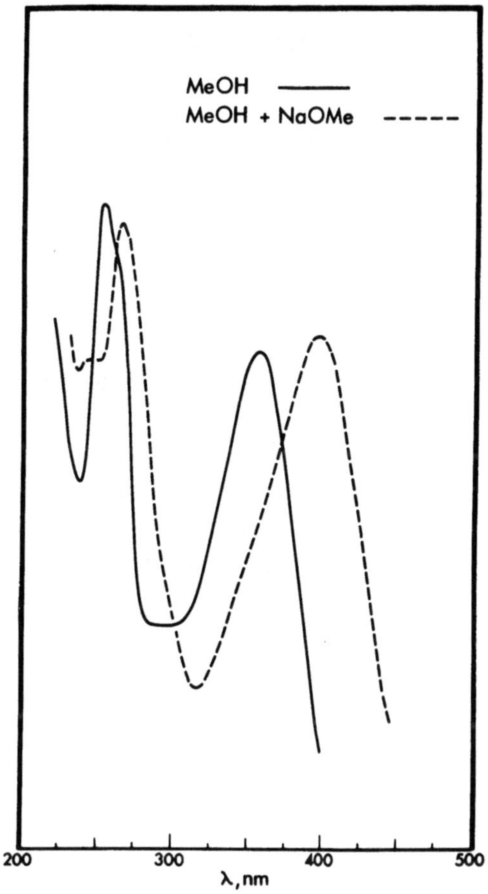

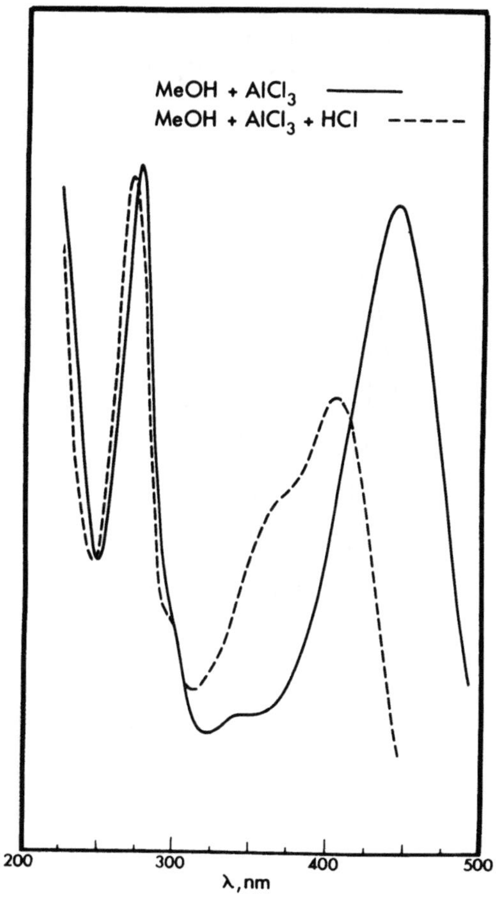

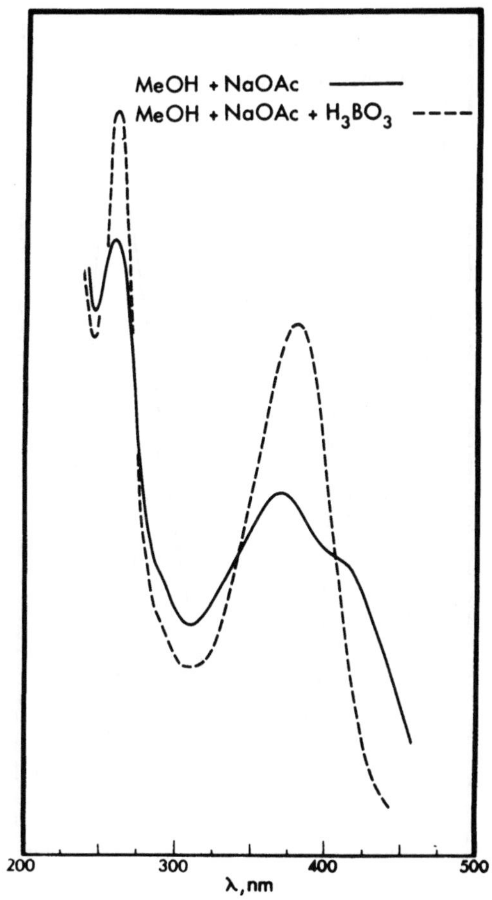

73
QUERCETIN 3-METHYL ETHER

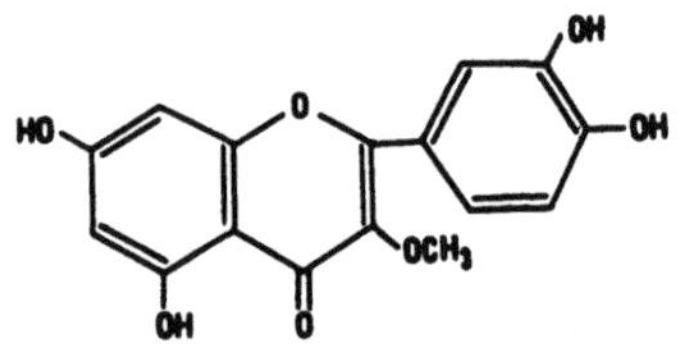

CHROMATOGRAPHIC DATA

Spot Appearance: (UV) deep purple
(UV/NH_3) yellow

R_f Values: 0.80 (TBA), 0.10 (HOAc)

UV SPECTRAL DATA (λ_{max},nm)

MeOH	257, 269sh, 294sh, 358
NaOMe	272, 329, 407
$AlCl_3$	277, 303sh, 336, 443
$AlCl_3$/HCl	268, 277sh, 299sh, 360, 402
NaOAc	273, 323, 383
NaOAc/H_3BO_3 (Proc. I)	262, 298sh, 378

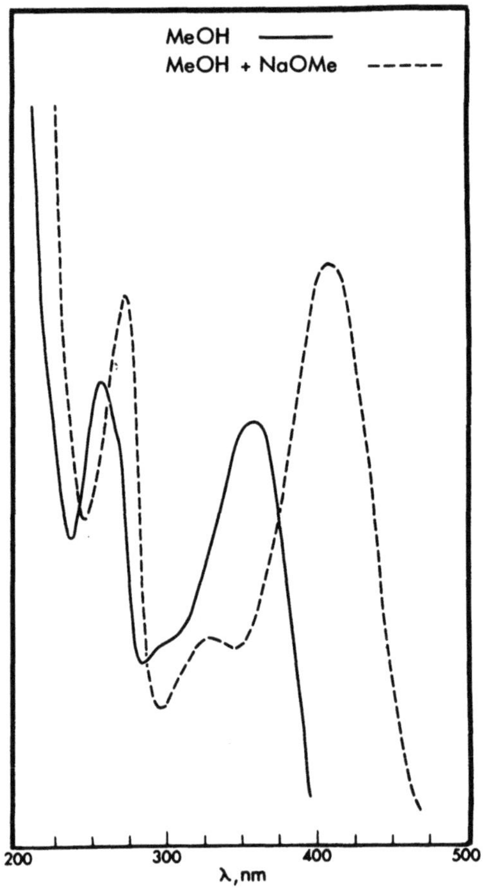

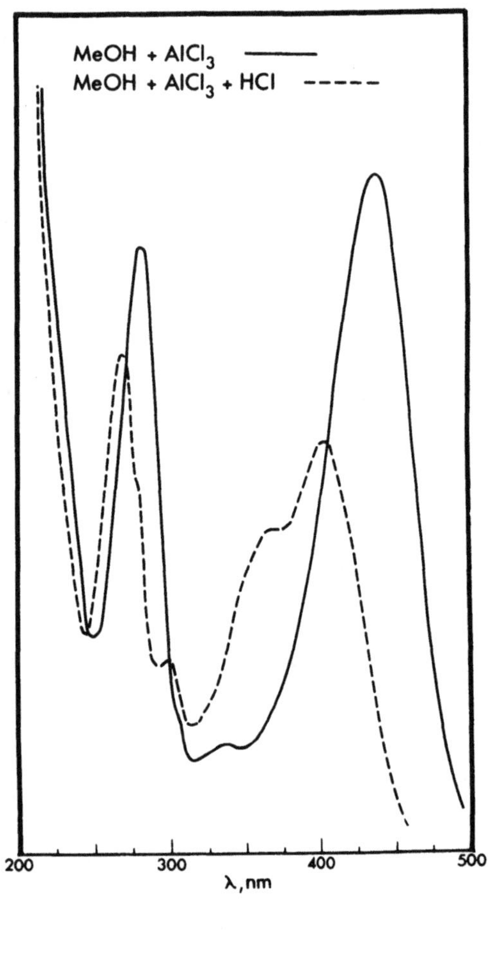

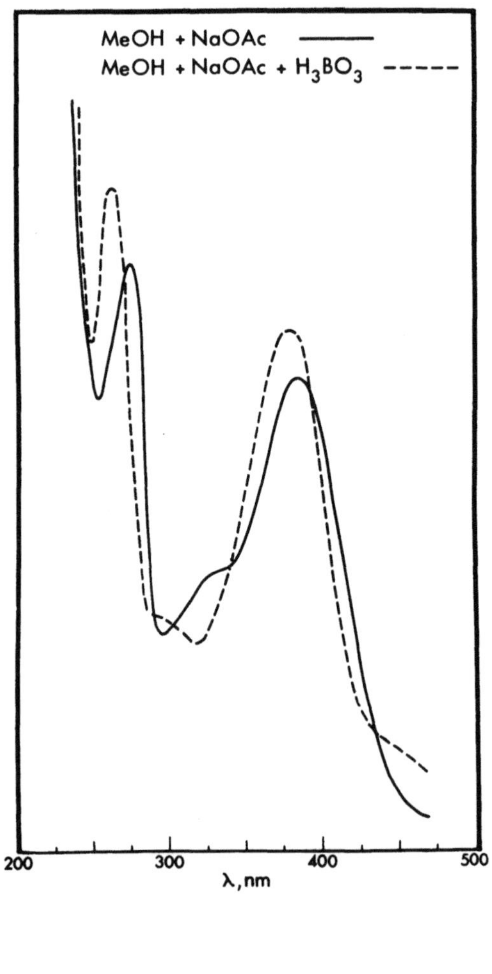

QUERCETIN 3-METHYL ETHER 4'-0-GLUCOSIDE 7-0-DIGLUCOSIDE

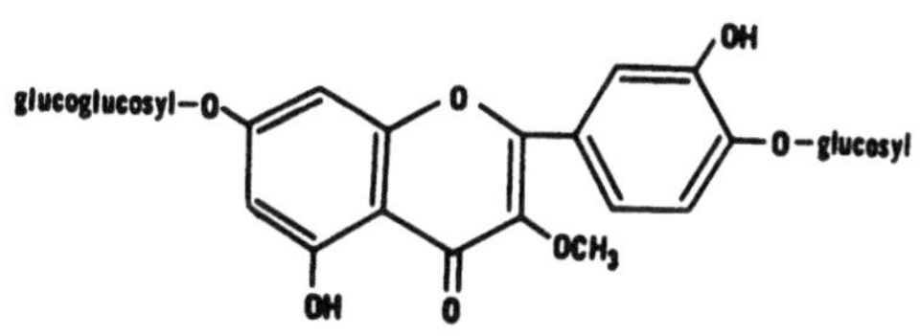

CHROMATOGRAPHIC DATA

Spot Appearance: (UV) deep purple
(UV/NH_3) deep purple

R_f Values: 0.29 (TBA), 0.52 (HOAc)

UV SPECTRAL DATA (λ_{max},*nm*)

MeOH	254, 269, 349
NaOMe	268, 376
$AlCl_3$	275, 298sh, 355, 400
$AlCl_3$/HCl	265sh, 279, 297sh, 348, 399
NaOAc	261, 350
NaOAc/H_3BO_3 (Proc. I)	254, 267, 350

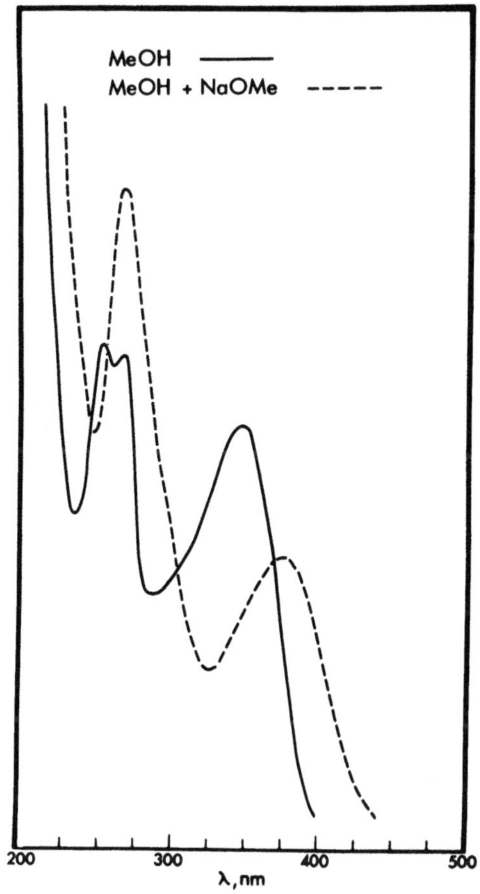

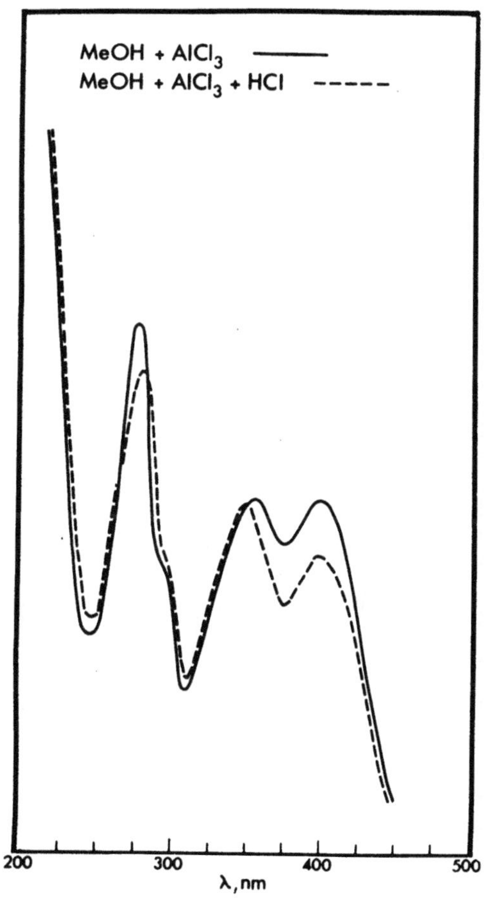

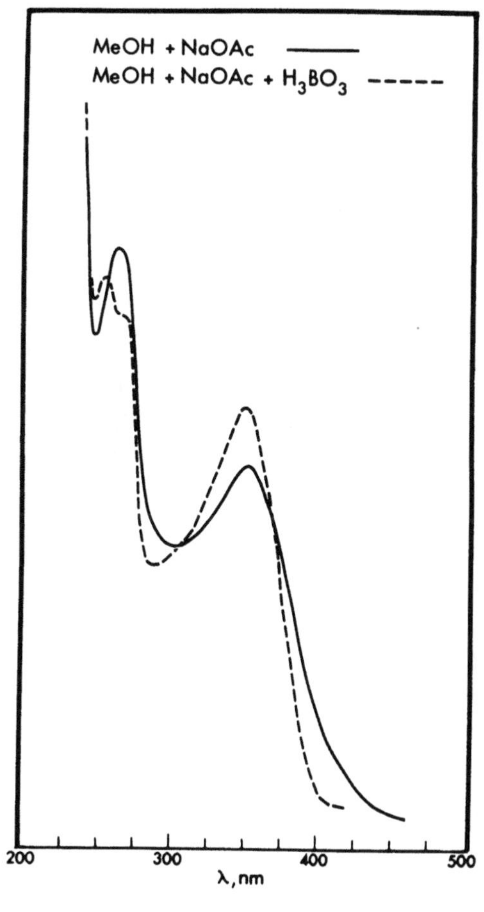

75

QUERCETIN 3′,4′,5,7-TETRAMETHYL ETHER

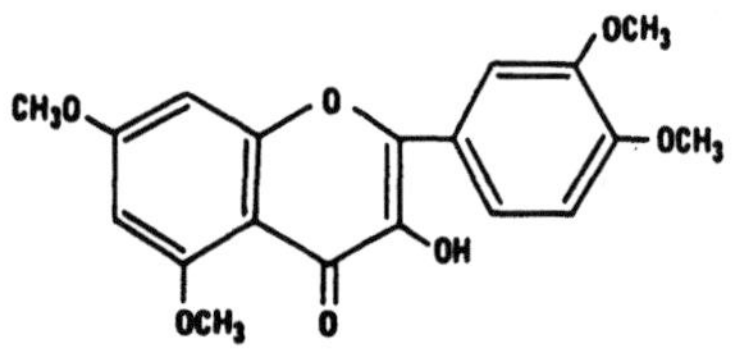

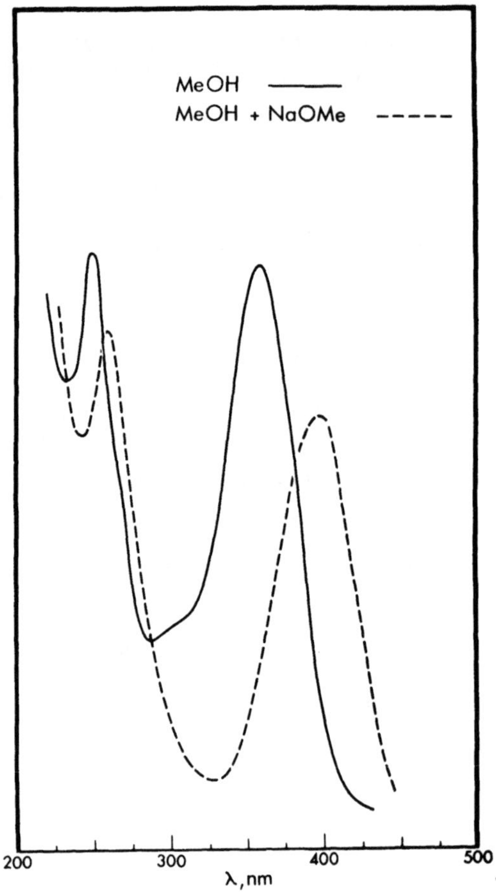

CHROMATOGRAPHIC DATA

Spot Appearance: (UV) fluorescent yellow
(UV/NH_3) fluorescent yellow

R_f Values: 0.57 (TBA), 0.41 (HOAc)

UV SPECTRAL DATA (λ_{max}, nm)

MeOH	252, 270sh, 304sh, 362
NaOMe	263, 400
$AlCl_3$	262, 269sh, 303sh, 343sh, 421
$AlCl_3$/HCl	260, 268sh, 303sh, 342sh, 420
NaOAc	250, 268sh, 365, 414sh
NaOAc/H_3BO_3 (Proc. I)	249sh, 269sh, 304sh, 361, 424sh

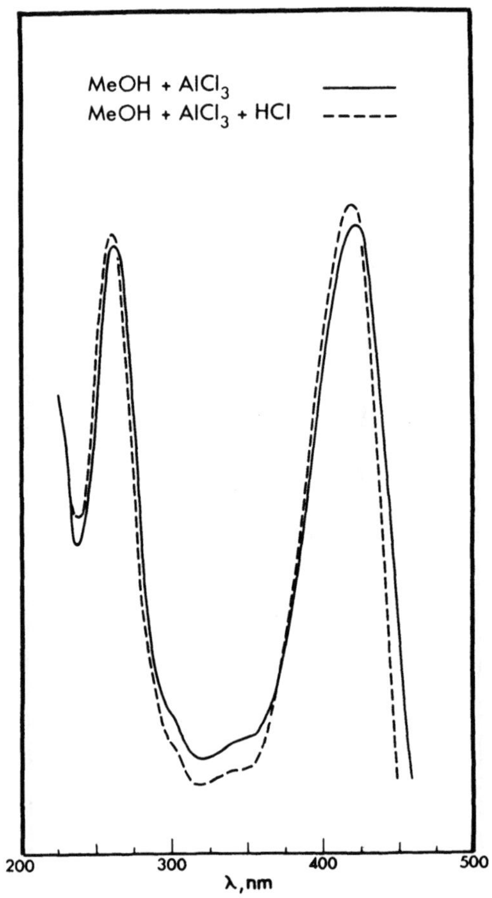

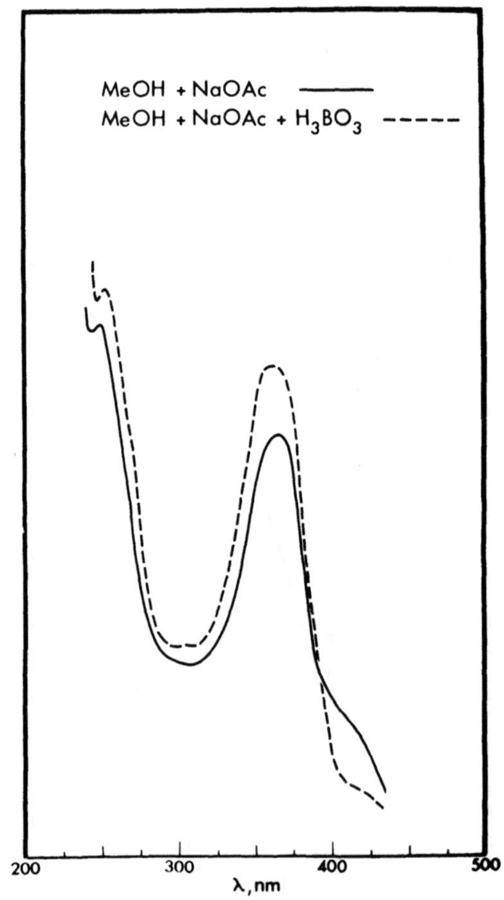

RHAMNETIN

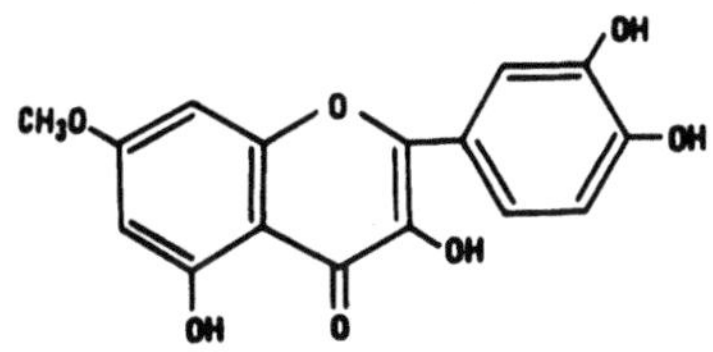

CHROMATOGRAPHIC DATA

Spot Appearance: (UV) dull yellow
(UV/NH_3) dull yellow

R_f Values: 0.63 (TBA), 0.04 (HOAc)

UV SPECTRAL DATA (λ_{max},nm)

MeOH	256, 270sh, 295sh, 371
NaOMe	242, 286, 331, 432 (dec.)
$AlCl_3$	273, 302sh, 330sh, 451
$AlCl_3$/HCl	268, 299sh, 363sh, 423
NaOAc	255, 292sh, 387, 422sh, (dec.)
NaOAc/H_3BO_3 (Proc. II)	260, 389

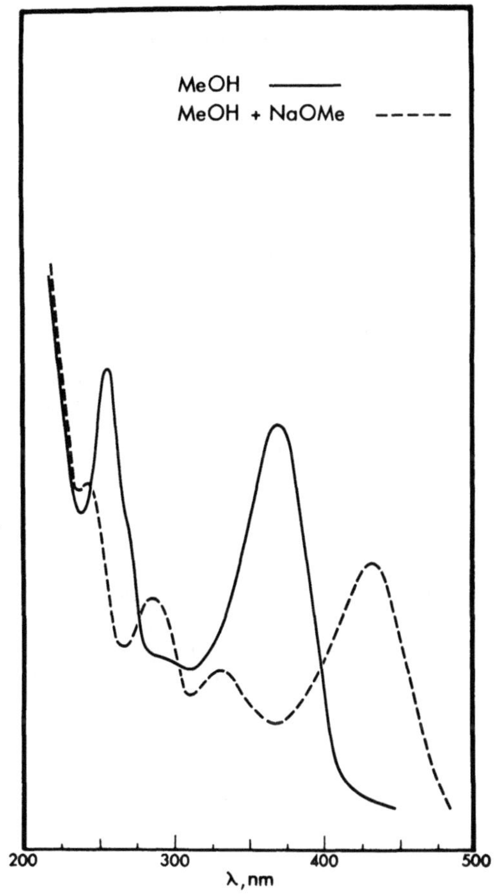

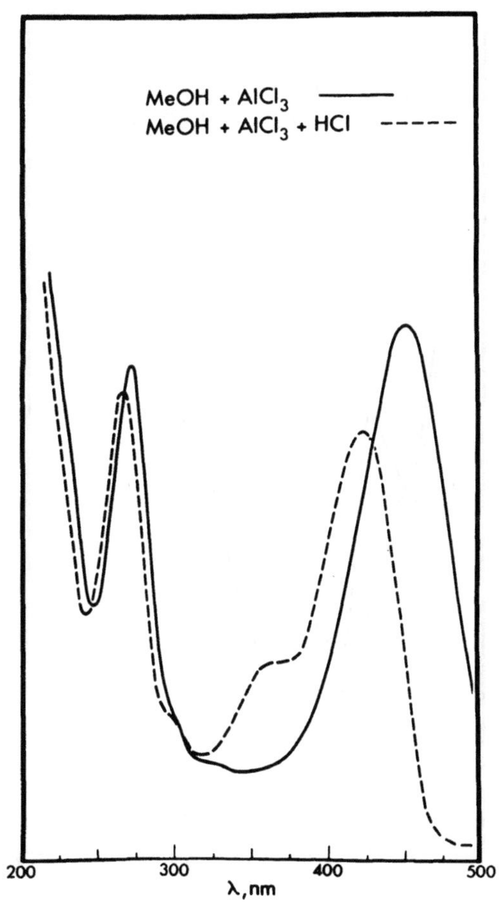

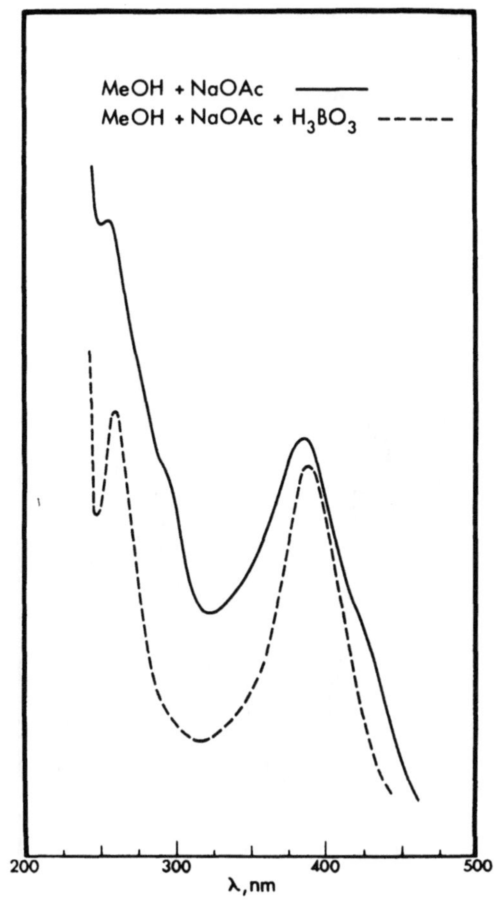

77

ISORHAMNETIN

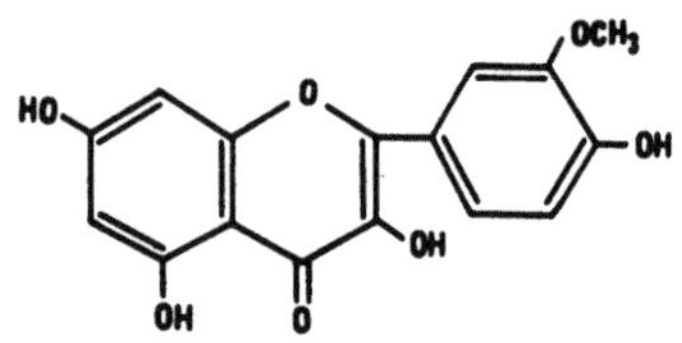

CHROMATOGRAPHIC DATA

Spot Appearance: (UV) dull yellow
(UV/NH_3) dull yellow

R_f Values: 0.68 (TBA), 0.02 (HOAc)

UV SPECTRAL DATA (λ_{max},nm)

MeOH	253, 267sh, 306sh, 326sh, 370
NaOMe	240sh, 271, 328, 435 (dec)
$AlCl_3$	264, 304sh, 361sh, 431
$AlCl_3$/HCl	242sh, 262, 271sh, 302sh, 357, 428
NaOAc	260sh, 274, 320, 393 (dec.)
NaOAc/H_3BO_3 (Proc. II)	255, 270sh, 306sh, 326sh, 377

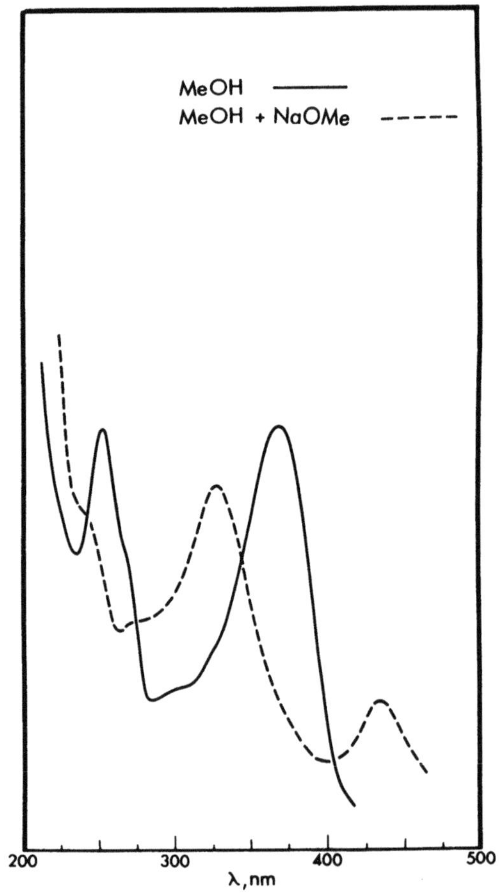

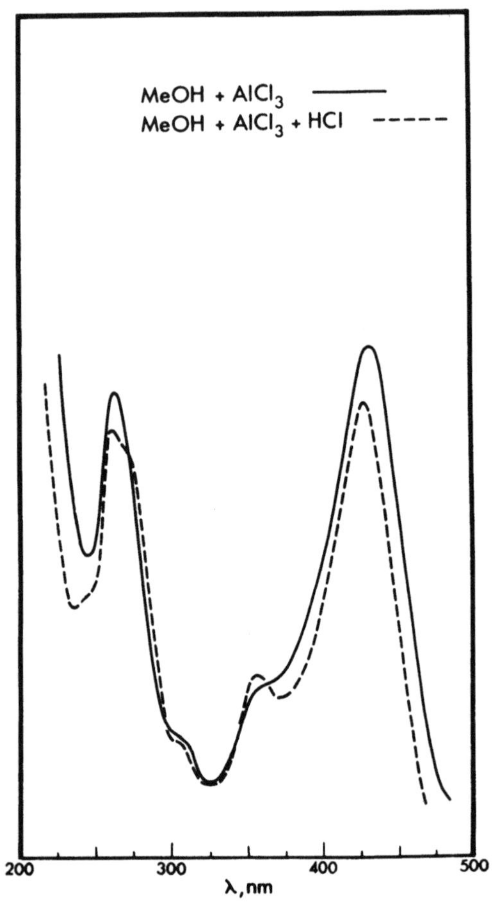

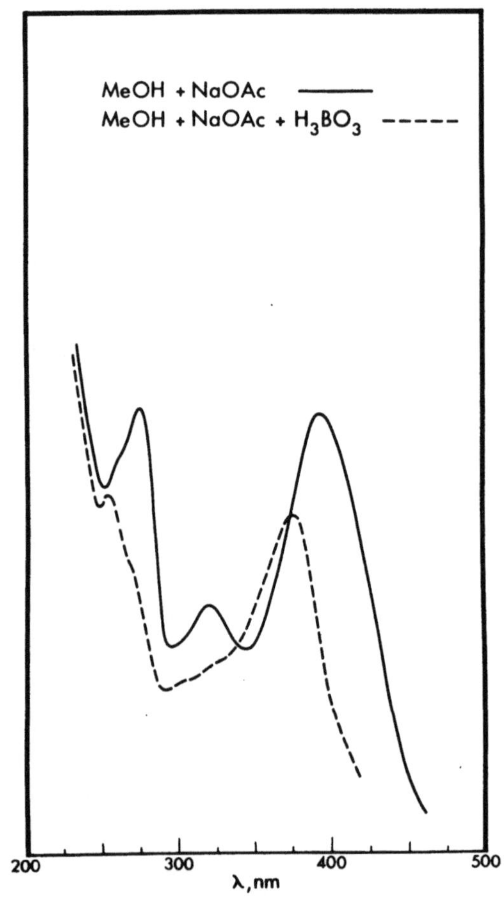

78
ISORHAMNETIN 3-0-GALACTOSIDE

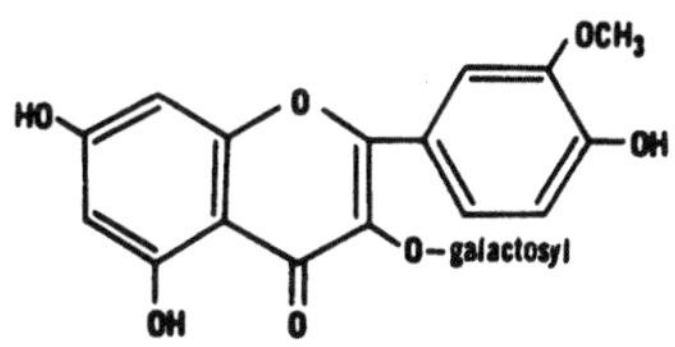

CHROMATOGRAPHIC DATA

Spot Appearance: (UV) deep purple
(UV/NH_3) yellow

R_f Values: 0.60 (TBA), 0.48 (HOAc)

UV SPECTRAL DATA (λ_{max},nm)

MeOH	255, 268sh, 303sh, 357
NaOMe	272, 327, 415
$AlCl_3$	269, 299sh, 365sh, 407
$AlCl_3$/HCl	267, 298sh, 357, 403
NaOAc	274, 316, 387
NaOAc/H_3BO_3 (Proc. I)	257, 267sh, 307sh, 361

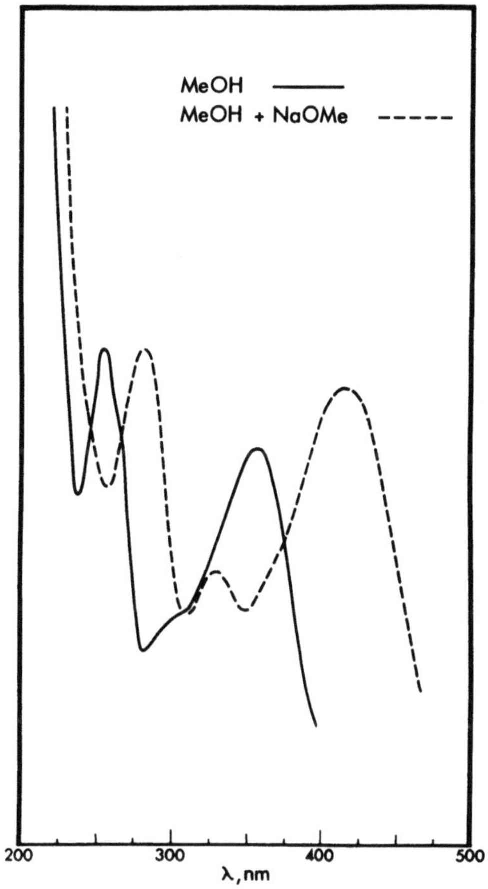

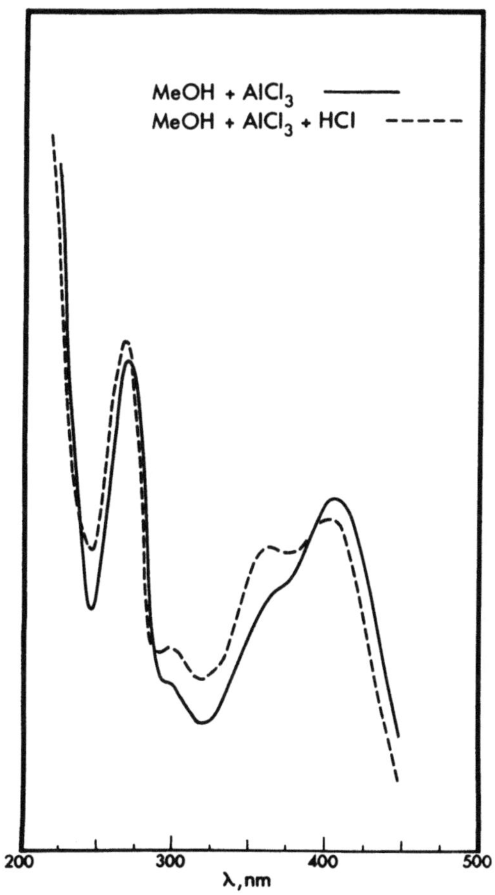

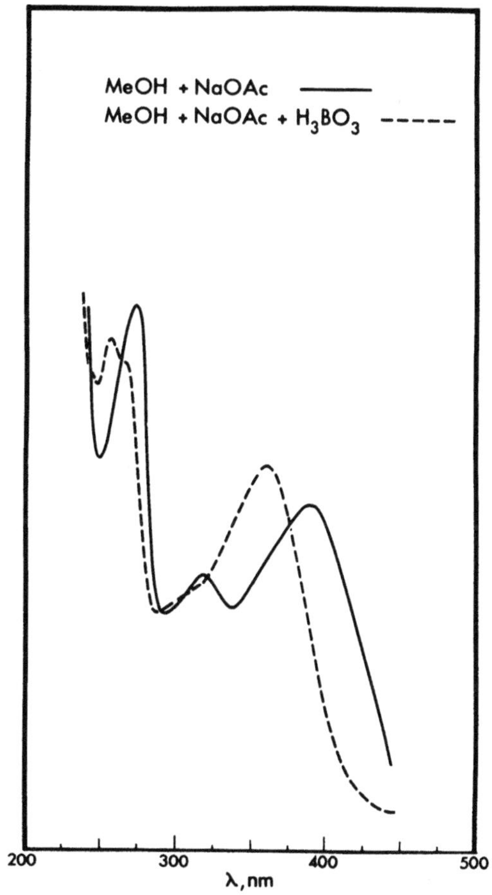

79
ISORHAMNETIN 3-0-RUTINOSIDE

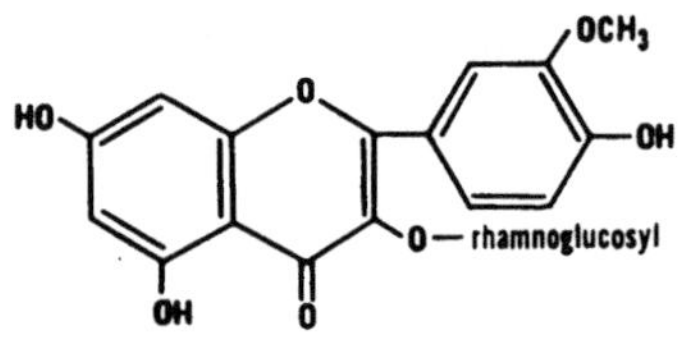

CHROMATOGRAPHIC DATA

Spot Appearance: (UV) deep purple
(UV/NH_3) yellow

R_f Values: 0.45 (TBA), 0.61 (HOAc)

UV SPECTRAL DATA (λ_{max},nm)

MeOH	254, 265sh, 305sh, 356
NaOMe	271, 328, 414
$AlCl_3$	268, 278sh, 300sh, 369sh, 402
$AlCl_3$/HCl	267, 275sh, 300sh, 359sh, 399
NaOAc	271, 320, 396
NaOAc/H_3BO_3 (Proc. I)	254, 267sh, 304sh, 360

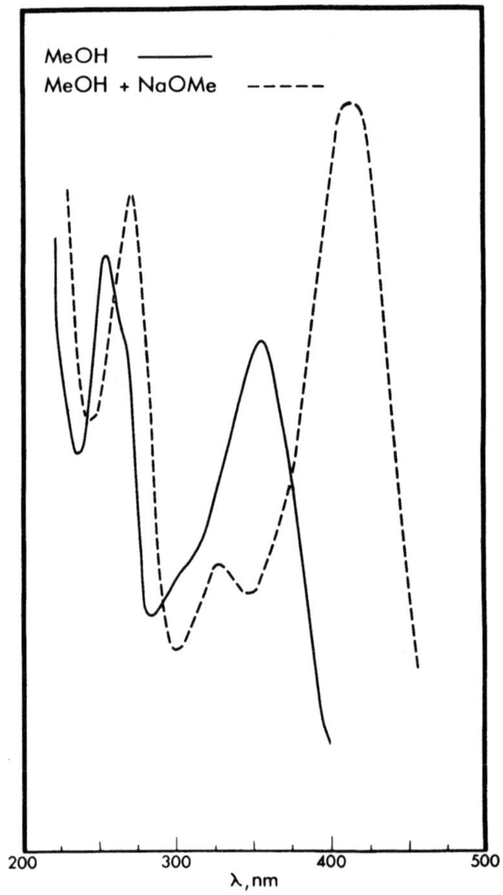

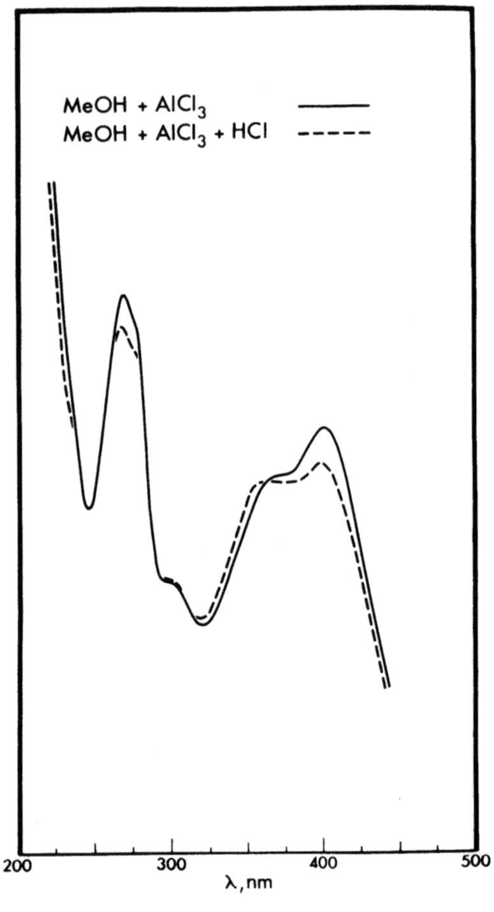

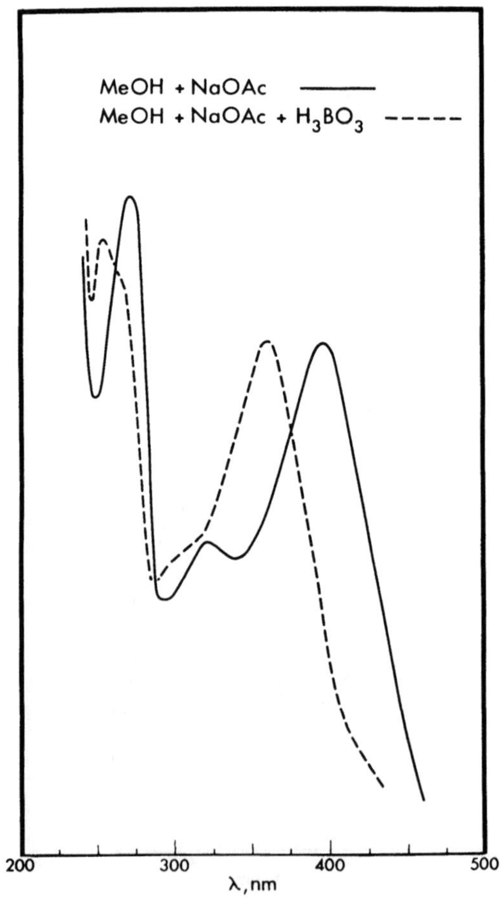

TAMARIXETIN 7-0-NEOHESPERIDOSIDE

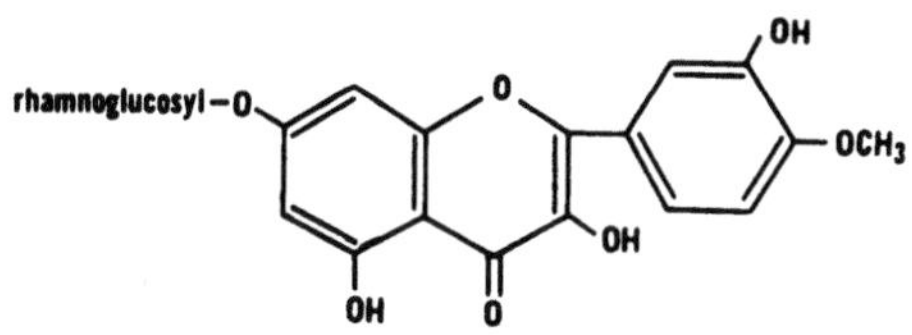

CHROMATOGRAPHIC DATA

Spot Appearance: (UV) yellow
(UV/NH_3) yellow

R_f Values: 0.17 (TBA), 0.29 (HOAc)

UV SPECTRAL DATA (λ_{max},nm)

MeOH	255, 269sh, 369
NaOMe	243, 268, 420
$AlCl_3$	266, 301sh, 360sh, 429
$AlCl_3$/HCl	242, 266, 301sh, 361, 427
NaOAc	257, 266sh, 328sh, 386, 419sh
NaOAc/H_3BO_3 (Proc. I)	255, 272sh, 372

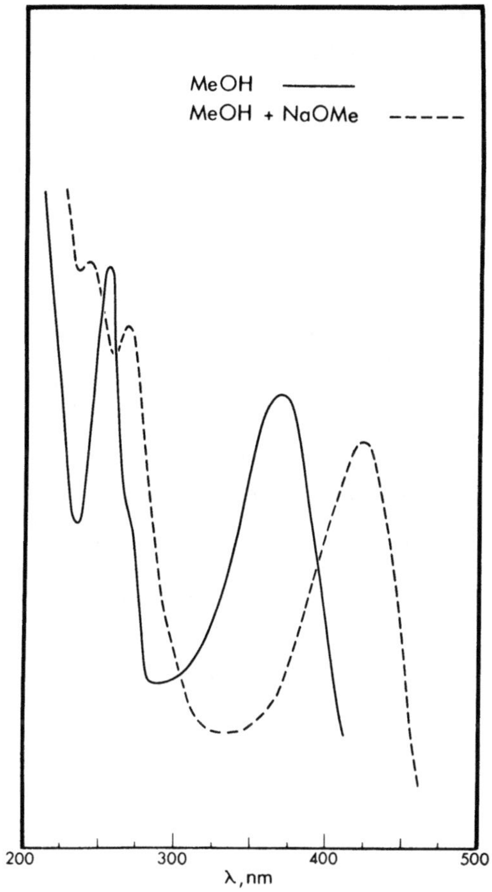

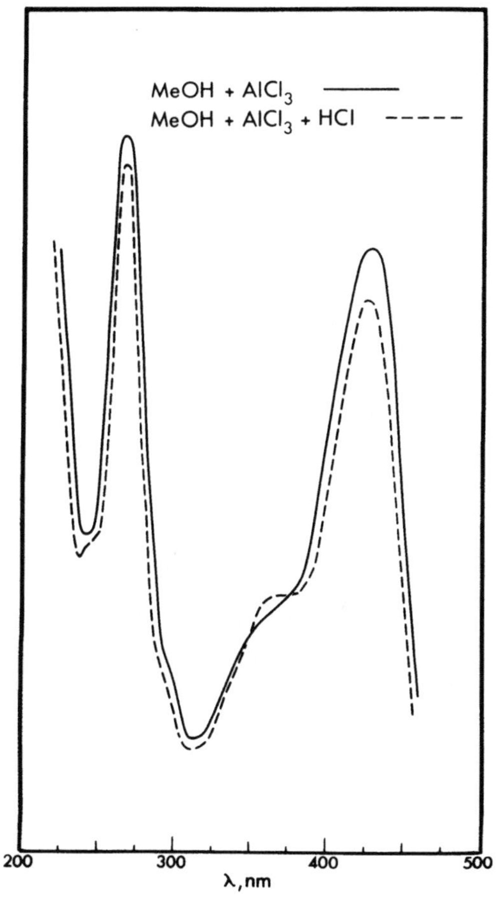

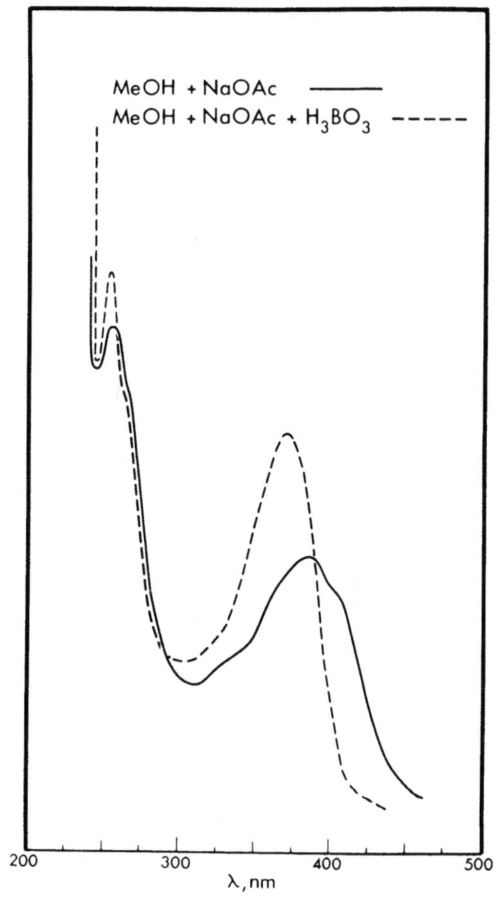

81

TAMARIXETIN 7-0-RUTINOSIDE

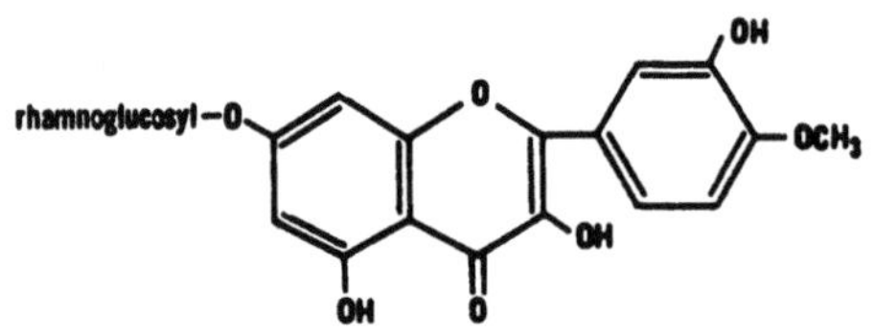

CHROMATOGRAPHIC DATA

Spot Appearance: (UV) pale yellow
(UV/NH_3) pale yellow

R_f Values: 0.16 (TBA), 0.32 (HOAc)

UV SPECTRAL DATA (λ_{max},*nm)*

MeOH	255, 271sh, 291sh, 367
NaOMe	242, 268, 415
$AlCl_3$	266, 303sh, 365sh, 427
$AlCl_3$/HCl	266, 301sh, 359, 423
NaOAc	256, 265sh, 327, 388, 415sh
NaOAc/H_3BO_3 (Proc. I)	255, 269sh, 292sh, 371

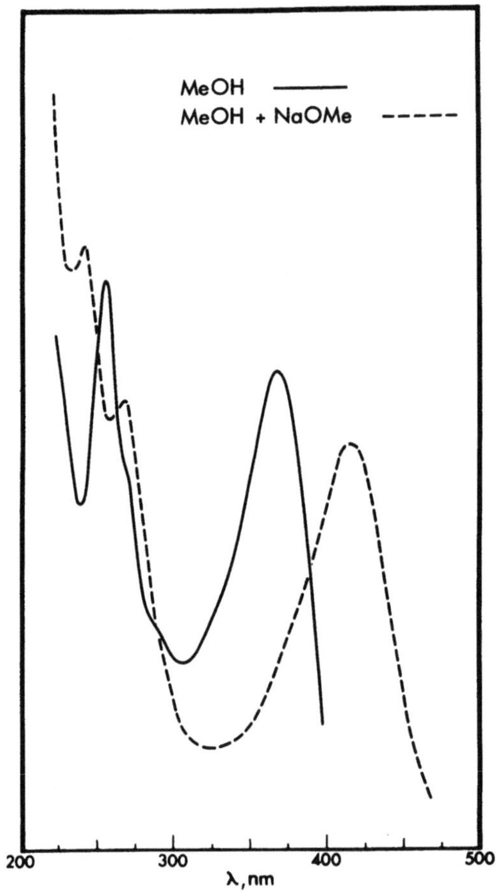

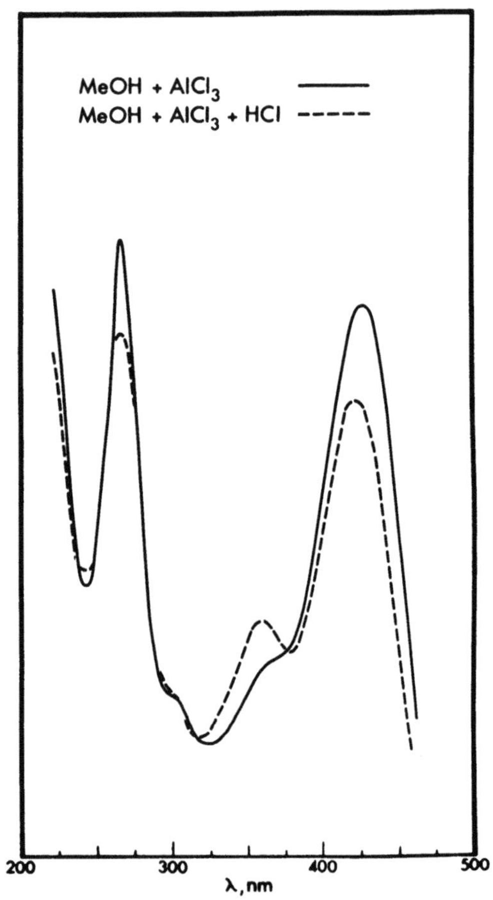

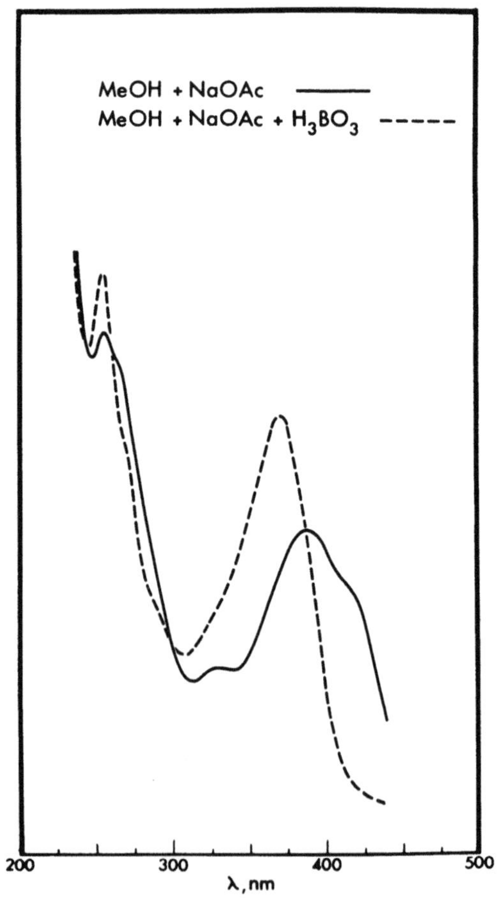

MORIN

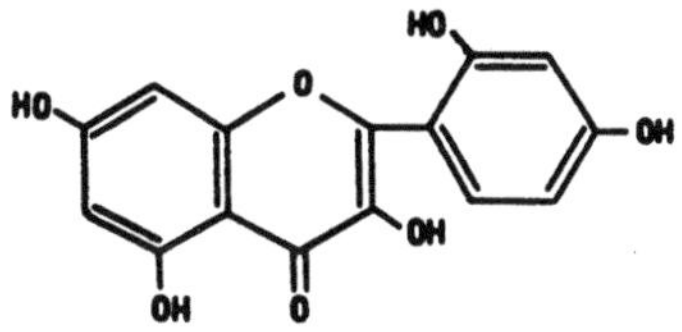

CHROMATOGRAPHIC DATA

Spot Appearance: (UV) yellow
(UV/NH_3) yellow

R_f Values: 0.76 (TBA), 0.22 (HOAc)

UV SPECTRAL DATA (λ_{max},nm)

MeOH	254sh, 264, 370
NaOMe	278, 314, 418 (dec.)
$AlCl_3$	268, 299sh, 352, 421
$AlCl_3/HCl$	267, 298sh, 349, 419
NaOAc	272, 315sh, 399
$NaOAc/H_3BO_3$ (Proc. II)	259sh, 267, 301sh, 374

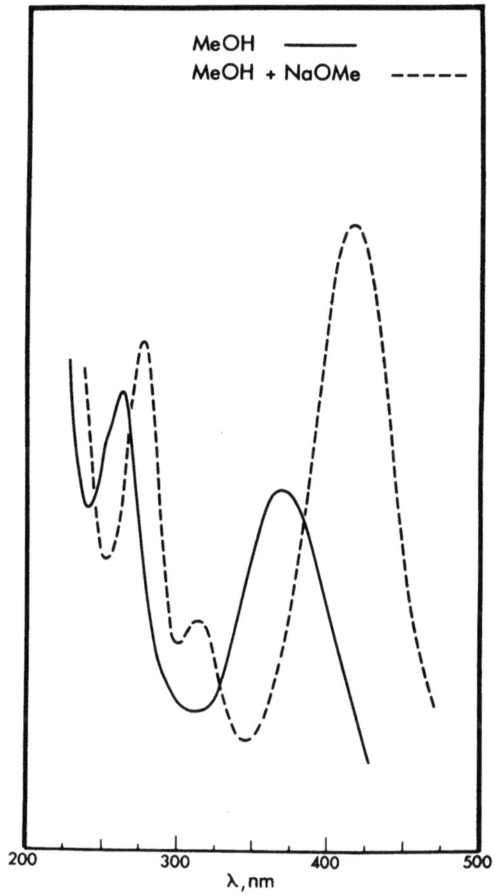

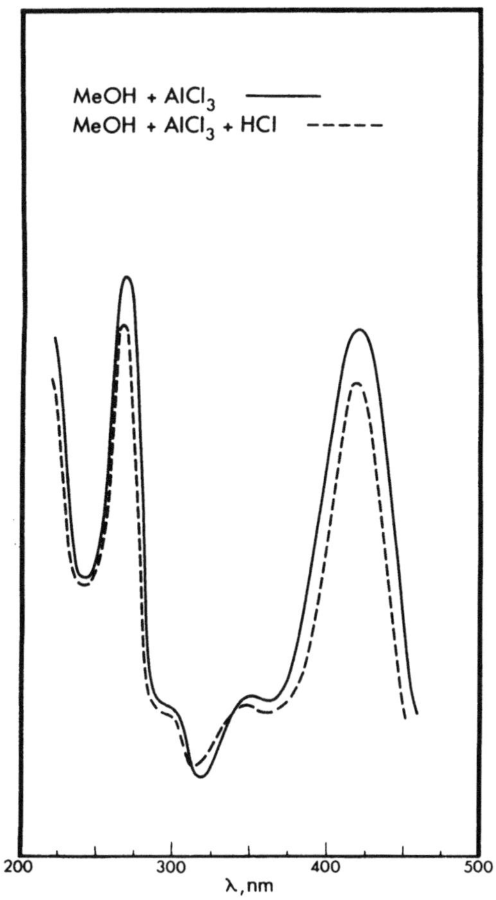

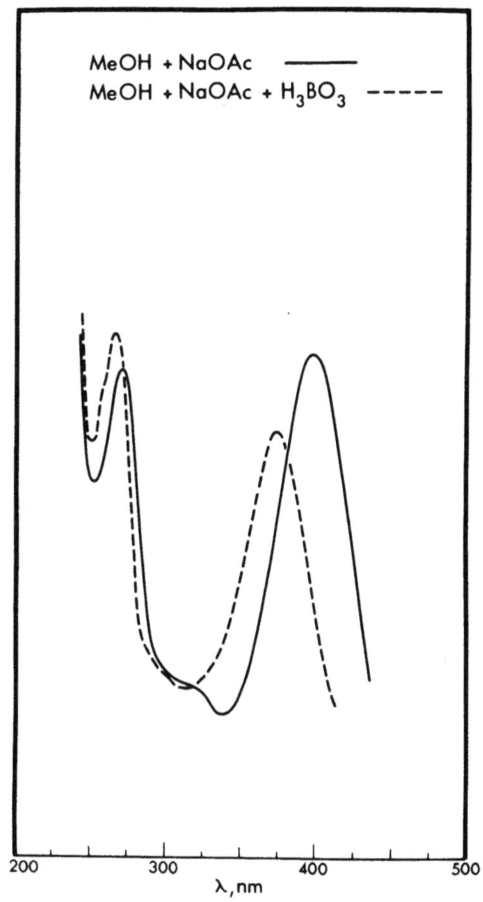

83

ROBINETIN

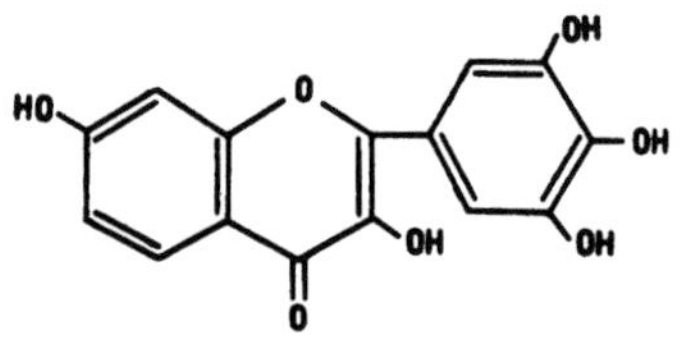

CHROMATOGRAPHIC DATA

Spot Appearance: (UV) fluorescent yellow
(UV/NH_3) fluorescent yellow

R_f Values: 0.25 (TBA), 0.03 (HOAc)

UV SPECTRAL DATA (λ_{max},*nm*)

MeOH	252, 266sh, 320, 367
NaOMe	264sh, 333, 475 (dec.)
$AlCl_3$	273, 281sh, 313, 447
$AlCl_3$/HCl	267, 275sh, 318, 426
NaOAc	257sh, 307sh, 346 (dec.)
NaOAc/H_3BO_3 (Proc. II)	256sh, 316, 385, 462sh

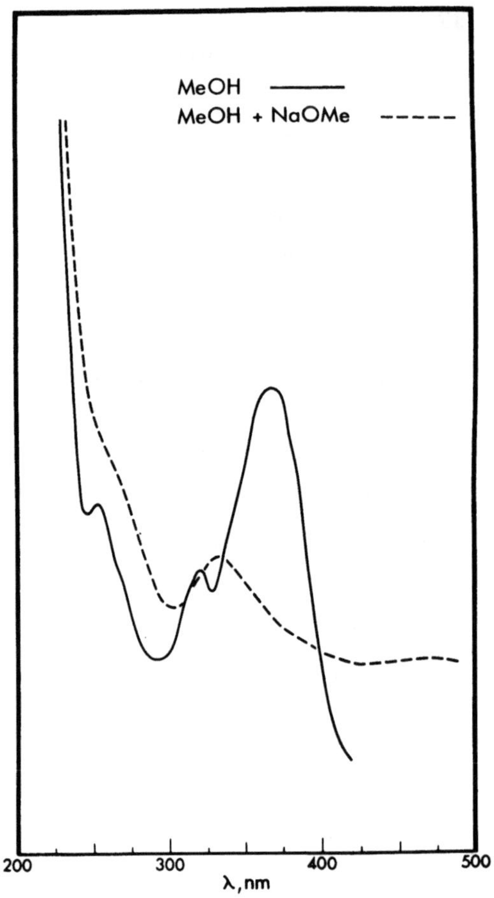

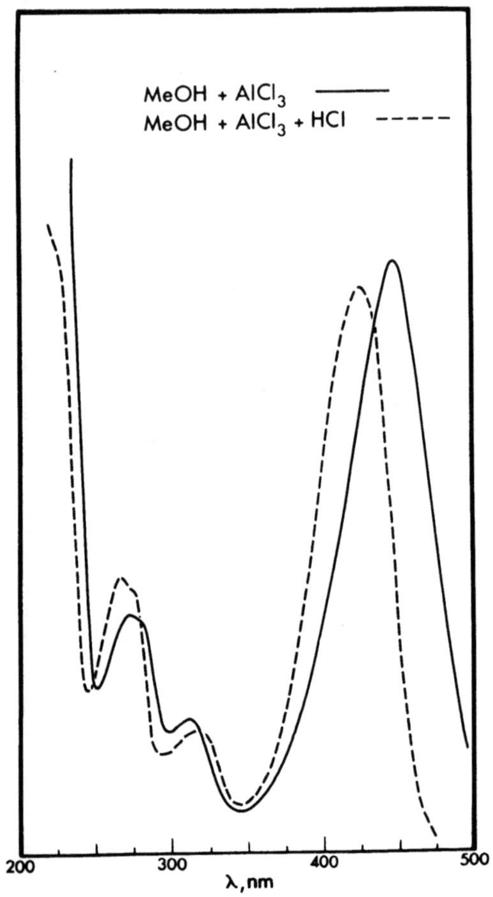

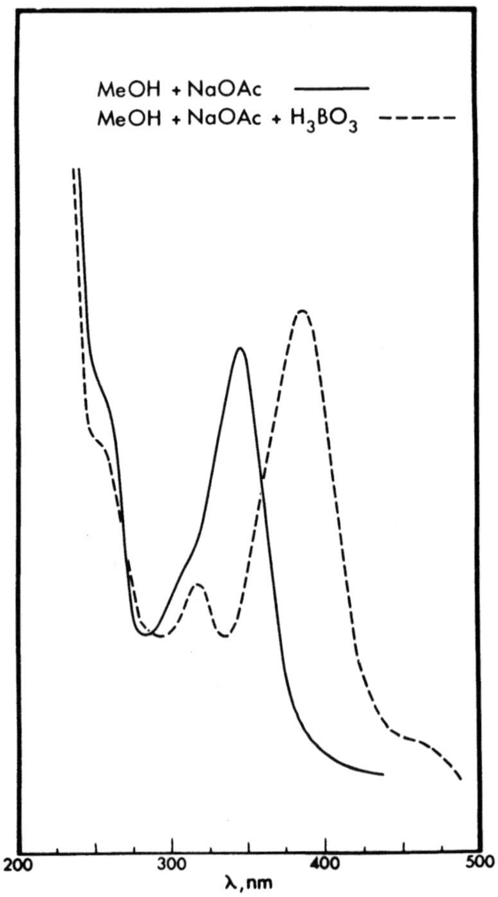

84

PENDULETIN

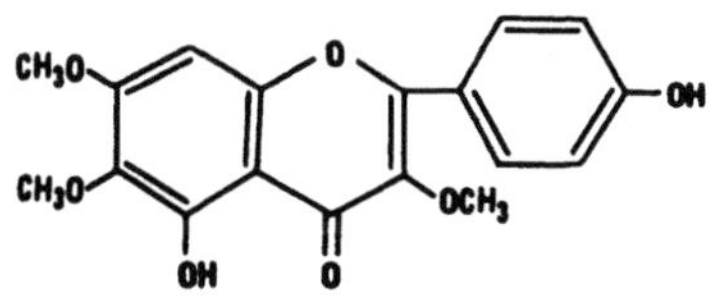

CHROMATOGRAPHIC DATA

Spot Appearance: (UV) deep purple
(UV/NH_3) greenish purple

R_f Values: 0.87 (TBA), 0.35 (HOAc)

UV SPECTRAL DATA (λ_{max},*nm*)

MeOH	271, 340
NaOMe	245sh, 274, 302sh, 350sh, 388
$AlCl_3$	268sh, 280, 302sh, 369, 396sh
$AlCl_3$/HCl	265sh, 283, 302sh, 359, 402sh
NaOAc	273, 294sh, 348, 396sh
NaOAc/H_3BO_3 (Proc. I)	271, 343

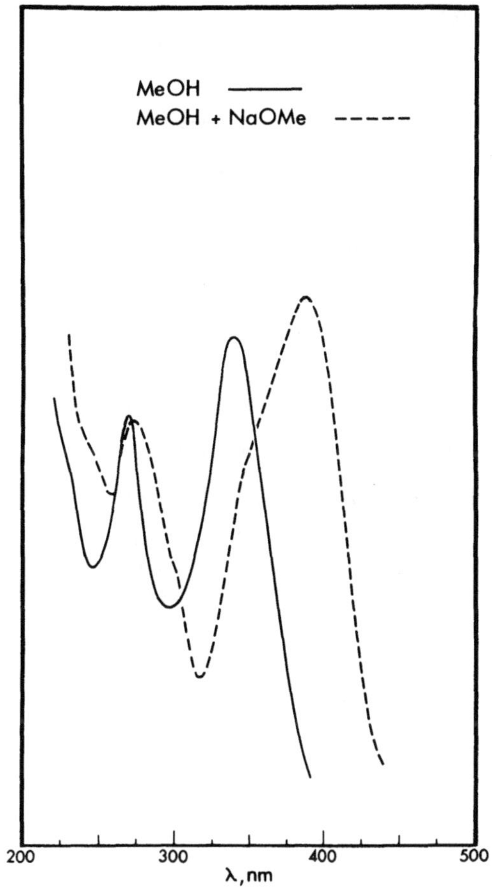

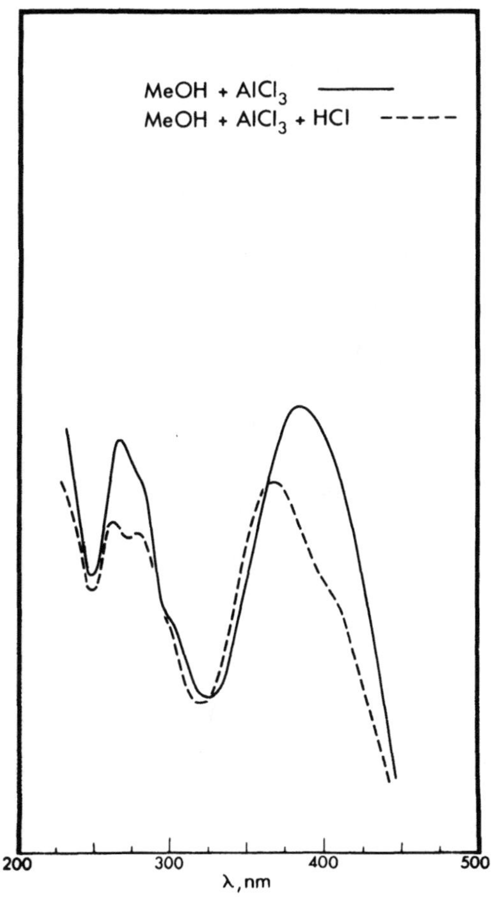

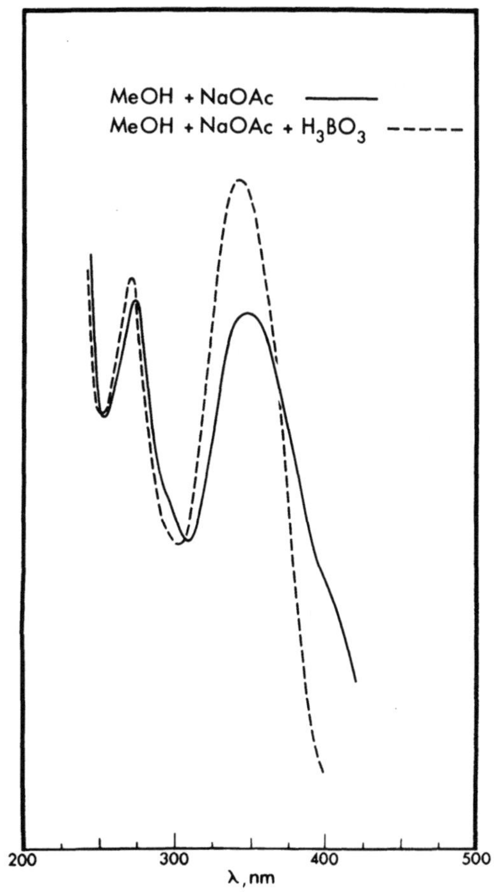

85

PENDULIN

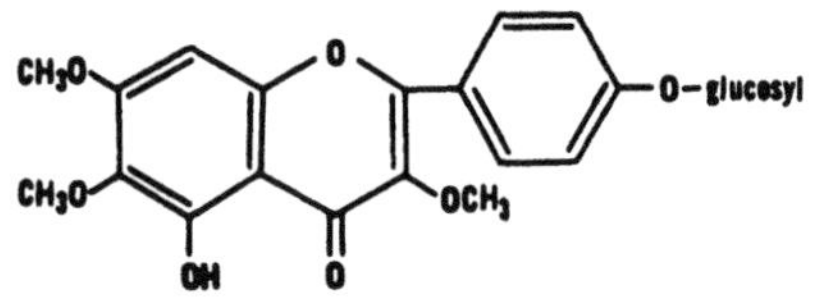

CHROMATOGRAPHIC DATA

Spot Appearance: (UV) deep purple
(UV/NH_3) deep purple

R_f Values: 0.61 (TBA), 0.64 (HOAc)

UV SPECTRAL DATA (λ_{max},nm)

MeOH	253sh, 273, 330
NaOMe	246sh, 290, 371sh
$AlCl_3$	262, 287, 303sh, 359, 403sh
$AlCl_3$/HCl	262, 288, 301sh, 356, 402sh
NaOAc	275, 328
NaOAc/H_3BO_3 (Proc. I)	273, 332

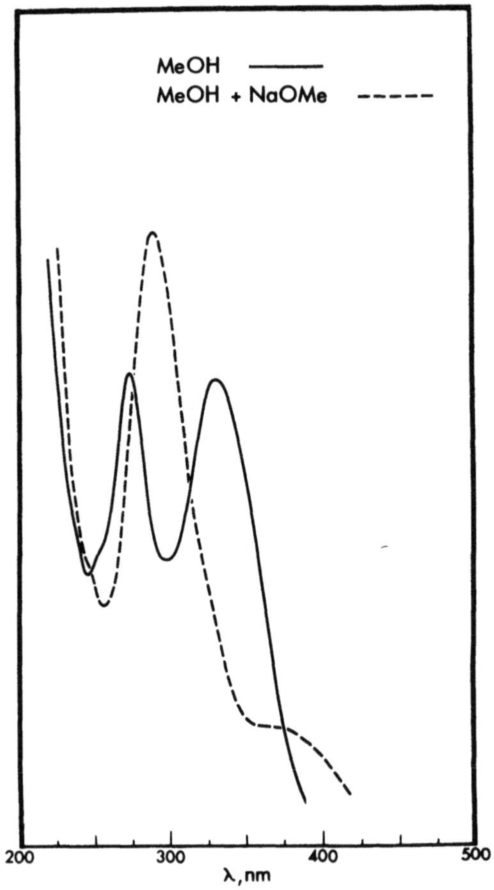

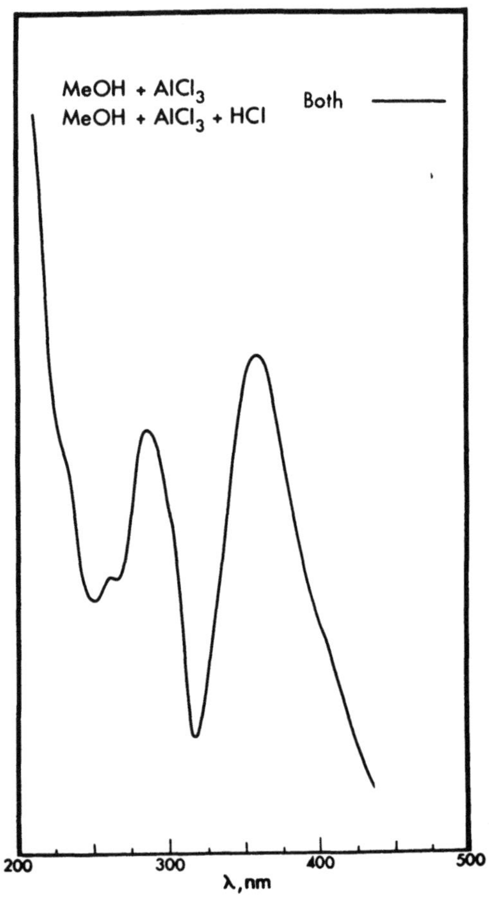

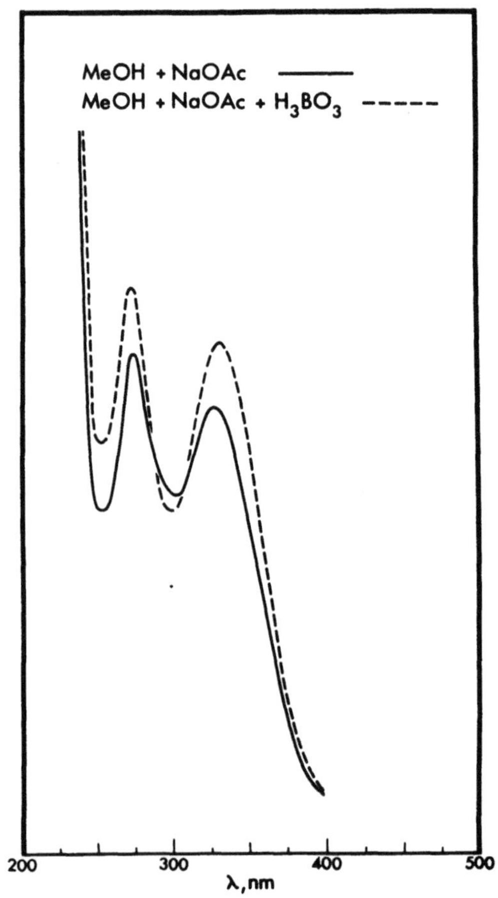

86

3,5,6,7,8-PENTAMETHOXYFLAVONE

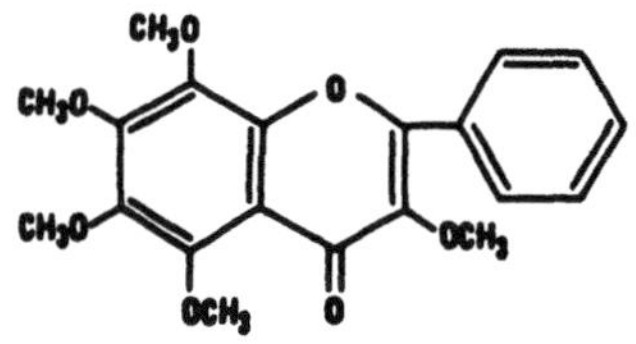

CHROMATOGRAPHIC DATA

Spot Appearance: (UV) deep purple
(UV/NH_3) deep purple

R_f Values: 0.91 (TBA), 0.44 (HOAc)

UV SPECTRAL DATA (λ_{max}, nm)

MeOH	268, 309, 338sh
NaOMe	268, 310, 335sh
$AlCl_3$	268, 309, 338sh
$AlCl_3$/HCl	268, 310, 340sh
NaOAc	268, 310, 334sh
NaOAc/H_3BO_3 (Proc. I)	268, 310, 335sh

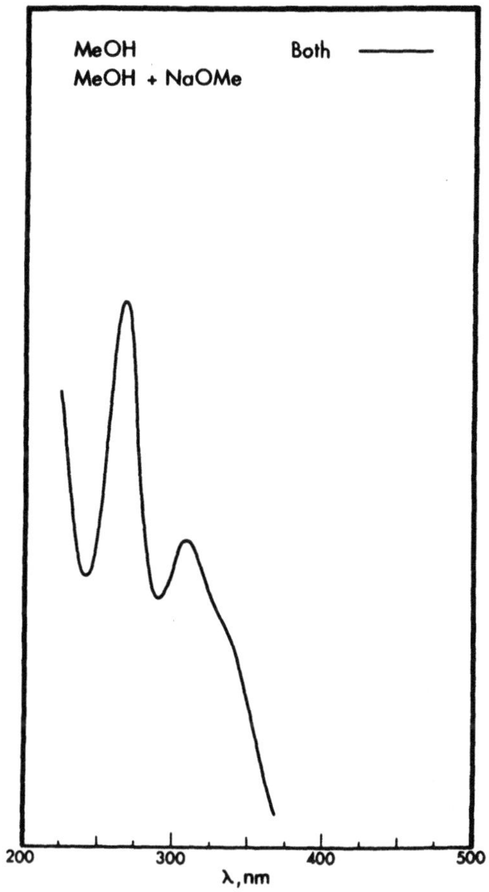

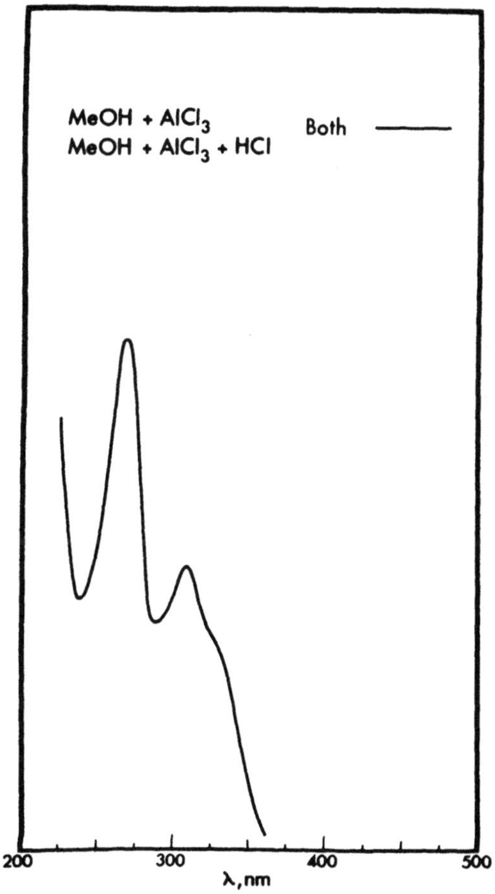

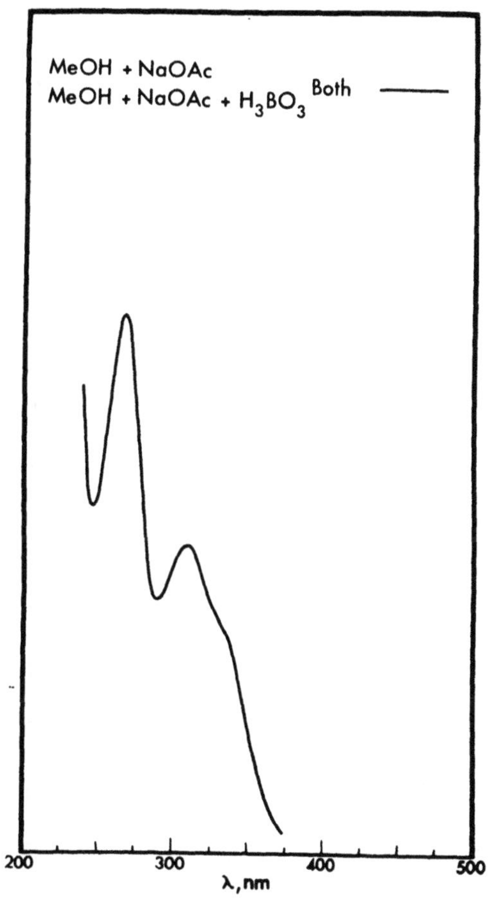

87

JACEIDIN

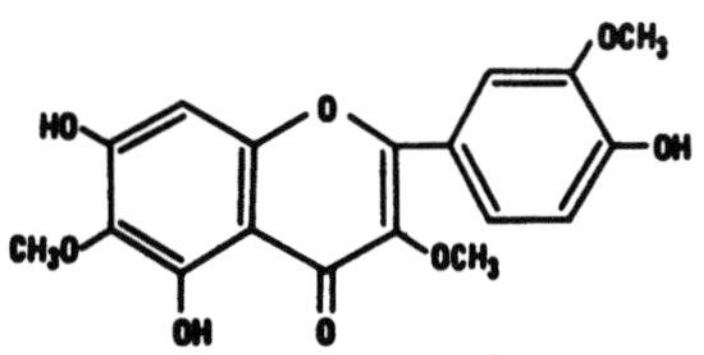

CHROMATOGRAPHIC DATA

Spot Appearance: (UV) deep purple
(UV/NH_3) greenish purple

R_f Values: 0.82 (TBA), 0.19 (HOAc)

UV SPECTRAL DATA (λ_{max},nm)

MeOH	256, 271, 351
NaOMe	272, 334, 412
$AlCl_3$	267, 281sh, 301sh, 384
$AlCl_3$/HCl	263, 279, 300sh, 368, 411sh
NaOAc	273, 322, 394
NaOAc/H_3BO_3 (Proc. I)	266, 288sh, 363

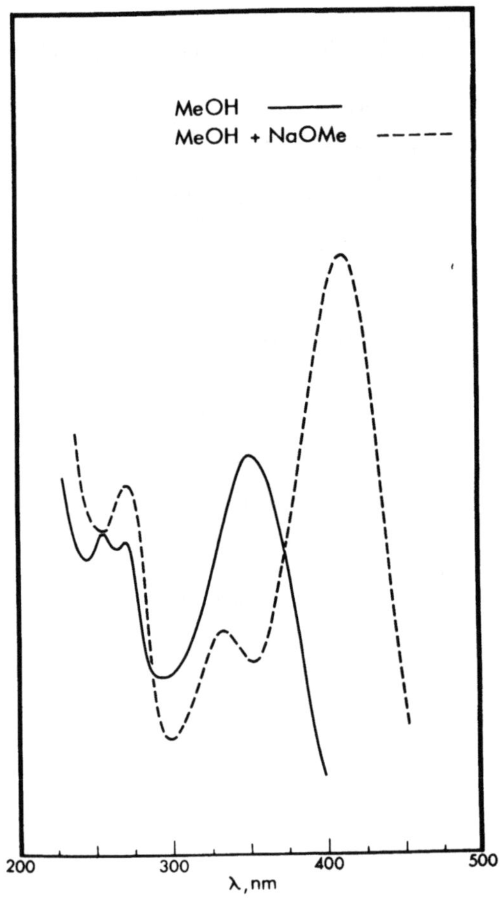

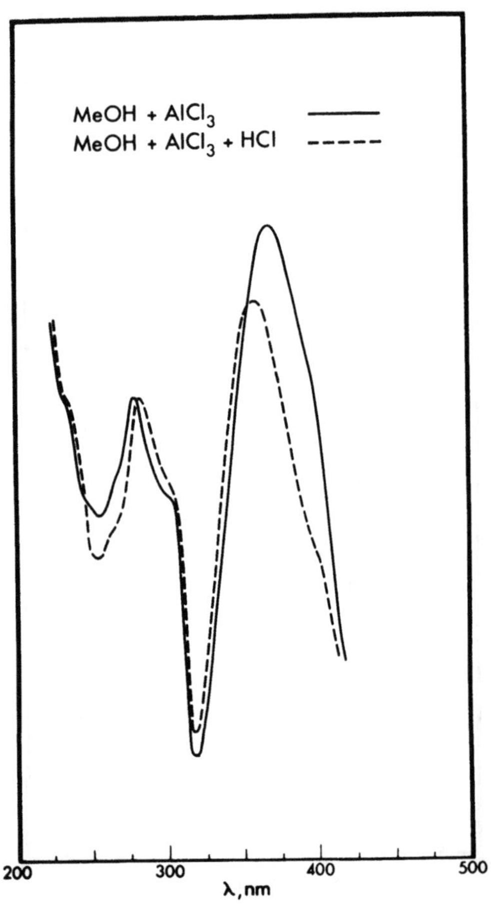

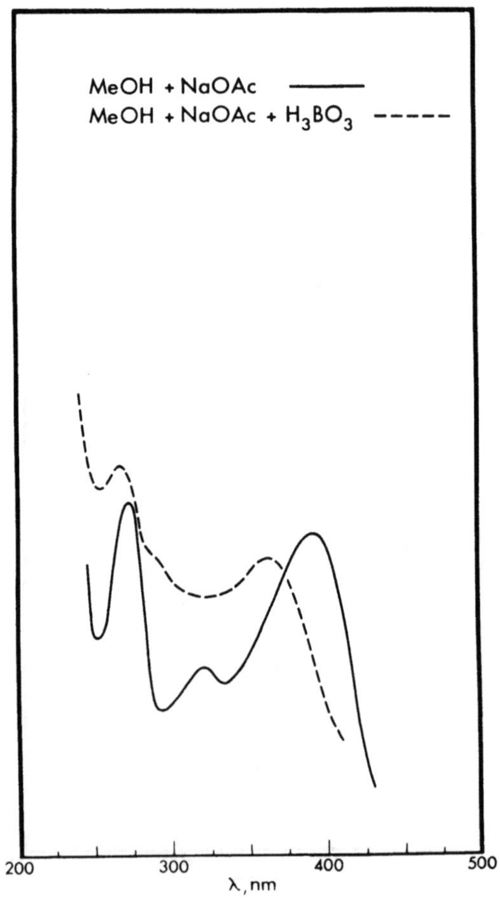

88

JACEIN

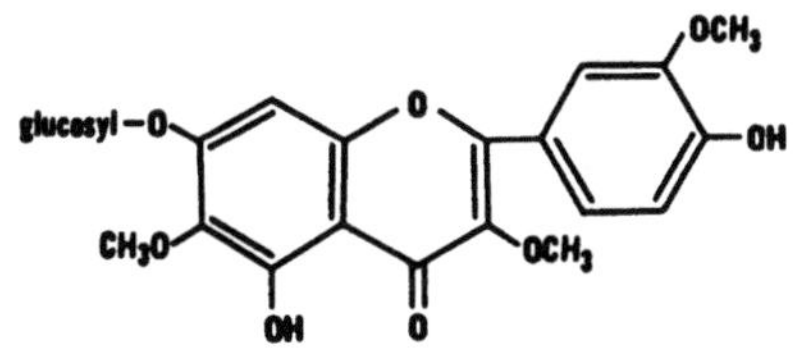

CHROMATOGRAPHIC DATA

Spot Appearance: (UV) deep purple
(UV/NH_3) yellow-green

R_f Values: 0.52 (TBA), 0.47 (HOAc)

UV SPECTRAL DATA (λ_{max},*nm*)

MeOH	257, 272sh, 352
NaOMe	248, 270, 401
$AlCl_3$	270, 280sh, 296sh, 387
$AlCl_3$/HCl	267, 280sh, 299sh, 369, 407sh
NaOAc	262, 373sh, 416
NaOAc/H_3BO_3 (Proc. I)	257, 271sh, 356

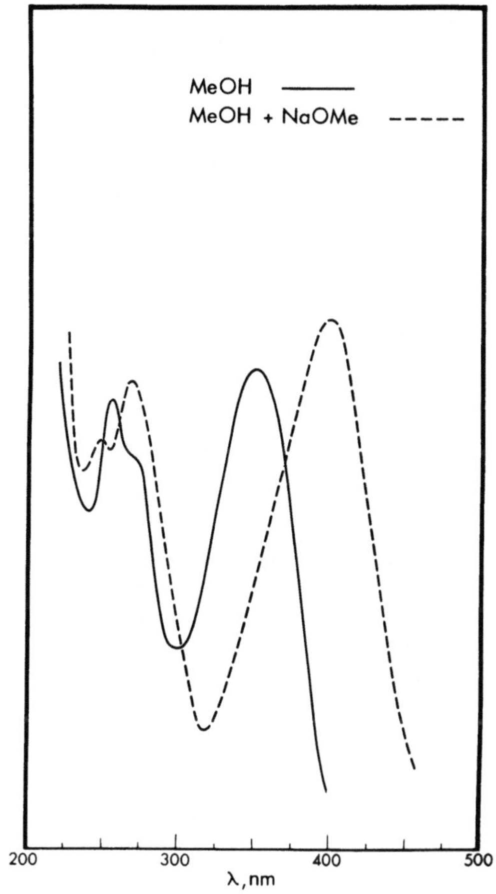

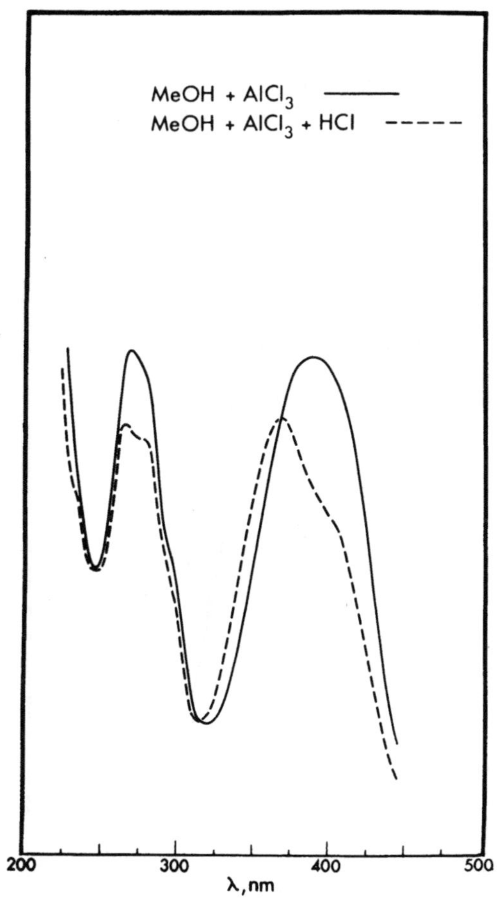

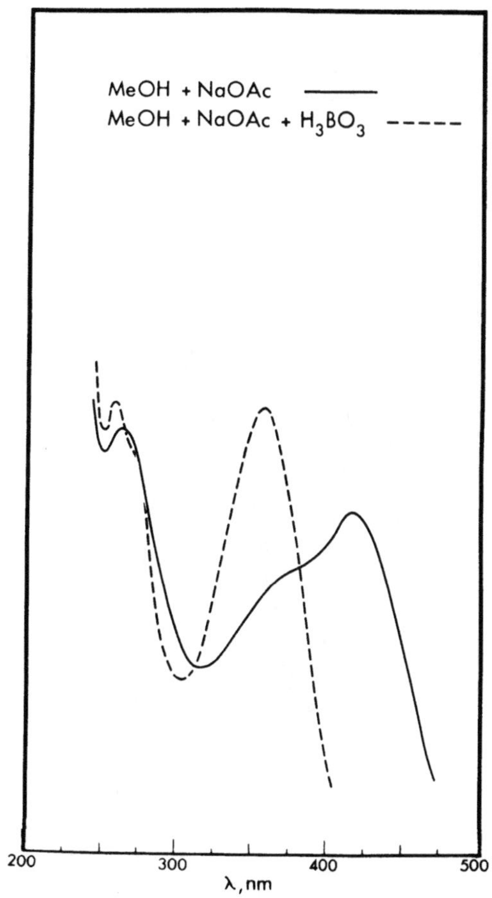

89

CENTAUREIN

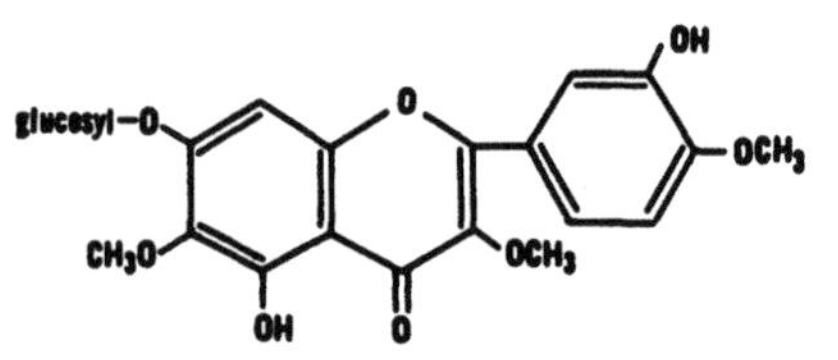

CHROMATOGRAPHIC DATA

Spot Appearance: (UV) deep purple
(UV/NH_3) deep purple

R_f Values: 0.49 (TBA), 0.54 (HOAc)

UV SPECTRAL DATA (λ_{max},*nm*)

MeOH	256, 270sh, 350
NaOMe	273, 381
$AlCl_3$	271, 281sh, 297sh, 381
$AlCl_3$/HCl	268, 281, 299sh, 368, 404sh
NaOAc	257, 272, 348
NaOAc/H_3BO_3 (Proc. I)	257, 271, 353

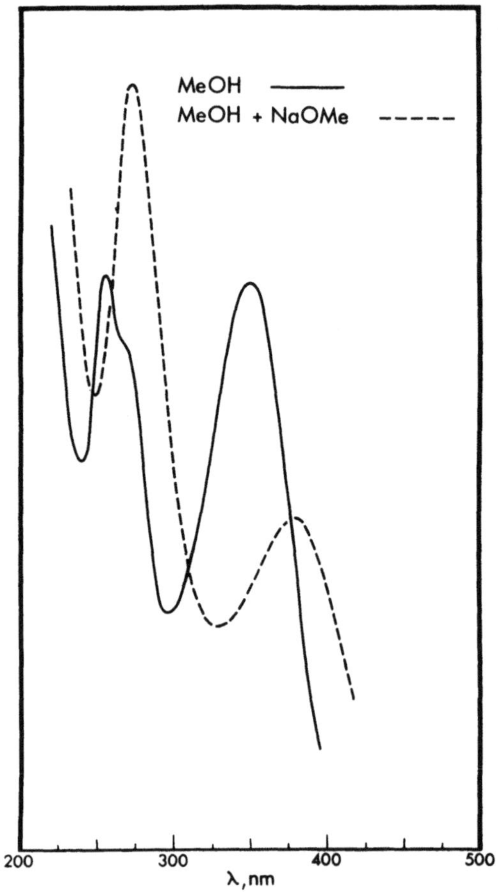

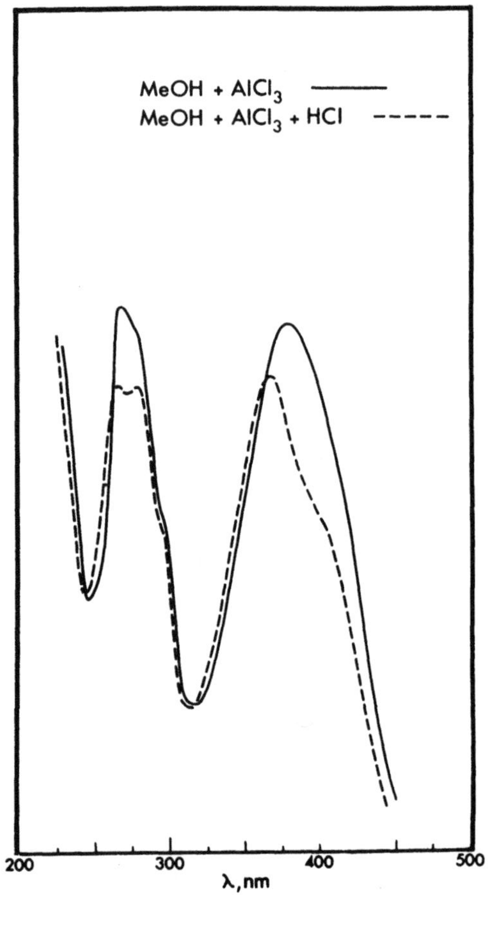

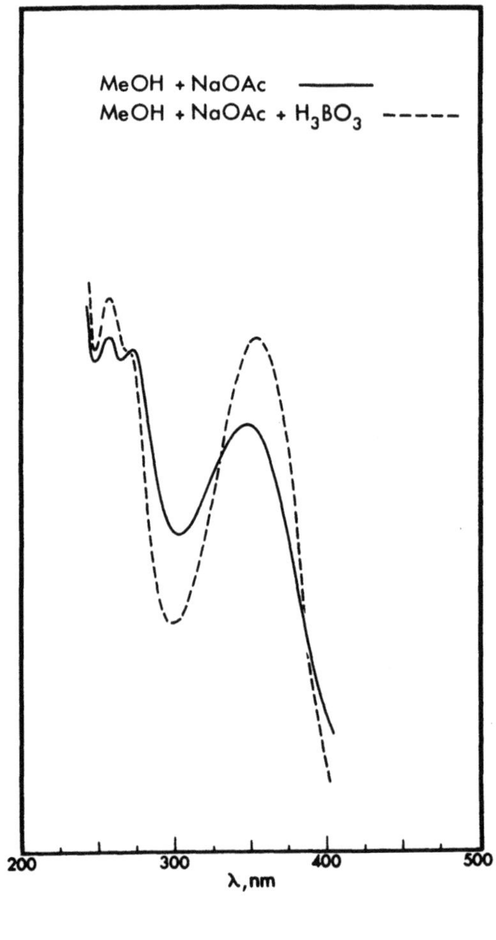

PATULETIN

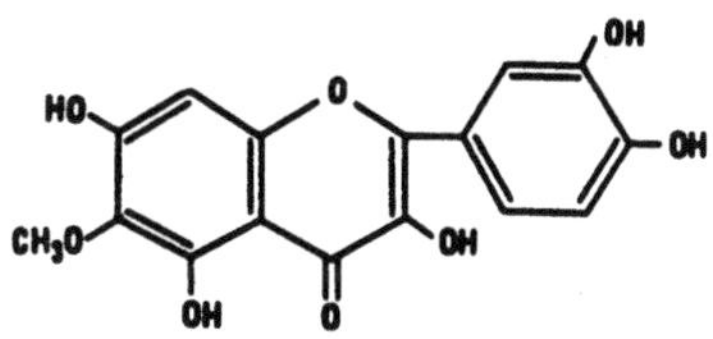

CHROMATOGRAPHIC DATA

Spot Appearance: (UV) yellow
(UV/NH_3) yellow

R_f Value: 0.54 (TBA), 0.04 (HOAc)

UV SPECTRAL DATA (λ_{max}, nm)

MeOH	258, 272sh, 293sh, 371
NaOMe	251sh, 296sh, 336, 411sh (dec.)
$AlCl_3$	238, 275, 308sh, 327sh, 459
$AlCl_3$/HCl	240, 268, 302sh, 381sh, 427
NaOAc	258sh, 274sh, 340, 394sh (dec.)
NaOAc/H_3BO_3 (Proc. II)	264, 393

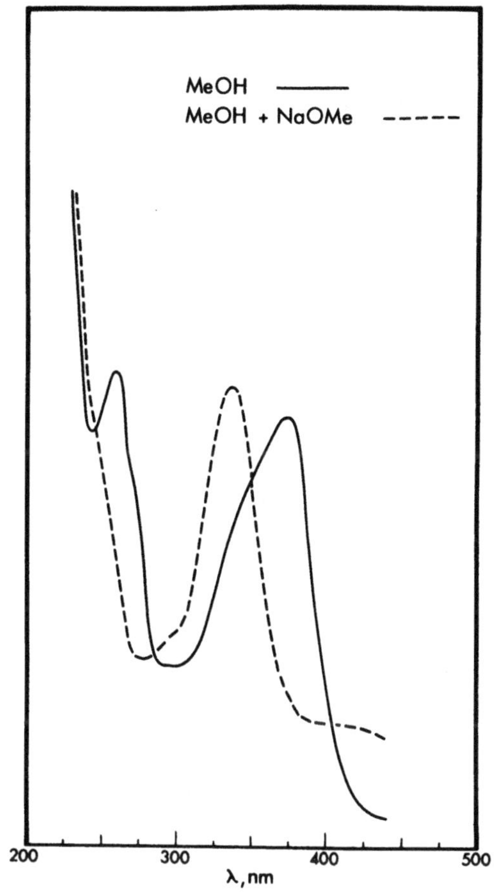

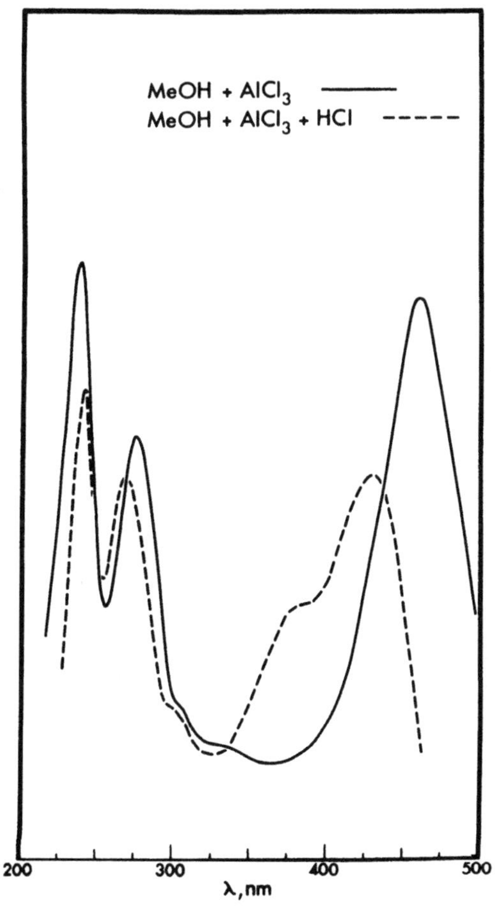

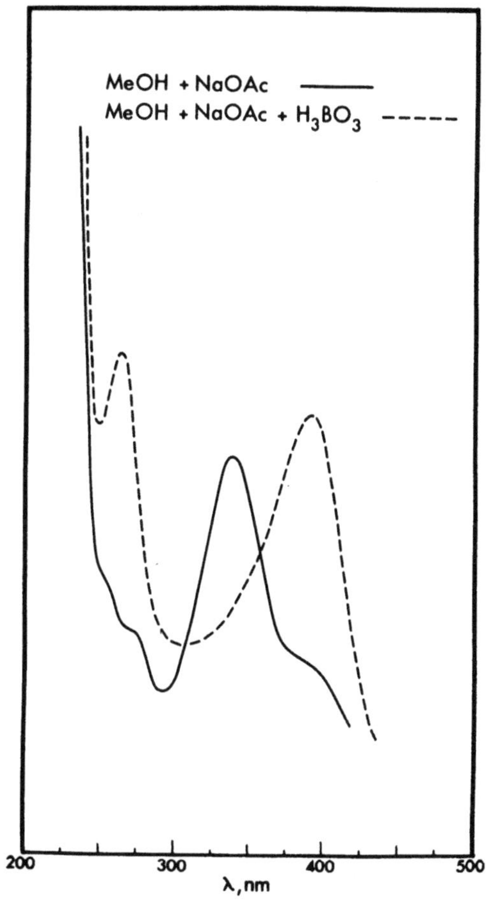

91

PATULETIN 3-0-GLUCOSIDE

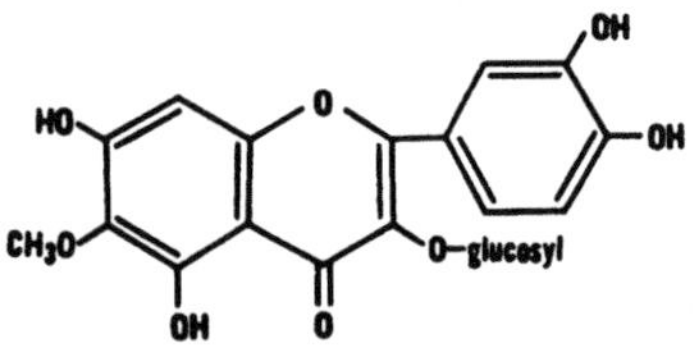

CHROMATOGRAPHIC DATA

Spot Appearance: (UV) deep purple
(UV/NH_3) green

R_f Values: 0.54 (TBA), 0.50 (HOAc)

UV SPECTRAL DATA (λ_{max},*nm*)

MeOH	261, 270sh, 355
NaOMe	273, 338, 410
$AlCl_3$	278, 302sh, 339sh, 438
$AlCl_3$/HCl	269, 281sh, 301sh, 373, 407sh
NaOAc	273, 326sh, 385
NaOAc/H_3BO_3 (Proc. I)	265, 382

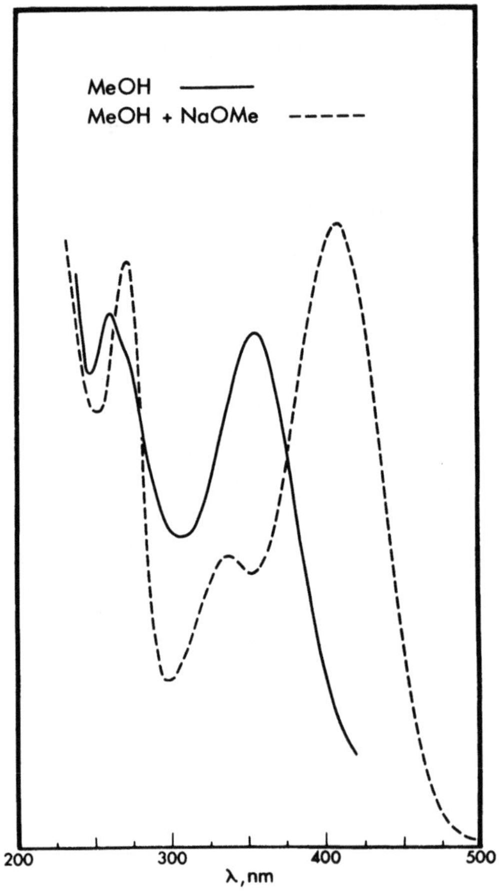

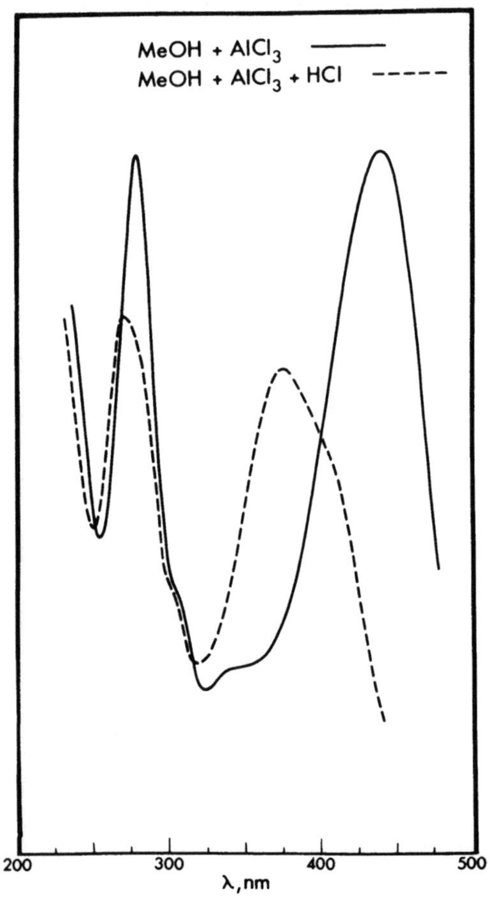

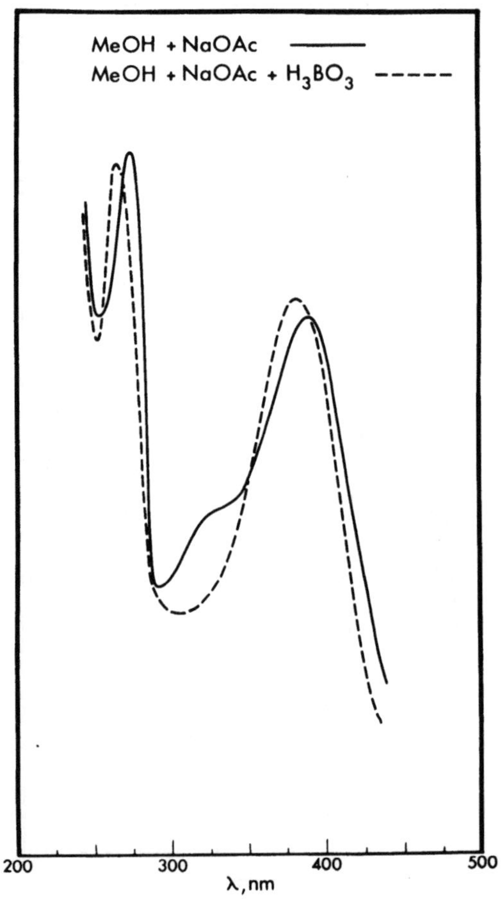

92

PATULETIN 3-0-RUTINOSIDE

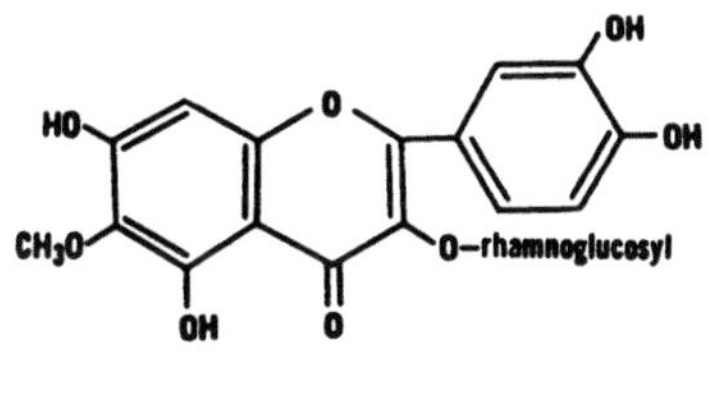

CHROMATOGRAPHIC DATA

Spot Appearance: (UV) deep purple
(UV/NH_3) dark green

R_f Values: 0.31 (TBA), 0.61 (HOAc)

UV SPECTRAL DATA (λ_{max},nm)

MeOH	259, 269sh, 356
NaOMe	273, 337, 411
$AlCl_3$	278, 310sh, 341sh, 435
$AlCl_3$/HCl	269, 279sh, 301sh, 375, 404sh
NaOAc	272, 328sh, 392
NaOAc/H_3BO_3 (Proc. I)	265, 381

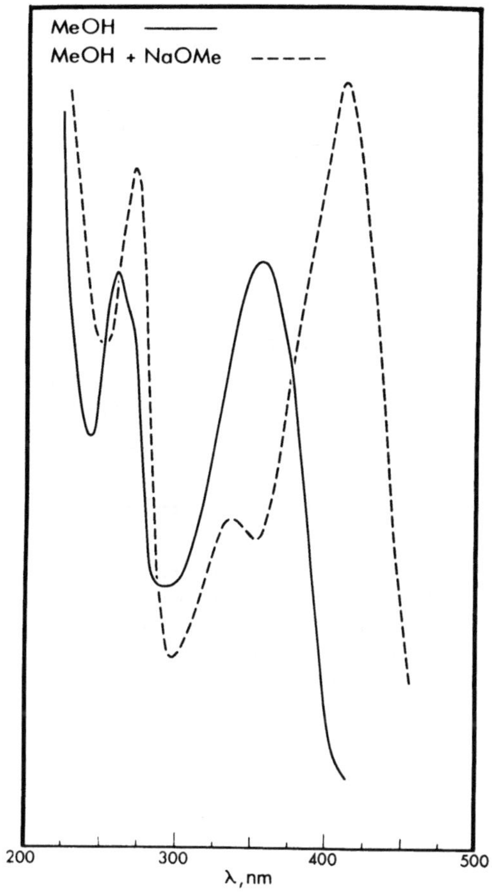

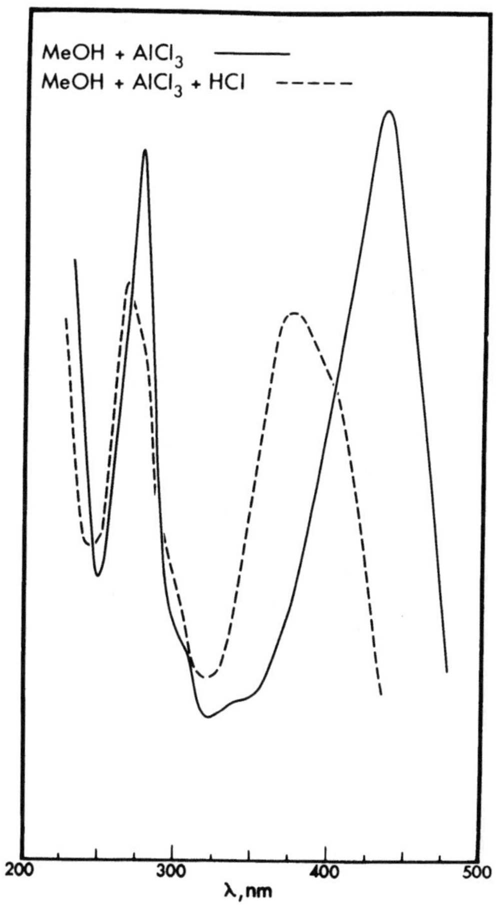

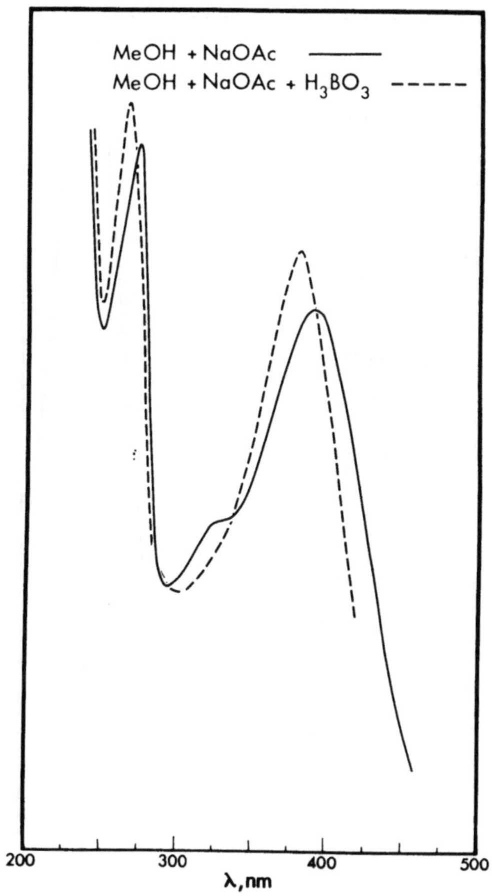

93

PATULITRIN

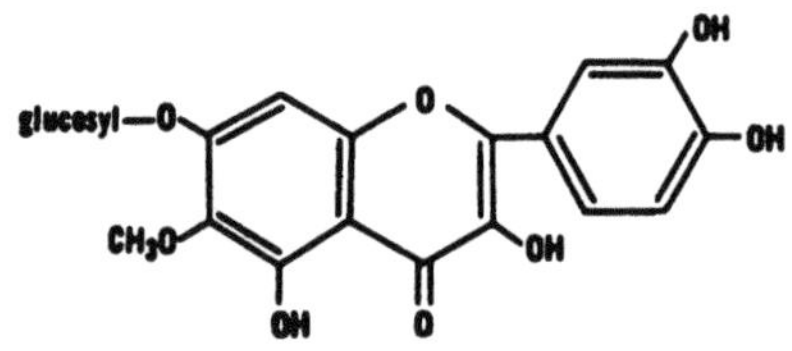

CHROMATOGRAPHIC DATA

Spot Appearance: (UV) dull yellow
(UV/NH_3) dull yellow

R_f Values: 0.25 (TBA), 0.15 (HOAc)

UV SPECTRAL DATA (λ_{max}, nm)

MeOH	259, 273sh, 338sh, 373
NaOMe	242, 292, 382, 467 (dec.)
$AlCl_3$	276, 349, 462
$AlCl_3$/HCl	269, 302sh, 380sh, 431
NaOAc	258, 343, 397, 417sh (dec.)
NaOAc/H_3BO_3 (Proc. II)	265, 394

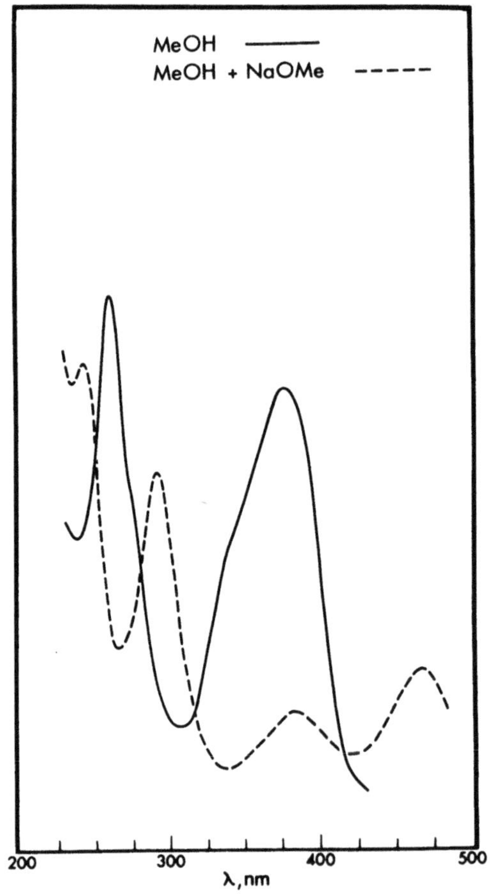

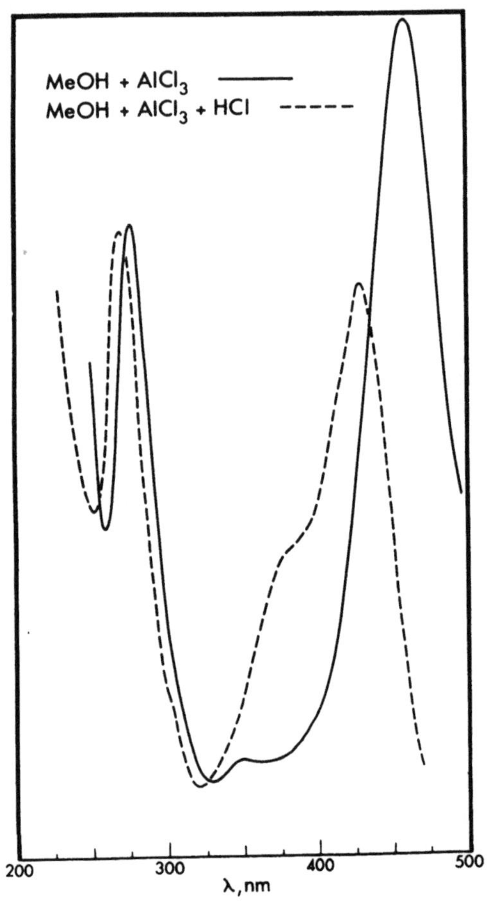

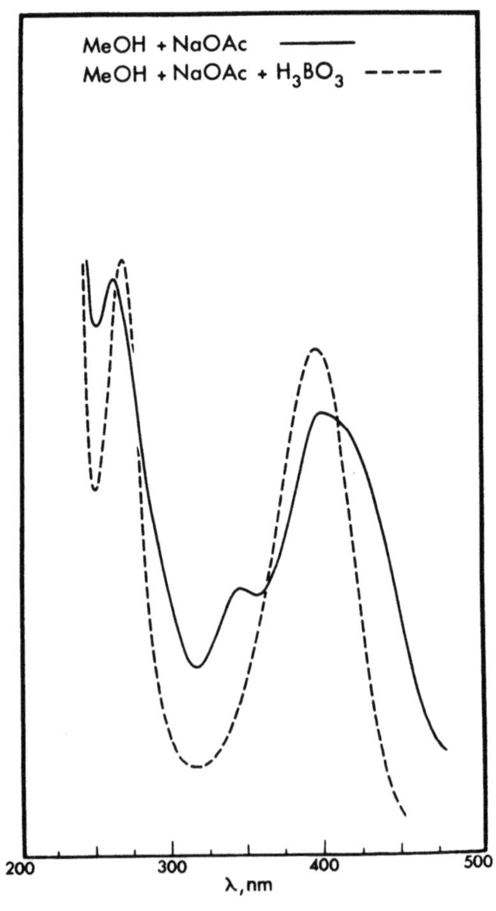

ARTEMETIN

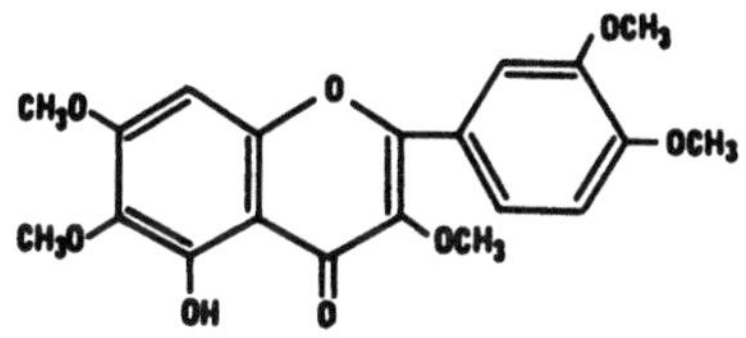

CHROMATOGRAPHIC DATA

Spot Appearance: (UV) deep purple
(UV/NH_3) deep purple

R_f Values: 0.86 (TBA), 0.34 (HOAc)

UV SPECTRAL DATA (λ_{max}, nm)

MeOH	254, 273, 345
NaOMe	250sh, 289, 325sh, 384sh
$AlCl_3$	266, 280sh, 299sh, 377
$AlCl_3$/HCl	264, 284, 366, 403sh
NaOAc	253sh, 274, 341
NaOAc/H_3BO_3 (Proc. I)	254, 272, 346

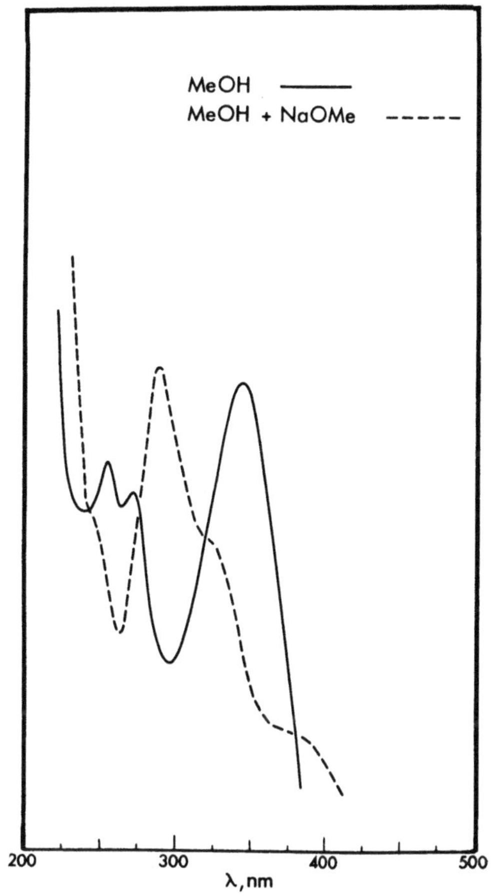

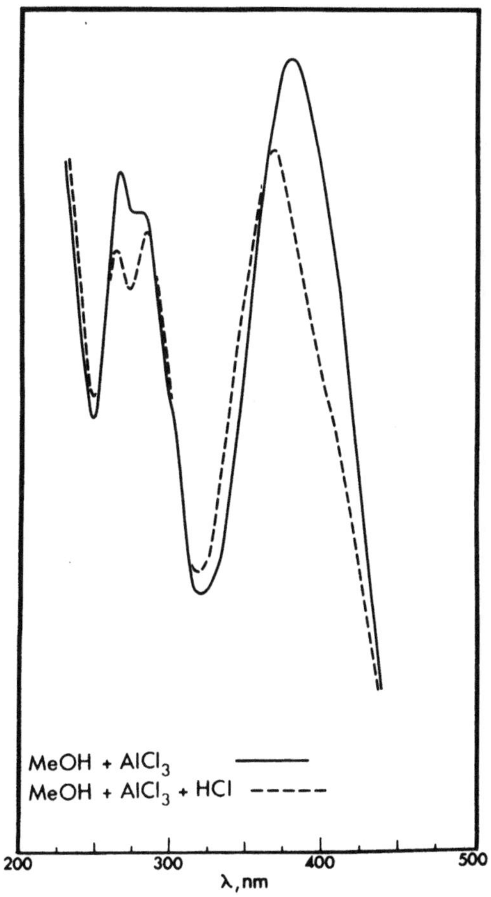

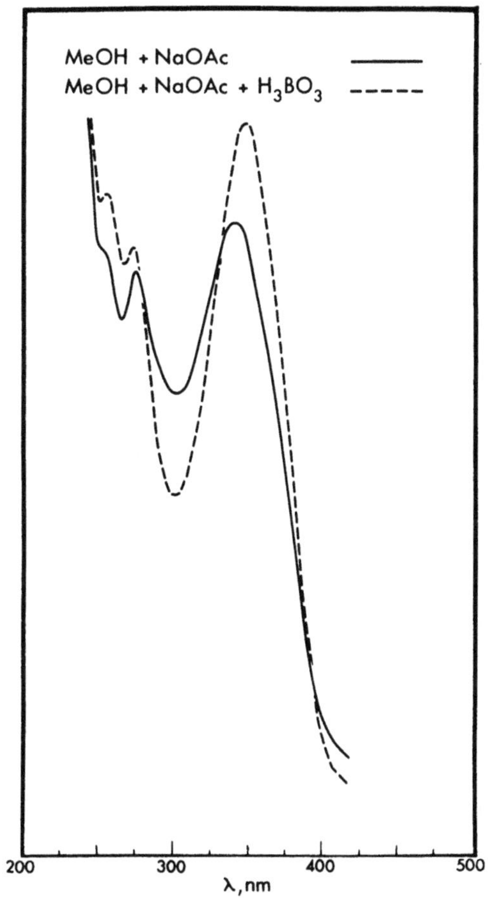

95

QUERCETAGETIN
3′,4′,5,6,7-PENTAMETHYL ETHER

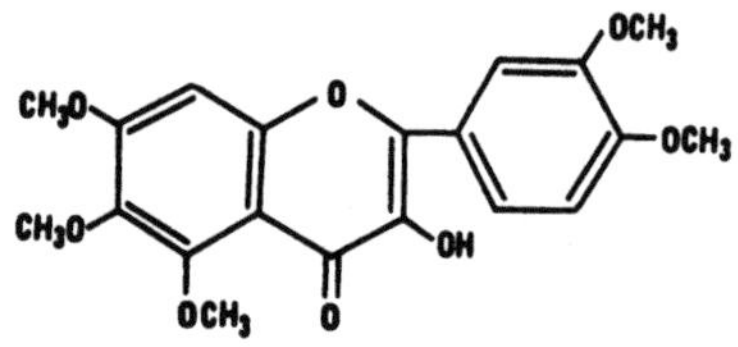

CHROMATOGRAPHIC DATA

Spot Appearance: (UV) fluorescent yellow
(UV/NH_3) fluorescent yellow

R_f Values: 0.78 (TBA), 0.14 (HOAc)

UV SPECTRAL DATA (λ_{max}, nm)

MeOH	254, 354
NaOMe	268, 327, 403
$AlCl_3$	268, 354sh, 421
$AlCl_3$/HCl	268, 354sh, 421
NaOAc	254sh, 361, 420sh
NaOAc/H_3BO_3	252sh, 360, 416sh
(Proc. I)	

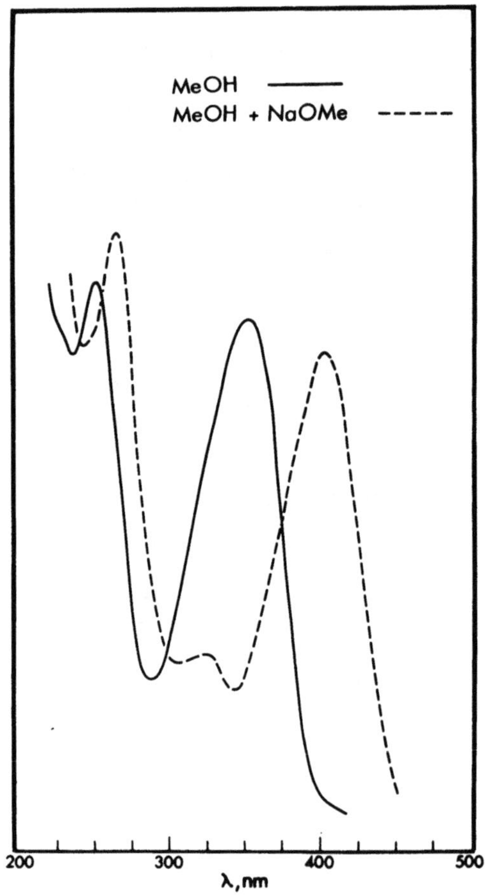

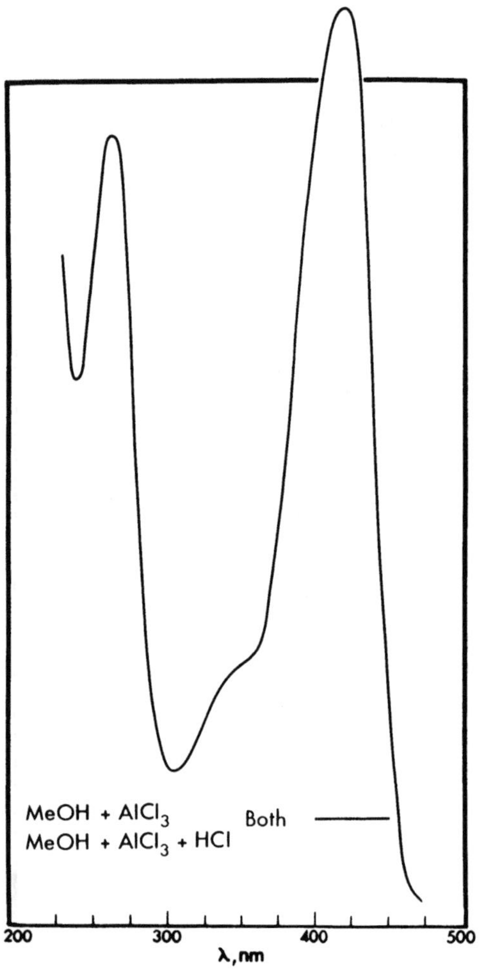

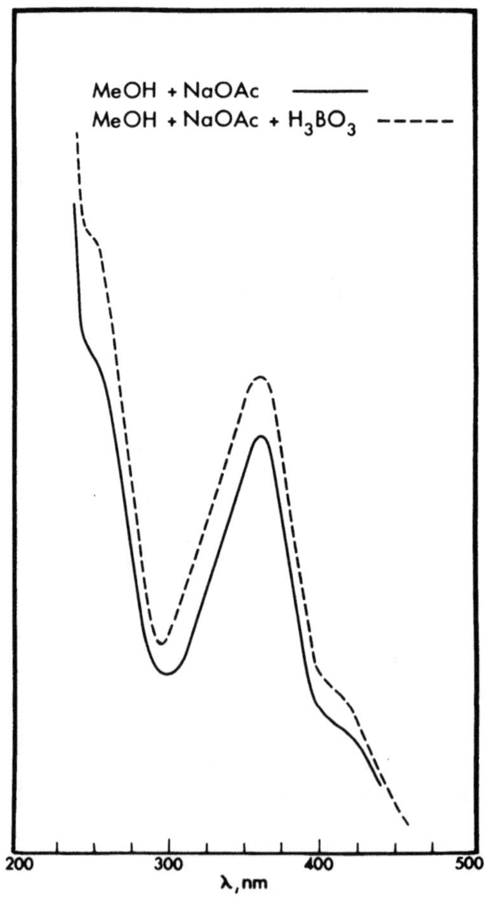

QUERCETAGETIN 3,3′,4′,5,6-PENTAMETHYL ETHER

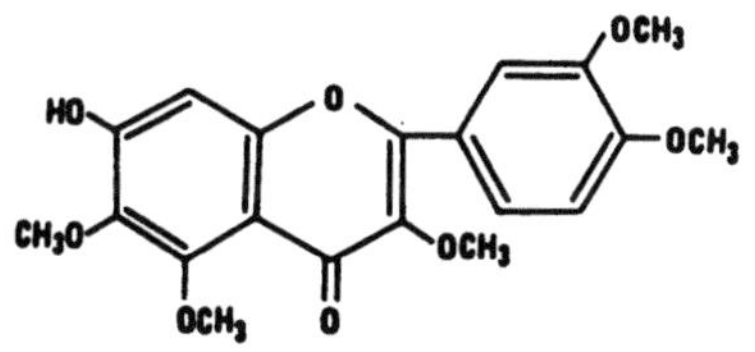

CHROMATOGRAPHIC DATA

Spot Appearance: (UV) fluorescent light blue

(UV/NH_3) fluorescent yellow-green

R_f Values: 0.82 (TBA), 0.34 (HOAc)

UV SPECTRAL DATA (λ_{max}, nm)

MeOH	251sh, 265sh, 338
NaOMe	269, 317, 363
$AlCl_3$	245sh, 267sh, 335
$AlCl_3$/HCl	245sh, 265sh, 336
NaOAc	269, 317, 365
NaOAc/H_3BO_3 (Proc. I)	265sh, 339

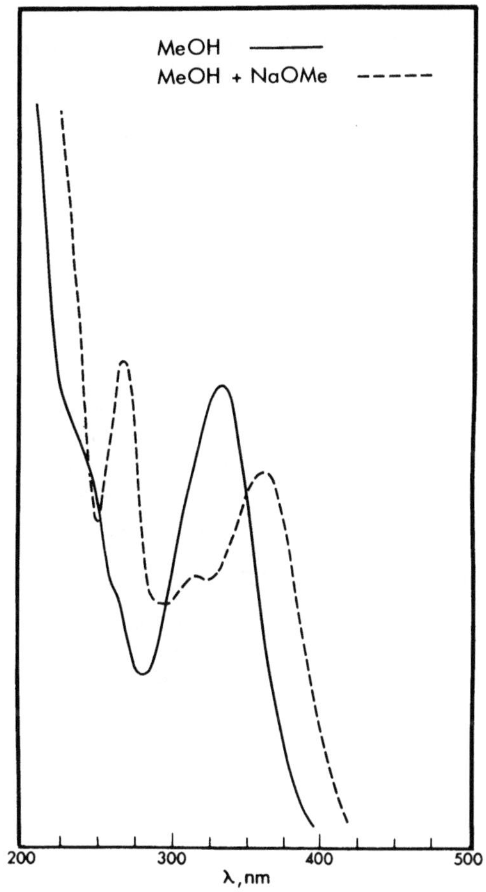

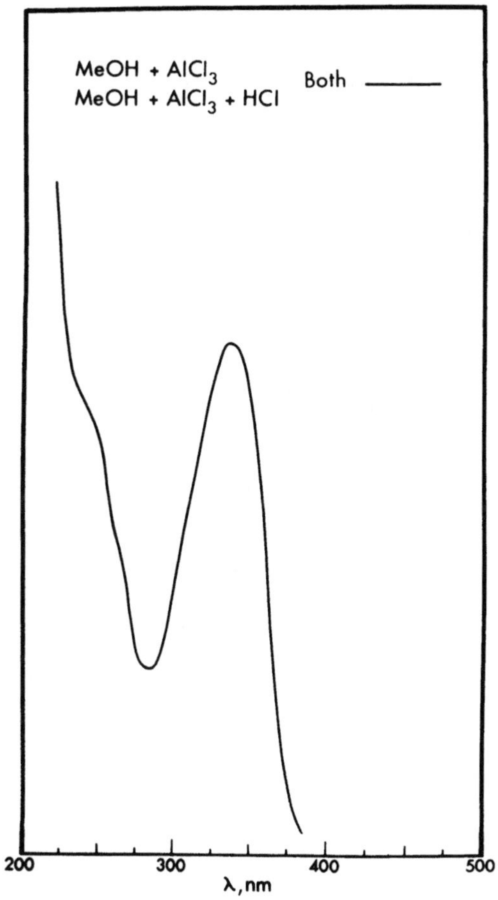

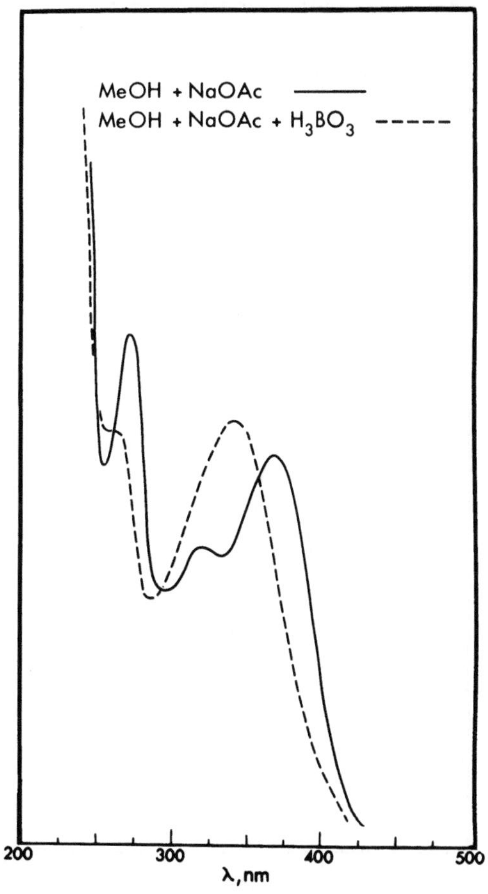

97

QUERCETAGETIN HEXAMETHYL ETHER

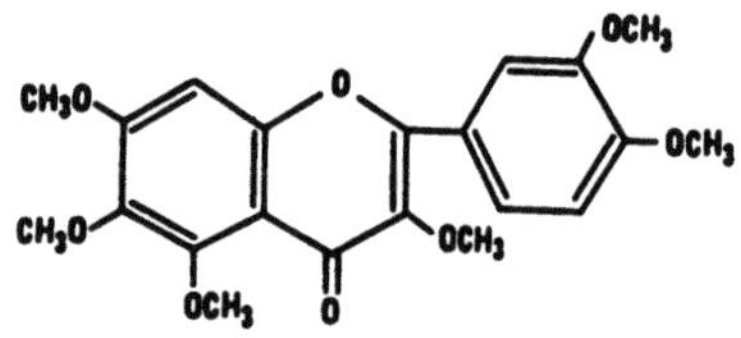

CHROMATOGRAPHIC DATA

Spot Appearance: (UV) fluorescent light blue

(UV/NH_3) fluorescent light blue

R_f Values: 0.86 (TBA), 0.44 (HOAc)

UV SPECTRAL DATA (λ_{max},*nm*)

MeOH	242, 252sh, 266sh, 333
NaOMe	254sh, 267sh, 334
$AlCl_3$	251sh, 264sh, 331
$AlCl_3$/HCl	255sh, 267sh, 329
NaOAc	264sh, 333
NaOAc/H_3BO_3 (Proc. I)	267sh, 331

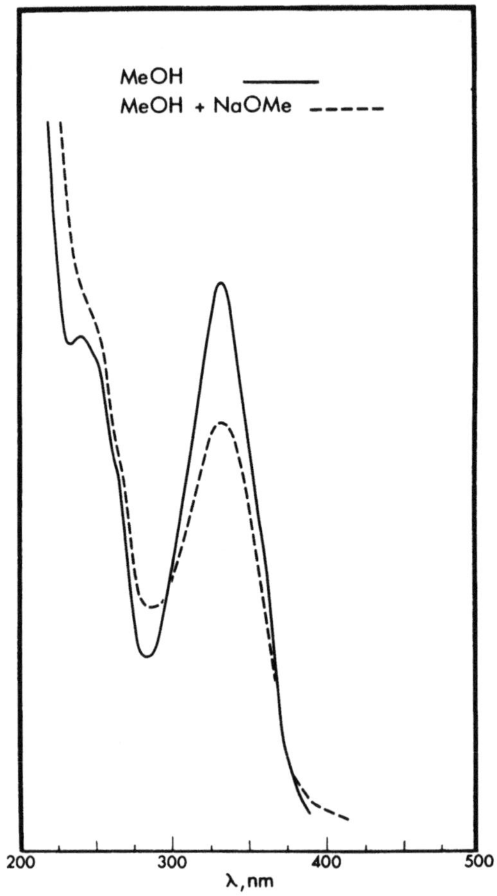

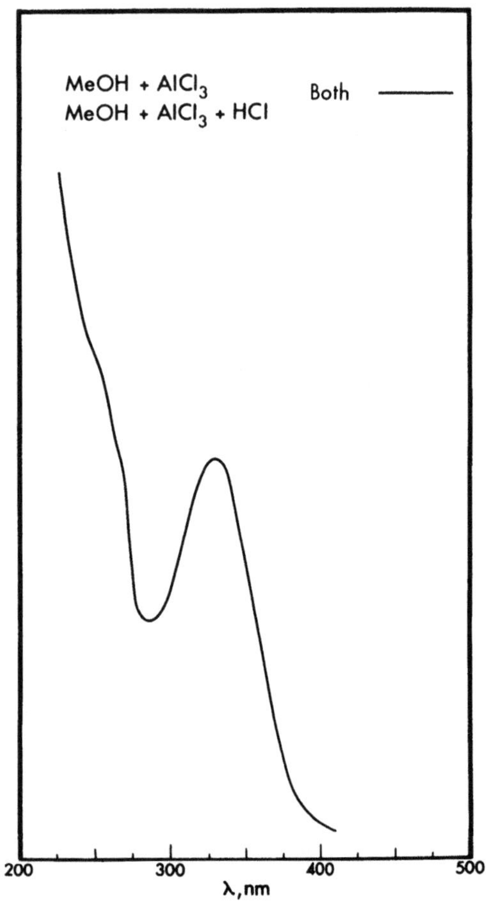

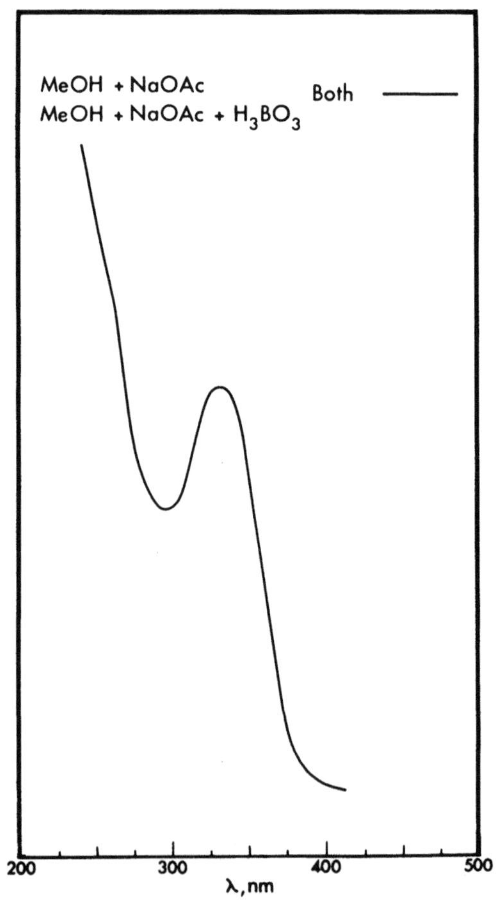

GOSSYPETIN

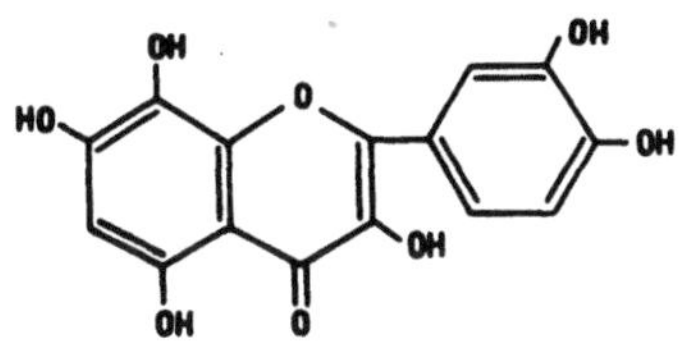

CHROMATOGRAPHIC DATA

Spot Appearance: (UV) yellow
(UV/NH_3) yellow

R_f Values: 0.28 (TBA), 0.04 (HOAc)

UV SPECTRAL DATA (λ_{max}, nm)

MeOH	261, 276, 309, 339, 385
NaOMe	251, 287, 366 (dec.)
$AlCl_3$	290, 327, 401, 492
$AlCl_3$/HCl	274, 292sh, 313, 372, 447
NaOAc	282, 366 (dec.)
NaOAc/H_3BO_3 (Proc. II)	273, 282sh, 314sh, 358, 406

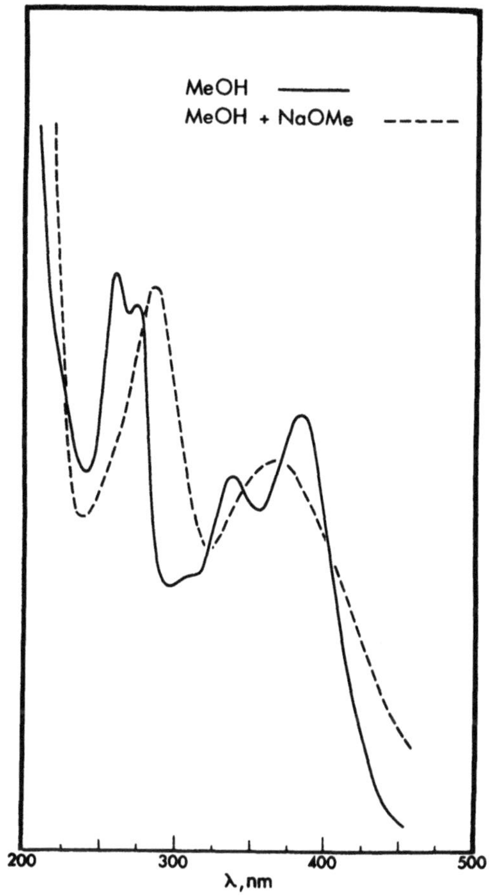

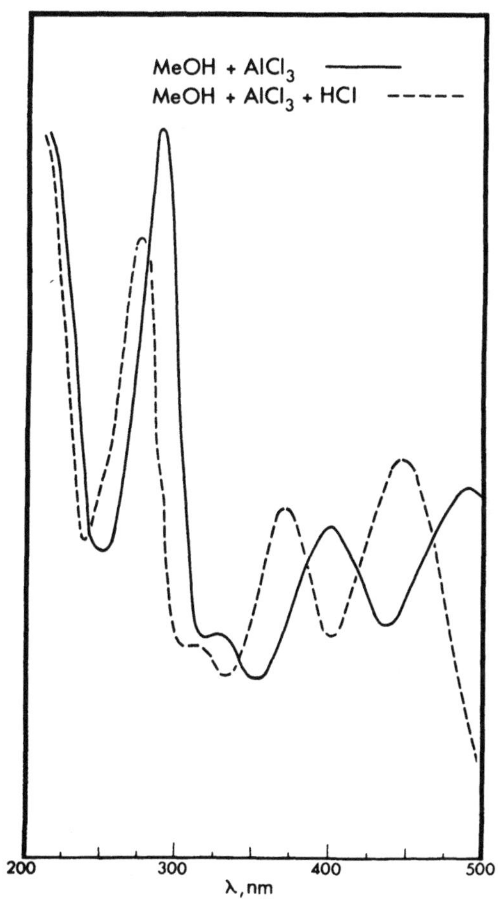

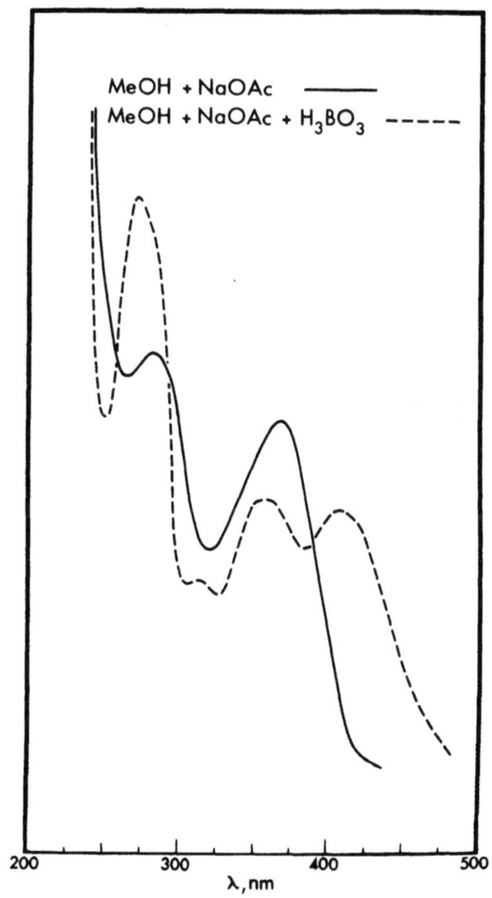

GOSSYPIN

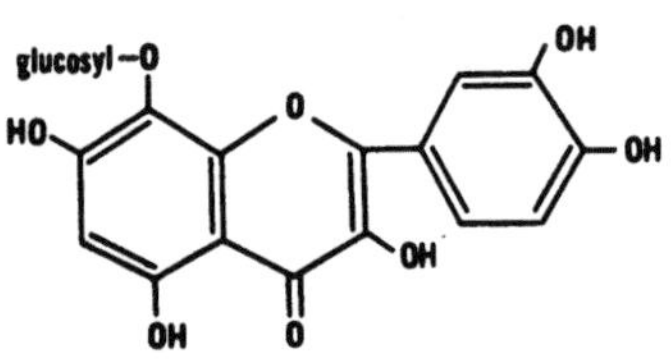

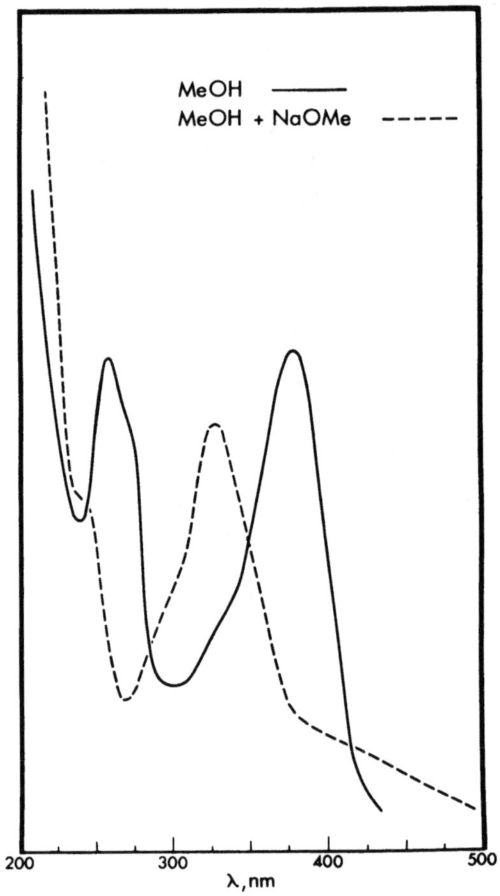

CHROMATOGRAPHIC DATA

Spot Appearance: (UV) yellow
(UV/NH_3) deep yellow

R_f Values: 0.22 (TBA), 0.09 (HOAc)

UV SPECTRAL DATA (λ_{max},nm)

MeOH	260, 273sh, 328sh, 380
NaOMe	245sh, 295sh, 331, 430sh (dec.)
$AlCl_3$	260sh, 275, 309sh, 364sh, 452
$AlCl_3$/HCl	269, 307sh, 367, 441
NaOAc	281, 328, 400 (dec.)
NaOAc/H_3BO_3 (Proc. II)	267, 277sh, 325, 400

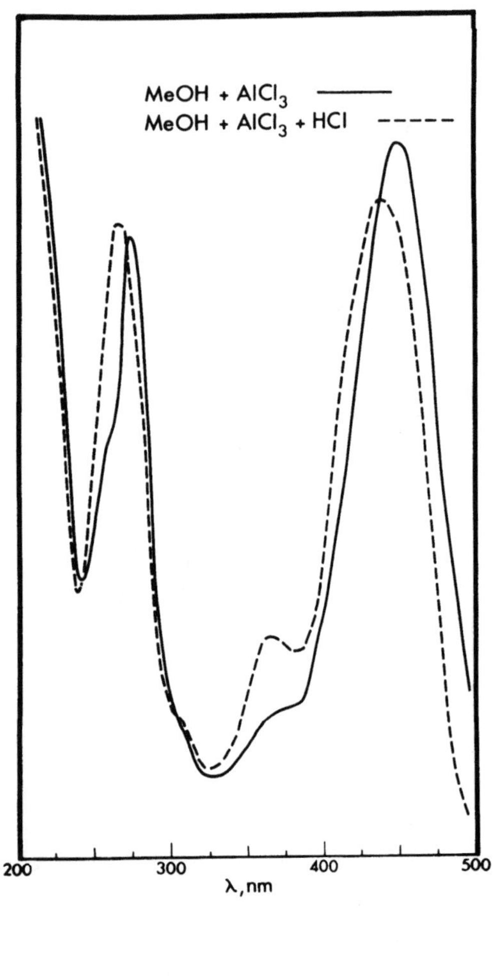

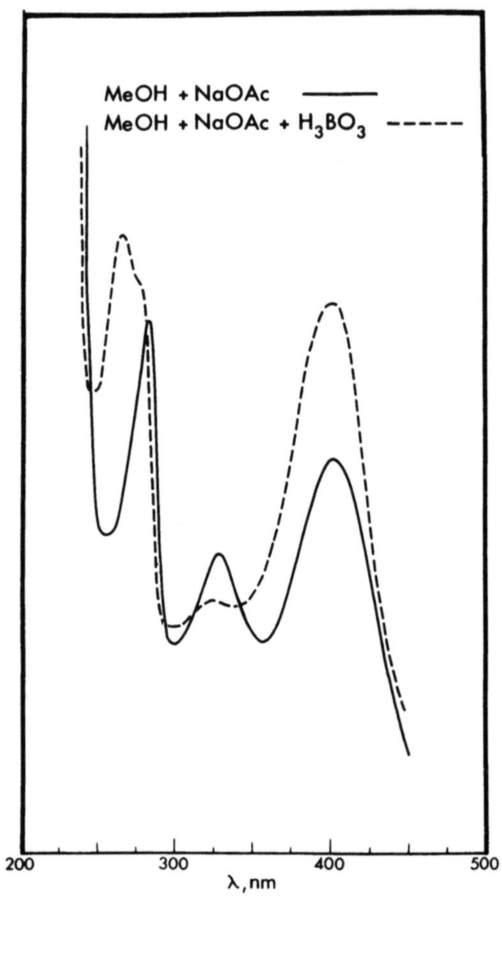

GOSSYPITRIN

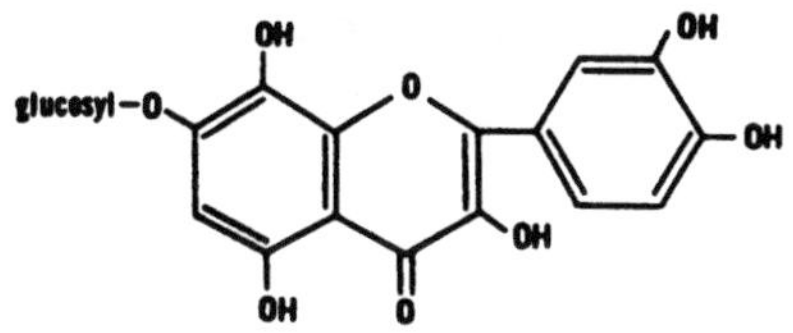

CHROMATOGRAPHIC DATA

Spot Appearance: (UV) yellow-green
(UV/NH_3) yellow-green

R_f Values: 0.15 (TBA), 0.08 (HOAc)

UV SPECTRAL DATA (λ_{max},*nm*)

MeOH	261, 279sh, 307sh, 343, 385
NaOMe	278, 371 (dec.)
$AlCl_3$	266sh, 277, 321sh, 475
$AlCl_3$/HCl	257sh, 272, 289sh, 316sh, 373, 454
NaOAc	273, 390, 450sh (dec.)
NaOAc/H_3BO_3 (Proc. II)	266, 399

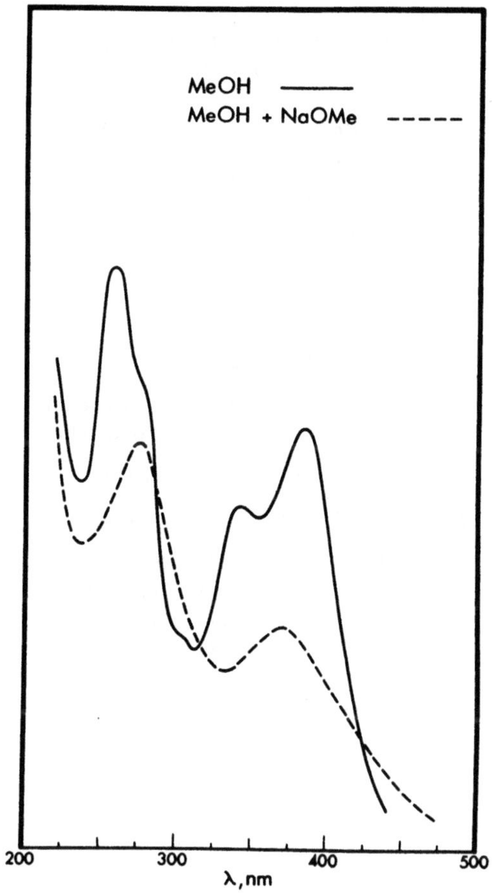

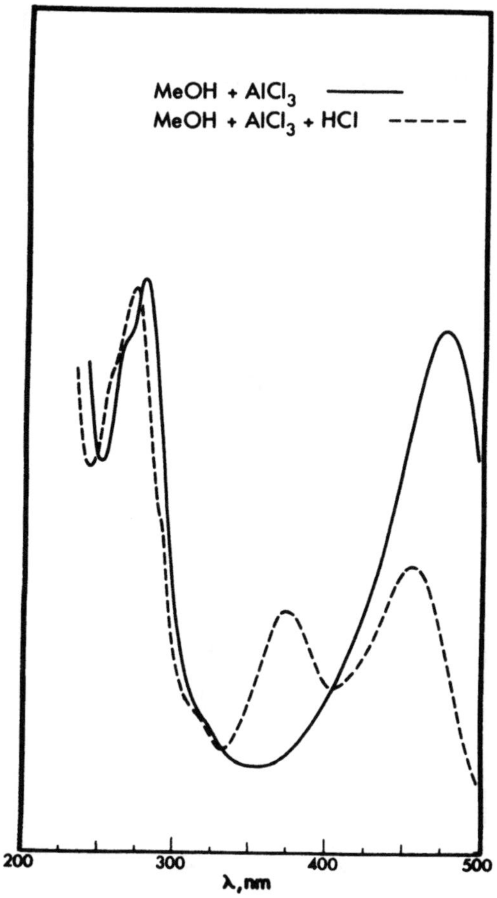

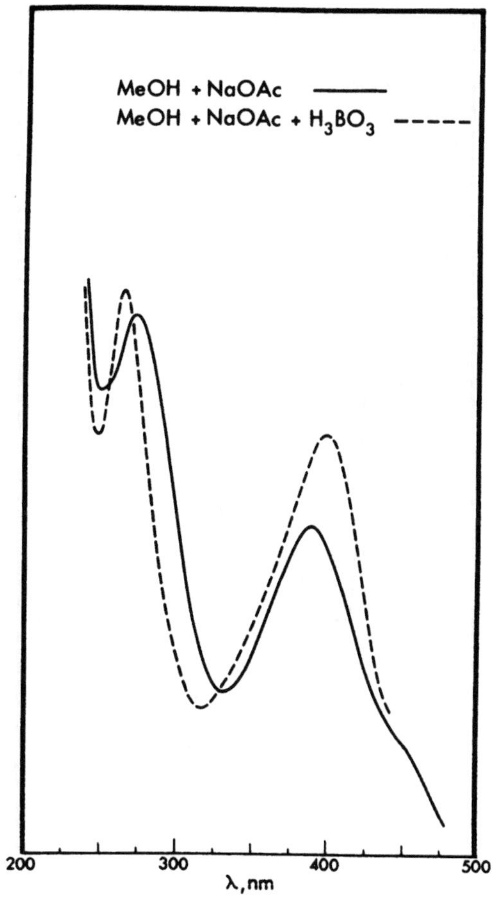

101

GOSSYPETIN HEXAMETHYL ETHER

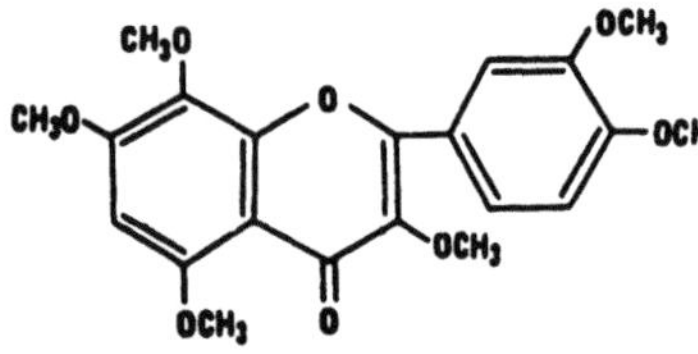

CHROMATOGRAPHIC DATA

Spot Appearance: (UV) dull yellow
(UV/NH_3) dull yellow

R_f Values: 0.79 (TBA), 0.30 (HOAc)

UV SPECTRAL DATA (λ_{max}, nm)

MeOH	252, 271, 301sh, 351
NaOMe	252, 271, 301sh, 351
$AlCl_3$	252, 271, 351
$AlCl_3$/HCl	252, 271, 351
NaOAc	252, 270, 301sh, 353
NaOAc/H_3BO_3 (Proc. I)	252sh, 271, 353

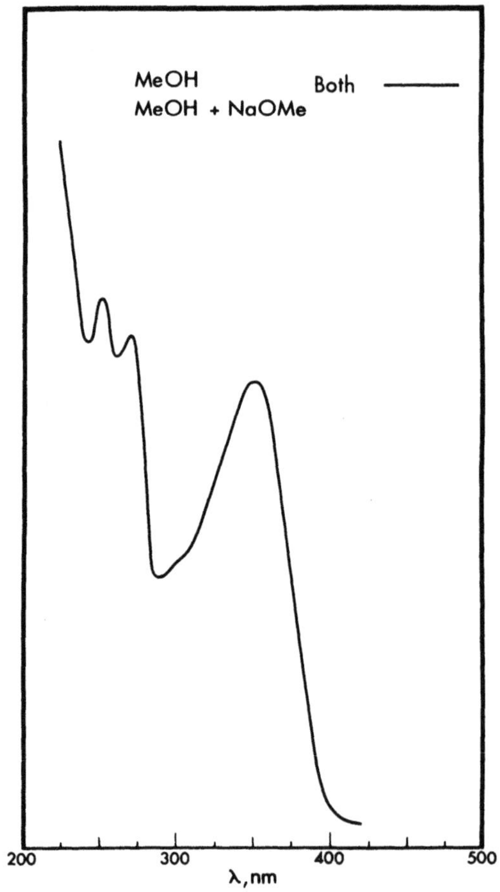

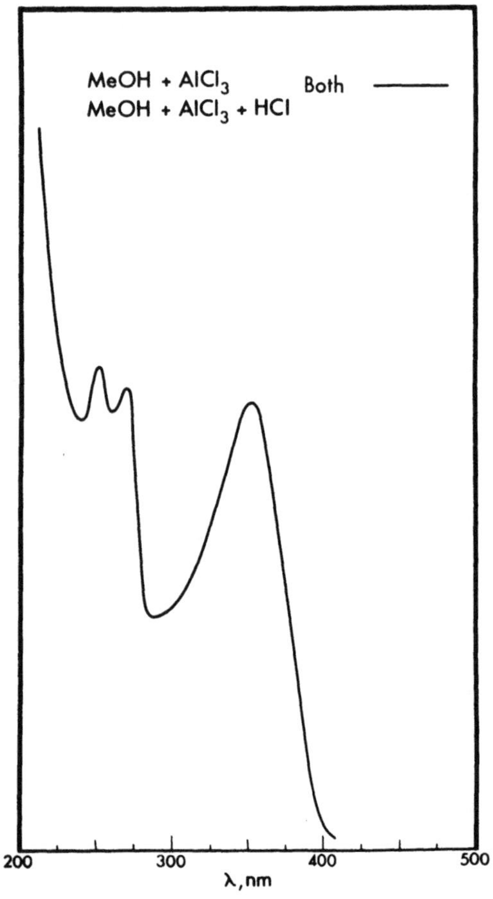

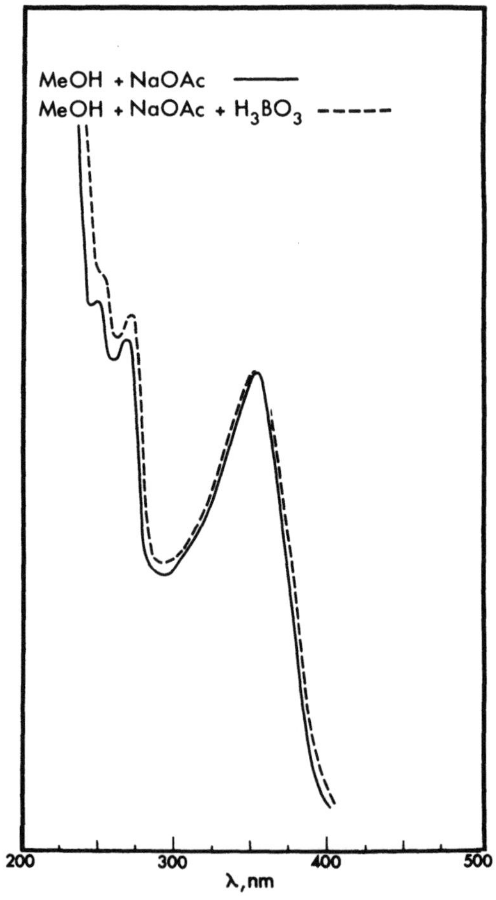

MYRICETIN

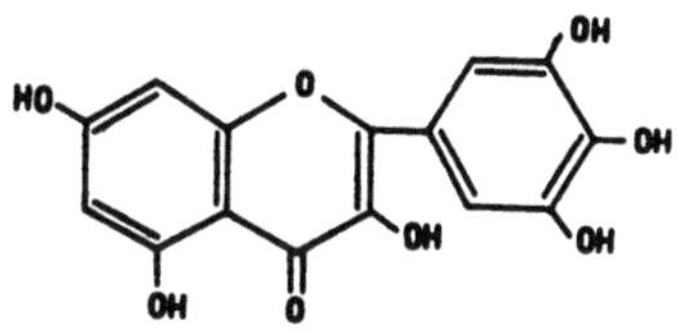

CHROMATOGRAPHIC DATA

Spot Appearance: (UV) yellow
(UV/NH_3) yellow

R_f Values: 0.29 (TBA), 0.01 (HOAc)

UV SPECTRAL DATA (λ_{max}, *nm*)

MeOH	254, 272sh, 301sh, 374
NaOMe	262sh, 285sh, 322, 423 (dec.)
$AlCl_3$	271, 316sh, 450
$AlCl_3$/HCl	266, 275sh, 308sh, 360sh, 428
NaOAc	269, 335 (dec.)
NaOAc/H_3BO_3 (Proc. II)	258, 304sh, 392

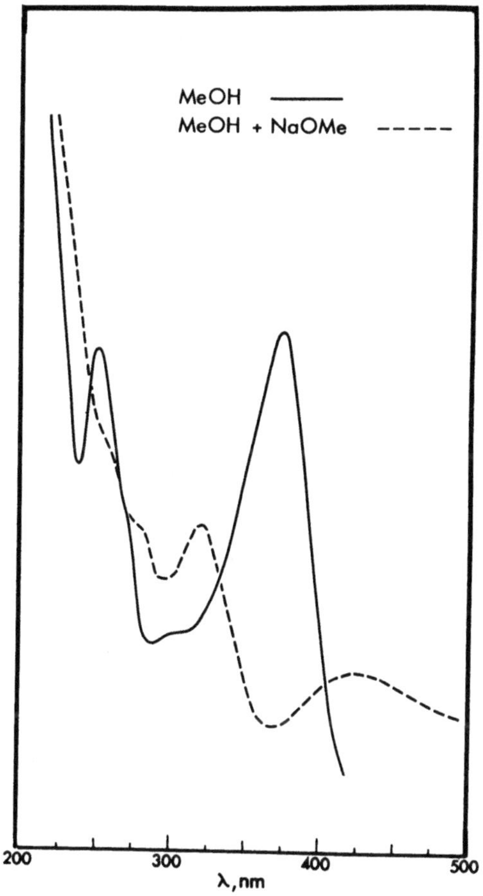

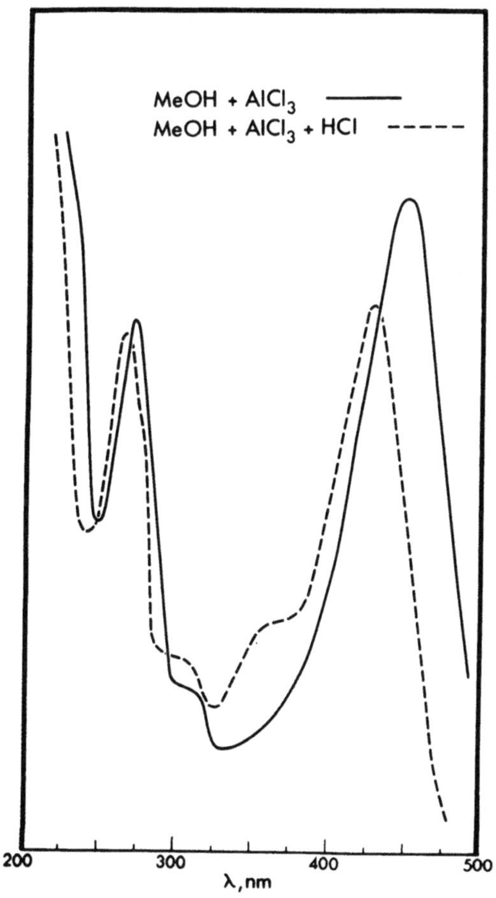

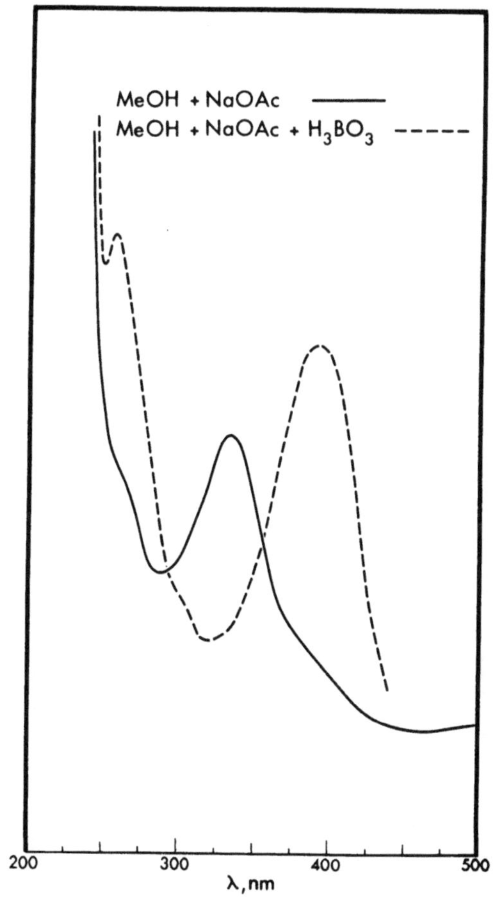

103

3',5,5'-TRIHYDROXY-3,4',6,7-TETRAMETHOXYFLAVONE

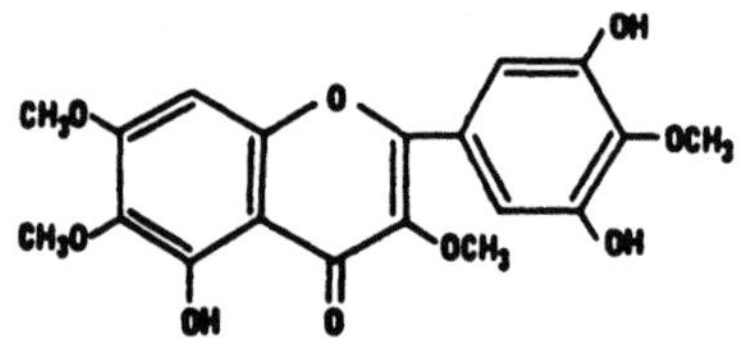

CHROMATOGRAPHIC DATA

Spot Appearance: (UV) deep purple
(UV/NH_3) deep purple

R_f Values: 0.82 (TBA), 0.42 (HOAc)

UV SPECTRAL DATA (*λ_{max},nm*)

MeOH	270, 335
NaOMe	269, 327, 366
$AlCl_3$	280, 306sh, 367, 398sh
$AlCl_3$/HCl	283, 306sh, 355, 403sh
NaOAc	270, 336
NaOAc/H_3BO_3 (Proc. I)	269, 339

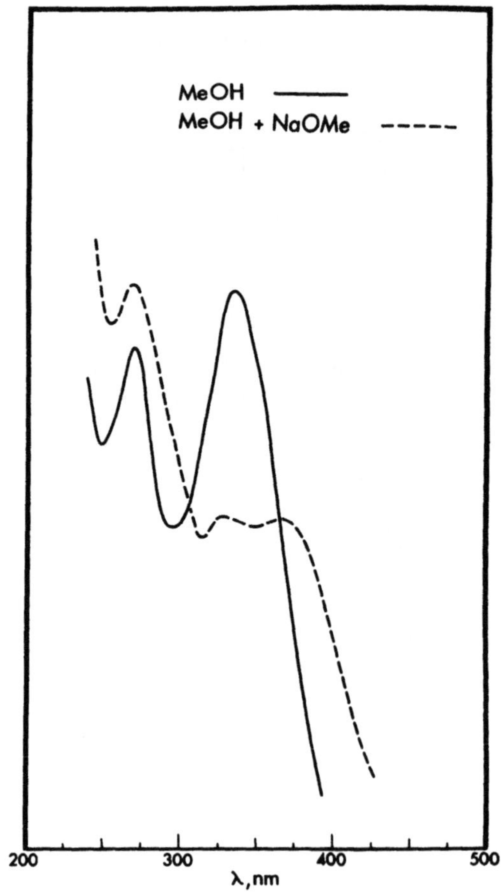

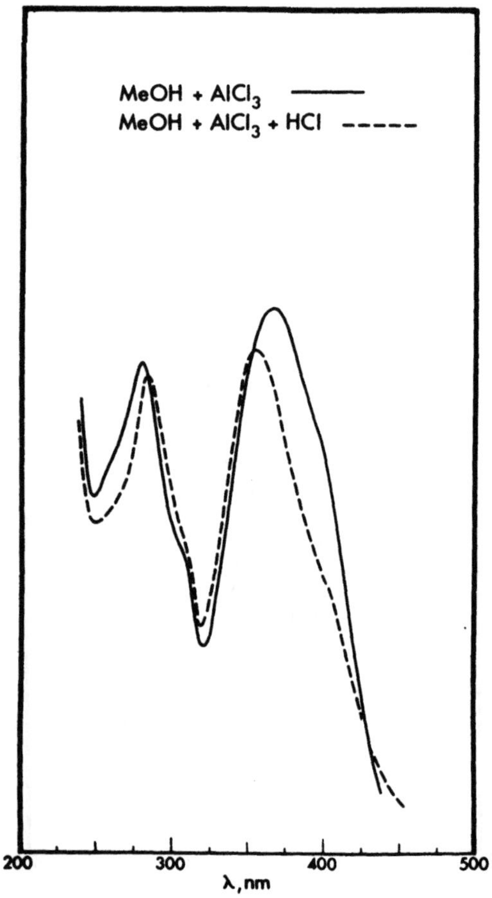

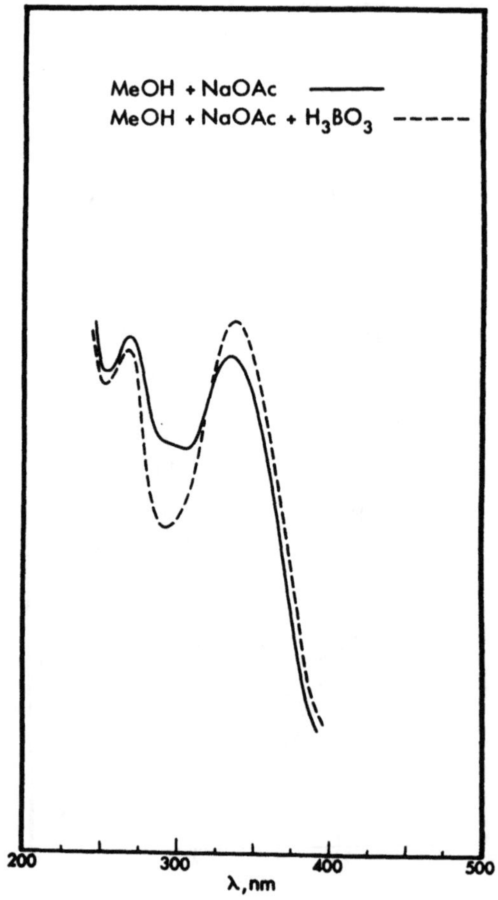

Chapter VI

The Ultraviolet Spectra of Isoflavones, Flavanones and Dihydroflavonols

VI-1. The UV Spectra of Isoflavones, Flavanones and Dihydroflavonols in Methanol . . . 165

VI-2. The UV Spectra of Isoflavones, Flavanones and Dihydroflavonols in the Presence of NaOMe . . . 167

2a. The Detection of 3′,4′-Dihydroxyl Groups in Isoflavones by the Effect of NaOMe on the UV Spectrum . . . 167

2b. The UV Spectra of Flavanones and Dihydroflavonols in the Presence of NaOMe . . . 168

VI-3. The UV Spectra of Isoflavones, Flavanones and Dihydroflavonols in the Presence of NaOAc . . . 169

3a. The Detection of 7-Hydroxyl Groups in Isoflavones by the Effect of NaOAc on the UV Spectrum . . . 169

3b. The Detection of 7-Hydroxyl Groups in Flavanones and Dihydroflavonols by the Effect of NaOAc on the UV Spectrum . . . 170

VI-4. The Detection of A-Ring Ortho-dihydroxyl Groups in Isoflavones, Flavanones and Dihydroflavonols by the Effect of $NaOAc/H_3BO_3$ on the UV Spectrum . . . 170

VI-5. The UV Spectra of Isoflavones, Flavanones and Dihydroflavonols in the Presence of $AlCl_3$ and $AlCl_3/HCl$ . . . 171

5a. The Detection of A-Ring Ortho-dihydroxyl Groups in Isoflavones, Flavanones and Dihydroflavonols by the Effect of $AlCl_3$ and $AlCl_3/HCl$ on the UV Spectrum . . . 171

5b. The Detection of 5-Hydroxyl Groups in Isoflavones, Flavanones and Dihydroflavonols by the Effect of $AlCl_3/HCl$ on the UV Spectrum . . . 171

VI-6. Index of Ultraviolet Absorption Spectra of Isoflavones, Flavanones and Dihydroflavonols . . . 172

References . . . 174

VI-1. The UV Spectra of Isoflavones (I), Flavanones (II) and Dihydroflavonols (III) in Methanol

Isoflavones, flavanones and dihydroflavonols all give similar UV spectra as a result of their having little or no conjugation between the A- and B-rings. They are all readily distinguished from flavones and flavonols by their UV spectra, which typically exhibit an intense Band II absorption with only a shoulder or low intensity peak representing Band I.

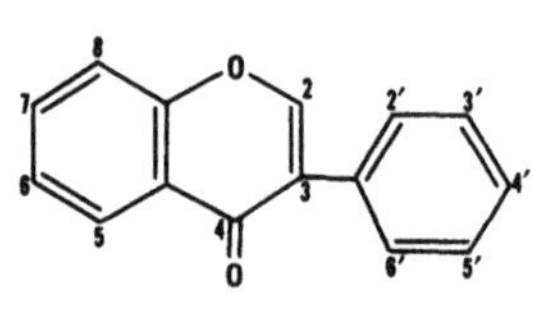
I. Isoflavone skeleton

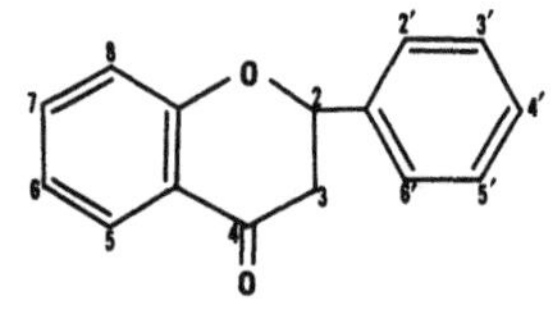
II. Flavanone skeleton

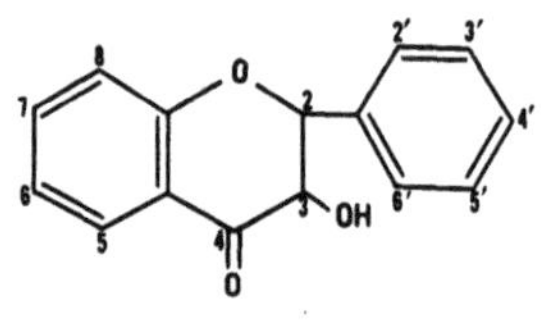
III. Dihydroflavonol skeleton

The UV Spectra of Isoflavones in MeOH

The Band II absorption of isoflavones usually occurs in the region 245–270 nm, and is relatively unaffected by increased hydroxylation of the B-ring, cf. the Band II position in: 5,7-dihydroxyisoflavone (259 nm); 5,7,4′-trihydroxyisoflavone (261 nm) and 5,7,3′,4′-tetrahydroxyisoflavone (262 nm). Band II is, however, shifted bathochromically by increased oxygenation in the A-ring (Table VI-1).

Table VI-1. *Band II in the UV spectra of isoflavones differing in their A-ring oxidation pattern*

Spectrum No.	Isoflavone	Oxidation pattern		Band II (nm)
		A-ring	B-ring	
108	Daidzein	7	4′	249
113	Genistein	5,7	4′	261
129	6-Hydroxygenistein	5,6,7	4′	270

Some specific effects of isoflavone oxidation and substitution pattern on the UV spectra are:

(1) Simple polyoxygenated isoflavones which have their Band II absorption in the 265–270 nm range are usually trioxygenated in the A-ring (cf. spectra 129, 130, 131, 132, 138 and 139).

(2) The UV spectra of 6,7-dioxygenated isoflavones, such as texasin and afrormosin (spectra 121 and 124) are unusual [*1*, *2*] in that Band I is abnormally intense; thus the spectra appear similar to those observed for flavones (however, paper chromatography or NMR spectroscopy clearly identifies them as isoflavones).

(3) Methylation or glycosylation of either 7- or 4′-hydroxyl groups in isoflavones has little or no effect on the UV spectrum (e.g. spectra pairs: 108, 109; 108, 120; 110, 111; 113, 114; 113, 115; 113, 116; 113, 118; 126, 127; 130, 131 and 133, 134) while substitution of the 5-hydroxyl groups causes a 5–10 nm hypsochromic shift of Band II (e.g. spectra pairs: 105, 106 and 113, 117).

(4) Loss of a 5-hydroxyl group causes a 7–17 nm hypsochromic shift in Band II (see spectra pairs: 104, 105; 108, 113; 109, 114; 110, 119; 111, 120; 121, 129; 124, 132; 125, 133; 126, 133 and 127, 135).

The UV Spectra of Flavanones and Dihydroflavonols in MeOH

The UV spectra of dihydroflavonols are almost identical with those obtained for the equivalent flavanones (cf. spectra 141, 148 and 146, 153); thus the presence or absence of the C-3 hydroxyl group in flavonoids which do not have a C_2-C_3 double bond makes little difference to the UV spectra. Both flavanones and dihydroflavonols have their major absorption peak (Band II) in the range 270–295 nm and are therefore clearly distinguished from the spectra of isoflavones (which have their Band II peaks between 245 and 270 nm). Removal of the 5-hydroxyl group from a flavanone or dihydroflavonol causes a 10–15 nm hypsochromic shift of the major absorption band, i.e., Band II (cf. spectra pairs 148, 151; 149, 153 and 141, 144). Increasing oxygenation in the B-ring of flavanones and dihydroflavonols has no noticeable effect on their UV spectra (cf. spectra 141, 149 and 155).

VI-2. The UV Spectra of Isoflavones, Flavanones and Dihydroflavonols in the Presence of NaOMe

2a. The Detection of 3′, 4′-Dihydroxyl Groups in Isoflavones by the Effect of NaOMe on the UV Spectrum

The spectra of isoflavones containing A-ring hydroxyl groups usually show bathochromic shifts of both Band I and Band II in the presence of NaOMe. In addition, the peaks in the UV spectrum of a 3′,4′-dihydroxyisoflavone show reduced intensity within a few minutes with added NaOMe (see spectra 125, 128, 133, 134, 135 and 137).

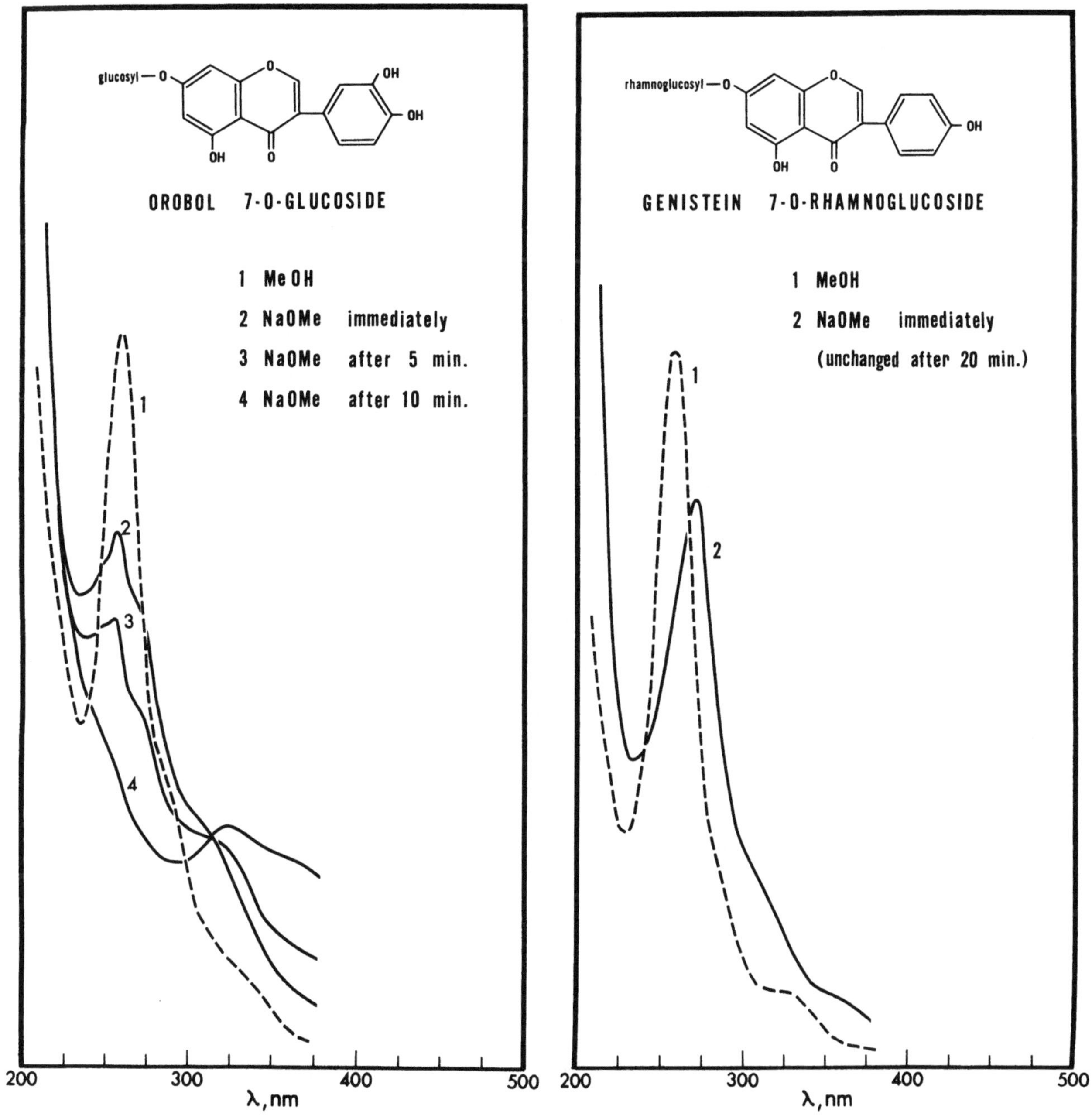

Fig. VI-1. The different effects of NaOMe on the spectra of isoflavones having a 3′,4′-dihydroxyl grouping (orobol 7-O-glucoside) and a 4′-hydroxyl group (genistein 7-O-rhamnoglucoside)

A comparison of the decomposition rates of a 4′-monohydroxyisoflavone and a 3′,4′-dihydroxyisoflavone (Fig. VI-1) suggests that the rapid decomposition of the latter may be diagnostic for the 3′,4′-dihydroxyl system in isoflavones. The only other isoflavone showing signs of decomposition in NaOMe was 6-hydroxygenistein (4′,5,6,7-tetrahydroxyisoflavone), which contains the alkali sensitive 5,6,7-hydroxylation pattern. Other methods for detecting ortho-dihydroxyl groups in flavonoids (e.g. $AlCl_3$ and $NaOAc/H_3BO_3$) do not distinguish the ortho-dihydroxyl system in the B-ring of isoflavones.

2b. The UV Spectra of Flavanones and Dihydroflavonols in the Presence of NaOMe

The UV spectra of all flavanones and dihydroflavonols with A-ring hydroxylation show bathochromic shifts of Band II with NaOMe. For dihydroflavonols, the magnitude

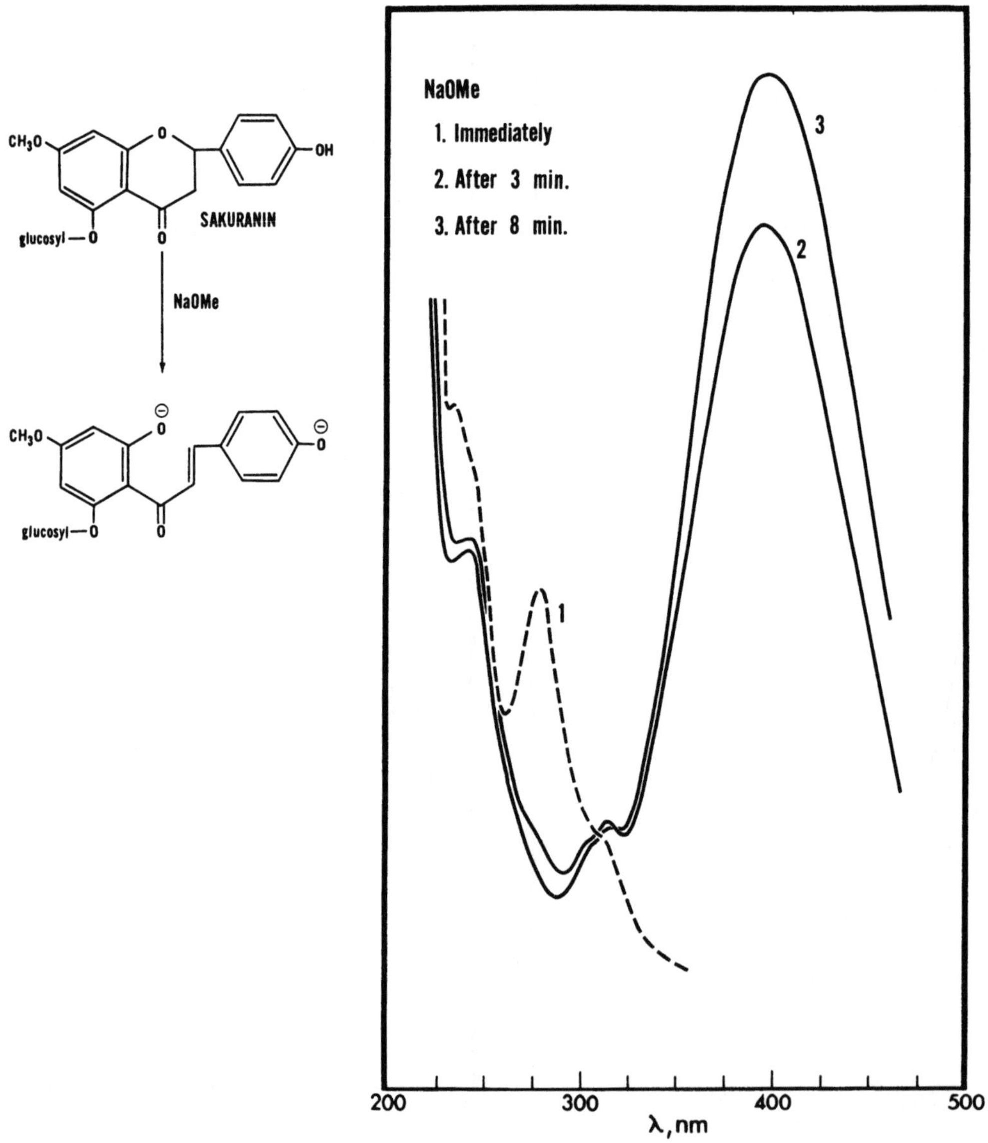

Fig. VI-2. The UV spectrum of sakuranin in the presence of NaOMe; the spectra illustrate the conversion of sakuranin to the equivalent ionized chalcone

of the bathochromic shift depends upon the presence or absence of a free 5-hydroxyl group. Spectra of 5,7-dihydroxy-dihydroflavonols (spectra 151, 152, 153 and 154) exhibit a consistent 34–40 nm shift of the major absorption peak, whereas spectra of the 7-hydroxy-dihydroflavonols lacking a free 5-hydroxyl group (spectra 148, 149, 150 and 155) show a 55–60 nm shift. In both cases, an increase in the Band II peak intensity is observed.

The UV spectra of flavanones in the presence of NaOMe also exhibit bathochromic shifts of the main absorption band (Band II) of about 35 nm for 5,7-dihydroxyflavanones and 60 nm for 7-hydroxyflavanones. Again, the shifts are accompanied by an increase in the intensity of Band II. Under alkaline conditions, however, some flavanones (in particular those lacking a free 5-hydroxyl group [*3*, *4*] will isomerize to chalcones, which have an entirely different UV spectrum (see Fig. VI-2).

Flavanones with 5,6,7 or 6,7,8 hydroxylation patterns decompose in the presence of NaOMe and, as a consequence, their UV spectra degenerate (e.g. spectrum 142). Horowitz and Jurd [*3*] have reported that 3′,4′-dihydroxyflavanones decompose rapidly in the presence of NaOH; however, this effect was not noticed with NaOMe in the present compilation.

VI-3. The UV Spectra of Isoflavones, Flavanones and Dihydroflavonols in the Presence of NaOAc

3a. The Detection of 7-Hydroxyl Groups in Isoflavones by the Effect of NaOAc on the UV Spectrum

NaOAc specifically ionizes the 7-hydroxyl group in isoflavones [*3*]. Unlike many flavones and flavonols, isoflavones do not contain ionizable 3- or 4′-hydroxyl groups, and therefore the interpretation of the spectral shifts is simplified. NaOAc causes Band II of the UV spectrum of a 7-hydroxyisoflavone to shift 6–20 nm bathochromically (see

Table VI-2. *The shift of Band II in the UV spectra of 7-hydroxyisoflavones with NaOAc*

Spectrum No.	Isoflavone	Bathochromic shift (nm)
104	7-Hydroxyisoflavone	21
105	5,7-Dihydroxyisoflavone	14
107	2-Carboxy-5,7-dihydroxyisoflavone	14
108	Daidzein	4
110	Formononetin	6
113	Genistein	10
116	Sophoricoside	11
117	Genistein 5-methyl ether	8
119	Biochanin A	11
121	Texasin	−2
123	2-Carboxy-6,7-dihydroxy-4′-methoxyisoflavone	13
124	Afrormosin	−2
125	3′,4′,7-Trihydroxyisoflavone	8
126	Pseudobaptigenin	8
128	Baptigenin	8
129	6-Hydroxygenistein	dec
130	Tectorigenin	6
132	Irisolidone	8
133	Orobol	8
136	Pratensein	9
138	Irigenin	5

Table VI-2); however, as was previously observed for flavones little or no shift occurs when there is an oxygen substituent at position 6 (e.g. spectra 121 and 124). Degeneration of the UV spectrum with time was observed for 6-hydroxygenistein (4',5,6,7-tetrahydroxyisoflavone, spectrum 129).

3b. The Detection of 7-Hydroxyl Groups in Flavanones and Dihydroflavonols by the Effect of NaOAc on the UV Spectrum

The presence or absence of a free 7-hydroxyl group in flavanones and dihydroflavonols may readily be determined from their UV spectra by comparing the positions of the major absorption peak (Band II) in the methanol spectrum with that of the same peak in the NaOAc spectrum. The shift is about 34 – 37 nm for the 5,7-dihydroxyflavanones and 5,7-dihydroxy-dihydroflavonols and 51 – 58 nm for their 5-deoxy-equivalents (see Table VI-3). Degeneration of the UV spectrum with time was observed for 5,6,7-trihydroxyflavanone (spectrum 142).

Table VI-3. *The shift of Band II in the UV spectra of 7-hydroxyflavanones and 7-hydroxydihydroflavonols in the presence of NaOAc*

Spectrum No.	Flavonoid	Bathochromic shift (nm) of Band II[a]
	Flavanones	
140	Pinocebrin	34
141	Liquiritigenin	51
142	5,6,7-Trihydroxyflavanone	dec
144	Naringenin	34
146	Eriodictyol	36
	Dihydroflavonols	
148	Garbanzol	58
149	Dihydrofisetin	57
150	(+)-Fustin 3-O-glucoside	58
151	Dihydrokaempferol	36
152	Engeletin	36
153	Taxifolin	37
154	Astilbin	37
155	Dihydrorobinetin	58

[a] Major absorption band.

VI-4. The Detection of A-Ring Ortho-dihydroxyl Groups in Isoflavones, Flavanones and Dihydroflavonols by the Effect of $NaOAc/H_3BO_3$ on the UV Spectrum

B-Ring ortho-dihydroxyl groups are not detectable by the effect of $NaOAc/H_3BO_3$ on the UV spectra of isoflavones, flavanones and dihydroflavonols because the B-rings in these flavonoids lack effective conjugation with the major chromophore. However, there is evidence (spectra 121 and 123) that 6,7-dihydroxyl groups in the A-ring of isoflavones and flavanones (and presumably dihydroflavonols) are detectable by a 10 to 15 nm bathochromic shift of Band I on the addition of $NaOAc/H_3BO_3$.

VI-5. The UV Spectra of Isoflavones, Flavanones and Dihydroflavonols in the Presence of $AlCl_3$ and $AlCl_3/HCl$

5a. The Detection of A-Ring Ortho-dihydroxyl Groups in Isoflavones, Flavanones and Dihydroflavonols by the Effect of $AlCl_3$ and $AlCl_3/HCl$ on the UV Spectrum

The presence of a 3′,4′-dihydroxyl group in isoflavones, flavanones and dihydroflavonols is not detectable by means of the $AlCl_3$ UV spectrum, because the B-ring has little or no conjugation with the major chromophore. In contrast, from limited data (spectra 121, 129, and 142) it appears that A-ring ortho-dihydroxyl groups which do not involve the C-5 hydrogen-bonded hydroxyl group are detectable in flavanones and isoflavones (and presumably also in dihydroflavonols, although no examples were available). When these compounds contain ortho-dihydroxyl groups at positions 6, 7 or 7, 8 the $AlCl_3$ spectrum exhibits bathochromic shifts (usually for both Band I and Band II) with respect to the $AlCl_3/HCl$ spectrum (cf., for example, spectra 121 and 142 with 122 and 143, respectively).

5b. The Detection of 5-Hydroxyl Groups in Isoflavones, Flavanones and Dihydroflavonols by the Effect of $AlCl_3/HCl$ on the UV Spectrum

Band II in the UV spectra of all 5-hydroxyisoflavones examined in the present investigation, with the exception of 2-carboxy-5,7-dihydroxyisoflavone (see Table VI-4, spectrum 107), undergoes a consistent 10 – 14 nm bathochromic shift (relative to the spectrum in methanol) in the presence of $AlCl_3/HCl$ (Table VI-4). The spectra of isoflavones lacking a free 5-hydroxyl group are unaffected by this reagent.

With 5-hydroxyflavanones and 5-hydroxydihydroflavonols the $AlCl_3/HCl$ reagent also causes a consistent bathochromic shift (20 – 26 nm) of Band II (Table VI-4).

Table VI-4. *The shift of Band II in the UV spectra of 5-hydroxyisoflavones, 5-hydroxyflavanones, and 5-hydroxydihydroflavonols in the presence of $AlCl_3/HCl$*

Spectrum No.	Flavonoid	Bathochromic shift of Band II (nm)	Spectrum No.	Flavonoid	Batochromic shift of Band II (nm)
	Isoflavones		136	Pratensein	11
105	5,7-Dihydroxyisoflavone	14	137	Pomiferin	11
107	2-Carboxy-5,7-dihydroxy-isoflavone	10	138	Irigenin	10
			139	Iridin	10
113	Genistein	21		*Flavanones*	
114	Genistin	12			
115	Sphaerobioside	11	140	Pinocebrin	20
116	Sophoricoside	10	142	5,6,7-Trihydroxyflavanone	22
118	Prunetin	12	143	5,6,7-Trihydroxyflavanone 7-O-glucuronide	26
119	Biochanin A	12			
120	Lanceolarin	12	144	Naringenin	22
129	6-Hydroxygenistein	11	146	Eriodictyol	20
130	Tectorigenin	11	147	Hesperidin	23
131	Tectoridin	10		*Dihydroflavonols*	
132	Irisolidone	12			
133	Orobol	12	151	Dihydrokaempferol	21
134	Orobol 7-O-glucoside	11	152	Engeletin	21
135	Orobol 7-O-rhamno glucoside	10	153	Taxifolin	22
			154	Astilbin	22

VI-6. Index[a] of Ultraviolet Absorption Spectra of Isoflavones, Flavanones and Dihydroflavonols

Spectrum No.		Oxidation pattern									
		2	3	5	6	7	8	2′	3′	4′	5′
	Isoflavones										
104	7-Hydroxyisoflavone					OH					
105	5,7-Dihydroxyisoflavone			OH		OH					
106	5,7-Dimethoxyisoflavone			OCH_3		OCH_3					
107	2-Carboxy-5,7-dihydroxyisoflavone	COOH		OH		OH					
108	Daidzein					OH				OH	
109	Daidzein 7-O-glucoside (Daidzin)					O-glu				OH	
110	Formononetin					OH				OCH_3	
111	Formononetin 7-O-glucoside					O-glu				OCH_3	
112	Formononetin 7-O-glucoside tetraacetate					[O-glu (Ac)$_4$]				OCH_3	
113	Genistein			OH		OH				OH	
114	Genistin			OH		O-glu				OH	
115	Sphaerobioside			OH		O-rut				OH	
116	Sophoricoside			OH		OH				O-glu	
117	Genistein 5-methyl ether			OCH_3		OH				OH	
118	Prunetin			OH		OCH_3				OH	
119	Biochanin A			OH		OH				OCH_3	
120	Lanceolarin			OH		[O-apio-syl-glu]				OCH_3	
121	Texasin				OH	OH				OCH_3	
122	Texasin 7-O-glucoside				OH	O-glu				OCH_3	
123	2-Carboxy-6,7-dihydroxy-4′-methoxyisoflavone	COOH			OH	OH				OCH_3	
124	Afrormosin				OCH_3	OH				OCH_3	
125	3′,4′,7-Trihydroxyisoflavone					OH			OH	OH	
126	Pseudobaptigenin					OH			-O-CH_2-O-		
127	Pseudobaptisin					O-rut			-O-CH_2-O-		
128	Baptigenin					OH			OH	OH	OH
129	6-Hydroxygenistein			OH	OH	OH				OH	
130	Tectorigenin			OH	OCH_3	OH				OH	
131	Tectoridin			OH	OCH_3	O-glu				OH	
132	Irisolidone			OH	OCH_3	OH				OCH_3	
133	Orobol			OH		OH			OH	OH	
134	Orobol 7-O-glucoside			OH		O-glu			OH	OH	

Spectrum No.		Oxidation pattern									
		2	3	5	6	7	8	2′	3′	4′	5′
135	Orobol 7-O-rhamnoglucoside			OH		[O-rh-glu]			OH	OH	
136	Pratensein			OH		OH			OH	OCH_3	
137	Pomiferin			OH	C_5H_9	O-C_5H_8			OH	OH	
138	Irigenin			OH	OCH_3	OH			OCH_3	OCH_3	OH
139	Iridin			OH	OCH_3	O-glu			OCH_3	OCH_3	OH
	Flavanones										
140	Pinocembrin			OH		OH					
141	Liquiritigenin					OH				OH	
142	5,6,7-Trihydroxyflavanone			OH	OH	OH					
143	5,6,7-Trihydroxyflavanone 7-O-glucuronide			OH	OH	O-gluc					
144	Naringenin			OH		OH				OH	
145	Sakuranin			O-glu		OCH_3				OH	
146	Eriodictyol			OH		OH			OH	OH	
147	Hesperidin			OH		O-rut			OH	OCH_3	
	Dihydroflavonols										
148	Garbanzol		OH			OH				OH	
149	Dihydrofisetin		OH			OH			OH	OH	
150	(+)-Fustin 3-O-glucoside		O-glu			OH			OH	OH	
151	Dihydrokaempferol		OH	OH		OH				OH	
152	Engeletin		O-rh	OH		OH				OH	
153	Taxifolin		OH	OH		OH			OH	OH	
154	Astilbin		O-rh	OH		OH			OH	OH	
155	Dihydrorobinetin		OH			OH			OH	OH	OH

[a] Abbreviations: glu = glucosyl; gluc = glucuronosyl; gly = glycosyl; rh-glu = rhamnoglucosyl; rut = rutinosyl.

References

1. Dyke, S.F., W.D. Ollis, M. Sainsbury, and J.S.P. Schwarz: Tetrahedron **20**, 1331 (1964).
2. Markham, K.R., W.T. Swift III, and T.J. Mabry: J. Org. Chem. **33**, 462 (1968).
3. Jurd, L., and R.M. Horowitz: J. Org. Chem. **26**, 2561 (1961).
4. Narasimhachari, N., and T.R. Seshadri: Proc. Indian Acad. Sci. **27 A**, 223 (1948); **30 A**, 271 (1959); Chem. Abstr. **44**, 1493 (1950).

7-HYDROXYISOFLAVONE

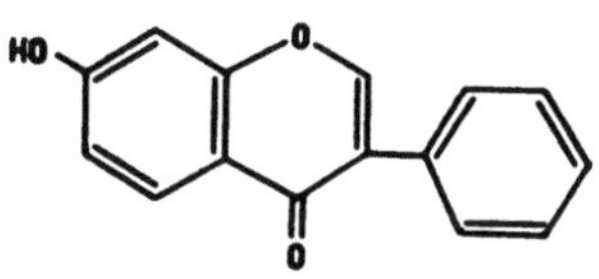

CHROMATOGRAPHIC DATA

Spot Appearance: (UV) invisible
(UV/NH_3) fluorescent light blue

R_f Values: 0.90 (TBA), 0.38 (HOAc)

UV SPECTRAL DATA (λ_{max}, *nm*)

MeOH	242, 299, 305sh
NaOMe	264, 336
$AlCl_3$	243, 299, 305sh
$AlCl_3$/HCl	243, 299, 305sh
NaOAc	263, 311sh, 336
NaOAc/H_3BO_3 (Proc. I)	252sh, 301

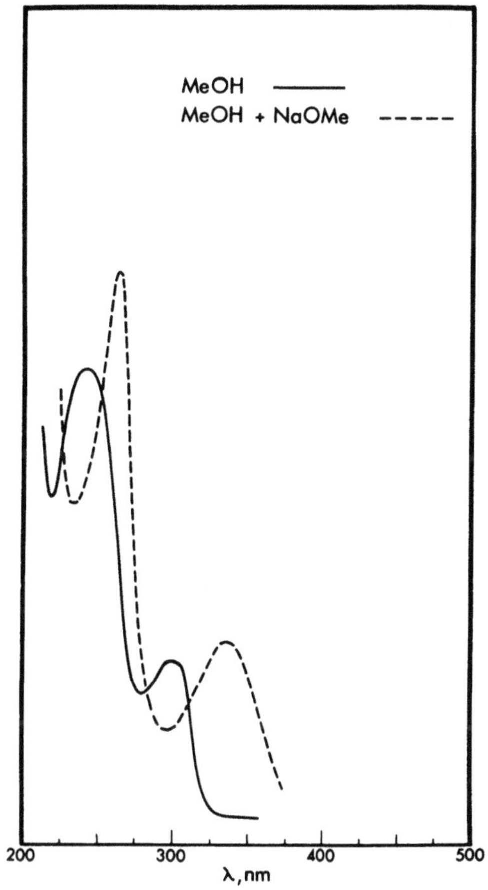

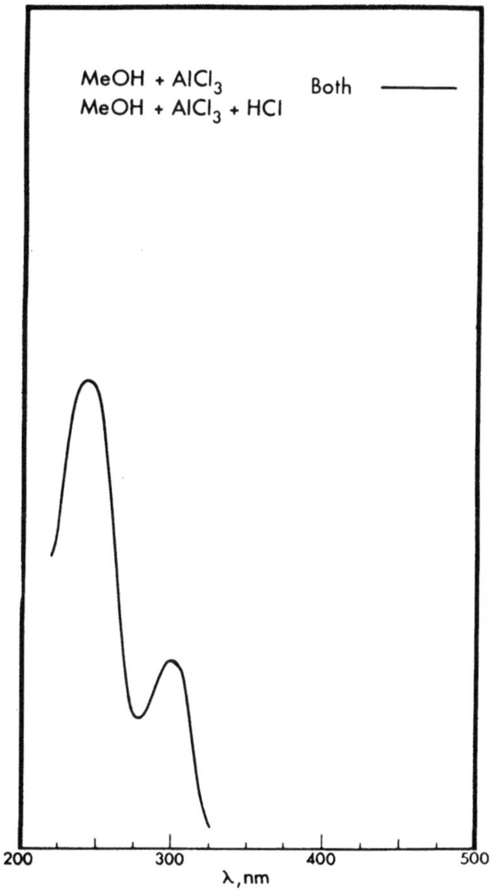

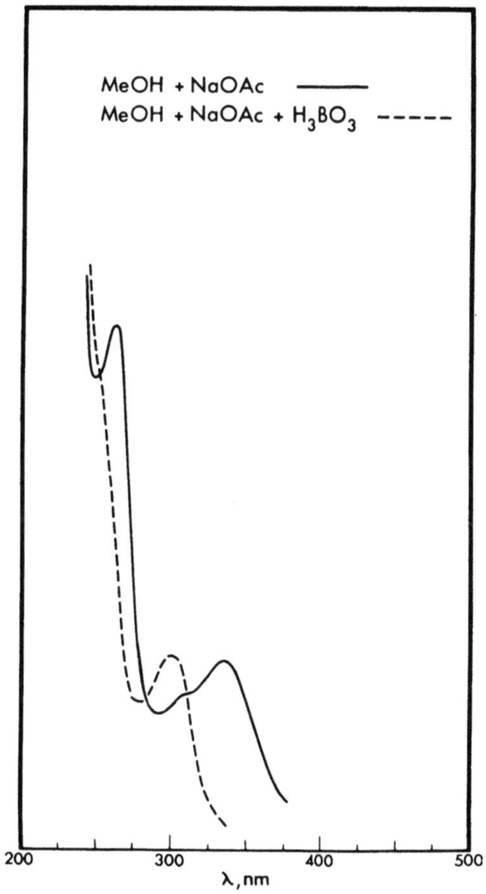

105

5,7-DIHYDROXYISOFLAVONE

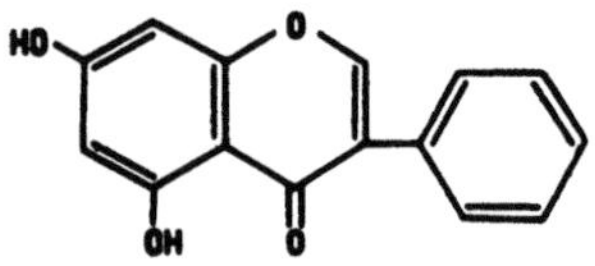

CHROMATOGRAPHIC DATA

Spot Appearance: (UV) deep purple
(UV/NH_3) deep purple

R_f Values: 0.93 (TBA), 0.33 (HOAc)

UV SPECTRAL DATA (λ_{max}, *nm*)

MeOH	259, 303sh, 315sh
NaOMe	274, 329
$AlCl_3$	272, 311, 367
$AlCl_3$/HCl	273, 313sh, 367
NaOAc	273, 327
NaOAc/H_3BO_3 (Proc. I)	260, 317sh

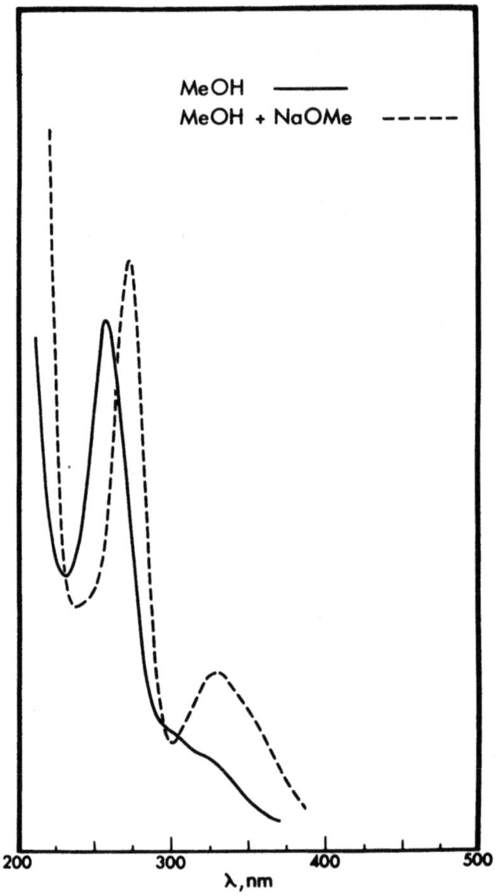

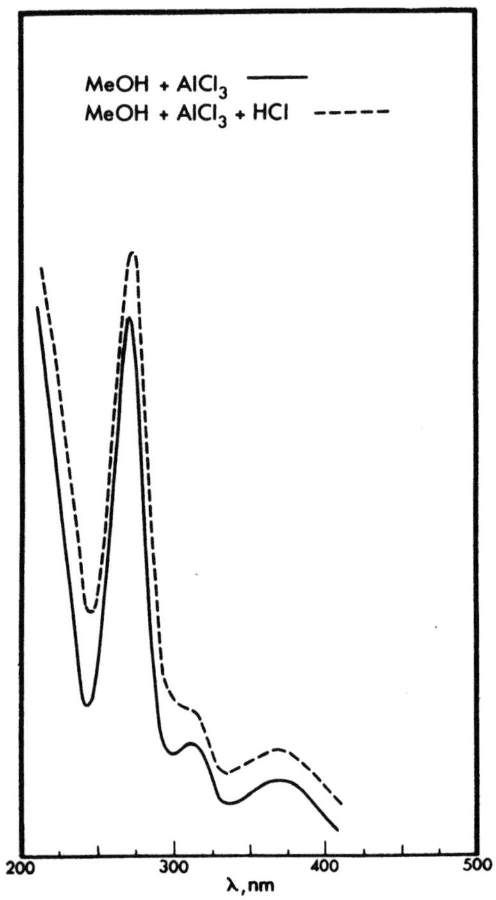

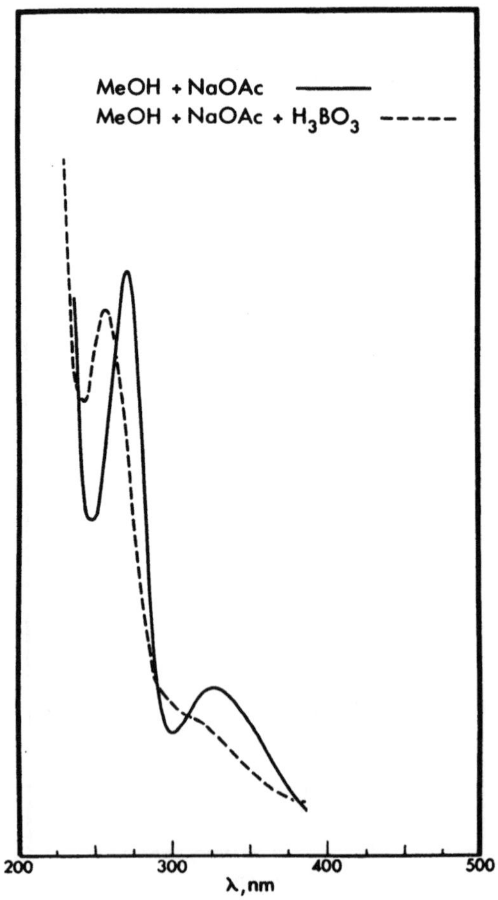

106

5,7-DIMETHOXYISOFLAVONE

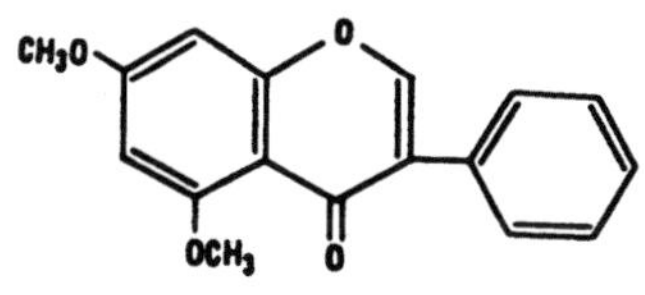

CHROMATOGRAPHIC DATA

Spot Appearance: (UV) light blue
(UV/NH_3) light blue

R_f Values: 0.90 (TBA), indef. (HOAc)

UV SPECTRAL DATA (λ_{max}, nm)

MeOH	251, 308sh
NaOMe	251, 309sh
$AlCl_3$	250, 305sh
$AlCl_3$/HCl	250, 305sh
NaOAc	252sh, 306sh
NaOAc/H_3BO_3	252sh, 306sh
(Proc. I)	

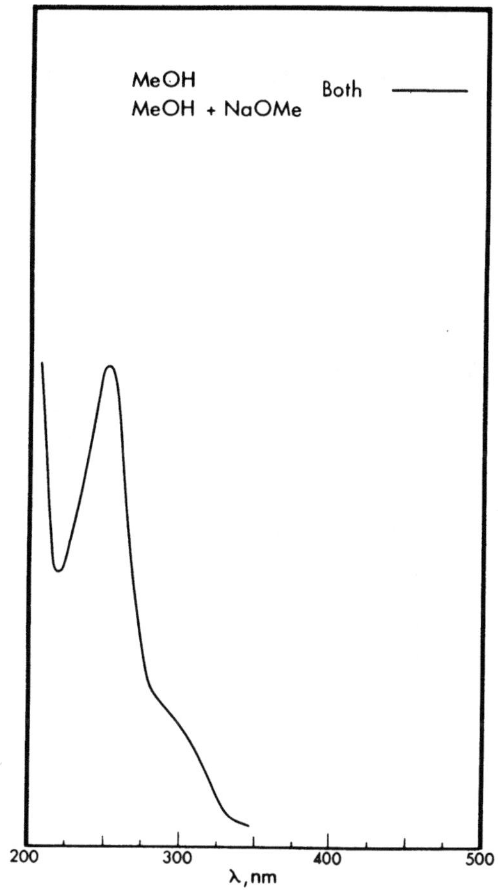

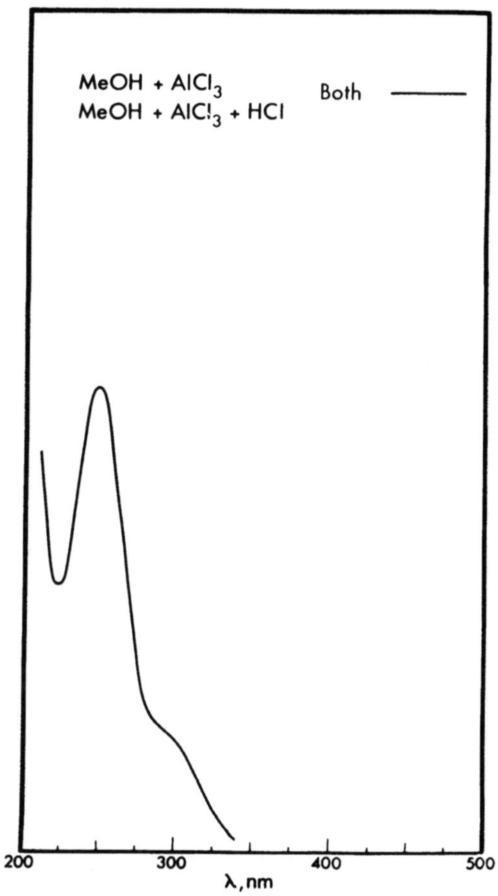

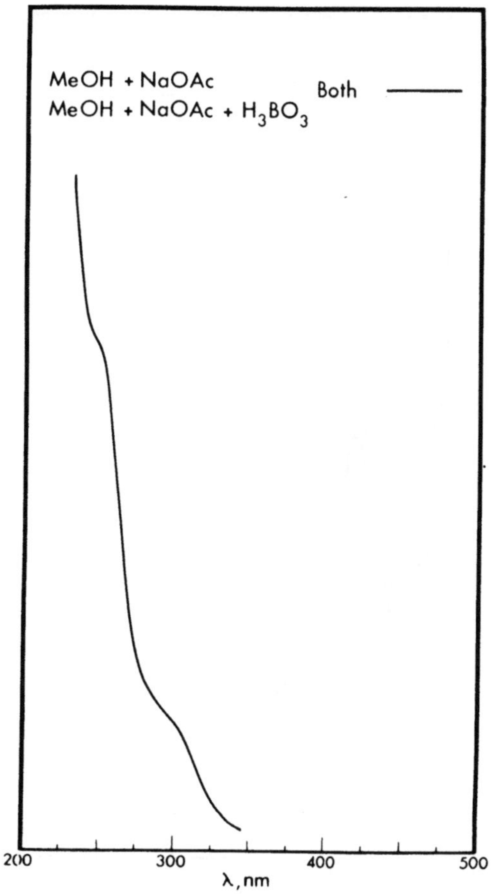

107

2-CARBOXY-
5,7-DIHYDROXYISOFLAVONE

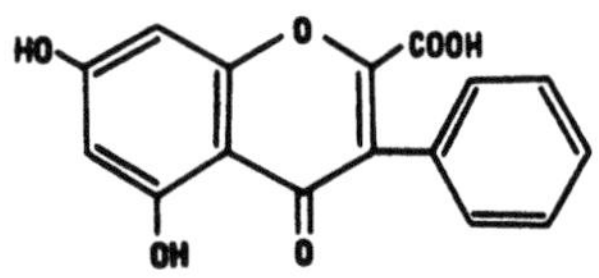

CHROMATOGRAPHIC DATA

Spot Appearance: (UV) deep purple
(UV/NH_3) deep purple

R_f Values: 0.75 (TBA), 0.74 (HOAc)

UV SPECTRAL DATA (λ_{max}, nm)

MeOH	257, 298sh, 323sh
NaOMe	272, 332
$AlCl_3$	243sh, 281, 324
$AlCl_3$/HCl	278, 317sh
NaOAc	271, 331
NaOAc/H_3BO_3 (Proc. I)	258, 309sh

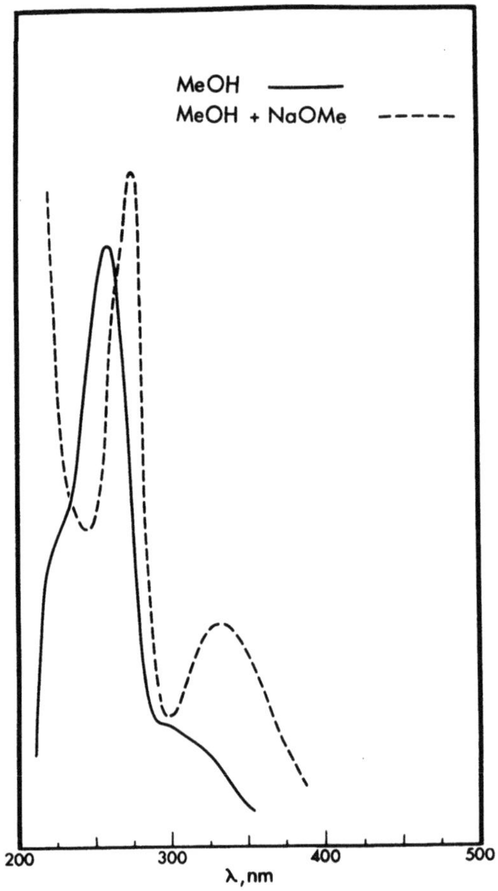

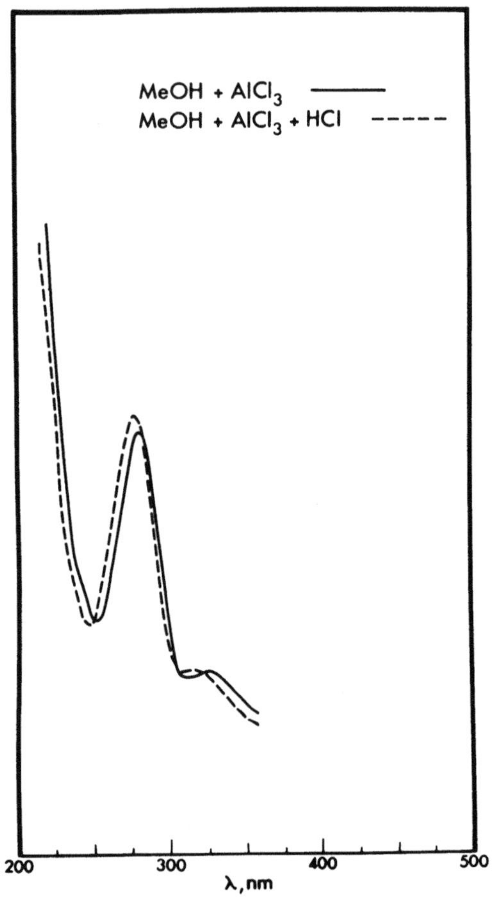

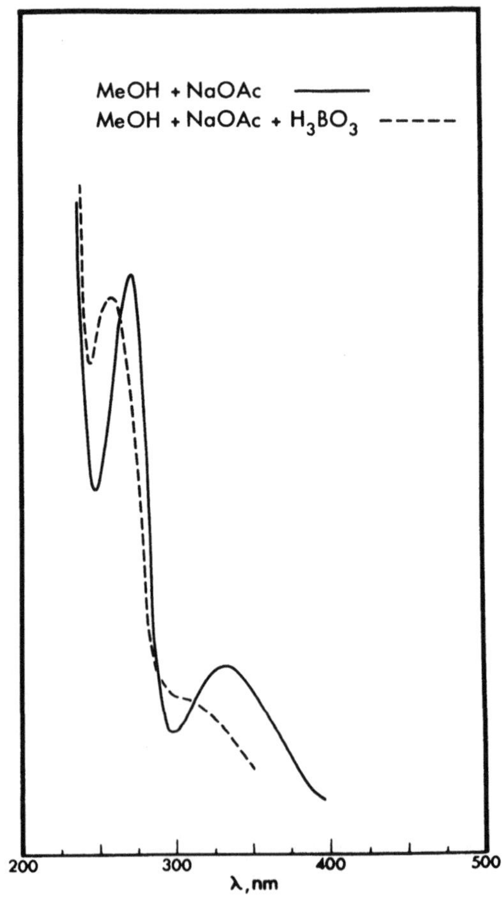

DAIDZEIN

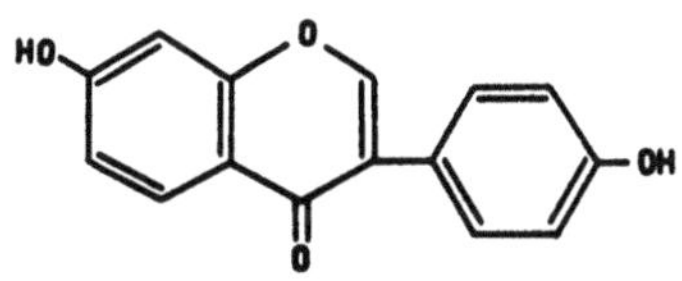

CHROMATOGRAPHIC DATA

Spot Appearance: (UV) invisible
(UV/NH_3) fluorescent light blue

R_f Values: 0.87 (TBA), 0.36 (HOAc)

UV SPECTRAL DATA (λ_{max},nm)

MeOH	238sh, 249, 259sh, 303sh
NaOMe	259, 289sh, 328
$AlCl_3$	240sh, 249, 260sh, 300sh
$AlCl_3$/HCl	240sh, 249, 262sh, 302sh
NaOAc	253, 272sh, 310, 330sh
NaOAc/H_3BO_3 (Proc. I)	261sh, 303

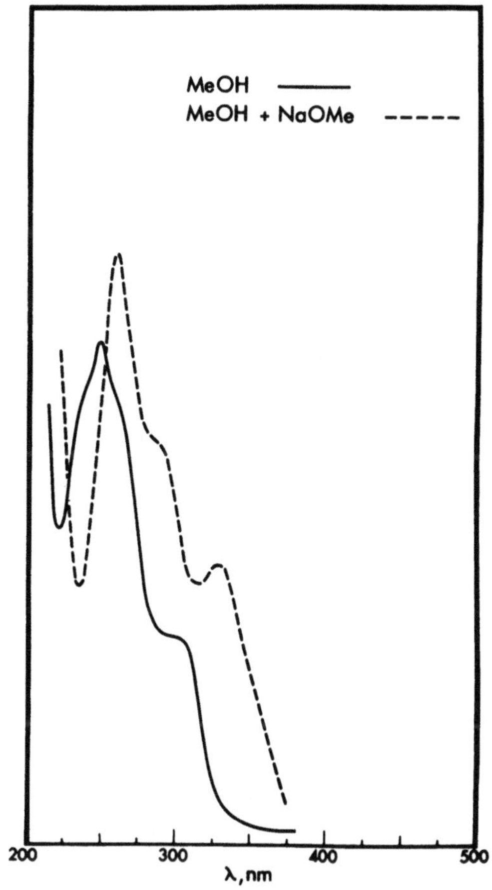

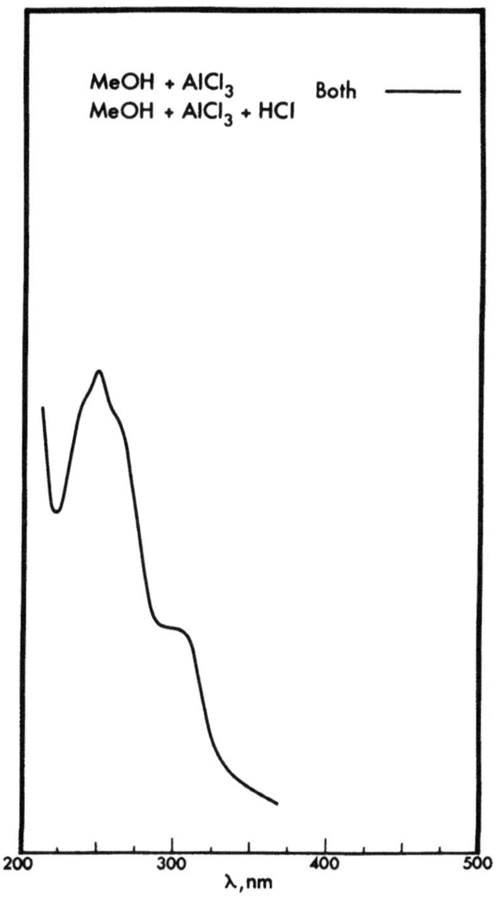

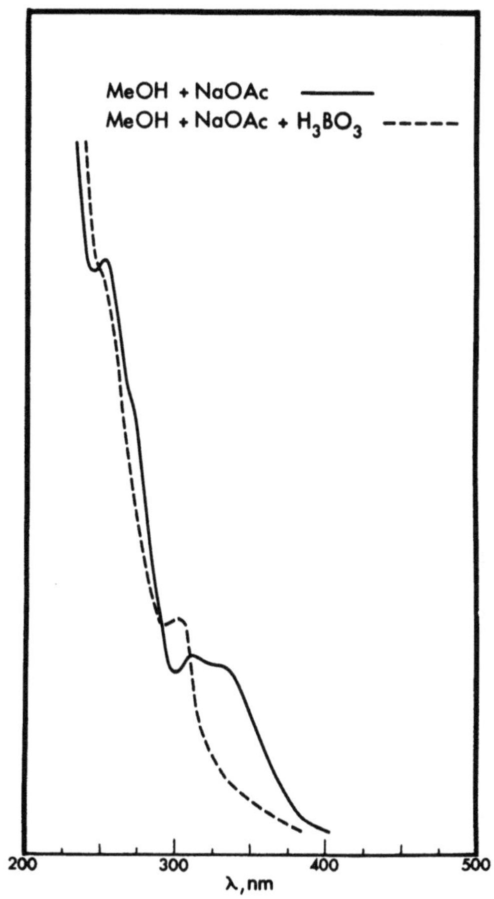

109
DAIDZEIN 7-0-GLUCOSIDE (DAIDZIN)

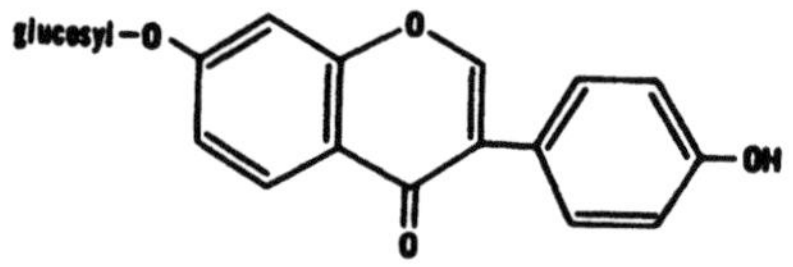

CHROMATOGRAPHIC DATA

Spot Appearance: (UV) invisible
(UV/NH_3) fluorescent light blue

R_f Values: 0.63 (TBA), 0.65 (HOAc)

UV SPECTRAL DATA (λ_{max}, *nm*)

MeOH	256, 313sh
NaOMe	256, 272sh, 320sh
$AlCl_3$	258, 304sh
$AlCl_3$/HCl	257, 303sh, 262sh
NaOAc	256, 322sh
NaOAc/H_3BO_3	254, 318sh
(Proc. I)	

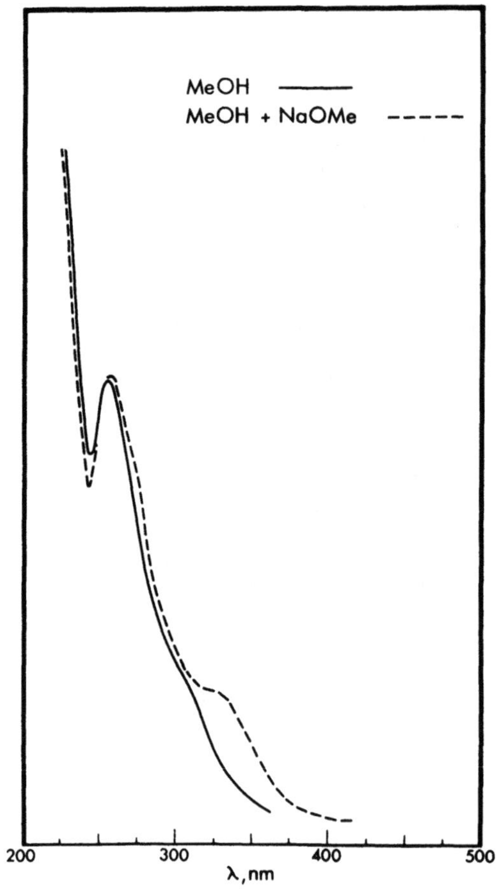

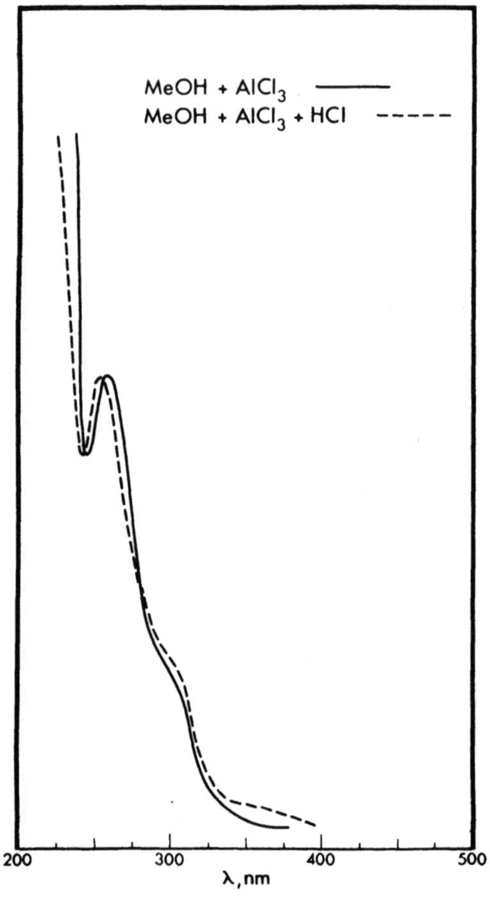

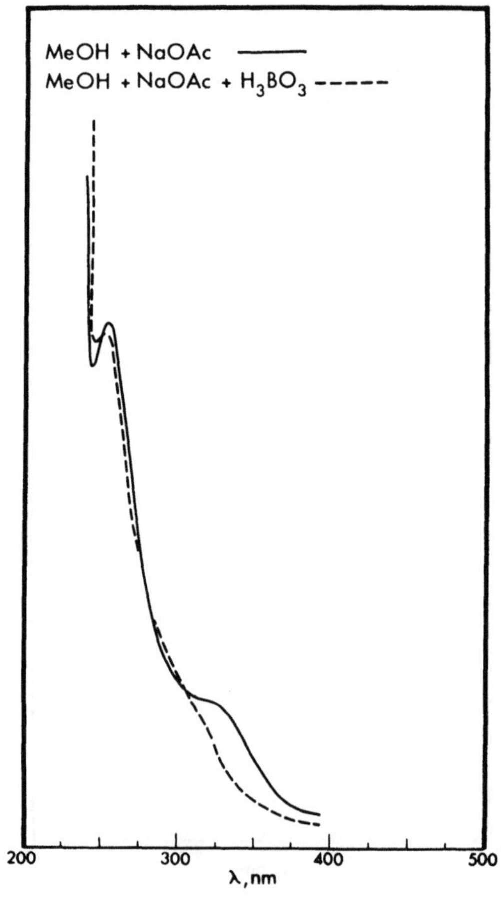

FORMONONETIN

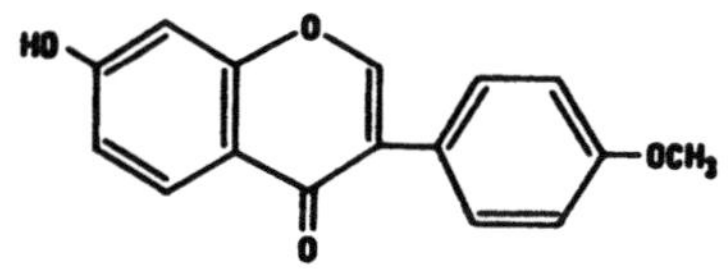

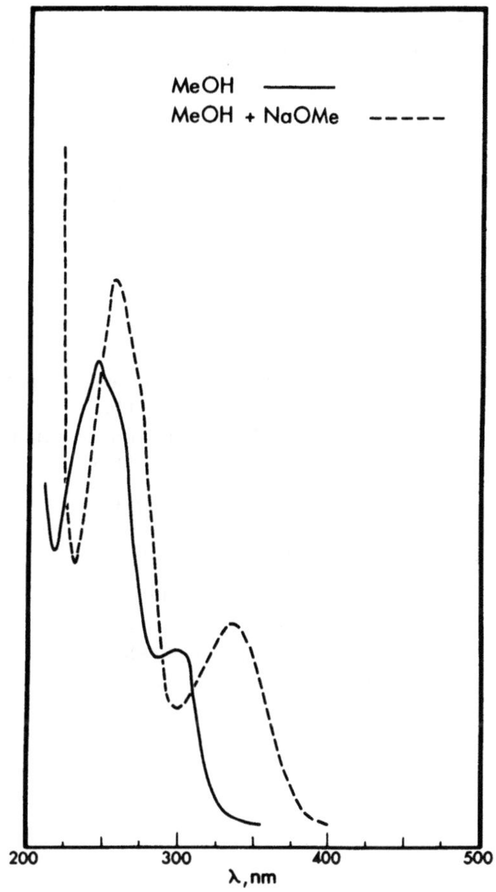

CHROMATOGRAPHIC DATA

Spot Appearance: (UV) invisible
(UV/NH_3) fluorescent light blue

R_f Values: 0.88 (TBA), 0.38 (HOAc)

UV SPECTRAL DATA (λ_{max}, nm)

MeOH	240sh, 248, 259sh, 311
NaOMe	255, 273sh, 335
$AlCl_3$	239sh, 248, 261sh, 301
$AlCl_3/HCl$	240sh, 249, 261sh, 301
NaOAc	254, 312sh, 334
$NaOAc/H_3BO_3$ (Proc. I)	264sh, 303

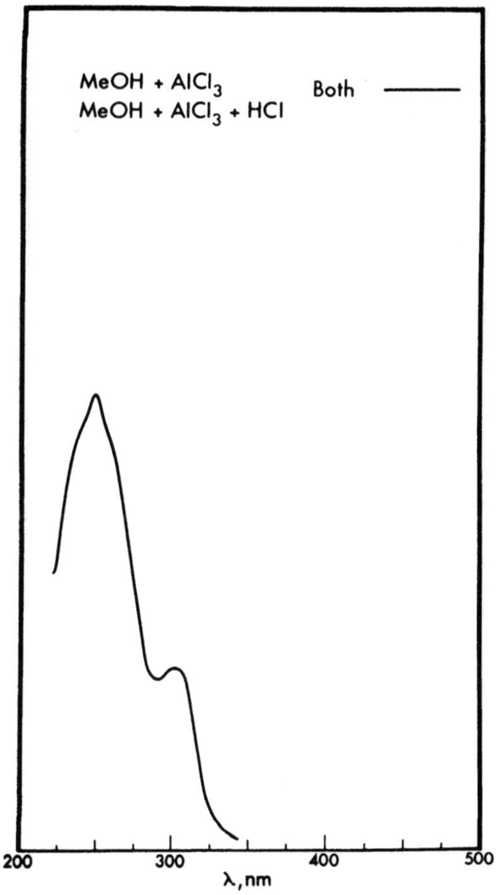

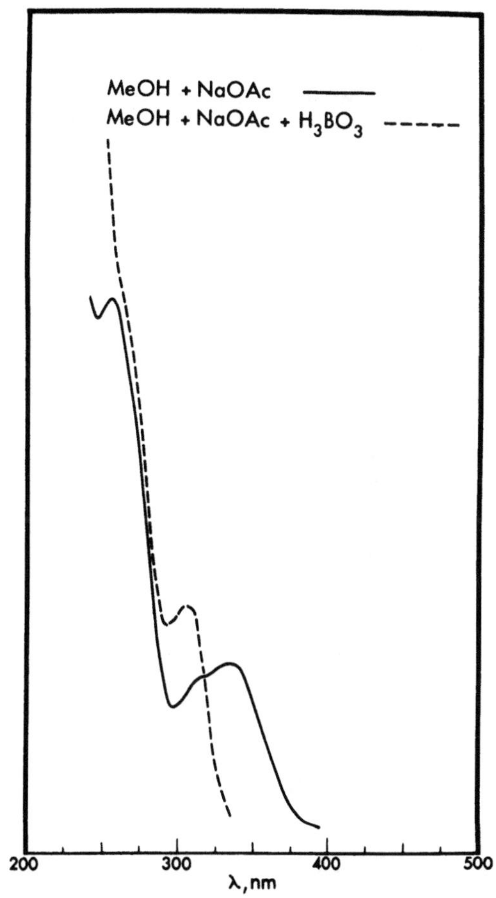

111
FORMONONETIN 7-O-GLUCOSIDE

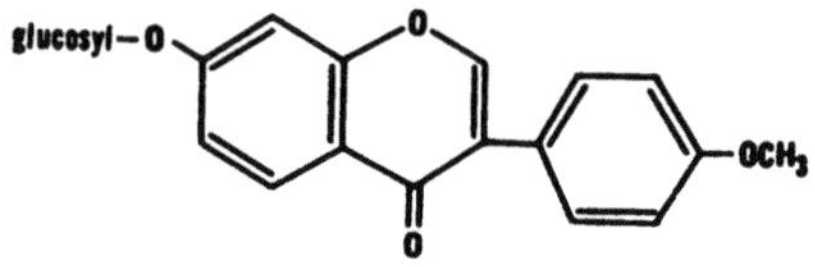

CHROMATOGRAPHIC DATA

Spot Appearance: (UV) fluorescent light blue
(UV/NH_3) fluorescent light blue

R_f Values: 0.66 (TBA), 0.74 (HOAc)

UV SPECTRAL DATA (λ_{max},*nm*)

MeOH	251sh, 258, 301sh
NaOMe	250sh, 258, 301sh
$AlCl_3$	251sh, 259, 300sh
$AlCl_3$/HCl	250sh, 257, 301sh
NaOAc	257, 304sh
NaOAc/H_3BO_3 (Proc. I)	255, 302sh

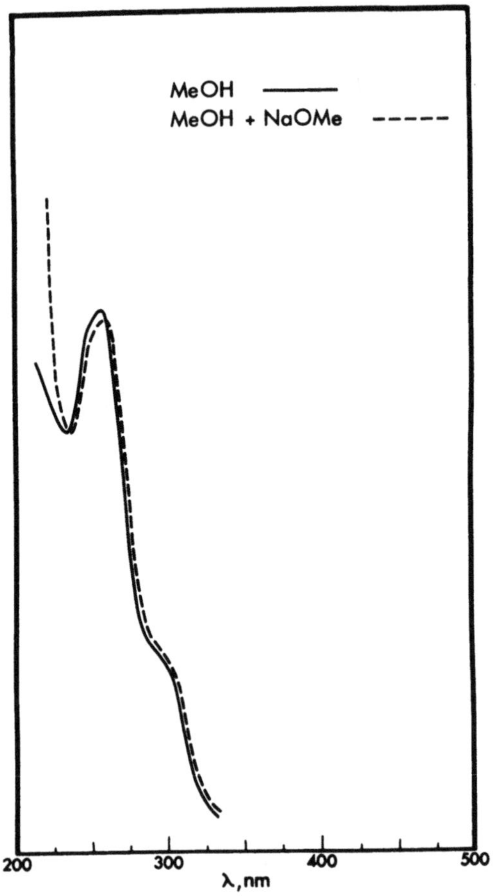

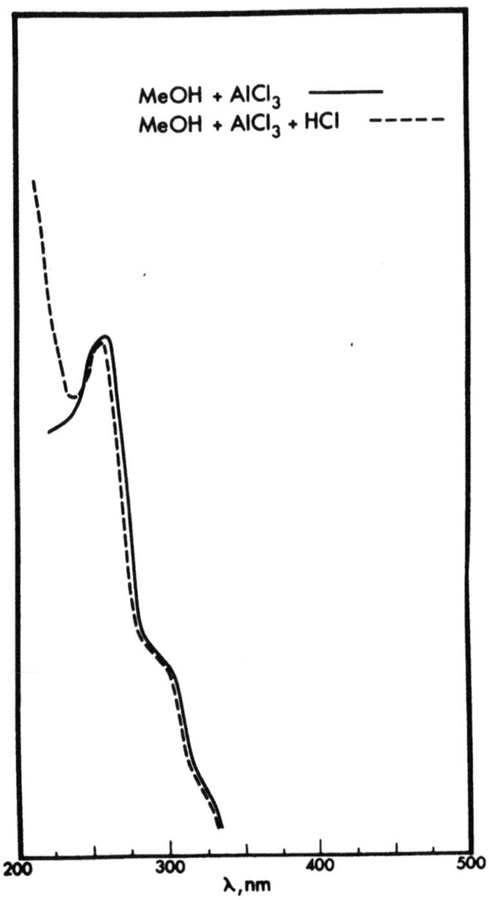

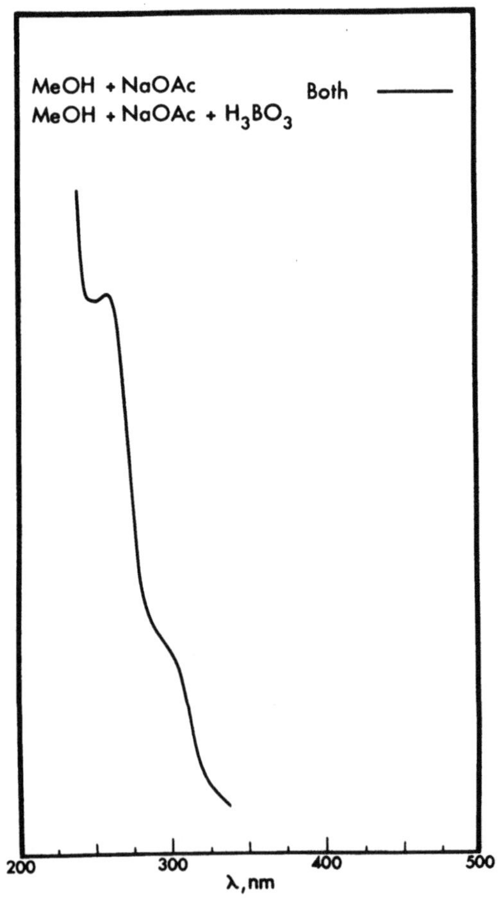

112

FORMONONETIN 7-0-GLUCOSIDE TETRAACETATE

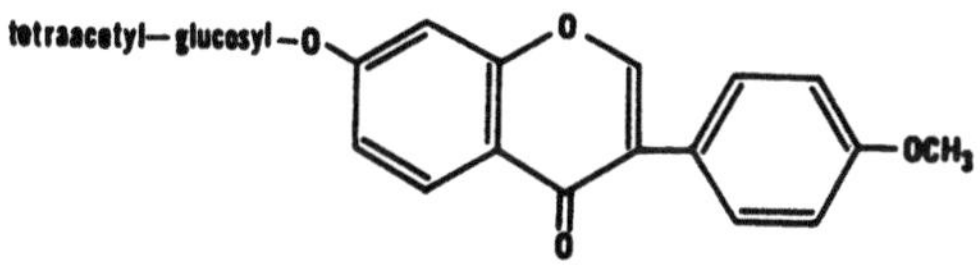

CHROMATOGRAPHIC DATA

Spot Appearance: (UV) fluorescent light blue

(UV/NH_3) fluorescent light blue

R_f Values: 0.93 (TBA), 0.01 (HOAc)

UV SPECTRAL DATA (λ_{max},*nm*)

MeOH	250sh, 258, 302sh
NaOMe	251sh, 259, 302sh
$AlCl_3$	251sh, 259, 302sh
$AlCl_3$/HCl	251sh, 260, 304sh
NaOAc	259, 305sh
NaOAc/H_3BO_3 (Proc. I)	259, 302sh

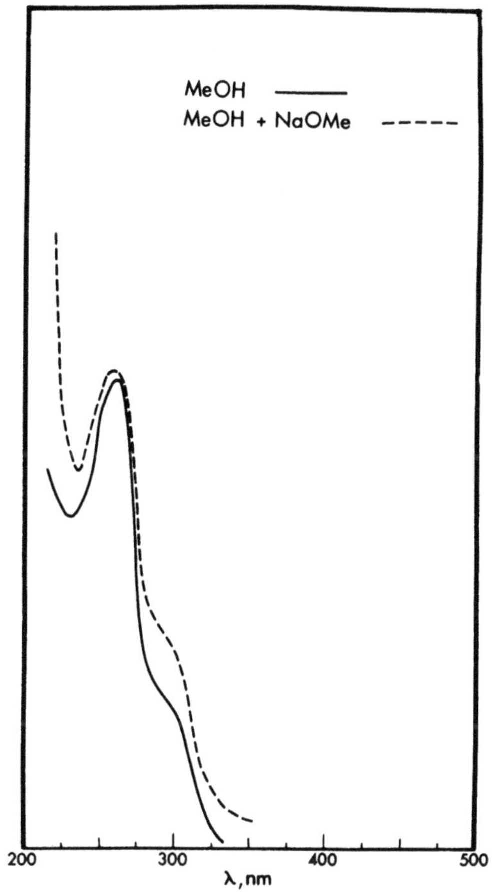

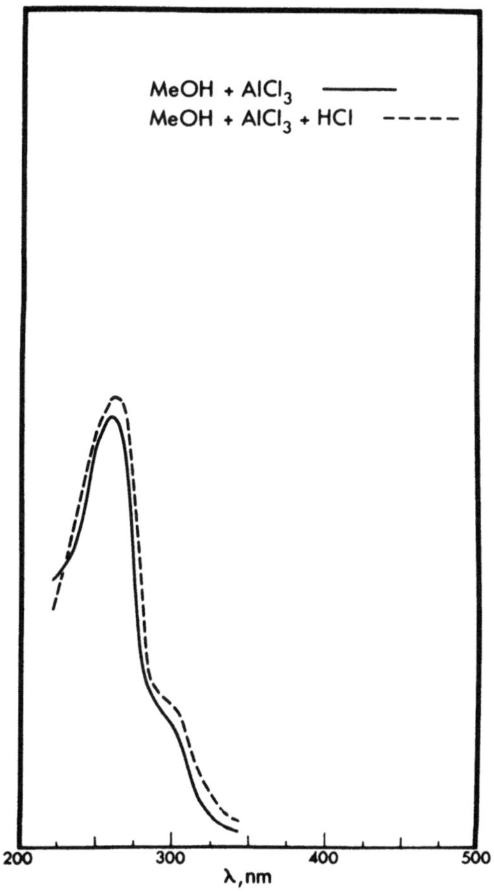

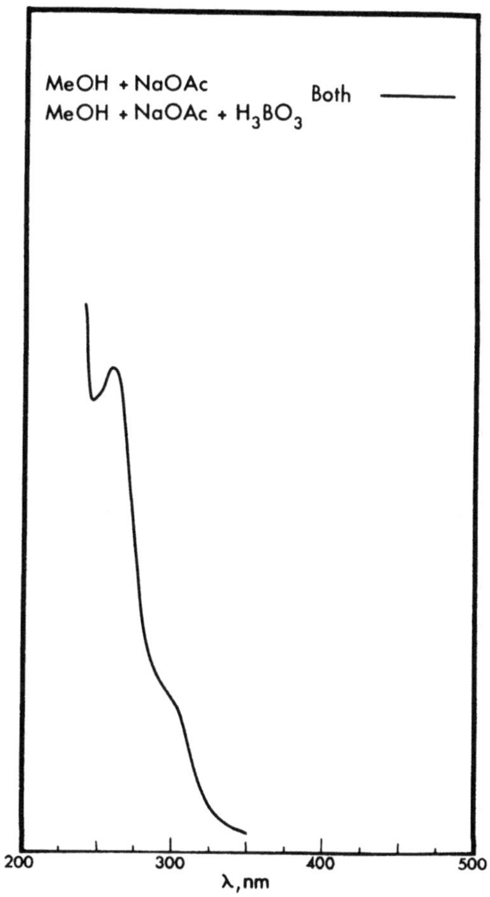

GENISTEIN

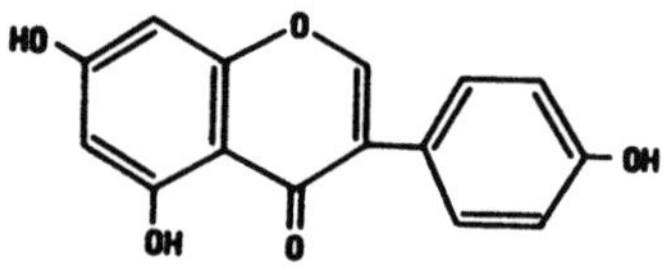

CHROMATOGRAPHIC DATA

Spot Appearance: (UV) deep purple
(UV/NH_3) deep purple

R_f Values: 0.85 (TBA), 0.30 (HOAc)

UV SPECTRAL DATA (λ_{max},*nm*)

MeOH	261, 328sh
NaOMe	276, 327sh
$AlCl_3$	272, 307sh, 372
$AlCl_3/HCl$	273, 309sh, 372
NaOAc	271, 325
$NaOAc/H_3BO_3$ (Proc. I)	262, 336sh

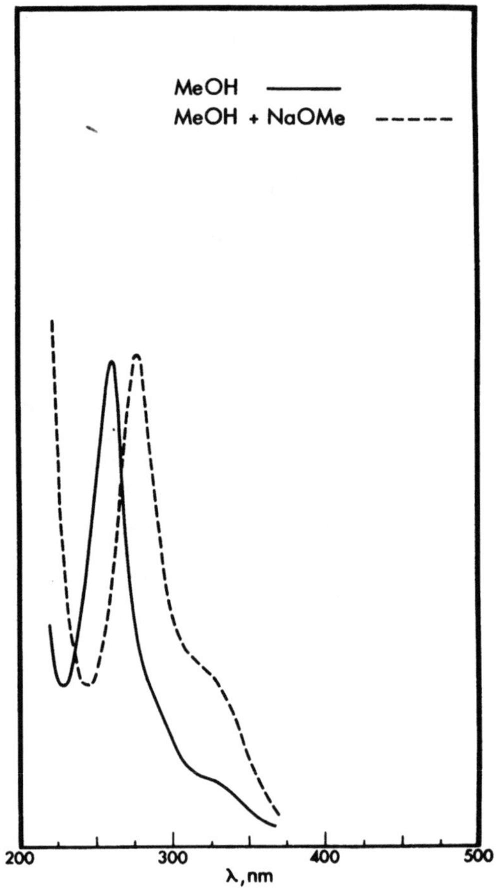

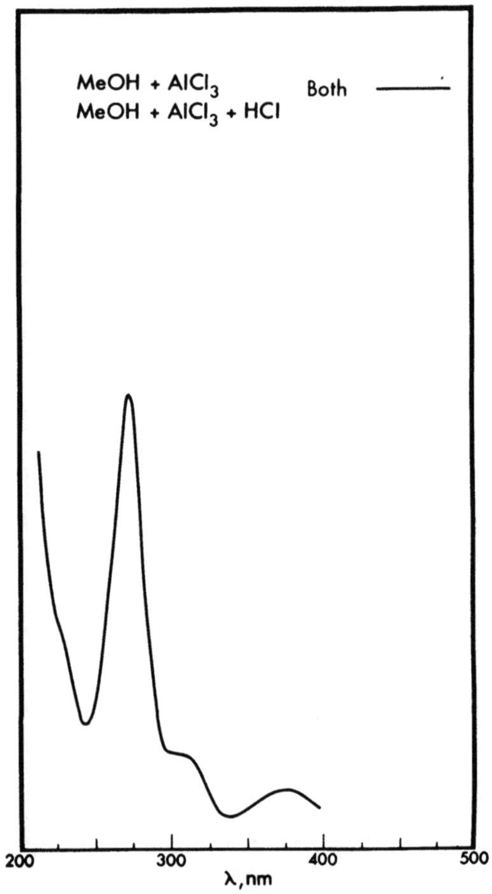

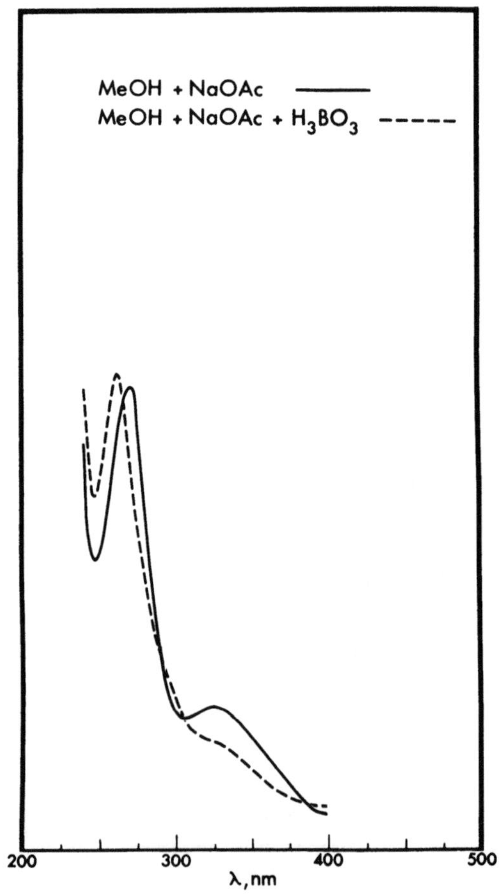

GENISTIN

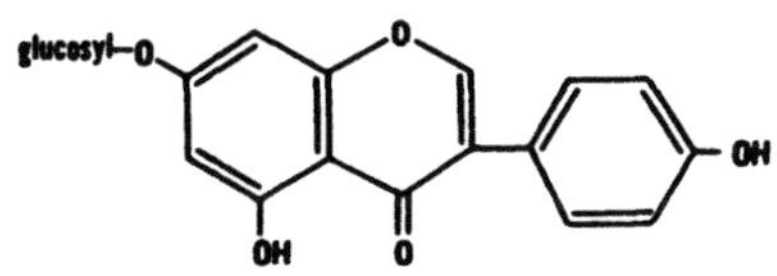

CHROMATOGRAPHIC DATA

Spot Appearance: (UV) deep purple
(UV/NH_3) deep purple

R_f Values: 0.55 (TBA), 0.63 (HOAc)

UV SPECTRAL DATA (λ_{max}, nm)

MeOH	261, 330sh
NaOMe	271, 356sh
$AlCl_3$	272, 308sh, 375
$AlCl_3$/HCl	272, 307sh, 374
NaOAc	261, 331sh
NaOAc/H_3BO_3 (Proc. I)	261, 328sh

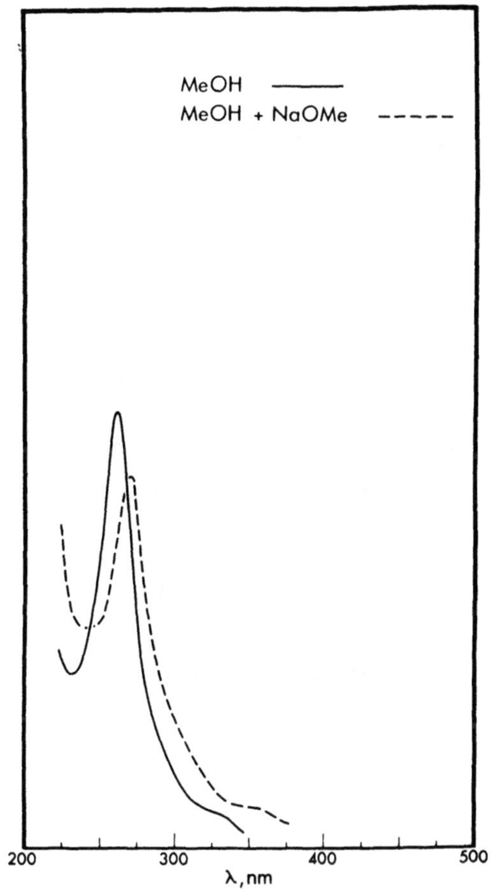

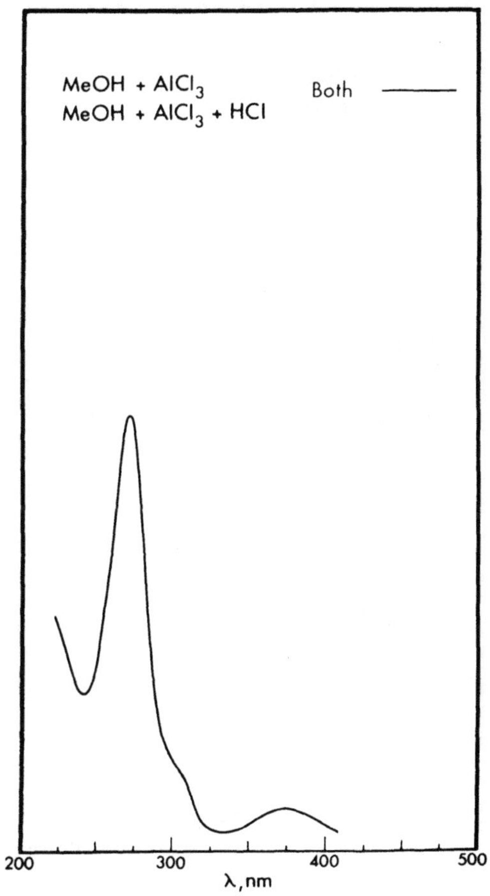

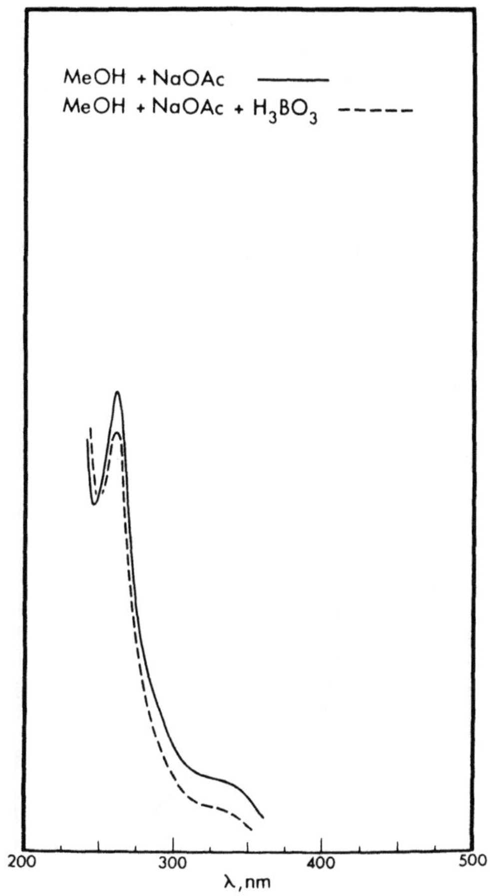

115

SPHAEROBIOSIDE

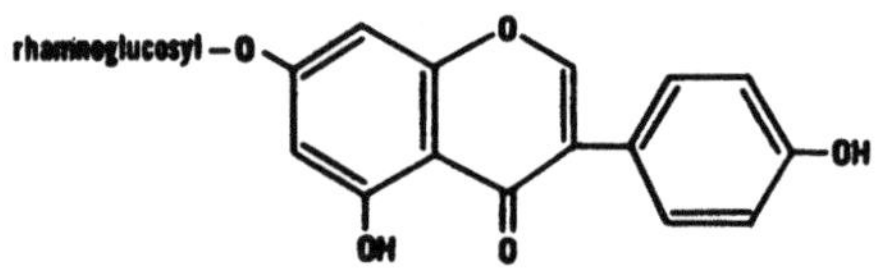

CHROMATOGRAPHIC DATA

Spot Appearance: (UV) deep purple
(UV/NH_3) deep purple

R_f Values: 0.57 (TBA), 0.74 (HOAc)

UV SPECTRAL DATA (λ_{max},*nm*)

MeOH	262, 327sh
NaOMe	270, 307sh, 351sh
$AlCl_3$	271, 308sh, 378
$AlCl_3$/HCl	272, 307sh, 378
NaOAc	262, 289sh, 325sh
NaOAc/H_3BO_3 (Proc. I)	262, 289sh, 325sh

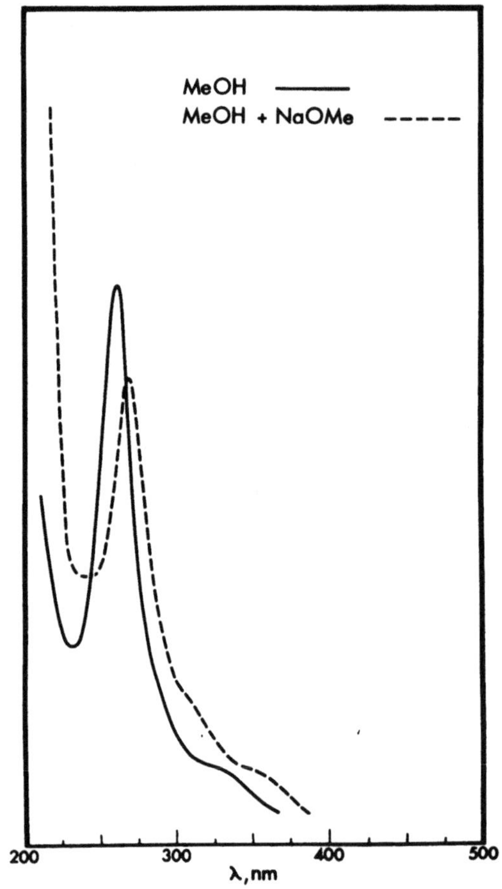

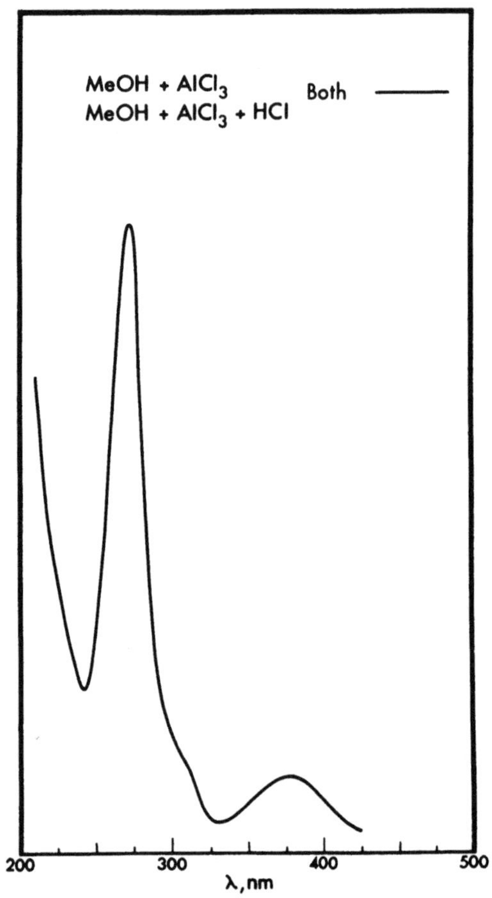

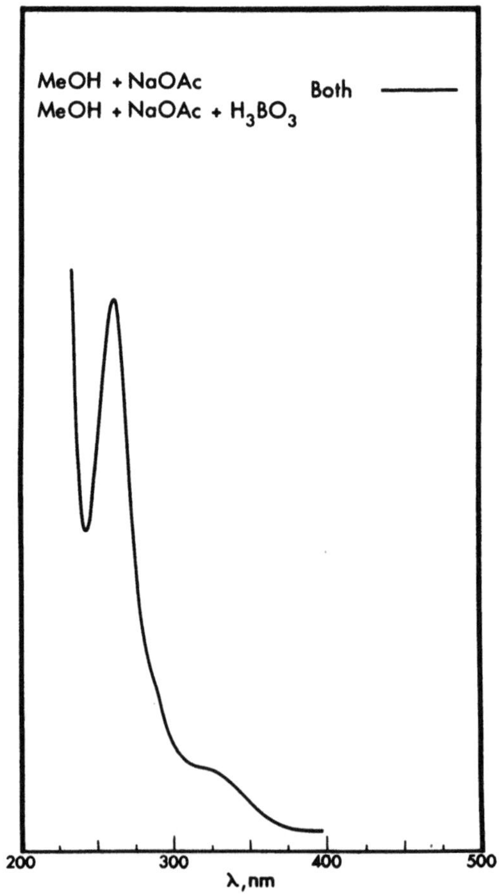

SOPHORICOSIDE

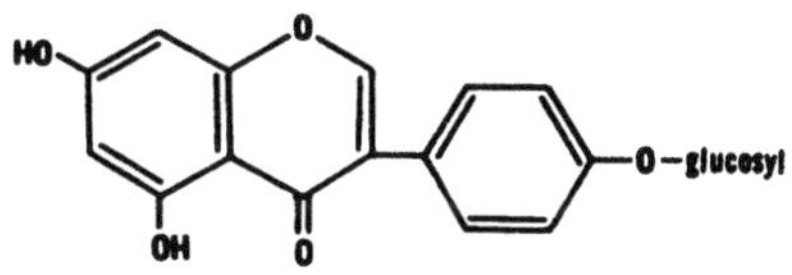

CHROMATOGRAPHIC DATA

Spot Appearance: (UV) deep purple
(UV/NH_3) deep purple

R_f Values: 0.60 (TBA), 0.58 (HOAc)

UV SPECTRAL DATA (λ_{max},*nm*)

MeOH	261, 324sh
NaOMe	248sh, 274, 326
$AlCl_3$	273, 311sh, 371
$AlCl_3$/HCl	273, 312sh, 371
NaOAc	272, 326
NaOAc/H_3BO_3 (Proc. I)	262, 327sh

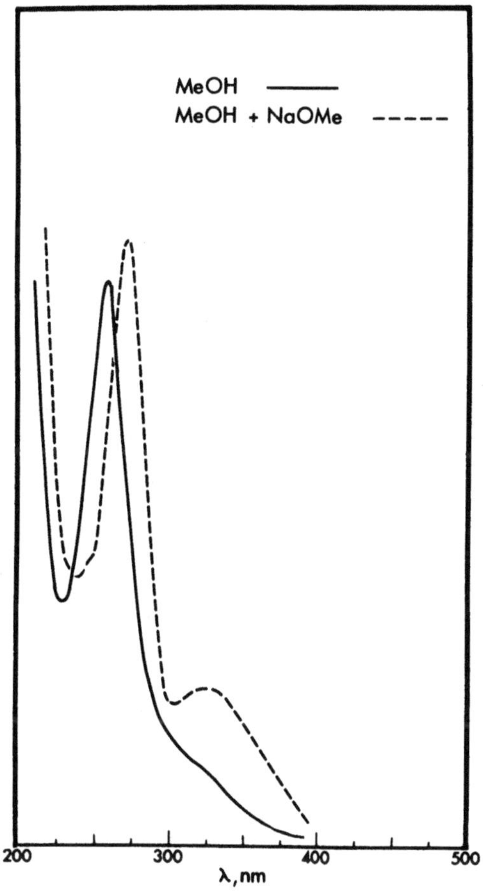

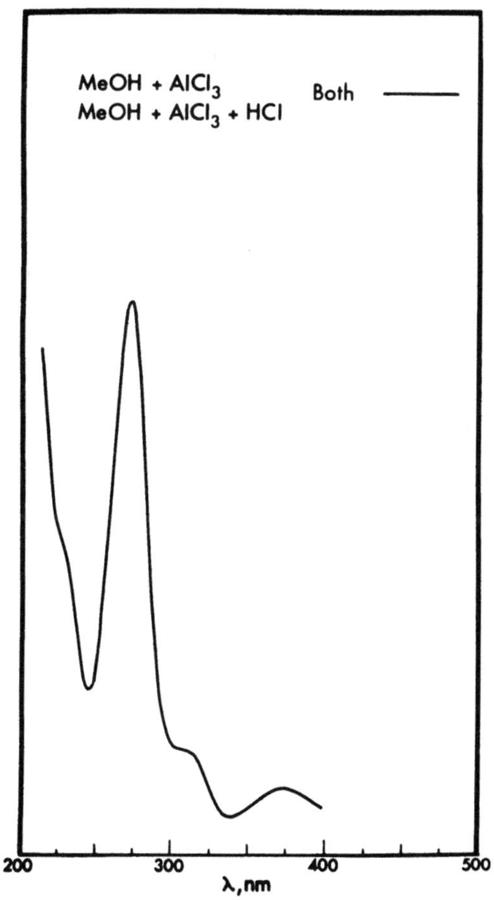

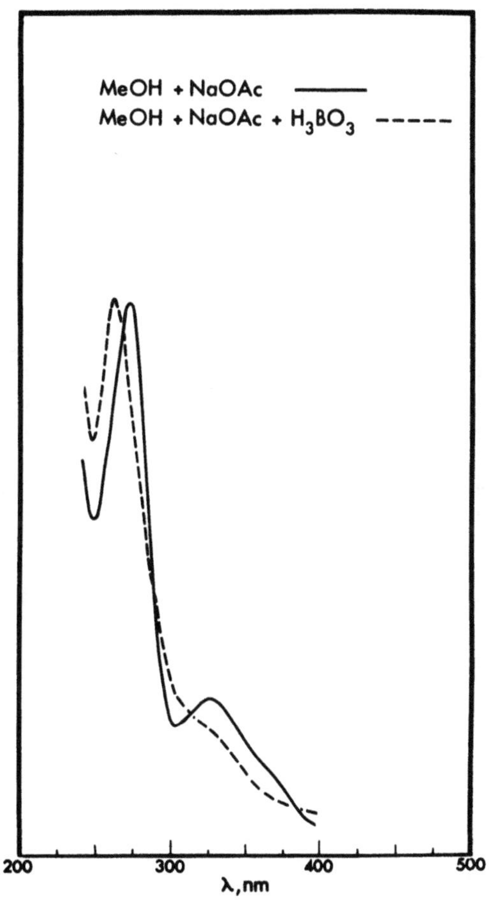

117
GENISTEIN 5-METHYL ETHER

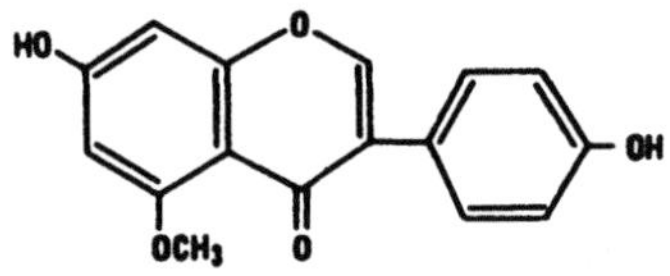

CHROMATOGRAPHIC DATA

Spot Appearance: (UV) invisible
(UV/NH_3) fluorescent light blue

R_f Values: 0.79 (TBA), 0.41 (HOAc)

UV SPECTRAL DATA (λ_{max},*nm*)

MeOH	256, 283sh, 317sh
NaOMe	266, 295sh
$AlCl_3$	256, 286sh, 317sh
$AlCl_3$/HCl	256, 284sh, 316sh
NaOAc	264, 315
NaOAc/H_3BO_3 (Proc. I)	256, 321sh

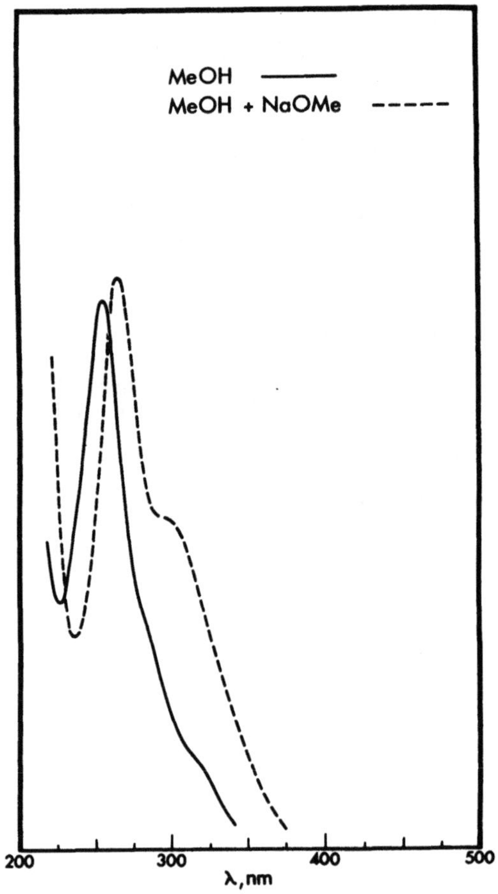

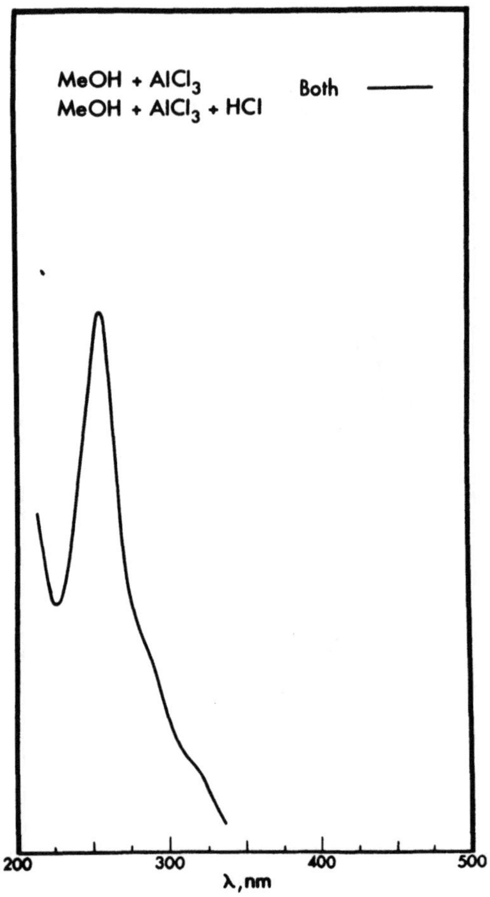

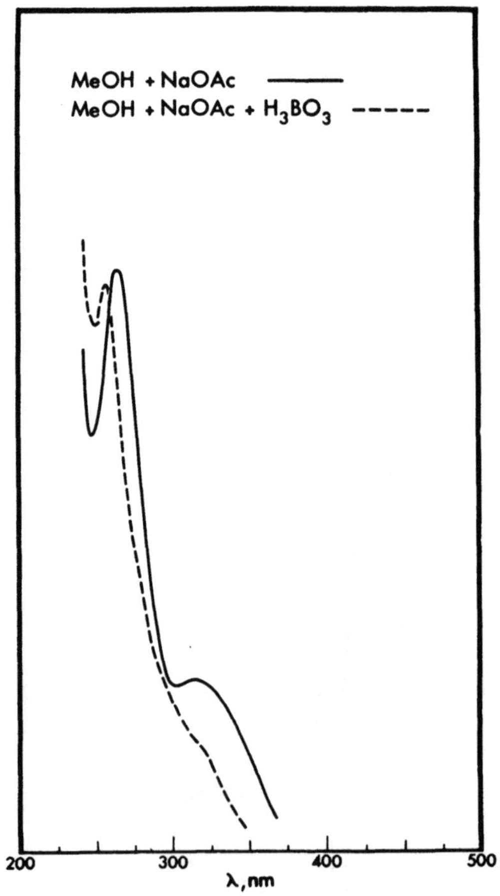

PRUNETIN

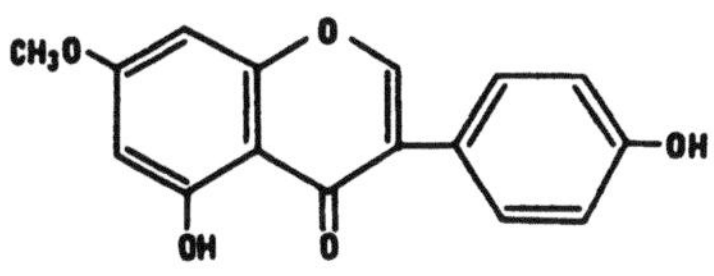

CHROMATOGRAPHIC DATA

Spot Appearance: (UV) deep purple
(UV/NH_3) deep purple

R_f Values: 0.86 (TBA), 0.35 (HOAc)

UV SPECTRAL DATA (λ_{max},*nm*)

MeOH	262, 327sh
NaOMe	272, 353sh
$AlCl_3$	273, 309sh, 374
$AlCl_3$/HCl	274, 310sh, 370
NaOAc	262, 330sh
NaOAc/H_3BO_3	262, 332sh
(Proc. I)	

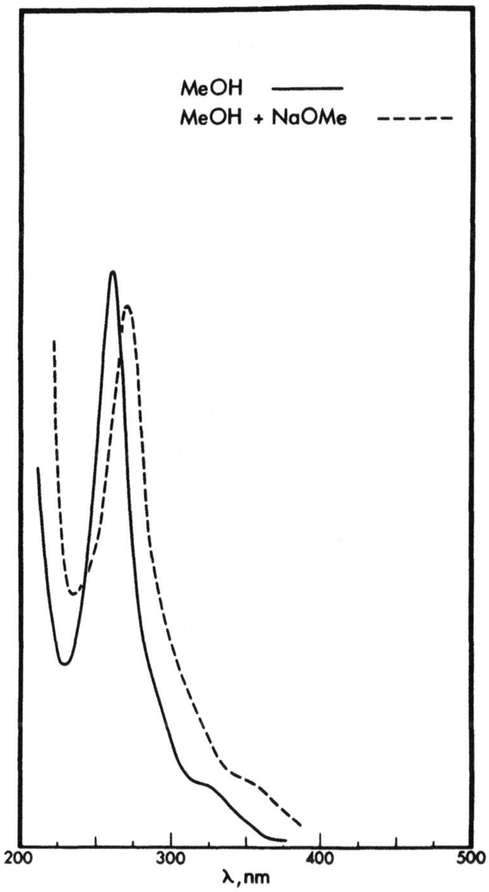

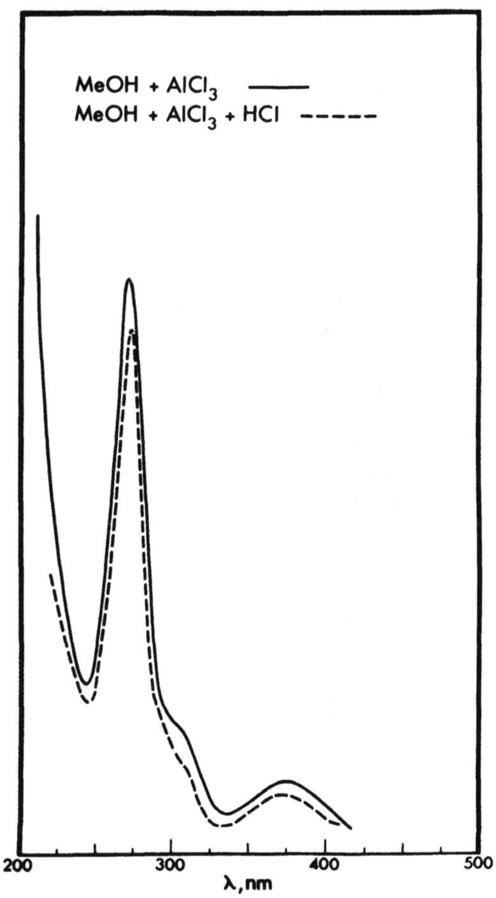

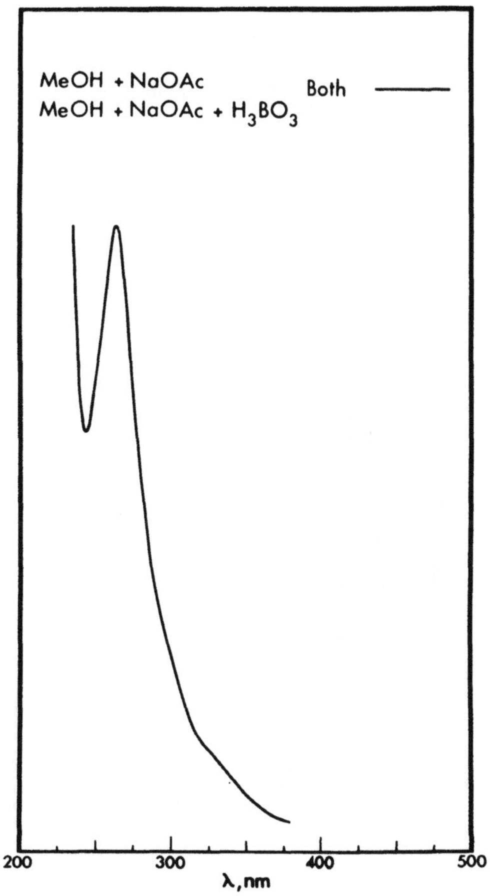

119

BIOCHANIN A

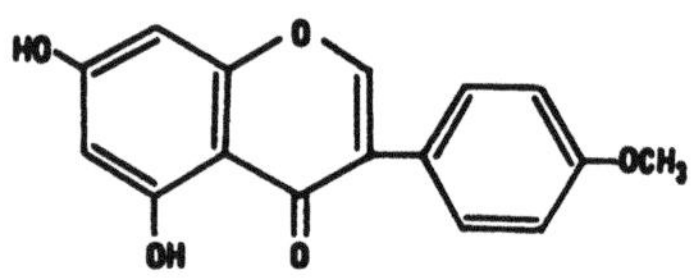

CHROMATOGRAPHIC DATA

Spot Appearance: (UV) deep purple
(UV/NH_3) deep purple

R_f Values: 0.89 (TBA), 0.30 (HOAc)

UV SPECTRAL DATA (λ_{max}, *nm*)

MeOH	261, 330sh
NaOMe	249sh, 273, 327
$AlCl_3$	273, 310sh, 375
$AlCl_3$/HCl	273, 310sh, 373
NaOAc	272, 327
NaOAc/H_3BO_3 (Proc. I)	262. 330sh

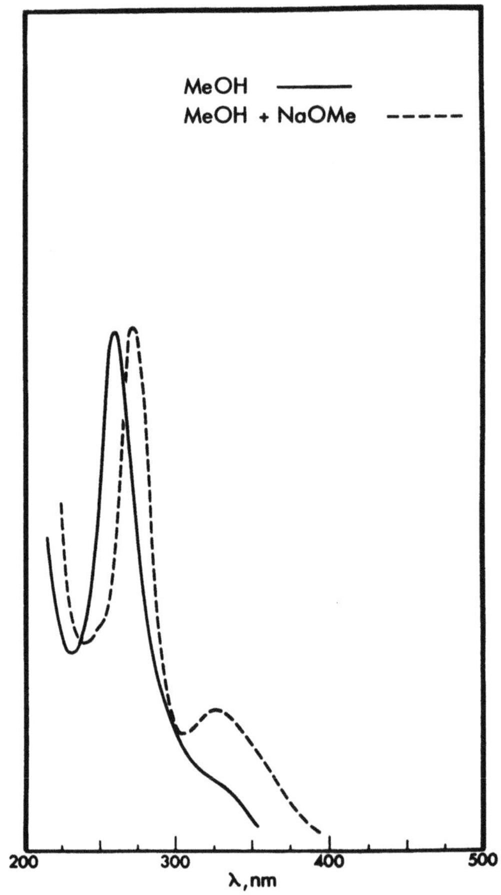

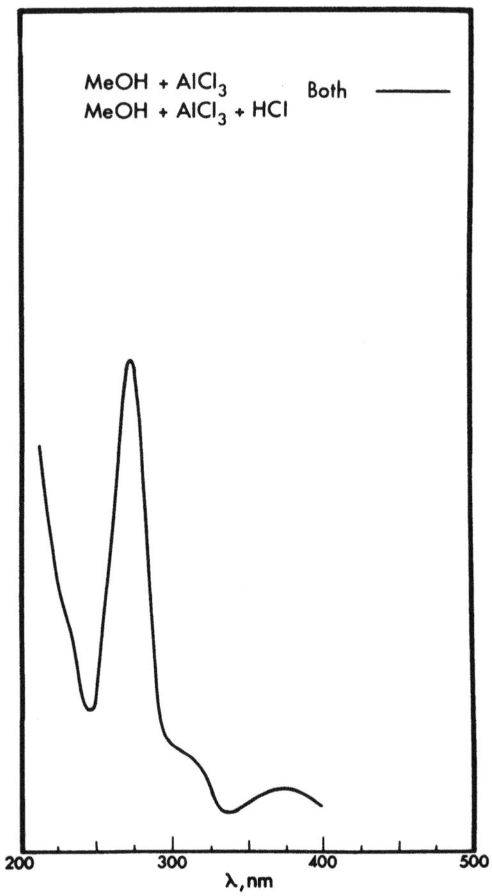

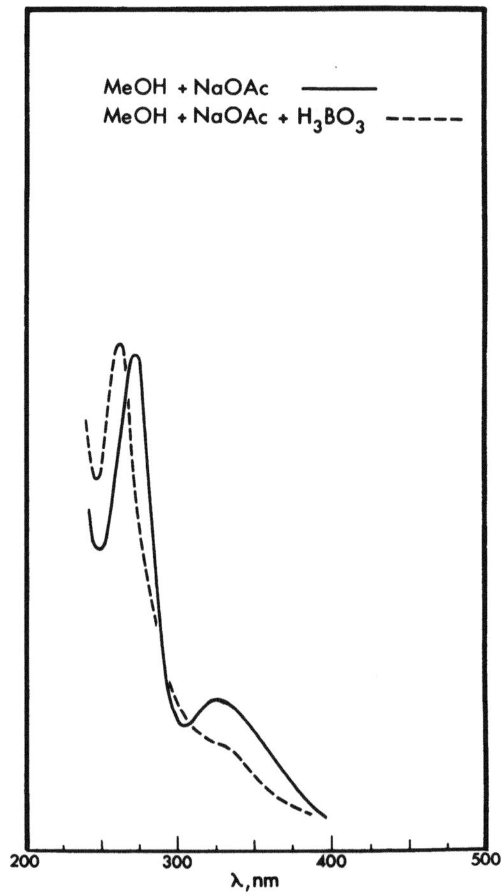

120

LANCEOLARIN

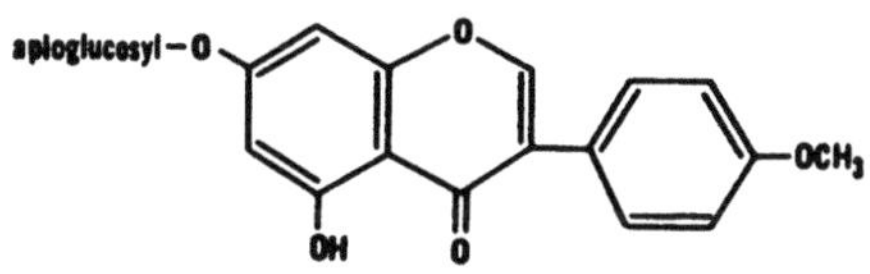

CHROMATOGRAPHIC DATA

Spot Appearance: (UV) deep purple
(UV/NH_3) deep purple

R_f Values: 0.68 (TBA), 0.67 (HOAc)

UV SPECTRAL DATA (λ_{max},nm)

MeOH	262, 325sh
NaOMe	244sh, 267, 368
$AlCl_3$	273, 305sh, 382
$AlCl_3$/HCl	273, 304sh, 380
NaOAc	261, 321sh
NaOAc/H_3BO_3 (Proc. I)	261, 320sh

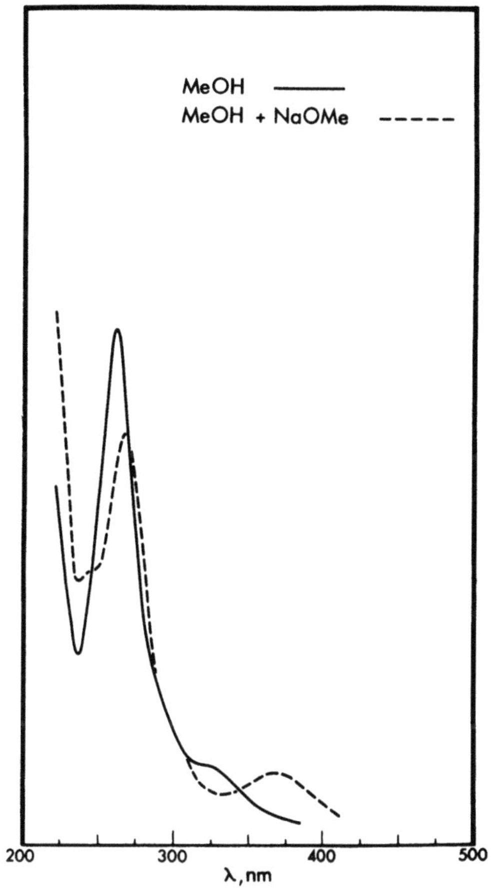

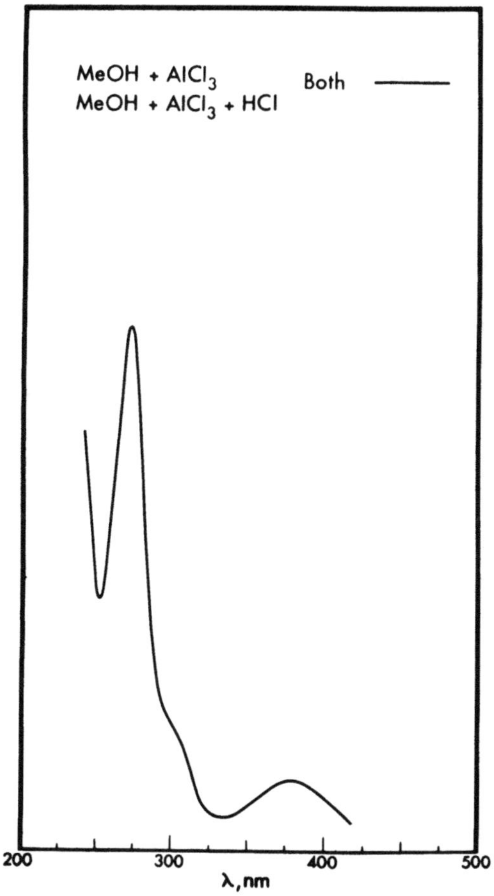

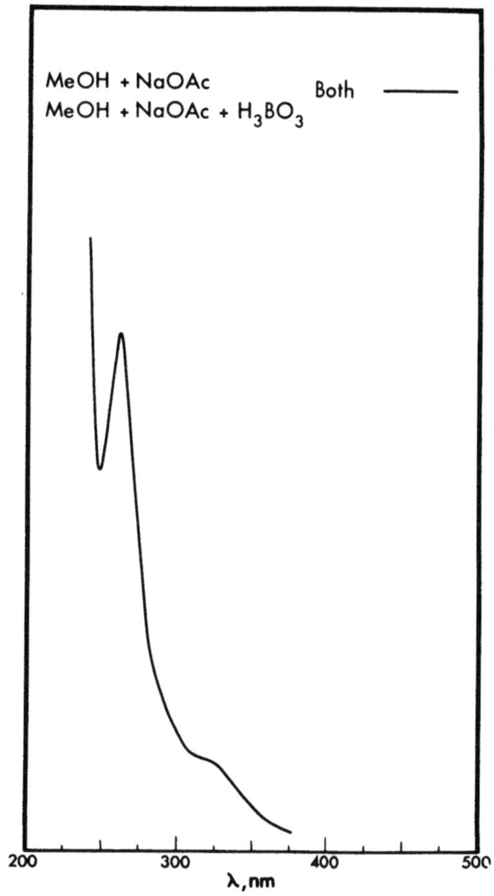

121

TEXASIN

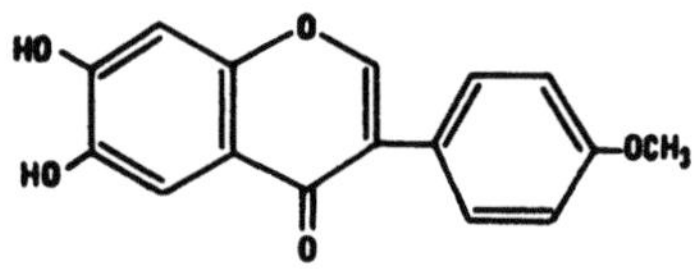

CHROMATOGRAPHIC DATA

Spot Appearance: (UV) fluorescent light blue

(UV/NH_3) fluorescent pale yellow

R_f Values: 0.80 (TBA), 0.31 (HOAc)

UV SPECTRAL DATA (*λ_{max},nm*)

MeOH	255, 325
NaOMe	254, 351
$AlCl_3$	237sh, 251, 344
$AlCl_3$/HCl	257, 325
NaOAc	253sh, 339
NaOAc/H_3BO_3 (Proc. I)	253sh, 338

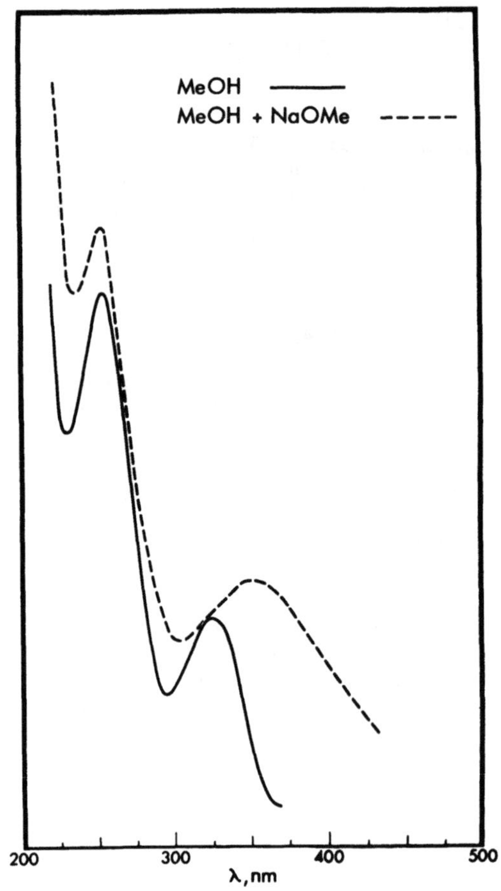

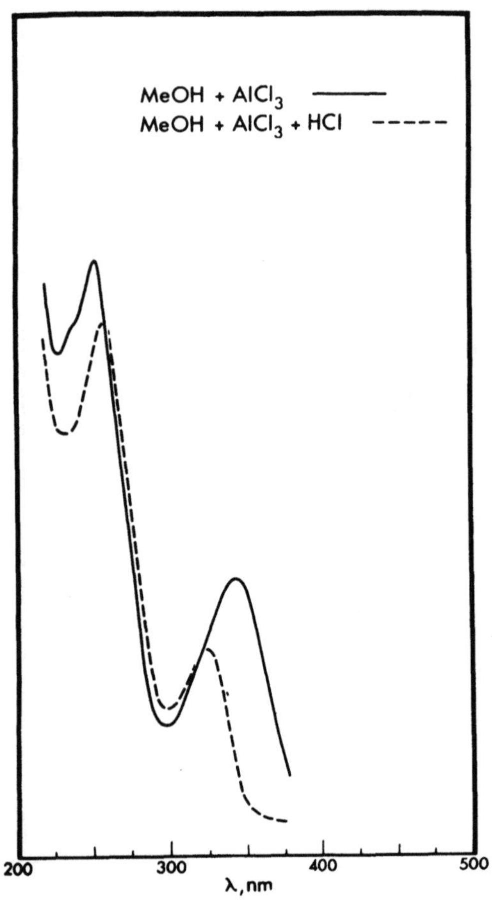

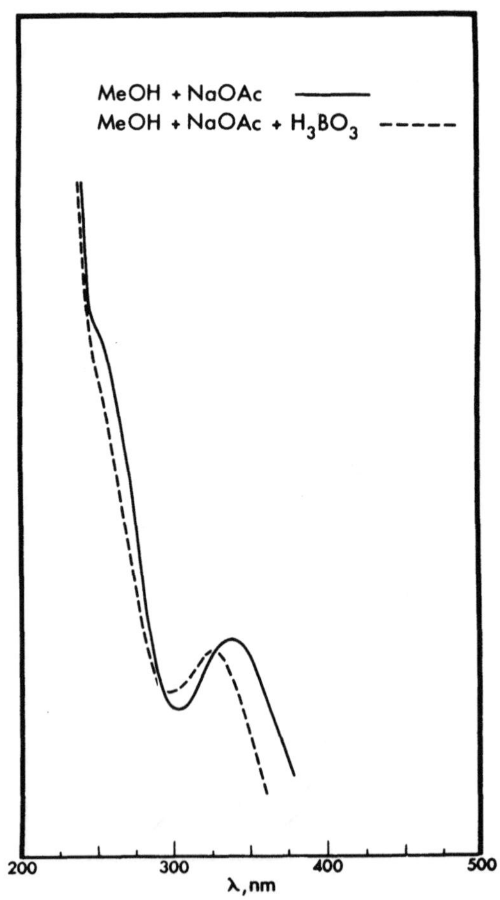

TEXASIN 7-0-GLUCOSIDE

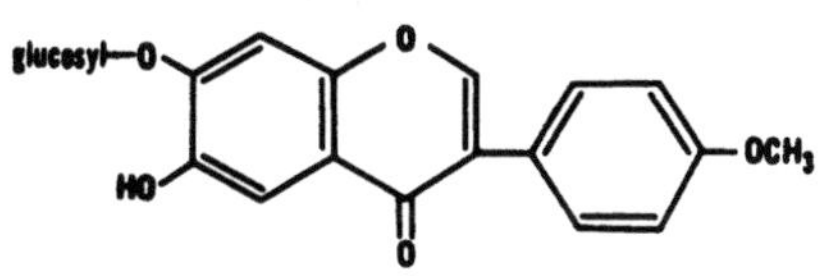

CHROMATOGRAPHIC DATA

Spot Appearance: (UV) fluorescent light blue
(UV/NH_3) fluorescent pale yellow

R_f Values: 0.56 (TBA), 0.66 (HOAc)

UV SPECTRAL DATA (λ_{max},nm)

MeOH	259, 326
NaOMe	255, 278sh, 368
$AlCl_3$	260, 325
$AlCl_3$/HCl	259, 325
NaOAc	257, 333, 366sh
NaOAc/H_3BO_3 (Proc. I)	259, 328

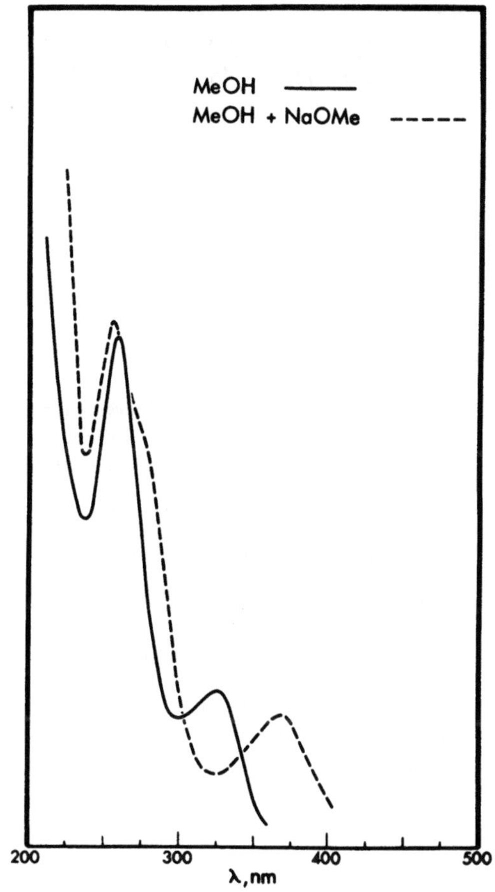

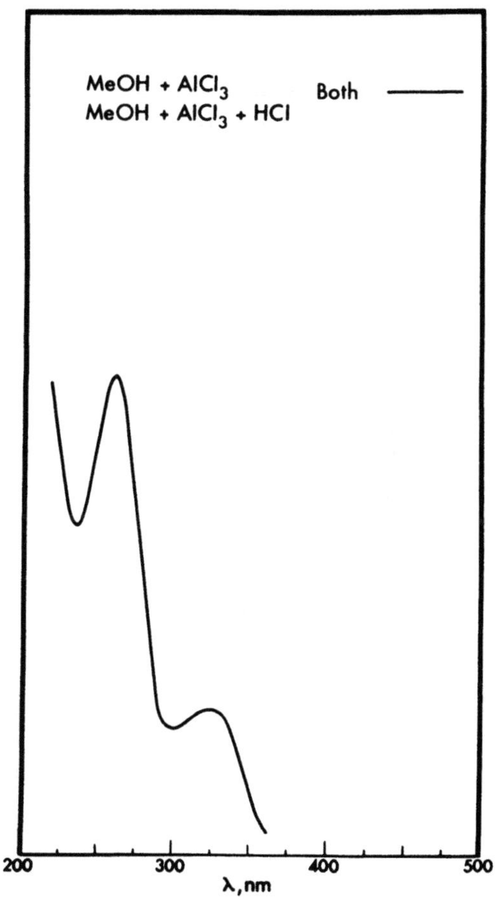

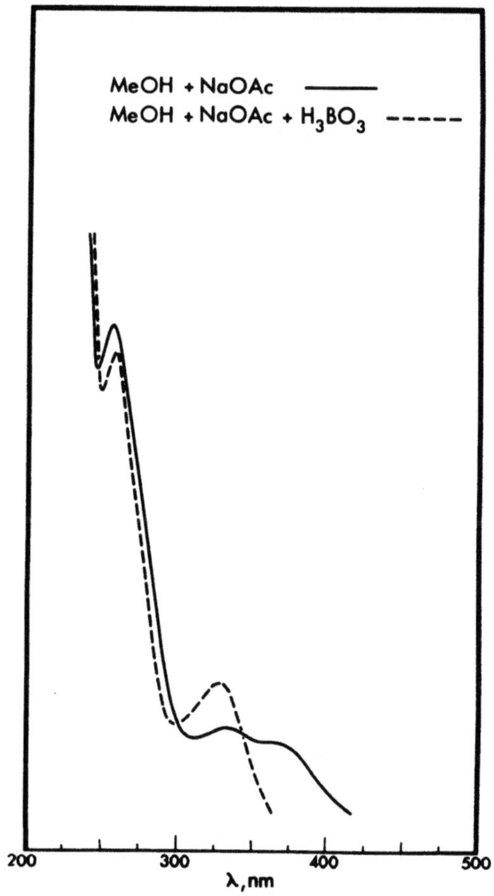

123

2-CARBOXY-6,7-DIHYDROXY-4'-METHOXYISOFLAVONE

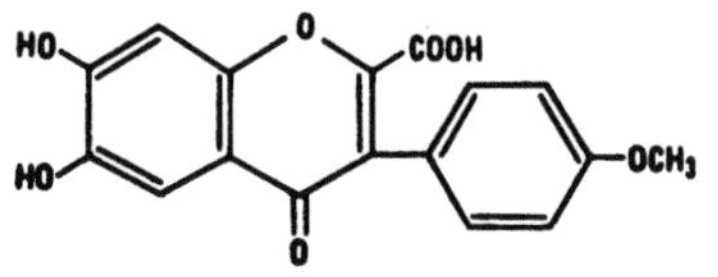

CHROMATOGRAPHIC DATA

Spot Appearance: (UV) fluorescent pale yellow

(UV/NH_3) fluorescent yellow

R_f Values: 0.63 (TBA), 0.71 (HOAc)

UV SPECTRAL DATA (λ_{max},*nm*)

MeOH	238, 254sh, 323
NaOMe	249, 349
$AlCl_3$	246sh, 289sh, 363
$AlCl_3$/HCl	238, 276sh, 336
NaOAc	251sh, 343
NaOAc/H_3BO_3 (Proc. I)	335

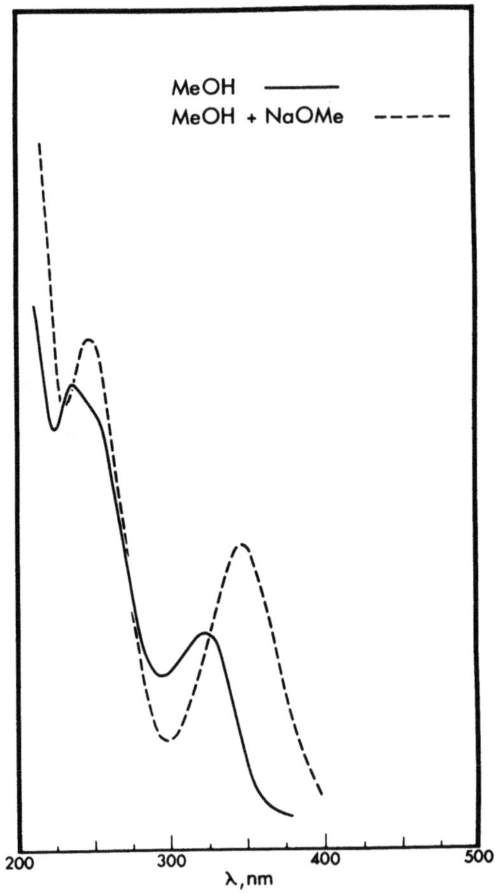

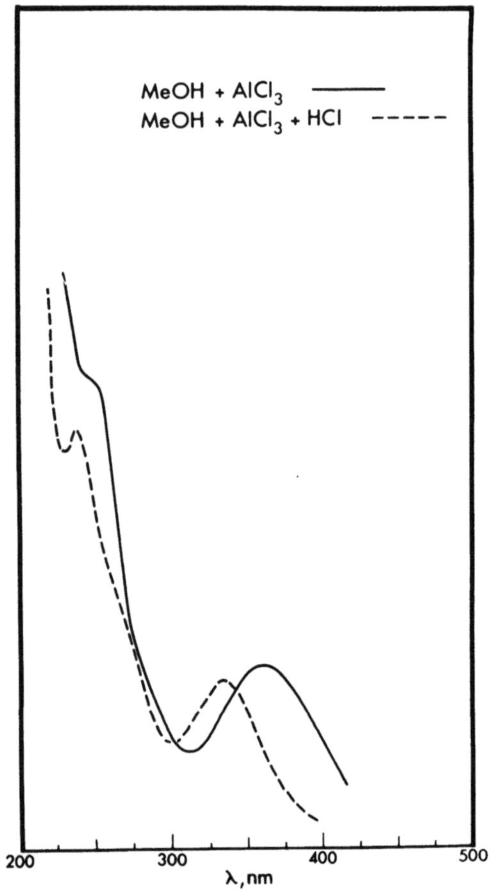

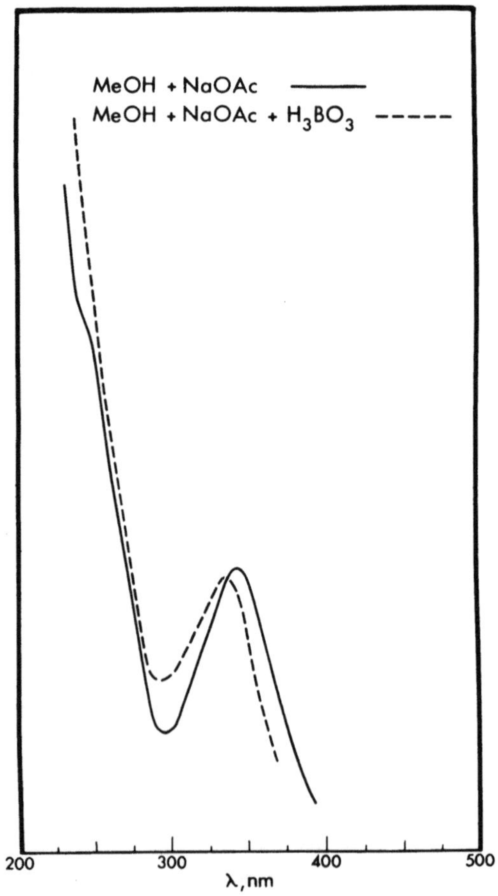

124

AFRORMOSIN

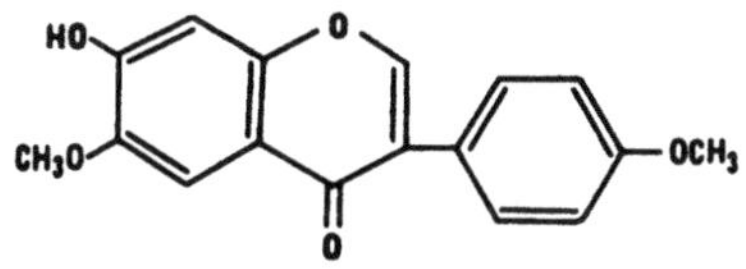

CHROMATOGRAPHIC DATA

Spot Appearance: (UV) fluorescent light blue

(UV/NH_3) brighter fluorescent blue

R_f Values: 0.86 (TBA), 0.39 (HOAc)

UV SPECTRAL DATA (λ_{max}, *nm*)

MeOH	258, 320
NaOMe	258, 349
$AlCl_3$	255, 319
$AlCl_3$/HCl	255, 318
NaOAc	256, 347
NaOAc/H_3BO_3 (Proc. I)	256, 325

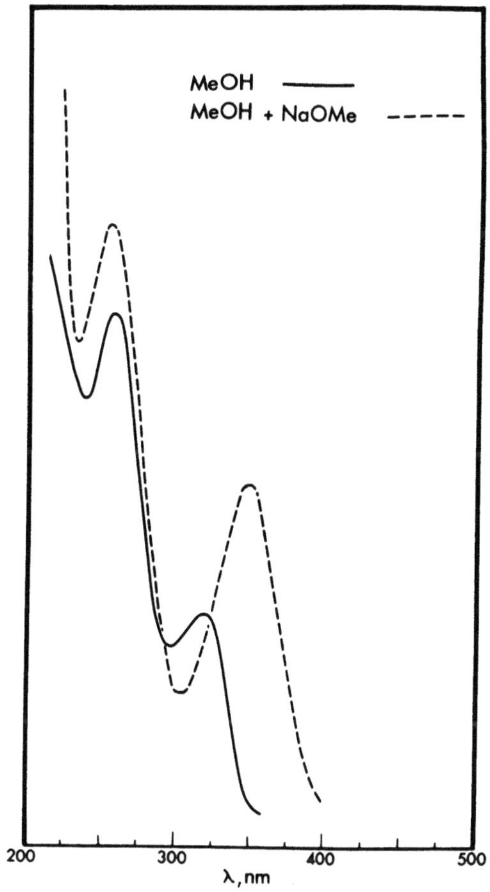

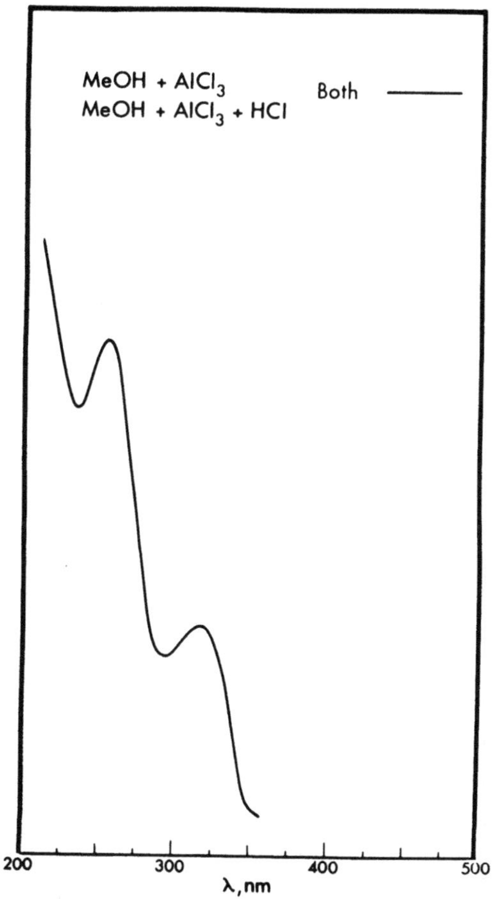

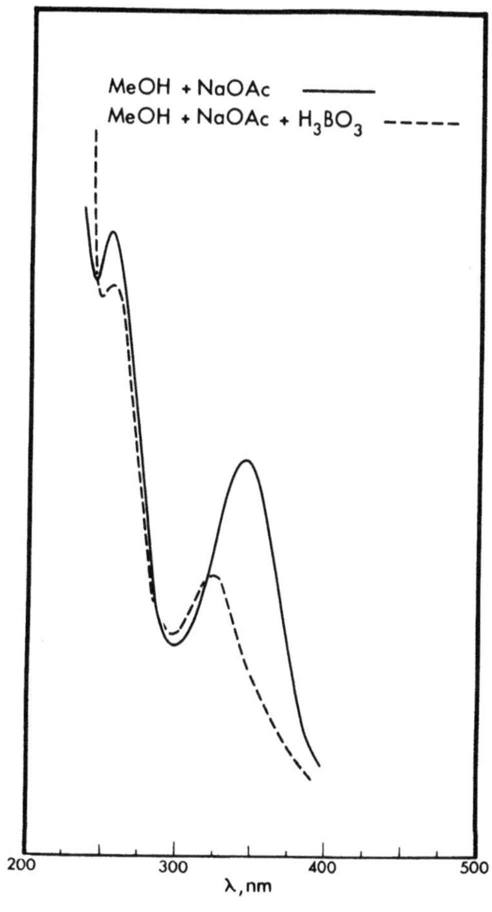

125

3',4',7-TRIHYDROXYISOFLAVONE

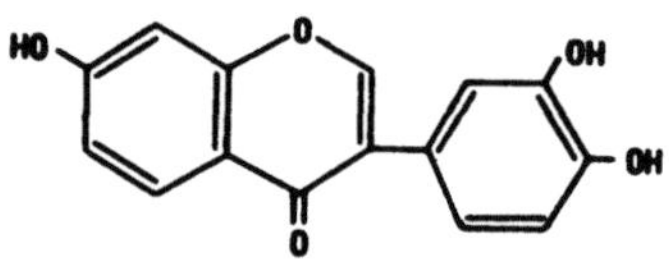

CHROMATOGRAPHIC DATA

Spot Appearance: (UV) invisible
(UV/NH_3) fluorescent light blue

R_f Values: 0.89 (TBA), 0.29 (HOAc)

UV SPECTRAL DATA (λ_{max}, nm)

MeOH	240, 249, 260sh, 293, 308sh
NaOMe	257, 336 (dec.)
$AlCl_3$	246sh, 275, 296, 364sh
$AlCl_3$/HCl	241, 249, 261sh, 292, 309sh
NaOAc	257, 291sh, 331
NaOAc/H_3BO_3 (Proc. II)	271, 297, 351sh

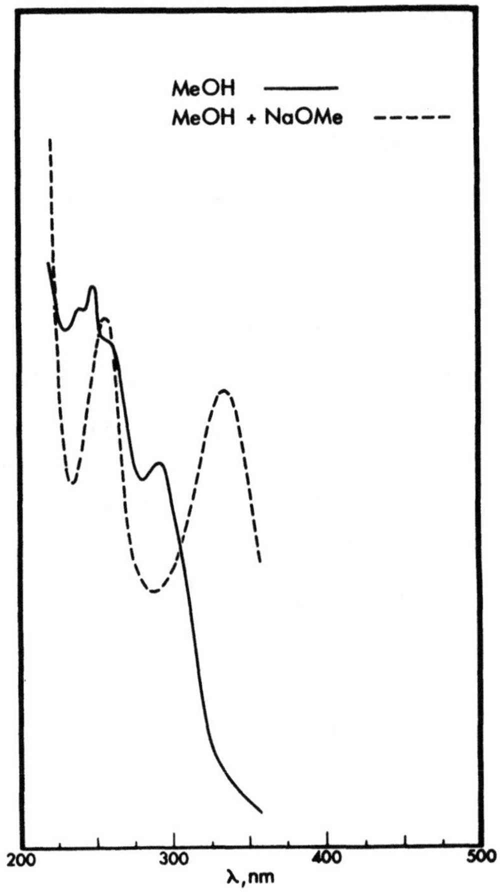

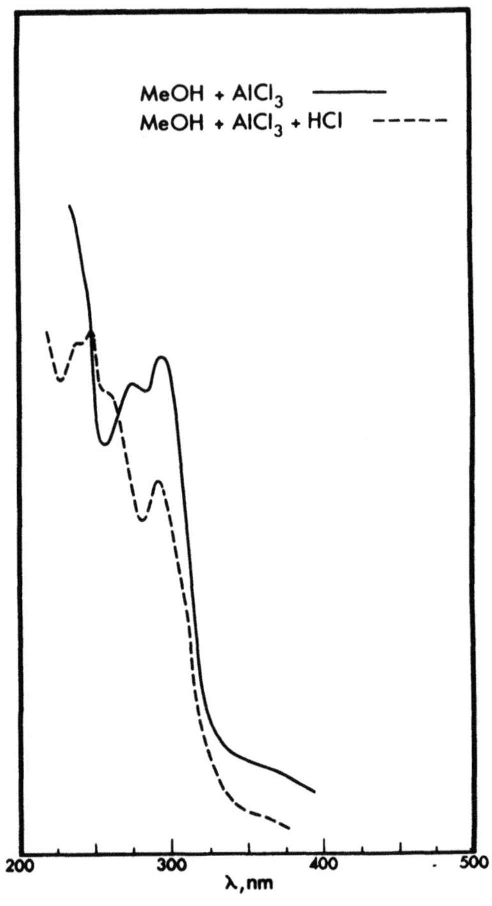

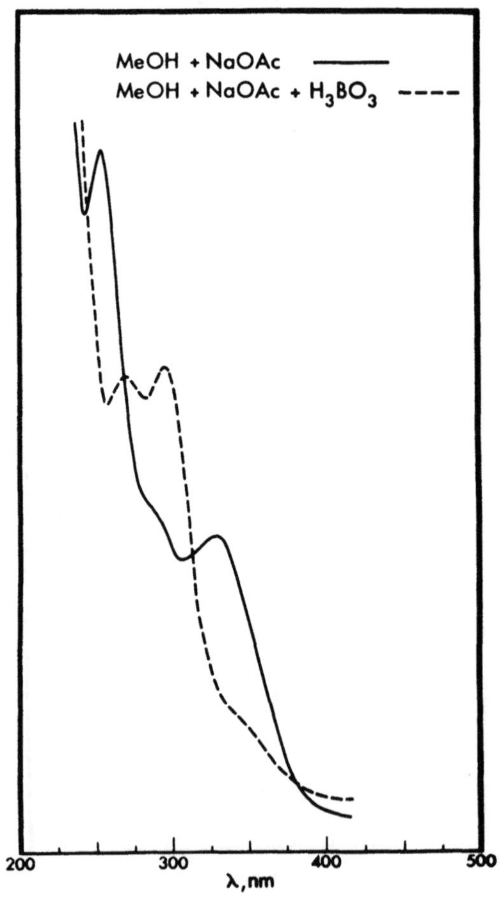

PSEUDOBAPTIGENIN

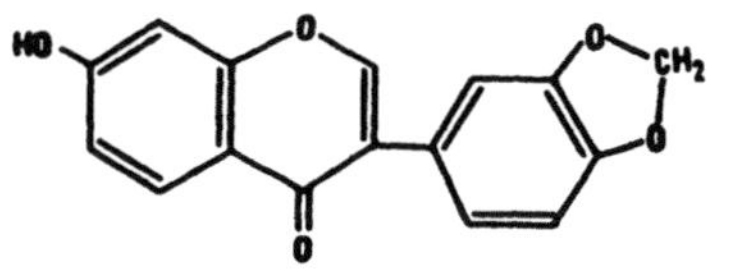

CHROMATOGRAPHIC DATA

Spot Appearance: (UV) invisible
(UV/NH_3) fluorescent light blue

R_f Values: 0.85 (TBA), 0.39 (HOAc)

UV SPECTRAL DATA (λ_{max}, nm)

MeOH	241sh, 250, 262sh, 295, 345sh
NaOMe	259, 293sh, 335 (dec.)
$AlCl_3$	242sh, 249, 264sh, 296
$AlCl_3$/HCl	242sh, 249, 262sh, 295
NaOAc	258, 297sh, 333
NaOAc/H_3BO_3 (Proc. II)	251, 262sh, 296

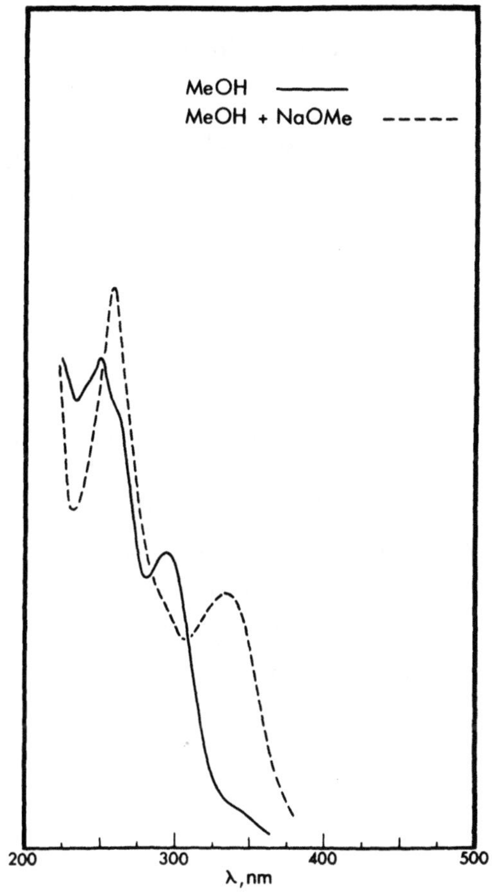

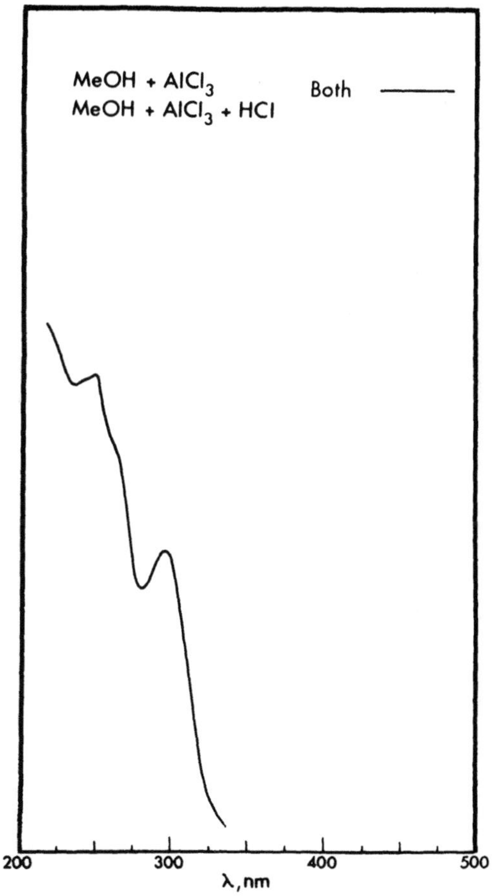

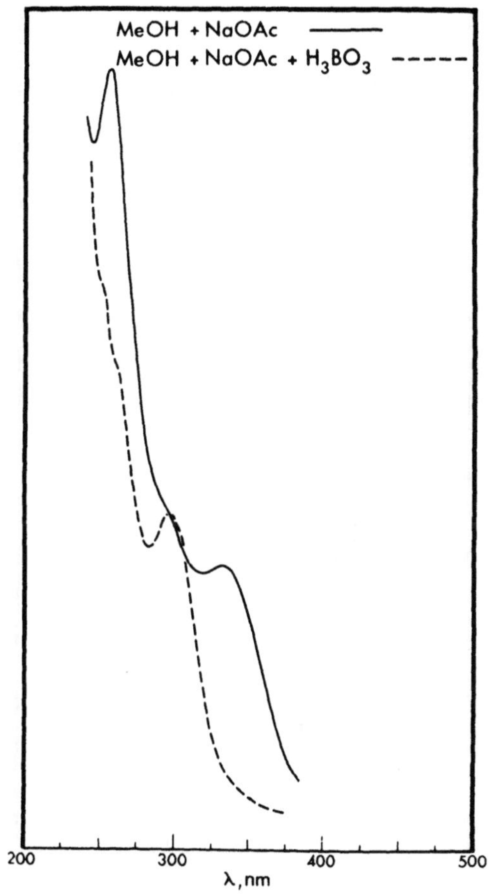

127

PSEUDOBAPTISIN

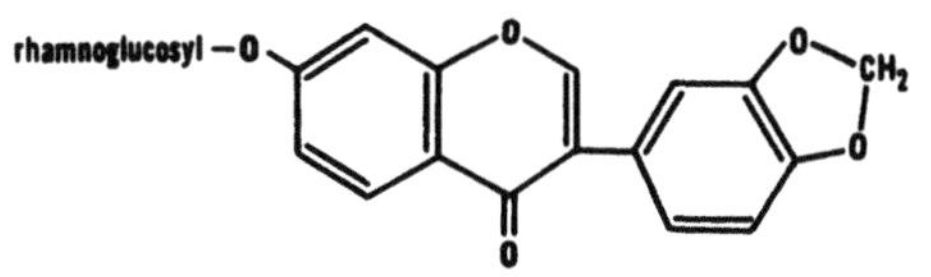

CHROMATOGRAPHIC DATA

Spot Appearance: (UV) fluorescent light blue

(UV/NH_3) fluorescent light blue

R_f Values: 0.55 (TBA), 0.75 (HOAc)

UV SPECTRAL DATA (λ_{max}, *nm*)

MeOH	249, 261, 292
NaOMe	249, 261, 292
$AlCl_3$	250, 262, 291
$AlCl_3$/HCl	249, 261, 291
NaOAc	261, 291
NaOAc/H_3BO_3	261, 291

(Proc. I)

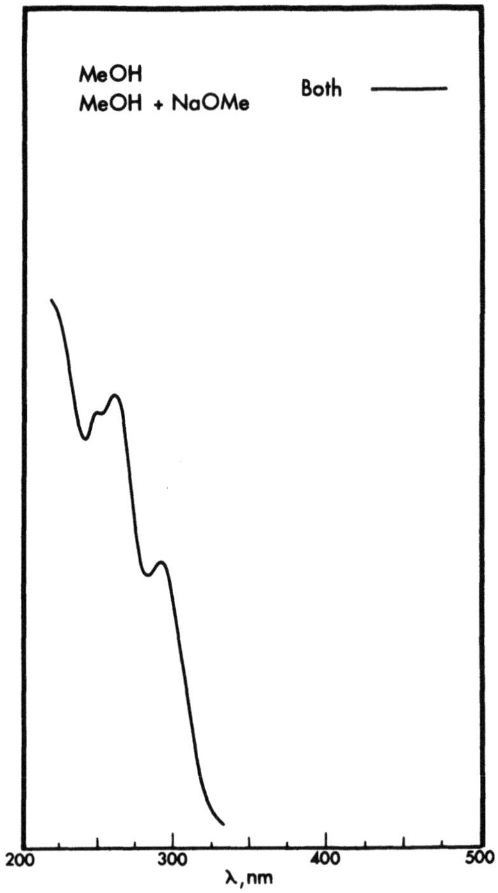

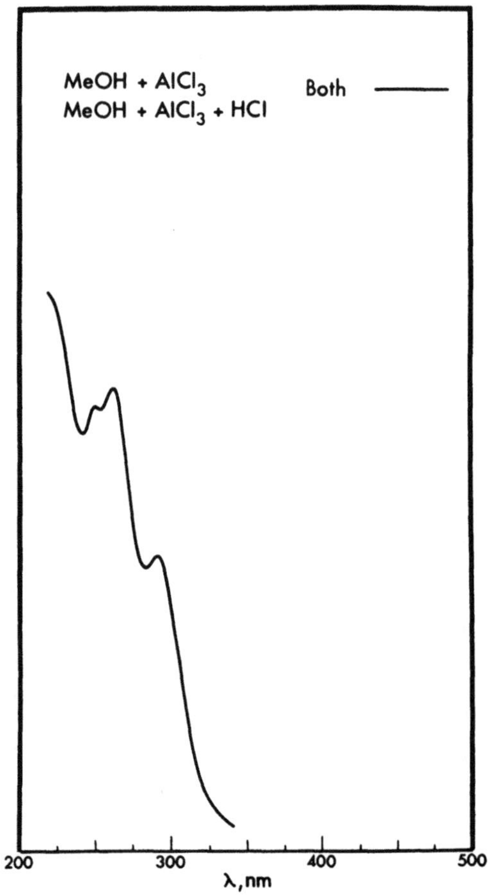

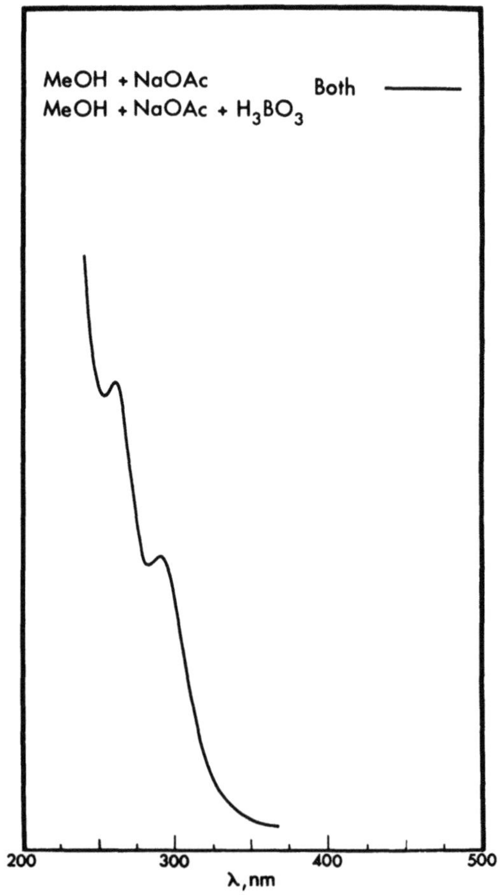

BAPTIGENIN

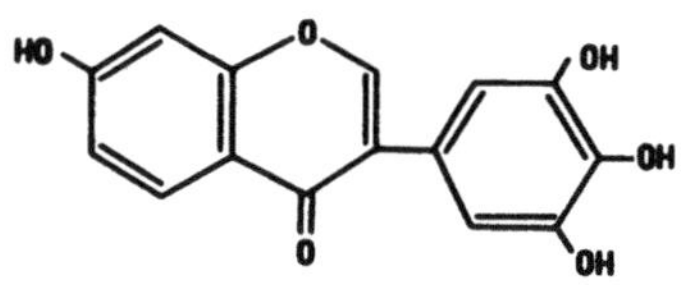

CHROMATOGRAPHIC DATA

Spot Appearance: (UV) fluorescent light blue
(UV/NH_3) brighter fluorescent light blue

R_f Values: 0.87 (TBA), 0.34 (HOAc)

UV SPECTRAL DATA (λ_{max}, nm)

MeOH	239, 247, 265, 304sh
NaOMe	245, 255sh, 286sh, 335 (dec.)
$AlCl_3$	238, 246sh, 283, 302sh
$AlCl_3$/HCl	238, 246, 266, 302sh
NaOAc	255, 285sh, 330
NaOAc/H_3BO_3 (Proc. II)	247sh, 258sh, 304sh

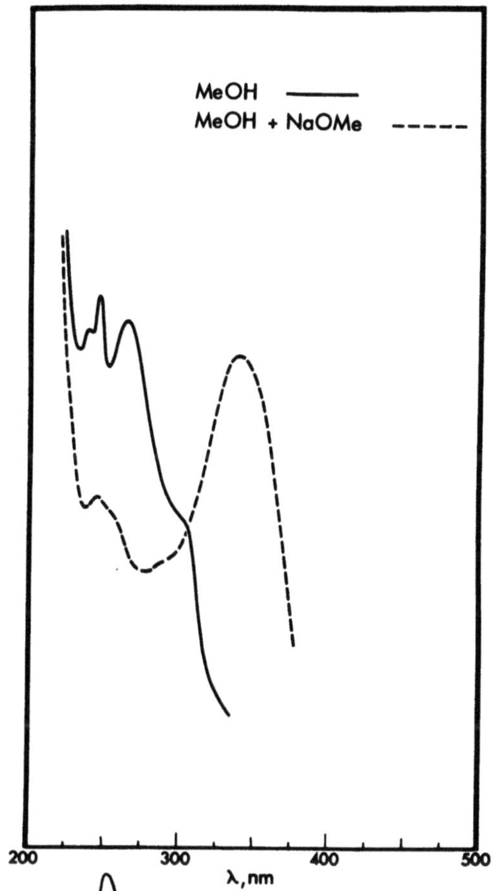

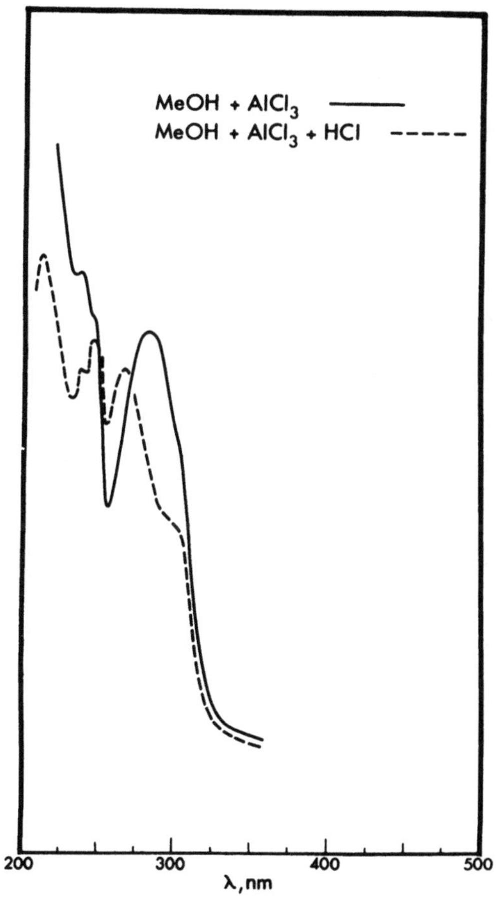

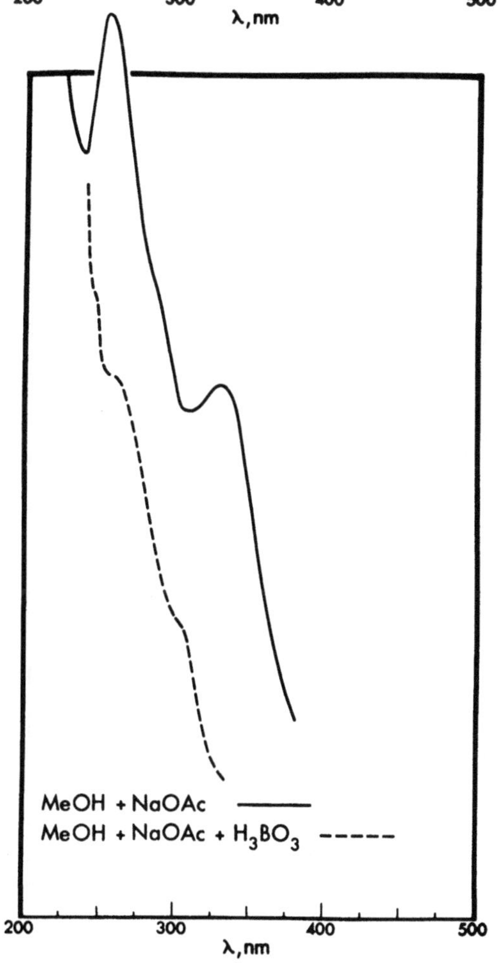

129

6-HYDROXYGENISTEIN

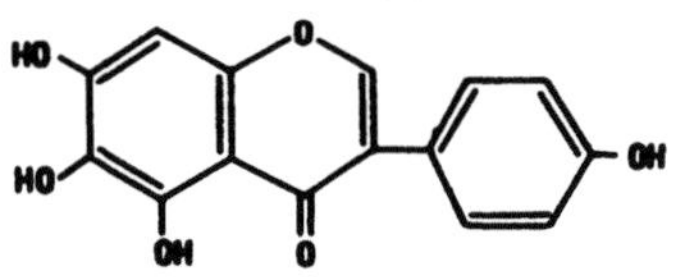

CHROMATOGRAPHIC DATA

Spot Appearance: (UV) deep purple
(UV/NH_3) deep purple

R_f Values: 0.79 (TBA), 0.36 (HOAc)

UV SPECTRAL DATA (λ_{max},*nm*)

MeOH	245sh, 270, 350sh
NaOMe	259, 307, 330sh (dec.)
$AlCl_3$	239, 248sh, 275, 295sh, 356
$AlCl_3$/HCl	281, 329
NaOAc	250sh, 303, 338sh, 418 (dec.)
NaOAc/H_3BO_3 (Proc. II)	275, 320

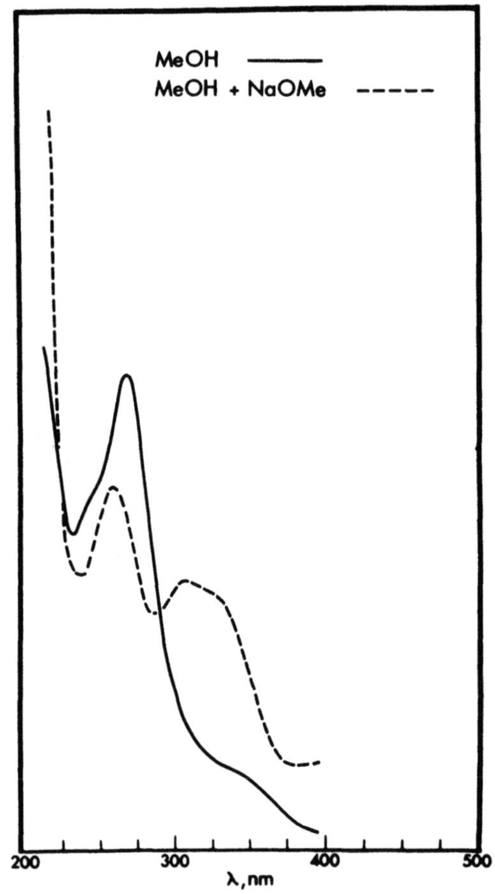

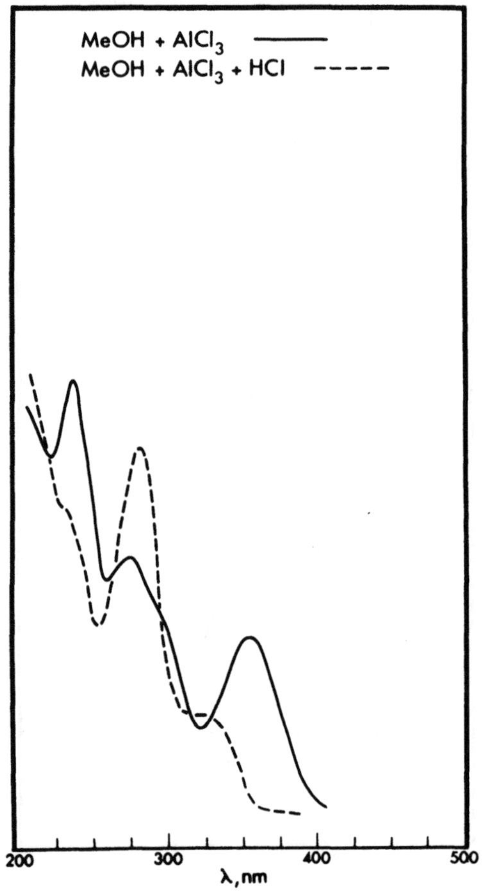

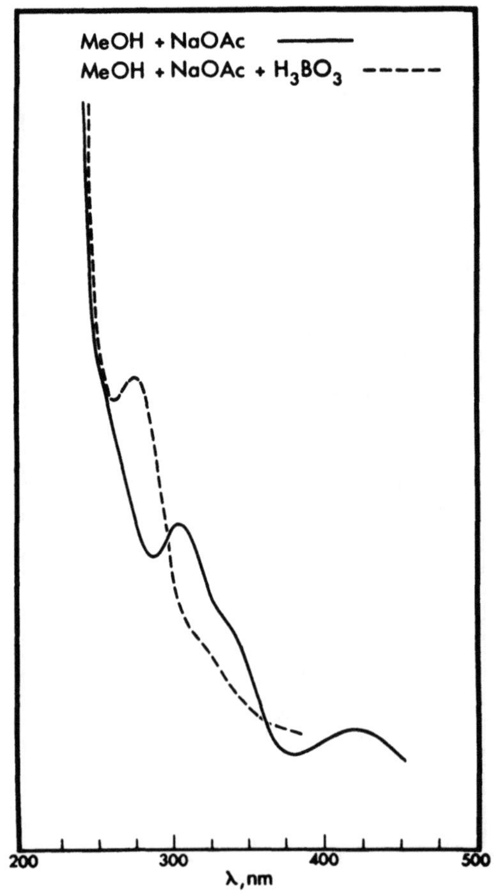

130

TECTORIGENIN

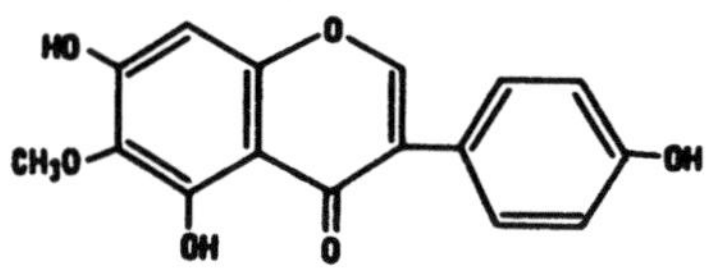

CHROMATOGRAPHIC DATA

Spot Appearance: (UV) deep purple
(UV/NH_3) deep purple

R_f Values: 0.83 (TBA), 0.38 (HOAc)

UV SPECTRAL DATA (λ_{max},*nm*)

MeOH	267, 330sh
NaOMe	278, 328
$AlCl_3$	276, 311, 378
$AlCl_3$/HCl	277, 309sh. 366
NaOAc	273, 339
NaOAc/H_3BO_3 (Proc. I)	268, 335sh

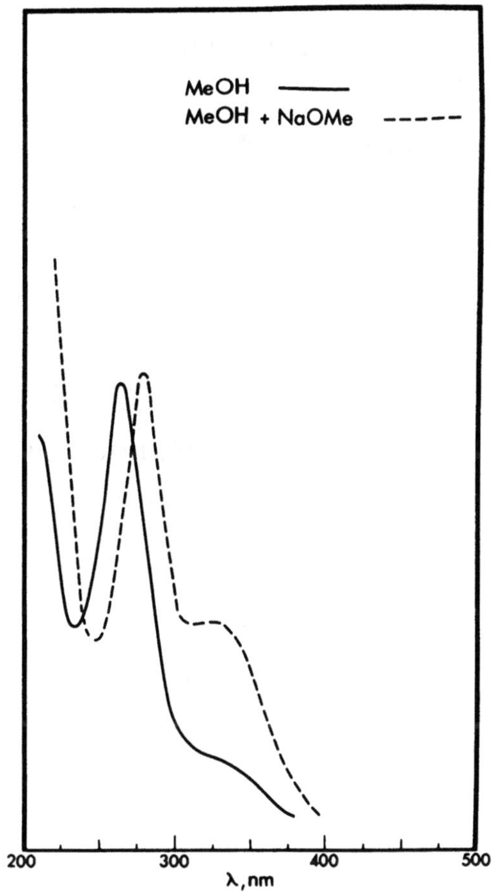

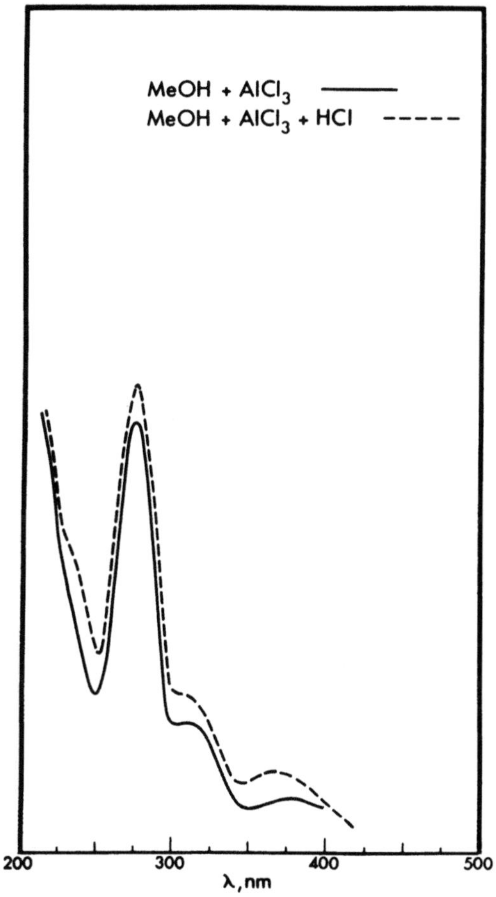

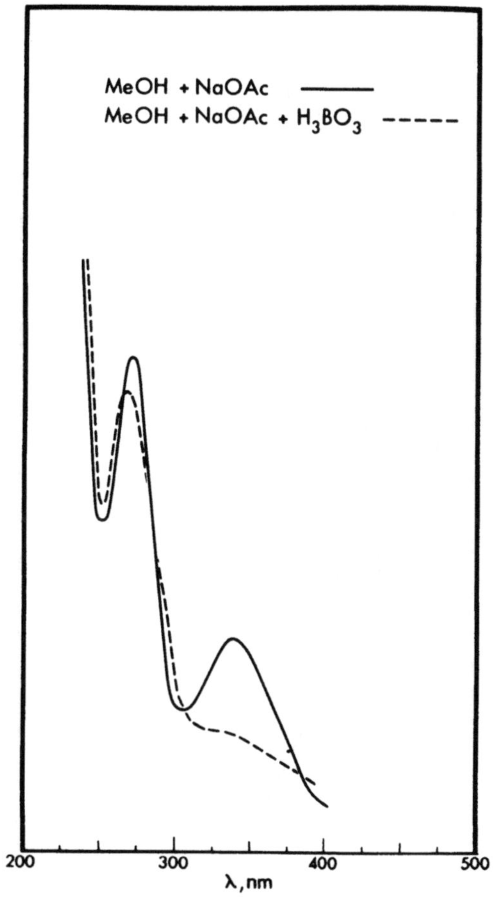

131

TECTORIDIN

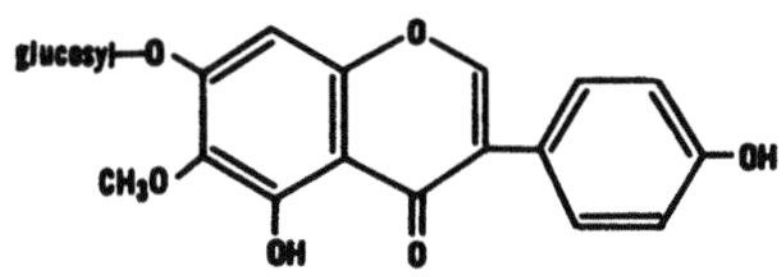

CHROMATOGRAPHIC DATA

Spot Appearance: (UV) deep purple
(UV/NH_3) deep purple

R_f Values: 0.59 (TBA), 0.74 (HOAc)

UV SPECTRAL DATA (λ_{max},*nm*)

MeOH	266, 331
NaOMe	274, 365
$AlCl_3$	277, 315sh, 380
$AlCl_3$/HCl	278, 322sh, 381
NaOAc	266, 331sh
NaOAc/H_3BO_3 (Proc. I)	266, 330sh

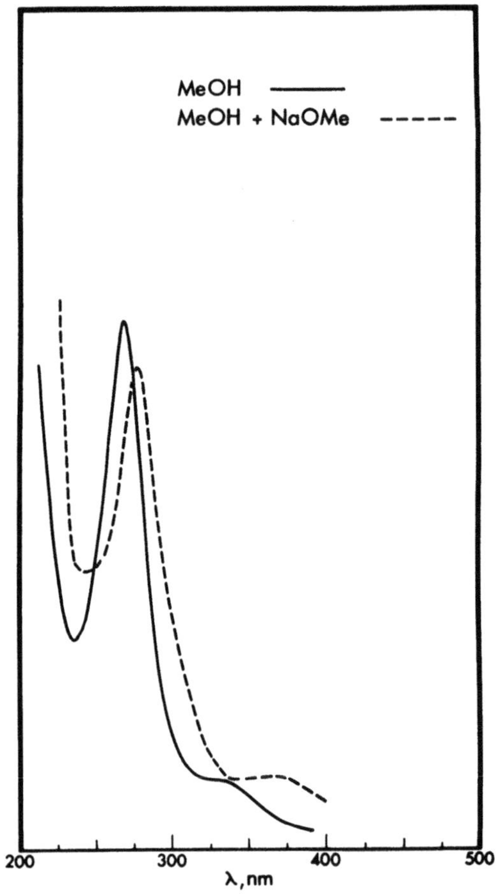

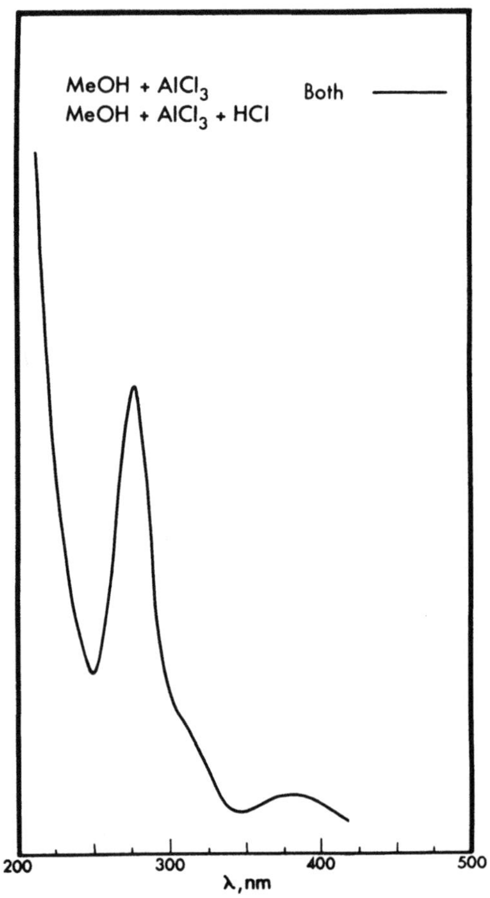

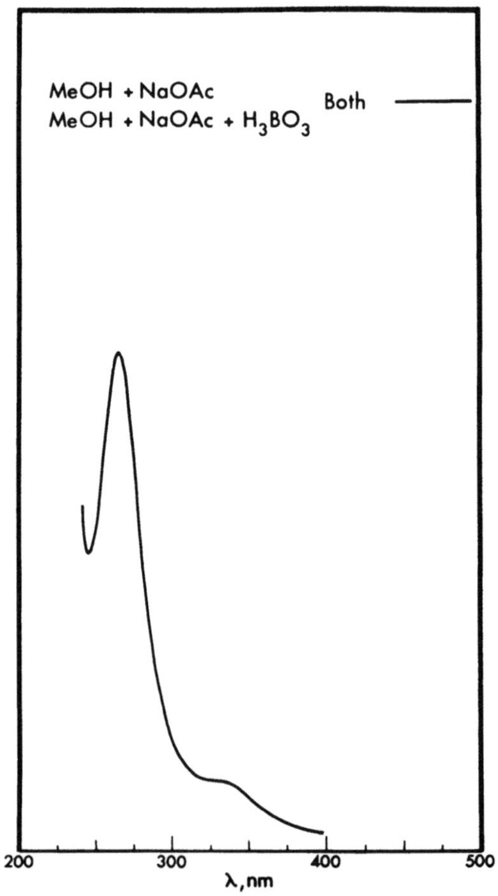

IRISOLIDONE

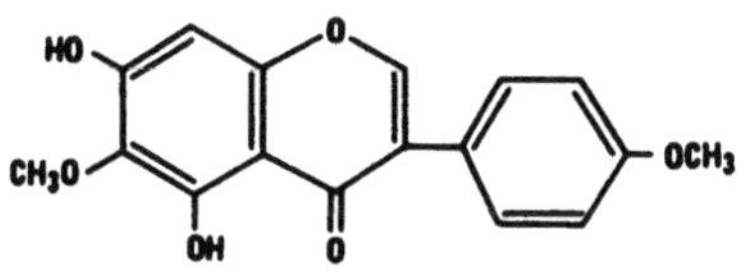

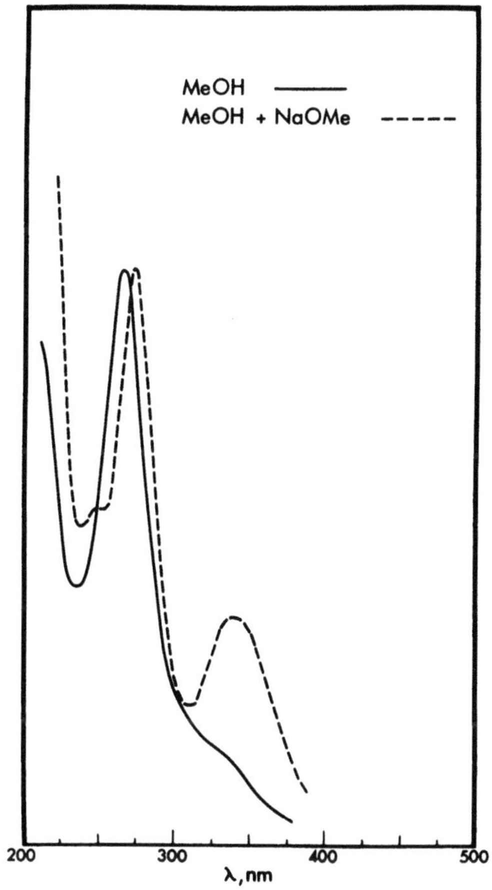

CHROMATOGRAPHIC DATA

Spot Appearance: (UV) deep purple
(UV/NH_3) deep purple

R_f Values: 0.88 (TBA). 0.34 (HOAc)

UV SPECTRAL DATA (λ_{max}, nm)

MeOH	265, 335sh
NaOMe	248, 273, 339
$AlCl_3$	276, 316, 378
$AlCl_3$/HCl	277, 312sh, 373
NaOAc	273, 339
NaOAc/H_3BO_3 (Proc. I)	271, 333

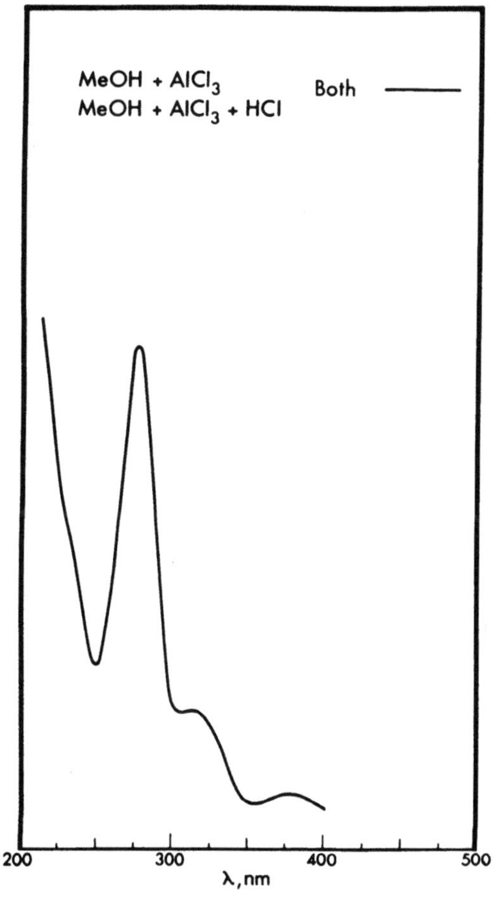

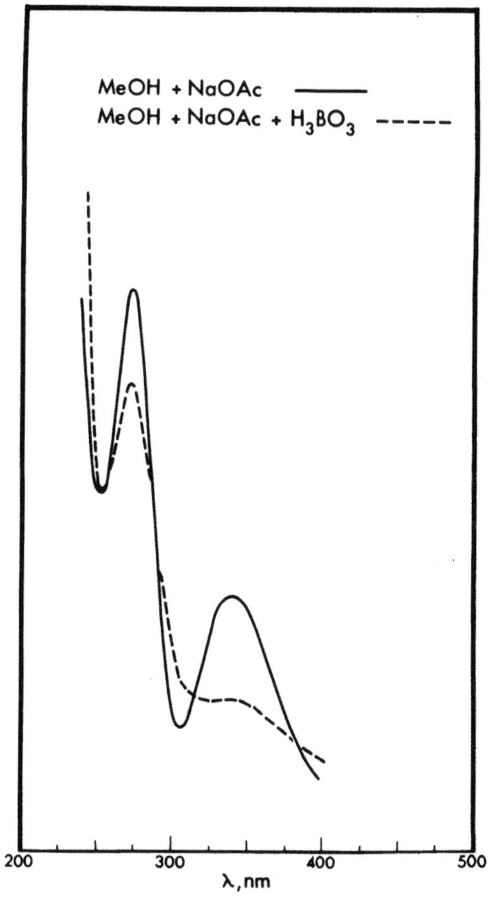

133

OROBOL

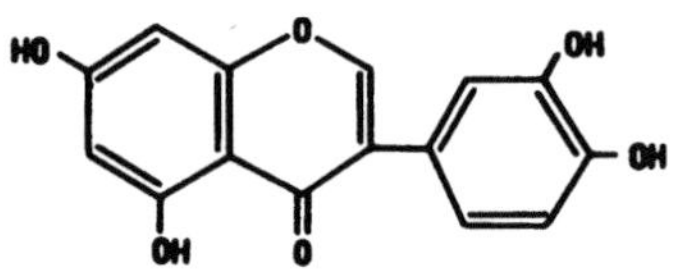

CHROMATOGRAPHIC DATA

Spot Appearance: (UV) deep purple
(UV/NH_3) deep purple

R_f Values: 0.80 (TBA), 0.24 (HOAc)

UV SPECTRAL DATA (λ_{max},*nm*)

MeOH	262, 294sh, 338sh
NaOMe	269, 334 (dec.)
$AlCl_3$	270, 298sh, 365
$AlCl_3$/HCl	273, 371
NaOAc	270, 322
NaOAc/H_3BO_3 (Proc. II)	266, 294sh

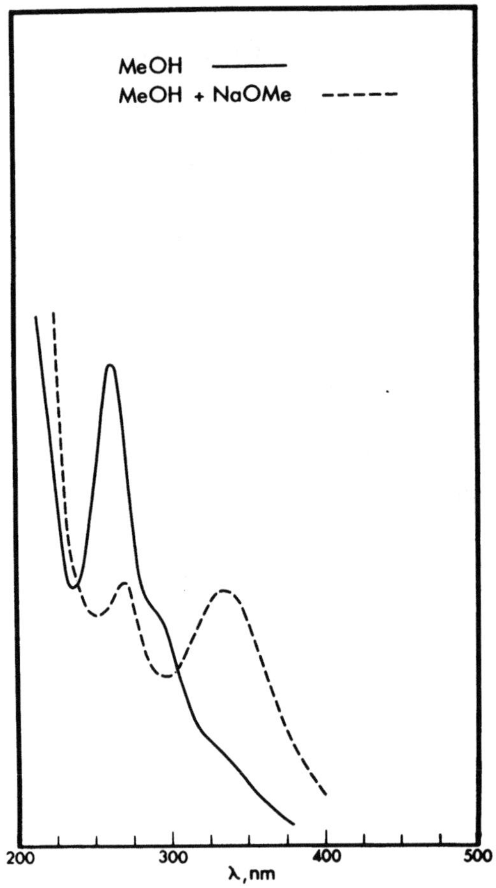

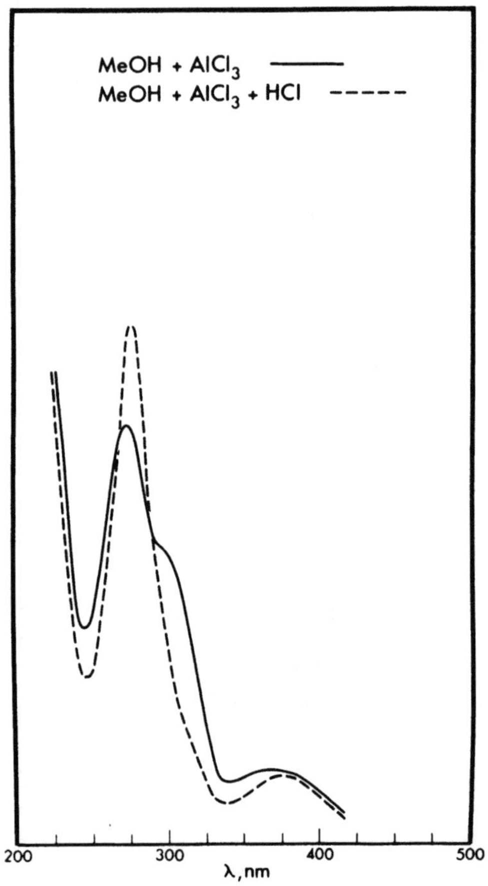

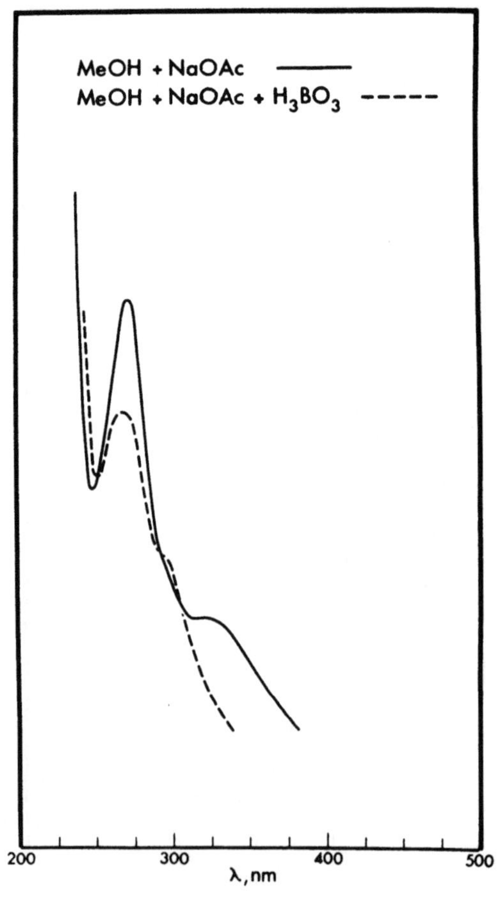

OROBOL 7-0-GLUCOSIDE

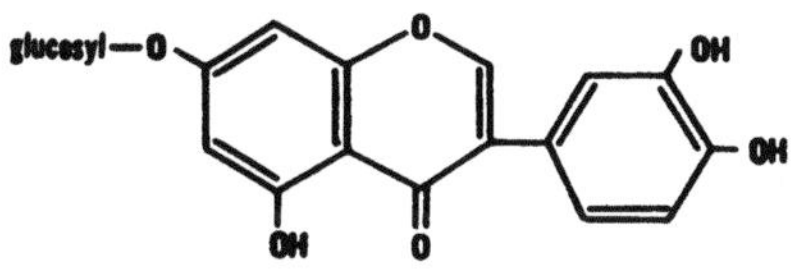

CHROMATOGRAPHIC DATA

Spot Appearance: (UV) deep purple
(UV/NH_3) deep purple

R_f Values: 0.39 (TBA), 0.54 (HOAc)

UV SPECTRAL DATA (λ_{max},*nm*)

MeOH	262, 290sh, 343sh
NaOMe	294sh, 337 (dec.)
$AlCl_3$	269, 297sh, 372
$AlCl_3$/HCl	272, 297sh, 376
NaOAc	261, 331sh
NaOAc/H_3BO_3 (Proc. II)	258, 269sh, 293sh, 322sh

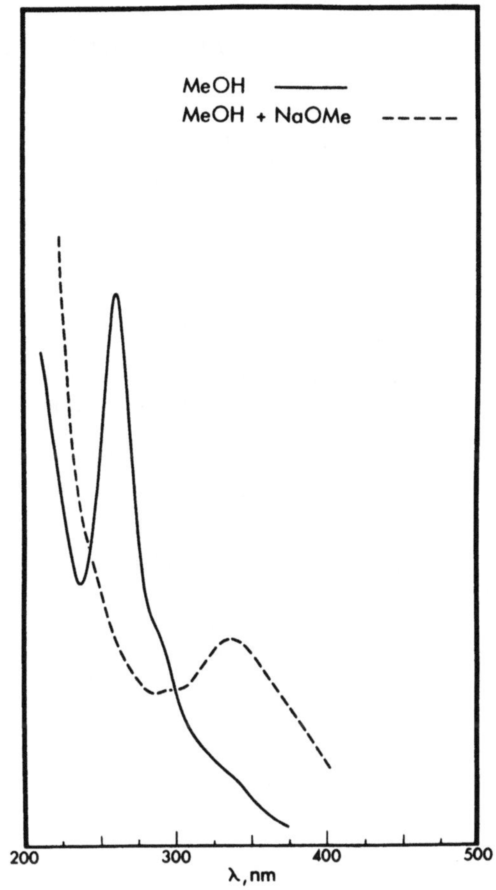

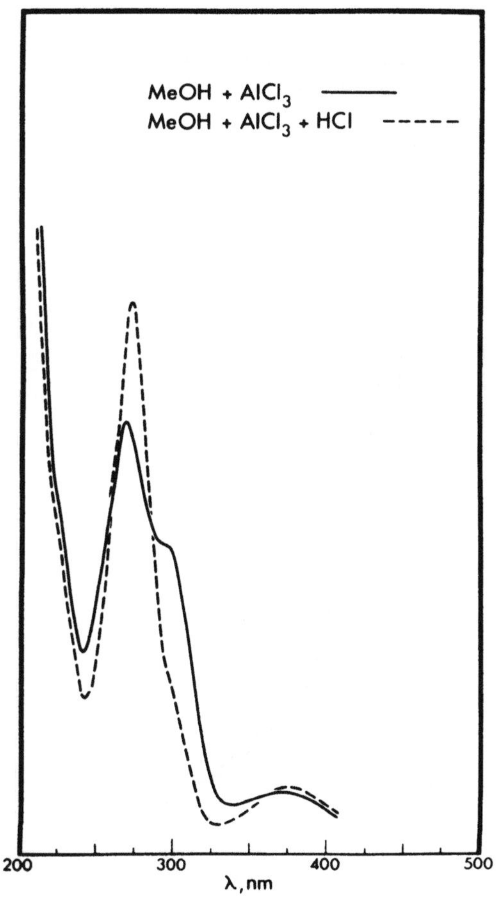

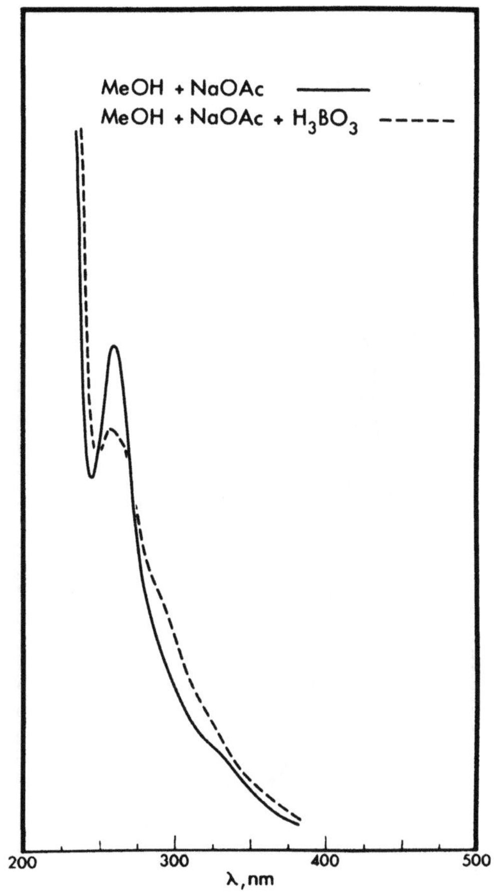

135

OROBOL 7-0-RHAMNOGLUCOSIDE

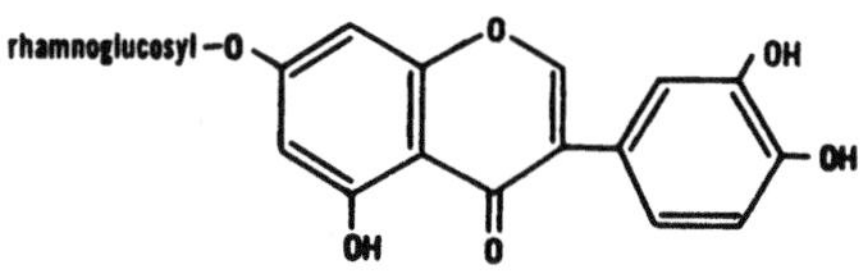

CHROMATOGRAPHIC DATA

Spot Appearance: (UV) deep purple
(UV/NH_3) deep purple

R_f Values: 0.33 (TBA), 0.65 (HOAc)

UV SPECTRAL DATA (λ_{max}, *nm*)

MeOH	262, 290sh, 343sh
NaOMe	294sh, 337 (dec.)
$AlCl_3$	269, 297sh, 372
$AlCl_3$/HCl	272, 297sh, 376
NaOAc	261, 331sh
NaOAc/H_3BO_3 (Proc. II)	258, 269sh, 293sh, 322sh

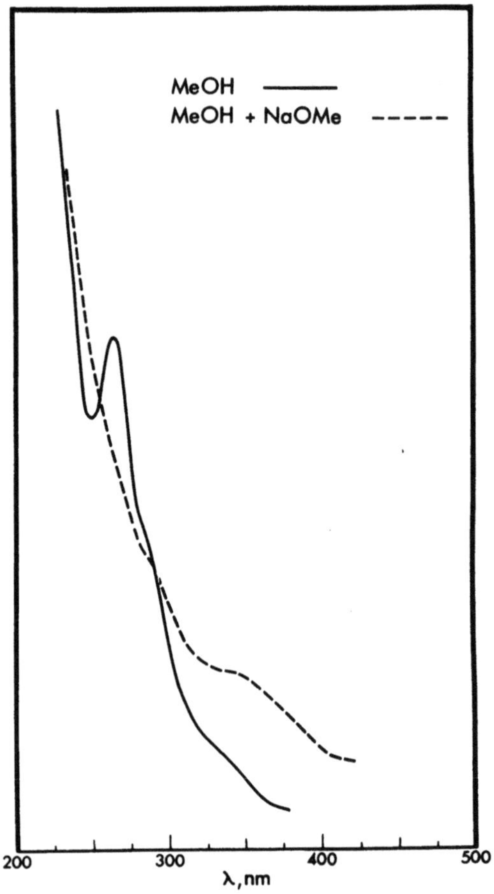

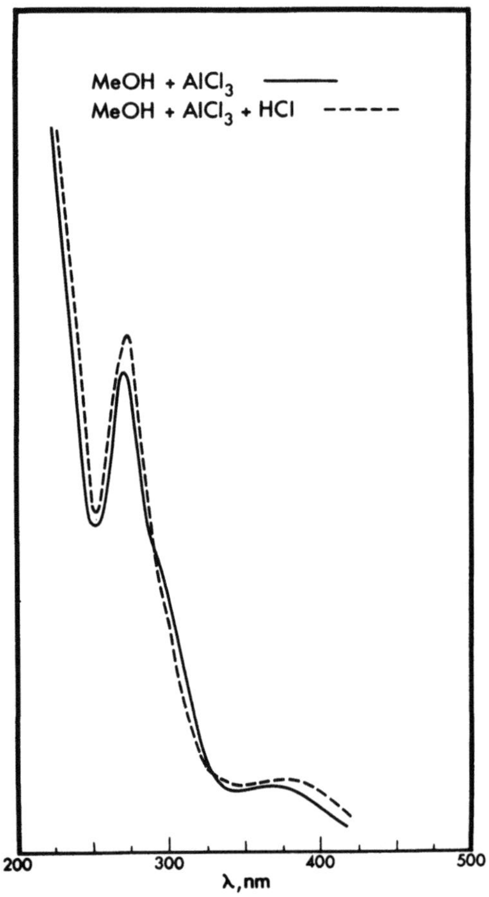

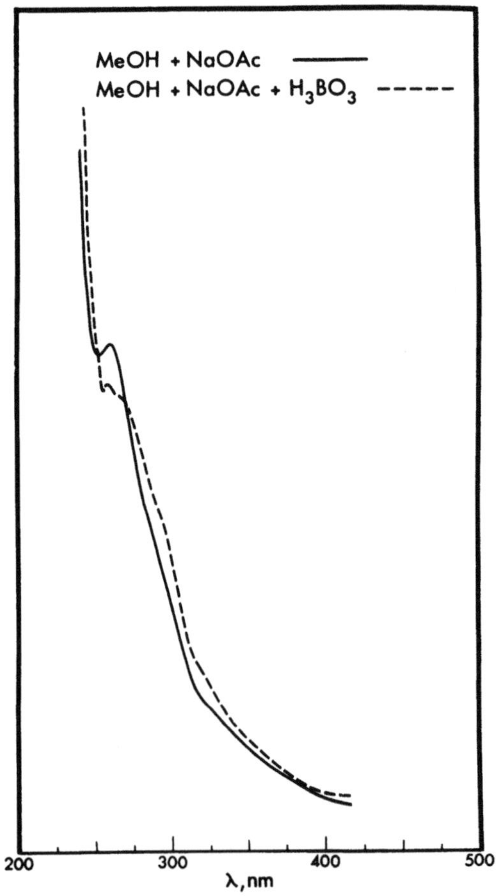

136

PRATENSEIN

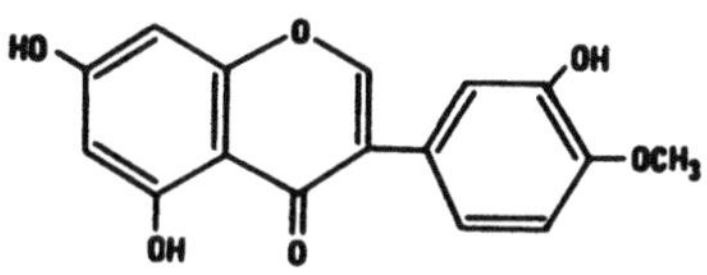

CHROMATOGRAPHIC DATA

Spot Appearance: (UV) deep purple
(UV/NH_3) deep purple

R_f Values: 0.81 (TBA), 0.31 (HOAc)

UV SPECTRAL DATA (λ_{max},*nm*)

MeOH	262, 292sh, 330sh
NaOMe	270, 321
$AlCl_3$	272, 311sh, 371
$AlCl_3$/HCl	273, 314sh, 371
NaOAc	271, 325sh
NaOAc/H_3BO_3 (Proc. I)	263, 295sh, 335sh

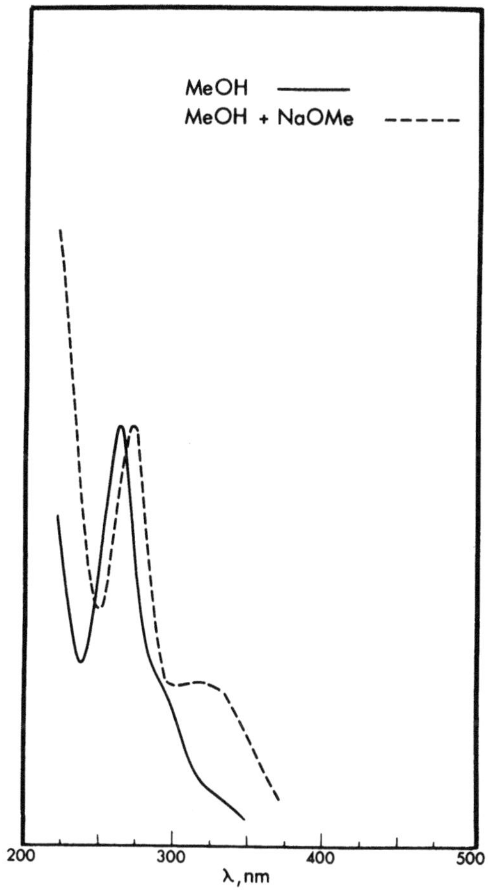

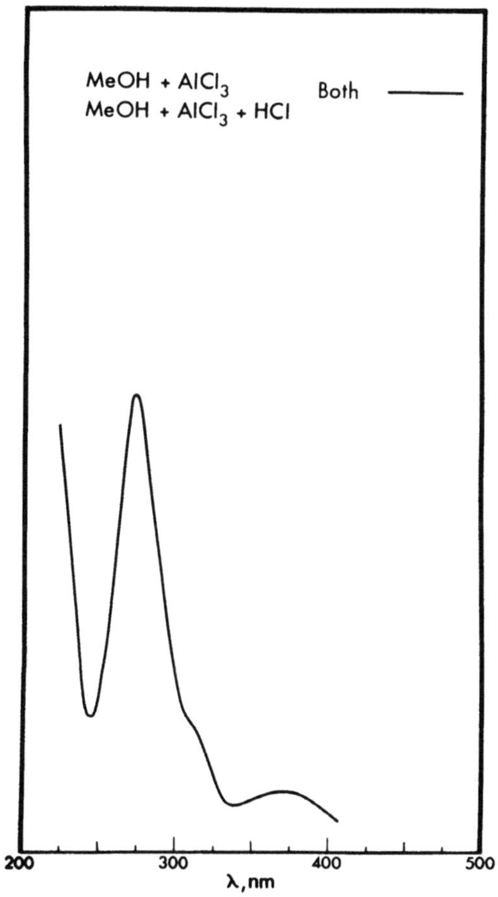

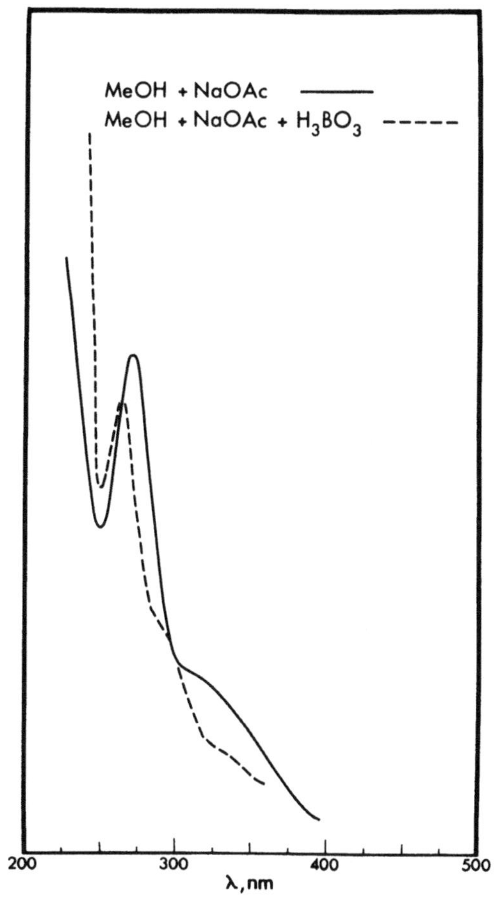

137

POMIFERIN

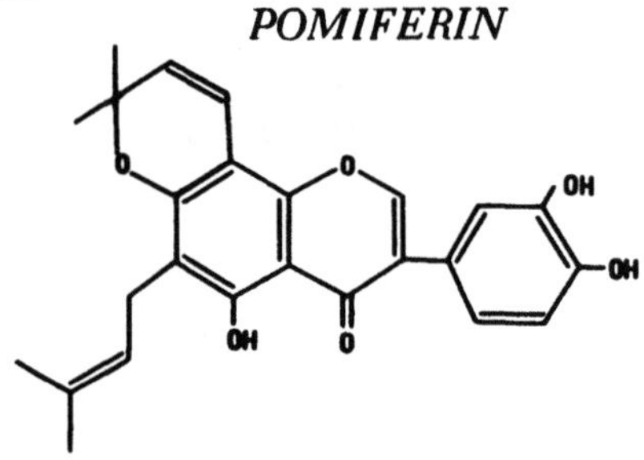

CHROMATOGRAPHIC DATA

Spot Appearance: (UV) deep purple
(UV/NH_3) deep purple

R_f Values: 0.93 (TBA), 0.02 (HOAc)

UV SPECTRAL DATA (λ_{max}, *nm*)

MeOH	274, 353sh
NaOMe	271 (dec.)
$AlCl_3$	284
$AlCl_3$/HCl	285
NaOAc	274, 352sh
NaOAc/H_3BO_3 (Proc. II)	276, 352sh

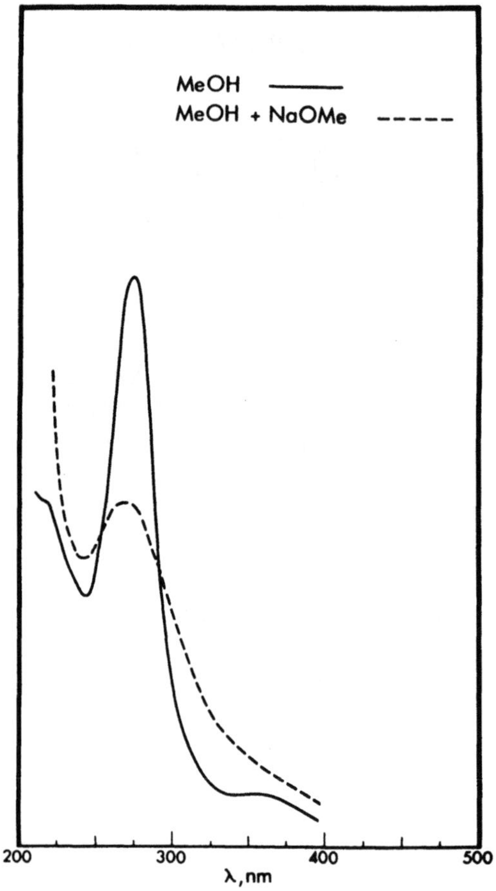

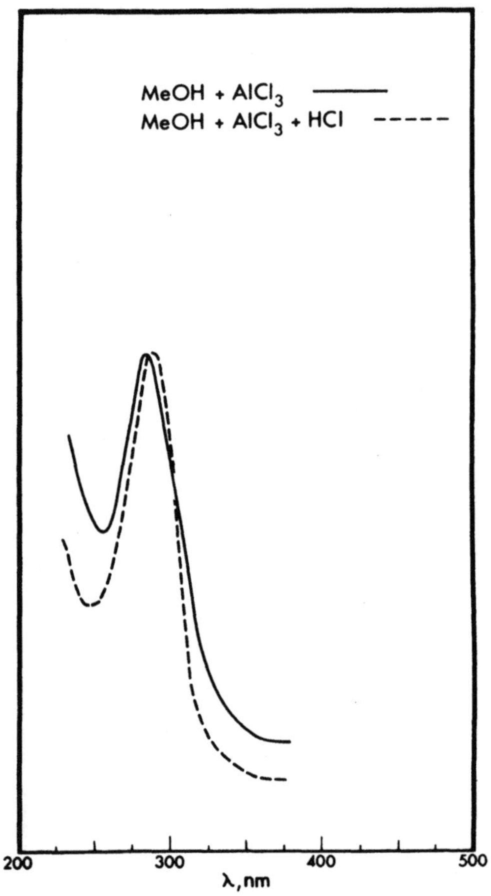

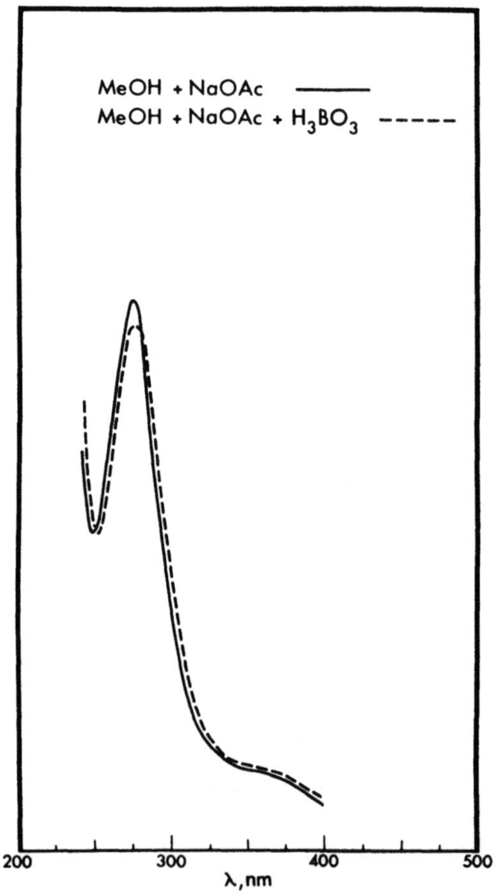

IRIGENIN

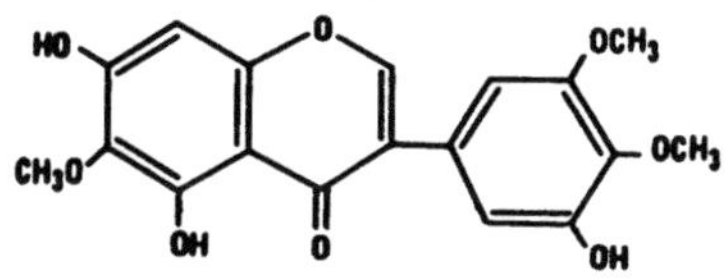

CHROMATOGRAPHIC DATA

Spot Appearance: (UV) deep purple
(UV/NH_3) deep purple

R_f Values: 0.85 (TBA), 0.48 (HOAc)

UV SPECTRAL DATA (λ_{max}, *nm*)

MeOH	268, 336sh
NaOMe	273, 336
$AlCl_3$	275, 316, 371
$AlCl_3$/HCl	278, 315sh, 374
NaOAc	273, 338
NaOAc/H_3BO_3 (Proc. I)	268, 339sh

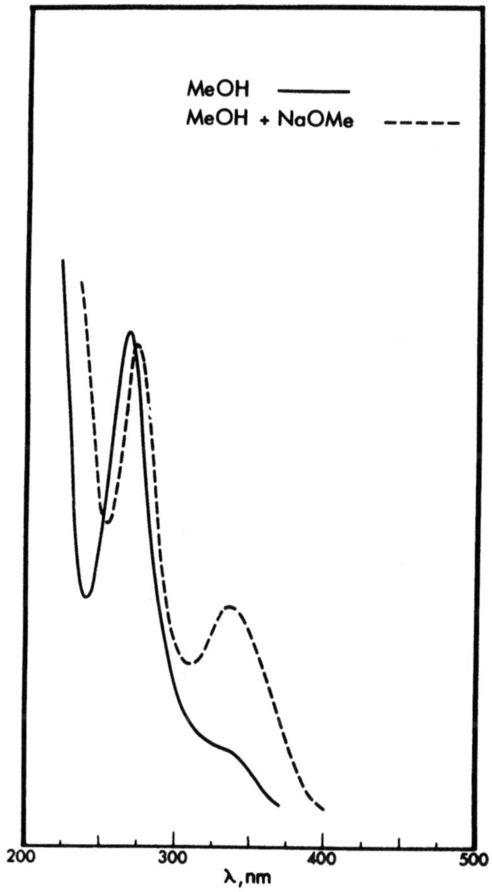

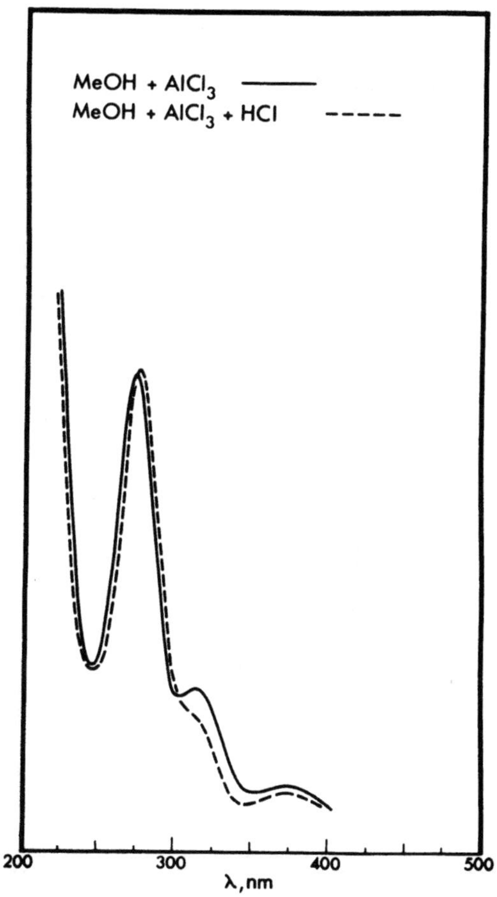

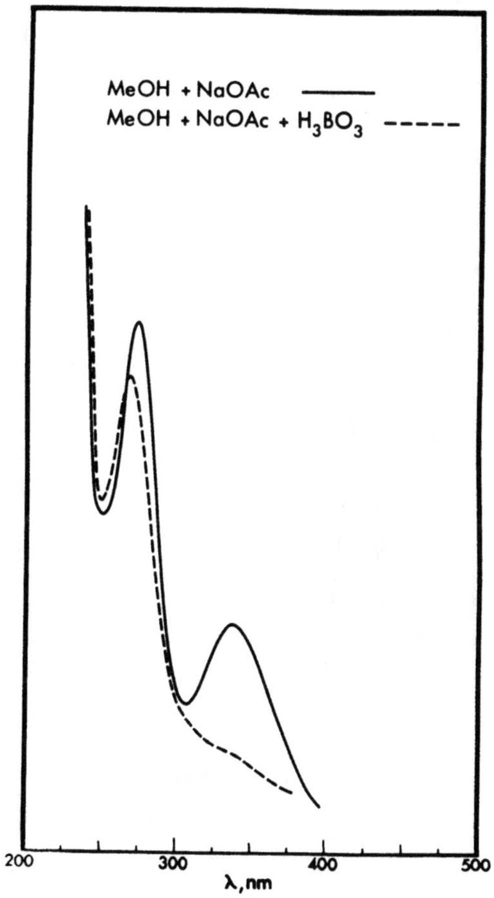

1.39

IRIDIN

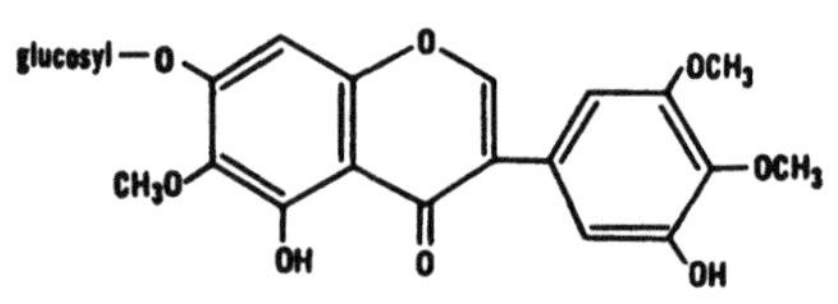

CHROMATOGRAPHIC DATA

Spot Appearance: (UV) deep purple
(UV/NH_3) deep purple

R_f Values: 0.61 (TBA), 0.78 (HOAc)

UV SPECTRAL DATA (λ_{max}, nm)

MeOH	268, 331sh
NaOMe	270, 356
$AlCl_3$	277, 319sh, 382
$AlCl_3$/HCl	278, 379
NaOAc	268, 335sh
NaOAc/H_3BO_3 (Proc. I)	268, 335sh

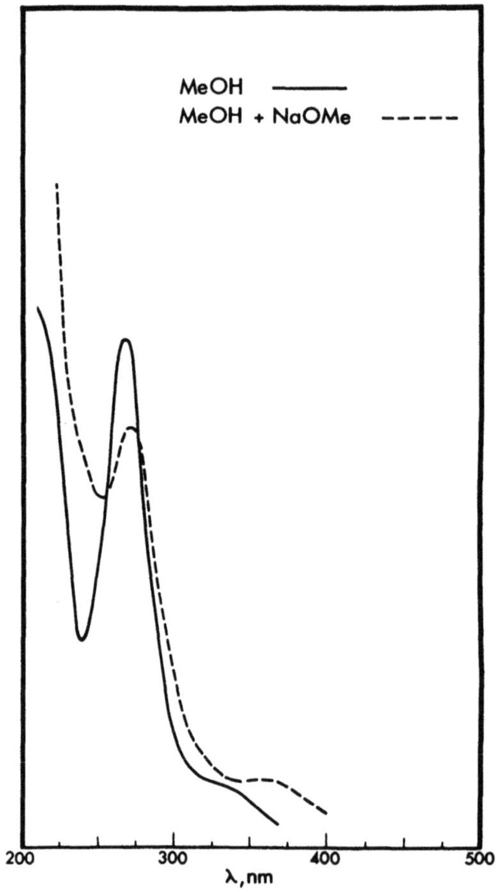

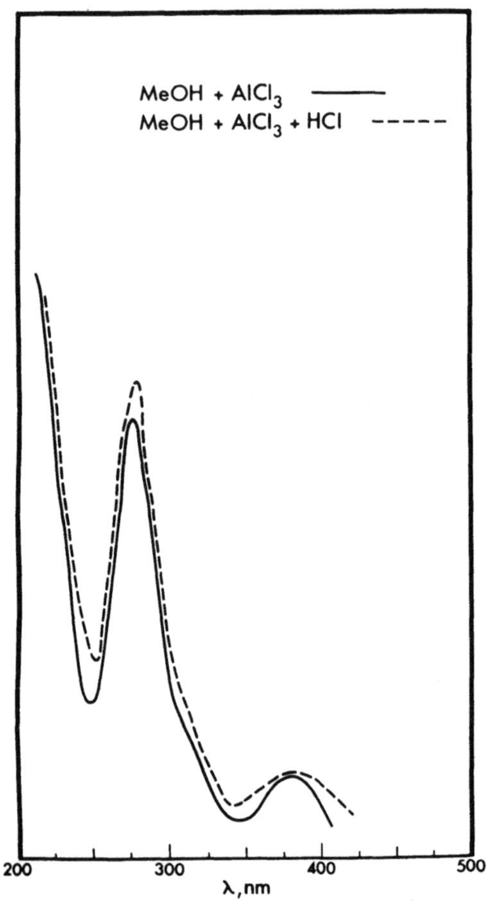

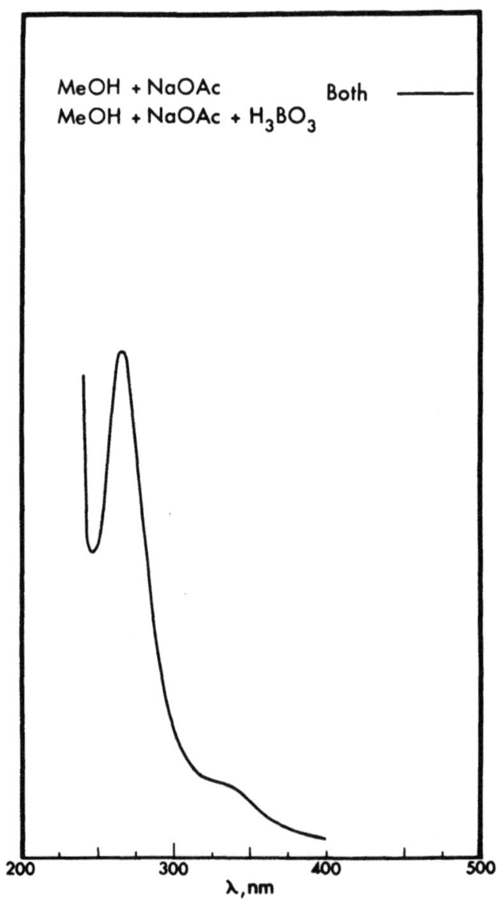

PINOCEMBRIN

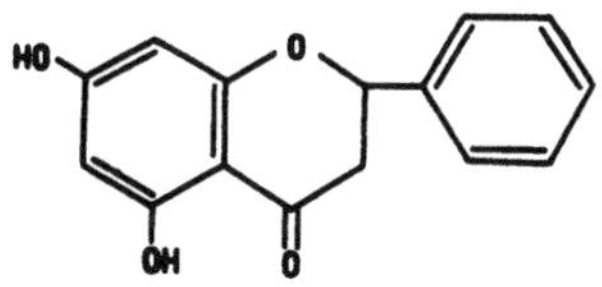

CHROMATOGRAPHIC DATA

Spot Appearance: (UV) deep purple
(UV/NH_3) deep purple

R_f Values: 0.92 (TBA), 0.29 (HOAc)

UV SPECTRAL DATA (λ_{max},*nm*)

MeOH	289, 325sh
NaOMe	245, 324
$AlCl_3$	311, 375
$AlCl_3$/HCl	309, 373
NaOAc	253sh, 323
NaOAc/H_3BO_3 (Proc. I)	291, 326sh

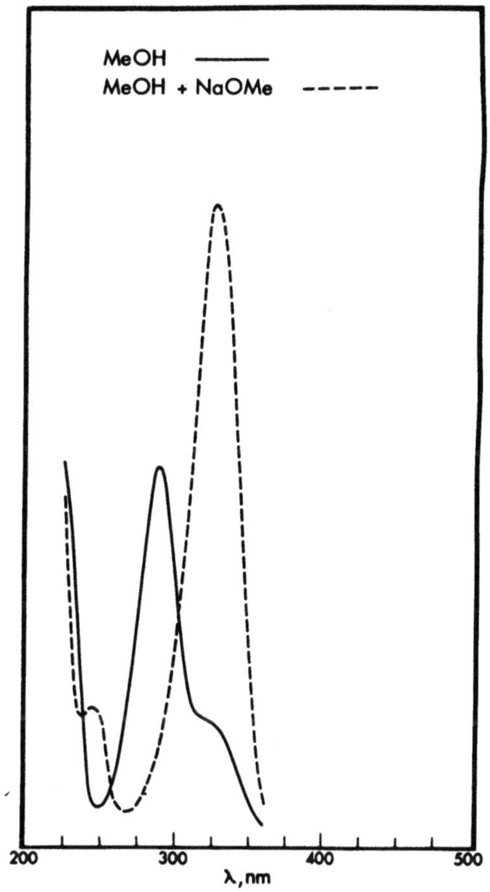

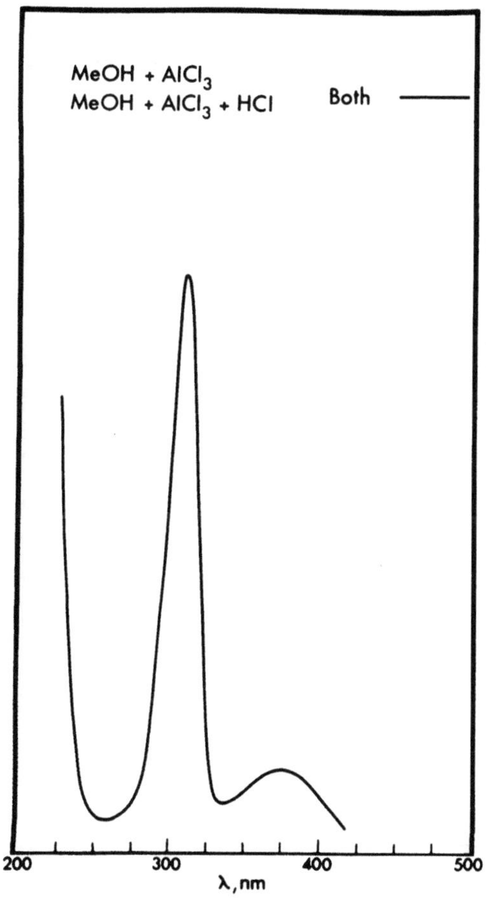

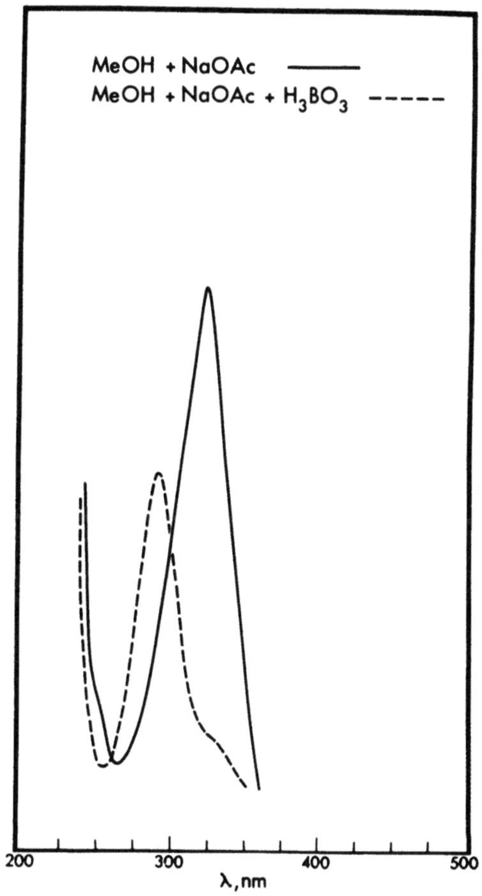

LIQUIRITIGENIN

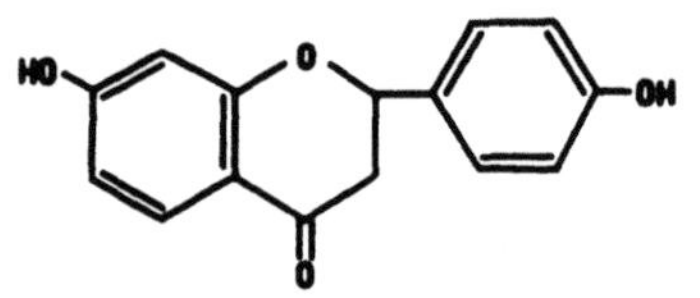

CHROMATOGRAPHIC DATA

Spot Appearance: (UV) fluorescent light blue
(UV/NH_3) fluorescent yellow

R_f Values: 0.87 (TBA), 0.39 (HOAc)

UV SPECTRAL DATA (λ_{max}, *nm*)

MeOH	276, 312
NaOMe	250, 298sh, 327sh, 335
$AlCl_3$	276, 311
$AlCl_3$/HCl	276, 311
NaOAc	255sh, 282, 327sh, 335
NaOAc/H_3BO_3 (Proc. I)	278, 312

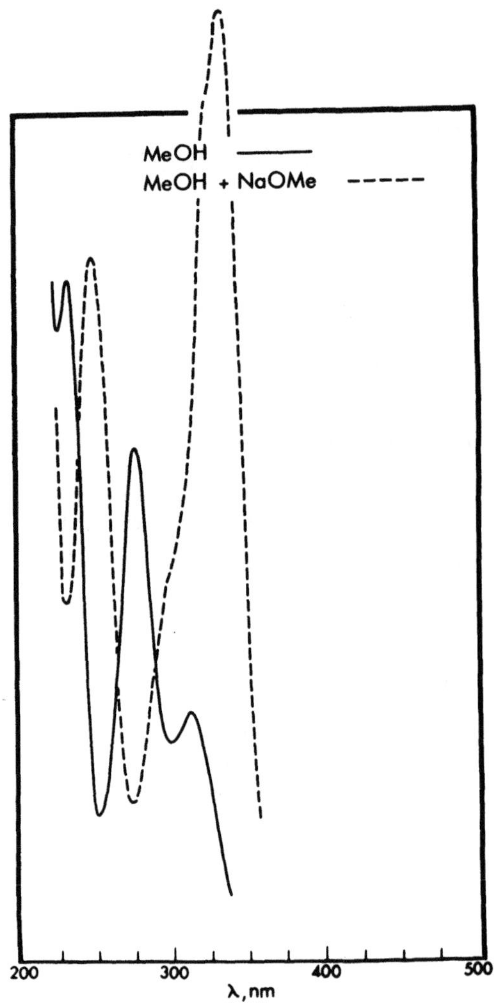

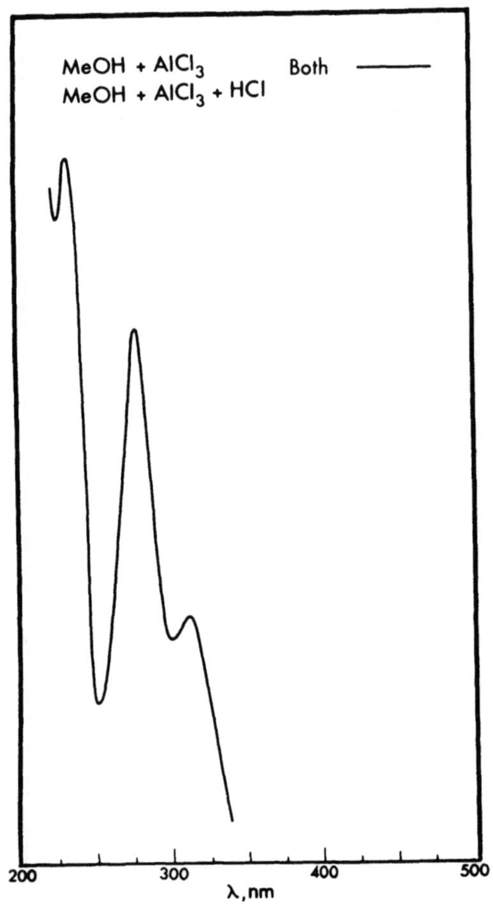

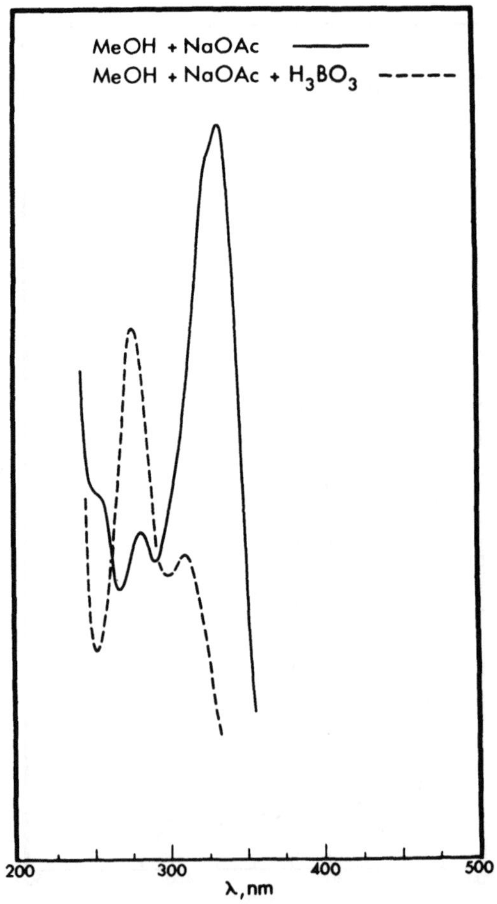

5,6,7-TRIHYDROXYFLAVANONE

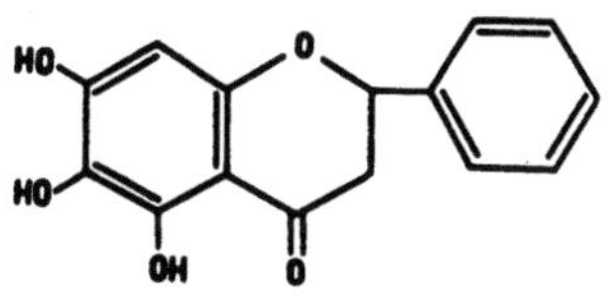

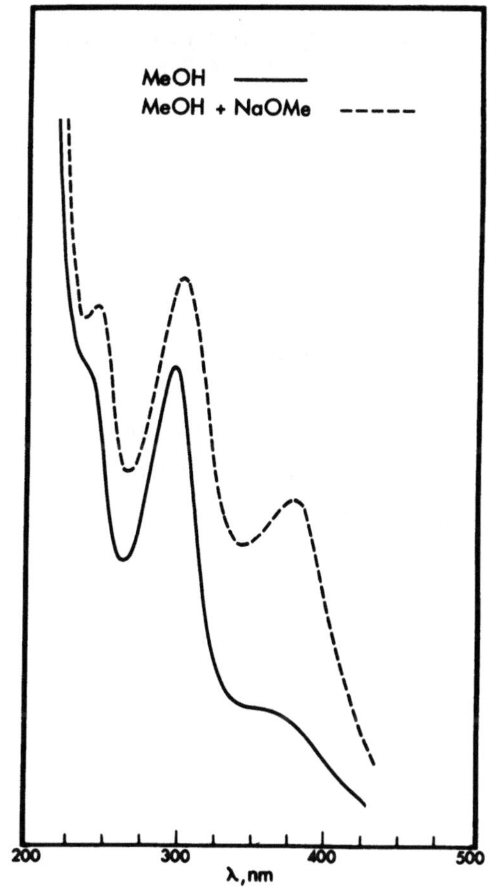

CHROMATOGRAPHIC DATA

Spot Appearance: (UV) deep purple
(UV/NH_3) deep purple

R_f Values: 0.83 (TBA), 0.47 (HOAc)

UV SPECTRAL DATA (λ_{max}, *nm*)

MeOH	242sh, 295, 362sh
NaOMe	245, 300, 377 (dec.)
$AlCl_3$	251sh, 328, 381sh
$AlCl_3$/HCl	251sh, 317, 373sh
NaOAc	248sh, 299, 385, (dec.)
NaOAc/H_3BO_3 (Proc. II)	249sh, 304, 371

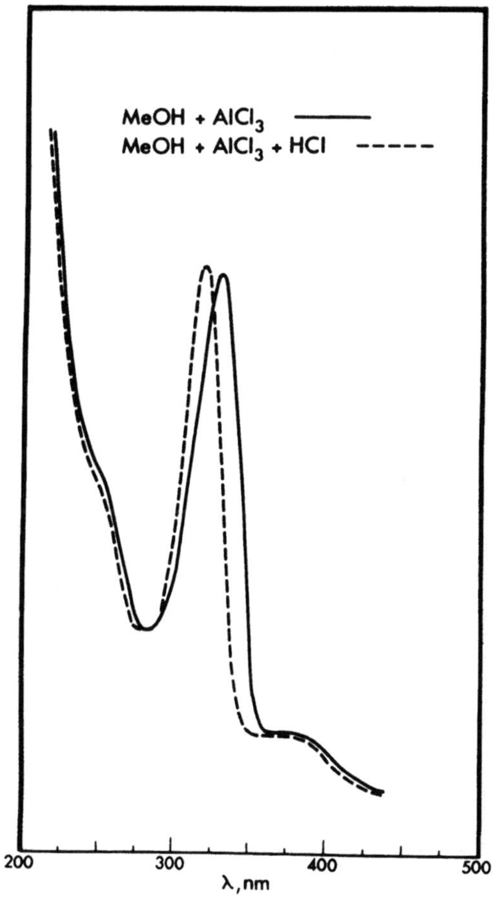

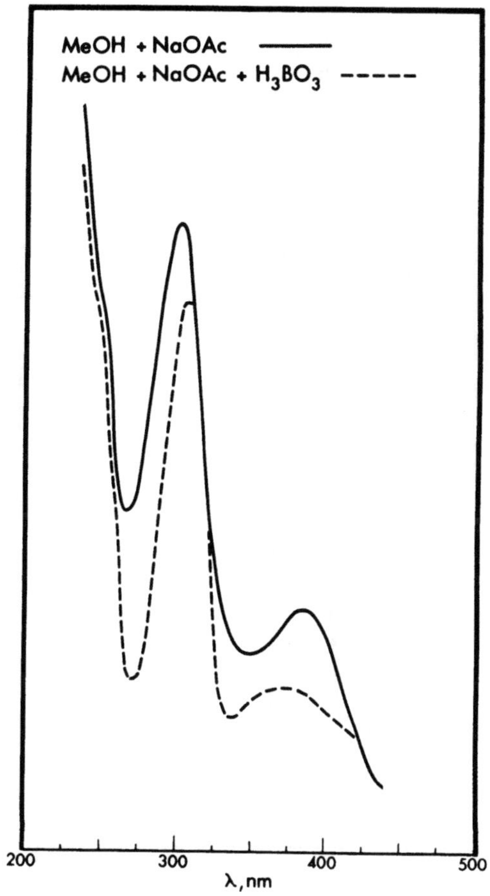

143

5,6,7-TRIHYDROXYFLAVANONE 7-0-GLUCURONIDE

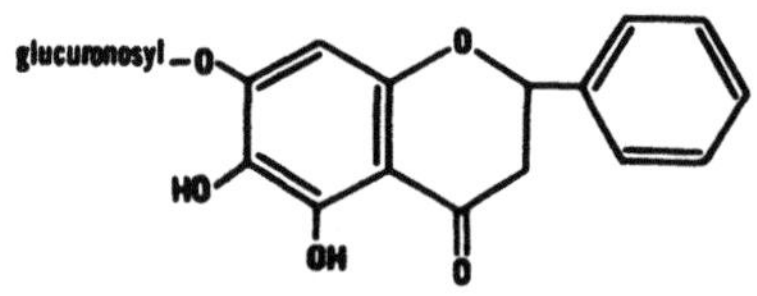

CHROMATOGRAPHIC DATA

Spot Appearance: (UV) deep purple
(UV/NH_3) greenish purple

R_f Values: 0.60 (TBA), 0.76 (HOAc)

UV SPECTRAL DATA (λ_{max}, nm)

MeOH	239, 288, 362
NaOMe	253sh, 295, 345sh (dec.)
$AlCl_3$	237sh, 316, 432
$AlCl_3$/HCl	237sh, 314, 427
NaOAc	287, 353 (dec.)
NaOAc/H_3BO_3 (Proc. II)	284, 377

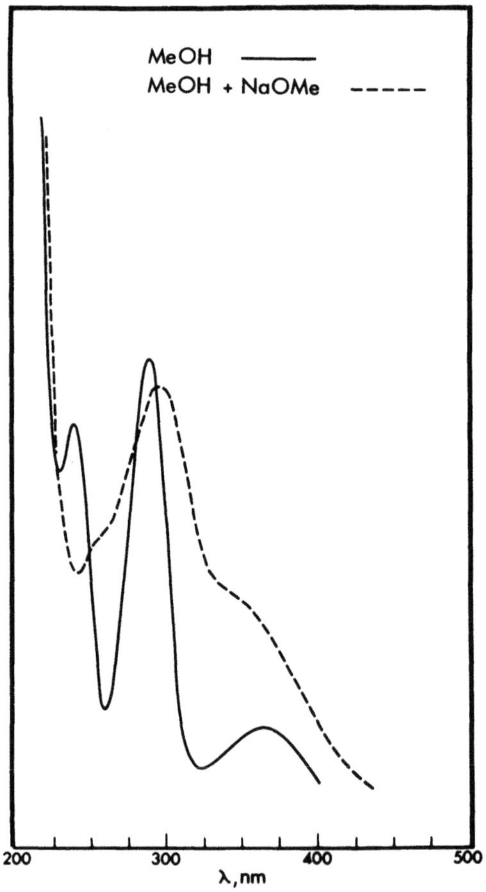

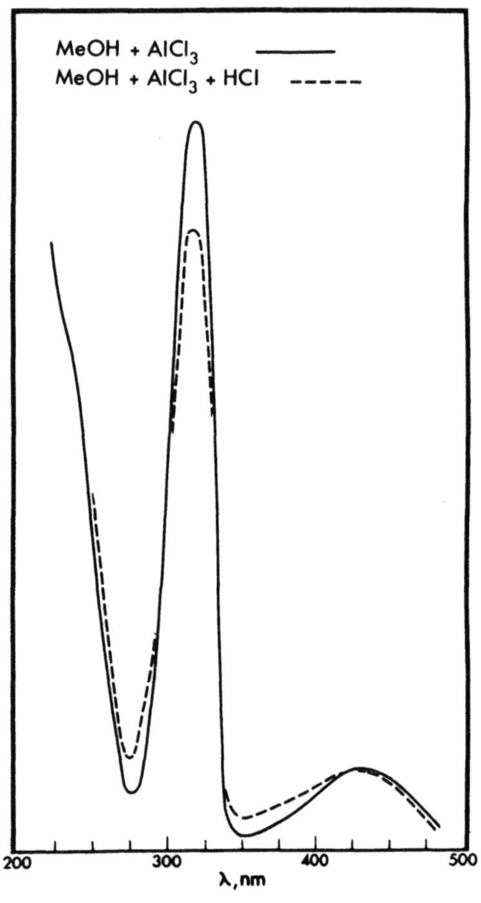

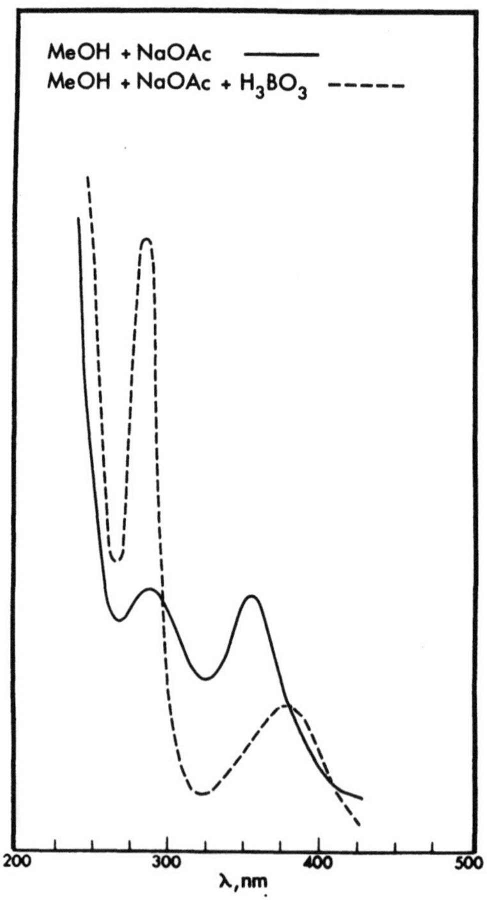

144

NARINGENIN

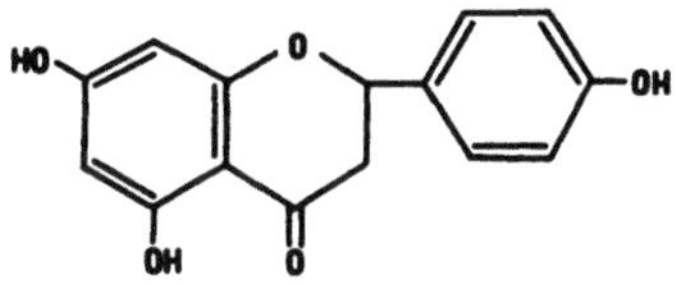

CHROMATOGRAPHIC DATA

Spot Appearance: (UV) deep purple
(UV/NH_3) greenish purple

R_f Values: 0.88 (TBA), 0.33 (HOAc)

UV SPECTRAL DATA (λ_{max}, nm)

MeOH	289, 326sh
NaOMe	245, 323
$AlCl_3$	312, 375
$AlCl_3$/HCl	311, 371
NaOAc	284sh, 323
NaOAc/H_3BO_3 (Proc. I)	290, 332sh

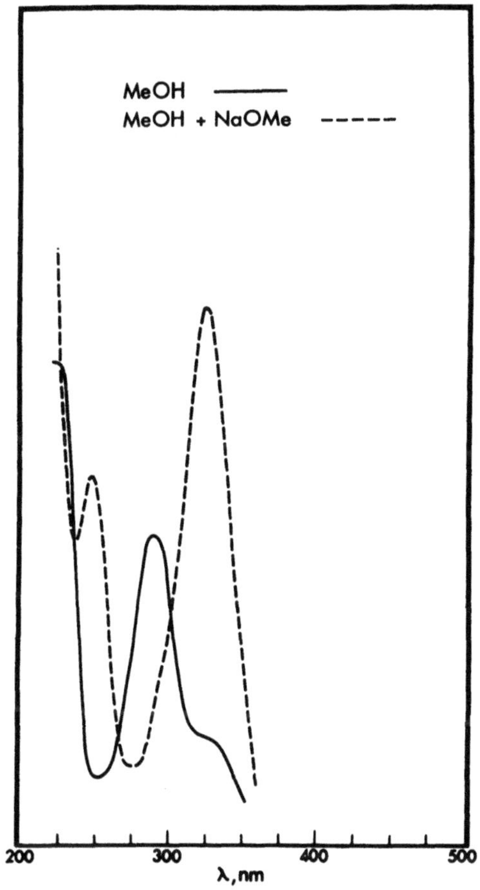

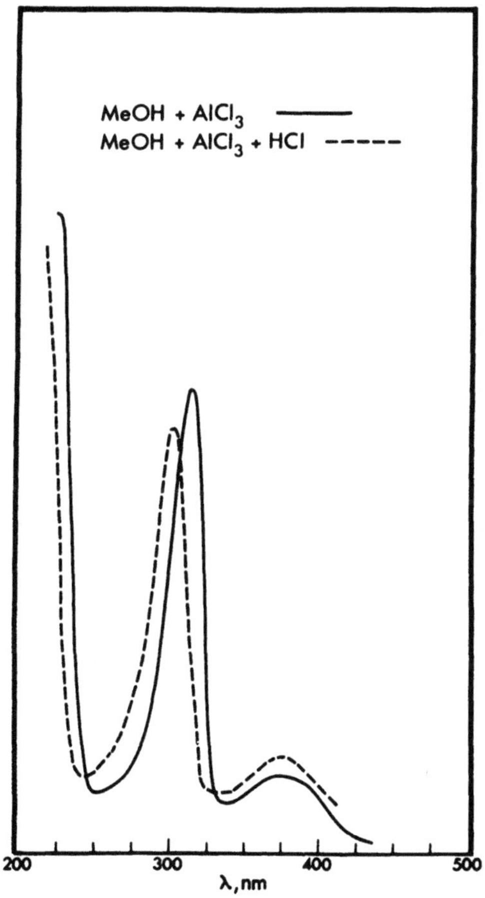

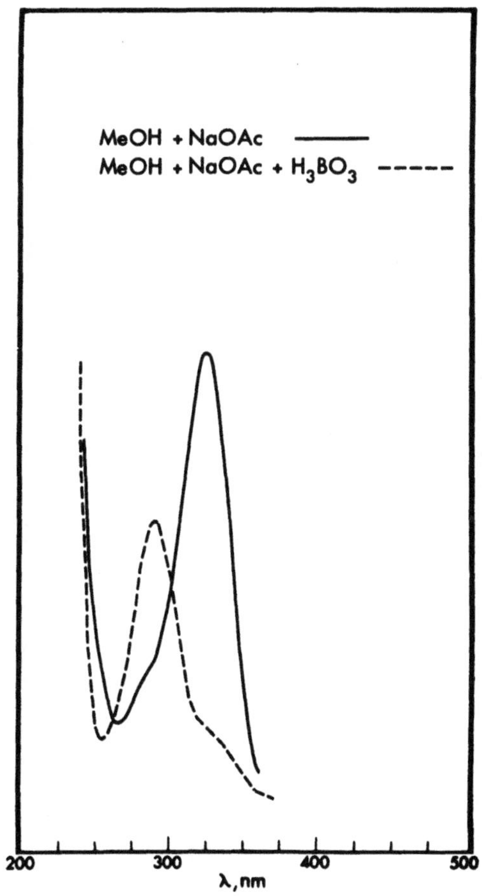

145

SAKURANIN

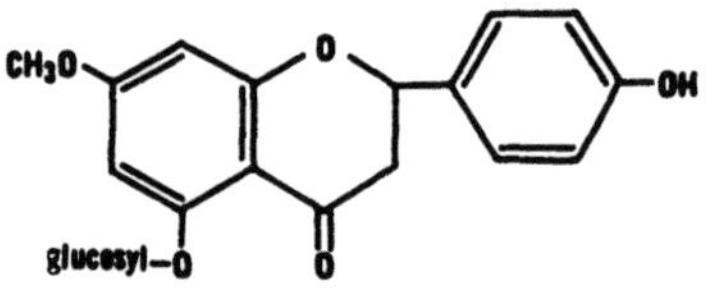

CHROMATOGRAPHIC DATA

Spot Appearance: (UV) fluorescent blue-green
(UV/NH_3) fluorescent green

R_f Values: 0.70 (TBA), 0.77 (HOAc)

UV SPECTRAL DATA (λ_{max}, *nm*)

MeOH	280, 317sh
NaOMe	315, 393
$AlCl_3$	280, 312sh
$AlCl_3$/HCl	280, 310sh
NaOAc	279, 314sh
NaOAc/H_3BO_3 (Proc. I)	279, 313sh

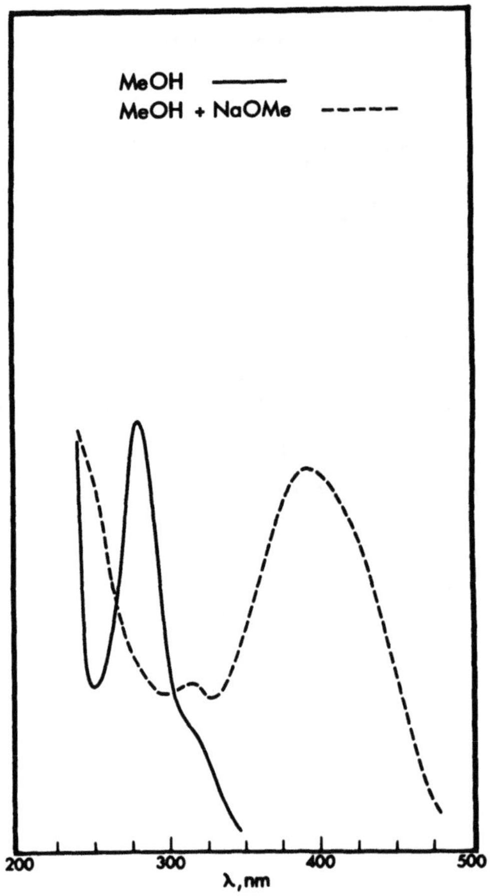

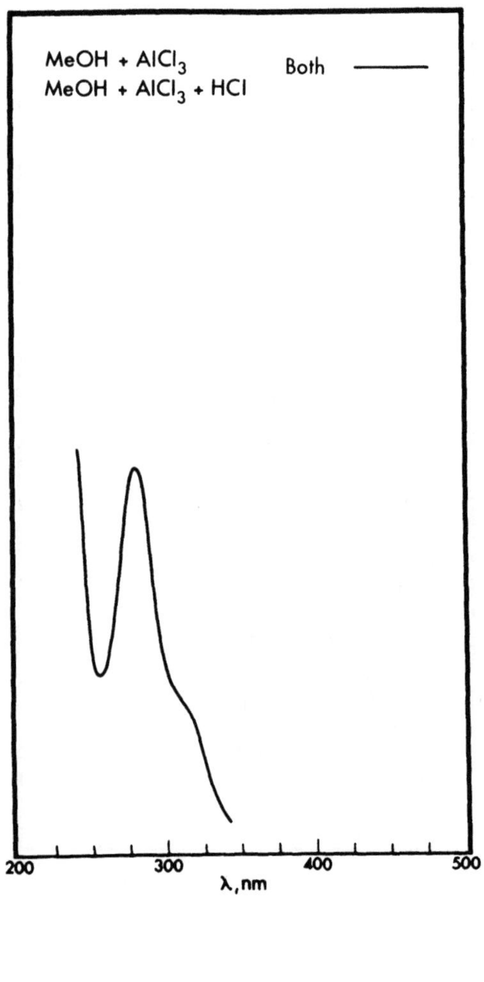

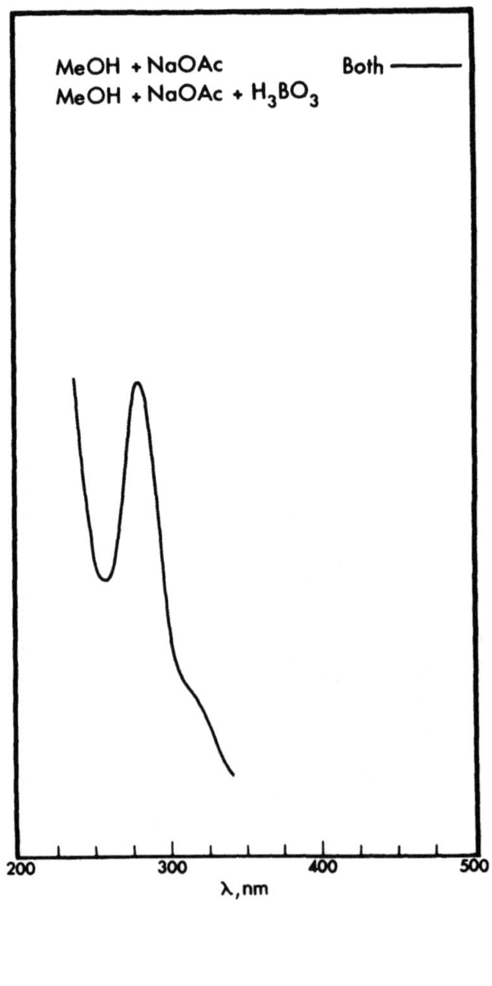

ERIODICTYOL

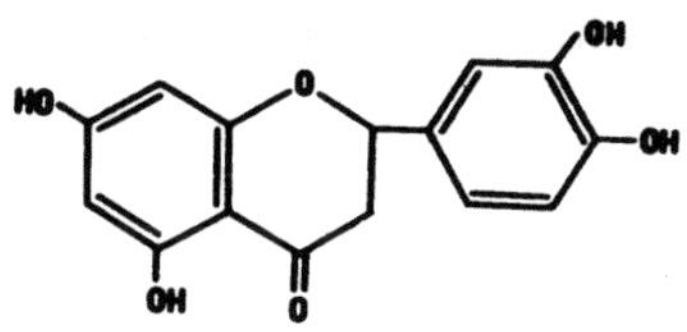

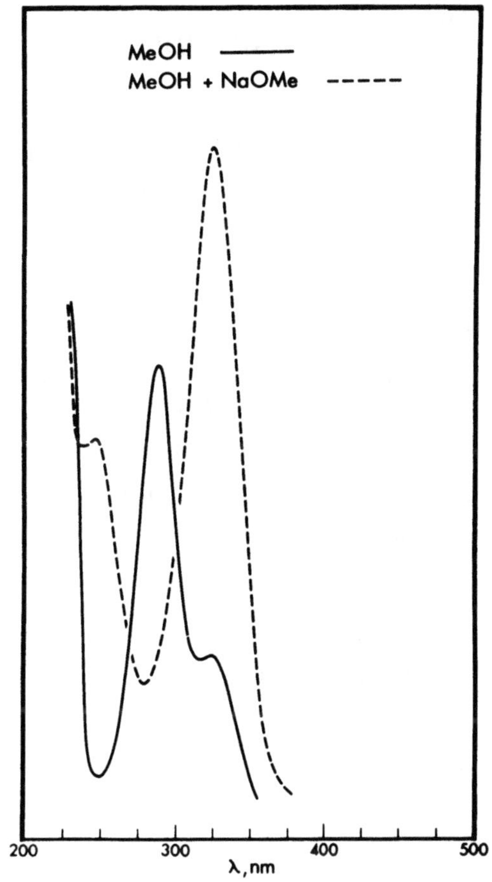

CHROMATOGRAPHIC DATA

Spot Appearance: (UV) deep purple
(UV/NH_3) deep purple

R_f Values: 0.84 (TBA), 0.39 (HOAc)

UV SPECTRAL DATA (λ_{max}, nm)

MeOH	289, 324sh
NaOMe	246, 324
$AlCl_3$	310, 378
$AlCl_3/HCl$	309, 373
NaOAc	289sh, 325
$NaOAc/H_3BO_3$ (Proc. I)	289, 333sh

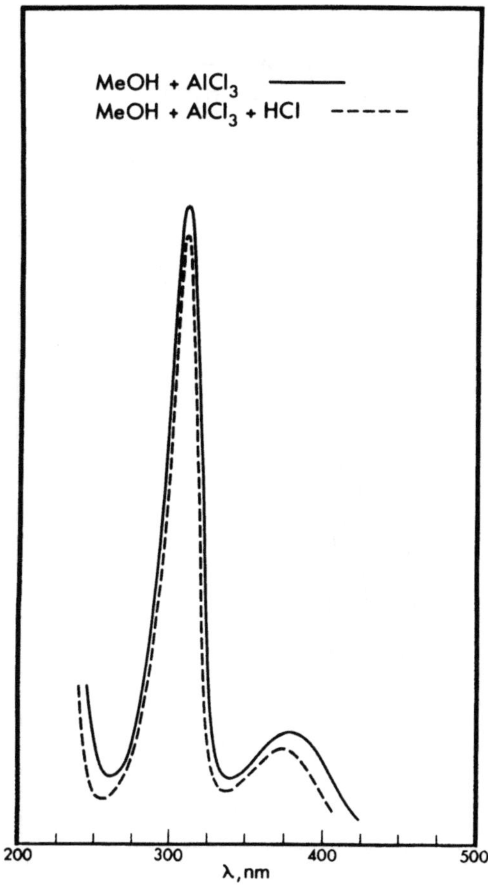

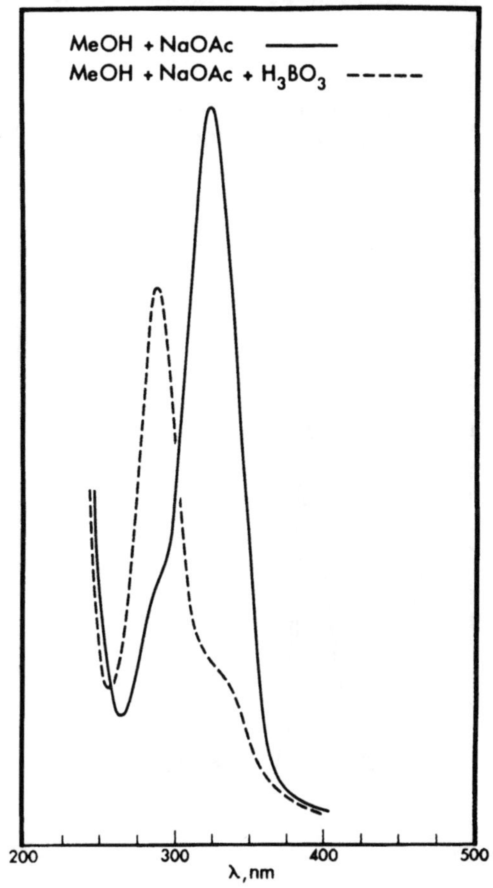

147

HESPERIDIN

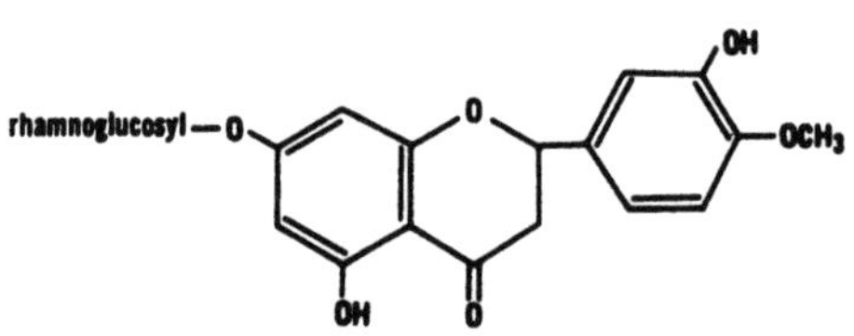

CHROMATOGRAPHIC DATA

Spot Appearance: (UV) deep purple
(UV/NH_3) light blue

R_f Values: 0.51 (TBA), 0.78 (HOAc)

UV SPECTRAL DATA (λ_{max},*nm*)

MeOH	283, 326
NaOMe	242, 286, 356
$AlCl_3$	308, 383
$AlCl_3$/HCl	306, 379
NaOAc	284, 328
NaOAc/H_3BO_3 (Proc. I)	284, 326

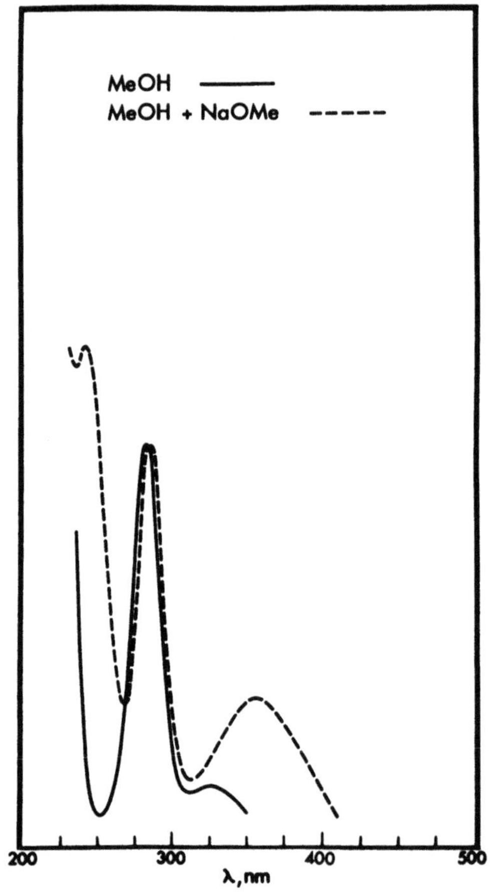

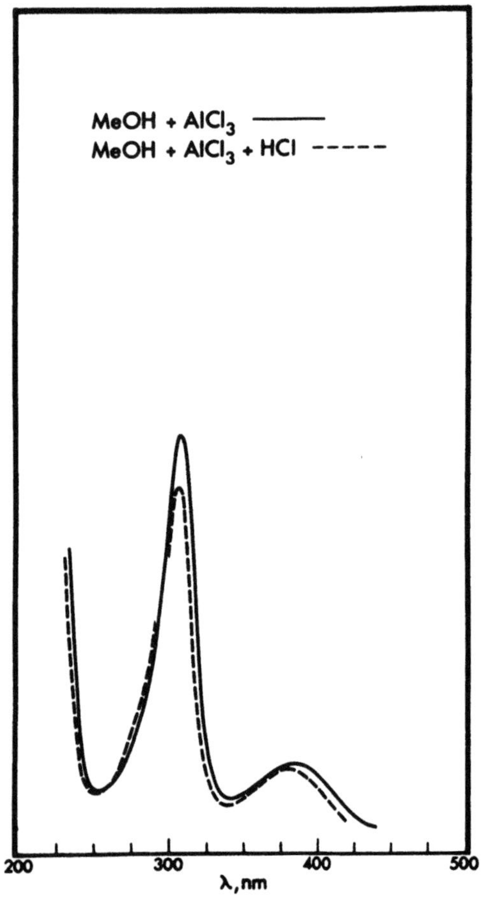

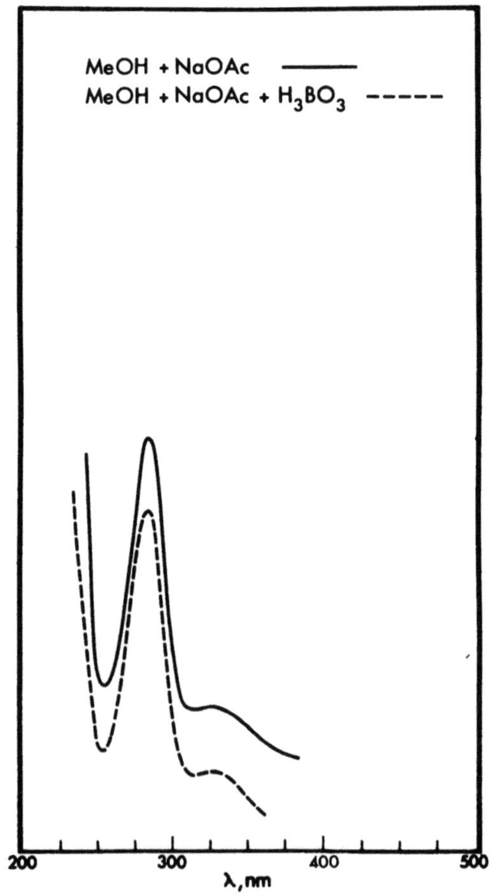

GARBANZOL

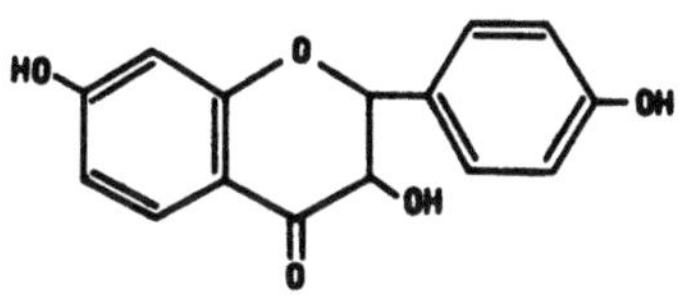

CHROMATOGRAPHIC DATA

Spot Appearance: (UV) pale yellow
(UV/NH_3) light yellow-purple

R_f Values: 0.87 (TBA), 0.52 (HOAc)

UV SPECTRAL DATA (λ_{max},nm)

MeOH	276, 311
NaOMe	250, 297sh, 334
$AlCl_3$	309, 347sh
$AlCl_3$/HCl	276, 309, 408sh
NaOAc	254, 282, 334
NaOAc/H_3BO_3 (Proc. I)	277, 312

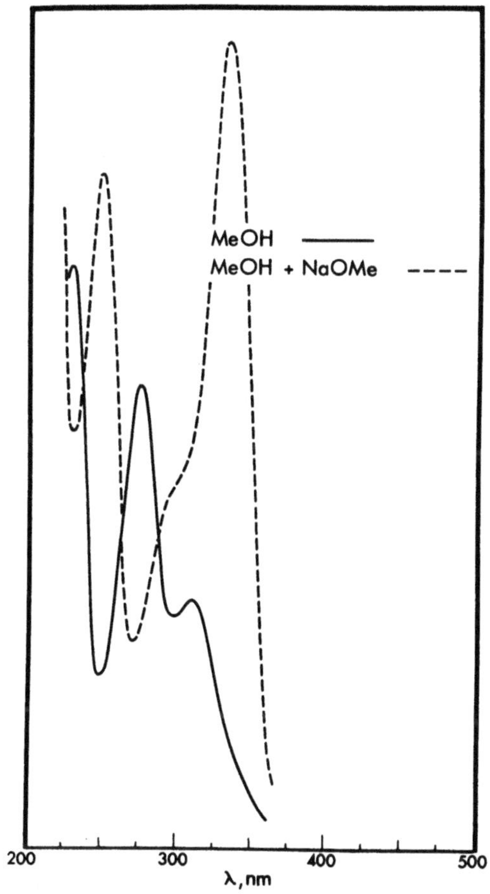

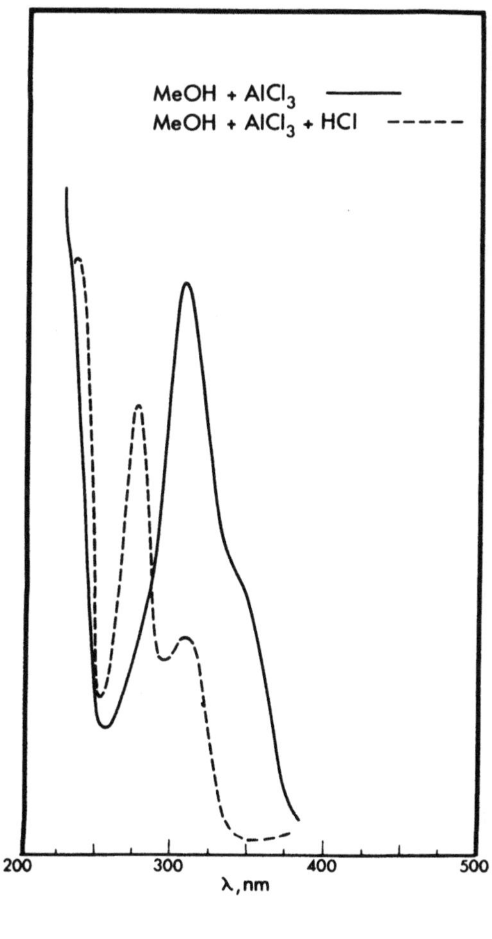

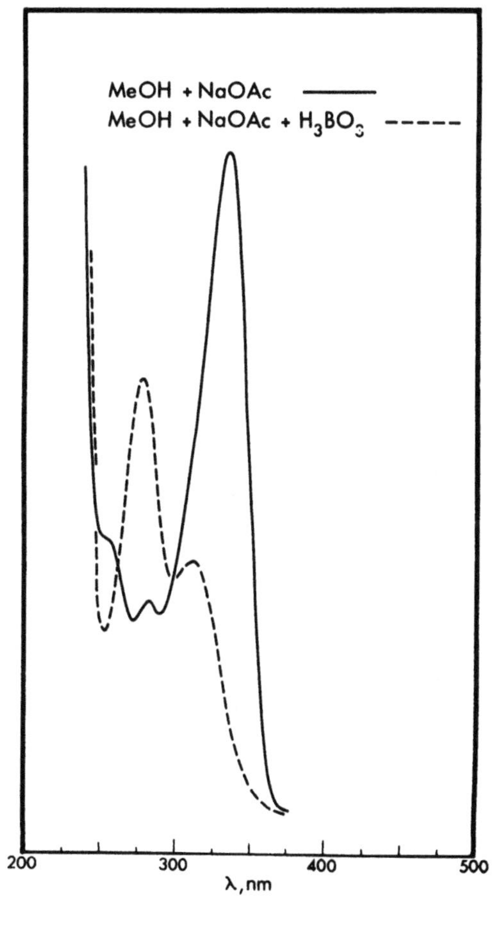

149

DIHYDROFISETIN

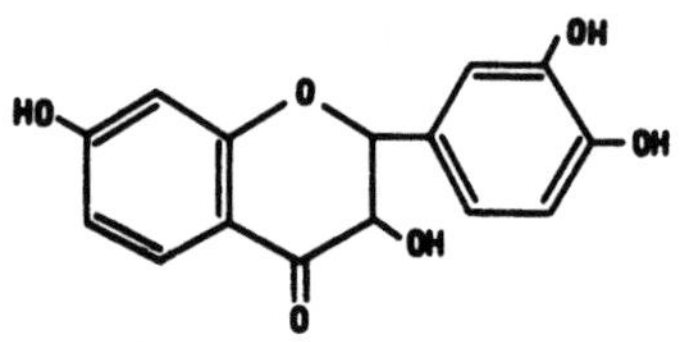

CHROMATOGRAPHIC DATA

Spot Appearance: (UV) pale yellow
(UV/NH_3) light yellow-purple

R_f Values: 0.75 (TBA), 0.63 (HOAc)

UV SPECTRAL DATA (λ_{max}, nm)

MeOH	277, 310
NaOMe	252, 297sh, 334 (dec.)
$AlCl_3$	235, 308, 349sh
$AlCl_3$/HCl	234, 278, 308
NaOAc	256sh, 285, 334
NaOAc/H_3BO_3 (Proc. II)	281, 314sh

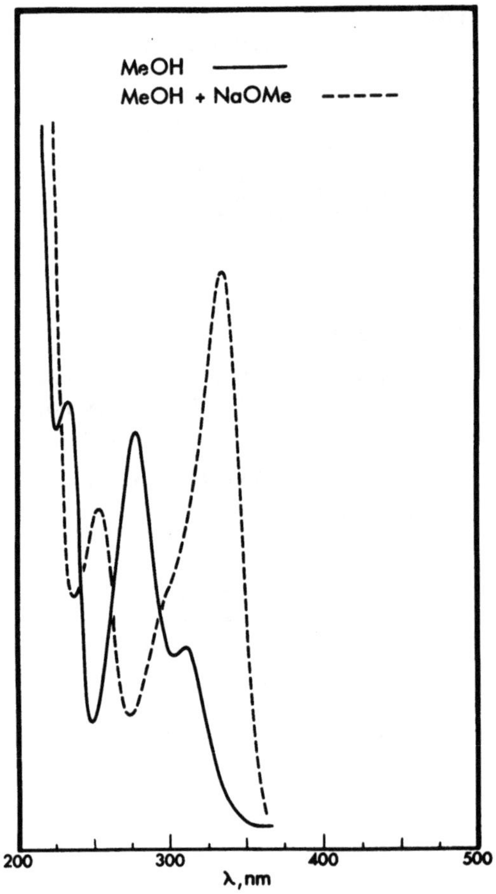

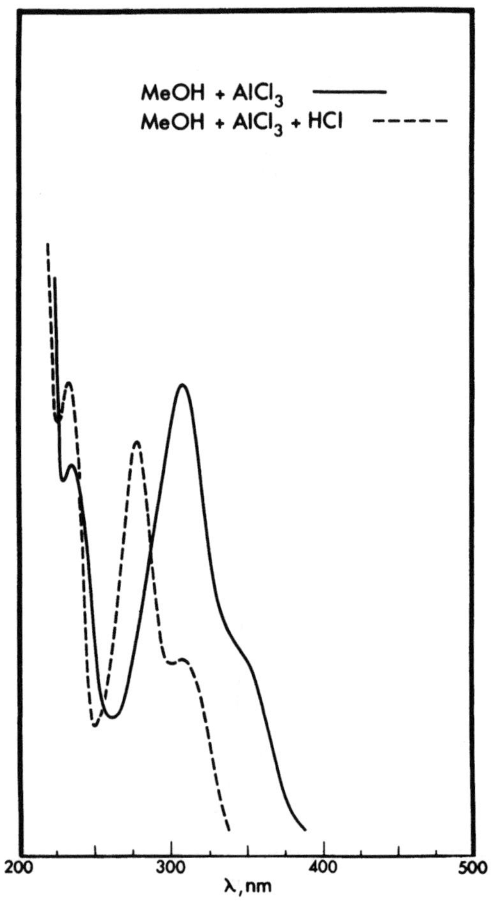

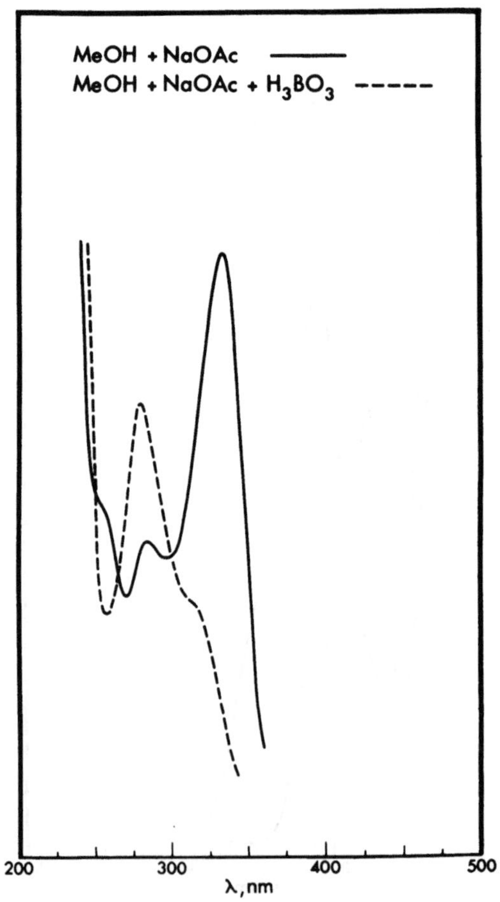

150

(+)-FUSTIN 3-0-GLUCOSIDE

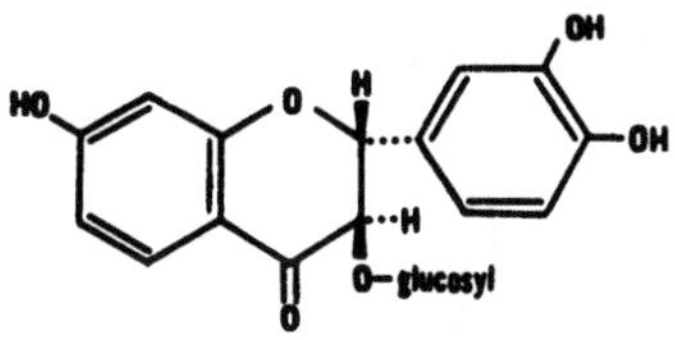

CHROMATOGRAPHIC DATA

Spot Appearance: (UV) pale yellow
(UV/NH_3) light yellow-purple

R_f Values: 0.57 (TBA), 0.78 (HOAc)

UV SPECTRAL DATA (λ_{max},nm)

MeOH	234sh, 280, 311sh
NaOMe	252, 296sh, 337
$AlCl_3$	237, 281, 318sh
$AlCl_3$/HCl	234sh, 280, 311sh, 394sh
NaOAc	254sh, 288, 338
NaOAc/H_3BO_3 (Proc. I)	284, 315sh

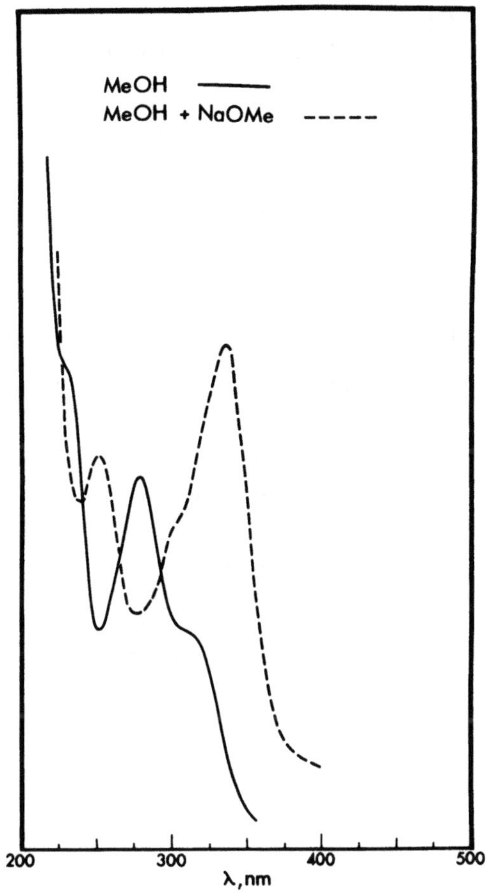

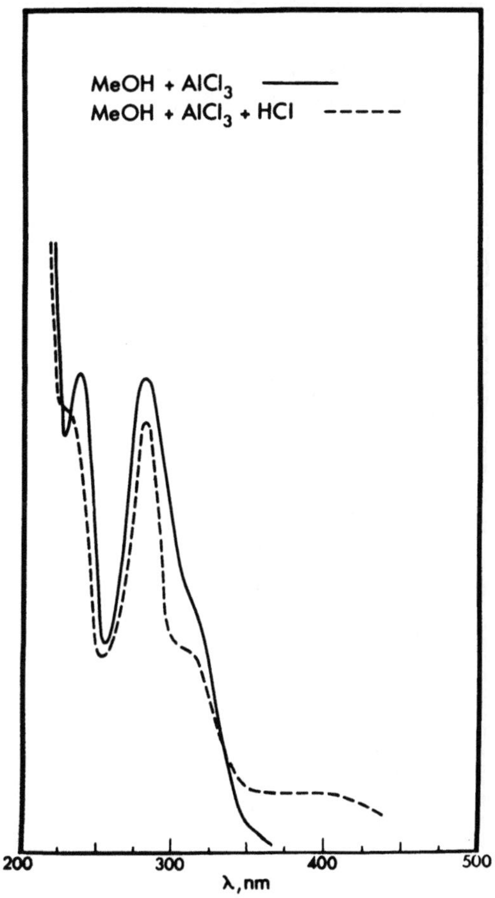

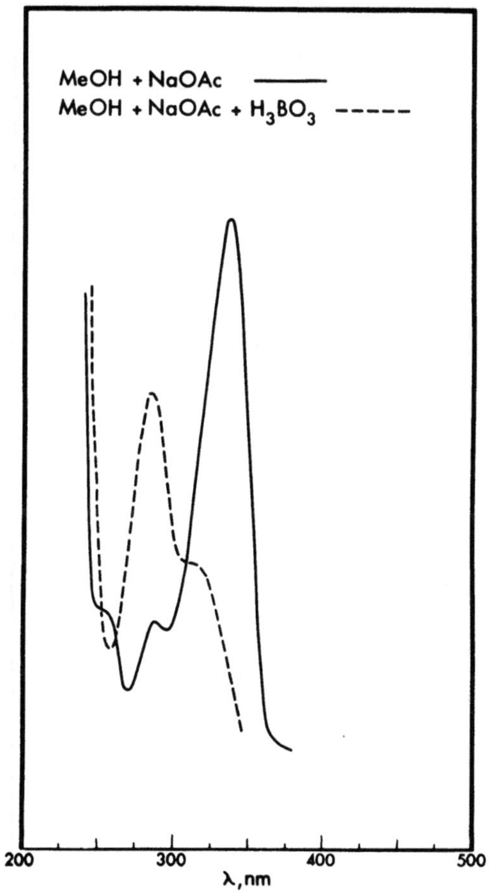

151

DIHYDROKAEMPFEROL

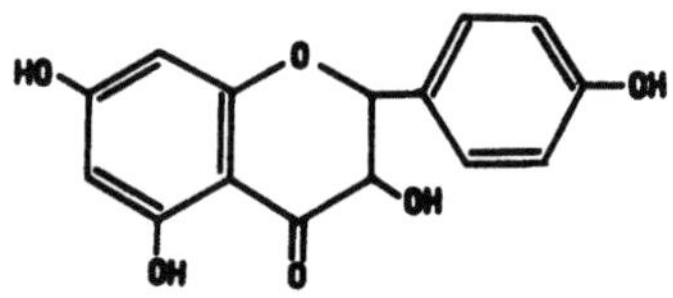

CHROMATOGRAPHIC DATA

Spot Appearance: (UV) deep purple
(UV/NH_3) deep purple

R_f Values: 0.86 (TBA), 0.48 (HOAc)

UV SPECTRAL DATA (λ_{max}, nm)

MeOH	291, 329sh
NaOMe	246, 325
$AlCl_3$	274sh, 316, 382
$AlCl_3$/HCl	280sh, 312, 378
NaOAc	254sh, 284sh, 327
NaOAc/H_3BO_3 (Proc. I)	296, 336sh

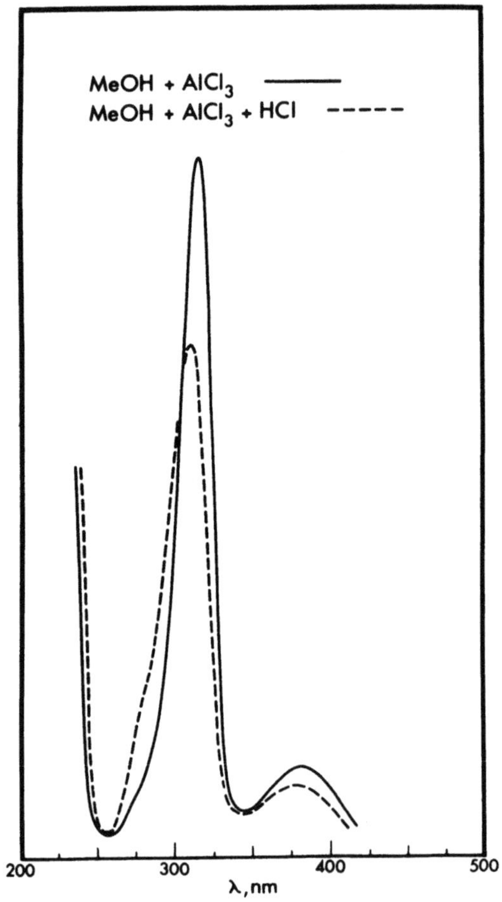

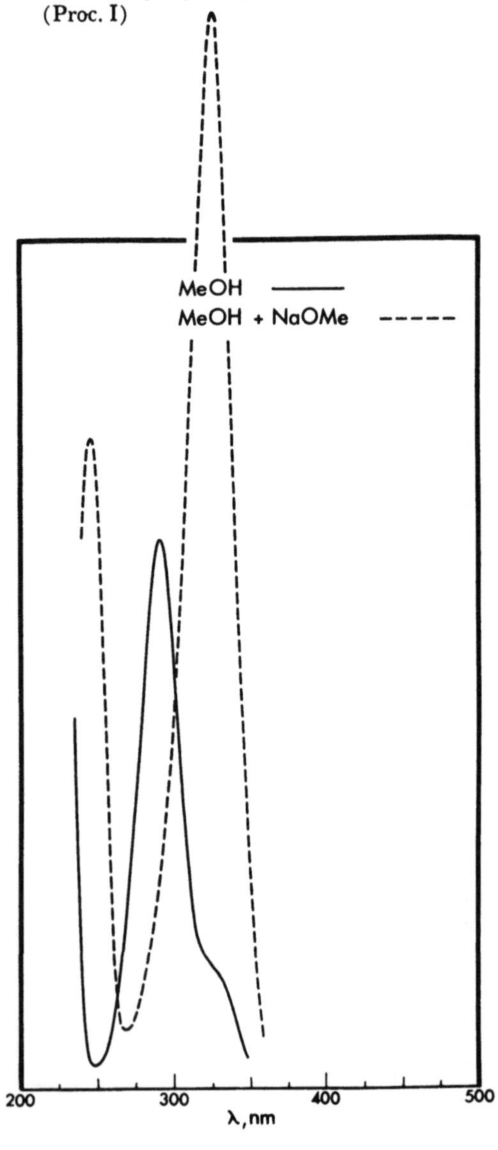

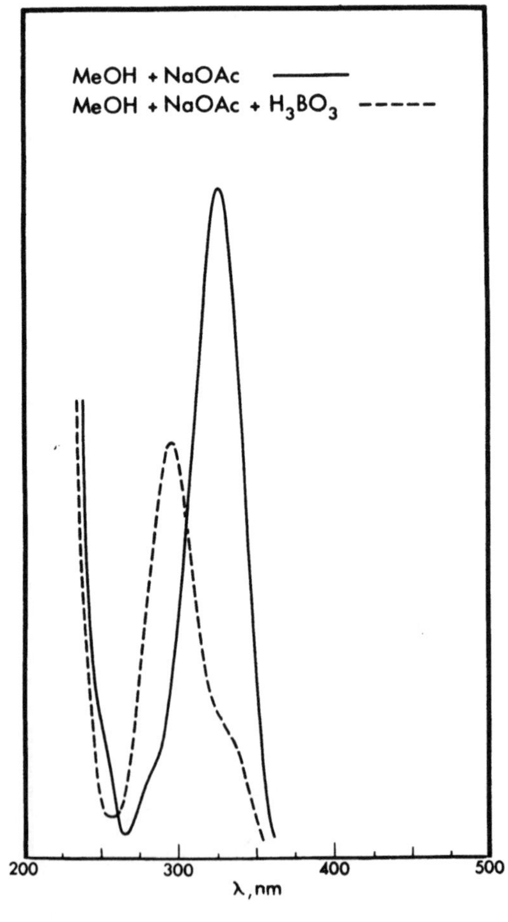

ENGELETIN

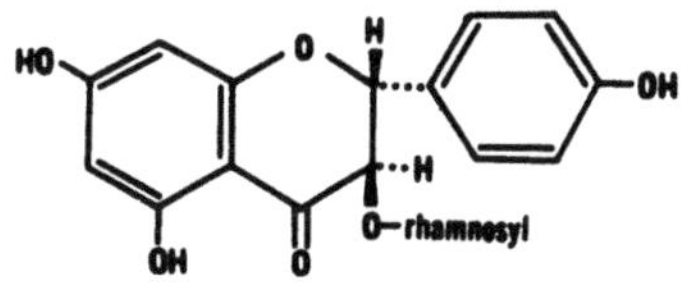

CHROMATOGRAPHIC DATA

Spot Appearance: (UV) deep purple
(UV/NH_3) deep purple

R_f Values: 0.76 (TBA), 0.74 (HOAc)

UV SPECTRAL DATA (λ_{max}, *nm*)

MeOH	293, 332sh
NaOMe	248, 327
$AlCl_3$	277sh, 329, 383sh
$AlCl_3$/HCl	269sh, 314, 379
NaOAc	283, 329
NaOAc/H_3BO_3 (Proc. I)	294, 338sh

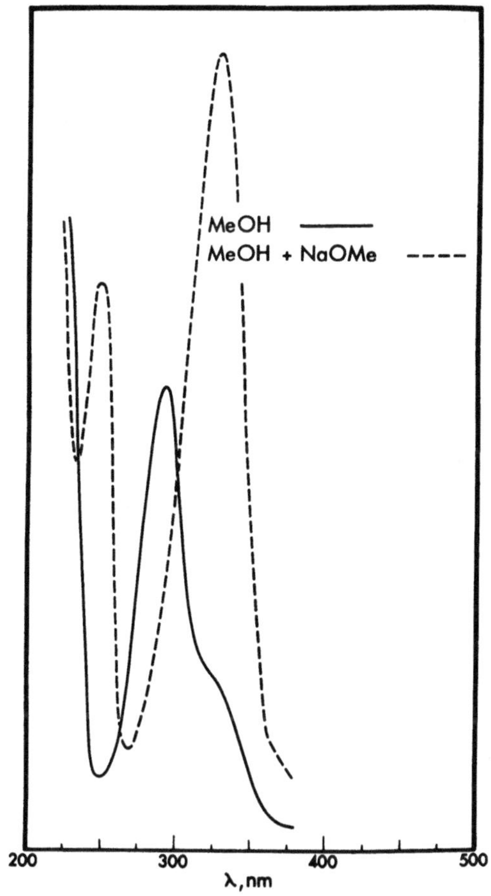

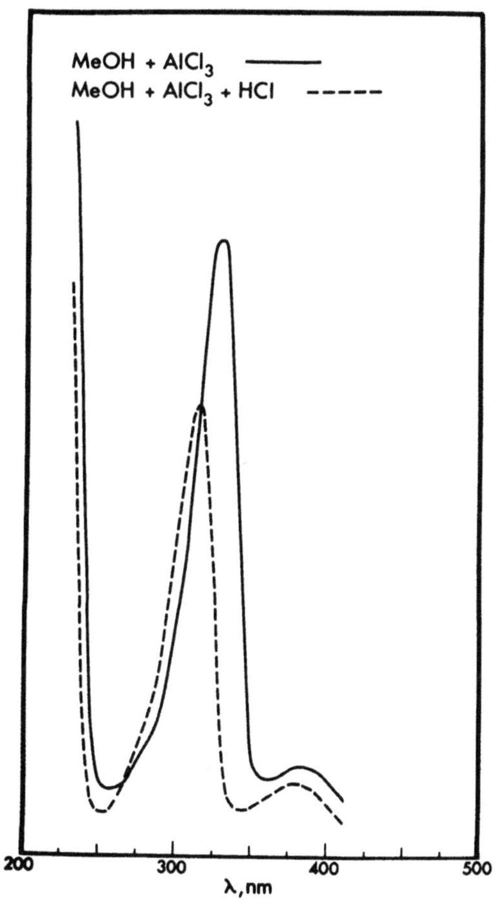

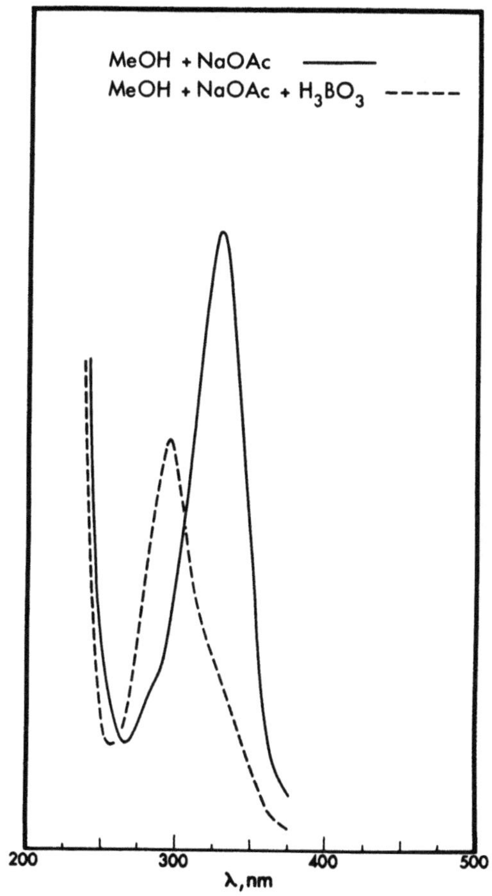

TAXIFOLIN

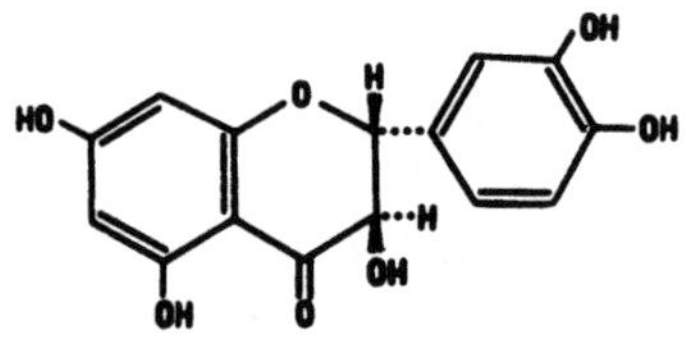

CHROMATOGRAPHIC DATA

Spot Appearance: (UV) deep purple
(UV/NH_3) deep purple

R_f Values: 0.72 (TBA), 0.59 (HOAc)

UV SPECTRAL DATA (λ_{max}, *nm*)

MeOH	290, 327sh
NaOMe	246sh, 326 (dec.)
$AlCl_3$	280sh, 312, 375
$AlCl_3/HCl$	312, 375
NaOAc	289sh, 327
$NaOAc/H_3BO_3$ (Proc. II)	292, 337sh

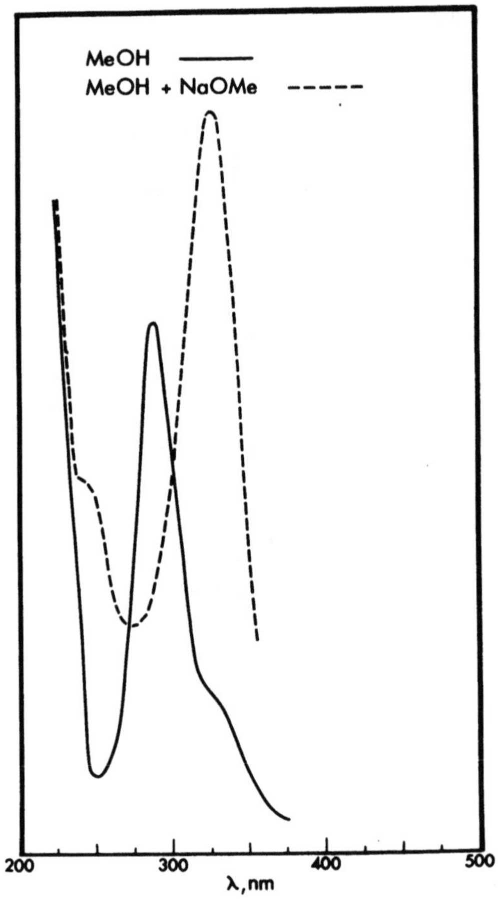

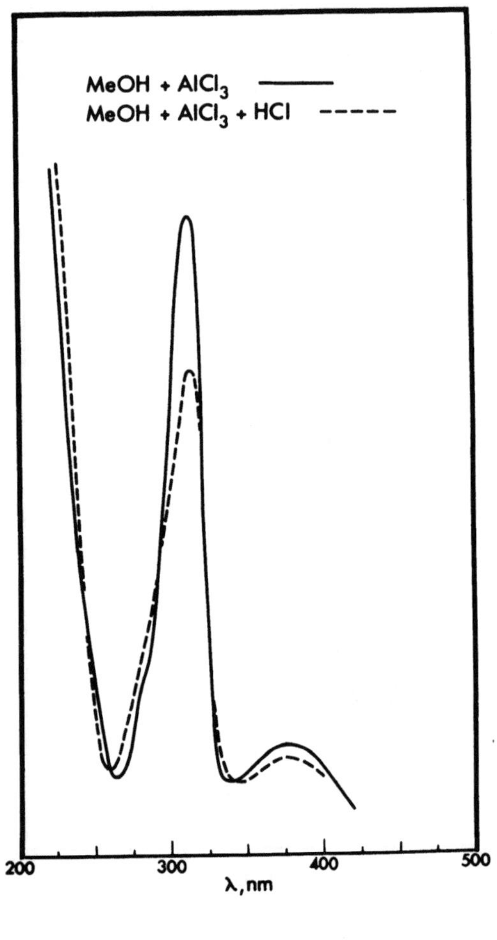

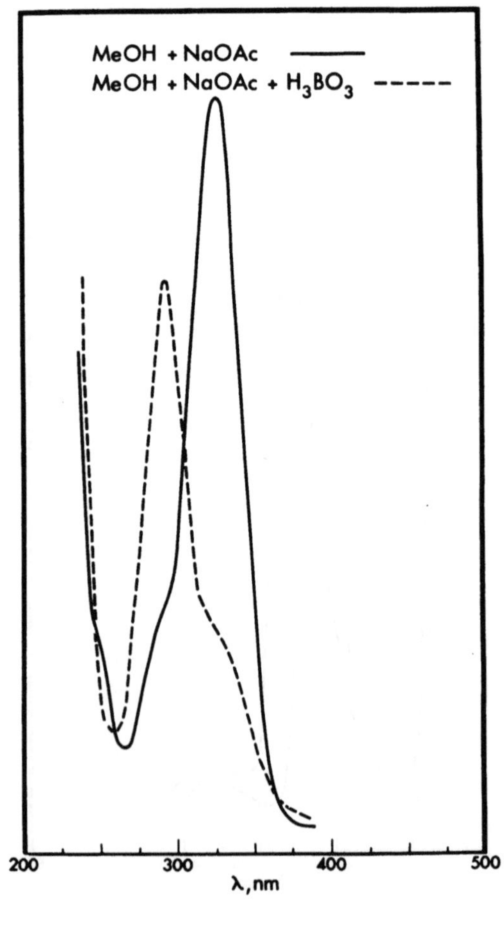

ASTILBIN

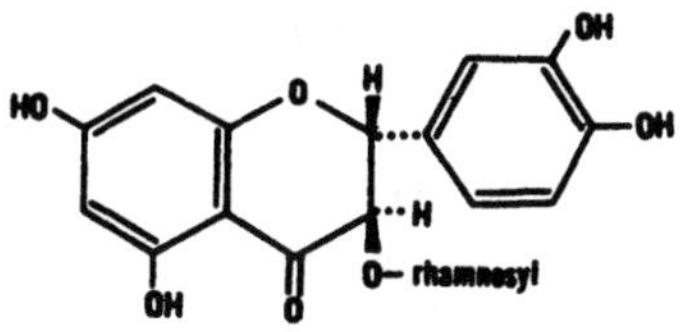

CHROMATOGRAPHIC DATA

Spot Appearance: (UV) deep purple
(UV/NH_3) deep purple

R_f Values: 0.66 (TBA), 0.71 (HOAc)

UV SPECTRAL DATA (λ_{max}, nm)

MeOH	292, 327sh
NaOMe	246, 328
$AlCl_3$	238, 316, 375sh
$AlCl_3$/HCl	287sh, 314, 378
NaOAc	290sh, 329
NaOAc/H_3BO_3 (Proc. I)	294, 335sh

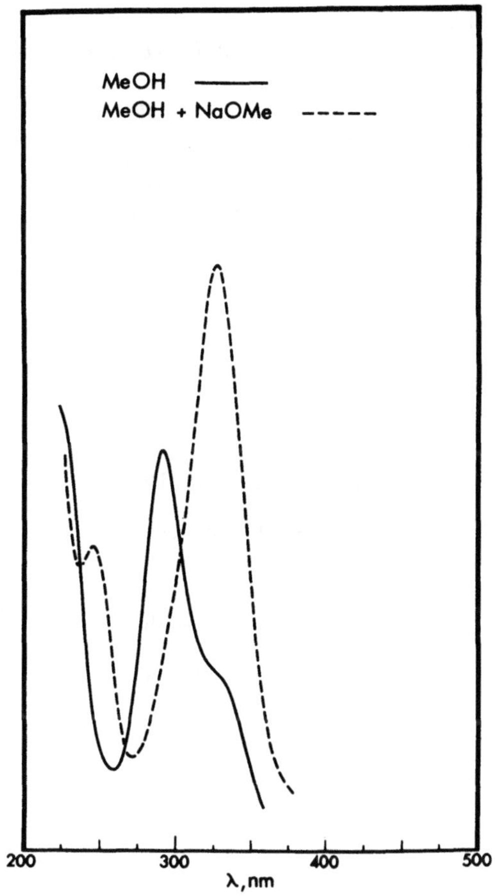

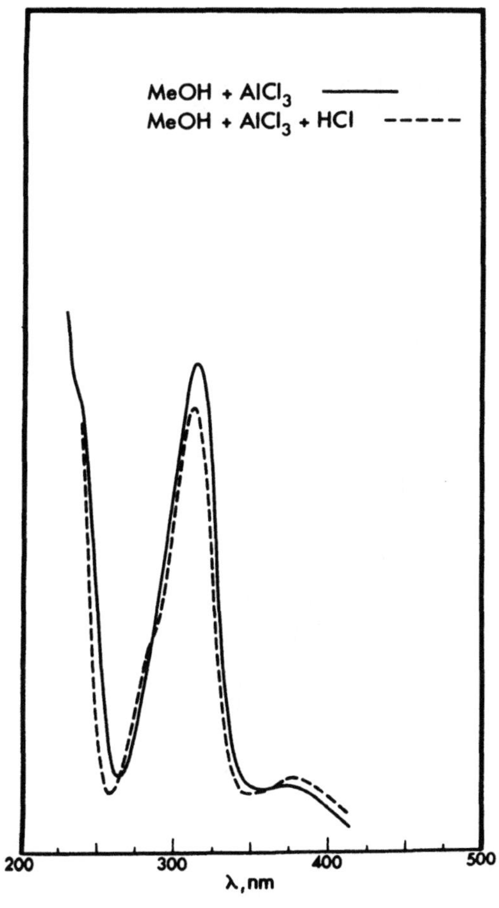

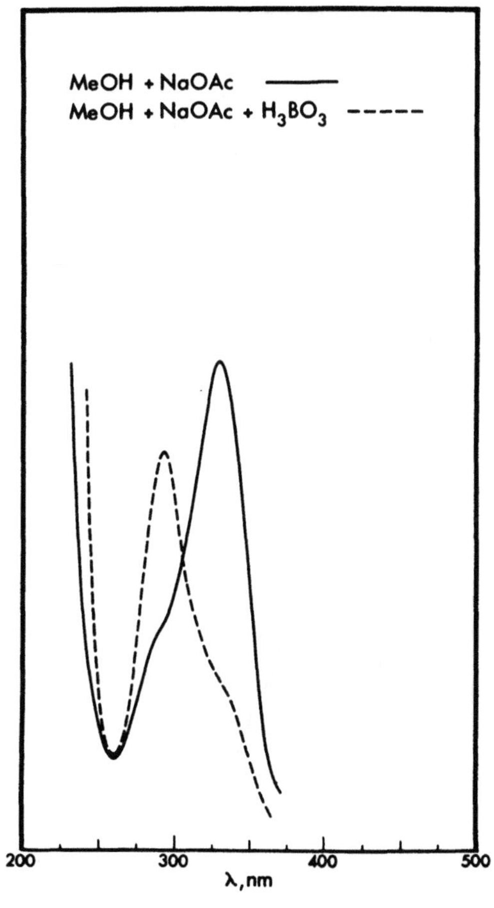

155

DIHYDROROBINETIN

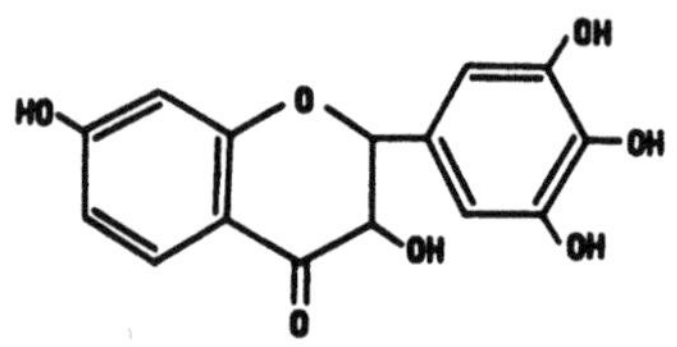

CHROMATOGRAPHIC DATA

Spot Appearance: (UV) pale yellow
(UV/NH_3) light yellow-purple

R_f Values: 0.36 (TBA), 0.58 (HOAc)

UV SPECTRAL DATA (λ_{max}, nm)

MeOH	275, 308
NaOMe	251, 334 (dec.)
$AlCl_3$	280, 307, 345sh
$AlCl_3$/HCl	275, 307
NaOAc	257sh, 280, 333
NaOAc/H_3BO_3 (Proc. II)	278, 312sh

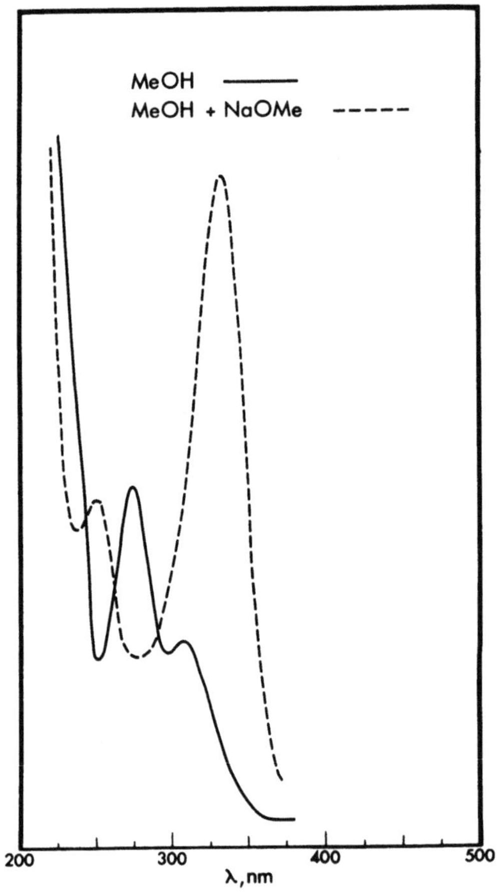

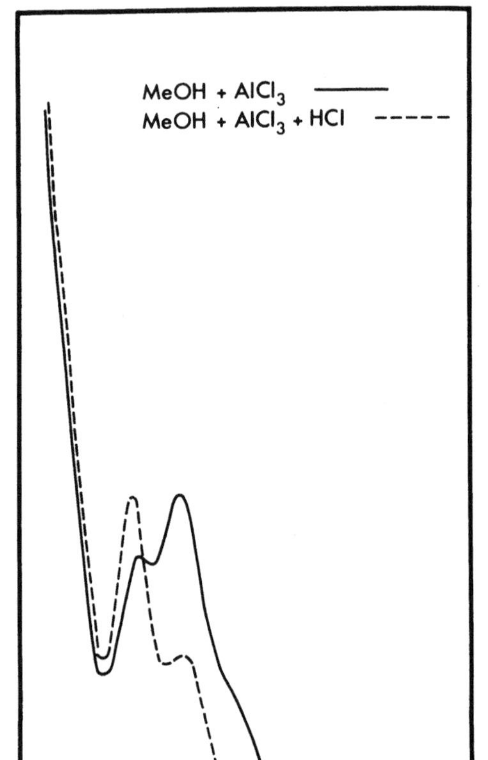

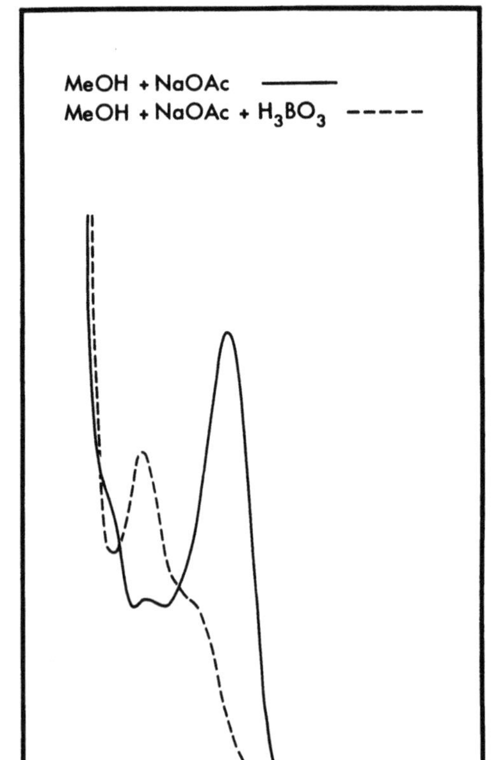

Chapter VII

The Ultraviolet Spectra of Chalcones and Aurones

VII-1. The UV Spectra of Chalcones and Aurones in MeOH 227
VII-2. The UV Spectra of Chalcones and Aurones in the Presence of NaOMe . 228
VII-3. The UV Spectra of Chalcones and Aurones in the Presence of NaOAc . . 228
VII-4. The Detection of Ortho-dihydroxyl Groups in Chalcones and Aurones by the Effect of $NaOAc/H_3BO_3$ on the UV spectrum 228
VII-5. The UV Spectra of Chalcones and Aurones in the Presence of $AlCl_3$ and $AlCl_3/HCl$ 229
5a. The Detection of Ortho-dihydroxyl Groups in Chalcones and Aurones by the Effect of $AlCl_3$ and $AlCl_3/HCl$ on the UV Spectrum 229
5b. The Detection of the 2′-Hydroxyl Group in Chalcones by the Effect of $AlCl_3/HCl$ on the UV Spectrum 229
VII-6. Index of Ultraviolet Absorption Spectra of Chalcones and Aurones . . . 230
References 230

Chalcones and aurones, which are often referred to as the "anthochlor" pigments because many turn red on contact with alkali, are relatively rare types of flavonoids [*1*]. Detailed spectral analyses, in particular of synthetic chalcones and aurones, have been published elsewhere [*2*–*5*], but for completeness a number have been included here. The UV spectra of both chalcones and aurones are characterized by an intense Band I and a diminished Band II absorption.

In connection with the discussion below, it should be noted that both aurones and chalcones have numbering schemes which differ from that used for other flavonoids.

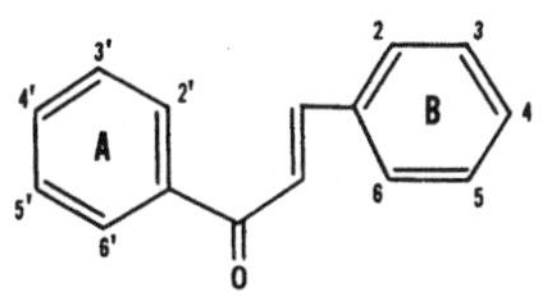

I. Chalcone skeleton

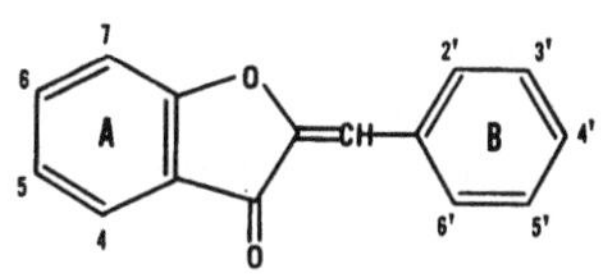

II. Aurone skeleton

VII-1. The UV Spectra of Chalcones (I) and Aurones (II) in MeOH

The major absorption band in chalcones (Band I) usually occurs in the range 340–390 nm, although chalcones lacking B-ring oxygenation may have their Band I absorption at considerably shorter wavelengths (see for example, spectra 157, 158 and 162). Band II is usually a minor peak in the 220–270 nm region. As with flavones and flavonols, increased oxygenation of either the A- or B-ring usually results in bathochromic shifts in Band I (see spectra pairs: 158, 162; 161, 164 and 165, 166).

The long wavelength absorption band in aurones is usually found in the 370–430 nm region, although some of the simpler aurones such as 6-hydroxyaurone [*2*] and 5,7-

dihydroxyaurone (spectrum 169) absorb at much shorter wavelengths. The position of the main absorption peak in naturally occurring aurones, however, ranges from 388 nm (hispidol) to 412 nm (maritimetin) [*1*].

Methylation or glycosylation of hydroxyl groups on the aurone nucleus has little effect on the UV spectrum, with the exception of the 18 nm hypsochromic shift observed when the 7-hydroxyl group in a 6,7-dihydroxyaurone is methylated [*2*]. Thus, the spectra of natural aurone glycosides closely resemble those of their aglycones. The lack of a hypsochromic shift on methylation of the 4-hydroxyl group is notable in that it is thought to indicate [*2*] that the 4-hydroxyl function is only weakly hydrogen-bonded to the carbonyl group (in contrast to the effect of methylation of the strongly hydrogen-bonded 5-hydroxyl group in other flavonoids).

VII-2. The UV Spectra of Chalcones and Aurones in the Presence of NaOMe

Characteristically, many chalcones and aurones turn red or orange in the presence of NaOMe. (As mentioned in Chapter I this color change in alkali distinguishes these flavonoids on a paper chromatogram.)

In the spectra of chalcones containing a free 4-hydroxyl group, the addition of NaOMe causes a 60 – 100 nm bathochromic shift of Band I with an increase in peak intensity (see spectra 159,161, 163, 165 and 166). Chalcones without a 4-hydroxyl group but with either a free 2- (see spectra 156 and 160) or 4'- (see spectrum 157) hydroxyl group also give, in the presence of NaOMe, a 60 – 100 nm bathochromic shift of Band I but without an increase in peak intensity [*2, 4*].

Aurones with a free 4'-hydroxyl group also exhibit a large (80 – 95 nm) diagnostic bathochromic shift of Band I [*3*] in the presence of NaOMe. Aurones which contain a free 6-hydroxyl group give a smaller (ca. 70 nm) shift with NaOMe. However, if both 6- and 4'-hydroxyl groups are present, these bathochromic shifts may be considerably reduced (cf. spectra 168 and 173 and 170, 171 and 174).

VII-3. The UV Spectra of Chalcones and Aurones in the Presence of NaOAc

The limited data (i.e. spectra 156 – 175) available from the present compilation suggest the following correlations regarding the effect of NaOAc on the UV spectra of chalcones and aurones: a bathochromic shift of Band I (or the appearance of a shoulder on the long wavelength side of Band I) can be related to the presence in chalcones of a free 4'- and/or 4-hydroxyl group, and in aurones to a free 4'- and/or 6-hydroxyl group. A free 2-hydroxyl group on the chalcone nucleus does not appear to be appreciably ionized by NaOAc. Chalcones with three adjacent hydroxyl groups may decompose in NaOAc (see spectrum 162).

VII-4. The Detection of Ortho-dihydroxyl Groups in Chalcones and Aurones by the Effect of NaOAc/H_3BO_3 on the UV Spectrum

B-Ring ortho-dihydroxyl groups are readily detected by the 28 – 36 nm bathochromic shift (Table VII-1) observed in Band I of the UV spectra of chalcones and aurones on the addition of NaOAc/H_3BO_3. A-Ring ortho-dihydroxyl groups also appear to be detectable by this procedure although a somewhat smaller shift (Table VII-1, spectra 162 and 170) is observed.

Table VII-1. *The shift of Band I in the UV spectra of ortho-dihydroxychalcones and ortho-dihydroxyaurones in the presence of $NaOAc/H_3BO_3$ and $AlCl_3$*

Spectrum No.	Flavonoid	Bathochromic shift with $NaOAc/H_3BO_3$ (nm) relative to MeOH spectrum	Bathochromic shift with $AlCl_3$ (nm) relative to $AlCl_3/HCl$ spectrum
	Chalcones		
159	3,4-Dihydroxychalcone	36	48
162	2',3',4'-Trihydroxychalcone	10	22
163	2',3,4-Trihydroxychalcone	30	67
166	2',3,4,4'-Tetrahydroxychalcone	36	63
	Aurones		
168	3',4'-Dihydroxyaurone	32	50
170	6,7-Dihydroxyaurone	22	39
172	3',4',6,7-Tetrahydroxyaurone (Maritimetin)	33	48
173	Leptosidin	28	44

VII-5. The UV Spectra of Chalcones and Aurones in the Presence of $AlCl_3$ and $AlCl_3/HCl$

5a. The Detection of Ortho-dihydroxyl Groups in Chalcones and Aurones by the Effect of $AlCl_3$ and $AlCl_3/HCl$ on the UV Spectrum

The presence of B-ring ortho-dihydroxyl groups in both chalcones and aurones can be detected by a 40 – 70 nm bathochromic shift of Band I (relative to the Band I position in the $AlCl_3/HCl$ UV spectrum) on the addition of $AlCl_3$ (see Table VII-2). A-Ring ortho-dihydroxyl groups can also be detected by this procedure; however, from the available data (spectra 162 and 170, Table VII-1) the shift appears to be somewhat smaller.

5b. The Detection of the 2'-Hydroxyl Group in Chalcones by the Effect of $AlCl_3/HCl$ on the UV Spectrum

Band I in the UV spectra of 2'-hydroxychalcones usually undergoes a large bathochromic shift of 48 – 64 nm in the presence of $AlCl_3/HCl$ (see Table VII-2); however, in the spectra of 2',3',4'-trihydroxychalcone (spectrum 162) and its derivatives [2] the shift is only about 40 nm. Lower wavelength bands also appear to shift bathochromically, but since these bands are often poorly defined, the shifts are difficult to determine.

Table VII-2. *The shift of Band I in the UV spectra of 2'-hydroxychalcones in the presence of $AlCl_3/HCl$*

Spectrum No.	Chalcone	Bathochromic shift of Band I (nm)
158	2'-Hydroxy-4'-methoxychalcone	64
160	2,2'-Dihydroxychalcone	64
161	2',4-Dihydroxychalcone	58
162	2',3',4'-Trihydroxychalcone	39
163	2',3,4-Trihydroxychalcone	63
164	2,2',4-Trihydroxychalcone	62
165	2',4,4'-Trihydroxychalcone	54
166	2',3,4,4'-Tetrahydroxychalcone	48

VII-6. Index of Ultraviolet Absorption Spectra of Chalcones and Aurones

Spectrum No.		Oxidation pattern					
		2′	3′	4′	2	3	4
	Chalcones						
156	2-Hydroxychalcone				OH		
157	4′-Hydroxychalcone			OH			
158	2′-Hydroxy-4′-methoxychalcone	OH		OCH_3			
159	3,4-Dihydroxychalcone					OH	OH
160	2,2′-Dihydroxychalcone	OH			OH		
161	2′,4-Dihydroxychalcone	OH					OH
162	2′,3′,4′-Trihydroxychalcone	OH	OH	OH			
163	2′,3,4-Trihydroxychalcone	OH				OH	OH
164	2,2′,4-Trihydroxychalcone	OH			OH		OH
165	2′,4,4′-Trihydroxychalcone	OH		OH			OH
166	2′,3,4,4′-Tetrahydroxychalcone	OH		OH		OH	OH
		3′	4′	4	5	6	7
	Aurones						
167	4′-Hydroxyaurone		OH				
168	3′,4′-Dihydroxyaurone	OH	OH				
169	5,7-Dihydroxyaurone				OH		OH
170	6,7-Dihydroxyaurone					OH	OH
171	6-Hydroxy-4′-methoxyaurone		OCH_3			OH	
172	6,7,3′,4′-Tetrahydroxyaurone (Maritimetin)	OH	OH			OH	OH
173	Leptosidin	OH	OH			OH	OCH_3
174	6-Hydroxy-4,3′,4′-trimethoxyaurone	OCH_3	OCH_3	OCH_3		OH	
175	3′-Hydroxy-4,4′,6-trimethoxyaurone	OH	OCH_3	OCH_3		OCH_3	

References

1. Harborne, J.B.: Comparative biochemistry of the flavonoids, p. 78, 83. London: Academic Press 1967.
2. Jurd, L., in: The chemistry of flavonoid compounds (edited by T.A. Geissman), p. 107–155. Oxford: Pergamon Press 1962.
3. Geissman, T.A., and J.B. Harborne: J. Am. Chem. Soc. **78**, 832 (1956).
4. Jurd, L., and R.M. Horowitz: J. Org. Chem. **26**, 2561 (1961).
5. Geissman, T.A., and J.B. Harborne: J. Am. Chem. Soc. **77**, 4622 (1955).

156

2-HYDROXYCHALCONE

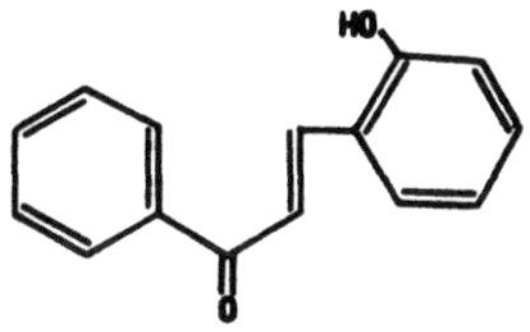

CHROMATOGRAPHIC DATA

Spot Appearance: (UV) dull green
(UV/NH_3) orange

R_f Values: 0.94 (TBA), 0.25 (HOAc)

UV SPECTRAL DATA (*λ_{max},nm*)

MeOH	243, 284, 344
NaOMe	242, 283sh, 322sh, 436
$AlCl_3$	238, 286sh, 348
$AlCl_3$/HCl	239, 281sh, 347
NaOAc	282sh, 345sh
NaOAc/H_3BO_3 (Proc. I)	281sh, 345

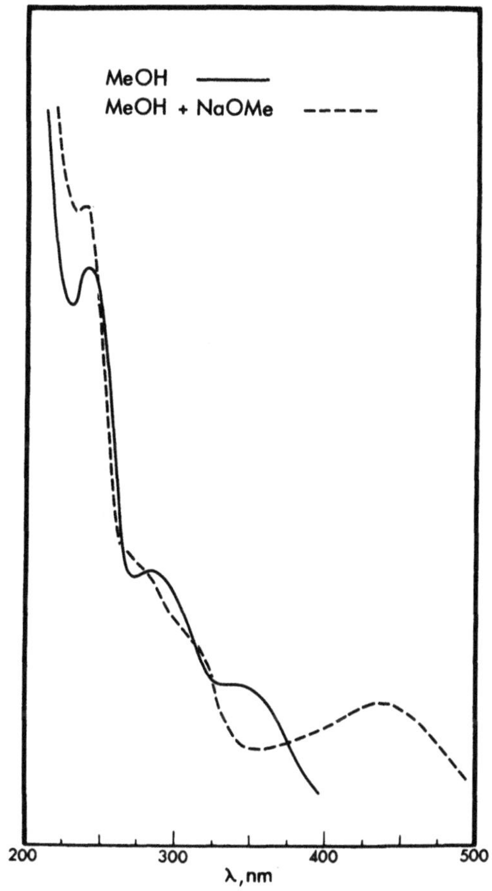

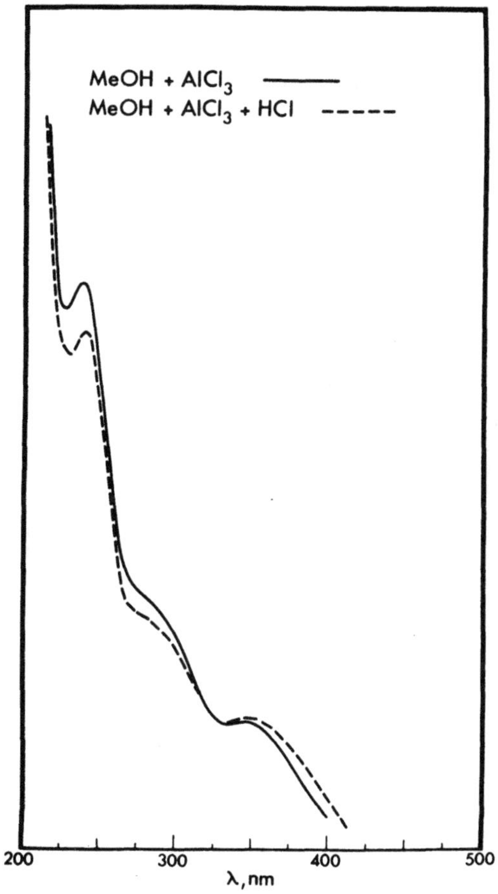

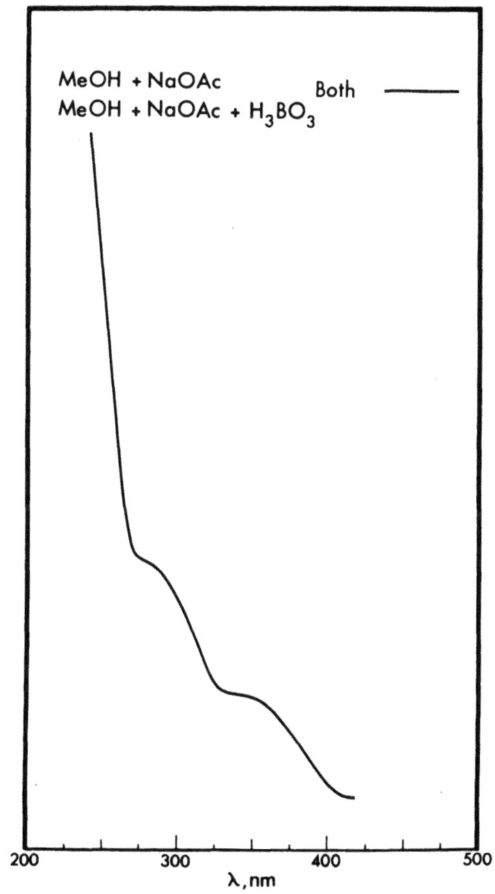

157

4'-HYDROXYCHALCONE

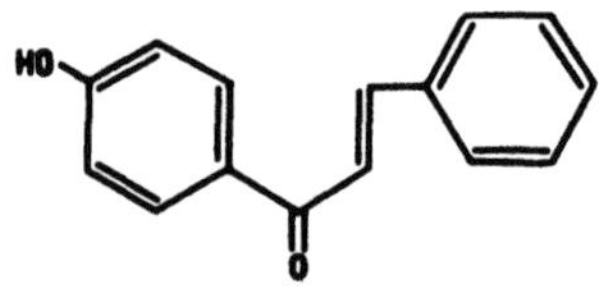

CHROMATOGRAPHIC DATA

Spot Appearance: (UV) deep purple
(UV/NH_3) yellow

R_f Values: 0.93 (TBA), 0.24 (HOAc)

UV SPECTRAL DATA (λ_{max}, nm)

MeOH	224, 318
NaOMe	267sh, 296, 380
$AlCl_3$	227, 318
$AlCl_3$/HCl	227, 318
NaOAc	267sh, 302, 375
NaOAc/H_3BO_3 (Proc. I)	320

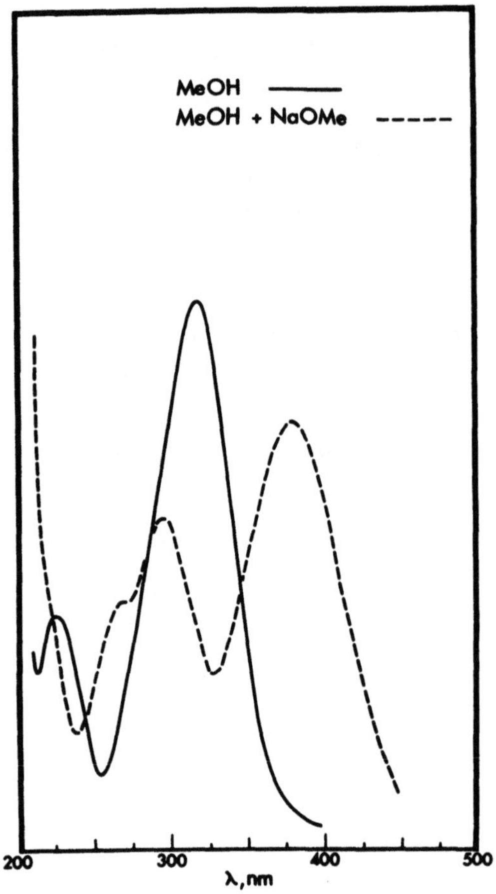

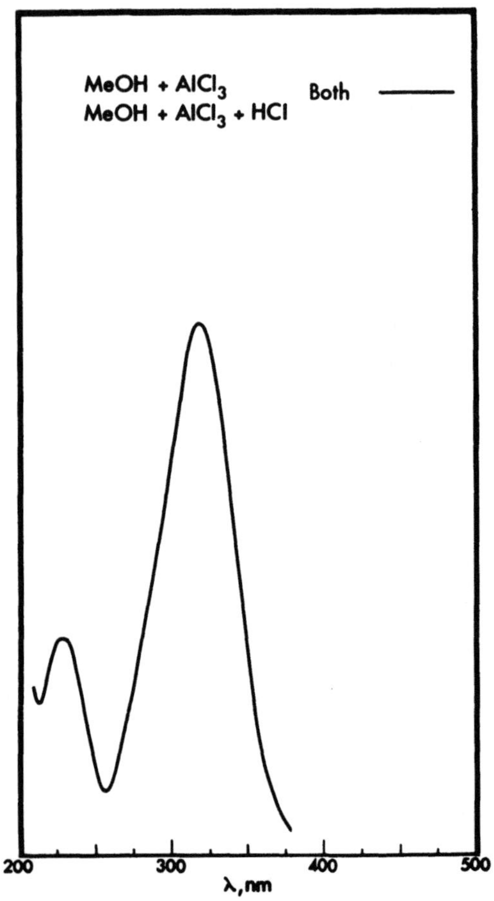

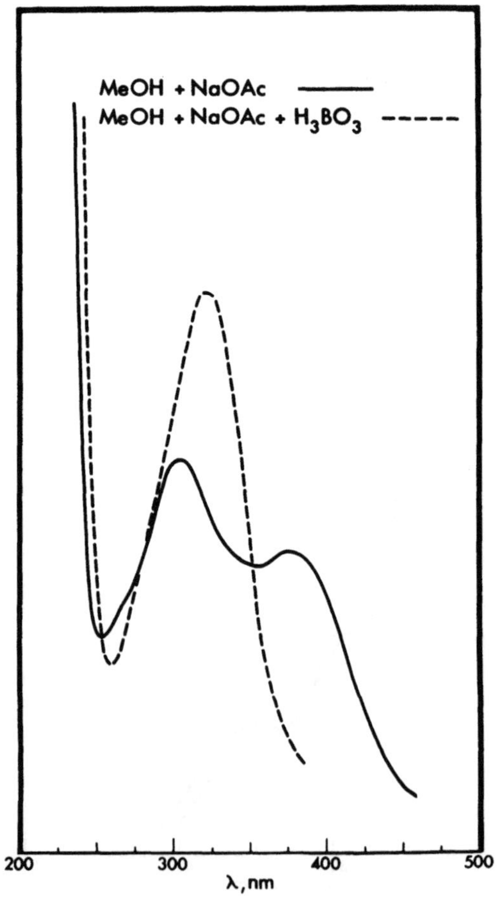

2′-HYDROXY-4′-METHOXYCHALCONE

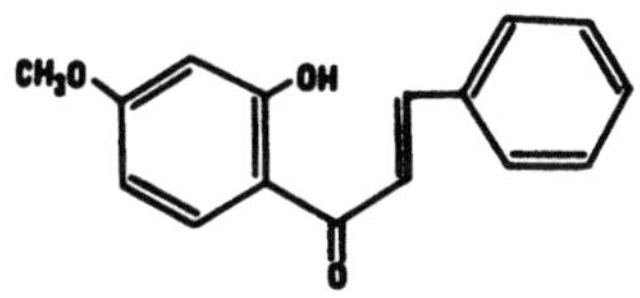

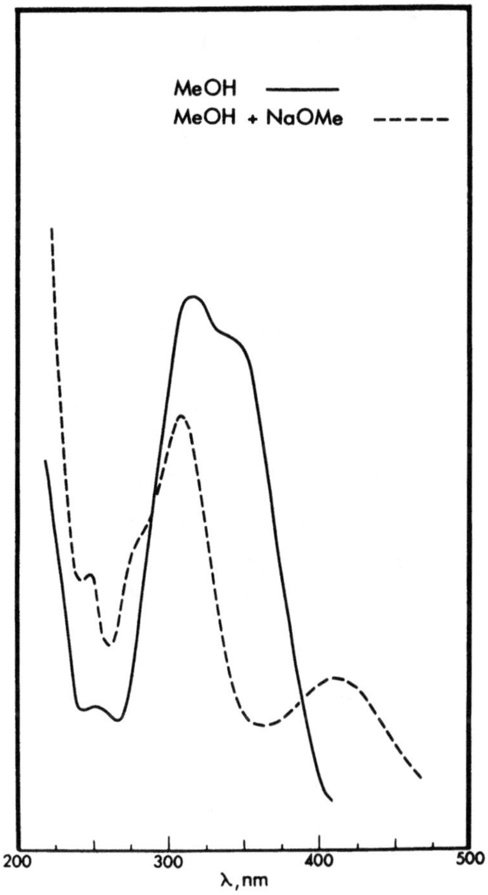

CHROMATOGRAPHIC DATA

Spot Appearance: (UV) deep purple
(UV/NH_3) deep purple

R_f Values: 0.93 (TBA), 0.09 (HOAc)

UV SPECTRAL DATA (λ_{max}, nm)

MeOH	252sh, 317, 342sh
NaOMe	249, 279sh, 309, 408
$AlCl_3$	231sh, 241sh, 304sh, 324sh, 357, 407
$AlCl_3$/HCl	231sh, 243sh, 272sh, 308sh, 323sh, 348, 406
NaOAc	256sh, 320, 343sh
NaOAc/H_3BO_3 (Proc. I)	260sh, 320, 343sh

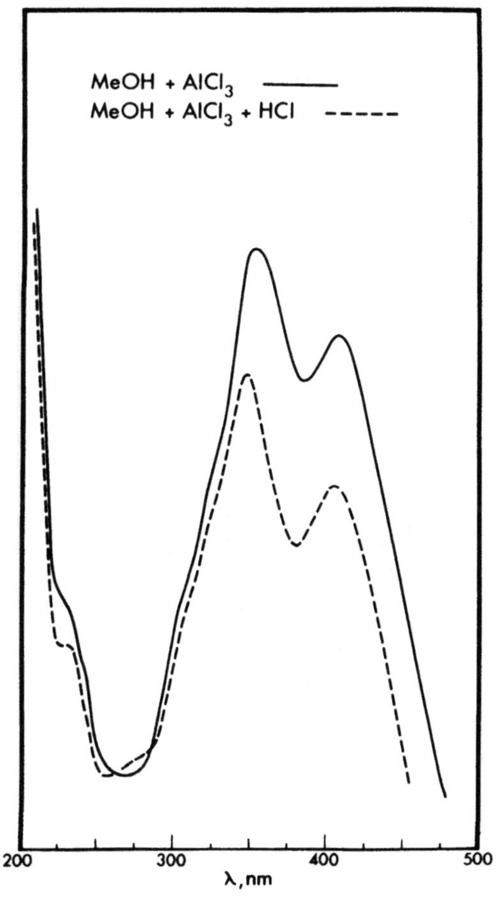

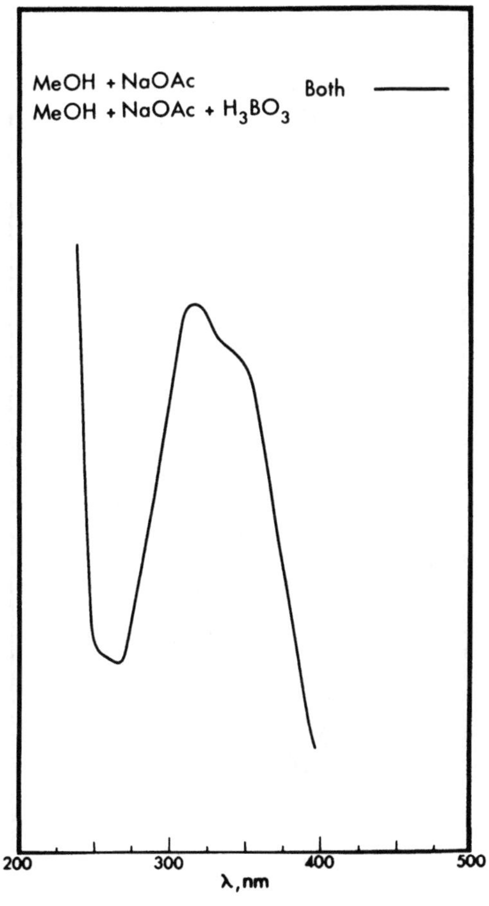

159

3,4-DIHYDROXYCHALCONE

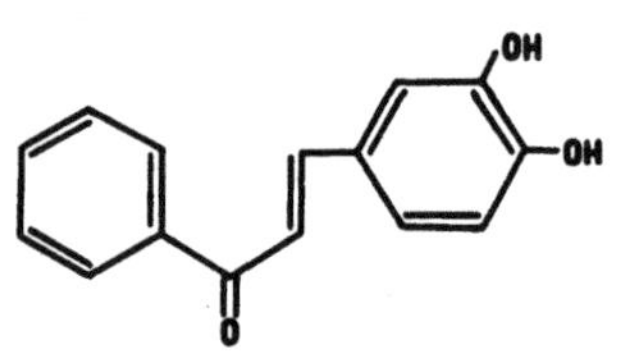

CHROMATOGRAPHIC DATA

Spot Appearance: (UV) fluorescent yellow-green
(UV/NH_3) orange

R_f Values: 0.87 (TBA), 0.26 (HOAc)

UV SPECTRAL DATA (λ_{max},*nm*)

MeOH	265, 316sh, 365
NaOMe	267, 341sh, 446
$AlCl_3$	263sh, 275, 332sh, 413
$AlCl_3$/HCl	265, 365
NaOAc	265, 377, 443sh
NaOAc/H_3BO_3 (Proc. I)	272, 327sh, 401

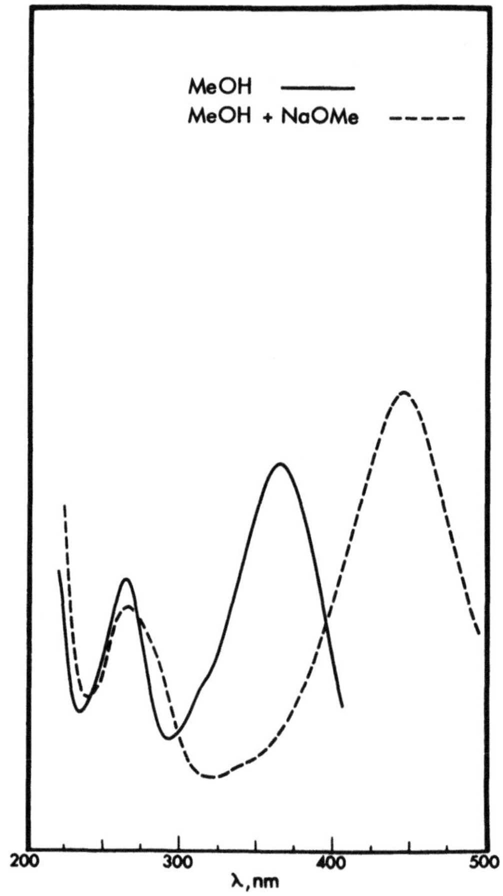

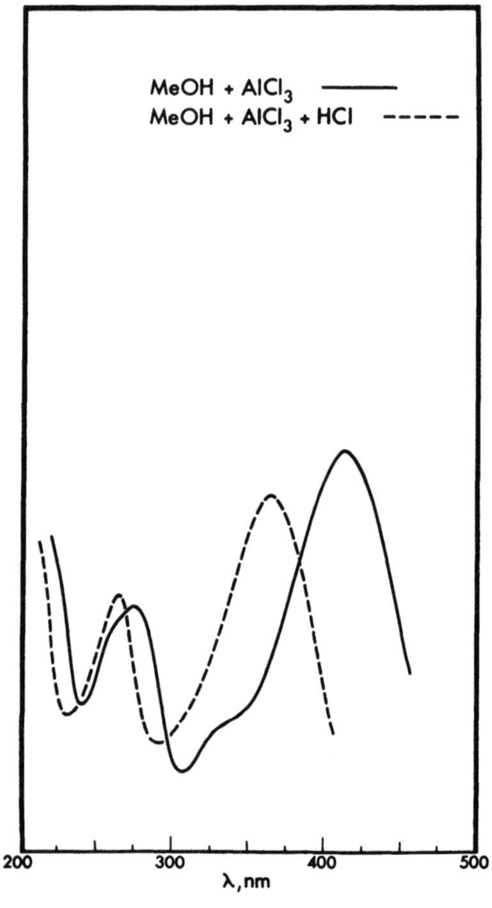

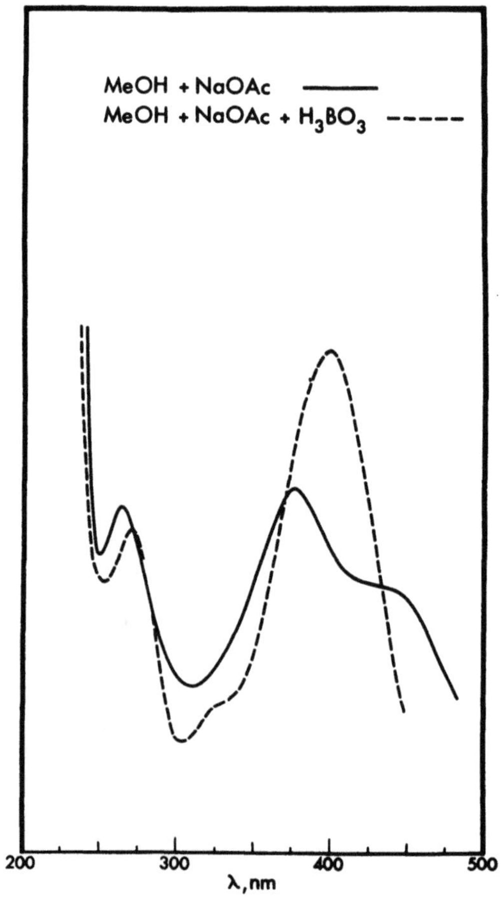

160

2,2'-DIHYDROXYCHALCONE

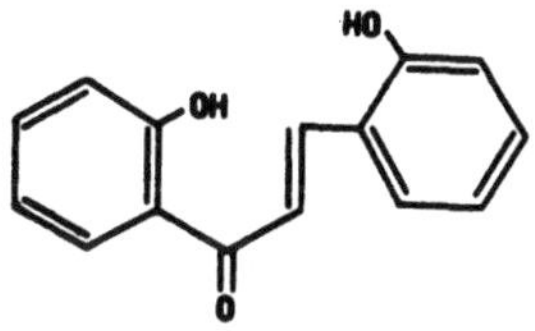

CHROMATOGRAPHIC DATA

Spot Appearance: (UV) deep purple
(UV/NH_3) red

R_f Values: 0.94 (TBA), 0.17 (HOAc)

UV SPECTRAL DATA (λ_{max},*nm*)

MeOH	240sh, 253, 309, 369
NaOMe	244sh, 276, 324, 444
$AlCl_3$	268, 303sh, 339, 392sh, 440
$AlCl_3$/HCl	263, 303sh, 335, 384, 433
NaOAc	256sh, 312, 371, 457
NaOAc/H_3BO_3	256sh, 311, 373
(Proc. I)	

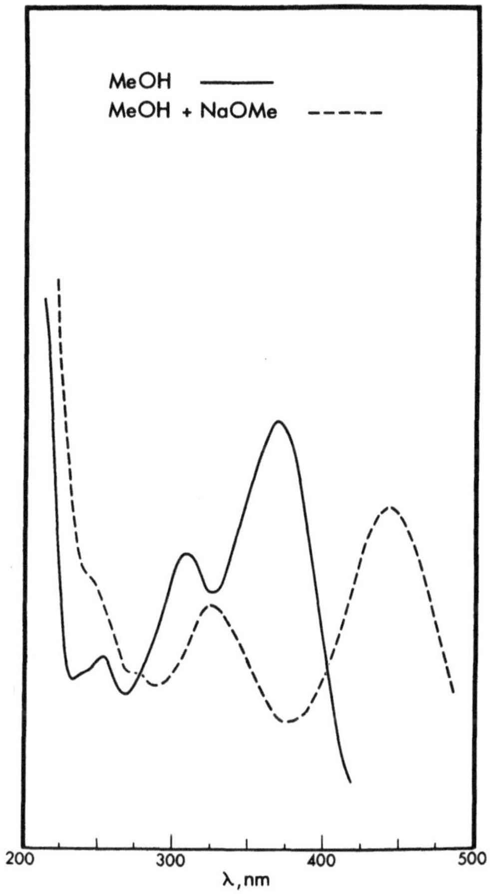

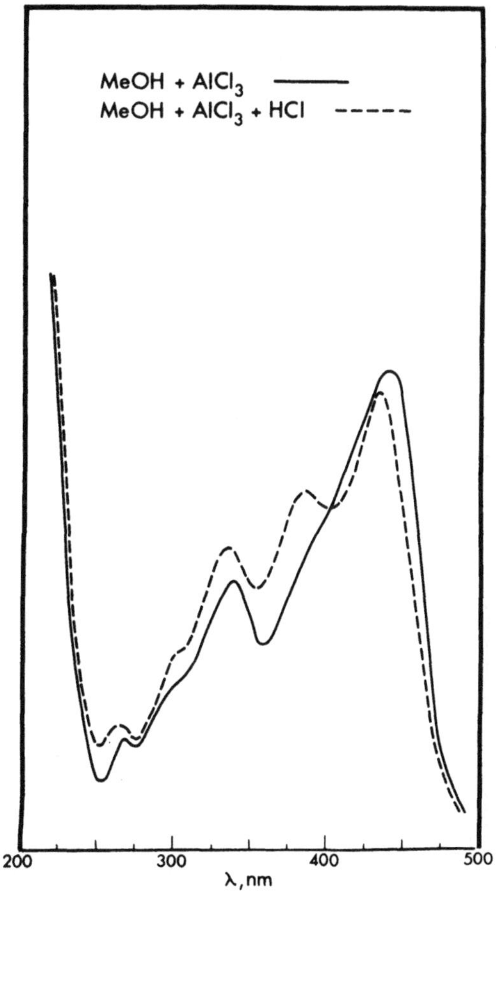

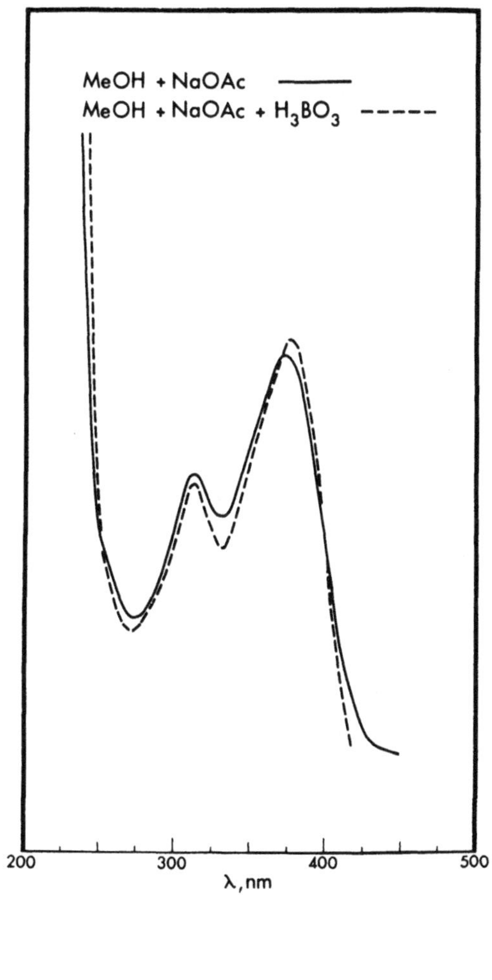

161

2′,4-DIHYDROXYCHALCONE

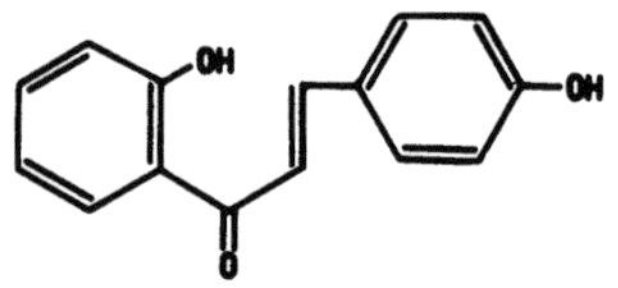

CHROMATOGRAPHIC DATA

Spot Appearance: (UV) deep purple
(UV/NH_3) orange

R_f Values: 0.93 (TBA), 0.11 (HOAc)

UV SPECTRAL DATA (λ_{max},*nm*)

MeOH	250, 278, 324sh, 369
NaOMe	249, 271, 320, 433
$AlCl_3$	247, 284, 301, 393, 443
$AlCl_3$/HCl	247, 282, 326sh, 383, 437
NaOAc	249sh, 275, 330sh, 373, 443sh
NaOAc/H_3BO_3 (Proc. I)	253sh, 277, 323sh, 372

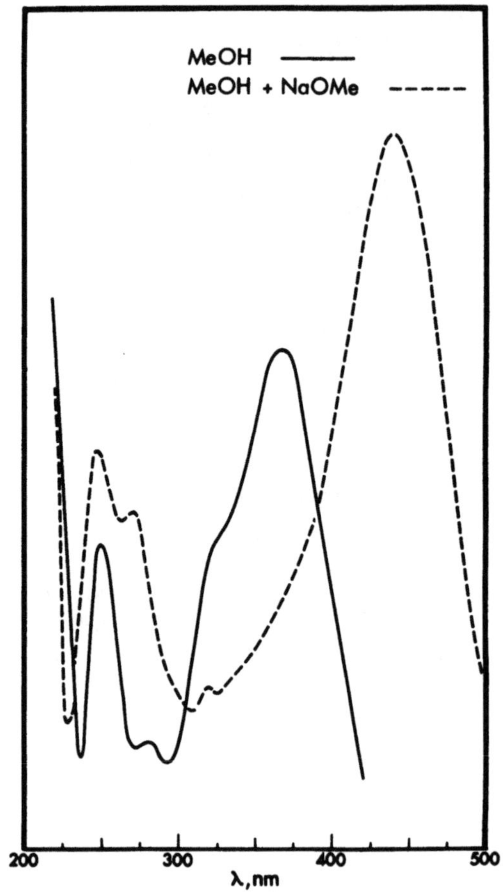

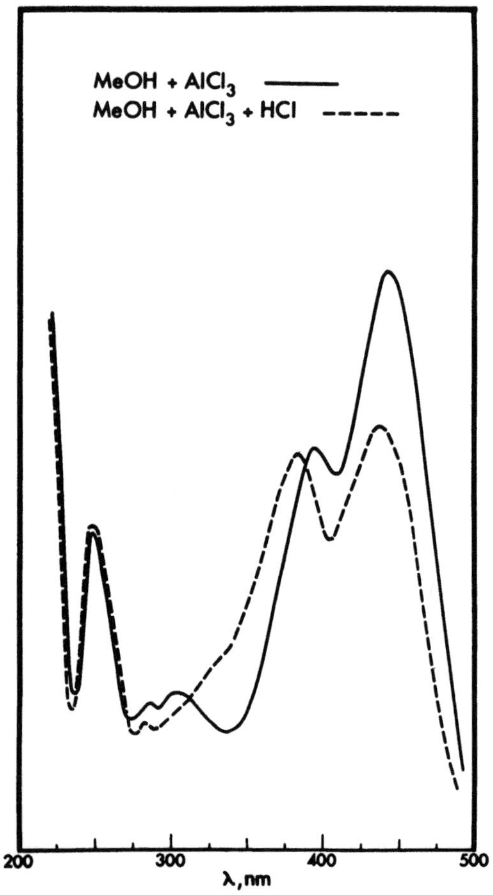

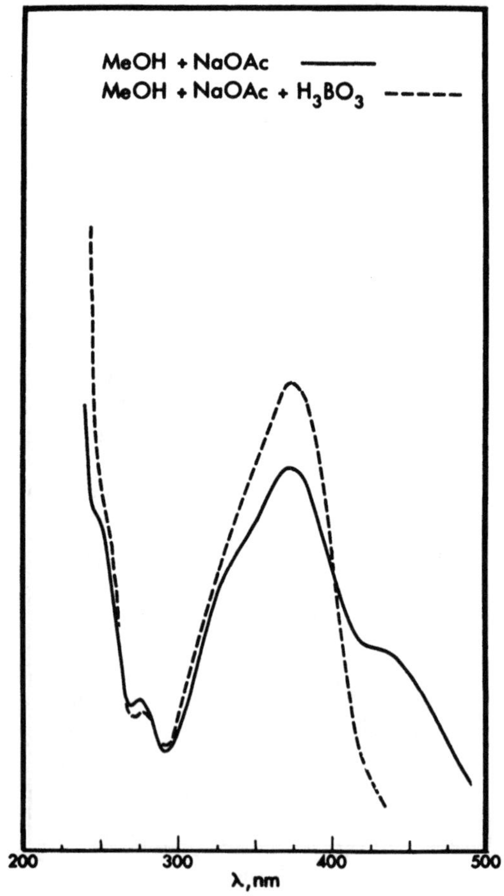

162

2′,3′,4′-TRIHYDROXYCHALCONE

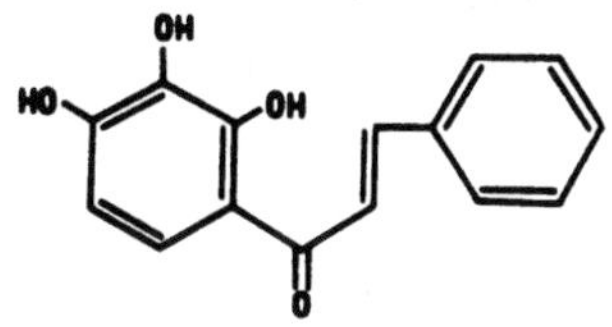

CHROMATOGRAPHIC DATA

Spot Appearance: (UV) yellow-green
(UV/NH_3) orange

R_f Values: 0.84 (TBA), 0.16 (HOAc)

UV SPECTRAL DATA (λ_{max}, nm)

MeOH	251sh, 309sh, 340
NaOMe	258, 298, 394 (dec.)
$AlCl_3$	236sh, 279sh, 306, 316sh, 332sh, 401
$AlCl_3$/HCl	238, 314, 328, 379
NaOAc	259, 287sh, 297, 389 (dec.)
NaOAc/H_3BO_3 (Proc. II)	308sh, 350

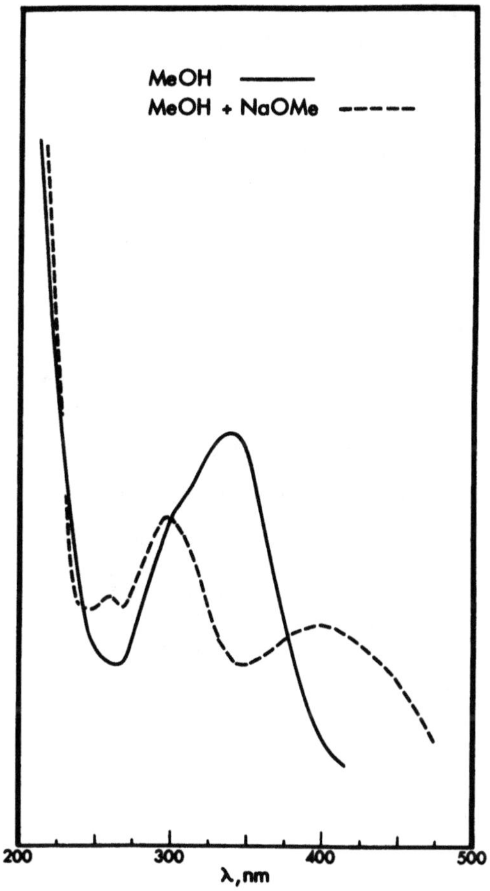

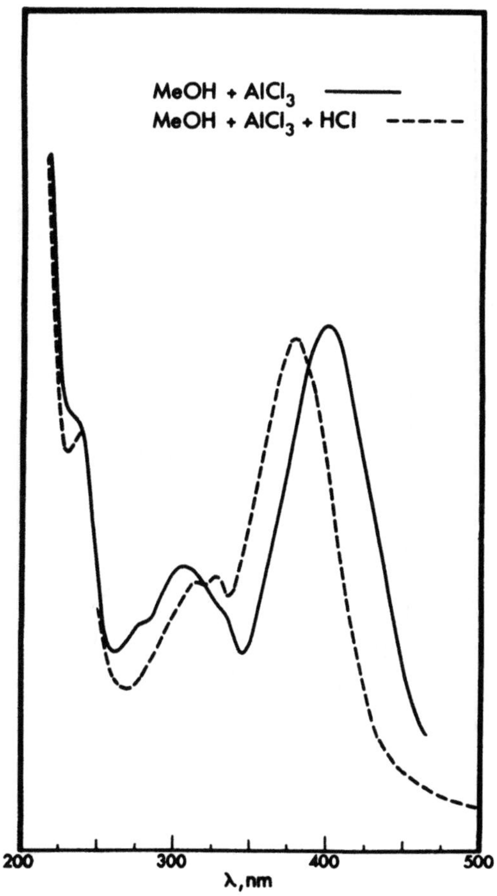

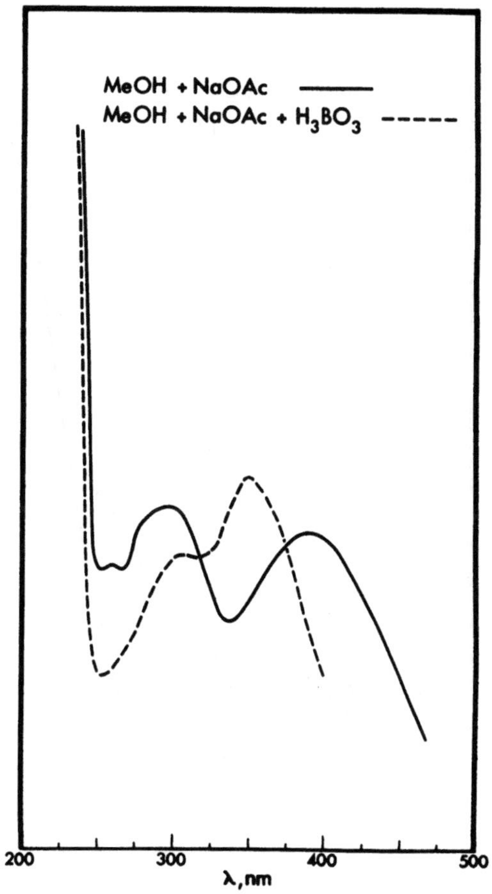

163

2′,3,4-TRIHYDROXYCHALCONE

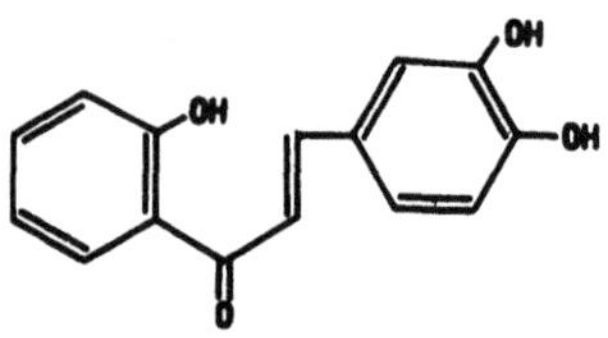

CHROMATOGRAPHIC DATA

Spot Appearance: (UV) dull yellow-green
(UV/NH_3) orange

R_f Values: 0.86 (TBA), 0.10 (HOAc)

UV SPECTRAL DATA (λ_{max},*nm*)

MeOH	246sh, 267, 320sh, 384
NaOMe	246sh, 275, 448
$AlCl_3$	288sh, 315sh, 375sh, 514
$AlCl_3$/HCl	273, 395, 447
NaOAc	273, 339, 402
NaOAc/H_3BO_3 (Proc. I)	277, 332, 414

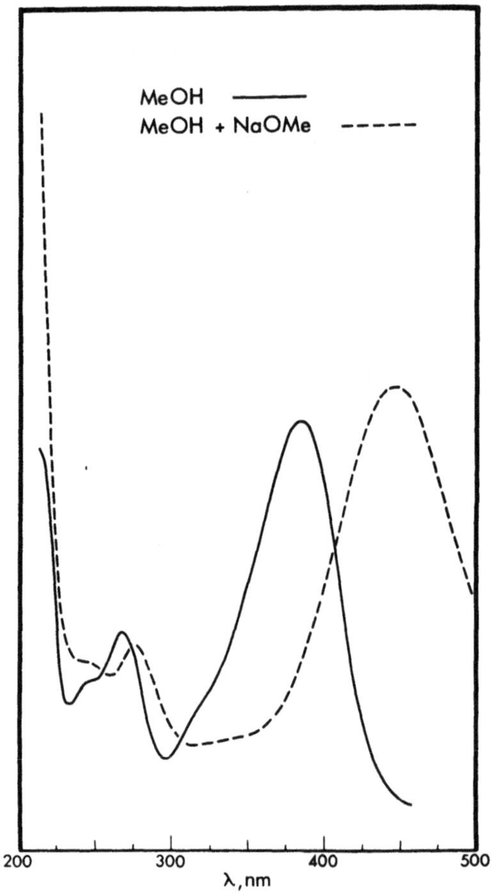

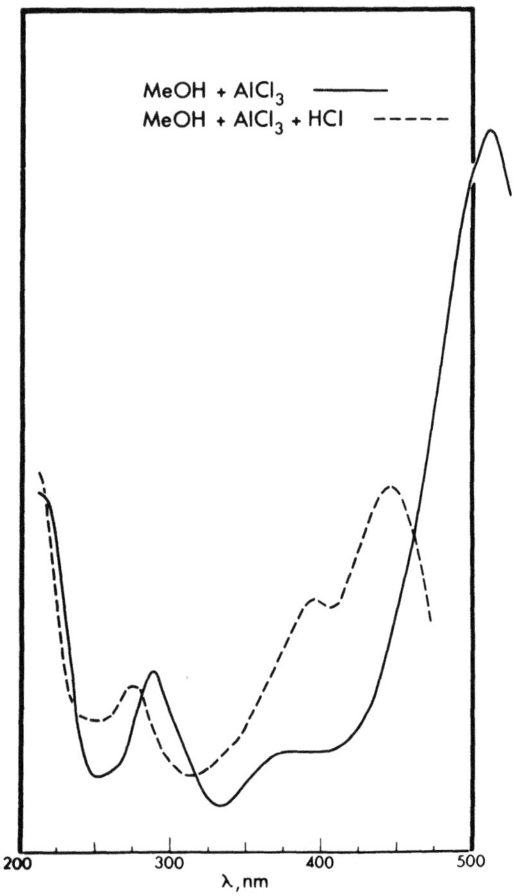

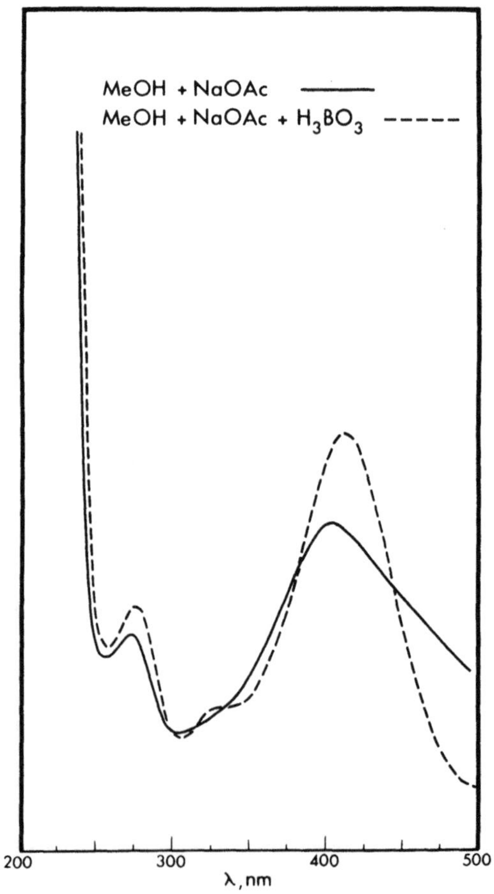

164

2,2′,4-TRIHYDROXYCHALCONE

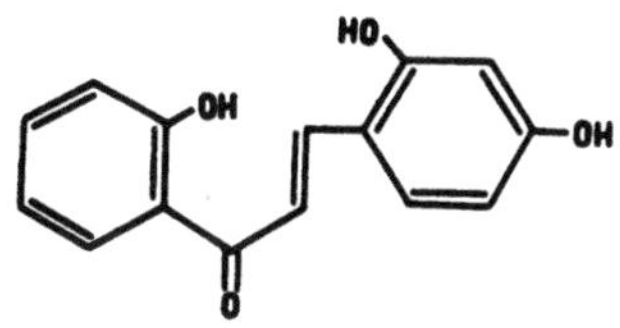

CHROMATOGRAPHIC DATA

Spot Appearance: (UV) fluorescent yellow-green

(UV/NH_3) red

R_f Values: 0.97 (TBA), 0.09 (HOAc)

UV SPECTRAL DATA (λ_{max},*nm*)

MeOH	253, 279sh, 322, 391
NaOMe	270, 302, 387sh, 501
$AlCl_3$	252, 281, 286sh, 321sh, 400sh, 465
$AlCl_3$/HCl	252, 279, 322, 399sh, 453
NaOAc	256sh, 275sh, 324, 402, 462sh
NaOAc/H_3BO_3 (Proc. I)	255sh, 277sh, 325, 398

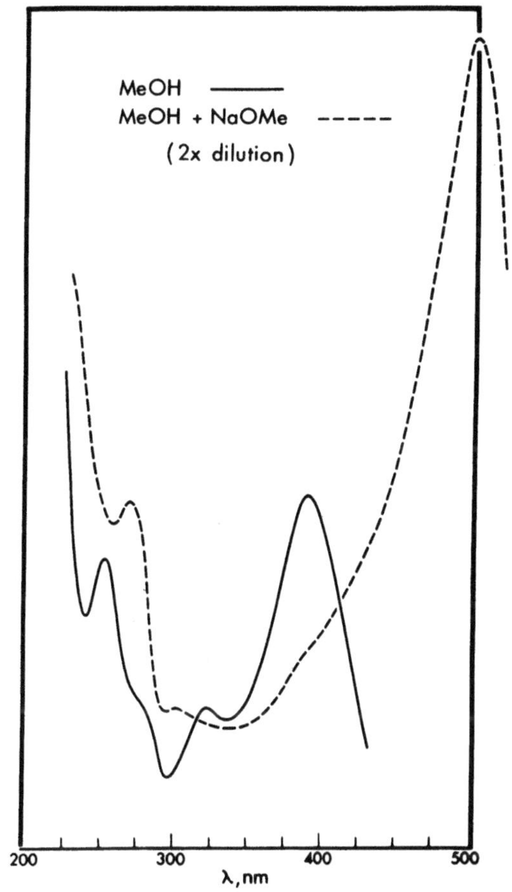

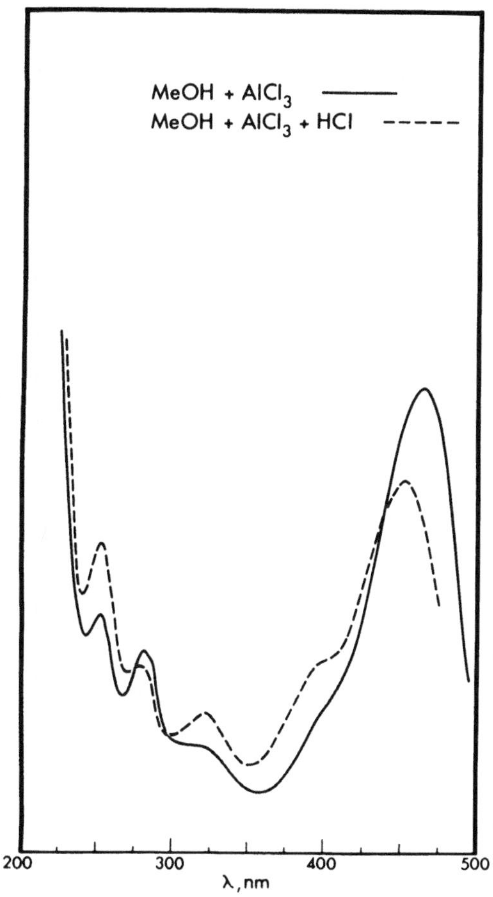

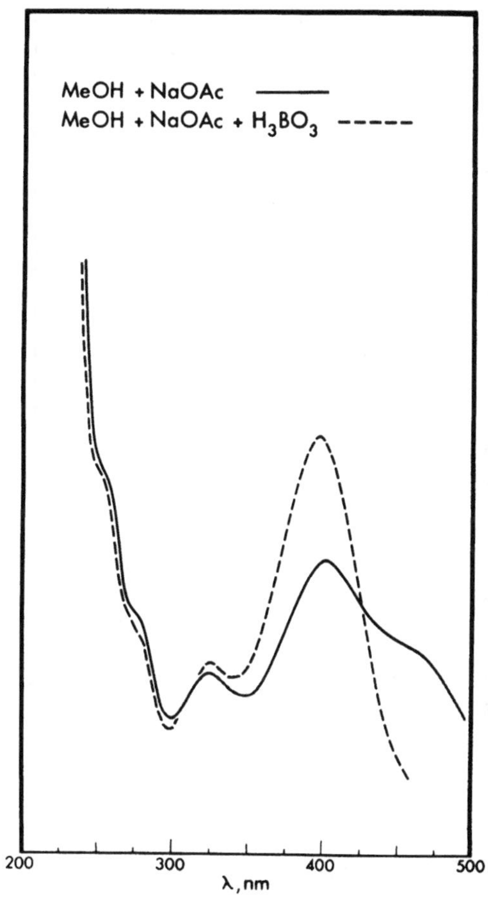

165

2',4,4'-TRIHYDROXYCHALCONE

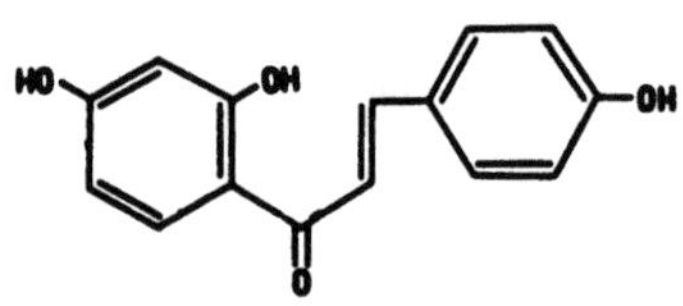

CHROMATOGRAPHIC DATA

Spot Appearance: UV) greenish purple
(UV/NH_3) orange

R_f Values: 0.87 (TBA), 0.07 (HOAc)

UV SPECTRAL DATA (λ_{max},*nm*)

MeOH	258sh, 298sh, 367
NaOMe	253sh, 280sh, 319sh, 349sh, 430
$AlCl_3$	258sh, 321, 382sh, 423
$AlCl_3$/HCl	319sh, 376sh, 421
NaOAc	281sh, 340, 350sh, 393
NaOAc/H_3BO_3	286, 353sh, 380, 443, 476sh
(Proc. I)	

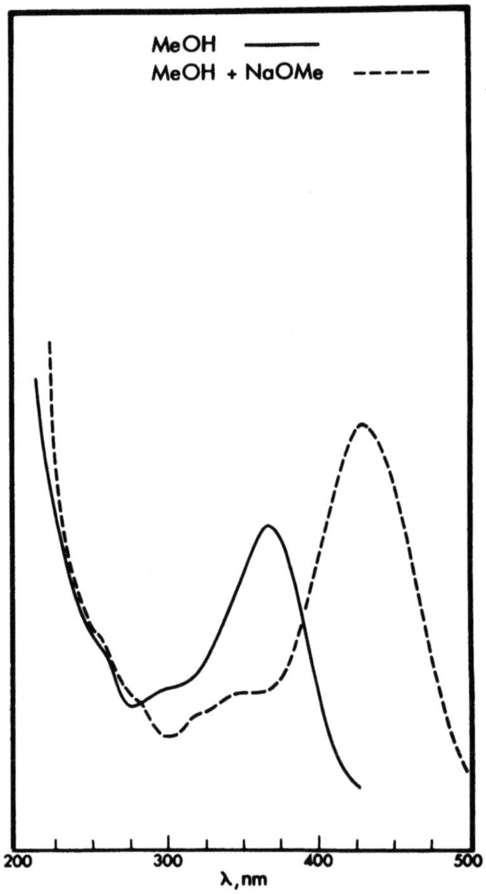

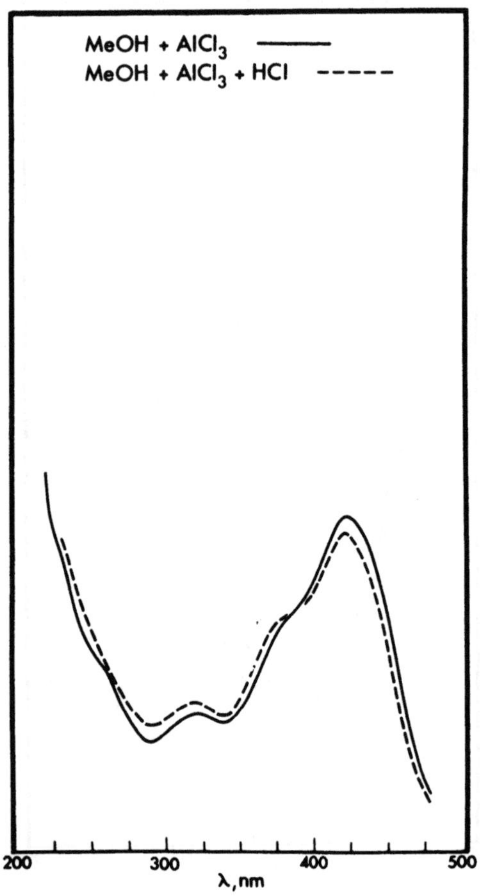

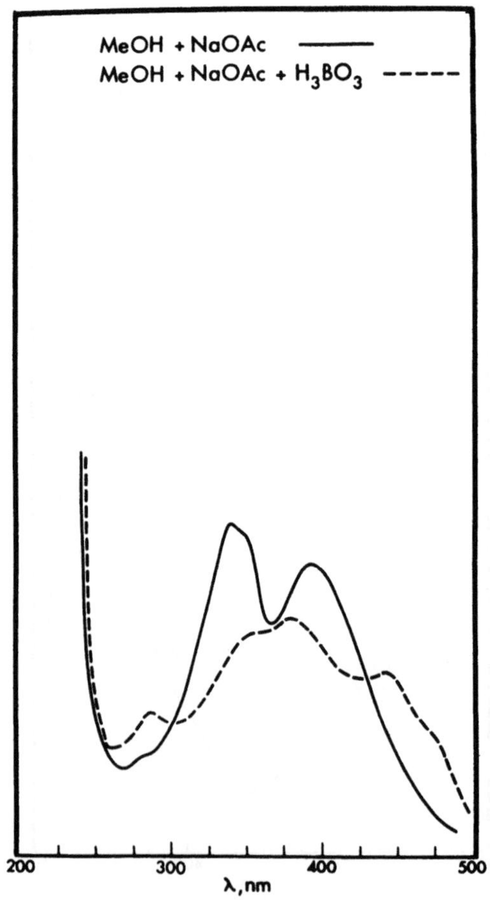

166

2′, 3,4,4′-TETRAHYDROXYCHALCONE

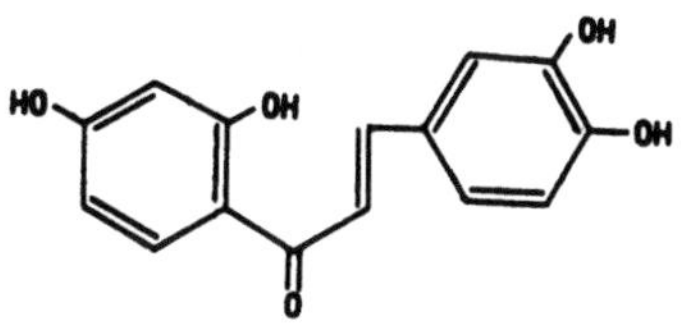

CHROMATOGRAPHIC DATA

Spot Appearance: (UV) green
(UV/NH_3) orange

R_f Values: 0.70 (TBA), 0.07 (HOAc)

UV SPECTRAL DATA (λ_{max}, *nm*)

MeOH	239sh, 266, 319sh, 379
NaOMe	251, 281, 344, 441
$AlCl_3$	254sh, 304sh, 318, 357sh, 490
$AlCl_3$/HCl	241sh, 275, 318, 384sh, 427
NaOAc	257sh, 279sh, 348, 397
NaOAc/H_3BO_3 (Proc. I)	282, 328, 415, 460sh, 489sh

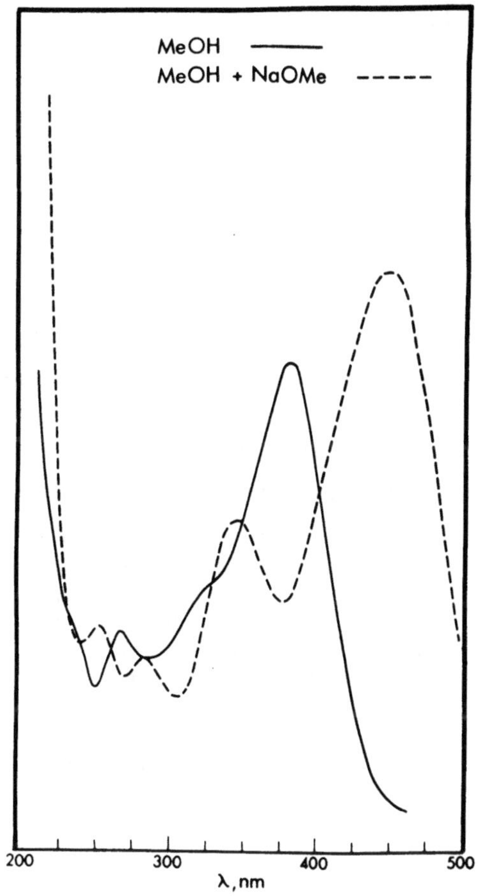

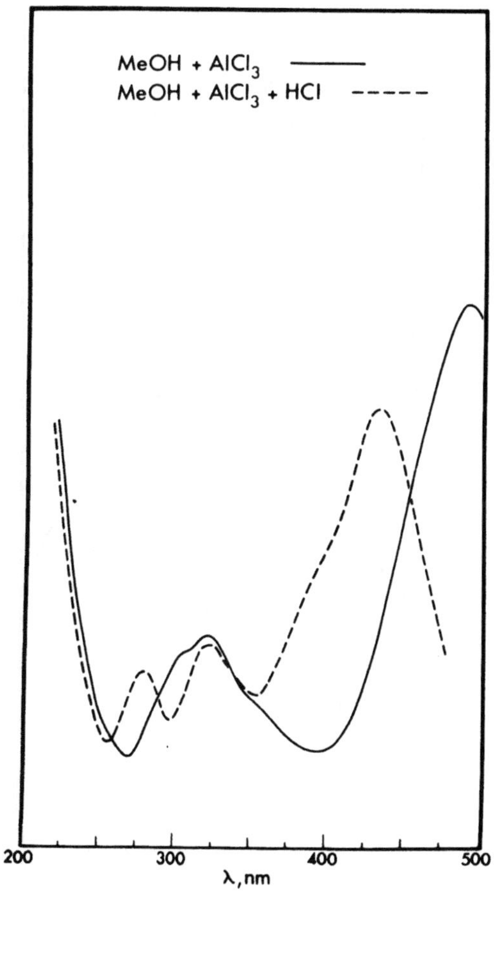

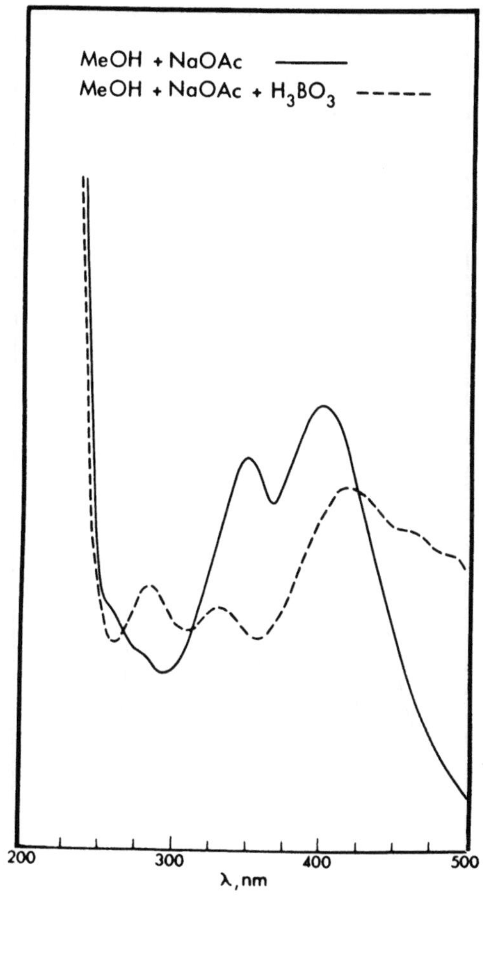

167

4'-HYDROXYAURONE

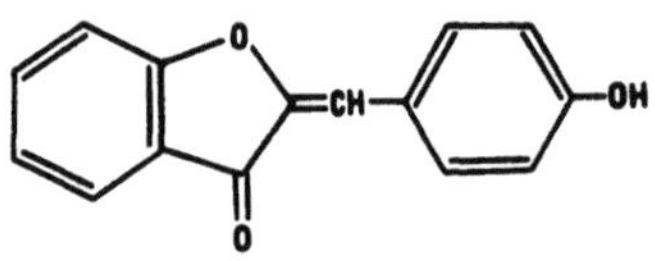

CHROMATOGRAPHIC DATA

Spot Appearance: (UV) fluorescent green
(UV/NH_3) orange

R_f Values: 0.89 (TBA), 0.10 (HOAc)

UV SPECTRAL DATA (λ_{max}, *nm*)

MeOH	255, 338sh, 397, 405sh
NaOMe	238sh, 277, 308sh, 350sh, 478
$AlCl_3$	255, 343sh, 396, 405sh
$AlCl_3$/HCl	255, 345sh, 396sh, 402
NaOAc	259, 277sh, 343sh, 410, 473
NaOAc/H_3BO_3 (Proc. I)	257sh, 344, 406

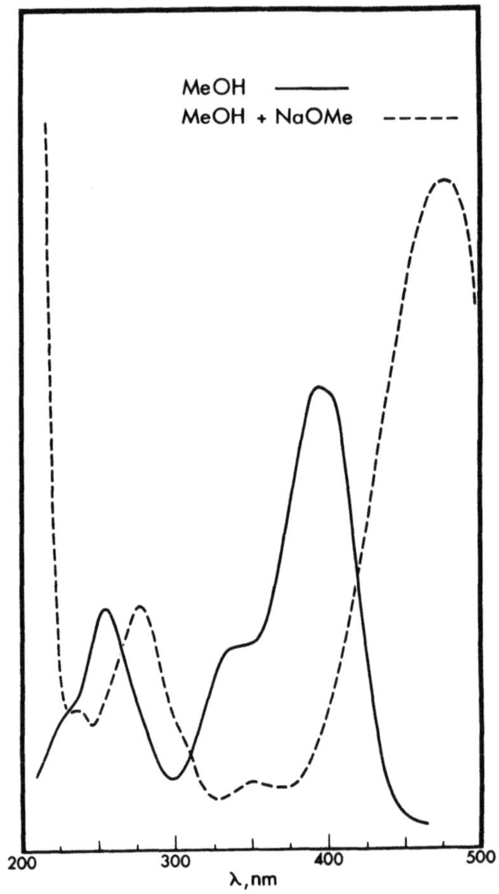

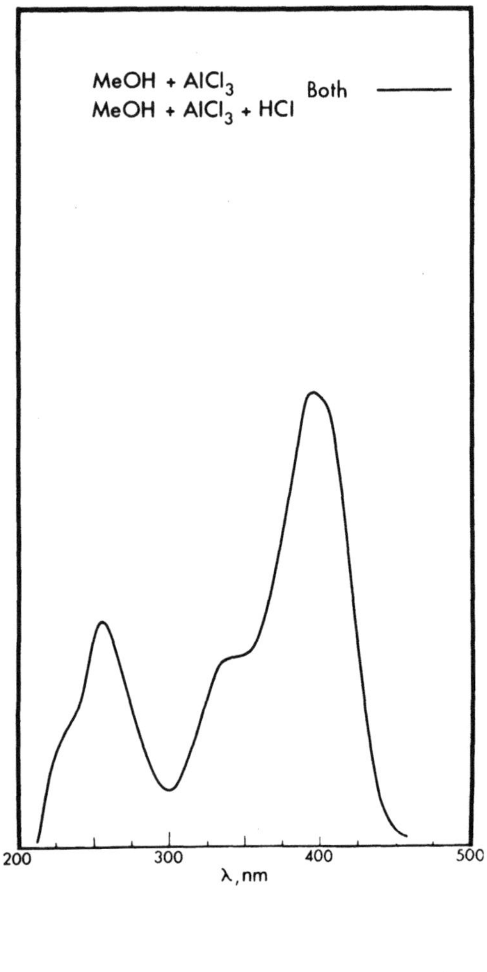

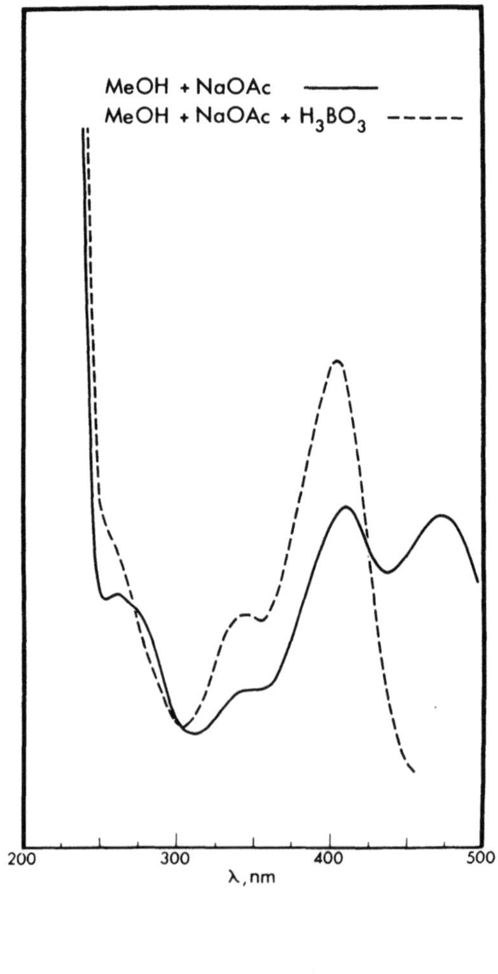

3',4'-DIHYDROXYAURONE

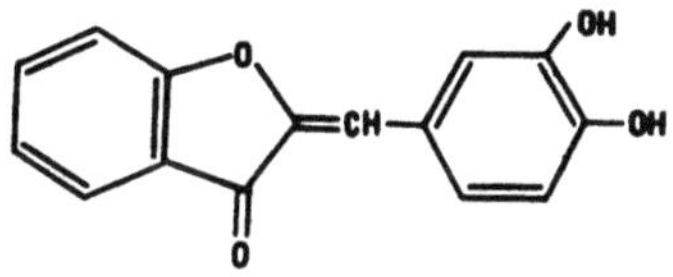

CHROMATOGRAPHIC DATA

Spot Appearance: (UV) fluorescent yellow-green

(UV/NH_3) red

R_f Values: 0.85 (TBA), 0.06 (HOAc)

UV SPECTRAL DATA (λ_{max}, *nm*)

MeOH	259, 277, 329sh, 413
NaOMe	279, 355sh, 502
$AlCl_3$	272sh, 287, 330sh, 463
$AlCl_3$/HCl	258, 277, 329sh, 413
NaOAc	260sh, 276, 313sh, 418, 502
NaOAc/H_3BO_3 (Proc. I)	265sh, 284, 332, 445

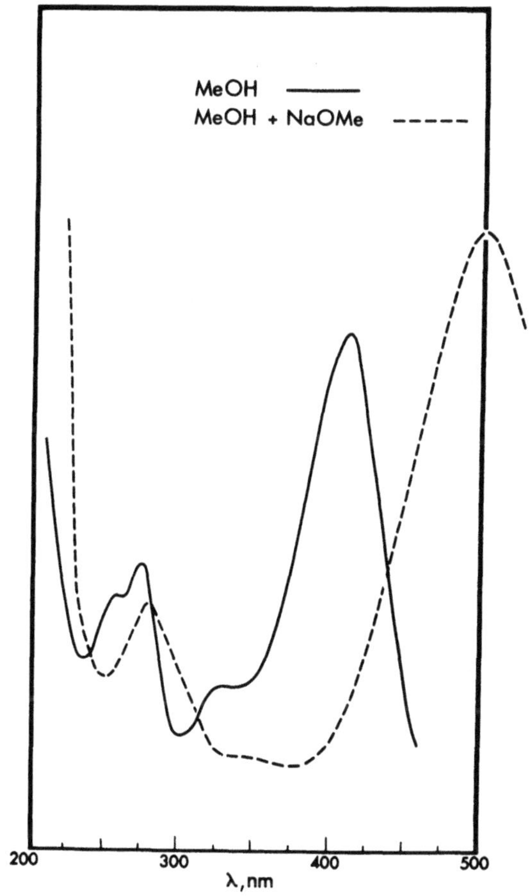

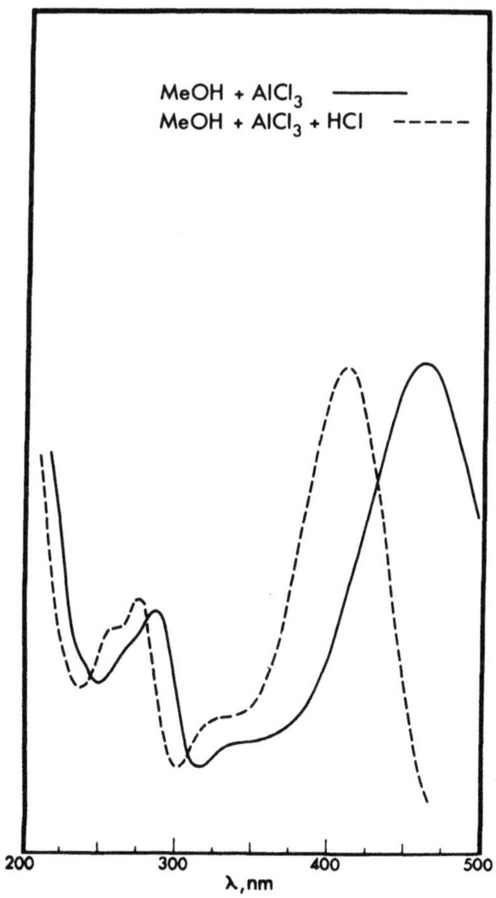

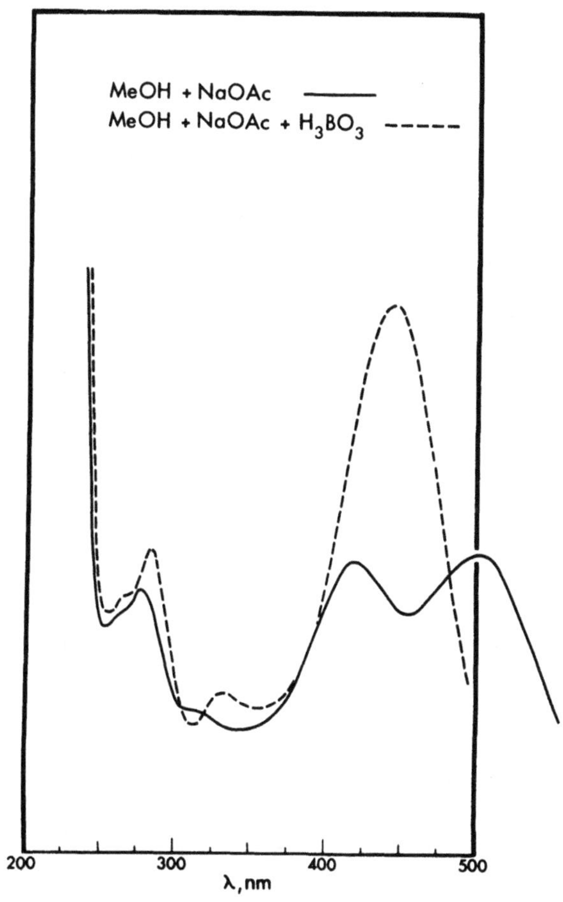

169

5,7-DIHYDROXYAURONE

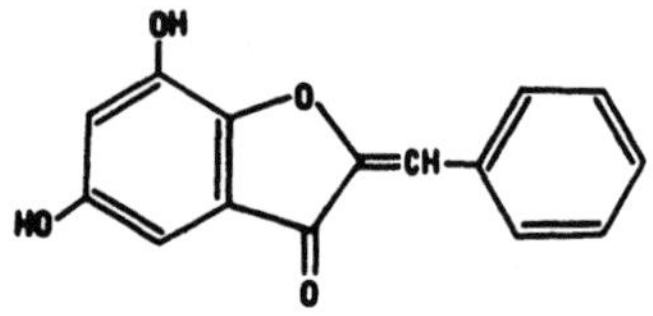

CHROMATOGRAPHIC DATA

Spot Appearance: (UV) blue
(UV/NH_3) fluorescent light blue

R_f Values: 0.76 (TBA), 0.63 (HOAc)

UV SPECTRAL DATA (λ_{max}, *nm*)

MeOH	283, 312sh
NaOMe	242sh, 308, 349sh
$AlCl_3$	301, 359
$AlCl_3$/HCl	282, 318sh
NaOAc	291, 313, 350sh
NaOAc/H_3BO_3 (Proc. I)	285, 314sh

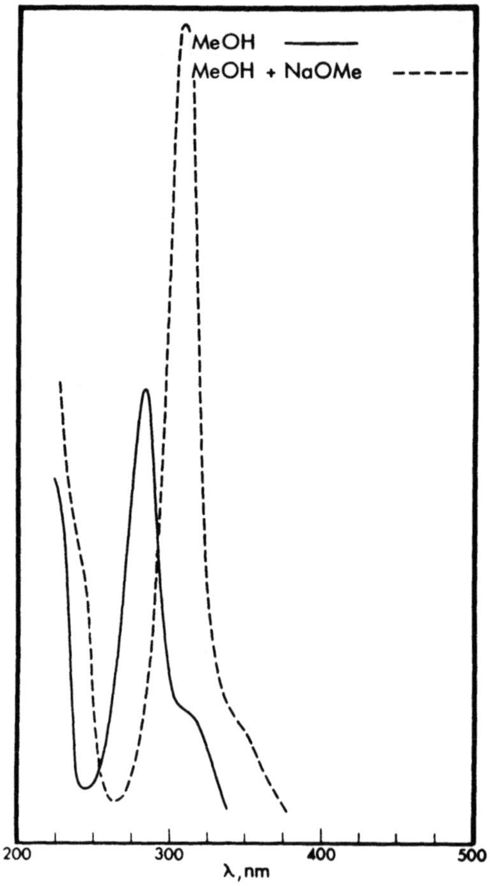

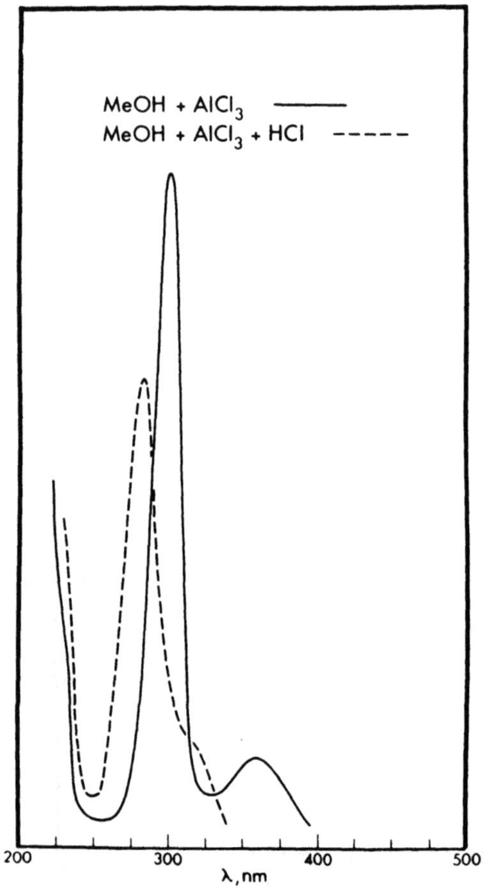

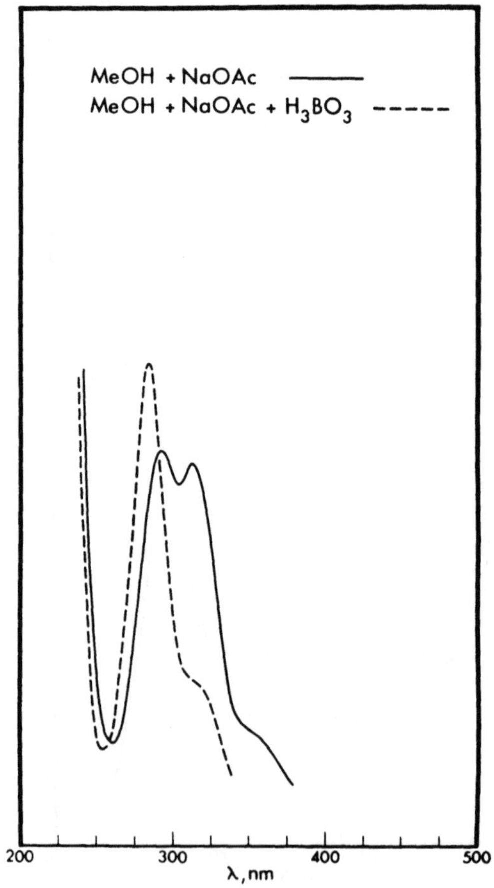

6,7-DIHYDROXYAURONE

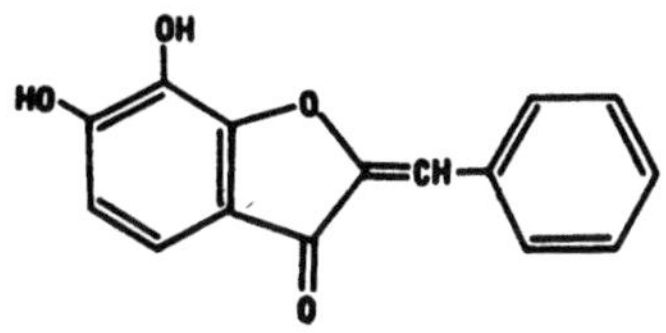

CHROMATOGRAPHIC DATA

Spot Appearance: (UV) greenish purple
(UV/NH_3) orange

R_f Values: 0.87 (TBA), 0.90 (HOAc)

UV SPECTRAL DATA (λ_{max},*nm*)

MeOH	242sh, 317, 379, 444sh
NaOMe	239sh, 279sh, 304, 320sh, 430
$AlCl_3$	261sh, 267, 318, 413
$AlCl_3$/HCl	244sh, 261sh, 320, 374
NaOAc	269, 311, 371sh, 431
NaOAc/H_3BO_3 (Proc. I)	264, 314, 401

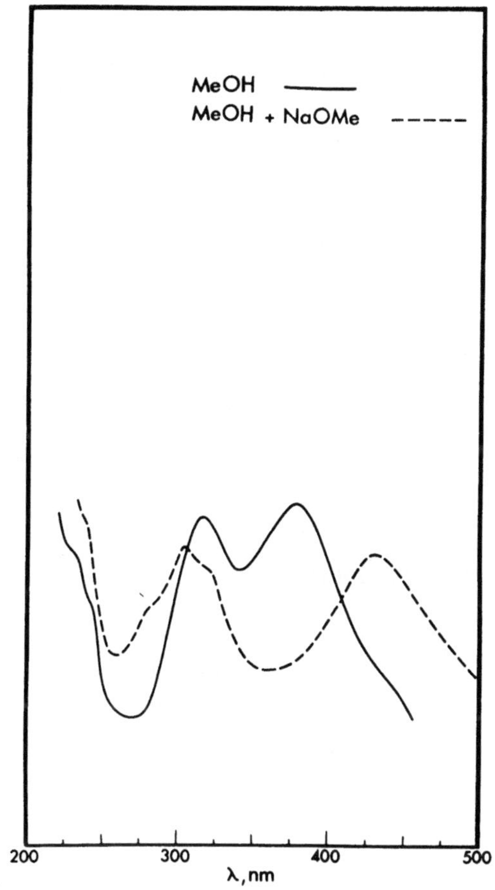

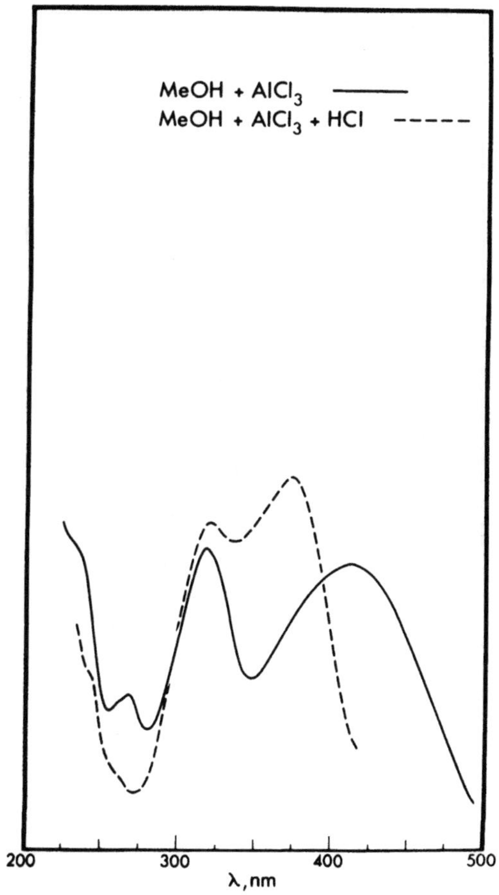

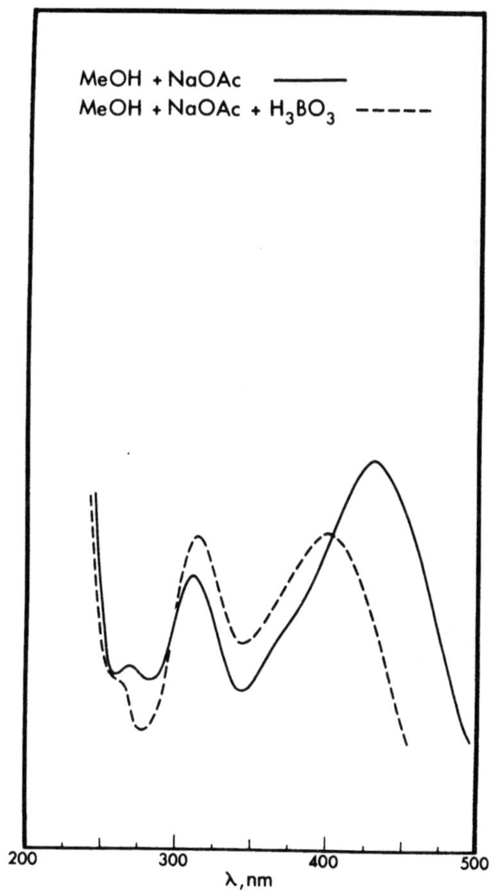

171

6-HYDROXY-4'-METHOXYAURONE

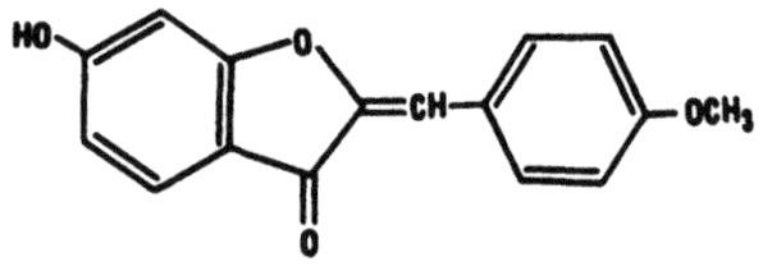

CHROMATOGRAPHIC DATA

Spot Appearance: (UV) fluorescent blue-green

(UV/NH_3) fluorescent yellow-green

R_f Values: 0.89 (TBA), 0.03 (HOAc)

UV SPECTRAL DATA (λ_{max},*nm*)

MeOH	252, 298sh, 373, 389sh
NaOMe	242, 303sh, 311, 379sh, 399
$AlCl_3$	254, 364, 389
$AlCl_3$/HCl	254, 301sh, 377
NaOAc	300sh, 311, 401
NaOAc/H_3BO_3 (Proc. I)	257sh, 301sh, 375

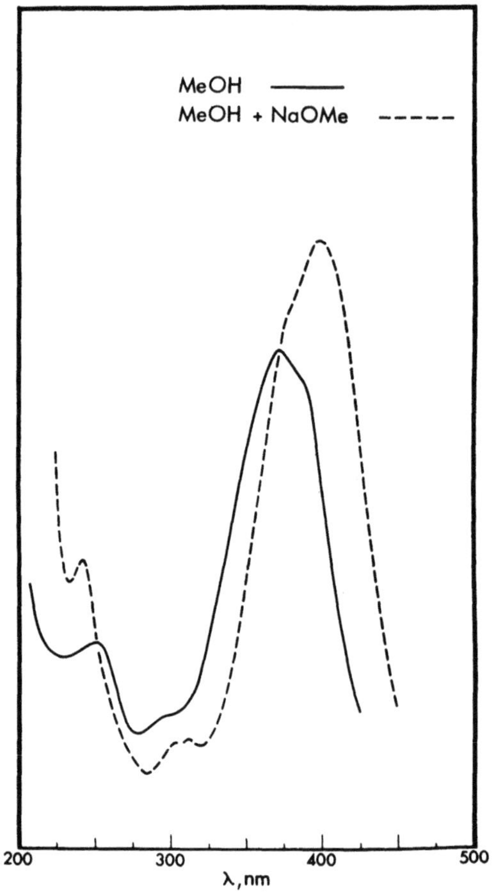

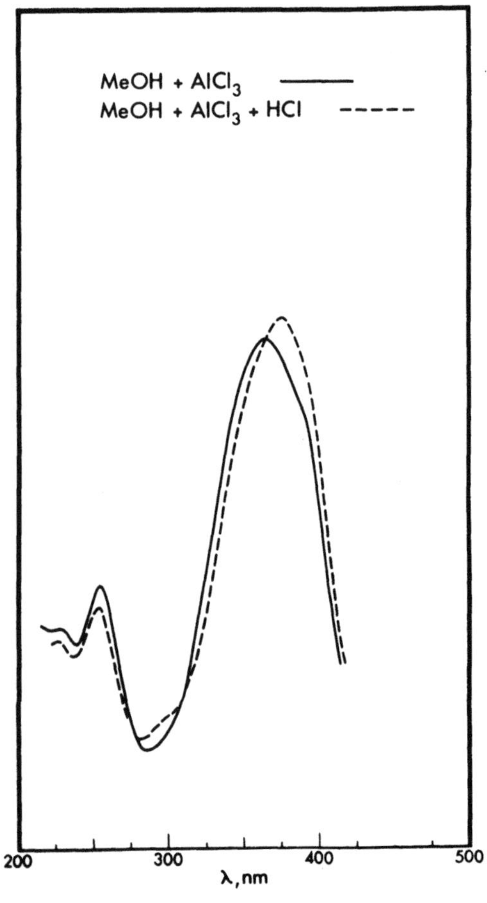

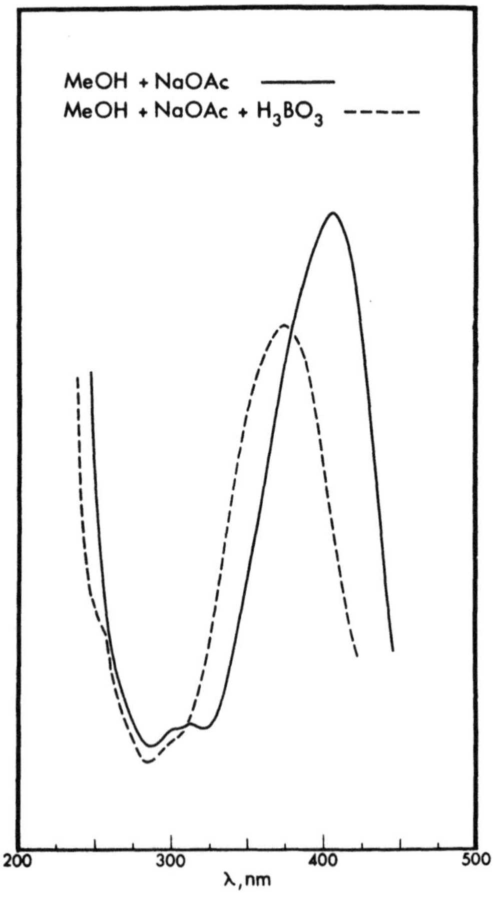

172

3′,4′,6,7-TETRAHYDROXYAURONE
(MARITIMETIN)

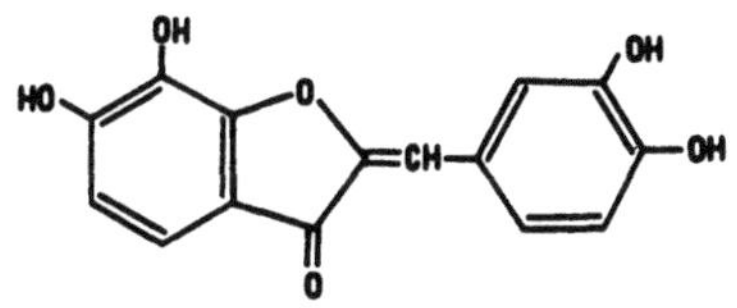

CHROMATOGRAPHIC DATA

Spot Appearance: (UV) fluorescent yellow-green

(UV/NH_3) pink

R_f Values: 0.39 (TBA), 0.02 (HOAc)

UV SPECTRAL DATA (λ_{max},*nm*)

MeOH	250sh, 271sh, 340sh, 412
NaOMe	247sh, 297sh, 409sh, 483
$AlCl_3$	267, 286sh, 383sh, 458, 603
$AlCl_3$/HCl	255sh, 272sh, 343sh, 410
NaOAc	266sh, 321sh, 385sh, 438
NaOAc/H_3BO_3 (Proc. I)	264sh, 280sh, 327sh, 369sh, 445

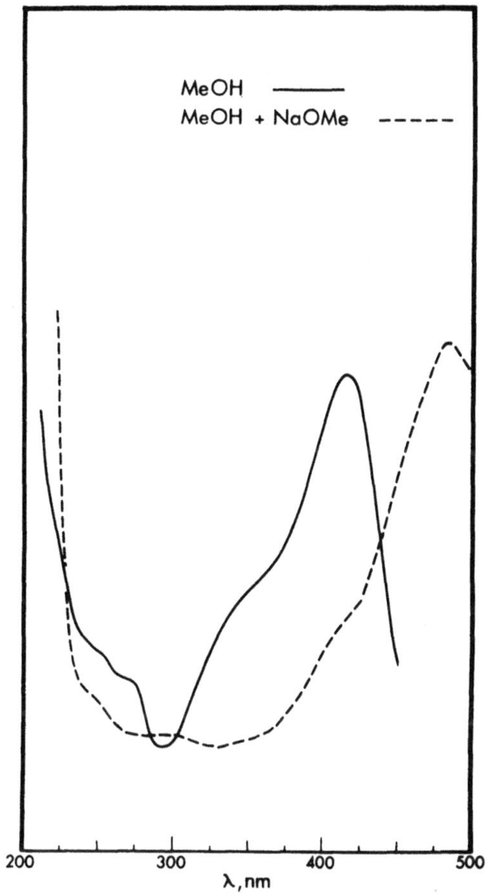

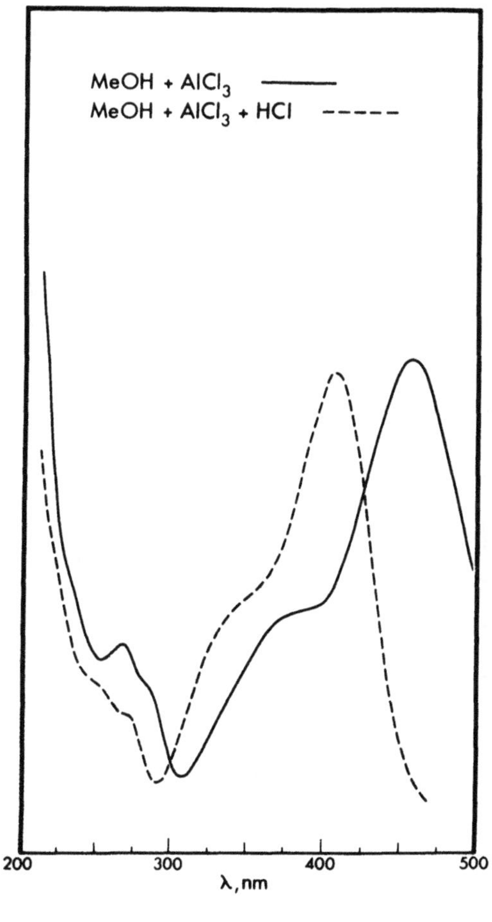

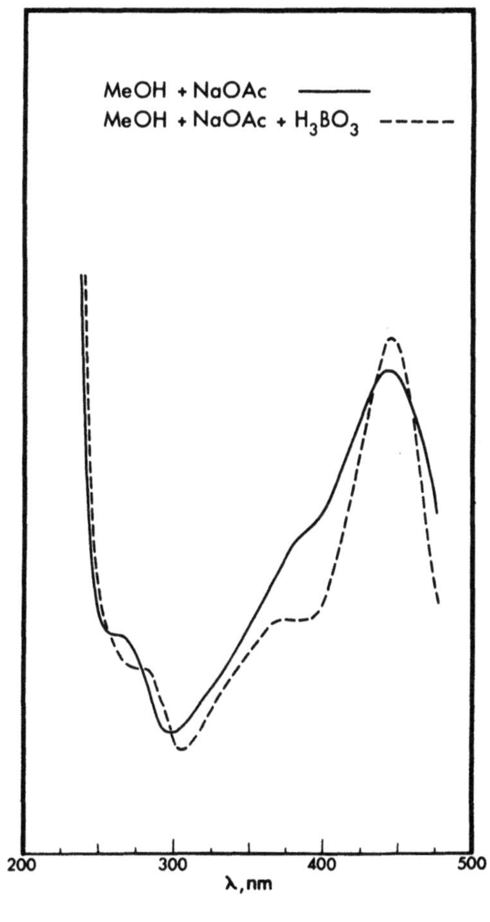

173

LEPTOSIDIN

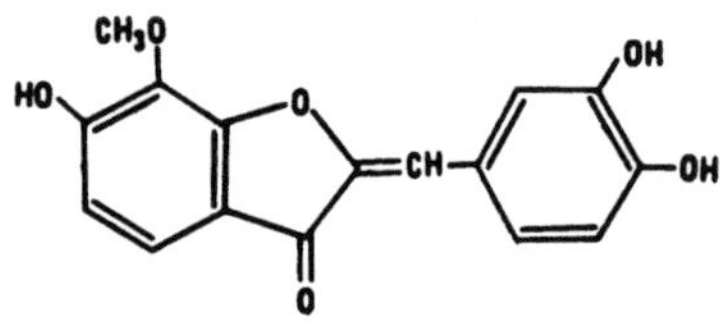

CHROMATOGRAPHIC DATA

Spot Appearance: (UV) fluorescent yellow-green

(UV/NH_3) orange

R_f Values: 0.64 (TBA), 0.02 (HOAc)

UV SPECTRAL DATA (λ_{max}, nm)

MeOH	244sh, 257sh, 269sh, 318sh, 392sh, 406
NaOMe	253, 273sh, 383sh, 402, 468
$AlCl_3$	259sh, 287, 342, 448
$AlCl_3$/HCl	255sh, 270sh, 325sh, 404
NaOAc	266, 318sh, 384sh, 426
NaOAc/H_3BO_3 (Proc. I)	262, 280, 346sh, 434

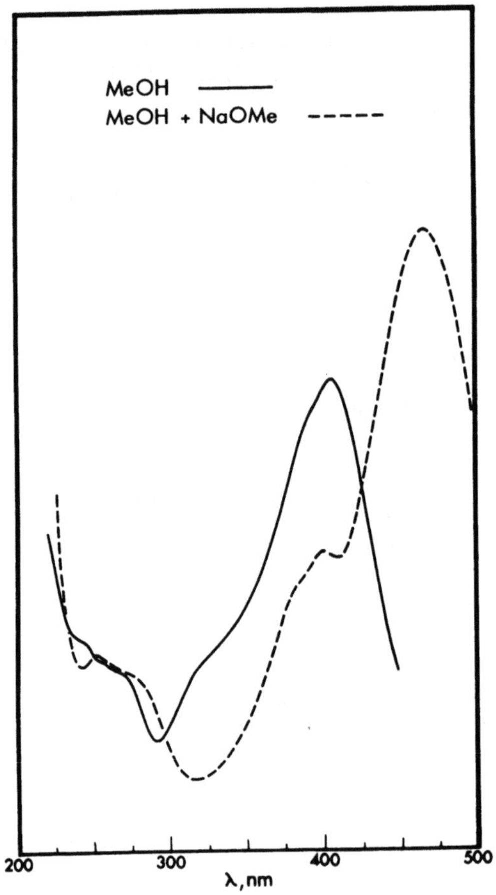

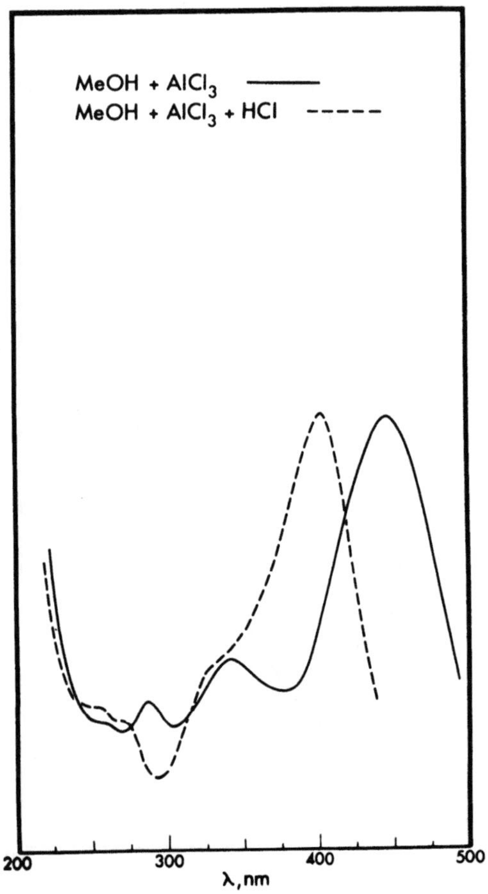

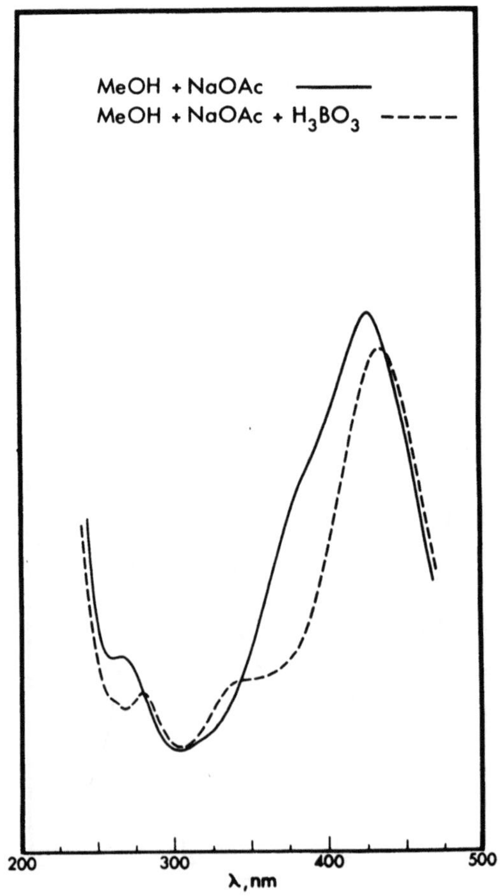

174

6-HYDROXY-3′,4,4′-TRIMETHOXYAURONE

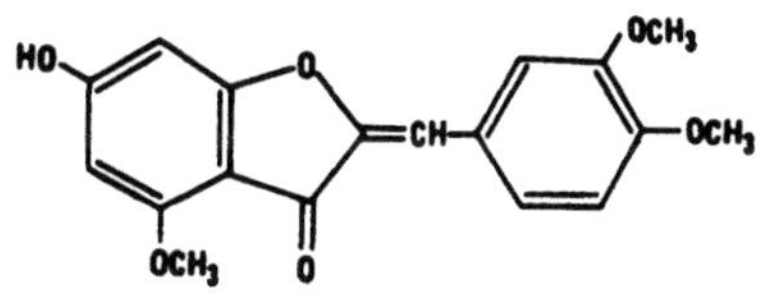

CHROMATOGRAPHIC DATA

Spot Appearance: (UV) fluorescent blue-green

(UV/NH_3) fluorescent blue-green

R_f Values: 0.61 (TBA), 0.02 (HOAc)

UV SPECTRAL DATA (λ_{max}, nm)

MeOH	251sh, 270sh, 320sh, 378sh, 395
NaOMe	242sh, 256sh, 311sh, 400
$AlCl_3$	253, 273sh, 336sh, 396
$AlCl_3$/HCl	251, 271sh, 327sh, 395
NaOAc	274sh, 311sh, 402
NaOAc/H_3BO_3 (Proc. I)	271sh, 313sh, 396

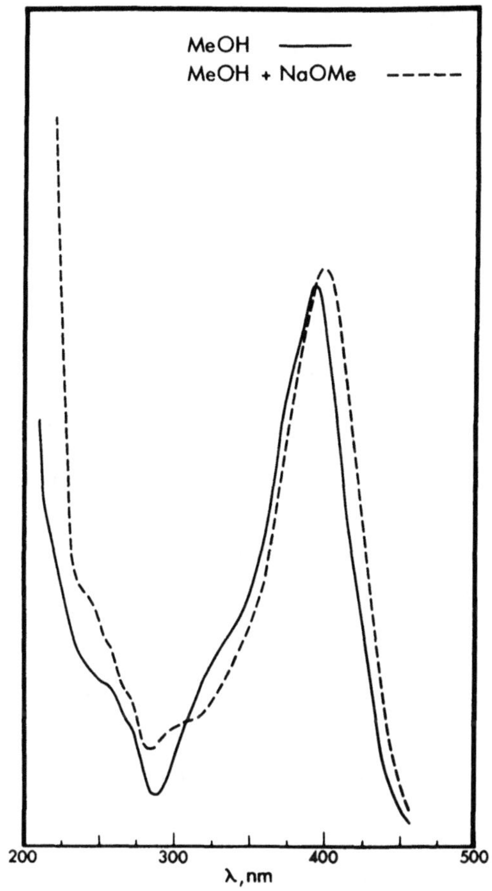

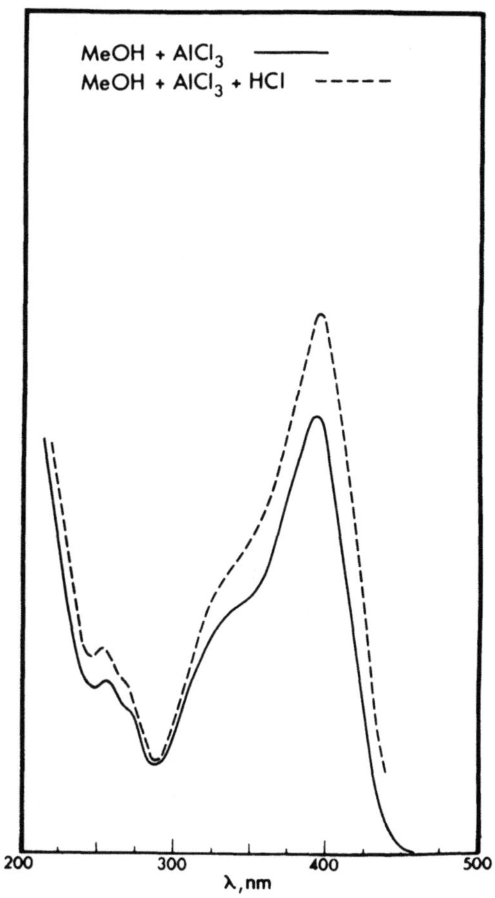

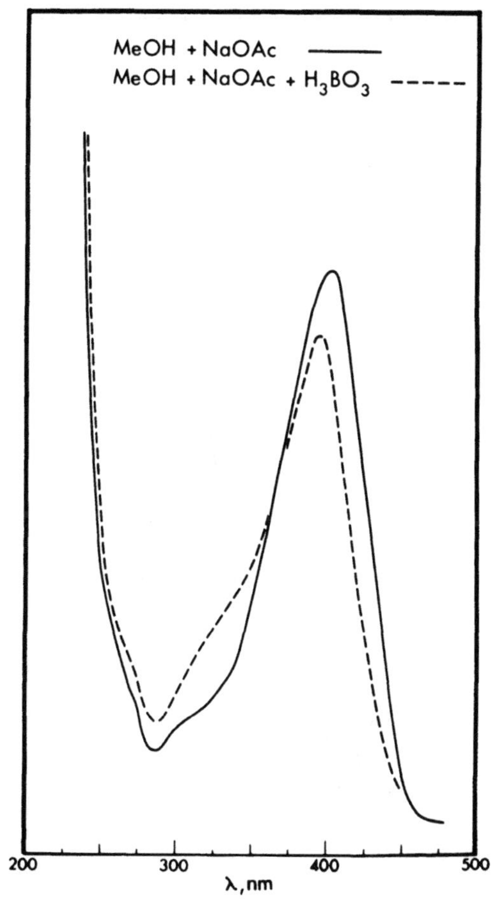

175

3'-HYDROXY-4,4',6-TRIMETHOXYAURONE

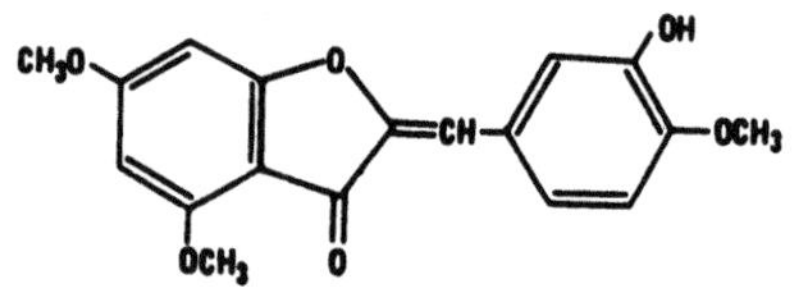

CHROMATOGRAPHIC DATA

Spot Appearance: (UV) fluorescent blue-green

(UV/NH_3) fluorescent blue-green

R_f Values: 0.74 (TBA), 0.05 (HOAc)

UV SPECTRAL DATA (λ_{max},*nm*)

MeOH	251sh, 268, 329sh, 396
NaOMe	243sh, 289, 337sh, 378, 435
$AlCl_3$	251sh, 268, 336sh, 397
$AlCl_3$/HCl	251sh, 268, 329sh, 397
NaOAc	269, 328, 396
$NaOAc/H_3BO_3$ (Proc. I)	269sh, 329sh, 398

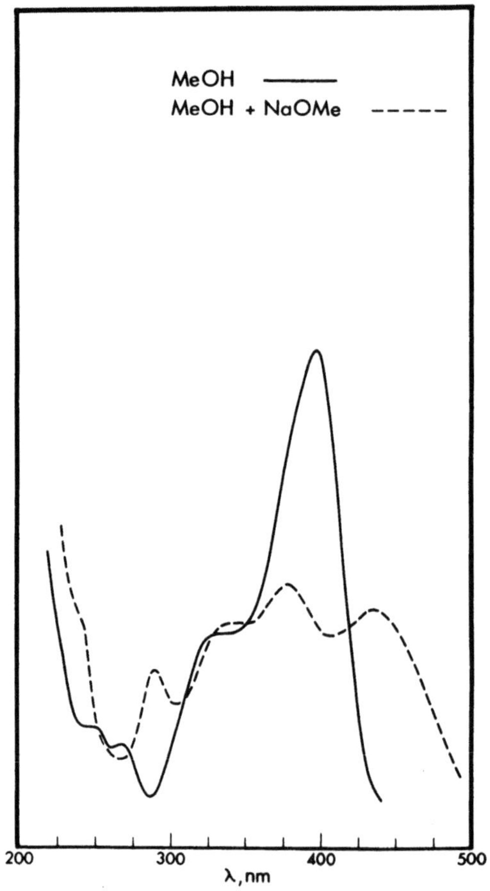

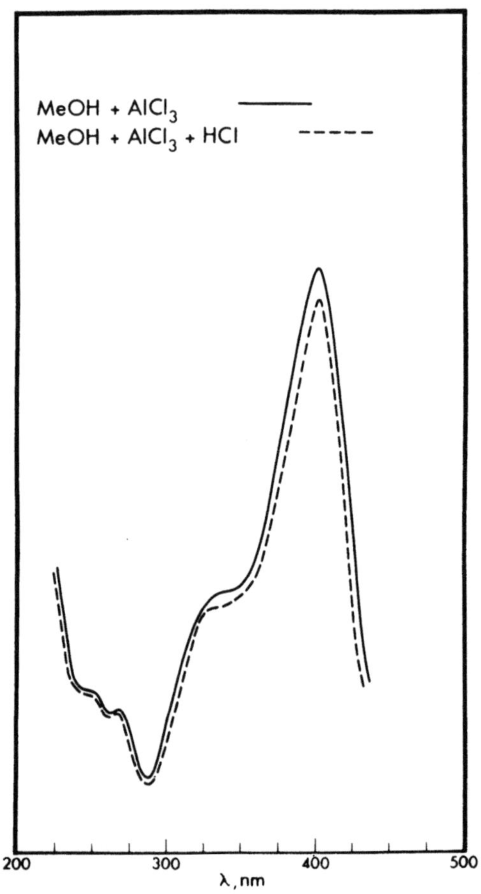

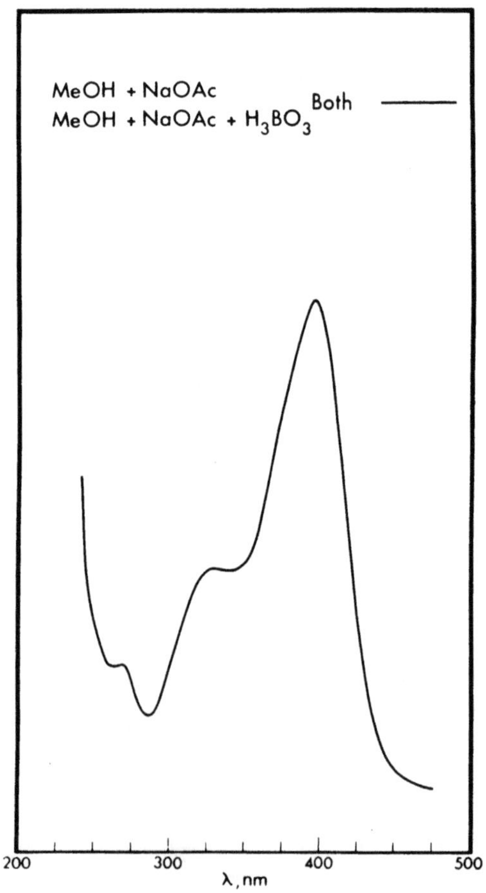

Part III

The Structure Analysis of Flavonoids by Proton Nuclear Magnetic Resonance Spectroscopy

Procedures for determining and interpreting the NMR spectra of flavonoids are discussed in Chapter VIII; Chapter IX presents 128 NMR spectra, most of which are for trimethylsilyl ether derivatives of flavonoids.

Chapter VIII

The Determination and Interpretation of NMR Spectra of Flavonoids

VIII-1. Introduction 254
VIII-2. The Use of DMSO-d_6 as Solvent for Flavonoid NMR Spectroscopy . . 254
VIII-3. Preparation of Trimethylsilyl Ether Derivatives of Flavonoids 255
3a. A Standard Procedure for the Preparation of TMS Ethers of Flavonoids 255
3b. Procedure for the Preparation of TMS Ethers of Flavonoids for Microcell Analysis 256
3c. Procedure for the Preparation of TMS Ethers of Flavonoids with Sterically Hindered C-5 Hydroxyl Groups 257
3d. Procedure for the Preparation of Partial TMS Ethers of Flavonoids having all Hydroxyl Groups Trimethylsilylated with the Exception of the C-5 Hydroxyl Group 258
3e. Procedure for the Complete De-Trimethylsilylation of TMS Ethers of Flavonoids 259
3f. Conversion of the TMS Ether Derivatives of Flavonoids to Methyl Ethers and Acetates 259
VIII-4. Interpretation of the NMR Spectra of Fully and Partially Trimethylsilylated Flavonoids 260
4a. A-Ring Protons 261
C-6 and C-8 Protons in Flavones, Flavonols and Isoflavones 261
C-6 and C-8 Protons in Flavanones and Dihydroflavonols 262
Distinguishing Between C-6, C-8 and C-3 Proton Signals 262
C-5, C-6 and C-8 Protons in 7-Hydroxyflavones, 7-Hydroxyisoflavones, 7-Hydroxyflavanones, 7-Hydroxyflavonols and 7-Hydroxydihydroflavonols 264
4b. B-Ring Protons 265
C-2′, C-3′, C-5′ and C-6′ Protons in 4′-Oxygenated Flavonoids . . . 265
C-2′, C-5′ and C-6′ Protons in 3′,4′-Oxygenated Flavonoids 265
C-2′ and C-6′ Protons in 3′,4′,5′-Oxygenated Flavonoids 266
4c. C-Ring Protons 267
C-3 Proton in Flavones; C-2 Proton in Isoflavones 267
α and β Protons in Chalcones; Benzylic Proton in Aurones 267
C-2 and C-3 Protons in Flavanones and Dihydroflavonols 267
4d. Sugar Protons 268
The C-1″ Proton in Flavonoid Monosaccharides 268
C-1″, C-1‴, and Rhamnose Methyl Protons (C-6‴) in Flavonoid Rhamnoglucosides 269
The Determination of the Sugar-Sugar Linkage in Xylosylvitexin . . 270
4e. Methoxyl and Acetoxyl Protons 271
4f. TMS Ether Protons 272
4g. 6- and 8-C-Methyl Protons 272
References 273

VIII-1. Introduction

The application of nuclear magnetic resonance (NMR) spectroscopy to the structure analysis of flavonoids is now well established. Many flavonoid aglycones, in particular isoflavones and highly methylated flavones and flavonols, are sufficiently soluble in the commonly used solvent, deuteriochloroform ($CDCl_3$) for direct NMR analysis. However, most naturally occurring flavonoids, including all of the flavonoid glycosides, have low solubility in $CDCl_3$; therefore, prior to 1964 most workers were limited to the NMR analysis of the more soluble methyl, ethyl and acetyl derivatives (see for example reference [*1*]). However, the signals observed for the substituent groups in these derivatives often obscure signals of other protons in the flavonoid.

In 1964/65, two groups of workers [*2, 3*] independently investigated the usefulness of trimethylsilyl ether derivatives for obtaining NMR spectra of flavonoids which were otherwise insoluble in $CDCl_3$. At about the same time [*4*] hexadeuteriodimethylsulfoxide (DMSO-d_6) was introduced as a solvent for the direct NMR analysis of flavonoids. We comment here on the relative merits of these two now widely used methods in addition to discussing in detail the flavonoid NMR spectra (mostly of trimethylsilyl ether derivatives) which are presented in Chapter IX.

VIII-2. The Use of DMSO-d_6 as Solvent for Flavonoid NMR Spectroscopy

DMSO-d_6 has been used as a solvent for a number of extensive investigations of flavonoid structures by NMR spectroscopy [*4, 5*]. Some of the advantages of this method relative to other available procedures include those listed below.

a) Most flavonoid aglycones and glycosides are sufficiently soluble in DMSO-d_6 for direct NMR analysis, thereby eliminating the necessity of preparing derivatives.

b) The proton signals resulting from the small amount of partially deuterated DMSO always present in DMSO-d_6 occur in a narrow band between 2.4 and 2.6 ppm, outside the region where most flavonoid protons absorb.

c) DMSO-d_6 (if it is anhydrous) can be used for observing protons on phenolic hydroxyl groups. For example, in the flavonoid aglycone galangin (3,5,7-trihydroxyflavone), the hydroxyl proton signals are readily distinguishable (Fig.VIII-1). Traces of

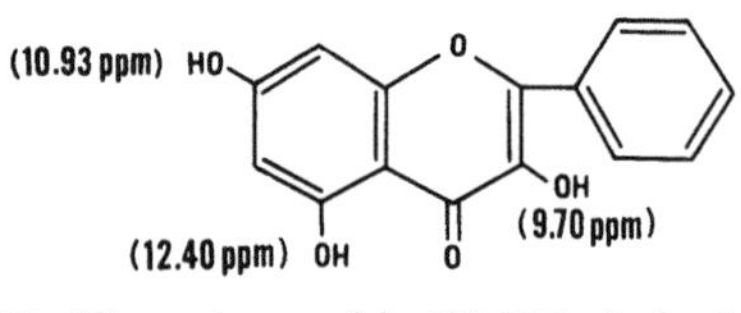

Fig. VIII-1. Chemical shifts (δ) as observed in DMSO-d_6 for hydroxyl protons in galangin

water in the DMSO-d_6 however, cause the flavonoid hydroxyl proton signals to broaden (as a result of rapid proton exchange), thus making their detection difficult. We have found it convenient to purchase anhydrous DMSO-d_6 in 1 g ampoules, which are only opened just before the solvent is needed for the NMR analyses. Once an ampoule is opened, any DMSO-d_6 not used immediately is kept dry by storage over molecular sieve[1].

There are however also a number of disadvantages associated with the use of DMSO-d_6 as solvent for flavonoid NMR spectroscopy.

[1] The molecular sieve (type 4a) was purchased from Fisher Scientific Co.

a) DMSO-d_6 has a boiling point of 189°, which makes the recovery of the flavonoid inconvenient.

b) DMSO-d_6 rapidly absorbs atmospheric moisture, and the signal obtained from the absorbed H_2O [variable around 3.5 ppm (δ)] often obscures NMR signals resulting from some of the flavonoid protons.

c) DMSO-d_6 must be handled carefully since it rapidly penetrates skin tissue carrying with it any dissolved substances.

d) Some flavonoids have been reported [*5 b*] to undergo decomposition in DMSO-d_6.

VIII-3. Preparation of Trimethylsilyl Ether Derivatives of Flavonoids

Waiss, Lundin, and Stern [*2*] and Mabry, Kagan, and Rösler [*3*] demonstrated the usefulness of trimethylsilyl ether derivatives for the NMR analysis of flavonoids. The method offers several important advantages.

a) Most flavonoid compounds can be converted to trimethylsilyl ether derivatives and prepared for NMR analysis in about 20 min, which is a considerably shorter preparation time than is required for most other derivatives.

b) Trimethylsilyl ether derivatives of flavonoids are miscible with carbon tetrachloride (CCl_4) in all proportions, and thus the use of more expensive deuterated solvents is avoided.

c) The signals of the trimethylsilyl protons occur at fields higher than 0.5 ppm (δ), and therefore are well out of the absorption region of flavonoid protons.

d) Both the preparation and the hydrolysis of trimethylsilyl ether derivatives are quantitative and may be carried out under mild conditions (see Fig. VIII-2).

e) By comparing the NMR spectra for fully and partially trimethylsilylated flavonoids, information can often be obtained about the substitution at C-3, C-6 and C-8.

In addition to the advantages noted above for using the trimethylsilyl ethers of flavonoids for NMR studies, it should be mentioned that both methylation (with dimethylsulfate [*3 c*] or diazomethane) and acetylation of flavonoids can conveniently be carried out with the trimethylsilyl ethers (see Section VIII-3 f).

3a. A Standard Procedure for the Preparation of TMS Ethers of Flavonoids

The trimethylsilylation of most flavonoids can be achieved using the procedure described below for the preparation of the TMS ether of hesperidin.

Preparation of Hesperidin Octa-Trimethylsilyl Ether (Fig. VIII-2). Hesperidin (50 mg) was dissolved in 2 ml of anhydrous pyridine (which had been stored over KOH) in a 50 ml round-bottom flask; 0.5 ml each of hexamethyldisilazane and trimethylchlorosilane (both reagents were obtained from Applied Science Laboratories, Inc., State College, Pa.) were subsequently added. The flask was stoppered and allowed to stand for 5 min at room temperature. The solvent and excess reagents were removed under high vacuum (oil pump) and the dry residue was extracted by the addition of 1 ml of CCl_4 (spectroscopic grade). The clear CCl_4 solution, which was obtained by filtering off the salts, was taken to dryness. (In some instances, the clear solution is concentrated to a volume suitable for NMR analysis.) The residue was dissolved in about 0.5 ml of CCl_4 to which was added a drop each of tetramethylsilane (used as an internal standard), hexamethyldisilazane and trimethylchlorosilane (the latter two reagents insure anhydrous conditions). The solution was then ready for NMR analysis.

TMS ether derivatives of flavonoids are hydrolyzed by water and for this reason their preparation must be carried out under anhydrous conditions. The addition of a small amount of each of the trimethylsilylating reagents to the CCl_4 solution just prior

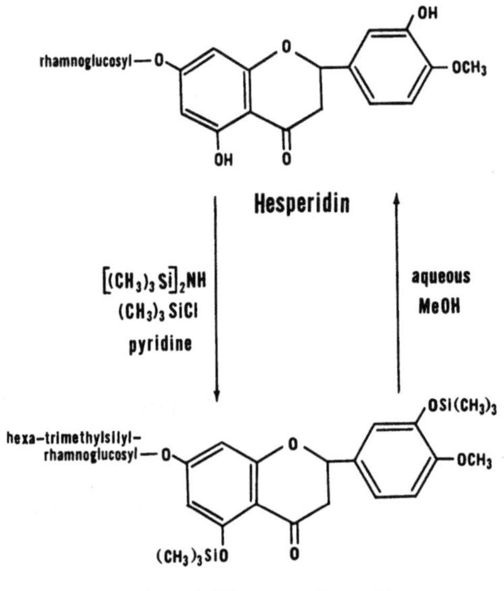

Fig. VIII-2. The reaction scheme for trimethylsilylation of hesperidin and the recovery of the flavonoid by hydrolysis of the derivative

to the NMR analysis was found to prevent hydrolysis during the spectral measurement. However, if an integration of the trimethylsilyl ether signals is required then these reagents should not be added. (An integration of the TMS ether signals can often be useful in the determination of the number of hydroxyl groups present in the original flavonoid [*3*]).

3b. Procedure for the Preparation of TMS Ethers of Flavonoids for Microcell Analysis

If a microcell is used for the NMR analysis, the amount of flavonoid and reagents used in procedure VIII-3a (above) may be reduced. We have found the 25 µl cell (obtained from NMR Specialities, Inc., New Kensington, Pa.) to be satisfactory for the NMR analysis of a 3 – 5 mg sample of flavonoid.

Preparation of Rutin Deca-Trimethylsilyl Ether for Microcell NMR Analysis. Rutin (I, 5 mg) was dissolved in pyridine (5 drops) in a 5 ml round-bottom flask, and then hexamethyldisilazane (2 drops) and trimethylchlorosilane (2 drops) were added. The reaction mixture, which was allowed to stand for 5 min, was worked up as in procedure VIII-3a (only 25 µl of CCl_4, which already contained TMS and the two trimethylsilylating reagents, were required for the NMR analysis).

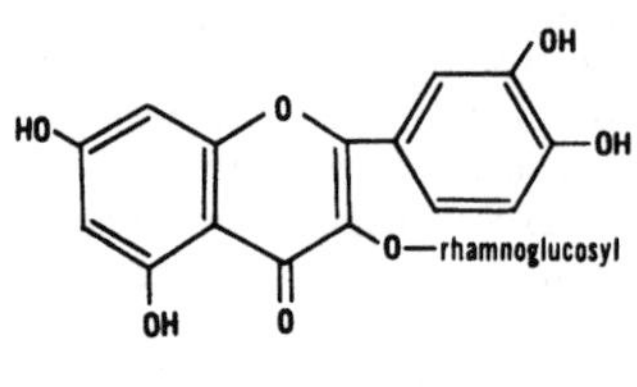

I. Rutin

Flavonoid samples of less than 1 mg may be analyzed by NMR spectroscopy if the spectrometer is equipped with a time averaging computer. For example, Fig. VIII-3 shows a time averaged spectrum (66 scans, about 5 hrs operating time) which was obtained from 0.5 mg of a flavonoid.

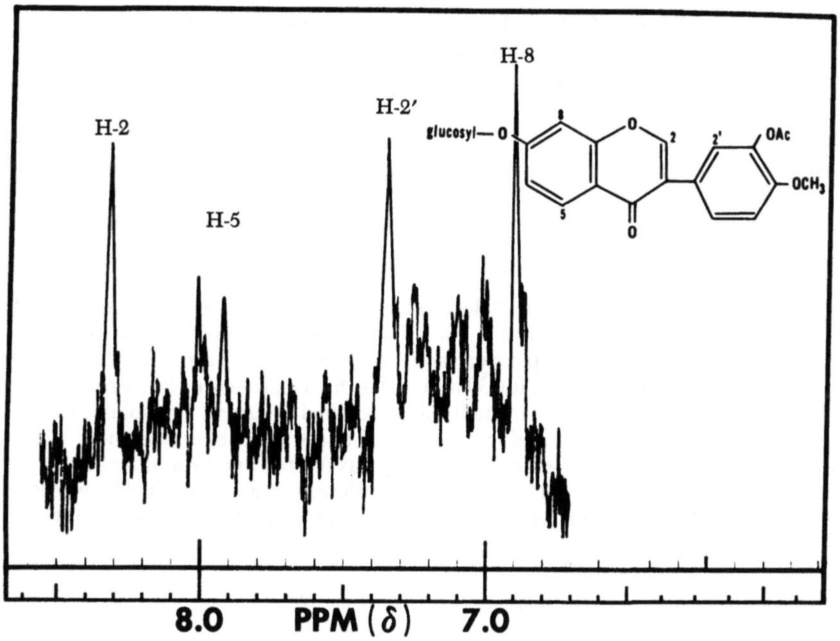

Fig. VIII-3. NMR spectrum (aromatic region after 66 scans) of the acetate of calycosin 7-O-β-D-glucoside (0.5 mg) in 25 μl of DMSO-d_6 using a 100 Mc spectrometer equipped with a time averaging computer

3c. Procedure for the Preparation of TMS Ethers of Flavonoids with Sterically Hindered C-5 Hydroxyl Groups

The strongly hydrogen-bonded hydroxyl group at C-5 in flavonoids reacts with the trimethylsilylating reagents less rapidly than other hydroxyl groups, and it has been observed that for the total trimethylsilylation of flavonoids with substituents *ortho* to the C-5 hydroxyl group, a longer reaction time is required.

Preparation of Isovitexin Hepta-Trimethylsilyl Ether. Isovitexin (II, 40 mg) was dissolved in pyridine (3 ml) and then hexamethyldisilazane (0.5 ml) and trimethylchlorosilane (0.5 ml) were added. The mixture was allowed to stand overnight in a stoppered flask and then worked up as described in procedure VIII-3a.

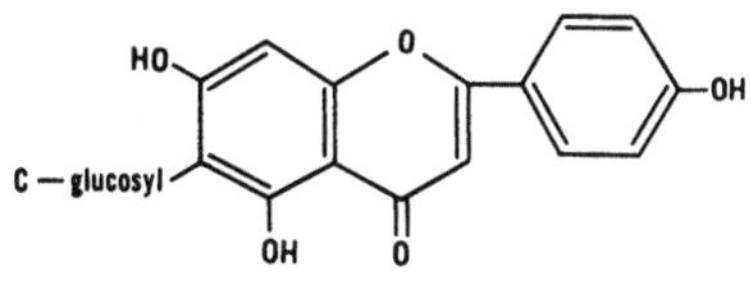

II. Isovitexin

The complete trimethylsilylation of another 6-C-glycoside, saponarin (spectrum 8) also required the longer reaction time described in procedure VIII-3c. A C-5 hydroxyl group *ortho* to a methoxyl group is less sterically hindered than one which is next to a C-glycosyl substituent; therefore, the complete trimethylsilylation of 5-hydroxy-6-methoxyflavonoids is usually achieved using a reaction time of about 30 min (spectra 21, 22, 23, 49, 50, 51, 53, 54, 55, 56 and 62).

3d. Procedure for the Preparation of Partial TMS Ethers of Flavonoids having all Hydroxyl Groups Trimethylsilylated with the Exception of the C-5 Hydroxyl Group

Useful information can often be gained by comparing the NMR spectrum of the fully trimethylsilylated compound with that of the same compound containing a free C-5 hydroxyl group (see Section VIII-4a, A-ring protons). Two methods for the selective de-trimethylsilylation of the C-5 TMS ether and one method for the direct preparation of a trimethylsilylated flavonoid with a free hydroxyl group are described below.

Complete Trimethylsilylation Followed by Selective De-Trimethylsilylation of Hymenoxin (III). Hymenoxin (6 mg) was dissolved in pyridine (5 drops) in a 5 ml round-bottomed flask and then hexamethyldisilazane (2 drops) and trimethylchlorosilane (2 drops) were added. The flask was stoppered and the reaction mixture was allowed to stand at room temperature for 30 min. After the usual work up (see procedure VIII-3a), NMR analysis (microcell) indicated that complete trimethylsilylation had occurred. The solution from the NMR tube was then evaporated to dryness, and the residue was exposed to the atmosphere for about 10 min. The solution which was obtained by dissolving the product in dry CCl_4 was filtered and evaporated to dryness. The residue was dissolved in CCl_4 (25 μl) containing 1% TMS and transferred to a microcell. The NMR spectrum of the product showed a signal near 12.5 ppm, typical for a C-5 hydrogen-bonded hydroxyl group.

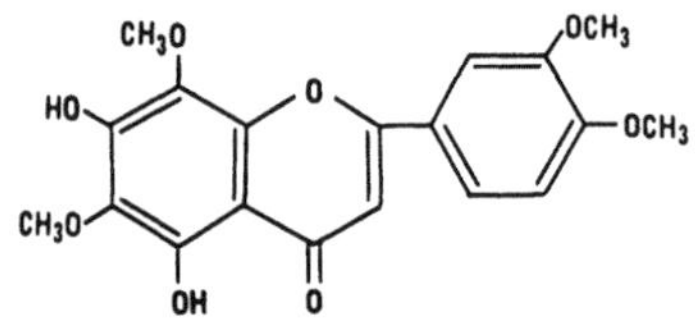

III. Hymenoxin

The conditions for the selective removal of the trimethylsilyl group from the oxygen at the 5-position vary from one compound to another, and a number of attempts may be required to achieve the desired result. For example, if the flavonoid is not substituted at C-6, the selective hydrolysis of the trimethylsilyl ether at C-5 may require a longer exposure to the atmosphere than indicated in procedure VIII-3d. Alternatively methanol may be used for the rapid hydrolysis of a C-5 TMS ether, which hydrolyzes in the presence of methanol somewhat faster than TMS ethers at other positions.

Selective De-Trimethylsilylation of 3,4′,7,8-Tetramethoxy-3′,5-Di-Trimethylsilylflavone. 3′,5-Dihydroxy-3,4′,7,8-tetramethoxyflavone (IV, 50 mg) was fully trimethylsilylated using procedure VIII-3a. The solution obtained after NMR analysis (0.5 ml) was mixed with methanol (0.5 ml), and the resulting solution was immediately evaporated under high vacuum. The residue was redissolved in CCl_4 (0.5 ml) which contained as internal reference 1% TMS. NMR analysis of the product showed the presence of one trimethylsilyl ether and one hydrogen-bonded hydroxyl group (12.3 ppm).

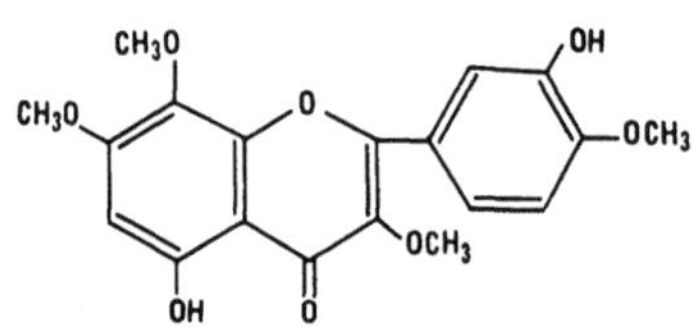

IV. 3′,5-Dihydroxy-3,4′,7,8-tetramethoxyflavone

Another method for preparing flavonoid derivatives with all the hydroxyl groups trimethylsilylated with the exception of the one at C-5 is to modify procedure VIII-3a by using a shorter reaction time. This procedure is particularly useful for compounds substituted at C-6.

Partial Trimethylsilylation of Patuletin 3-O-Rhamnoglucoside (V). Patuletin 3-O-rhamnoglucoside (7 mg) was dissolved in pyridine (6 drops) in a 5 ml round-bottom flask and then hexamethyldisilazane (2 drops) and trimethylchlorosilane (2 drops) were added. The mixture was immediately evaporated under high vacuum. The residue was worked up as described in procedure VIII-3b. NMR analysis of the material thus obtained showed that the main product was a TMS ether containing a free C-5 hydroxyl group. A small amount of the fully trimethylsilylated flavonoid was also present.

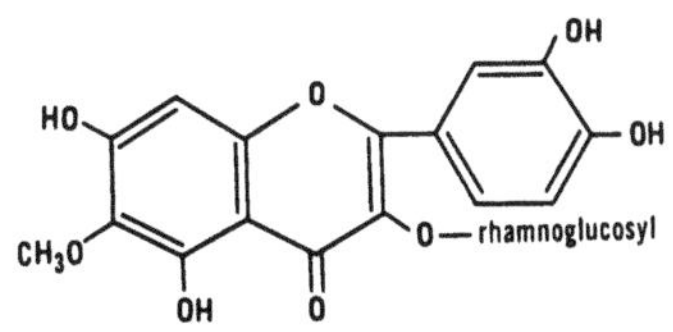

V. Patuletin 3-O-rhamnoglucoside

The success of each of the above trimethylsilylation and de-trimethylsilylation experiments can best be evaluated by observing the signal (near 13 ppm) of the hydrogen-bonded C-5 hydroxyl group.

3e. Procedure for the Complete De-Trimethylsilylation of TMS Ethers of Flavonoids

The original substance can often be regenerated unaltered after NMR analysis by allowing the trimethylsilyl ether derivative to stand overnight in 20% aqueous methanol containing a drop of acetic acid. Alternatively, the TMS ether may be refluxed for about 30 min or longer as required in an aqueous methanolic solution. Frequently the flavonoid crystallizes directly from the water/methanol solution, otherwise the hydrolyzed product may be isolated by chromatography on silica gel or polyamide (see Chapter II).

Complete De-Trimethylsilylation of Pseudobaptisin Hexatrimethylsilyl Ether. The solution obtained after NMR analysis of 50 mg of pseudobaptisin hexa-trimethylsilyl ether was evaporated to dryness and the residue was dissolved in methanol (5 ml). Water (about 1 ml) and acetic acid (1 drop) were added to the methanolic solution and the resulting solution was allowed to stand at room temperature overnight. White crystals of pseudobaptisin (VI, 35 mg) separated from the solution.

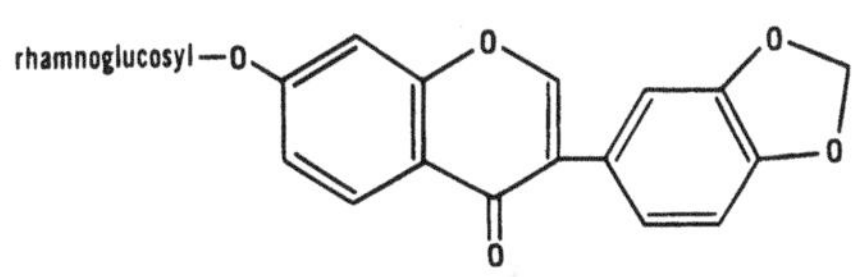

VI. Pseudobaptisin

3f. Conversion of the TMS Ether Derivatives of Flavonoids to Methyl Ethers and Acetates

In addition to the advantages noted previously for using the trimethylsilyl ethers of flavonoids for NMR studies, it should be mentioned that both methylation (with dimethylsulfate [*3c*] or diazomethane) and acetylation of flavonoids can be carried out with the trimethylsilyl ethers. The TMS-ethers of flavonoids, all of which are readily soluble in the organic solvents used for the preparation of the methyl and acetyl derivatives, undergo hydrolysis fast enough (in either mild acidic or basic media) to permit formation of the usual acetyl and methyl derivatives. Comparison of the NMR spectra for the trimethylsilyl, acetyl and methyl derivatives of flavonoids may be important for making structural assignments.

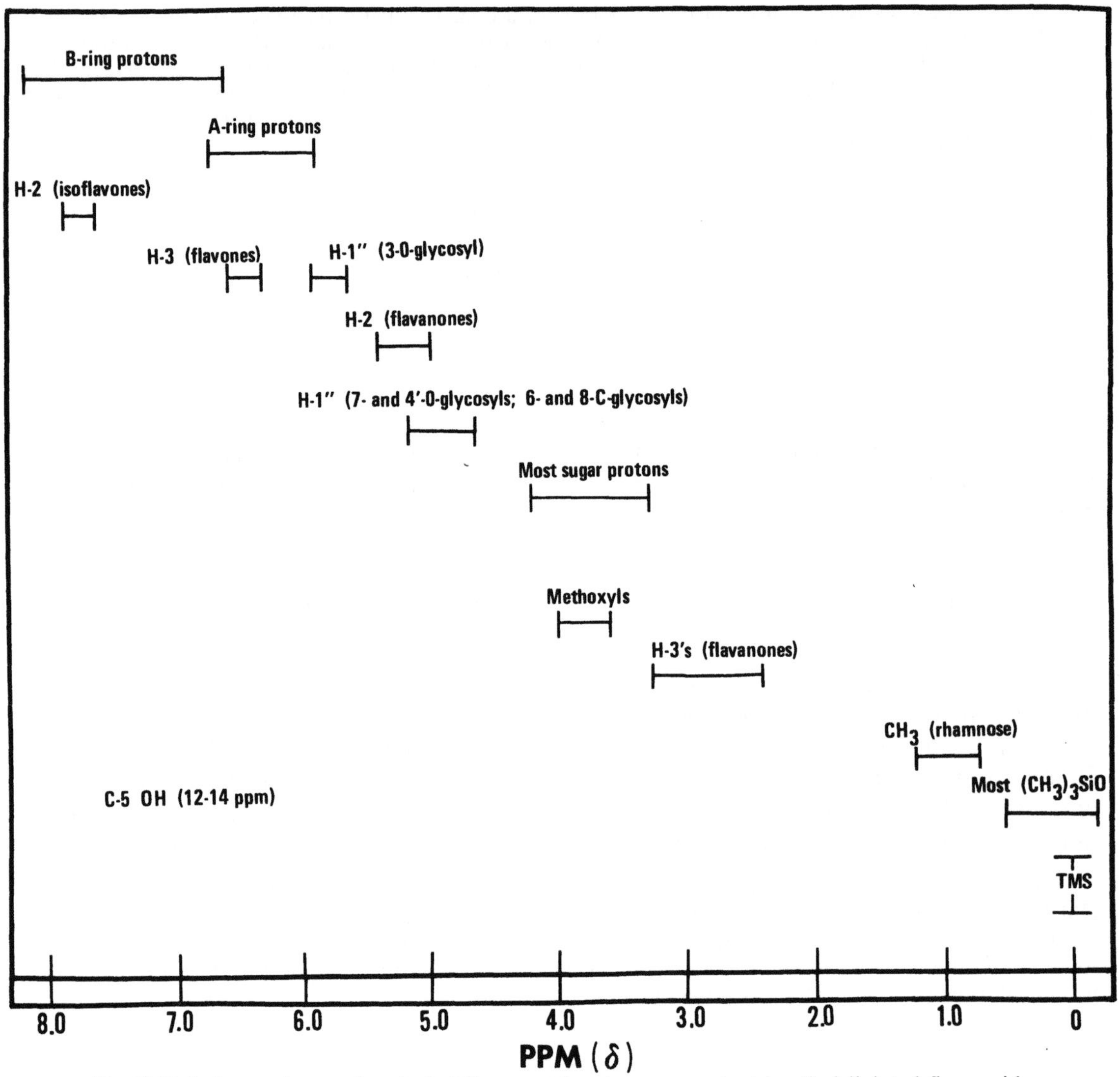

Fig. VIII-4. Approximate chemical shift ranges for the protons in trimethylsilylated flavonoids

VIII-4. Interpretation of the NMR Spectra of Fully and Partially Trimethylsilylated Flavonoids

Proton signals obtained in the NMR spectra of trimethylsilylated flavonoids generally occur in the range 0–9 ppm and most fall into a number of well separated groups (Fig. VIII-4). In the interpretation of the NMR spectra of flavonoids we have discussed the protons under the following headings: A-ring protons; B-ring protons; C-ring protons; sugar protons; methoxyl and acetoxyl protons; TMS ether protons; 6- and 8-C-methyl protons.

As described in the previous section all hydroxyl groups in both the aglycones and glycosides may be converted to trimethylsilyl ethers and in the discussion and Tables that follow, unless otherwise noted, *the expression "hydroxyl group" refers to the trimethylsilylated hydroxyl group.* In all cases, the NMR spectra were recorded relative to tetramethylsilane as a standard (usually internal) on a Varian A-60 spectrometer.

All chemical shift values reported in this chapter are given in ppm (δ-scale) relative to TMS (tetramethylsilane).

The numbering system used in the NMR discussions for flavonoids other than aurones and chalcones is presented below:

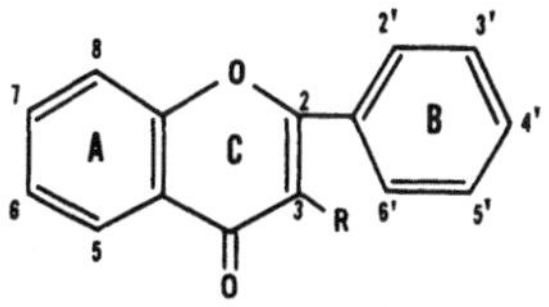

Flavones, R=H,
Flavonols, R=OH

4a. A-Ring Protons

C-6 and C-8 Protons in Flavones, Flavonols and Isoflavones

The protons at C-6 and C-8 of flavones, flavonols and isoflavones which contain the common 5,7-dihydroxy substitution pattern give rise to two doublets ($J=2.5$ cps) in the range 6.0–6.5 ppm. The H-6 doublet occurs consistently at higher field than the signal for the H-8 (Table VIII-1). When a sugar is attached to the oxygen at C-7 the signals for both H-8 and H-6 are shifted downfield (Table VIII-1).

Table VIII-1. *Chemical shifts of C-6 and C-8 protons in 5,7-dihydroxyflavones, 5,7-dihydroxyflavonols and 5,7-dihydroxyisoflavones and their 7-O-glycosides*

	H-6[a] (δ, ppm)	H-8[a] (δ, ppm)	Spectra No.
	6.0–6.2	6.3–6.5	2, 10, 13, 14, 15, 16, 25, 29, 30, 31, 33, 38, 39, 40, 46, 60, 70, 85
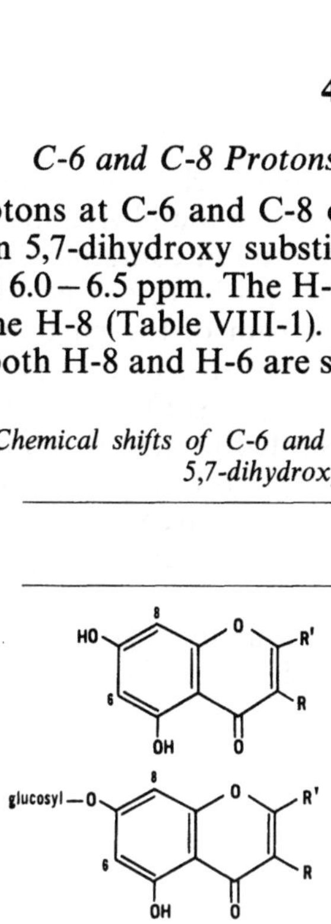	6.2–6.4	6.5–6.9	3, 5, 11, 12, 17, 26, 27, 32, 35, 36, 41, 43, 68, 86

[a] Doublet, $J=2.5$.

Table VIII-2. *Chemical shifts of C-6 and C-8 protons in 5,7-dihydroxyflavanones and 5,7-dihydroflavonols and their 7-O-glycosides*

Type of Flavonoid	H-6[a] (δ, ppm)	H-8[a] (δ, ppm)	Spectra No.
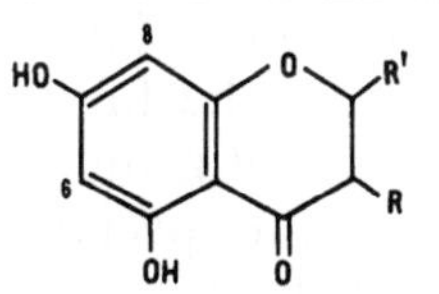	5.75–5.95	5.9–6.1	92, 93, 95, 96, 100 105, 106, 109, 110
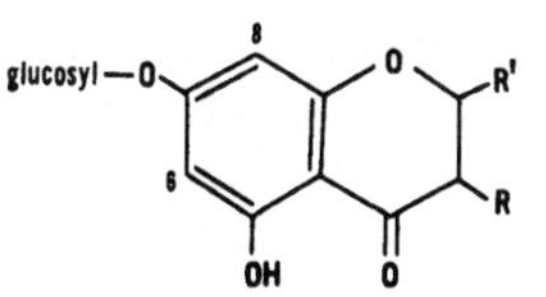	5.9–6.1	6.1–6.4	101, 107, 108

[a] Doublet, $J=2.5$ cps.

C-6 and C-8 Protons in Flavanones and Dihydroflavonols

In flavanones and dihydroflavonols which contain the 5,7-dihydroxy substitution pattern (Table VIII-2) the signals for the A-ring protons appear at higher field than in the corresponding flavones and flavonols.

Distinguishing Between C-6, C-8 and C-3 Proton Signals

The only other proton of the flavonoid nucleus which consistently gives a signal in the same region of the NMR spectrum as those of the C-6 and C-8 protons is the C-3 proton of flavones, which appears as a singlet near 6.3 ppm. Occasionally however, the C-3′ and C-5′ proton signals will also appear in this region of the spectrum. In flavones with the common 5,7-dihydroxylation pattern the C-6 and C-8 protons occur as doublets ($J=2.5$ cps) and are therefore readily distinguished from the C-3 proton singlet. However, many flavones have only one A-ring proton (e.g. 5,6,7- or 5,7,8-substitution patterns) and in these compounds the lone A-ring proton produces a singlet which is often in the same region as the C-3 proton signal. A useful technique for distinguishing the C-6, C-8 and C-3 protons in such cases is to compare the NMR spectrum of the fully trimethylsilylated compound with that of the compound containing trimethylsilyl substituents on all hydroxyl groups but the one at C-5 [*3b*, 6, 7] (see procedures in Section VIII-3d for experimental details). In the spectrum of the partially trimethylsilylated compound (5-hydroxyl group unsubstituted) relative to the spectrum of the fully trimethylsilylated flavonoid, the C-3 proton singlet is shifted downfield (usually more than 0.15 ppm) while the C-8 proton signal is shifted upfield by about 0.15 ppm. The signal of the C-6 proton is almost unaffected. These shifts are clearly observed when the spectrum of fully trimethylsilylated luteolin (VII) is compared with the spectrum of luteolin 3′,4′,7-trimethylsilyl (VIII) ether (spectra 13 and 14, the aromatic proton regions of which are reproduced in Fig. VIII-5).

As an example of the usefulness of the above technique for distinguishing C-6, C-8 and C-3 protons, the identification of patulitrin (IX) is described below.

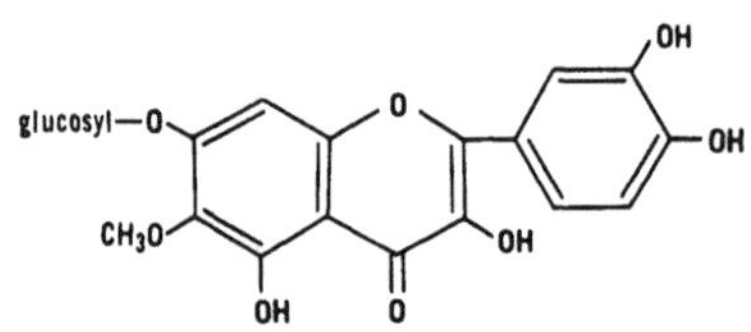

IX. Patulitrin

Patulitrin (IX), a flavonoid isolated from *Hymenoxys scaposa* [7], was trimethylsilylated by the standard procedure described in Section VIII-3a. The NMR spectrum obtained for the TMS ether (see Fig. VIII-6a) was typical of a flavonoid having the 3′,4′-substitution in ring B (see Section VIII-4b). However, the intensities of the low field signal at 12.35 ppm (C-5 hydrogen-bonded hydroxyl group) and at 6.52 ppm, (the intensities of both signals corresponded to 0.3 protons relative to the singlet at 6.65 ppm) suggested the presence of two compounds, i.e. the fully trimethylsilylated derivative (XI) and the TMS ether with a free C-5 hydroxyl group (X). For complete trimethylsilylation the mixture was therefore subjected to the special trimethylsilylating conditions for 6-methoxy substituted flavonoids (procedure VIII-3a, modified to a 30 min reaction time) and then the NMR analysis was repeated (see Fig. VIII-6b). The signals at 12.35 and 6.52 ppm were not present in the latter spectrum, thus indicating that the partially trimethylsilylated compound had a proton signal (6.62 ppm) which moved downfield more than 10 cps after complete trimethylsilylation. The only proton which has a downfield shift of this magnitude is a C-8 proton. Therefore the compound must be substituted at C-3 and C-6 (i.e. as in patulitrin, IX).

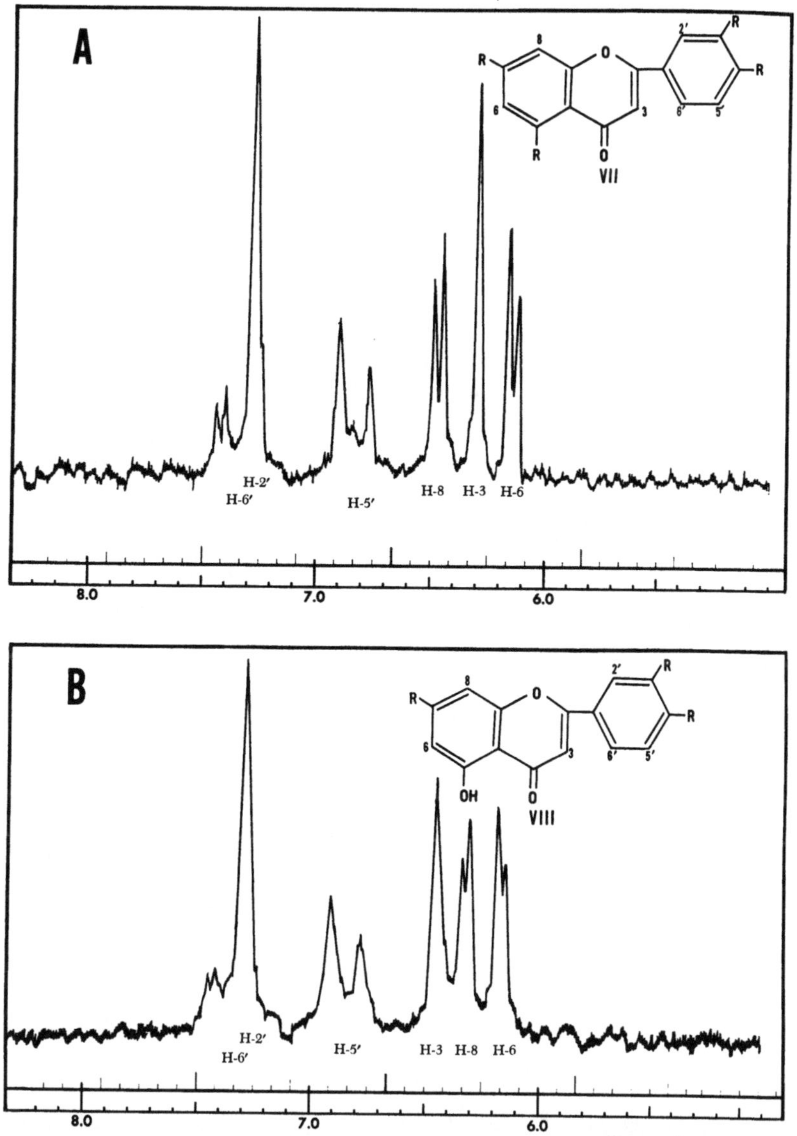

Fig. VIII-5. (A) NMR spectrum of luteolin 3',4',5,7-tetratrimethylsilyl ether (VII) [R = $OSi(CH_3)_3$] and (B) luteolin 3',4',7-trimethylsilyl ether (VIII) [R = $OSi(CH_3)_3$] in CCl_4

It appears that the TMS ethers of some 8-C-glycosylflavones exist in two forms (which may be conformers involving the rotational orientation of the C-glycosyl group with respect to the flavone nucleus); this conclusion is based on the fact that the NMR spectra for the fully trimethylsilylated derivative of vitexin (spectrum no. 6) and cytisoside (spectrum 9) exhibit two signals for H-3. The NMR spectrum of the TMS ether of vitexin shows a major H-3 signal at 6.48 ppm and a minor one upfield near 6.42 ppm; the spectrum of the TMS ether of cytisoside has a third H-3 signal downfield which corresponds to a small amount of material with a free C-5 hydroxyl group.

We recently observed that the signal for an H-6 proton in acetylated 8-C-glycosylflavones (which had hydroxyl groups at C-5 and C-7 before acetylation) appears in the region 6.5 – 6.7 ppm while the H-8 proton signal in acetylated 6-C-glycosylflavones falls between 7.25 – 7.4 ppm. This difference in chemical shift can be used to distinguish some 6- and 8-C-glycosylflavones.

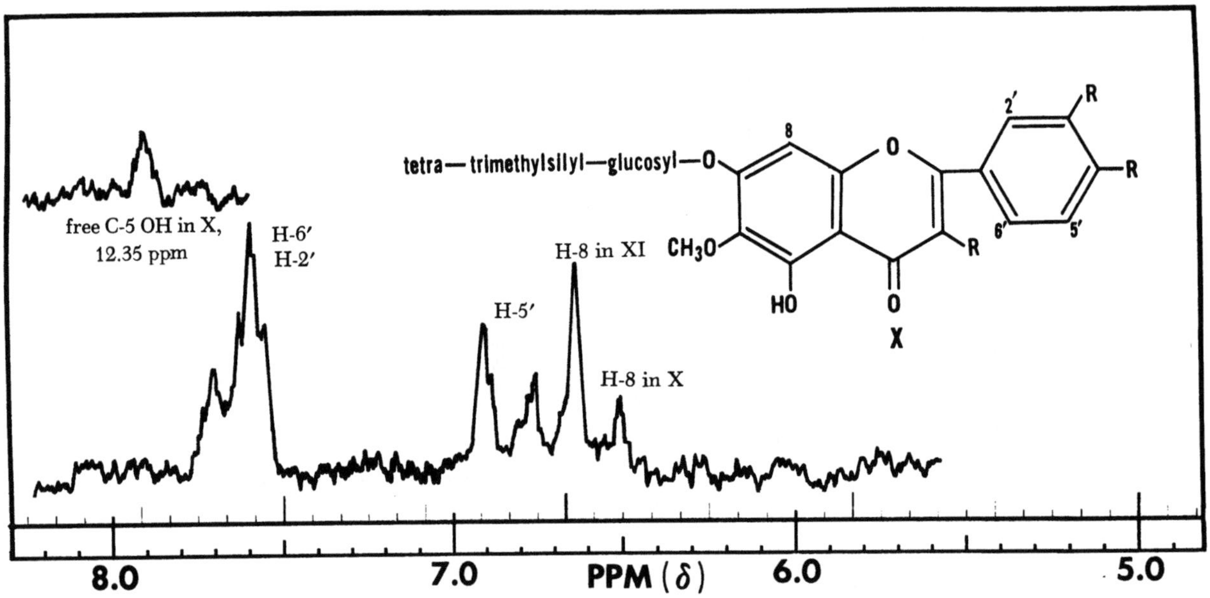

Fig. VIII-6a. A partial NMR spectrum of a 9:1 mixture of X [$R=OSi(CH_3)_3$] and XI [$R=OSi(CH_3)_3$], which was obtained by the standard trimethylsilylation reaction (see Section VIII-3a) of patulitrin (IX)

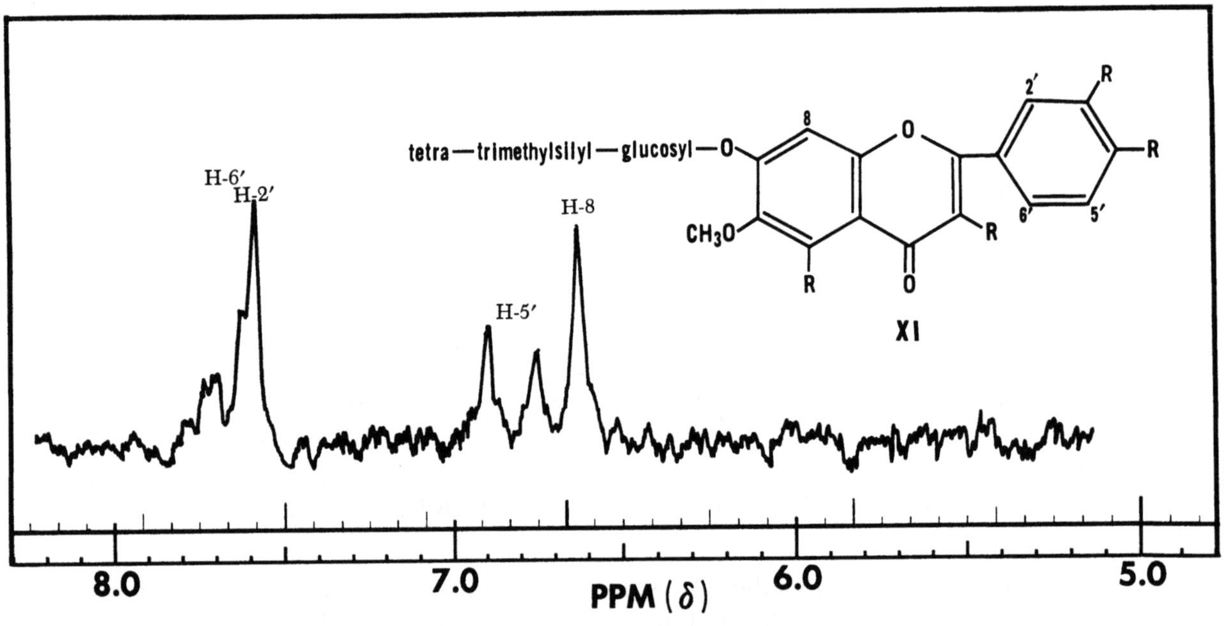

Fig. VIII-6b. A partial NMR spectrum of fully trimethylsilylated patulitrin (XI) [$R=OSi(CH_3)_3$], which was obtained by a 30 min trimethylsilylation reaction

C-5, C-6 and C-8 Protons in 7-Hydroxyflavones, 7-Hydroxyisoflavones, 7-Hydroxyflavanones, 7-Hydroxyflavonols and 7-Hydroxydihydroflavonols

Some naturally occurring flavonoids have oxygenation only at position 7 in the A-ring. In these compounds the C-5 proton is deshielded by the C-4 keto group and therefore absorbs near 8.0 ppm (Table VIII-3), and thus at lower field than most aromatic protons. The C-5 proton in these 7-oxygenated flavonoids appears as a doublet ($J = ca.$ 9 cps) as a result of *ortho* coupling between the C-5 and C-6 protons. The signals for the protons on C-6 and C-8 both occur at lower field than in the 5,7-dihydroxy-

Table VIII-3. *Chemical shifts of C-5, C-6, and C-8 protons in 7-oxygenated flavonoids*

Type of flavonoid	H-5[a] (δ, ppm)	H-6[b] (δ, ppm)	H-8[c] (δ, ppm)	Spectra No.
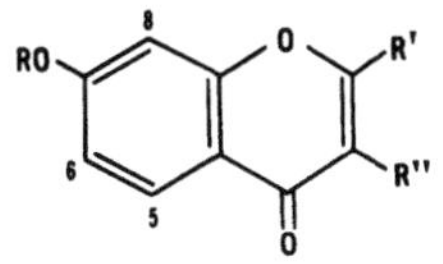	7.9–8.2	6.7–7.1	6.7–7.0	47, 66, 82
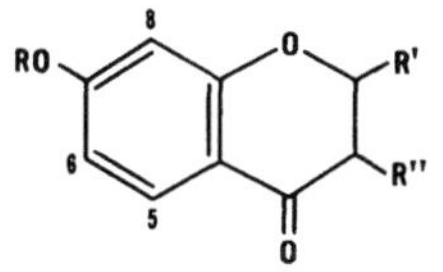	7.7–7.9	6.4–6.5	6.3–6.4	91, 94, 97, 99

[a] Doublet, $J=9$ cps.
[b] Quartet, $J=9$ and 2.5 cps.
[c] Doublet, $J=2.5$ cps.

flavonoids. Moreover, in 5-deoxyflavonoids the C-8 proton may absorb at either lower or higher field than the C-6 proton [in 5,7-dihydroxyflavonoids, as far as is known, the C-6 proton always absorbs at higher field than the C-8 proton (see Tables VIII-1 and VIII-2)].

4b. B-Ring Protons

C-2′, C-3′, C-5′ and C-6′ Protons in 4′-Oxygenated Flavonoids

The protons of ring B usually appear in the range 6.7–7.9 ppm, which is downfield from the region where the A-ring protons absorb. The signal pattern observed for the B-ring protons is characteristic for the substitution pattern of that ring and, in addition, suggests the oxidation level of ring C.

If ring B is oxygenated at C-4′, a typical four-peak pattern of two doublets (each $J=8.5$ cps) is observed. The doublet for the C-3′ and -5′ protons (which are shielded by the C-4′ oxygen substituent) always appears upfield from the C-2′ and -6′ protons and generally falls in the range 6.65–7.1 ppm for all types of flavonoids. The position of the C-2′, -6′ doublet depends to some extent on the oxidation level of ring C (see Table VIII-4); however, it consistently appears at lower field (7.1–8.1 ppm) than the C-3′, -5′ doublet.

C-2′, C-5′ and C-6′ Protons in 3′,4′-Oxygenated Flavonoids

The NMR spectra of flavonoids with the 3′,4′-oxygenation pattern in the B-ring are more complex than for compounds with the 4′-oxygenation. The C-5′ proton of 3′,4′-

Table VIII-4. *Chemical shifts of C-2′, -6′, and C-3′, -5′ protons in 4′-oxygenated flavonoids*

4′-Oxygenated flavonoids	H-2′, -6′[a] (δ, ppm)	H-3′, 5′[a] (δ, ppm)	Spectra No.
Flavanones	7.1–7.3	6.5–7.1	99, 100, 101, 102, 103, 104
Dihydroflavonols	7.2–7.4	6.5–7.1	91, 92, 93
Isoflavones	7.2–7.5	6.5–7.1	66, 68, 70, 71, 73, 74, 75, 80, 81, 87, 88
Chalcones[b]	7.4–7.6	6.5–7.1	111, 112, 113, 116, 117
Aurones	7.6–7.8	6.5–7.1	122, 123, 124
Flavones	7.7–7.9	6.5–7.1	2, 3, 5, 6, 7, 8, 9, 21, 22
Flavonols	7.9–8.1	6.5–7.1	25, 26, 27, 48, 49, 50, 59

[a] Doublet, *J ca.* 8.5 cps.
[b] Different numbering system (see Fig. VIII-7); values are quoted for H-2, -6 (first column) and H-3, -5 (second column).

Table VIII-5. *Chemical shifts of C-2′ and C-6′ protons in 3′,4′-oxygenated flavones*

Type of flavonoid	H-2′ [a] (δ, ppm)	H-6′ [b] (δ, ppm)	Spectra No.
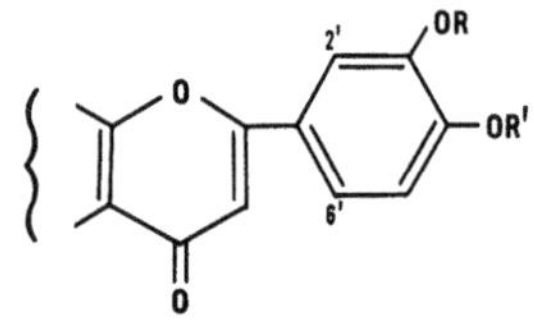 a) R=R′=H b) R=H, R′=CH_3	7.2–7.3	7.3–7.5	11, 12, 13, 14, 15, 16, 17, 23
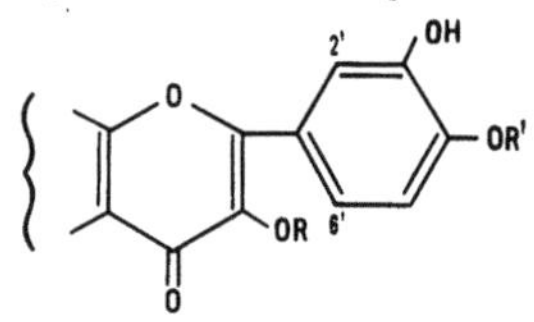 R=H or CH_3 R′=H or CH_3	7.5–7.7	7.6–7.9	29, 37, 40, 41, 43, 51, 56
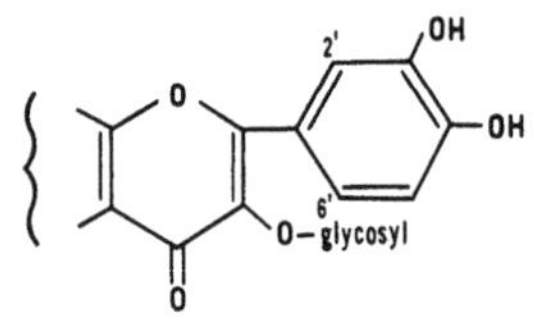	7.2–7.5	7.3–7.7	30, 31, 32, 33, 35, 36, 53, 54
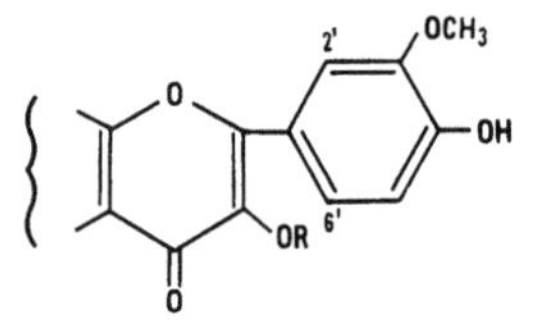 R=H, CH_3 or glycosyl	7.6–7.8	7.4–7.6	38, 39, 55

[a] Doublet, J=2.5 cps.
[b] Quartet, J=2.5 and 8.5 cps.

oxygenated flavones and flavonols appears as a doublet centered between 6.7 and 7.1 ppm (J=8.5 cps), and the C-2′ and -6′ proton signals, which often overlap, usually occur between 7.2 and 7.9 ppm. The relative positions of the signals for the C-2′ and -6′ protons may be used to distinguish the 3′-methoxy-4′-hydroxy from the 4′-methoxy-3′-hydroxy B-ring substitution pattern in flavonols. The C-2′ proton signal is usually centered at slightly higher field than the C-6′ proton signal in flavonoids containing the 4′-methoxyl group (Table VIII-5). In contrast, their positions are reversed when a 3′-methoxyl group is present in a 3′,4′-oxygenated flavonol (cf. spectra 55 and 56).

Different spectral patterns are observed for 3′,4′-oxygenated isoflavones, flavanones and dihydroflavonols. These compounds give a complex multiplet, usually two peaks, for the C-2′, -5′ and -6′ protons in the region 6.7–7.1 ppm (see spectra 82, 85, 86, 94, 96, 105, 106, 107, 108, and 109); the chemical shifts for B-ring protons in these flavonoids depend upon whether or not the proton is *ortho* or *para* to an oxygen function. In 3′,4′-dioxygenated isoflavones, flavanones and dihydroflavonols, the C-2′, -5′ and -6′ protons are either *ortho* or *para* to an oxygen substituent and therefore have similar chemical shifts.

C-2′ and C-6′ Protons in 3′,4′,5′-Oxygenated Flavonoids

The C-2′ and -6′ proton signals usually overlap in the region 6.5–7.5 ppm in flavonoids having the 3′,4′,5′-oxygenation pattern (see spectra 47, 60, 61, 89, 90, and 97).

4c. C-Ring Protons

C-3 Proton in Flavones; C-2 Proton in Isoflavones

Considerable variation is found in the chemical shifts of the C-ring protons among the different flavonoid classes depending upon the oxidation level of the C-ring. The C-3 proton in flavones gives a sharp singlet near 6.3 ppm (spectra 1 – 24), which usually overlaps the signals produced by the A-ring protons. On the other hand, the C-2 proton in isoflavones, which is in a beta position to the C-4 keto function, occurs in the range 7.6 – 7.8 ppm (in CCl_4), a region downfield from where most aromatic proton signals appear. The position of the signal obtained for the C-2 proton when the isoflavone (or its TMS ether) is examined in either CCl_4 or $CDCl_3$ shifts about 1 ppm downfield when the compound is analysed in DMSO-d_6 (Table VIII-6). In certain instances, this change in chemical shifts can be used for structure assignments [*8*].

Table VIII-6. *Chemical shift of the C-2 proton of isoflavones examined in CCl_4, $CDCl_3$, and DMSO-d_6*

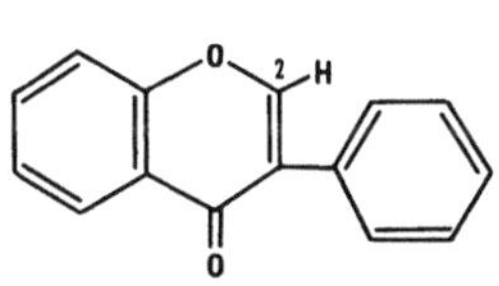

Solvent	C-2 singlet (δ, ppm)	Spectra No.
CCl_4	7.6 – 7.8	66, 68, 70, 71, 72, 73, 80, 82, 85, 86, 87, 88, 89, 90
$CDCl_3$	7.8 – 8.1	62, 64, 65, 67, 69, 74, 75, 81, 84
DMSO-d_6	8.5 – 8.7	63, 83

α and β Protons of Chalcones; Benzylic Proton in Aurones

The H-α and H-β protons of chalcones (Fig. VIII-7) occur as doublets (J = *ca.* 17 cps) in the ranges 6.7 – 7.4 ppm (H-α) and 7.3 – 7.7 ppm (H-β), while the aurone benzylic proton appears as a singlet at 6.5 – 6.7 ppm.

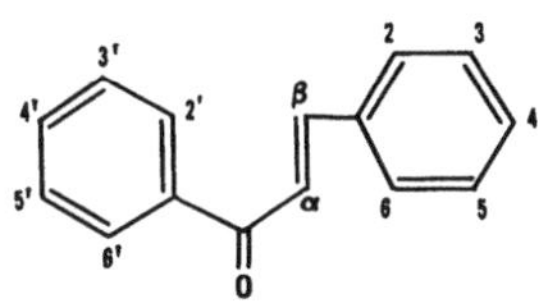

Chalcone

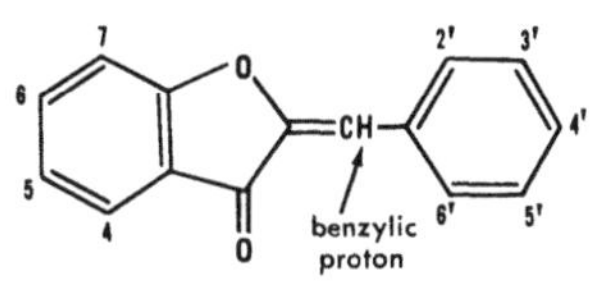

Aurone

Fig. VIII-7. Numbering systems for aurones and chalcones

C-2 and C-3 Protons in Flavanones and Dihydroflavonols

The signal for the C-2 proton of flavanones appears as a quartet (two doublets J_{cis} = *ca.* 5 cps, J_{trans} = *ca.* 11 cps) near 5.2 ppm as a result of the coupling of the C-2 proton with the two C-3 protons [*9*]. The C-3 protons couple with each other (J = 17 cps) in addition to their spin-spin interaction with the C-2 proton, thus giving rise to two overlapping quartets near 2.8 ppm. Two of the signals of each quartet, however, are weak and are often not observed (spectra 98 – 110).

In naturally occurring dihydroflavonols (spectra 91 – 97) the C-2 proton signal occurs as a doublet (J = *ca.* 11 cps) near 5.2 ppm, while the C-3 proton doublet appears further upfield at about 4.3 ppm (Table VIII-7). The 11 cps coupling constant is typical for 1,2-diaxial protons; thus, the naturally occurring dihydroflavonols, which are known to have the *R* absolute configuration at both C-2 and C-3 [*10*], can be represented by the structure shown [*9*] in Fig. VIII-8.

Table VIII-7. *Chemical shifts of C-2 and C-3 protons in flavanones and dihydroflavonols*

Type of flavonoid	H-2 (δ, ppm)	H-3 (δ, ppm)	Spectra No.
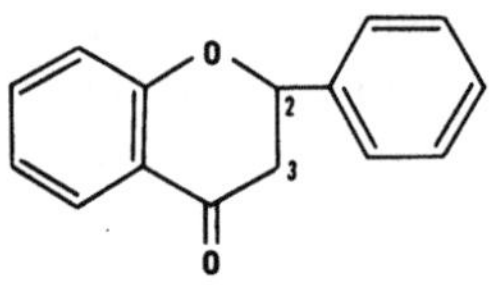	5.0 – 5.5[a]	near 2.8[b]	98 – 110
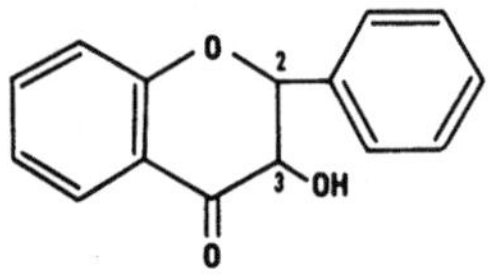	4.8 – 5.0[c]	4.1 – 4.3[c]	92, 94, 95, 97
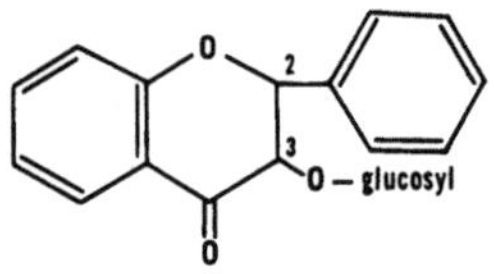	5.0 – 5.6[c]	4.3 – 4.6[c]	91, 93, 96

[a] Quartet, $J = ca.$ 5, 11 cps.
[b] Two quartets, $J = ca.$ 17, 5 cps; $J = ca.$ 17, 11 cps.
[c] Doublet, $J = ca.$ 11 cps.

In dihydroflavonol 3-O-glycosides the C-2 and C-3 proton signals are both at significantly lower field than in the equivalent aglycones (Table VIII-7). This observation is potentially useful for determining the sugar location in dihydroflavonol glycosides.

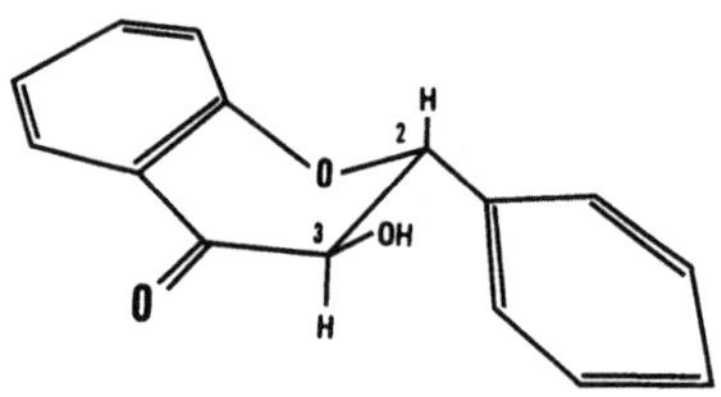

Fig. VIII-8. The absolute stereochemistry (2R, 3R) of naturally occurring dihydroflavonols

4d. Sugar Protons

The C-1″ Proton in Flavonoid Monosaccharides[2]

The chemical shift of the C-1″ proton of a sugar directly attached to the flavonoid hydroxyl group depends both on the nature of the flavonoid and on the position and stereochemistry of attachment to it.

With glucosides (and presumably some other glycosides) a sugar on the C-3 hydroxyl group can be readily distinguished from one at C-4′ (see spectrum 50), C-5 (see spectrum 103) or C-7 (see Table VIII-8). In the latter three types of flavonoid O-glucosides, the C-1″ proton signal occurs near 5.0 ppm, while in flavonol 3-O-glucosides the C-1″ proton signal appears further downfield at about 5.8 ppm.

Glucose commonly forms a β-linkage in flavonoid glycosides and the C-1″ proton of the β-linked sugar has a diaxial coupling with the C-2″ proton. Thus the C-1″ proton usually appears as a doublet with a coupling constant of about 7 cps. In flavonoid 7-O-glucosides, however, the glucosyl C-1″ proton does not appear as a sharp doublet but instead gives a complex multiplet. Molecular models indicate that the reason for this is that as the 7-O-glucosyl moiety rotates with respect to the flavonoid nucleus, the

[2] In flavonoid glycosides, the protons in the sugar attached to the flavonoid are denoted as C-1″, C-2″ and so on; in disaccharides the protons of the terminal sugar are designated C-1‴, C-2‴ and so on.

Table VIII-8. *Chemical shifts of the C-1″ protons of glucosyl and rhamnosyl substituents which are directly attached to flavonoid hydroxyl groups*

Flavonoid glycoside	Sugar C-1 proton (δ, ppm)	Spectra No.
Flavonoid 7-O-glucoside	4.8 – 5.2	3, 5, 8, 11, 17, 27, 32, 35, 41, 43, 51, 55, 56, 66, 68, 80, 82, 88, 90, 101, 107, 108
Flavonol 3-O-glucoside	5.7 – 6.0	32, 33, 35, 36, 53, 54
Flavonol 7-O-rhamnoside	5.1 – 5.3	26, 36
Flavonol 3-O-rhamnoside	5.0 – 5.1	30
Dihydroflavonol 3-O-glucoside	4.1 – 4.3	91
Dihydroflavonol 3-O-rhamnoside	4.0 – 4.2	93, 96

C-1″ proton of the sugar experiences different electronic environments. As rotation becomes more restricted by C-8 and/or C-6 substituents, the signal for the C-1″ proton tends to become the expected doublet [see flavonoid 6- and 8-C-glycosides (spectra 6, 7, 8, and 9)]. With 3-O-glucosides and galactosides (spectra 26, 31, 32, 33, 35, 39, 53, and 54) and 5-O-glucosides (spectrum 103), the rotation of the glycosyl moiety is restricted and in the spectra of these flavonoids the C-1″ proton appears as a doublet.

Flavonoid rhamnosides occur naturally as α-L-rhamnosides in which the rhamnosyl C-1″ proton has an equatorial-equatorial coupling with the C-2″ proton ($J = ca.$ 2 cps). In both 3- and 7-O-rhamnosides the C-1″ proton signal occurs in the range 5.0 – 5.3 ppm (Table VIII-8). In the NMR spectra of flavonoid rhamnosides, the signal for the rhamnose methyl group, which occurs at 0.8 – 1.2 ppm, is also a distinguishing feature. The rhamnose methyl group in quercetin 3-O-rhamnoside (spectrum 30) gives a sharp doublet at 0.85 ppm ($J = 6.5$ cps) while in quercetin 3-glucoside 7-rhamnoside (spectrum 36), the rhamnose methyl appears as a complex signal at 1.2 ppm. This difference observed in the coupling patterns probably reflects the freer rotation possible for a rhamnosyl moiety at C-7.

C-1″, C-1‴, and Rhamnose Methyl Protons (C-6‴) in Flavonoid Rhamnoglucosides

All the known flavonoid rhamnoglucosides contain either the rutinosyl or neohesperidosyl moieties (Fig. VIII-9A and -9B). These two substituents, which differ only in the point of attachment of the rhamnose to the glucose can be distinguished in flavonoid disaccharides by the NMR analysis of their TMS ethers [*11*]. In the trimethylsilyl ethers of 7- and 3-O-rutinosides, rhamnose is characterized by a C-1‴ proton signal near 4.2 – 4.4 ppm ($J = 2$ cps) and a broad peak for the C-6‴ protons (methyl group) at 0.7 – 1.0 ppm (see spectra 11, 17, 33, 43, 54, 68, 82, and 107), whereas in flavonoid 7- and 3-O-neohesperidosides, the rhamnose C-1‴ proton absorbs at 4.9 – 5.0 ppm ($J = 2$ cps) and the signal for the methyl group appears as a doublet at 1.1 – 1.3 ($J = 6$ cps) (see spectra 3, 27, 41, 101, and 108). These differences for the rhamnose signals are illustrated in Fig. VIII-9A and -9B, by spectra A and B, which represent the sugar

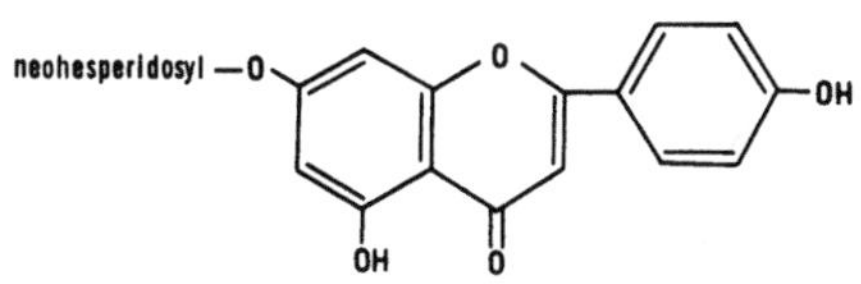

XII. Apigenin 7-O-neohesperidoside

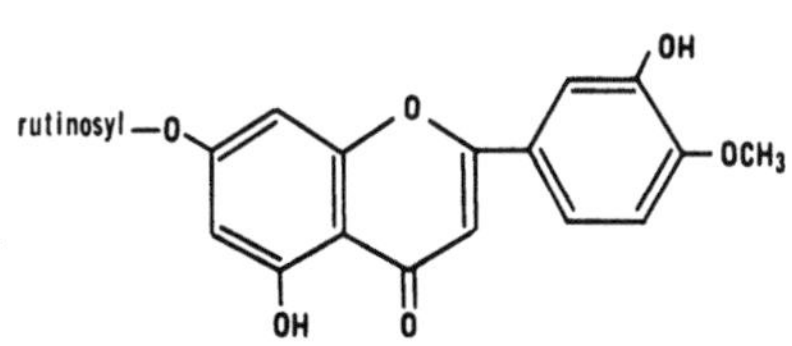

XIII. Diosmin

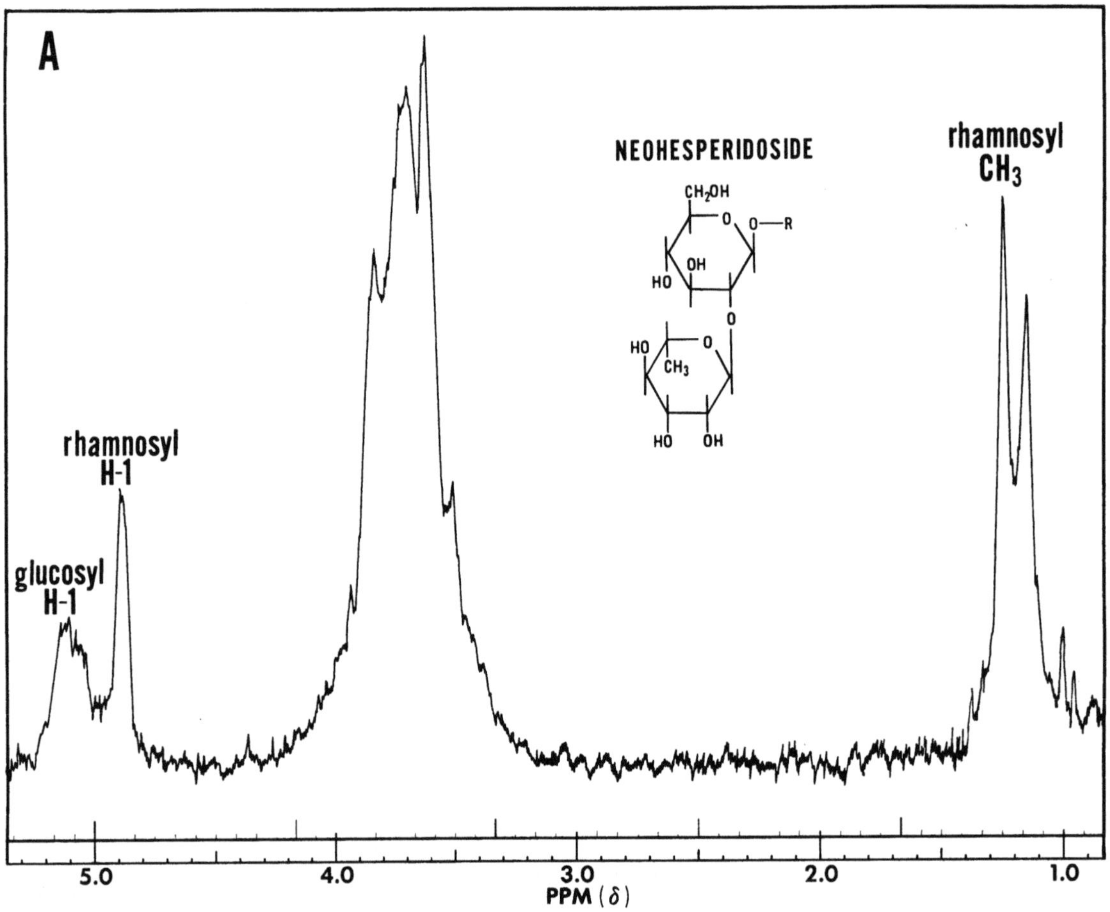

Fig. VIII-9A. NMR spectra of the sugar moiety in trimethylsilylated apigenin 7-O-neohesperidoside

moieties of the trimethylsilyl ethers of apigenin 7-neohesperidoside (XII) and diosmin (XIII), respectively. These interpretations can be confirmed by studying the NMR spectra of the flavonoids after acetylation [11].

The Determination of the Sugar-Sugar Linkage in Xylosylvitexin (XIV)

The sugar-sugar linkage in a different kind of flavonoid disaccharide was determined by Horowitz using NMR spectroscopy [12]. He showed that xylosylvitexin was a 2″-O-β-D-xylopyranoside of vitexin by an extensive NMR analysis of a number of acetyl and methyl derivatives. The NMR analysis of xylosylvitexin was based in part on the observation (recognized previously by Hillis and Horn [13]) that in the spectra of

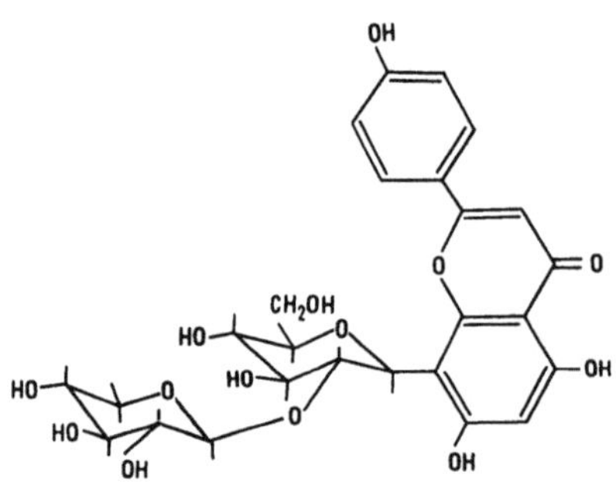

XIV. 2″-O-Xylosylvitexin

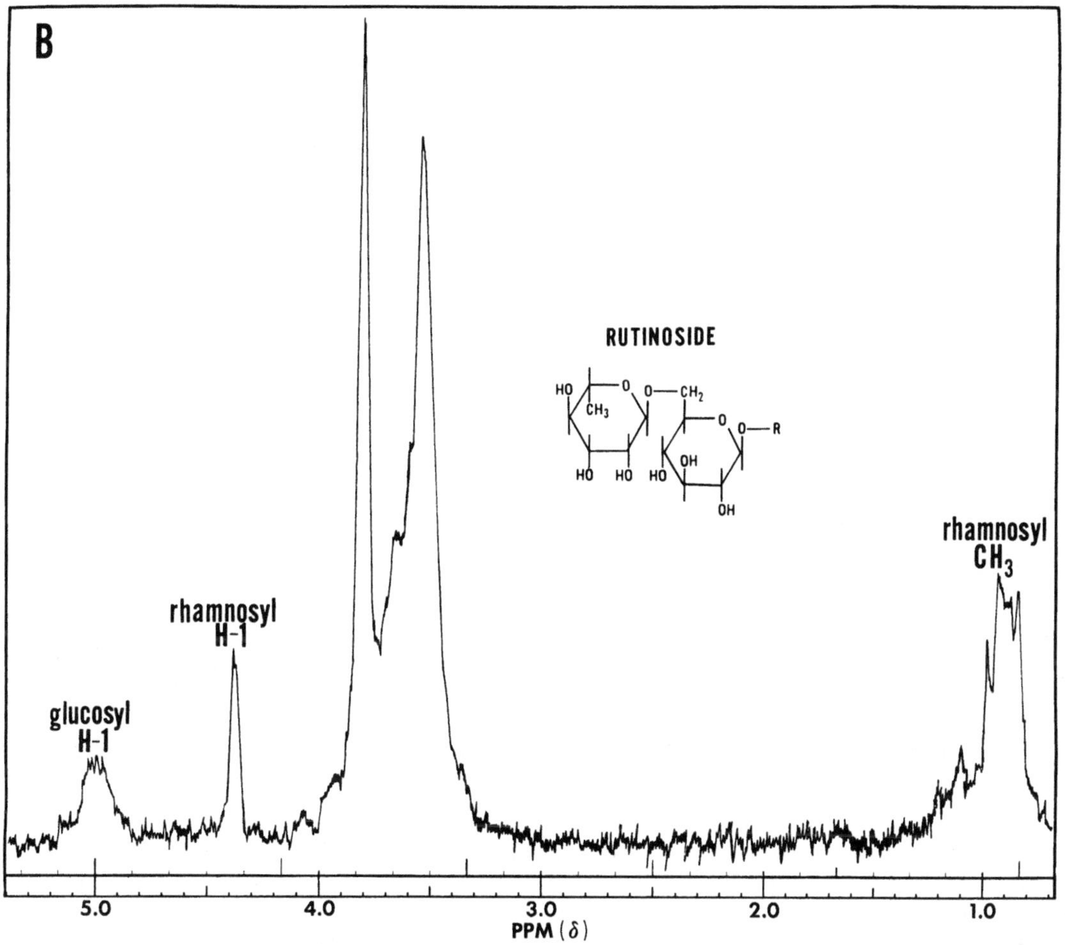

Fig. VIII-9 B. NMR spectra of the sugar moiety in trimethylsilylated diosmetin 7-O-rutinoside (diosmin)

acetylated 8- and 6-C-glucosylflavones the 2″-O-acetyl groups give signals near 1.75 and 1.80 ppm (δ), respectively. Moreover, the signal for a 6″-O-acetyl group in acetylated 8-C-glucosylflavones comes between 1.90 – 1.95 ppm. The signals for almost all other aliphatic acetyl groups appear in the range 2.0 – 2.10 ppm while the aromatic acetyl groups give signals in the range 2.25 – 2.50 ppm. Since the spectrum of acetylated xylosylvitexin did not exhibit an acetyl signal near 1.75 ppm, it was concluded that the xylosyl group was attached to the 2″-position of vitexin as in XIV. This conclusion was confirmed by comparison of the NMR spectra of hepta-O-methylvitexin and nona-O-methylxylosylvitexin. In the spectrum of hepta-O-methylvitexin, the signal for the 2″-O-methyl group, which is markedly shielded by the flavone nucleus, appears diagnostically upfield at 3.00 ppm. The signal for the 6″-O-methyl group comes near 3.4 ppm while all the other O-methyl groups give rise to signals between 3.5 – 4.1 ppm. The NMR spectrum for the nona-O-methyl derivative of xylosylvitexin did not exhibit a methyl signal near 3.00 ppm, thus supporting the previous conclusion that the xylosyl moiety is attached to the 2″-position of vitexin.

4e. Methoxyl and Acetoxyl Protons

In flavonoids, methoxyl proton signals, with few exceptions, appear in the region 3.5 – 4.1 ppm, while most aromatic acetoxyl proton signals occur in the range 2.25 –

2.50 ppm. Signals for the protons on the flavonoid nucleus *ortho* and *para* to acetoxyl groups are shifted downfield by about 0.3 and 0.5 ppm respectively relative to their positions in the spectrum of the flavonoid TMS ether; in contrast, *meta* proton signals are shifted only about 0.15 ppm (compare, for example, spectrum 3 with spectrum 4).

Horowitz has recently investigated [*14*] the applicability of determining the positions of methoxyl and acetoxyl groups on flavonoid nuclei by comparing the NMR spectra of the derivatives first in $CDCl_3$ and then in benzene. It appears from preliminary studies that when flavonoids are examined first in $CDCl_3$ and then in benzene, the C-3 and C-6 methoxyl and C-5 acetoxyl signals shift upfield less than 0.18 ppm (occasionally a downfield shift is observed) while all other methoxyl and acetoxyl signals are shifted upfield more than 0.3 ppm.

Other workers [*15*] also observed that methoxyl groups at C-2′, C-4′, C-5 and C-7 in flavones and flavonols exhibit large upfield benzene-induced solvent shifts (0.5 – 0.8 ppm) in the absence of *ortho* methoxyl or hydroxyl substituents. In contrast, a C-3 methoxyl group and methoxyl groups flanked by two *ortho* methoxyl functions (or one *ortho* methoxyl and one *ortho* hydroxyl group) show little or no shift (up- or downfield). These workers observed that a C-6 methoxyl adjacent to a C-5 methoxyl forces the C-5 group into the sphere of influence of the carbonyl function which causes a downfield solvent shift of the C-5 group.

A number of investigators [*1, 12, 13*] have observed that the aromatic and aliphatic acetyl signals of C-glycosylflavonoid acetates fall in characteristic regions of the NMR spectra: the aromatic acetate signals appear between 2.25 and 2.50 ppm while the aliphatic acetate signals occur between 1.65 and 2.10 ppm. The signals of the 4′- and 7-acetoxyl groups in simple flavone acetates usually appear in the range of 2.30 – 2.35 ppm, while the signal for a 5-acetyl group occurs around 2.45 ppm.

As mentioned in Section VIII-4d, in 8-C-glucosylflavones the 2″-O-acetyl signal consistently occurs in the range 1.70 – 1.75 ppm while the 6″-O-acetyl signal usually appears in the range 1.90 – 1.95 ppm; other acetyl signals associated with the C-glucosyl moiety come in the range 2.0 – 2.10 ppm. In 6-C-glycosylflavones, the signal for the 2″-O-acetyl group comes near 1.80 ppm and a band for the 6″-O-acetyl group appears near 2.0 ppm.

4f. TMS Ether Protons

Most of the trimethylsilyl proton signals of flavonoid TMS ethers occur between 0.1 and 0.5 ppm. However the signals for a few trimethylsilyloxy groups appear in the range 0 to 0.1 ppm upfield from tetramethylsilane. For example, the signal for the trimethylsilyloxy group at C-3 in dihydroflavonols occurs upfield from the signal for tetramethylsilane (see spectra 92, 94, 95 and 97). In addition, the NMR spectra of TMS ethers of C-glycosylflavonoids usually exhibit a signal upfield from tetramethylsilane.

The number of hydroxyl groups present in a flavonoid can be determined by integrating the signals for the trimethylsilyl protons. For this determination, the trimethylsilylating reagents must be excluded from the NMR tube (see Section VIII-3a).

4g. 6- and 8-C-Methyl Protons

An NMR study of eighteen synthetic 6- and 8-C-methylisoflavones [*8*] showed that the signal for the 6-C-methyl protons, which occurred in the range 2.04 – 2.27 ppm, was almost always about 0.2 ppm upfield from the signal for the 8-C-methyl group in the isomeric isoflavone (2.14 – 2.45 ppm). This same relative chemical shift difference was previously observed for the signals for the H-6 and H-8 protons in almost all flavonoids (see Section VIII-4a).

References

1. a) Massicot, J., and J.-P. Marthe: Bull. Soc. Chim. France (1962). b) Massicot, J., J.-P. Marthe, and S. Heitz: Bull. Soc. Chim. France 2712 (1963).
2. Waiss, A.C., R.E. Lundin, and D.J. Stern: Tetrahedron Letters 513 (1964).
3. a) Mabry, T.J., J. Kagan, and H. Rösler: Phytochemistry **177**, 487 (1965). b) Mabry, T.J., J. Kagan, and H. Rösler: Univ. of Texas Publ. No. 6418 (1964). c) Rösler, H., T. J. Mabry and J. Kagan: Chem. Ber. **98**, 2193 (1965).
4. Batterham, T.J., and R.J. Highet: Australian J. Chem. **17**, 428 (1964).
5. a) Grouiller, A.: Bull. Soc. Chim. France 2405 (1966). b) Grouiller, A., and H. Pacheco: Bull. Soc. Chim. France 1938 (1967).
6. a) Seikel, M.K., and T.J. Mabry: Tetrahedron Letters 1105 (1965). b) Hörhammer, L., H. Wagner, L. Rosprim, T.J. Mabry, and H. Rösler: Tetrahedron Letters 1707 (1965).
7. Thomas, M.B., and T.J. Mabry: Phytochemistry **7**, 787 (1968).
8. Markham, K.R., W. Rahman, S. Jehan, and T.J. Mabry: J. Heterocyclic Chem. **4**, 61 (1967).
9. Clark-Lewis, J.W., L.M. Jackman, and T.M. Spotswood: Australian J. Chem. **17**, 632 (1964).
10. Markham, K.R., and T.J. Mabry: Tetrahedron **24**, 823 (1968).
11. Rösler, H., T.J. Mabry, M.F. Cranmer, and J. Kagan: J. Org. Chem. **30**, 4346 (1965).
12. Horowitz, R.M., and B. Gentili: Chem. Ind. (London) 625 (1966).
13. a) Hillis, W.E., and D.H.S. Horn: Australian J. Chem. **18**, 531 (1965). b) Eade, R.A., W.E. Hillis, D.H.S. Horn, and J.J.H. Simes: Australian J. Chem. **18**, 715 (1965).
14. Horowitz, R.M.: Abstract and lecture presented at the Seventh Annual Meeting of the Phytochemical Society of North America, Madison, Wisconsin, August, 1967.
15. Wilson, R.G., J.H. Bowie, and Dudley H. Williams: Tetrahedron **24**, 1407 (1968).

Chapter IX

The NMR Spectra of Flavonoids

NMR spectra are presented in order of increasing oxidation and substitution of the flavonoid nucleus for flavones, flavonols, isoflavones, dihydroflavonols, flavanones, chalcones and aurones.

Index of NMR spectra[a] of flavonoids[b]

Spectrum No.	Flavonoids	Oxidation pattern											
		2	3	4	5	6	7	8	2′	3′	4′	5′	6′
	Flavones												
1	Tectochrysin				OH		OCH_3						
2	Apigenin				OH		OH				OH		
3	Apigenin 7-O-neohesperidoside				OH		O-neo-hesp				OH		
4	Apigenin 7-O-neohesperidoside acetate				OAc		O-hexa-acetyl-neohesp				OAc		
5	Apigenin 7-O-glucoside				OH		O-glu				OH		
6	Vitexin				OH		OH	C-glu			OH		
7	Isovitexin				OH	C-glu	OH				OH		
8	Saponarin				OH	C-glu	O-glu				OH		
9	Cytisoside				OH		OH	C-glu			OCH_3		
10	5,7-Dihydroxy-2′-methoxyflavone				OH		OH		OCH_3				
11	Luteolin 7-O-rutinoside				OH		O-rut			OH	OH		
12	Luteolin 7-O-glucoside				OH		O-glu			OH	OH		
13	Luteolin				OH		OH			OH	OH		
14	Luteolin (5-hydroxy-3′,4′,7-TMS ether)				OH		OH			OH	OH		
15	5,7-Dihydroxy-3′,4′-dimethoxy-flavone				OH		OH			OCH_3	OCH_3		
16	Diosmetin				OH		OH			OH	OCH_3		
17	Diosmin				OH		O-rut			OH	OCH_3		
18	Diosmin acetate				OAc		O-hexa-acetyl-rut			OAc	OCH_3		
19	Zapotinin				OH	OCH_3			OCH_3				OCH_3
20	Zapotin				OCH_3	OCH_3			OCH_3				OCH_3
21	Xanthomicrol				OH	OCH_3	OCH_3	OCH_3			OH		
22	Demethoxysudachitin				OH	OCH_3	OH	OCH_3			OH		
23	Hymenoxin				OH	OCH_3	OH	OCH_3		OCH_3	OCH_3		
24	Scaposin				OH	OCH_3	OH	OCH_3		OCH_3	OCH_3	OH	
	Flavonols												
25	Kaempferol		OH		OH		OH				OH		
26	Kaempferol 3-O-robinoside 7-O-rhamnoside		O-rh-gal		OH		O-rh				OH		

Spectrum No.	Flavonoids	Oxidation pattern											
		2	3	4	5	6	7	8	2′	3′	4′	5′	6′
27	Kaempferol 7-O-neohesperidoside		OH		OH		O-neo-hesp				OH		
28	Kaempferol 7-O-neohesperidoside acetate		OAc		OAc		O-hexa-acetyl-neohesp				OAc		
29	Quercetin		OH		OH		OH			OH	OH		
30	Quercitrin		O-rh		OH		OH			OH	OH		
31	Hyperin		O-gal		OH		OH			OH	OH		
32	Quercetin 3,7-O-diglucoside		O-glu		OH		O-glu			OH	OH		
33	Rutin		O-rut		OH		OH			OH	OH		
34	Rutin acetate		O-hexa-acetyl-rut		OAc		OAc			OAc	OAc		
35	Quercetin 3-O-glucoside 7-O-rutinoside		O-glu		OH		O-rut			OH	OH		
36	Quercetin 3-O-glucoside 7-O-rhamnoside		O-glu		OH		O-rh			OH	OH		
37	Rhamnetin		OH		OH		OCH_3			OH	OH		
38	Isorhamnetin		OH		OH		OH			OCH_3	OH		
39	Isorhamnetin 3-O-galactoside		O-gal		OH		OH			OCH_3	OH		
40	Tamarixetin		OH		OH		OH			OH	OCH_3		
41	Tamarixetin 7-O-neohesperidoside		OH		OH		O-neo-hesp			OH	OCH_3		
42	Tamarixetin 7-O-neohesperidoside acetate		OAc		OAc		O-hexa-acetyl-neohesp			OAc	OCH_3		
43	Tamarixetin 7-O-rutinoside		OH		OH		O-rut			OH	OCH_3		
44	Quercetin 3′,4′,5,7-tetramethyl ether		OH		OCH_3		OCH_3			OCH_3	OCH_3		
45	3,5,6,7,8-Pentamethoxyflavone		OCH_3		OCH_3	OCH_3	OCH_3	OCH_3					
46	Morin		OH		OH		OH		OH		OH		
47	Robinetin		OH				OH			OH	OH	OH	
48	Herbacetin 8-methyl ether		OH		OH		OH	OCH_3			OH		
49	Penduletin		OCH_3		OH	OCH_3	OCH_3				OH		
50	Pendulin		OCH_3		OH	OCH_3	OCH_3				O-glu		
51	Patulitrin (CCl_4)		OH		OH	OCH_3	O-glu			OH	OH		
52	Patulitrin (DMSO-d_6)		OH		OH	OCH_3	O-glu			OH	OH		
53	Patuletin 3-O-glucoside		O-glu		OH	OCH_3	OH			OH	OH		

54	Patuletin 3-O-rutinoside	O-rut	OH	OCH_3	OH		OH	OH	
55	Jacein	OCH_3	OH	OCH_3	O-glu		OCH_3	OH	
56	Centaurein	OCH_3	OH	OCH_3	O-glu		OH	OCH_3	
57	Artemetin	OCH_3	OH	OCH_3	OCH_3		OCH_3	OCH_3	
58	Quercetagetin hexamethyl ether	OCH_3	OCH_3	OCH_3	OCH_3		OCH_3	OCH_3	
59	3,4′,5,6,7,8-Hexamethoxyflavone	OCH_3	OCH_3	OCH_3	OCH_3	OCH_3		OCH_3	
60	Myricetin	OH	OH		OH		OH	OH	OH
61	3′,5,5′-Trihydroxy-3,4′,6,7-tetramethoxyflavone	OCH_3	OH	OCH_3	OCH_3		OH	OCH_3	OH
	Isoflavones								
62	8-Carbomethoxy-5-hydroxy-6-methylisoflavone ($CDCl_3$)		OH	CH_3		CO_2CH_3			
63	8-Carbomethoxy-5-hydroxy-6-methylisoflavone (DMSO-d_6)		OH	CH_3		CO_2CH_3			
64	5-Methoxy-8-methylisoflavone		OCH_3			CH_3			
65	7-Acetyloxy-6-carbomethoxy-isoflavone			$OCOCH_3$	CO_2CH_3				
66	Formononetin 7-O-glucoside				O-glu			OCH_3	
67	5,7-Dimethoxy-8-methylisoflavone		OCH_3		OCH_3	CH_3			
68	Sphaerobioside		OH		O-rut			OH	
69	Sphaerobioside acetate		OAc		O-hexa-acetyl-rut			OAc	
70	Biochanin A		OH		OH			OCH_3	
71	Genistein 5-methyl ether		OCH_3		OH			OH	
72	Genistein 4′,5-dimethyl ether		OCH_3		OH			OCH_3	
73	5,7-Dihydroxy-4′-methoxy-8-methylisoflavone		OH		OH	CH_3		OCH_3	
74	4′,5,7-Trimethoxy-6-methyl-isoflavone		OCH_3	CH_3	OCH_3			OCH_3	
75	4′,5,7-Trimethoxy-8-methyl-isoflavone		OCH_3		OCH_3	CH_3		OCH_3	
76	2-Carbethoxy-5,7-dihydroxy-4′-methoxy-8-methylisoflavone (CCl_4)	CO_2CH_2-CH_3	OH		OH	CH_3		OCH_3	
77	2-Carbethoxy-5,7-dihydroxy-4′-methoxy-8-methylisoflavone ($CDCl_3$)	CO_2CH_2-CH_3	OH		OH	CH_3		OCH_3	
78	2-Carbethoxy-5,7-dihydroxy-4′-methoxy-6-methylisoflavone (CCl_4)	CO_2CH_2-CH_3	OH	CH_3	OH			OCH_3	

Spectrum No.	Flavonoids	Oxidation pattern											
		2	3	4	5	6	7	8	2′	3′	4′	5′	6′
79	2-Carbethoxy-5,7-dihydroxy-4′-methoxy-6-methylisoflavone ($CDCl_3$)	CO_2CH_2-CH_3			OH	CH_3	OH				OCH_3		
80	Texasin 7-O-glucoside					OH	O-glu				OCH_3		
81	Afrormosin					OCH_3	OH				OCH_3		
82	Pseudobaptisin (CCl_4)						O-rut			O-CH_2-O			
83	Pseudobaptisin (DMSO-d_6)						O-rut			O-CH_2-O			
84	5-Hydroxy-7,8-dimethoxyisoflavone				OH		OCH_3	OCH_3					
85	Orobol				OH		OH			OH	OH		
86	Orobol 7-O-glucoside				OH		O-glu			OH	OH		
87	Irisolidone				OH	OCH_3	OH				OCH_3		
88	Tectoridin				OH	OCH_3	O-glu				OH		
89	Irigenin				OH	OCH_3	OH			OCH_3	OCH_3	OH	
90	Iridin				OH	OCH_3	O-glu			OCH_3	OCH_3	OH	
	Dihydroflavonols												
91	Lecontin		O-glu				OH				OH		
92	Dihydrokaempferol		OH		OH		OH				OH		
93	Engeletin		O-rh		OH		OH				OH		
94	Dihydrofisetin		OH				OH			OH	OH		
95	Dihydroquercetin		OH		OH		OH			OH	OH		
96	Astilbin		O-rh		OH		OH			OH	OH		
97	Dihydrorobinetin		OH				OH			OH	OH	OH	
98	6-Hydroxyflavanone					OH							
99	Liquiritigenin						OH				OH		
100	Naringenin				OH		OH				OH		
101	Naringin				OH		O-neohesp				OH		
102	Sakuranetin				OH		OCH_3				OH		
103	Sakuranin				O-glu		OCH_3				OH		
104	Naringenin 4′,7-dimethyl ether				OH		OCH_3				OCH_3		
105	Homoeroidicytol				OH		OH			OCH_3	OH		
106	Hesperetin				OH		OH			OH	OCH_3		
107	Hesperidin				OH		O-rut			OH	OCH_3		
108	Neohesperidin				OH		O-neohesp			OH	OCH_3		
109	5,7-Dihydroxy-3′,4′-dimethoxyflavanone				OH		OH			OCH_3	OCH_3		
110	Hesperetin quinone				OH		OH			=O	OCH_3		=O

	Chalcones									
111	4,4′-Dimethoxychalcone		OCH_3					OCH_3		
112	2′,4-Dihydroxy-4′-methoxy-chalcone		OH			OH		OCH_3		
113	2′-Hydroxy-4,4′-dimethoxy-chalcone		OCH_3			OH		OCH_3		
114	2′,3,4,4′-Tetrahydroxychalcone	OH	OH			OH		OH		
115	2′,4′-Dihydroxy-3,4-dimethoxychalcone	OCH_3	OCH_3			OH		OH		
116	2′-Hydroxy-4,4′,6′-trimethoxychalcone		OCH_3			OH		OCH_3		OCH_3
117	2′,4,4′,6′-Tetramethoxychalcone		OCH_3	OCH_3		OCH_3		OCH_3		
118	2′,3′,4′,5′,6′-Pentamethoxychalcone					OCH_3	OCH_3	OCH_3	OCH_3	OCH_3
119	2′,3,4,4′,6′-Pentamethoxychalcone	OCH_3	OCH_3			OCH_3		OCH_3		OCH_3
120	Chalcone α,β-epoxide									
121	2′,4,4′,6′-Tetramethoxychalcone α,β-epoxide		OCH_3			OCH_3		OCH_3		OCH_3
	Aurones									
122	4′,7-Dihydroxyaurone				OH			OH		
123	6-Hydroxy-4′-methoxyaurone			OH				OCH_3		
124	4,4′,6-Trimethoxyaurone		OCH_3					OCH_3		OCH_3
125	3′,4′,6,7-Tetrahydroxyaurone			OH	OH		OH	OH		
126	3′,4′,6,7-Tetramethoxyaurone			OCH_3	OCH_3		OCH_3	OCH_3		
127	4,6,7-Trimethoxy-4′-benzyloxyaurone		OCH_3	OCH_3	OCH_3			benzyl-oxy		
	Biflavonyls									
128	Amentoflavone			{OH OH}	{OH OH}			{OH OH}		

[a] Abbreviations: gal = galactosyl; glu = glucosyl; neohesp = neohesperidosyl; rh = rhamnosyl; rh-glu = rhamnoglucosyl; rut = rutinosyl.
[b] Most of the NMR spectra are for fully trimethylsilylated flavonoids, thus for these derivatives OH refers to $-O-Si(CH_3)_3$.

NMR of TMS Ether of Tectochrysin in CCl_4

1

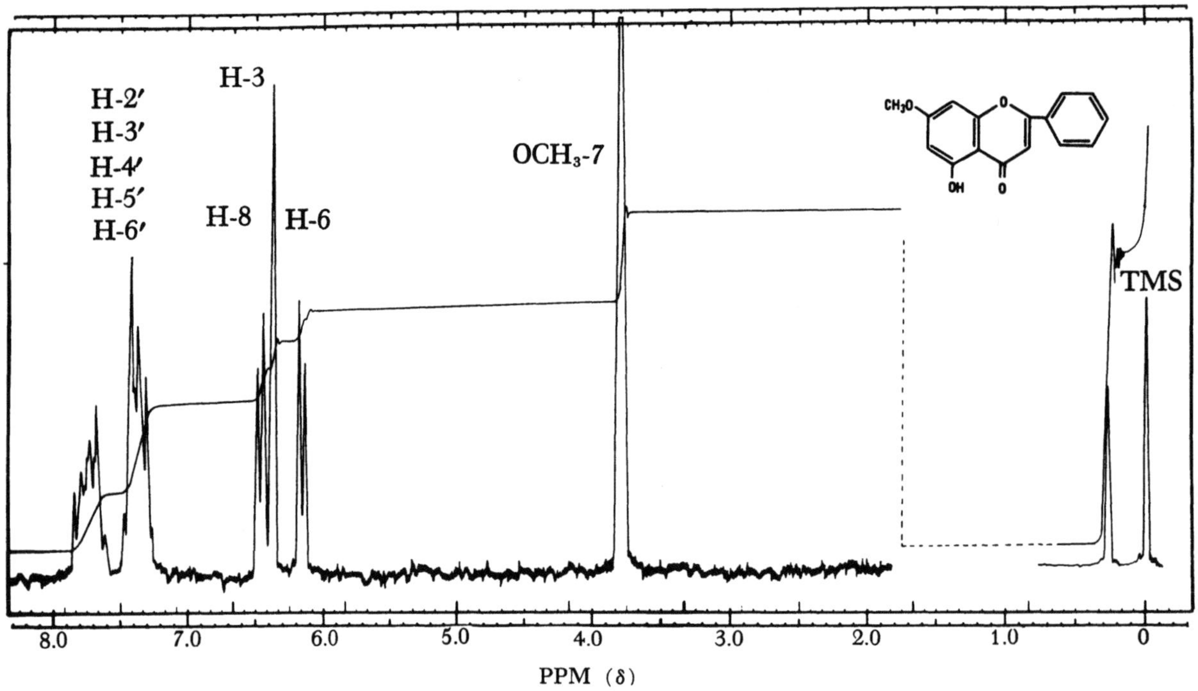

NMR of TMS Ether of Apigenin in CCl_4

2

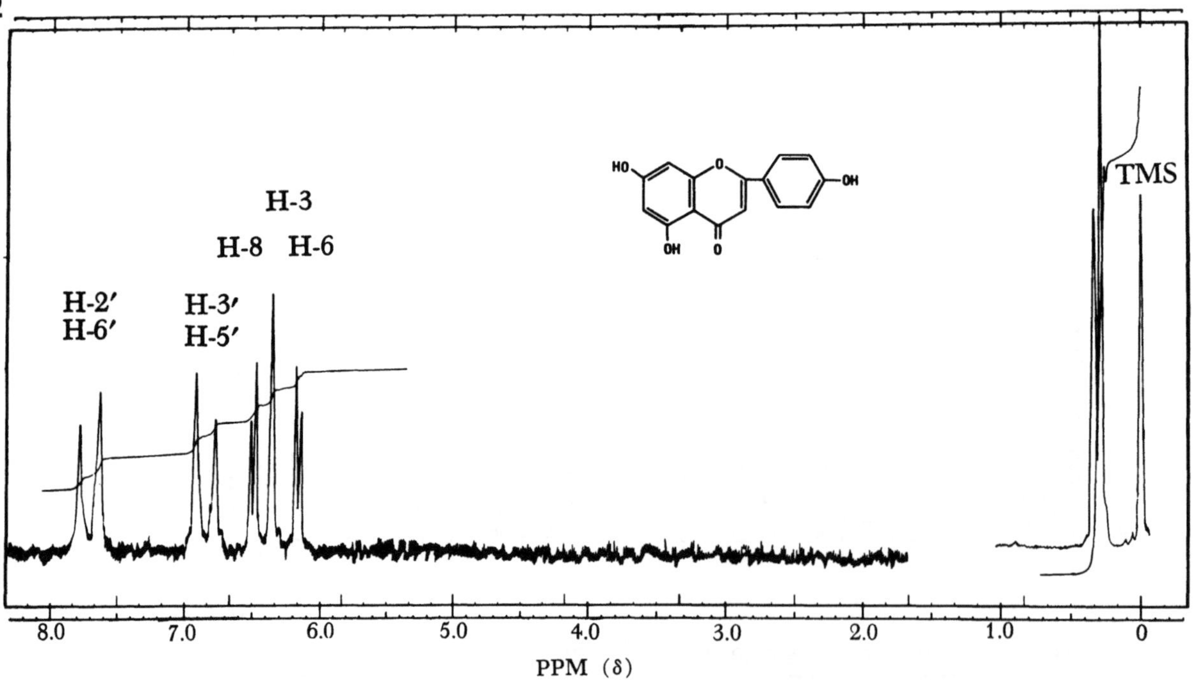

NMR of TMS Ether of Apigenin 7-O-Neohesperidoside in CCl_4

3

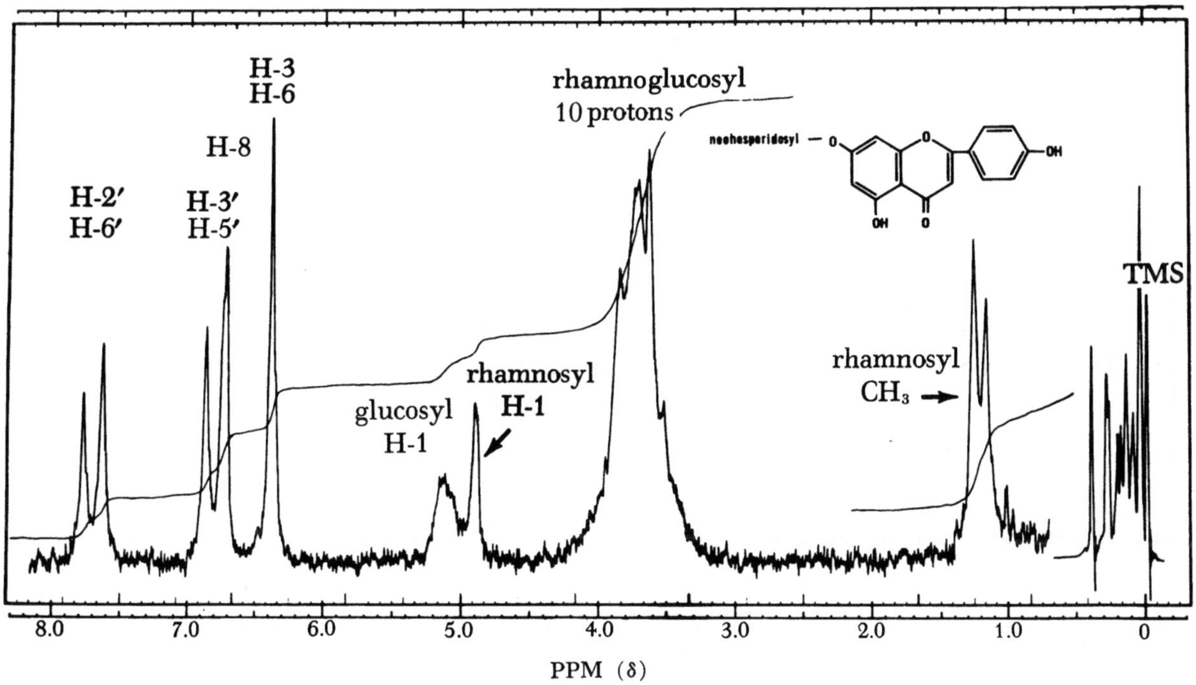

NMR of Apigenin 7-O-Neohesperidoside Acetate in $CDCl_3$

4

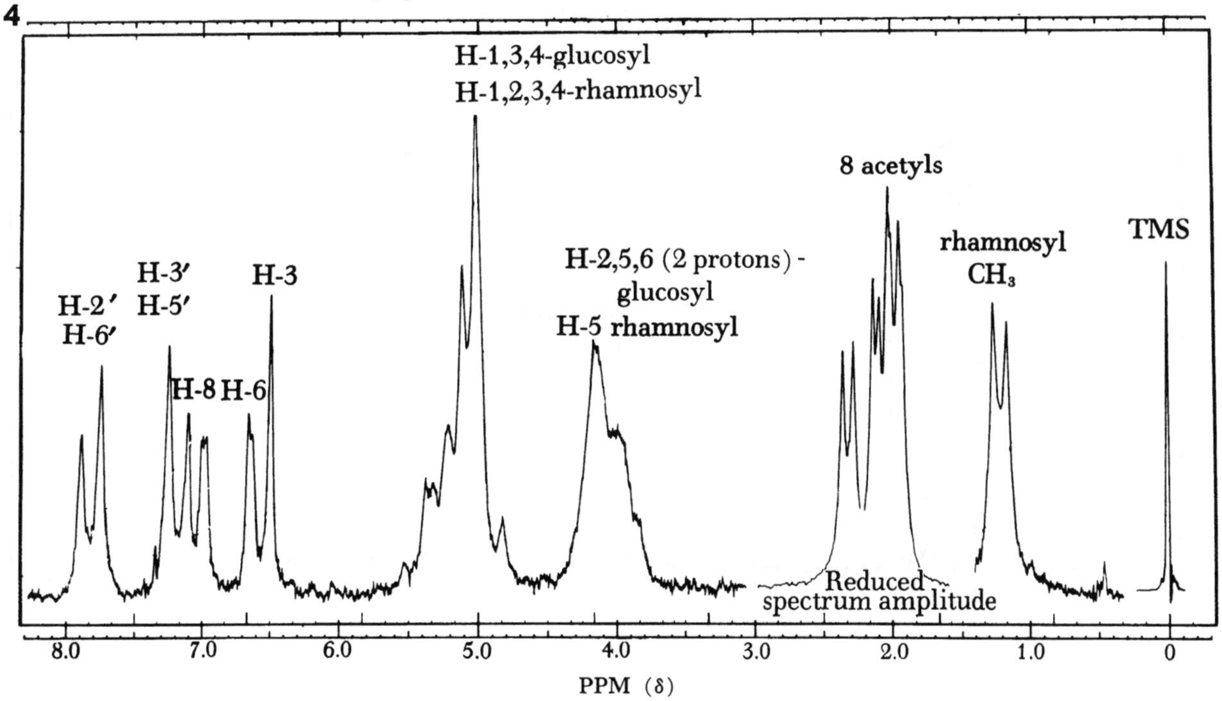

NMR of TMS Ether of Apigenin 7-O-Glucoside in CCl_4

5

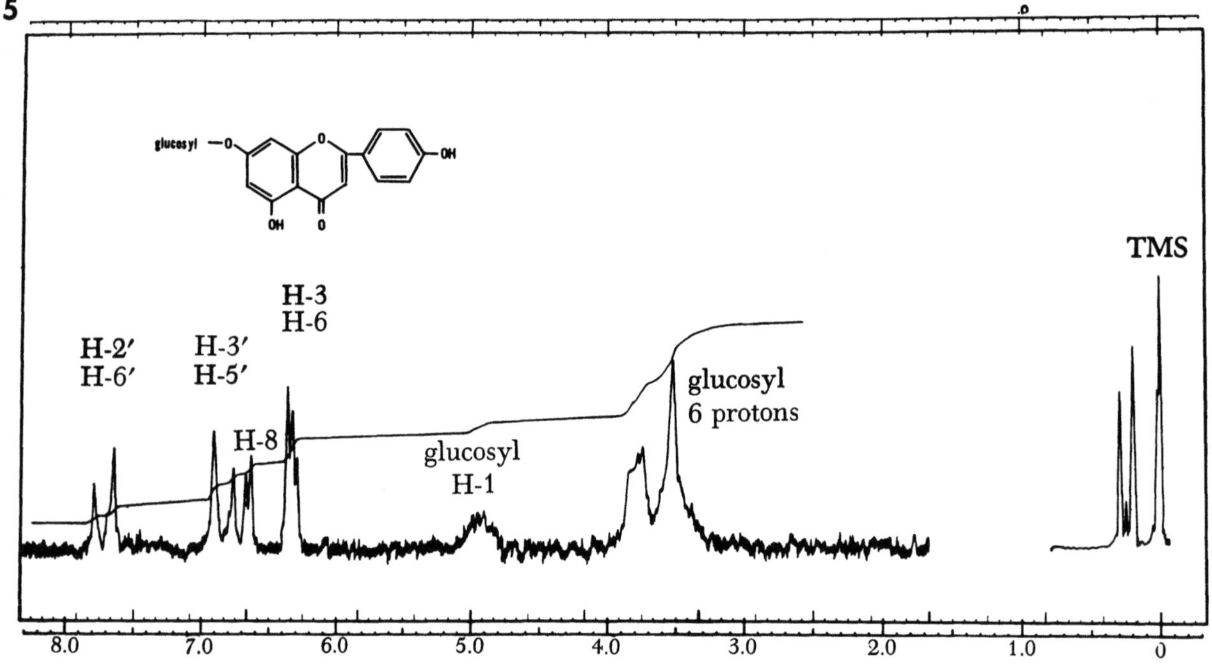

NMR of TMS Ether of Vitexin in CCl_4

6

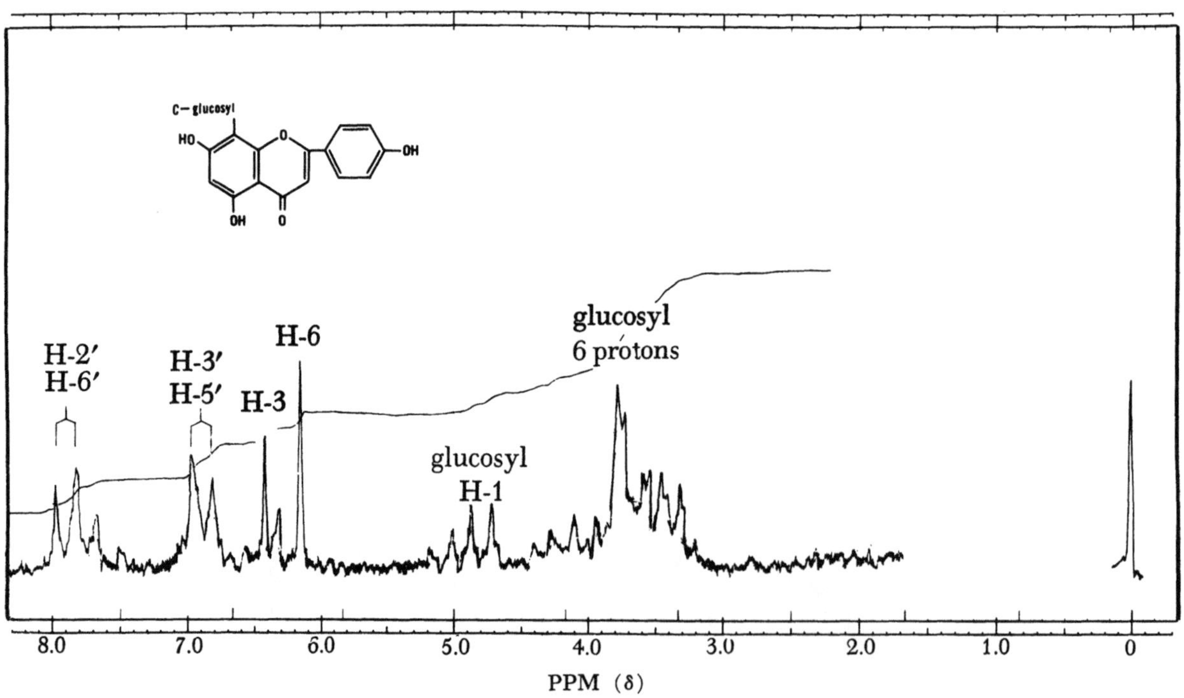

7

NMR of TMS Ether of Isovitexin in CCl_4

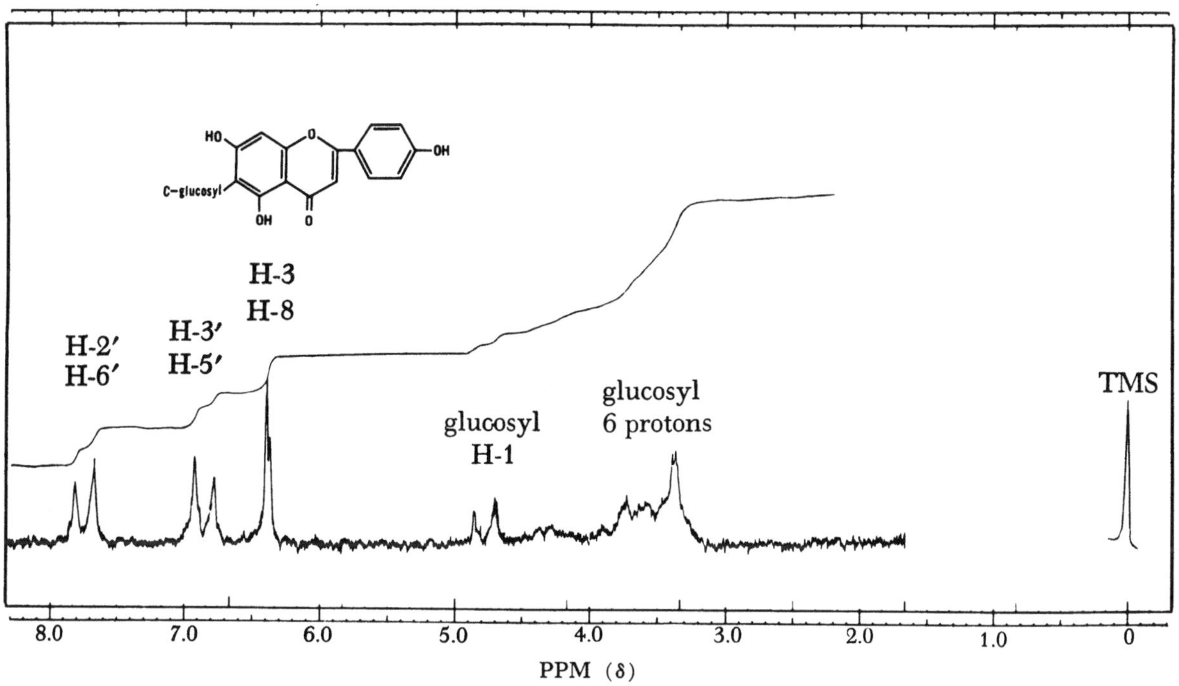

8

NMR of TMS Ether of Saponarin in CCl_4

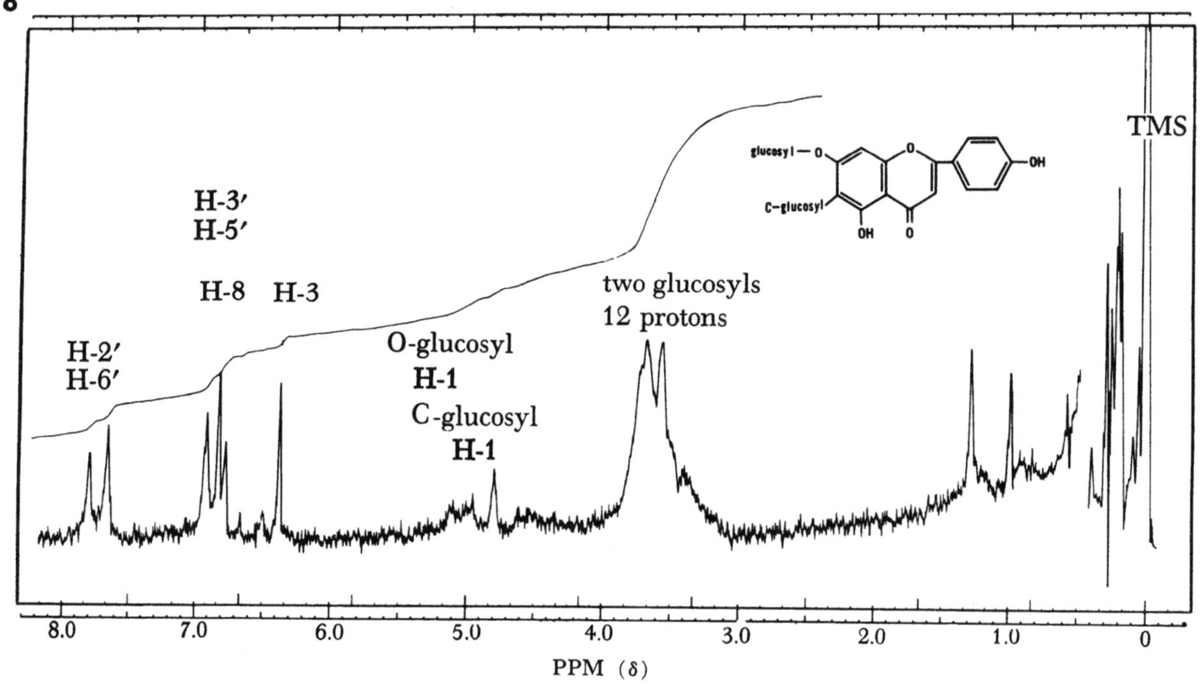

NMR of TMS Ether of Cytisoside in CCl_4

9

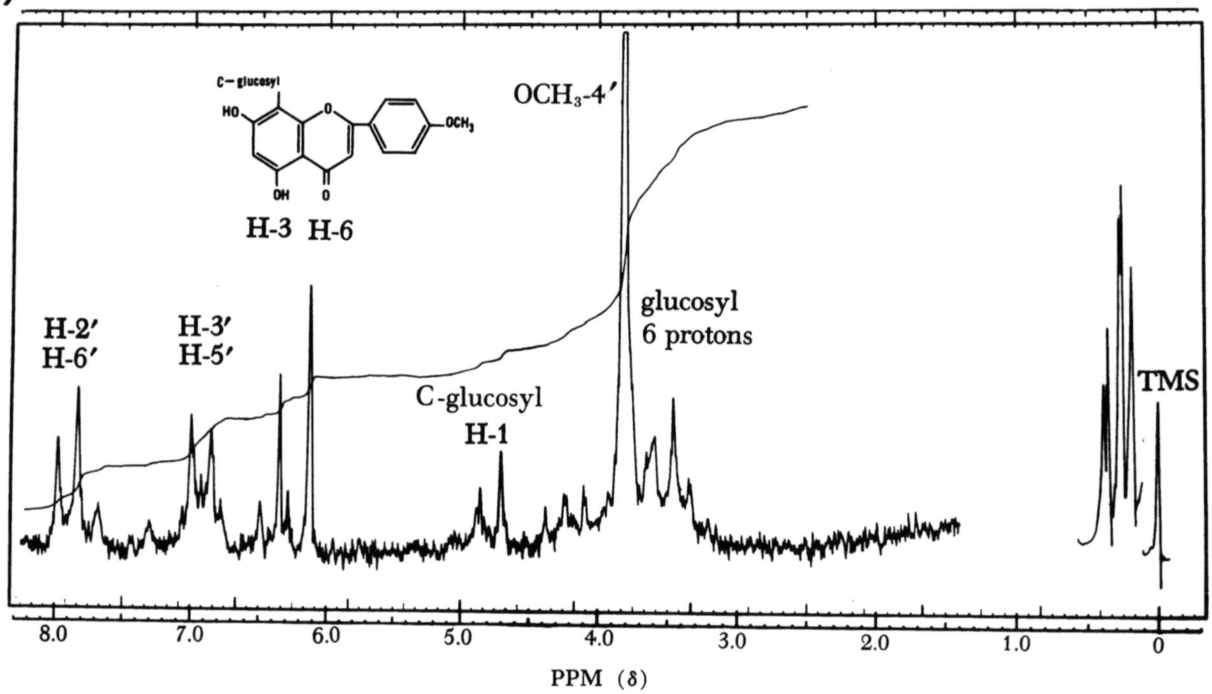

NMR of TMS Ether of 5,7-Dihydroxy-2′-methoxyflavone in CCl_4

10

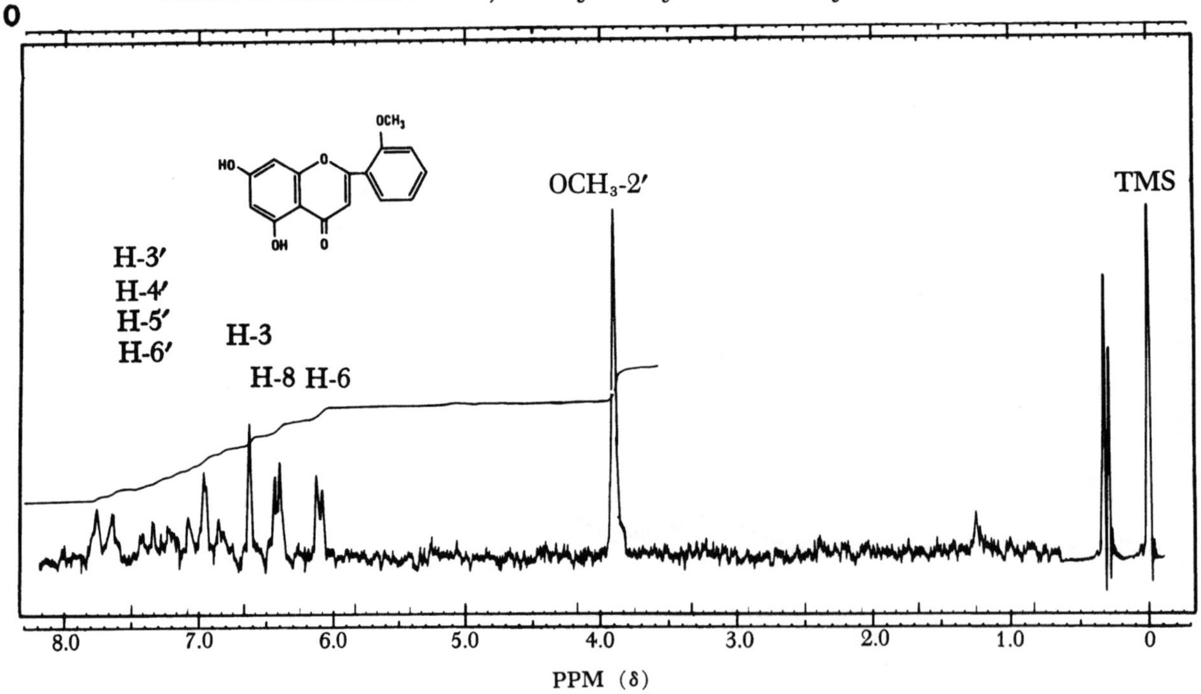

NMR of TMS Ether of Luteolin 7-O-Rutinoside in CCl_4

11

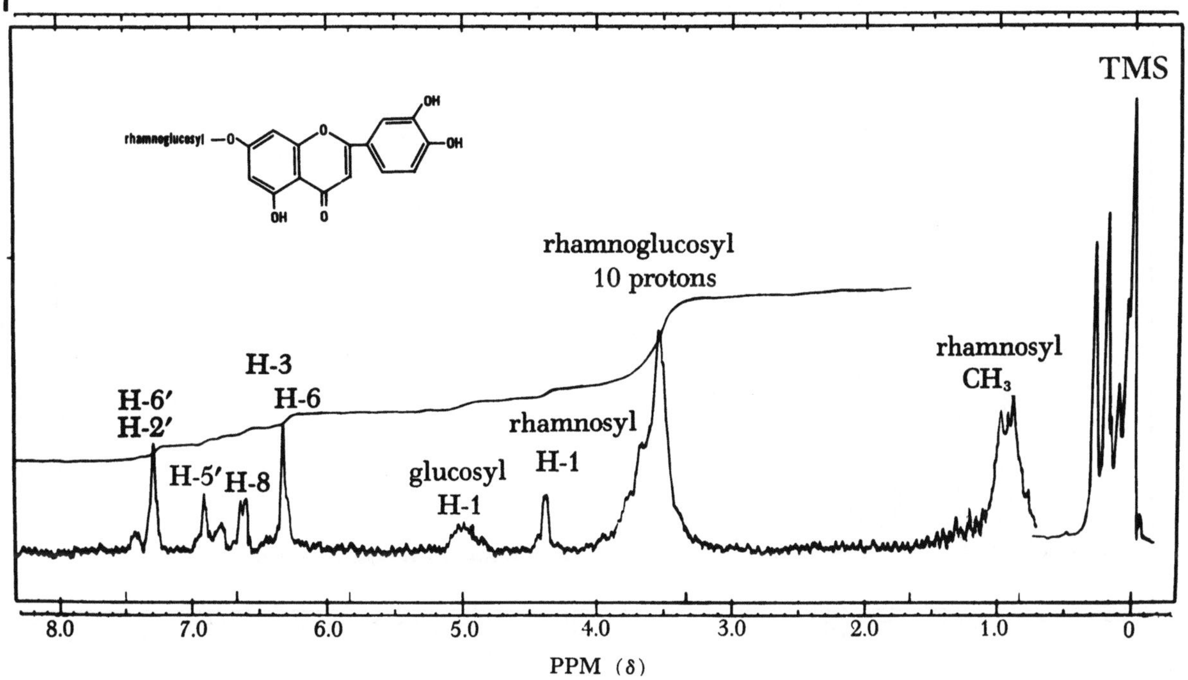

NMR of TMS Ether of Luteolin 7-O-Glucoside in CCl_4

12

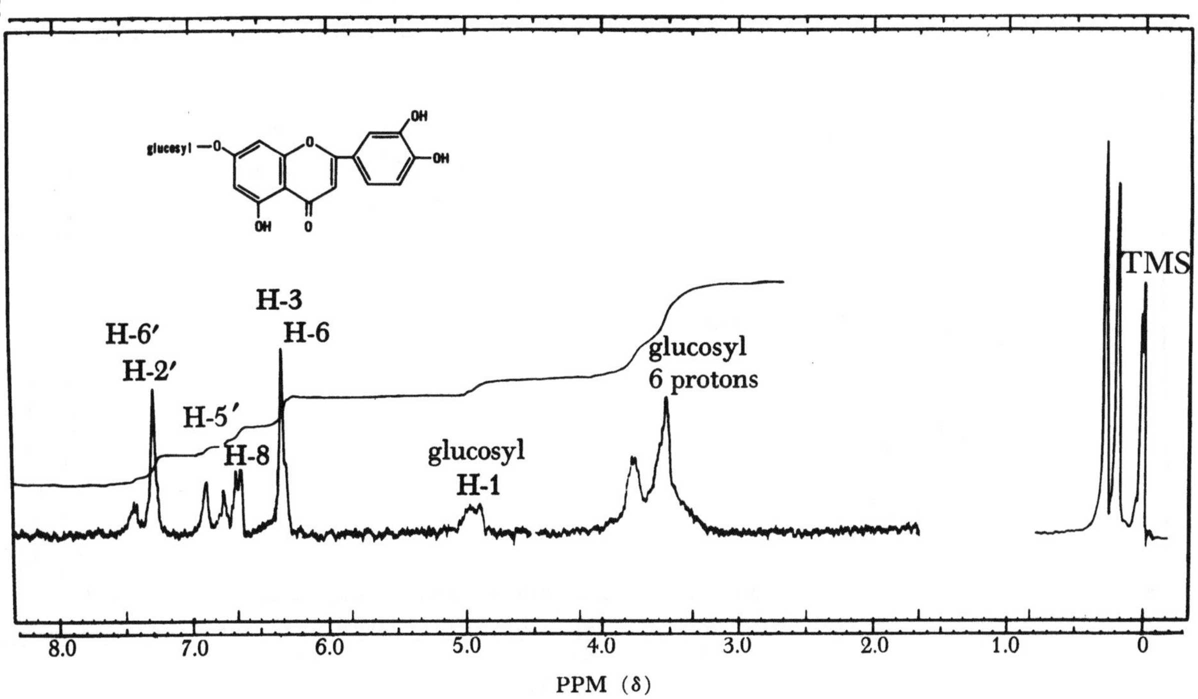

13 NMR of TMS Ether of Luteolin in CCl_4

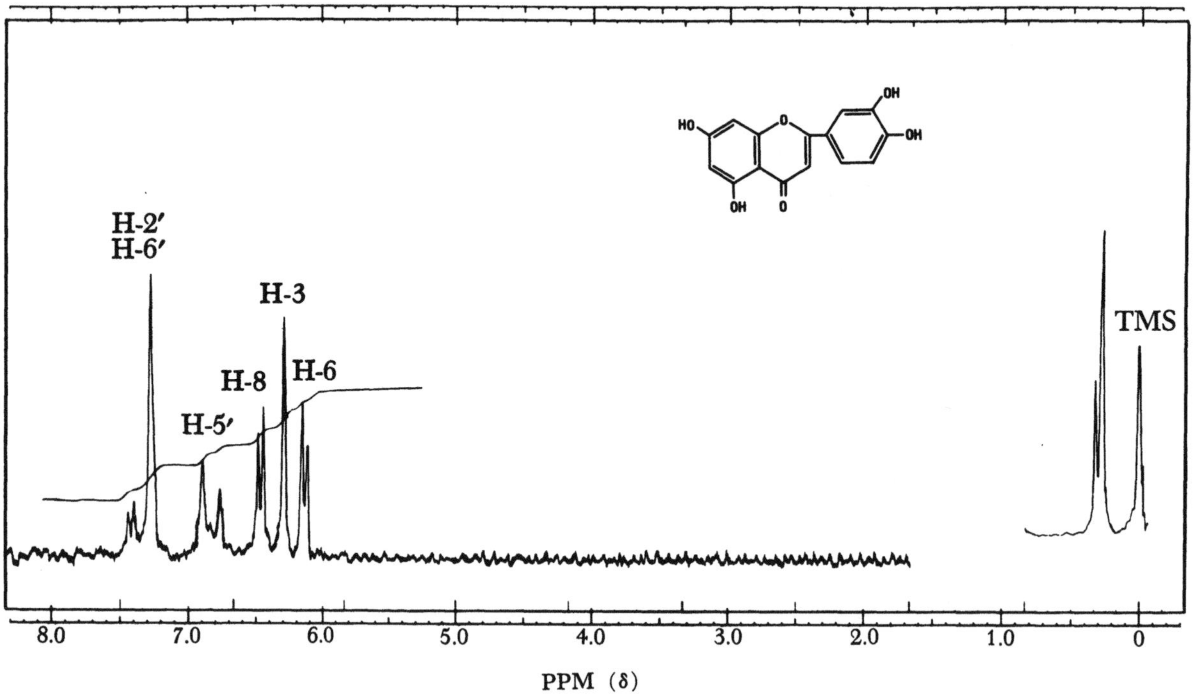

14 NMR of Luteolin 3′,4′,7 -TMS Ether in CCl_4

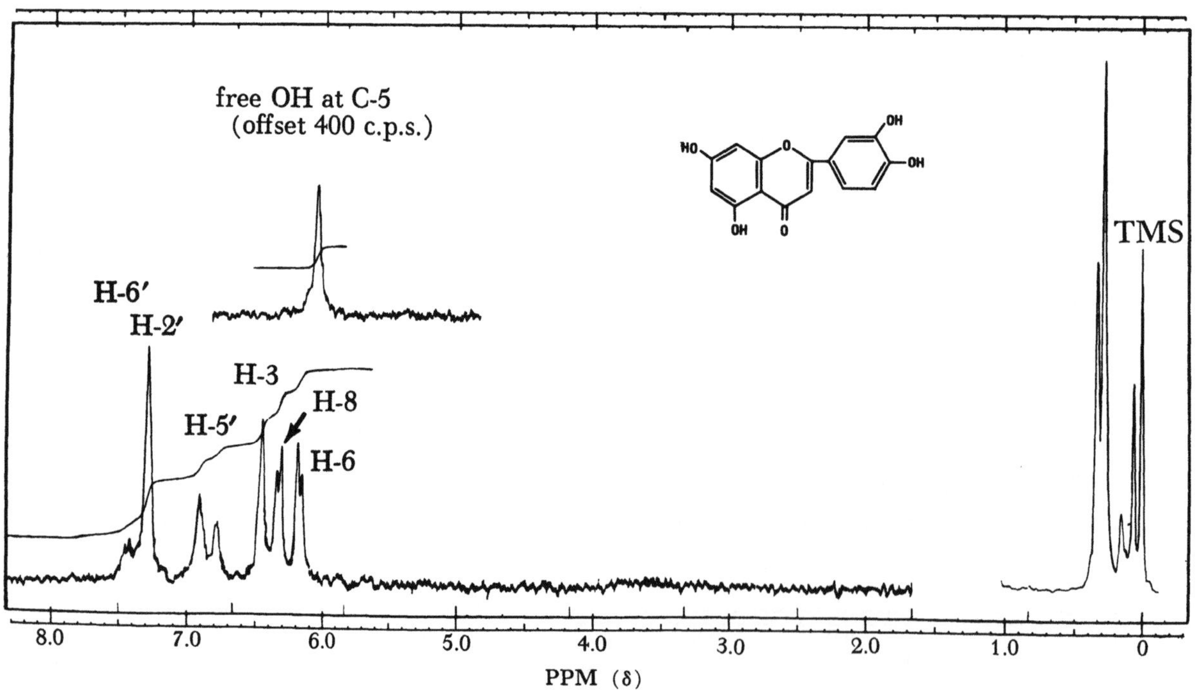

NMR of TMS Ether of 5,7-Dihydroxy-3′,4′-dimethoxyflavone in CCl_4

15

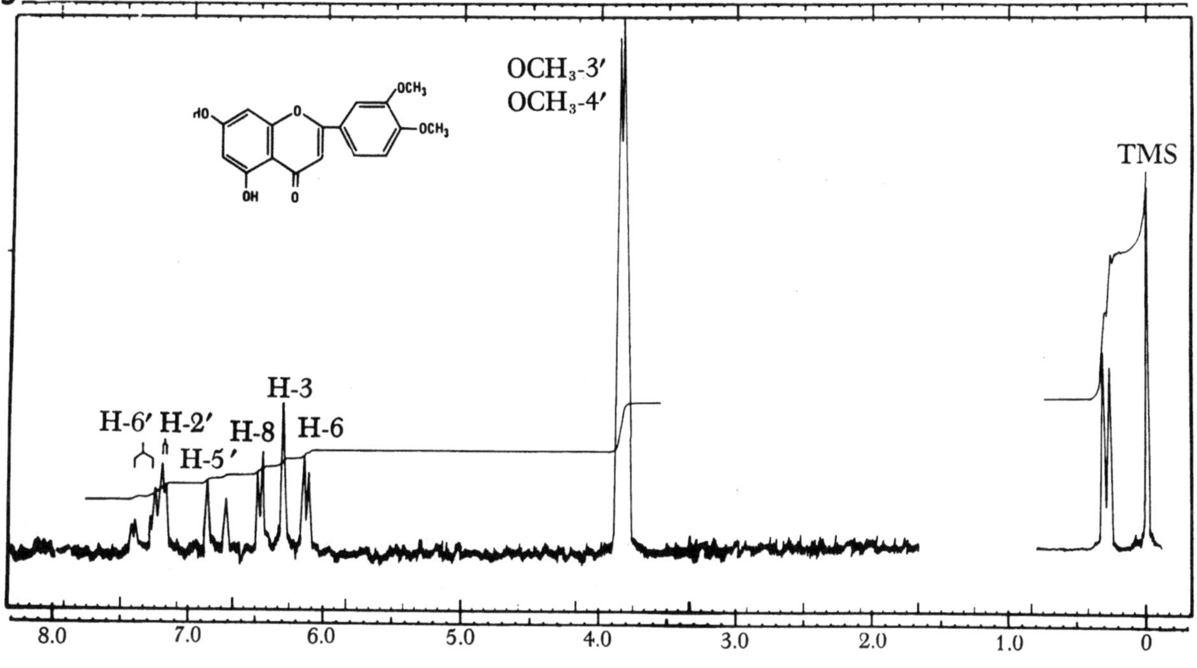

NMR of TMS Ether of Diosmetin in CCl_4

16

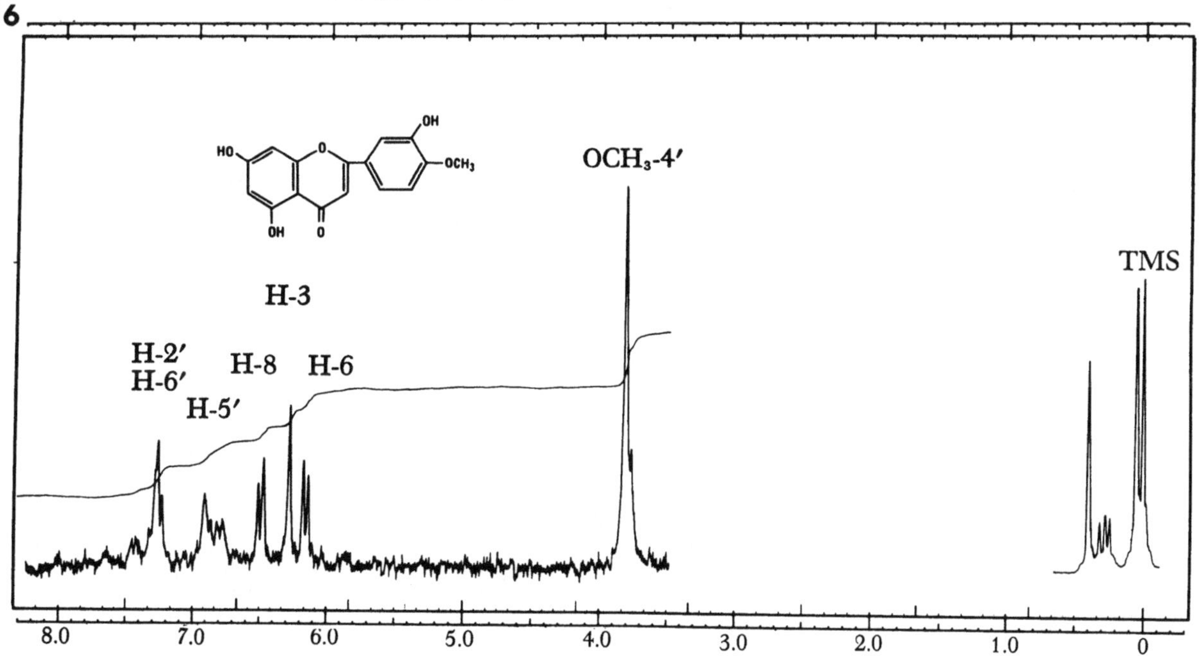

NMR of TMS Ether of Diosmin in CCl_4

17

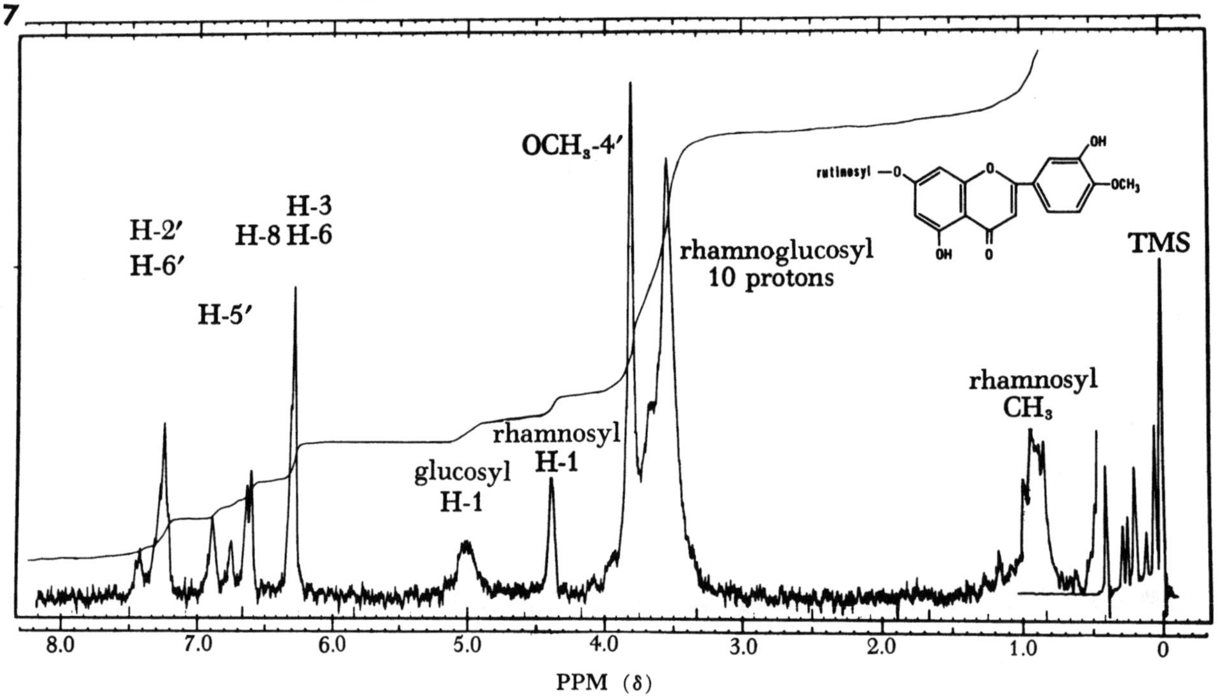

NMR of Diosmin Acetate in $CDCl_3$

18

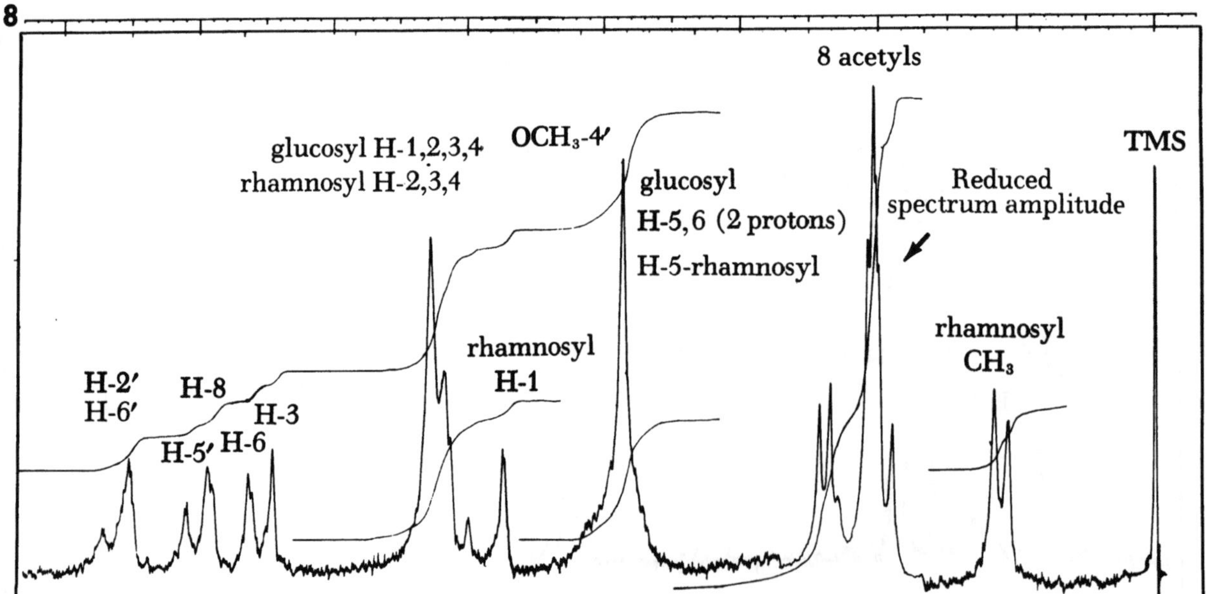

19

NMR of Zapotinin in $CDCl_3$

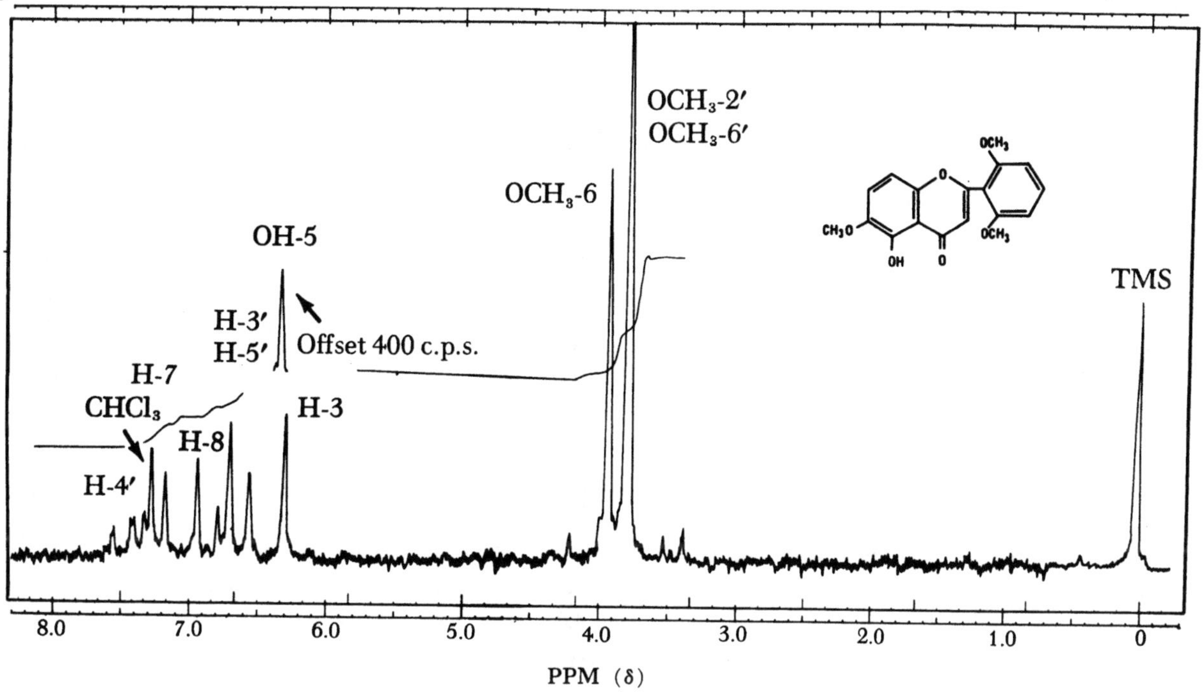

20

NMR of Zapotin in $CDCl_3$

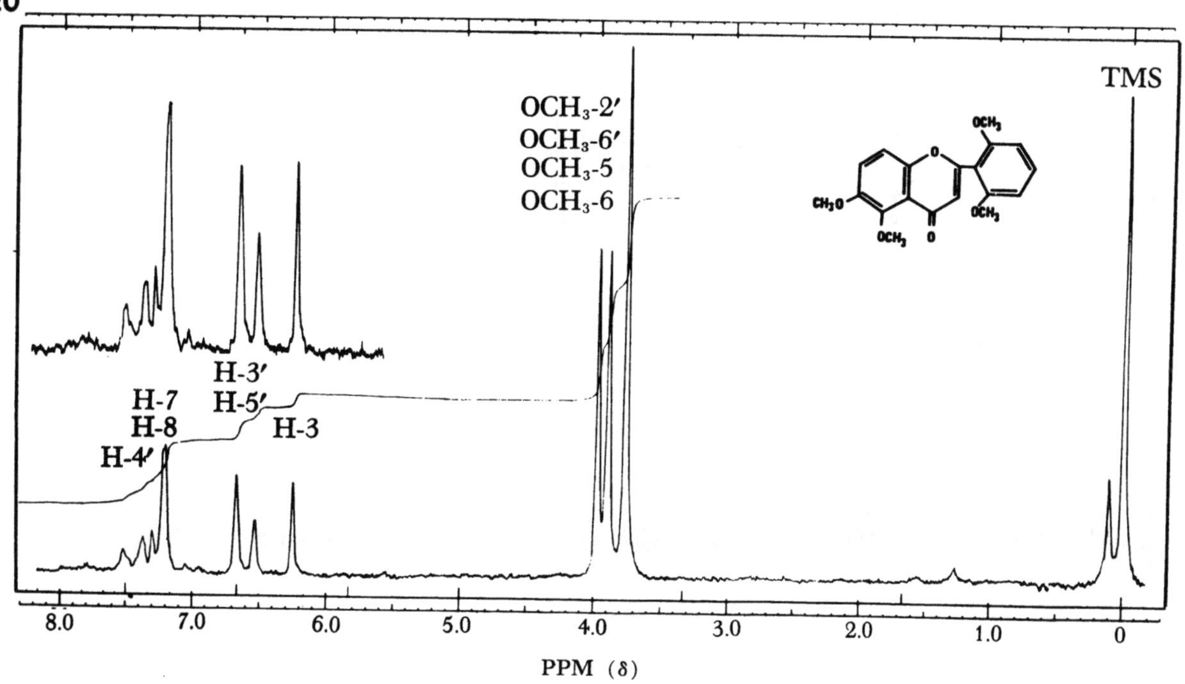

NMR of TMS Ether of Xanthomicrol in CCl_4

21

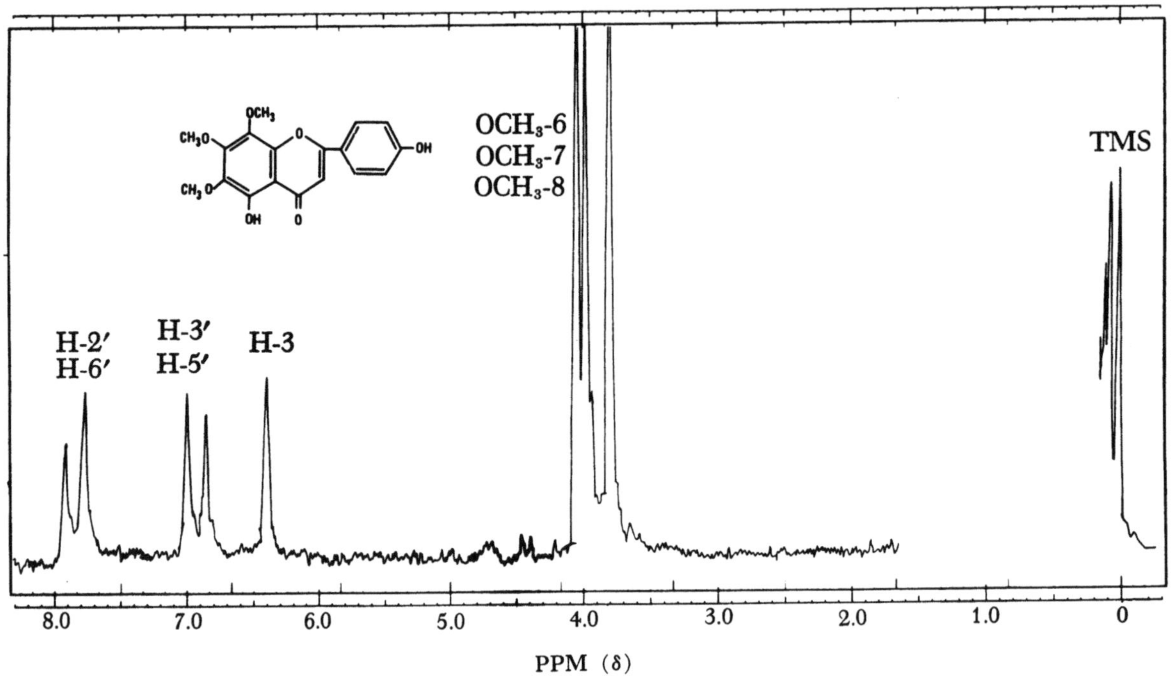

NMR of TMS Ether of Demethoxysudachitin in CCl_4

22

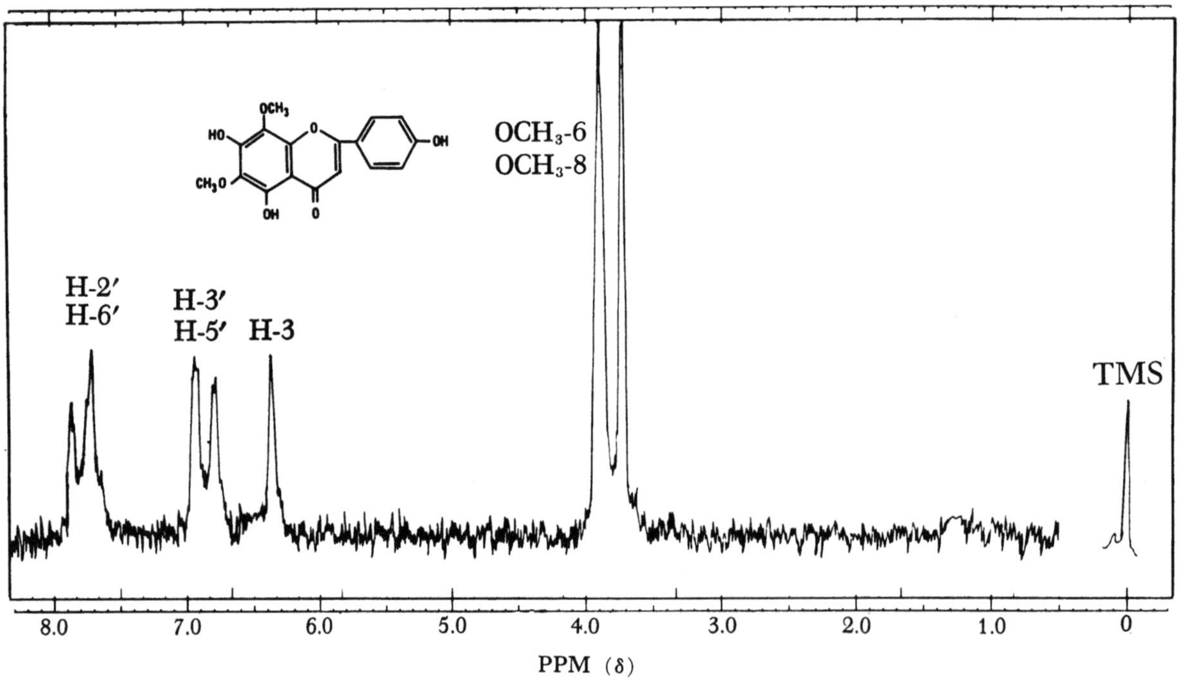

NMR of TMS Ether of Hymenoxin in CCl_4

23

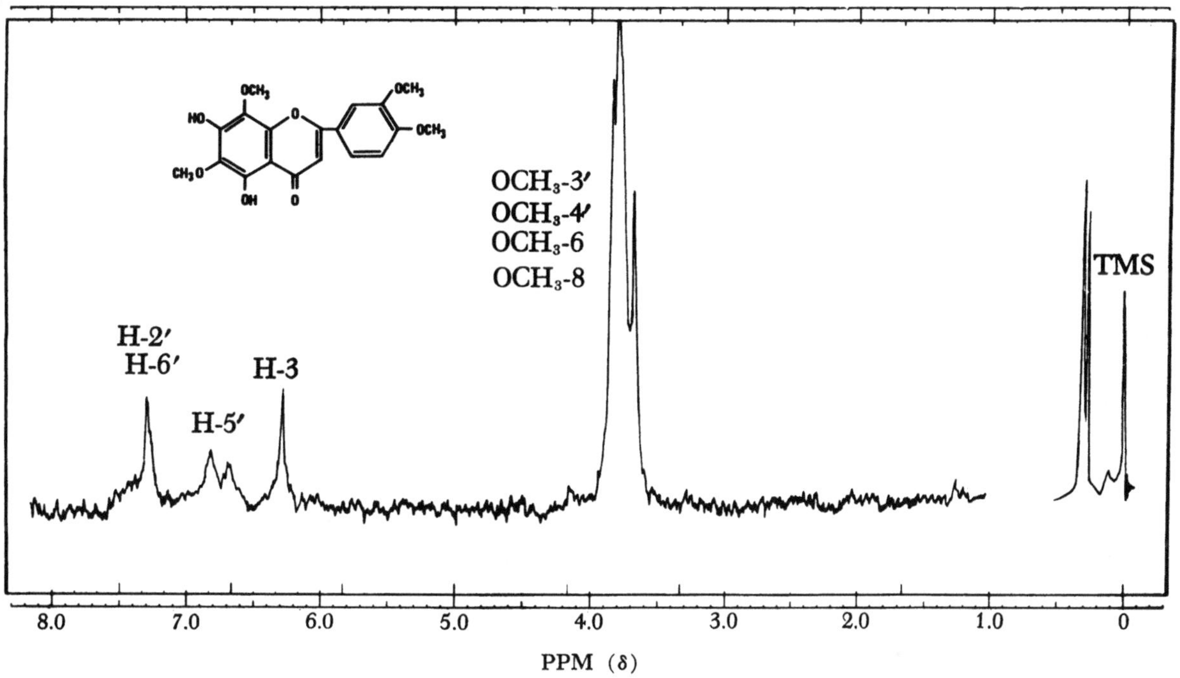

NMR of Scaposin in DMSO (Deuterated)

24

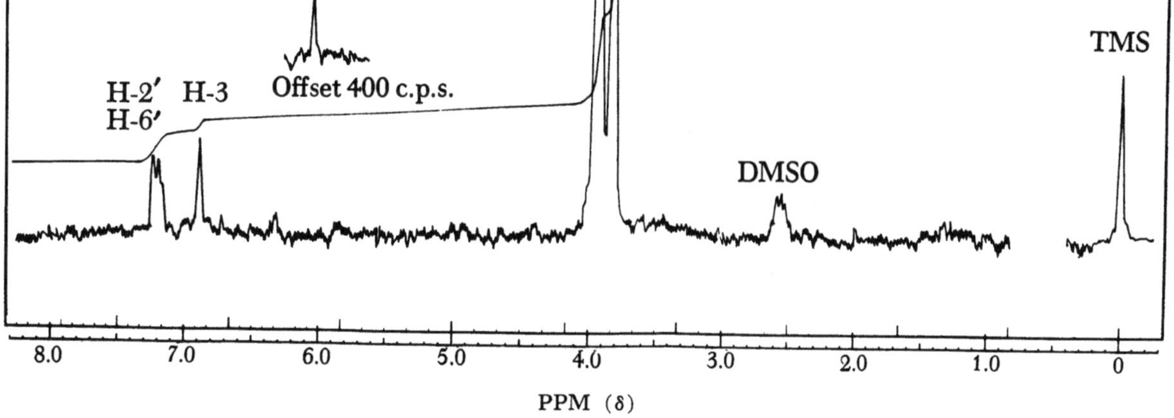

NMR of TMS Ether of Kaempferol in CCl_4

25

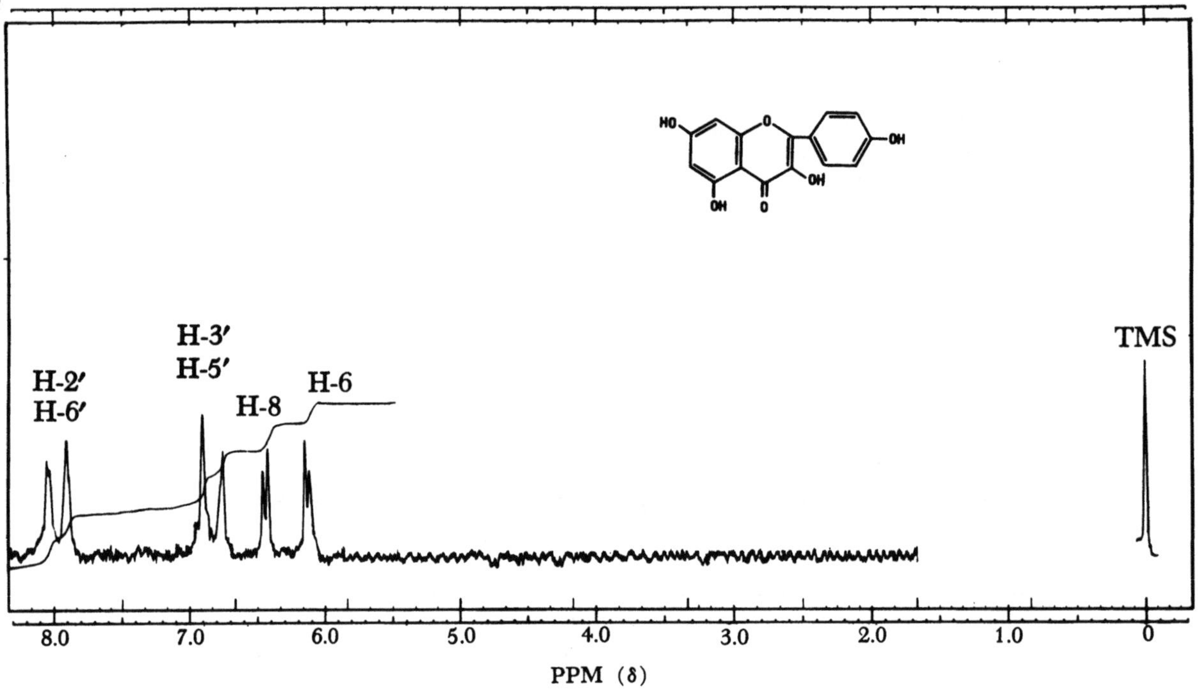

NMR of TMS Ether of Kaempferol 3-O-Robinoside 7-O-Rhamnoside in CCl_4

26

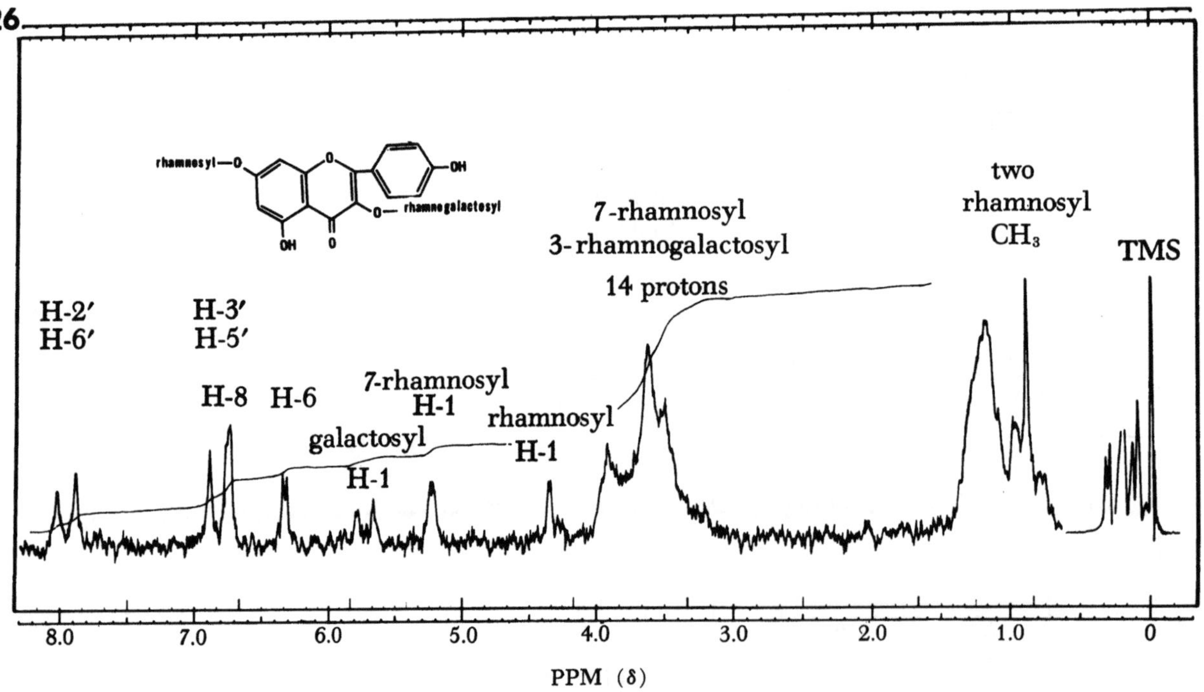

NMR of TMS Ether of Kaempferol 7-O-Neohesperidoside in CCl_4

27

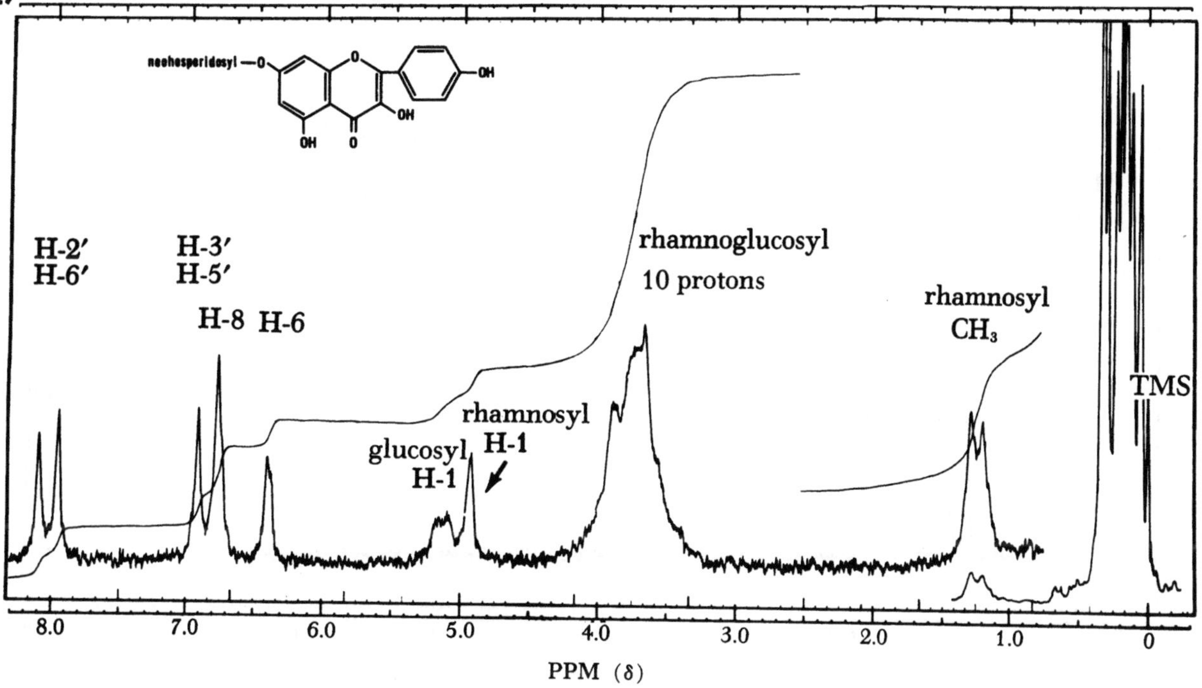

NMR of Kaempferol 7-O-Neohesperidoside Acetate in $CDCl_3$

28

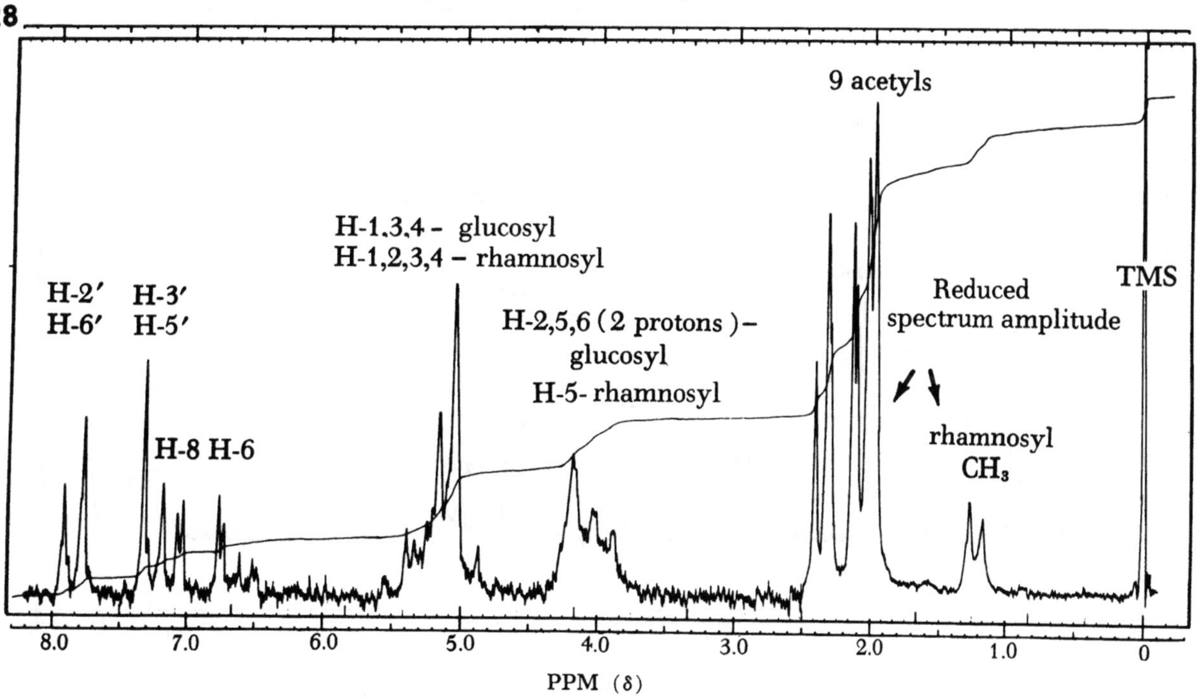

NMR of TMS Ether of Quercetin in CCl_4

29

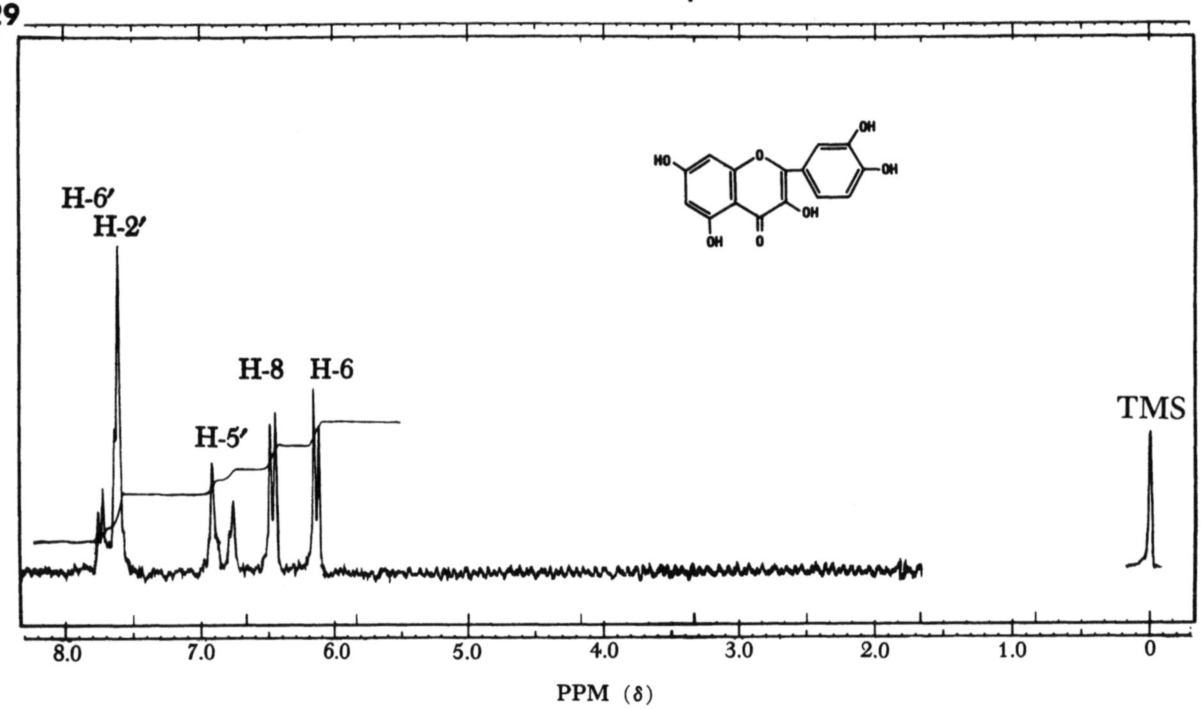

NMR of TMS Ether of Quercitrin in CCl_4

30

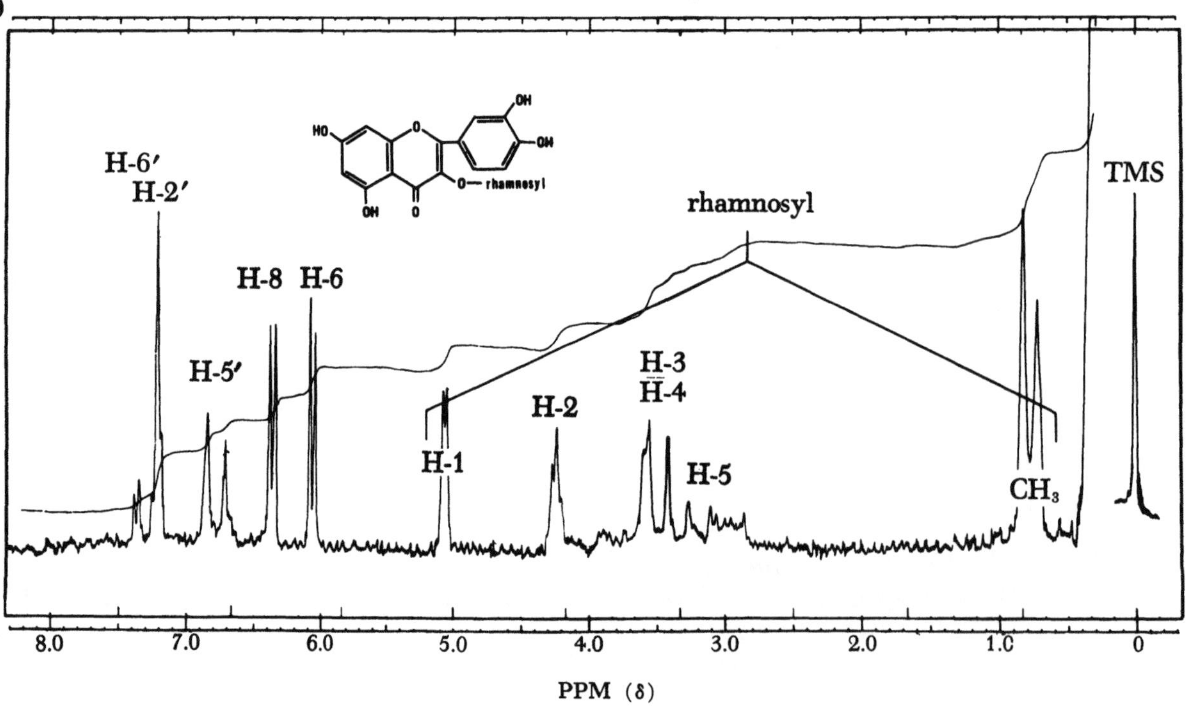

NMR of TMS Ether of Hyperin in CCl_4

31

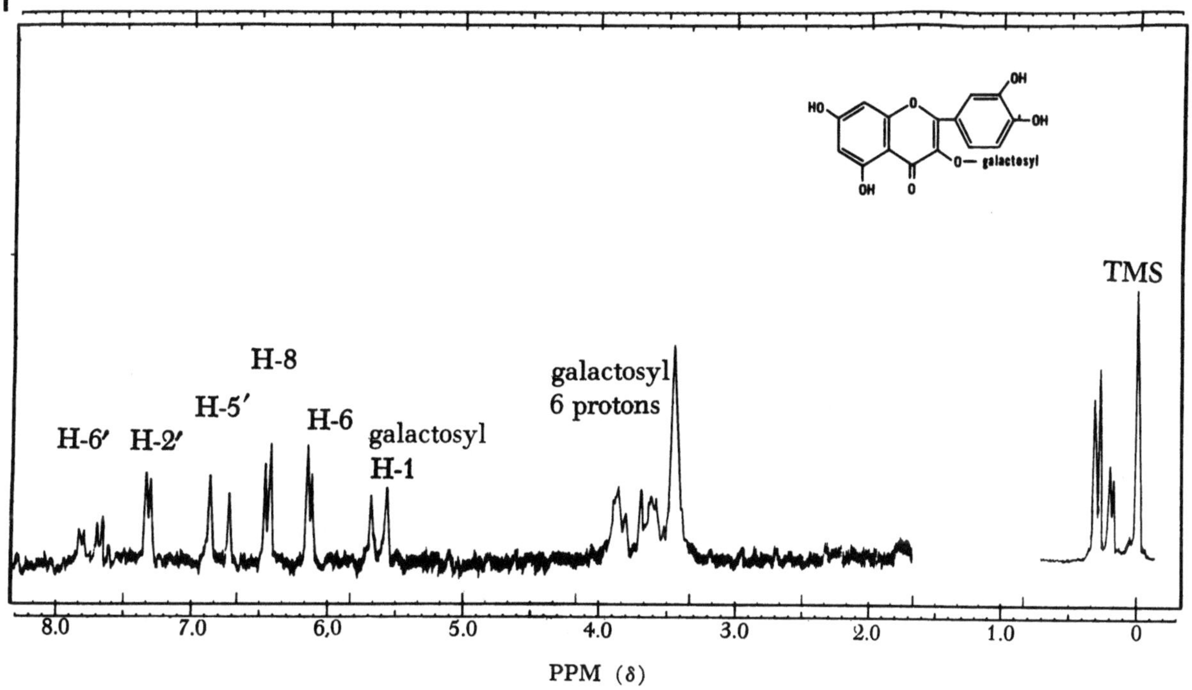

NMR of TMS Ether of Quercetin 3,7-O-Diglucoside in CCl_4

32

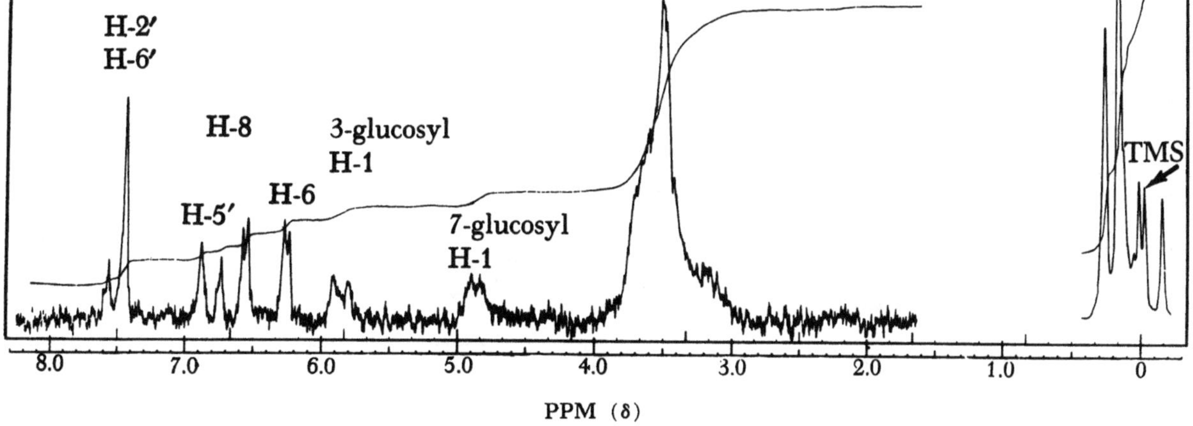

NMR of TMS Ether of Rutin in CCl_4

33

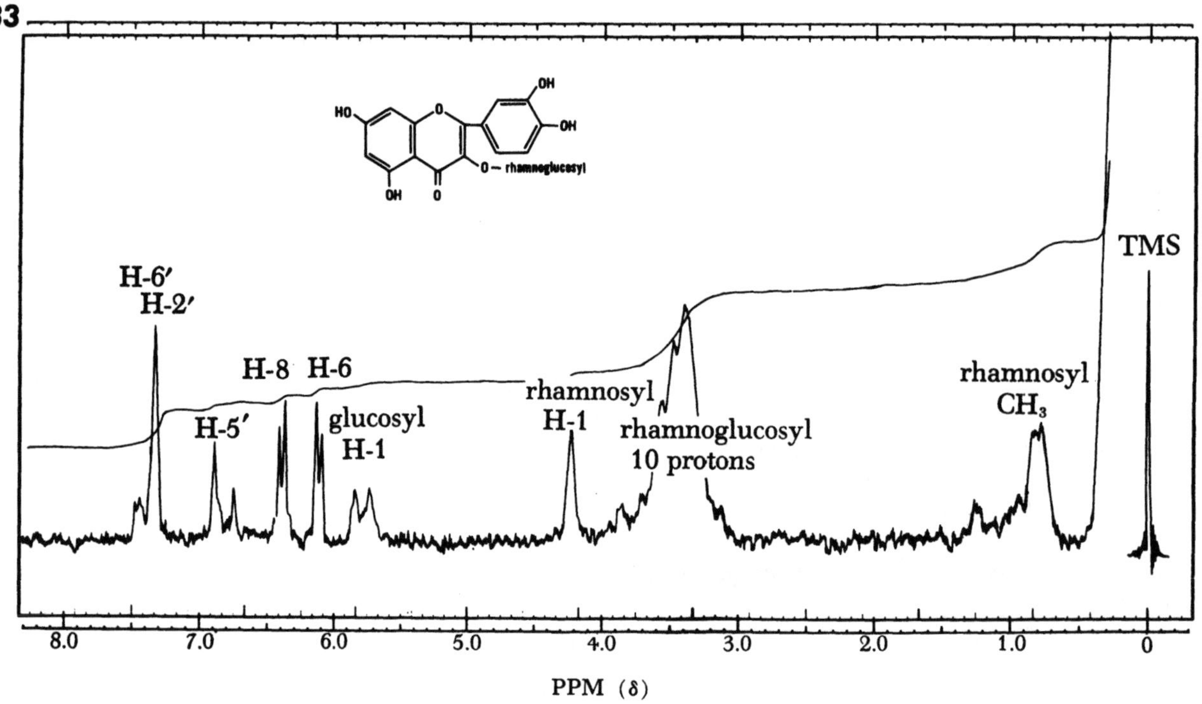

NMR of Rutin Acetate in $CDCl_3$

34

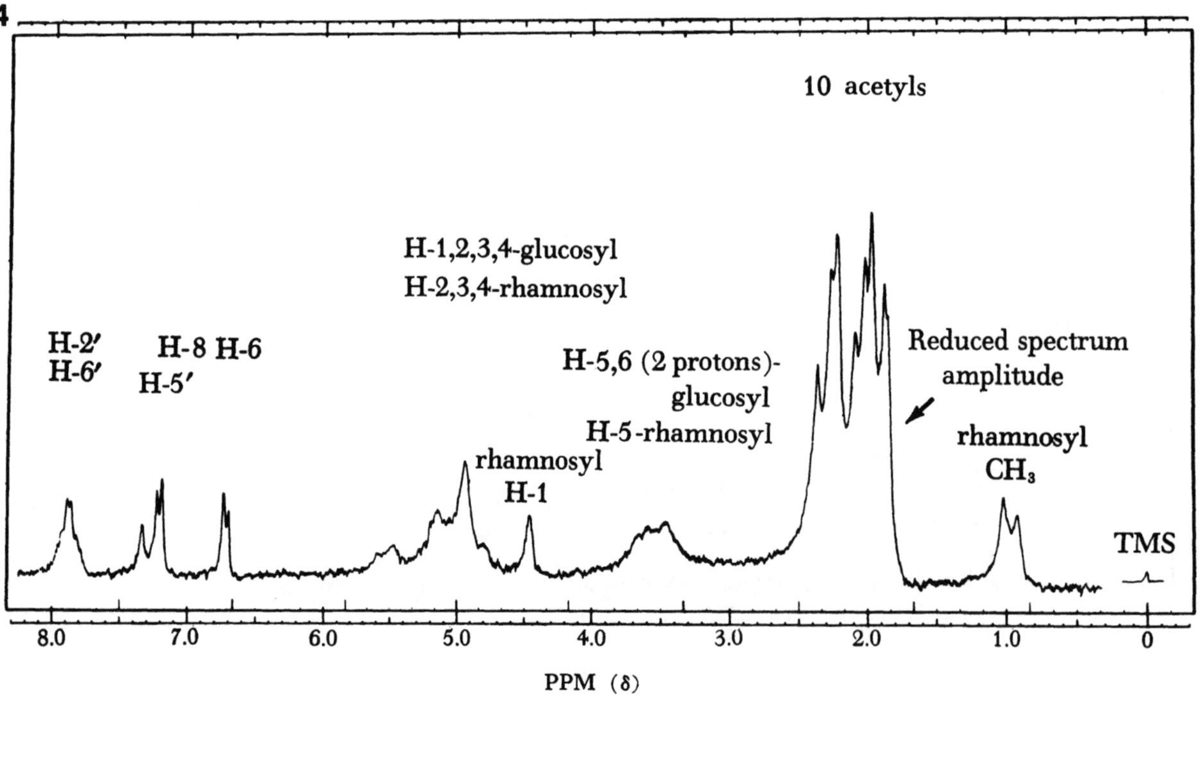

NMR of TMS Ether of Quercetin 3-O-Glucoside 7-O-Rutinoside in CCl_4

35

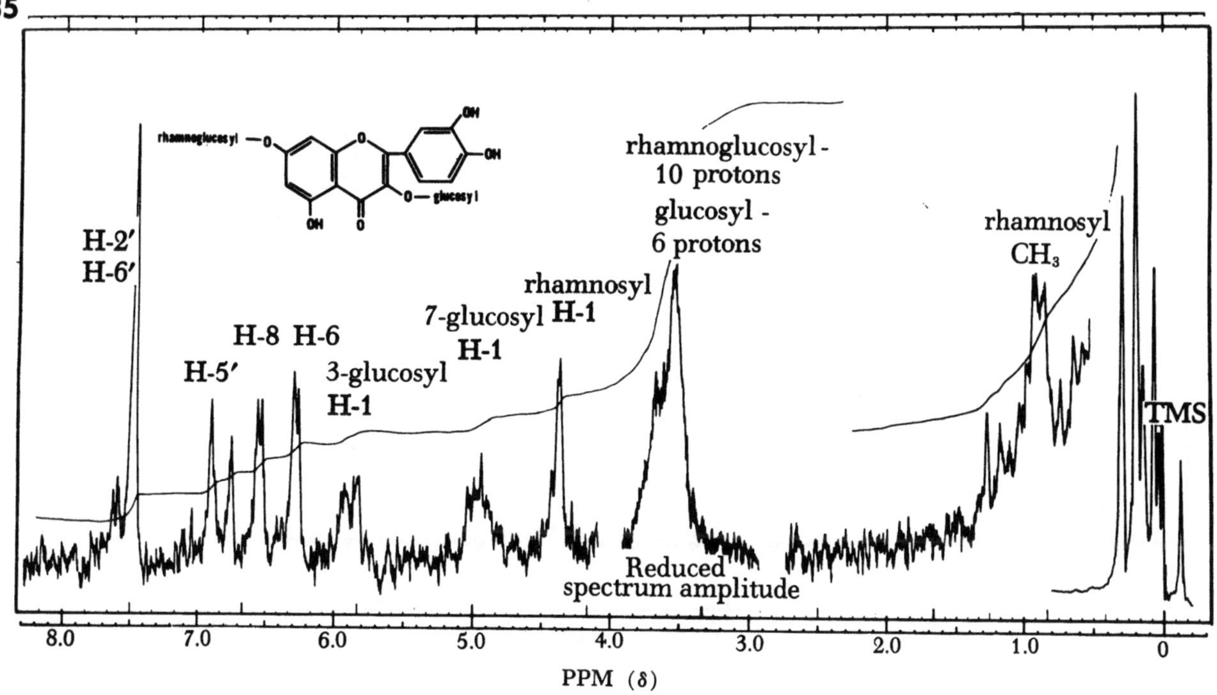

NMR of TMS Ether of Quercetin 3-O-Glucoside 7-O-Rhamnoside in CCl_4

36

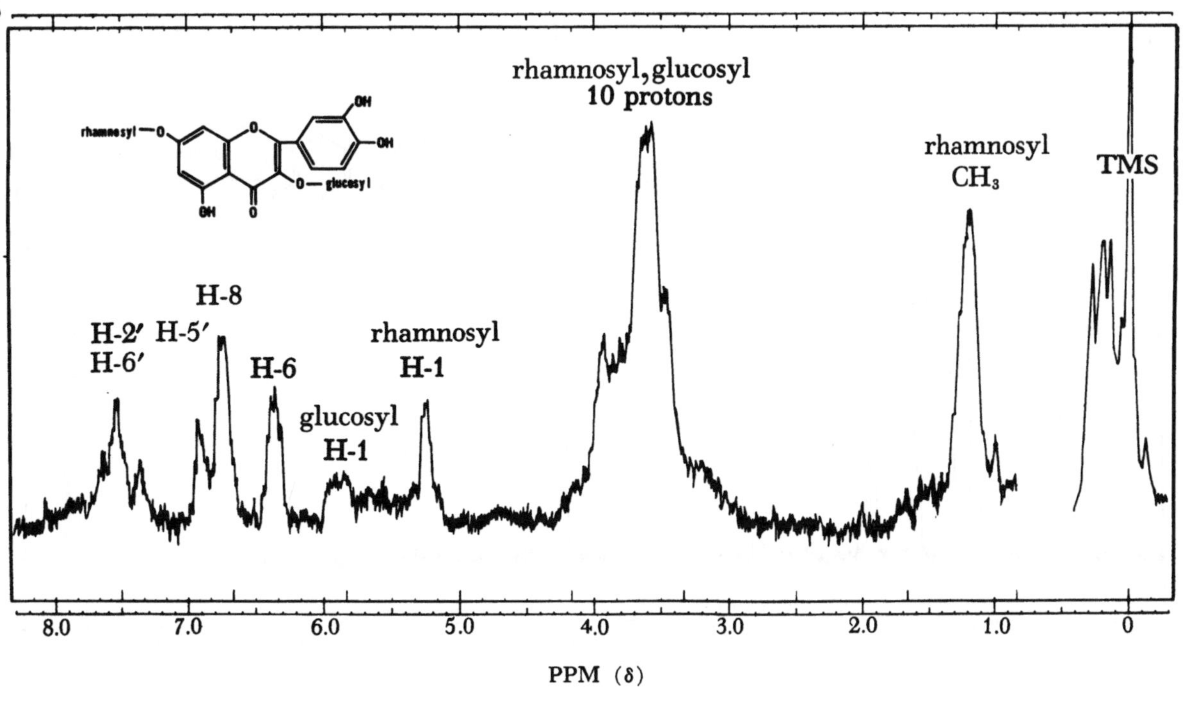

NMR of TMS Ether of Rhamnetin in CCl_4

37

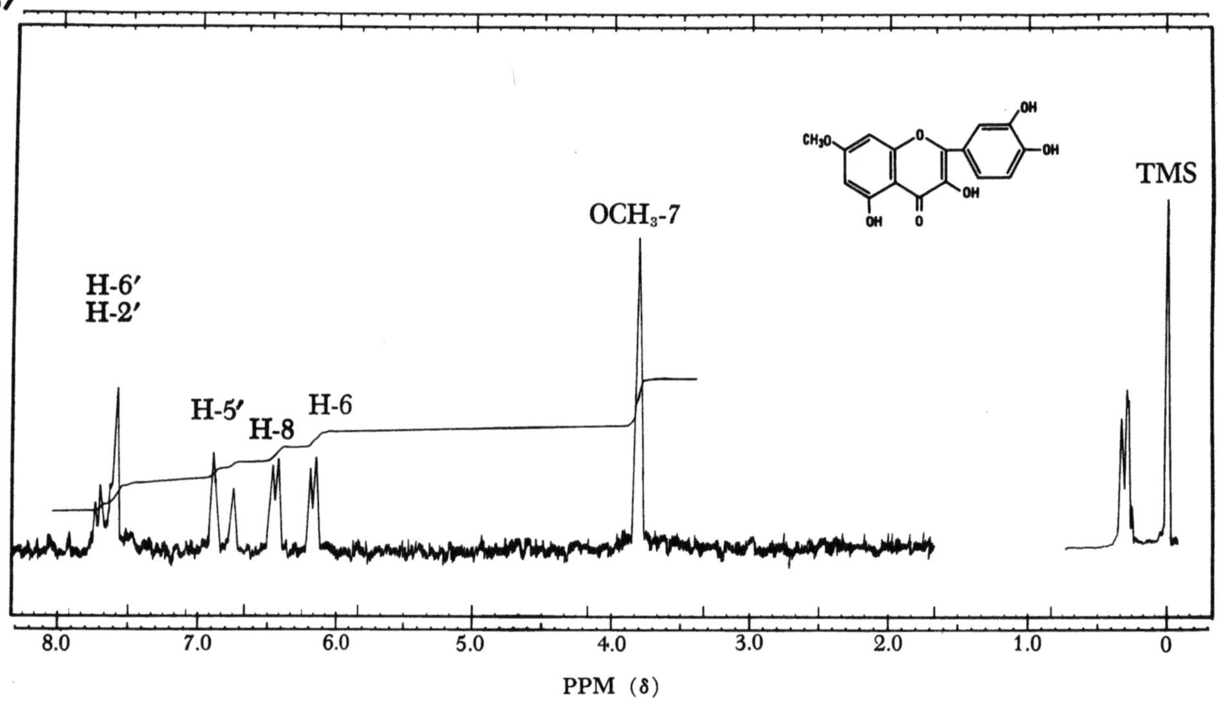

NMR of TMS Ether of Isorhamnetin in CCl_4

38

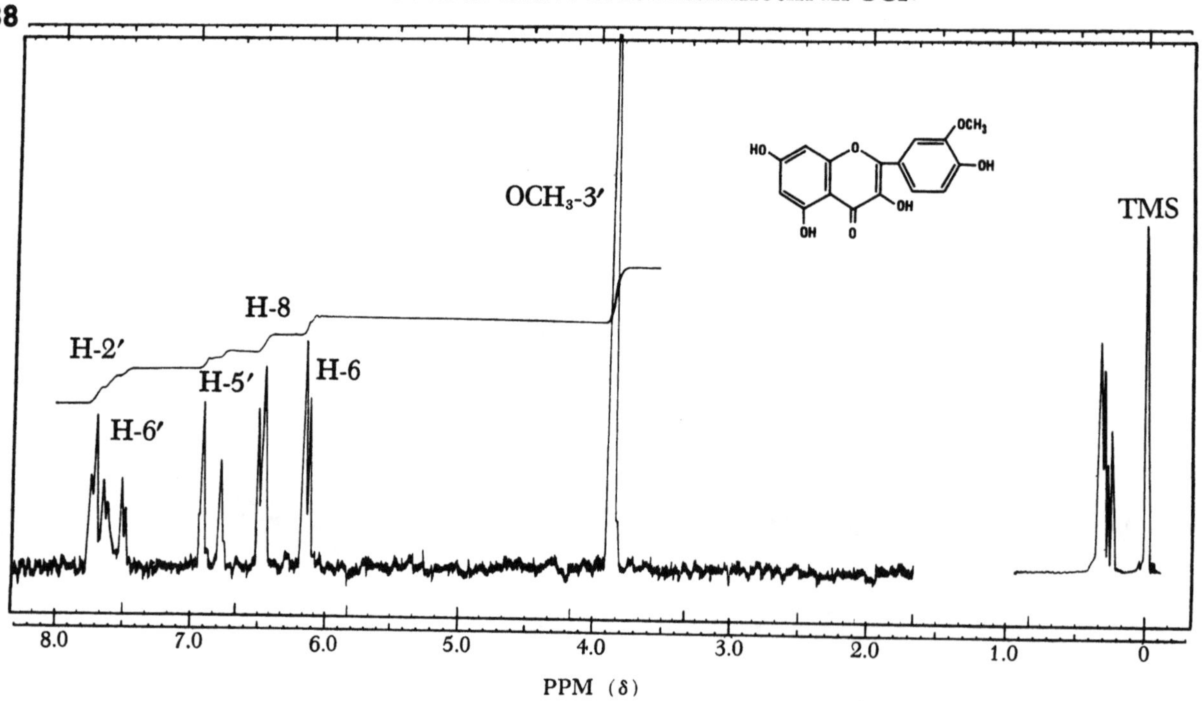

NMR of TMS Ether of Isorhamnetin 3-O-Galactoside in CCl_4

39

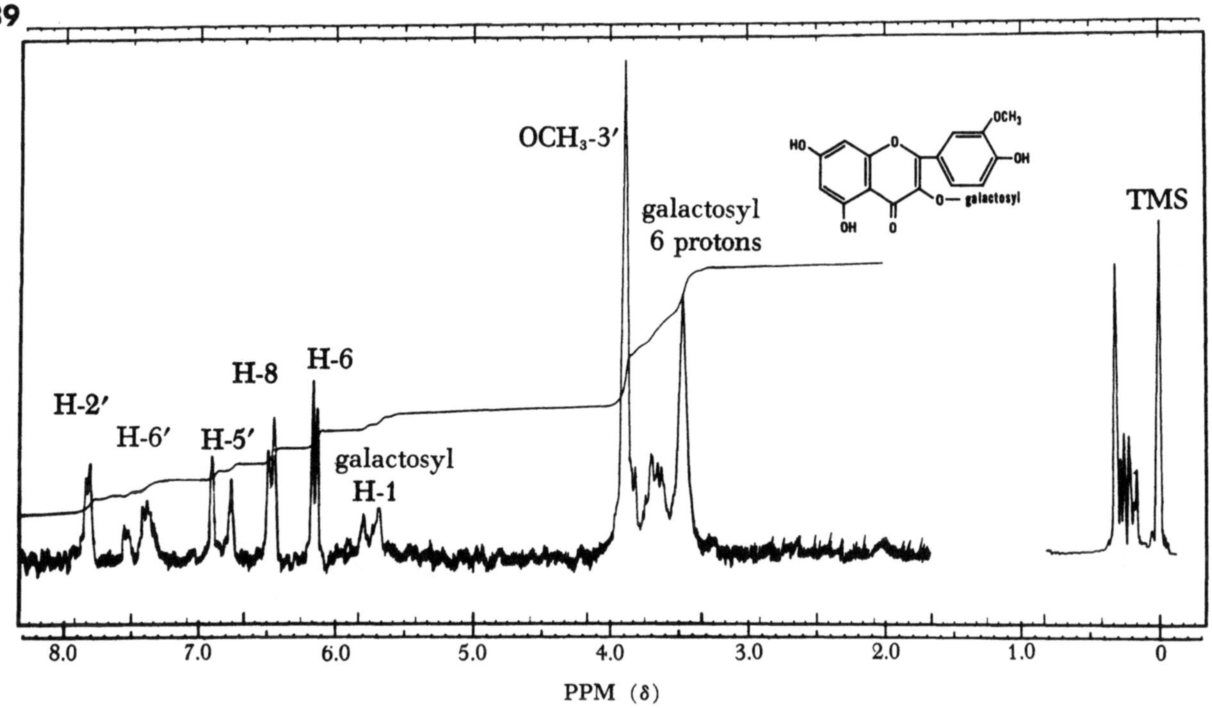

NMR of TMS Ether of Tamarixetin in CCl_4

40

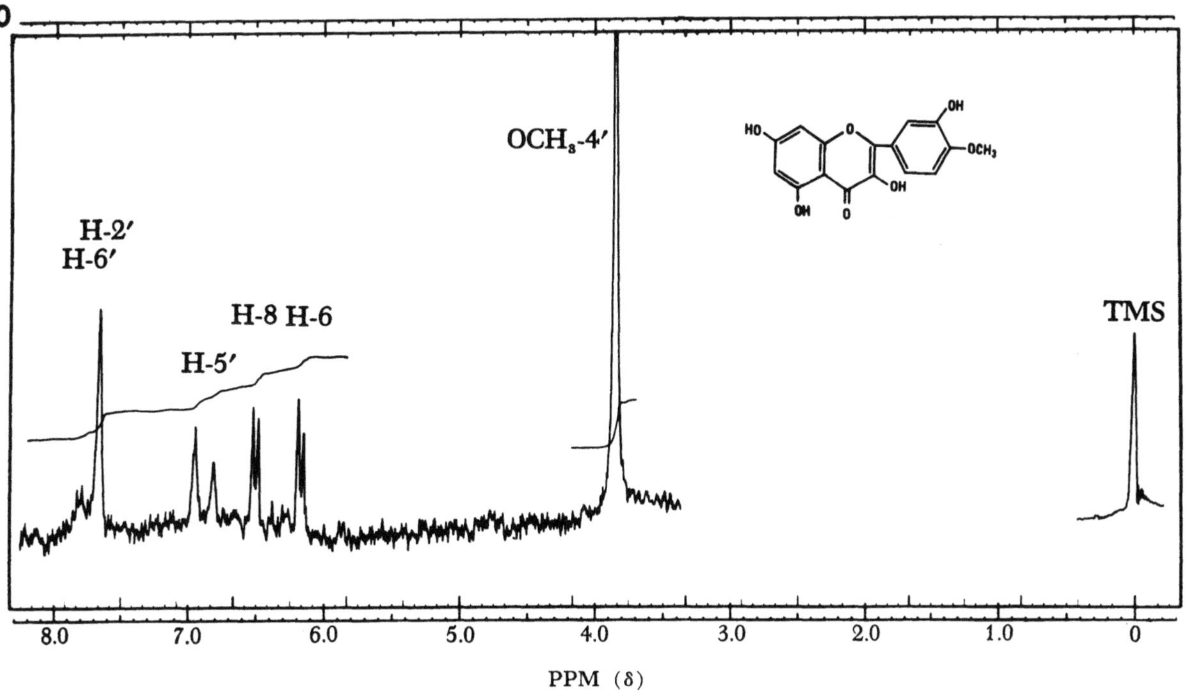

NMR of TMS Ether of Tamarixetin 7-O-Neohesperidoside in CCl_4

41

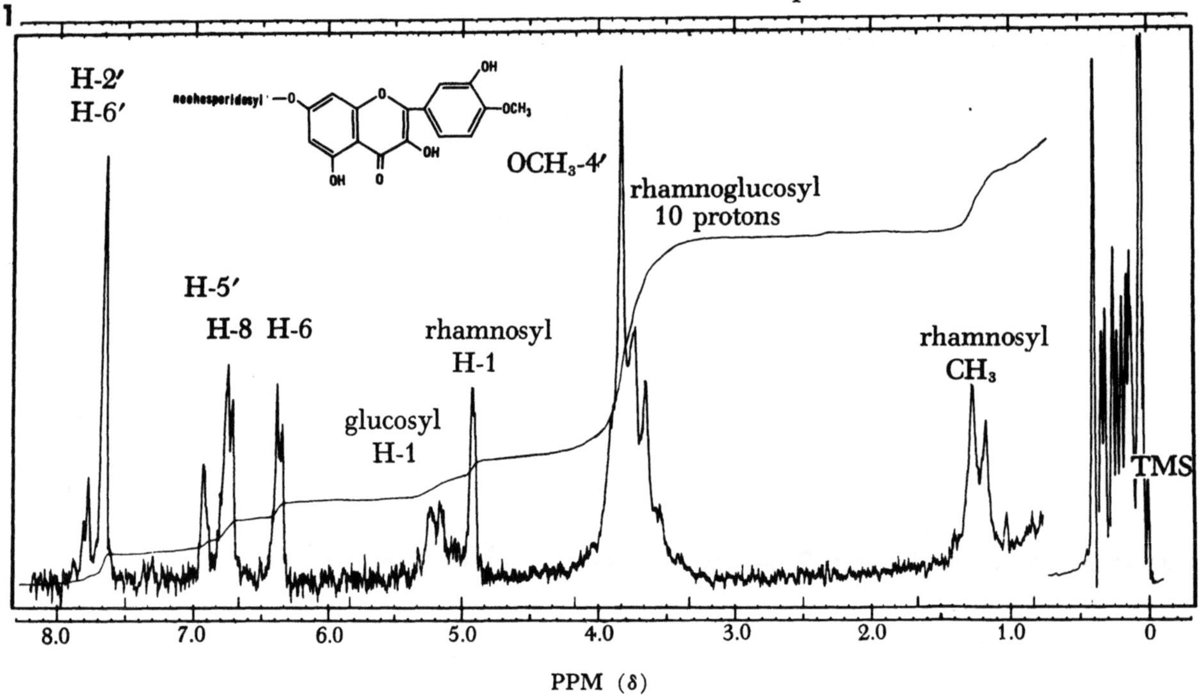

NMR of Tamarixetin 7-O-Neohesperidoside Acetate in $CDCl_3$

42

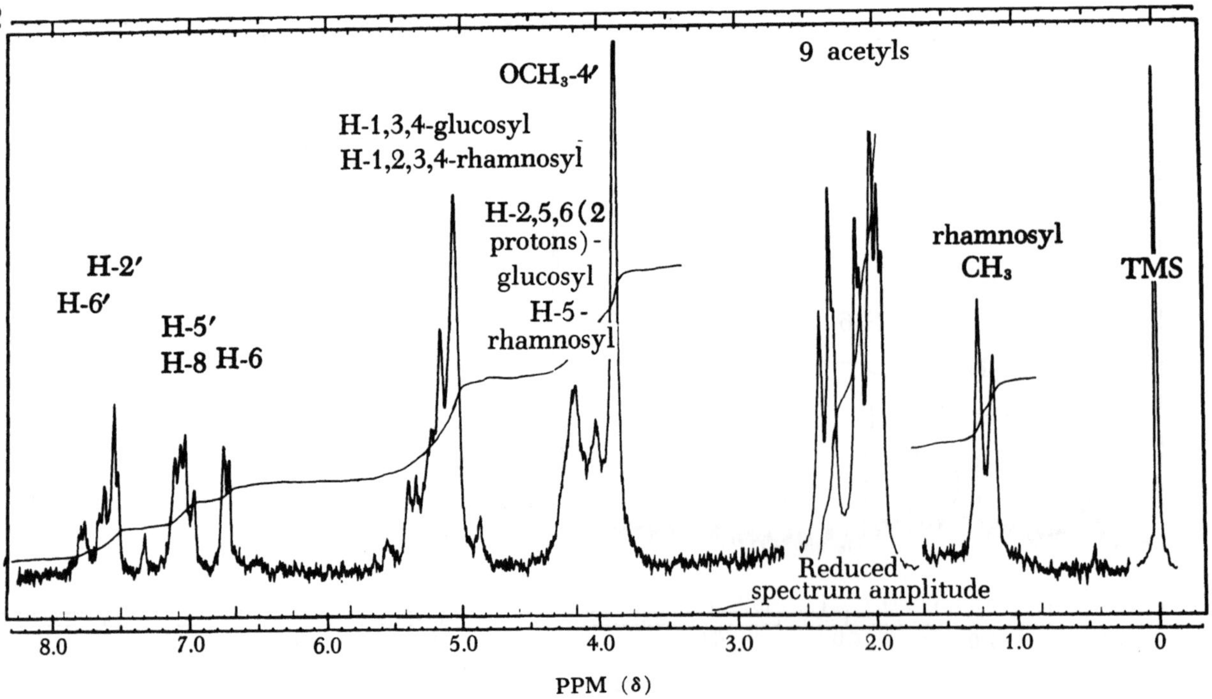

NMR of TMS Ether of Tamarixetin 7-O-Rutinoside in CCl_4

43

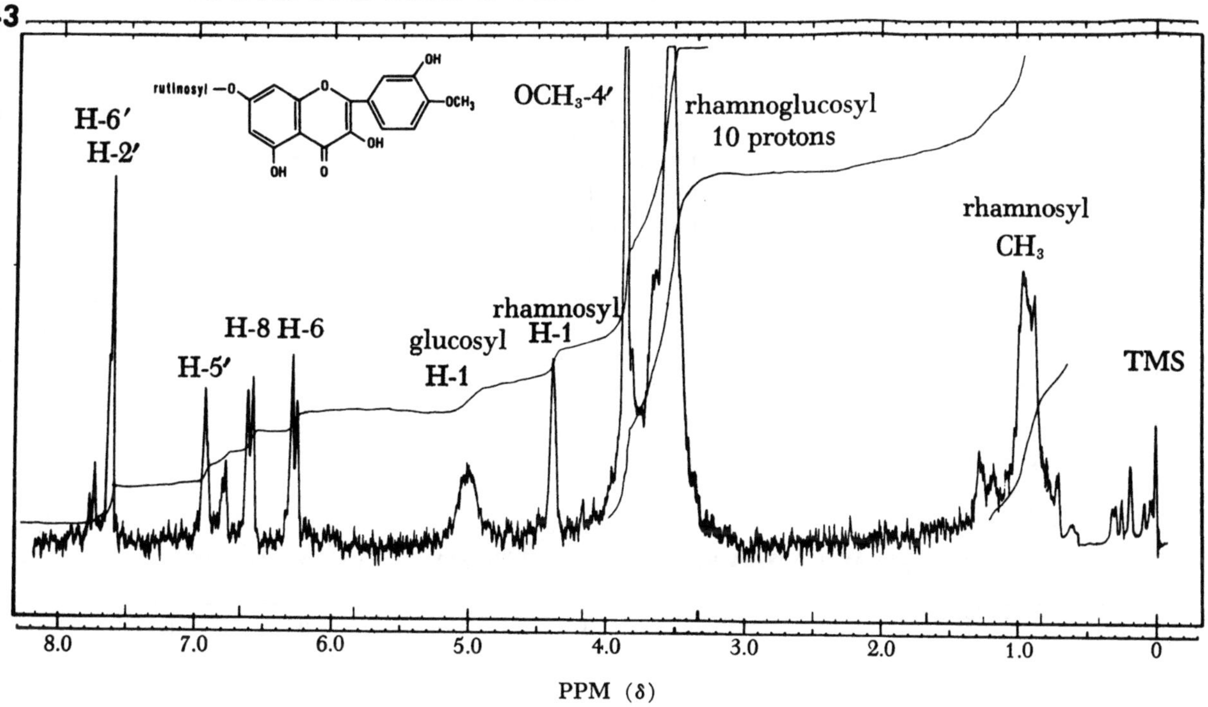

NMR of Quercetin 3′,4′,5,7-Tetramethyl ether in $CDCl_3$

44

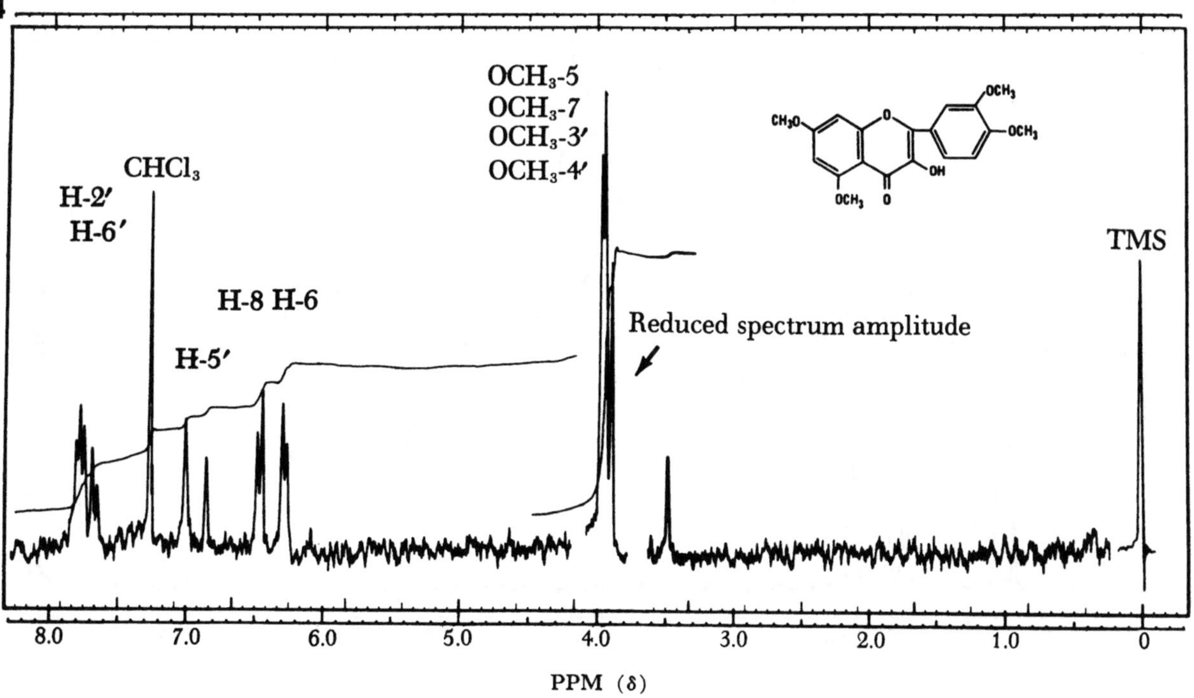

NMR of 3,5,6,7,8-Pentamethoxyflavone in $CDCl_3$

45

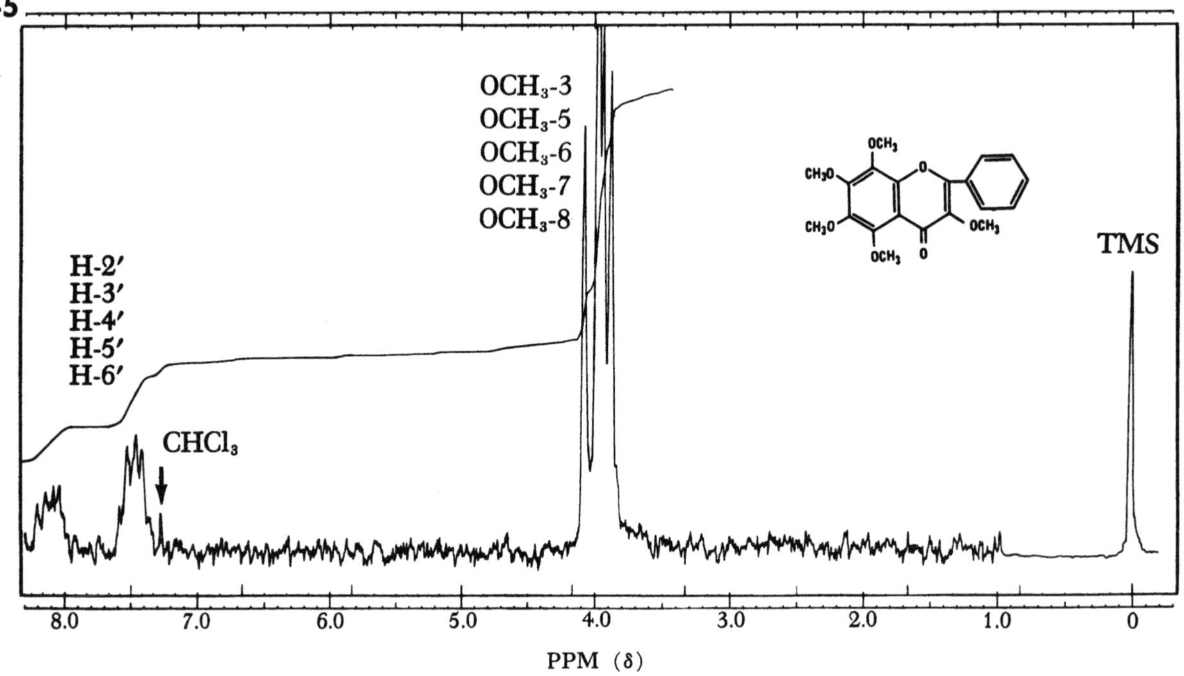

NMR of TMS Ether of Morin in CCl_4

46

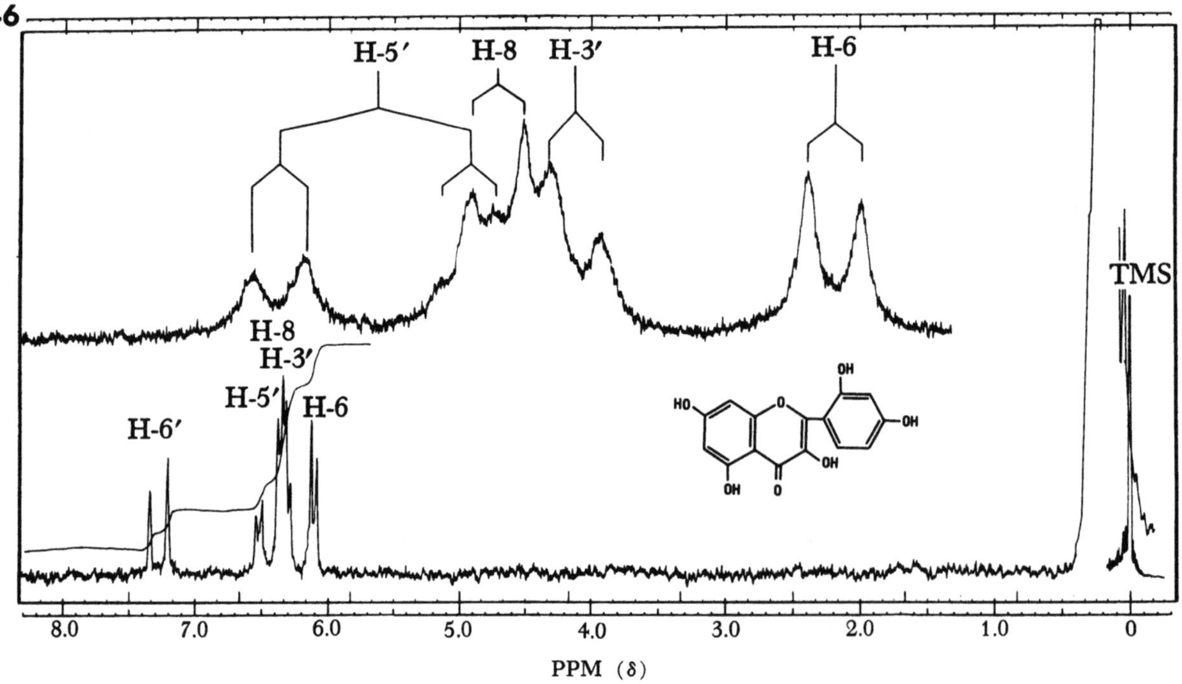

NMR of TMS Ether of Robinetin in CCl_4

47

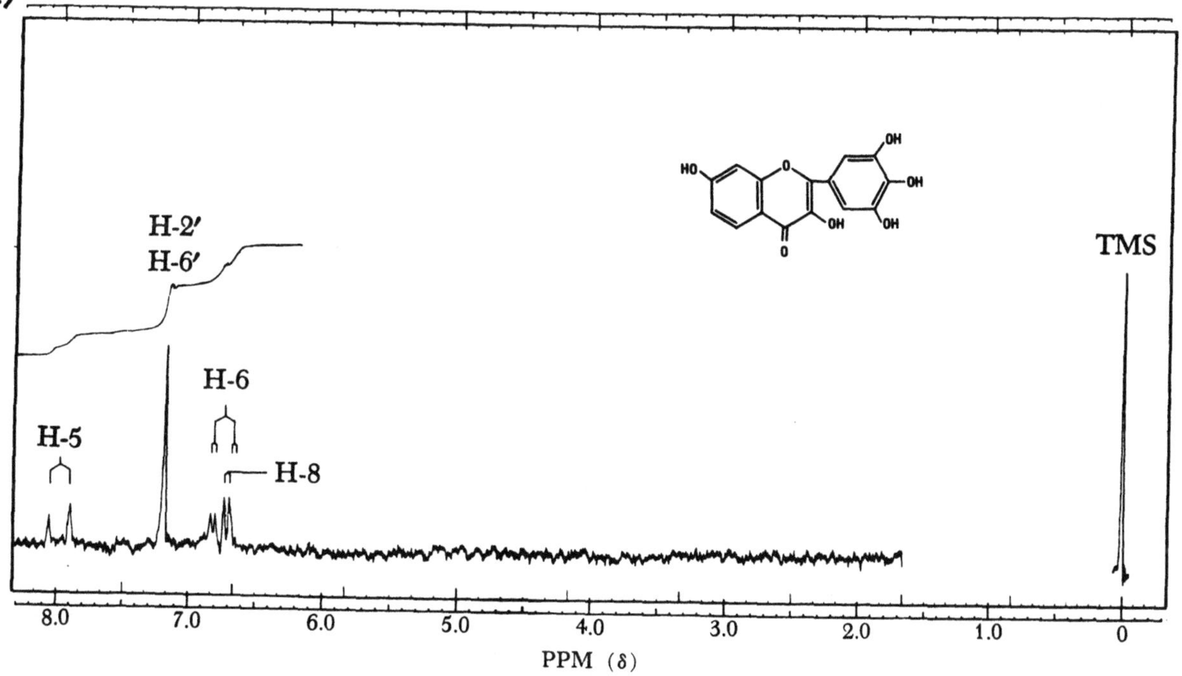

NMR of TMS Ether of Herbacetin 8-O-Methyl Ether in CCl_4

48

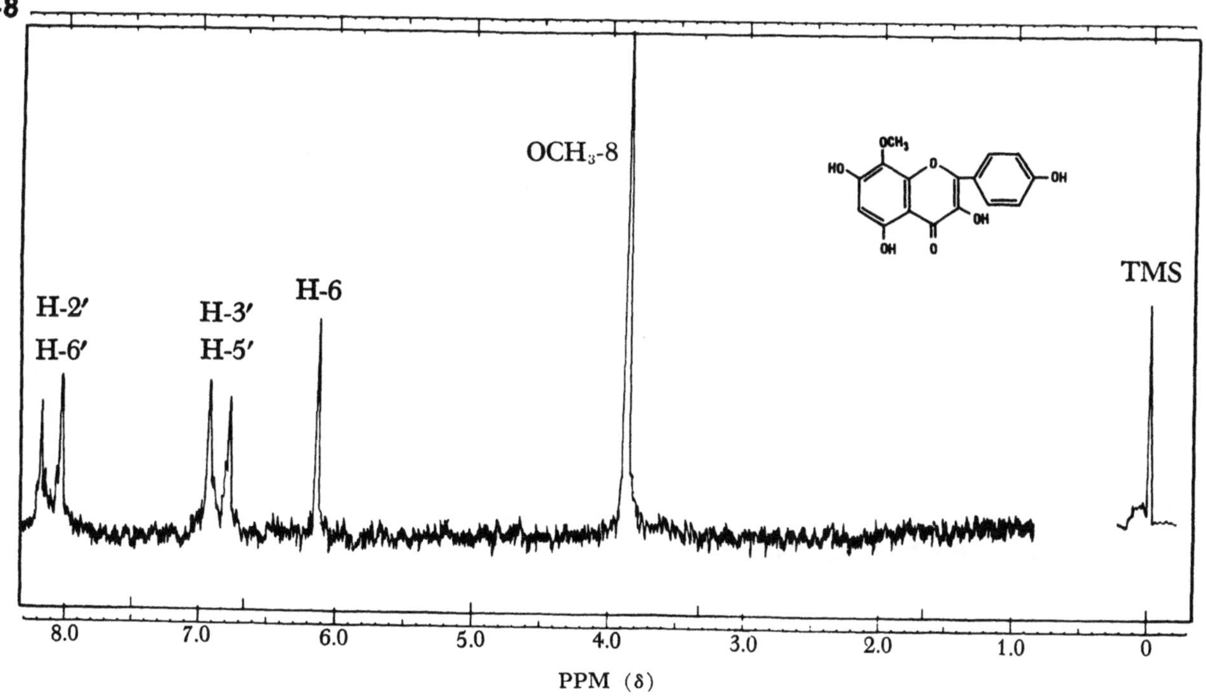

NMR of TMS Ether of Penduletin in CCl_4

49

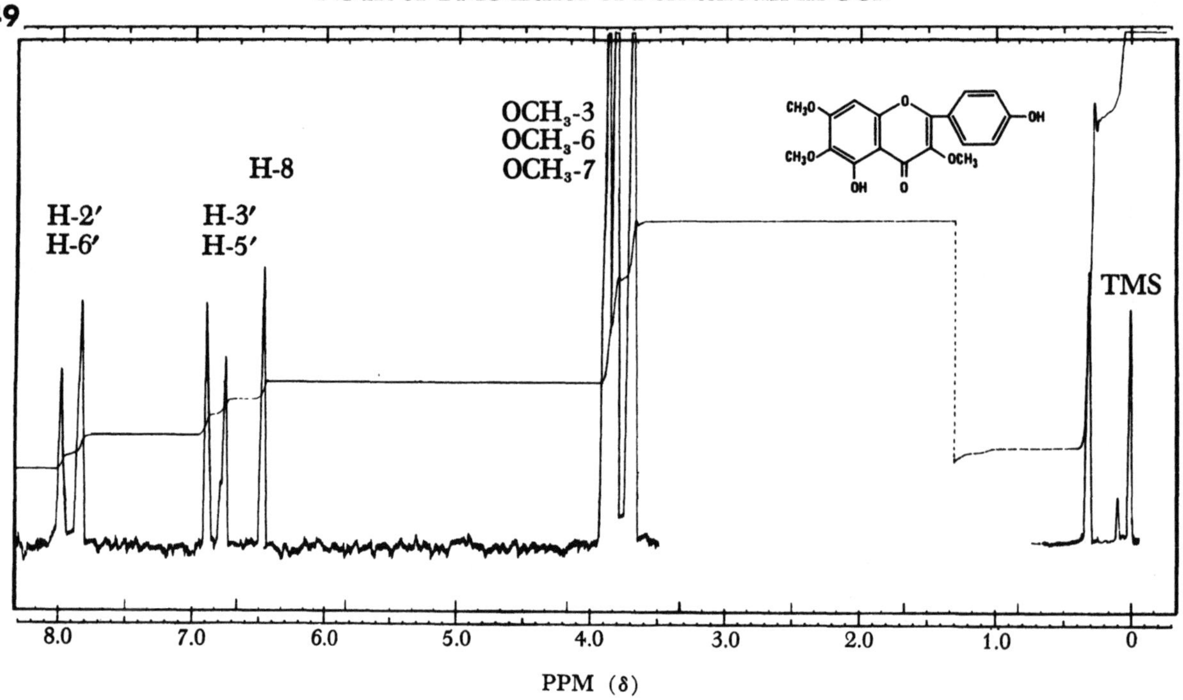

NMR of TMS Ether of Pendulin in CCl_4

50

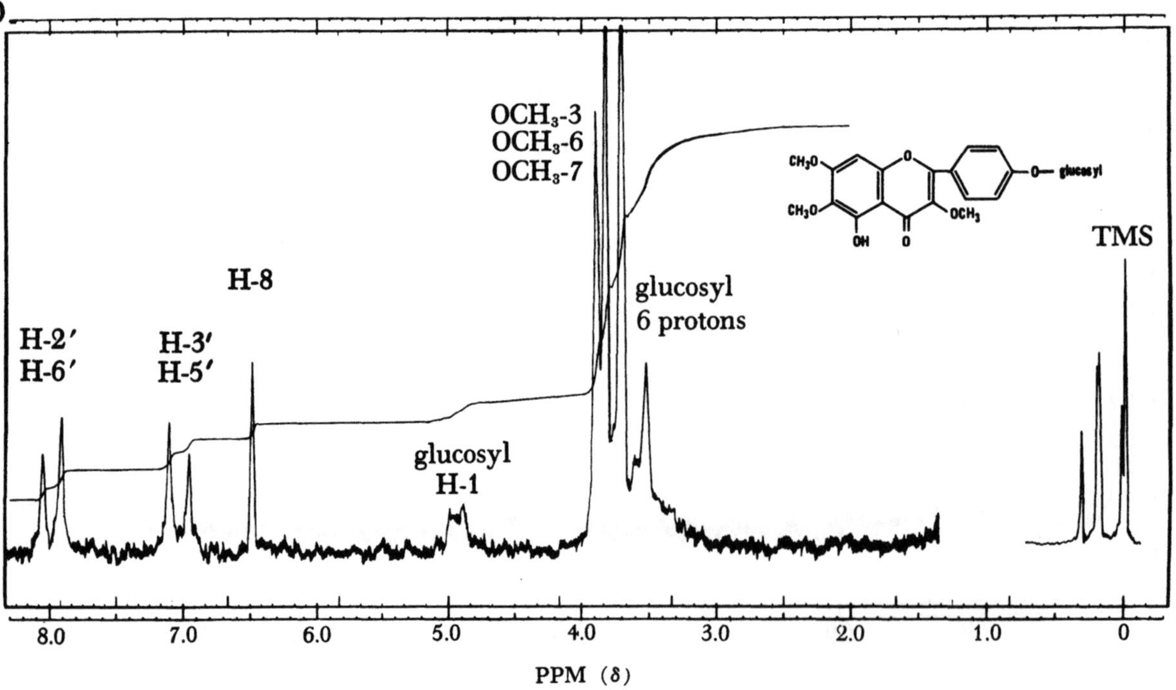

NMR of TMS Ether of Patulitrin in CCl_4

51

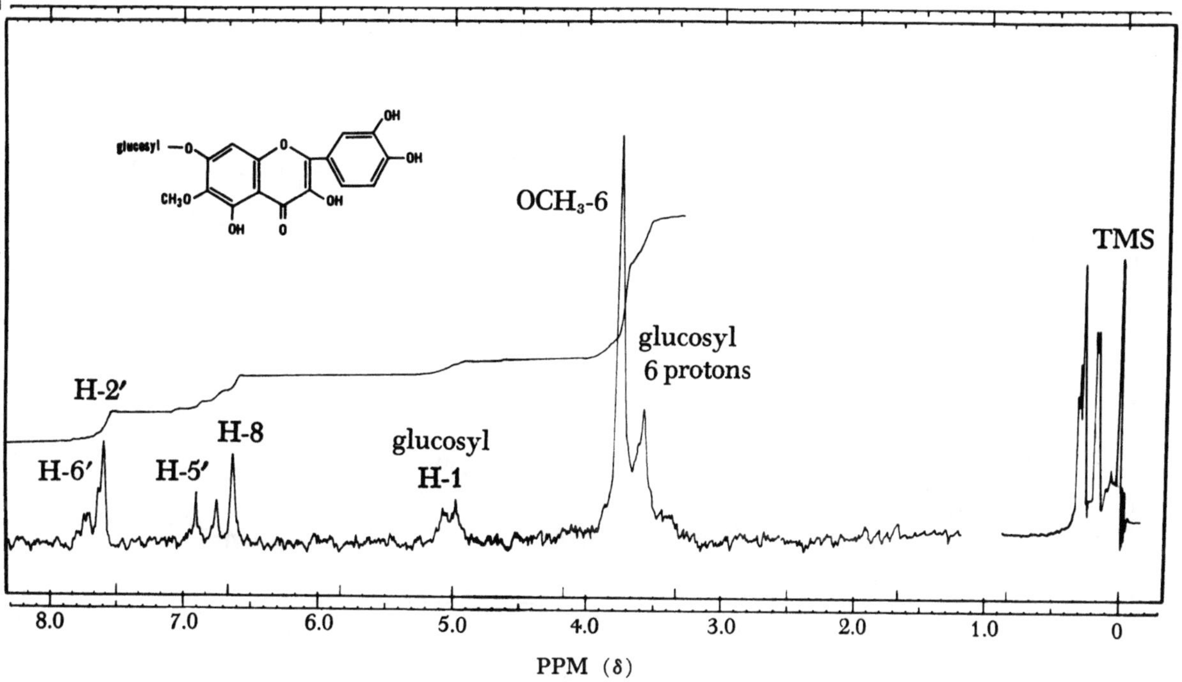

NMR of Patulitrin in DMSO (Deuterated)

52

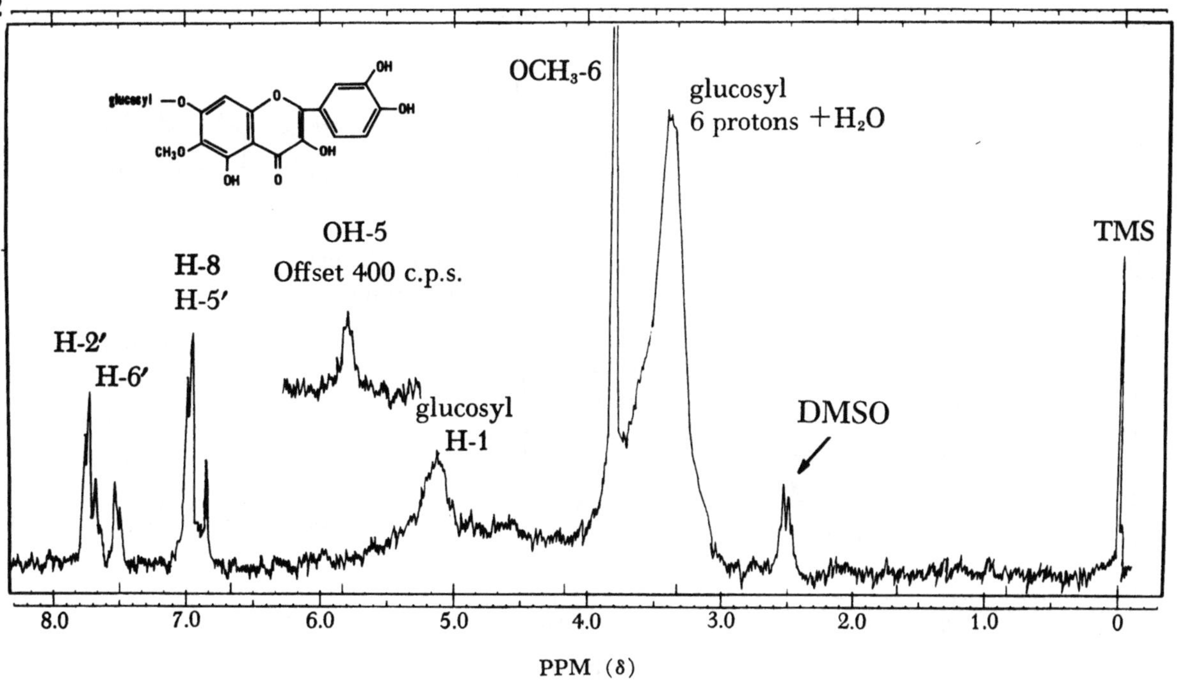

NMR of TMS Ether of Patuletin 3-O-Glucoside in CCl_4

53

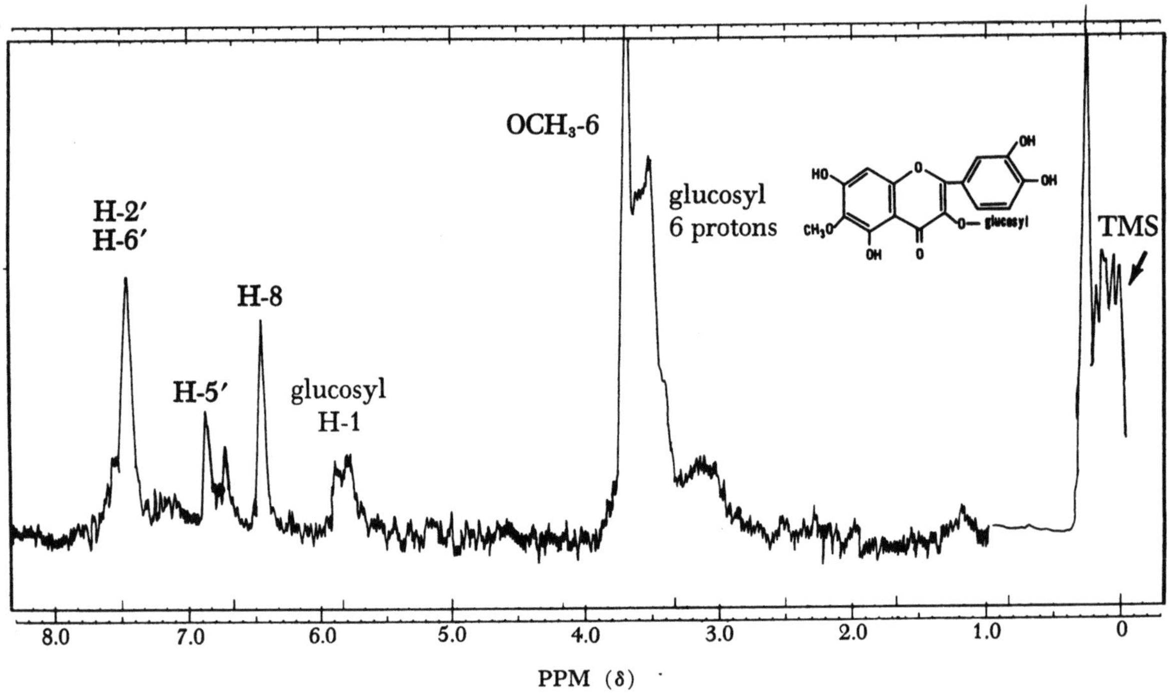

NMR of TMS Ether of Patuletin 3-O-Rutinoside in CCl_4

54

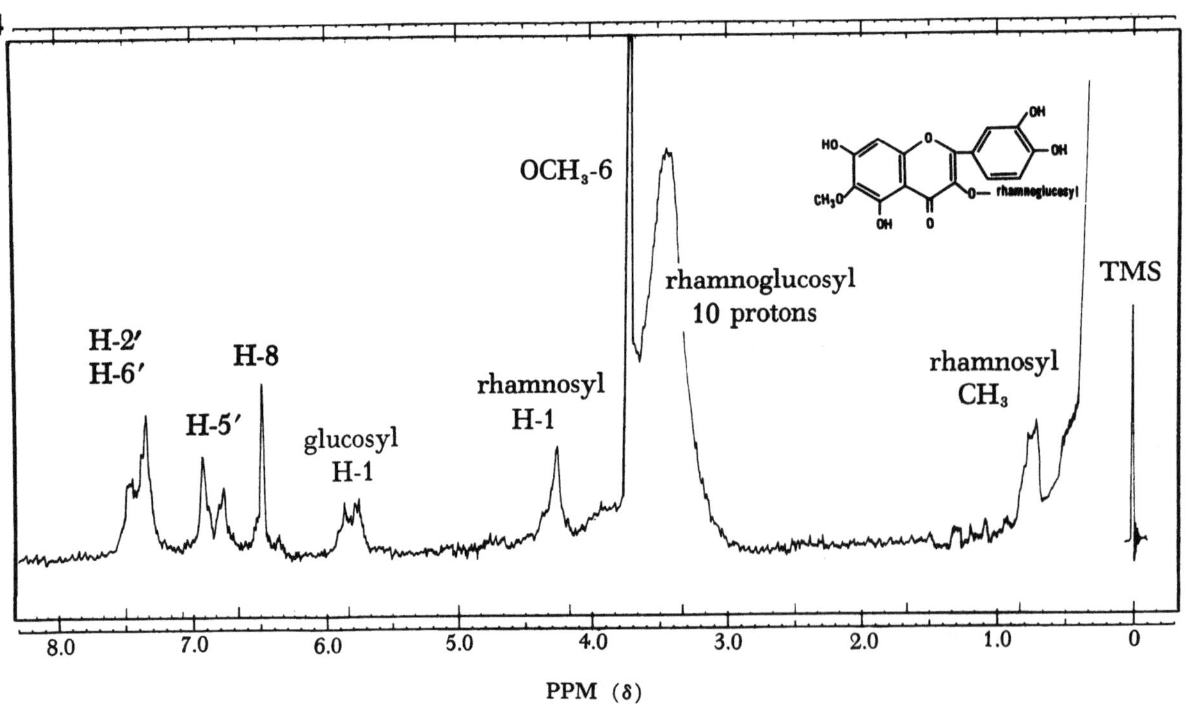

NMR of TMS Ether of Jacein in CCl_4

55

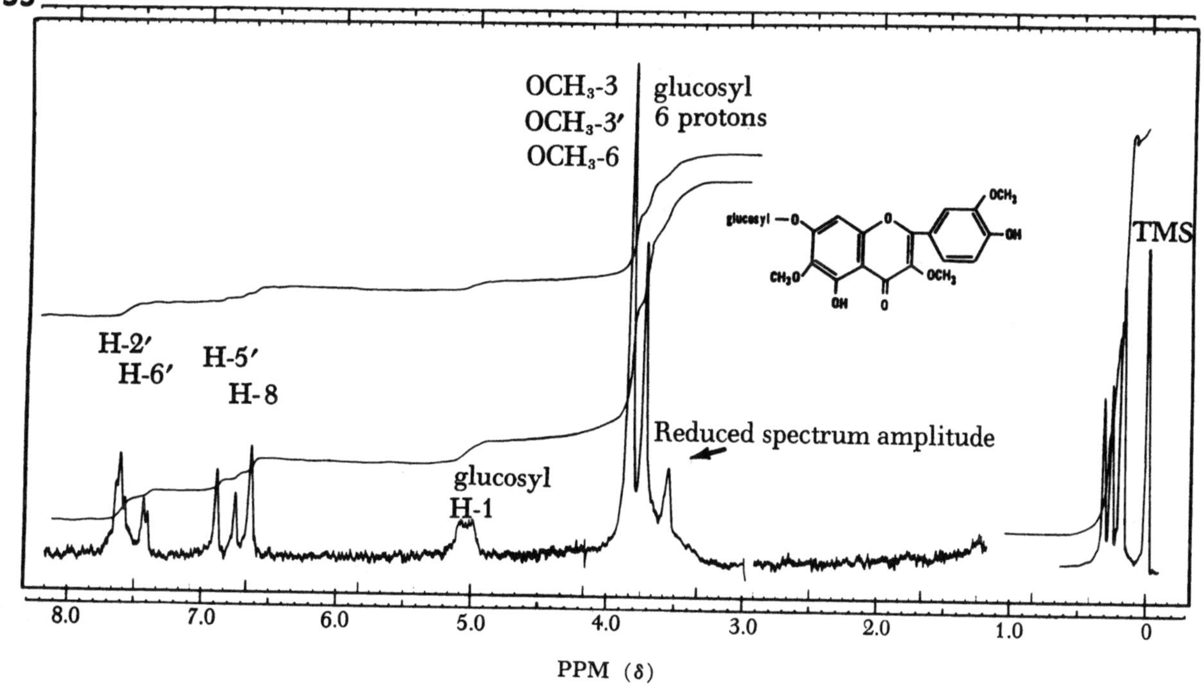

NMR of TMS Ether of Centaurein in CCl_4

56

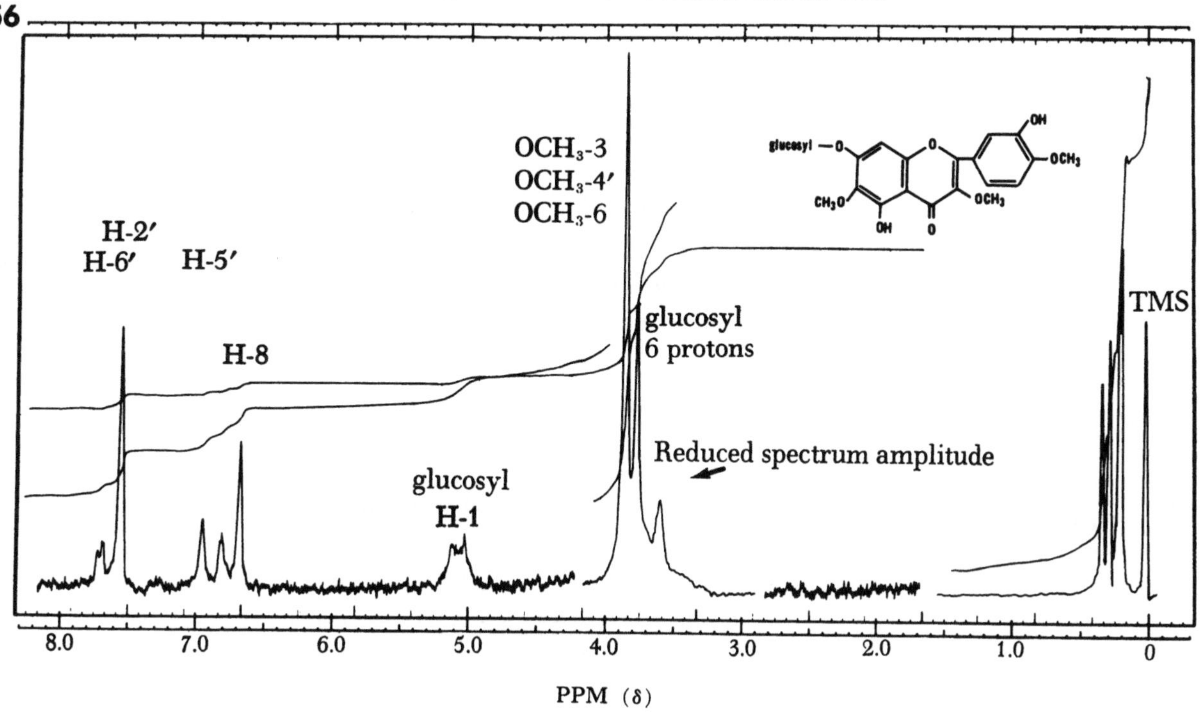

NMR of Artemetin (Artemisetin) in $CDCl_3$

57

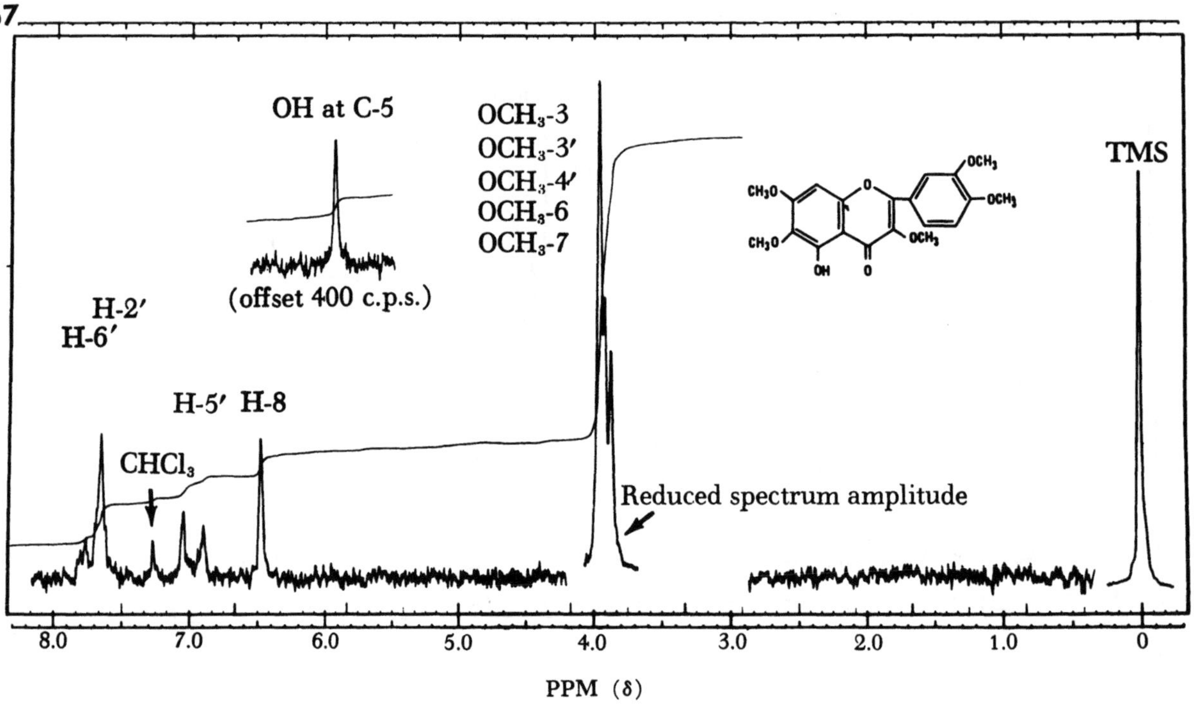

NMR of Quercetagetin Hexamethyl Ether in $CDCl_3$

58

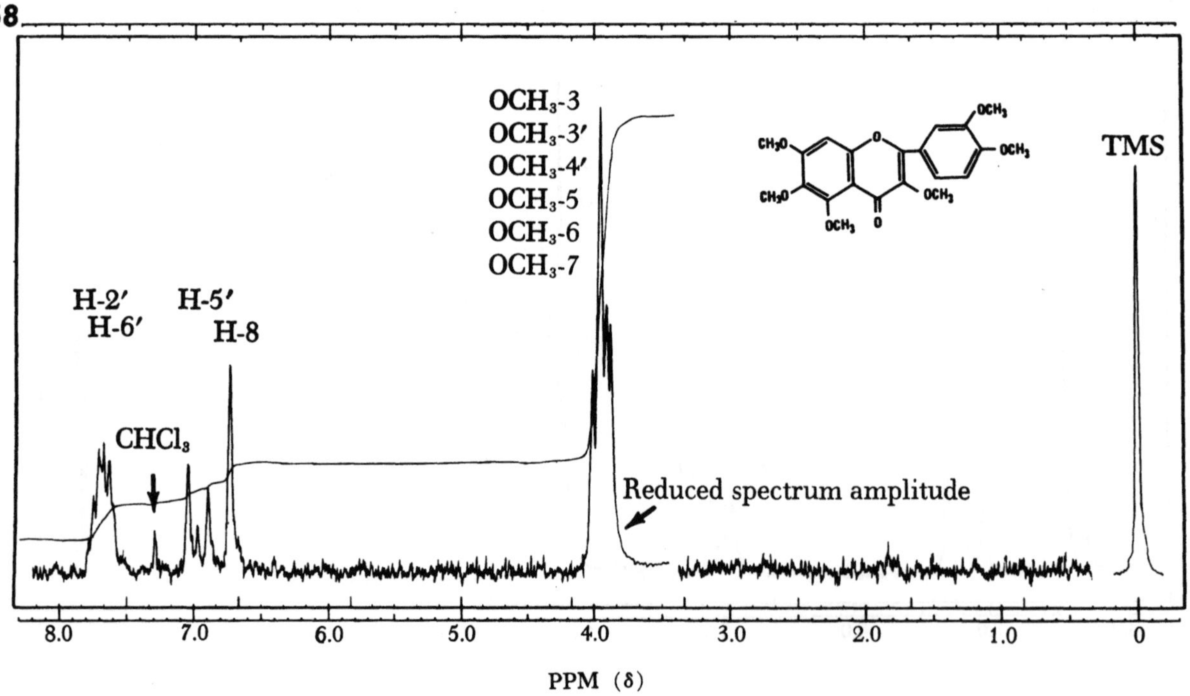

59

NMR of 3,4′,5,6,7,8-Hexamethoxyflavone in $CDCl_3$

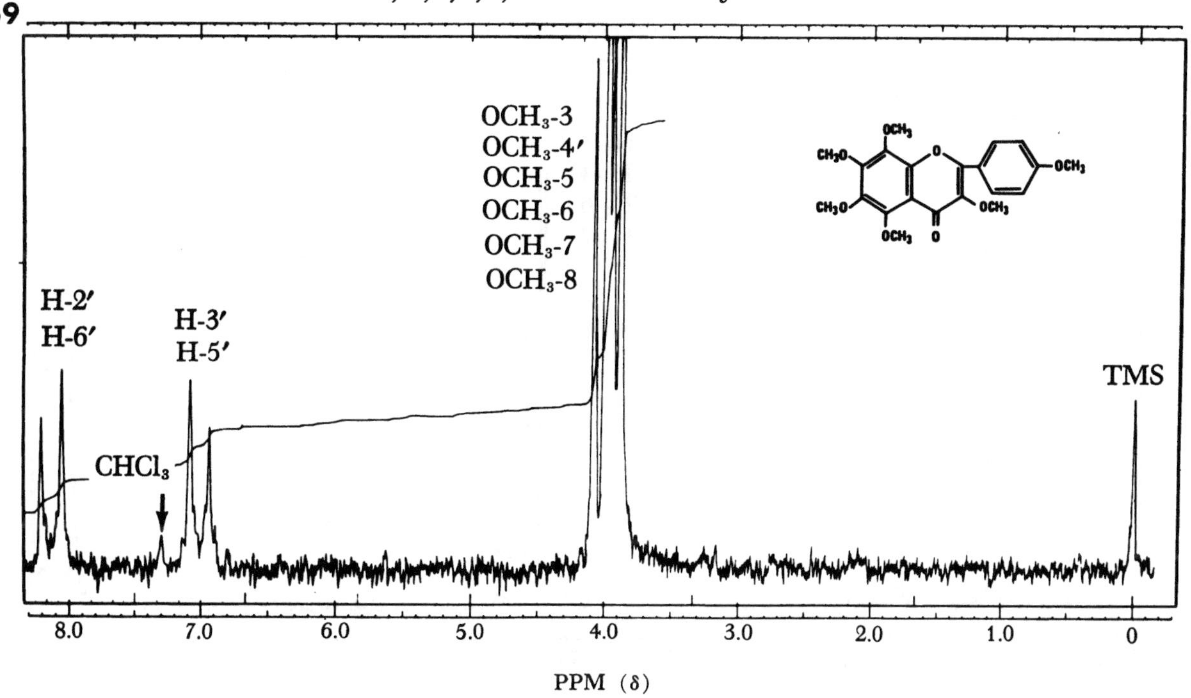

60

NMR of TMS Ether of Myricetin in CCl_4

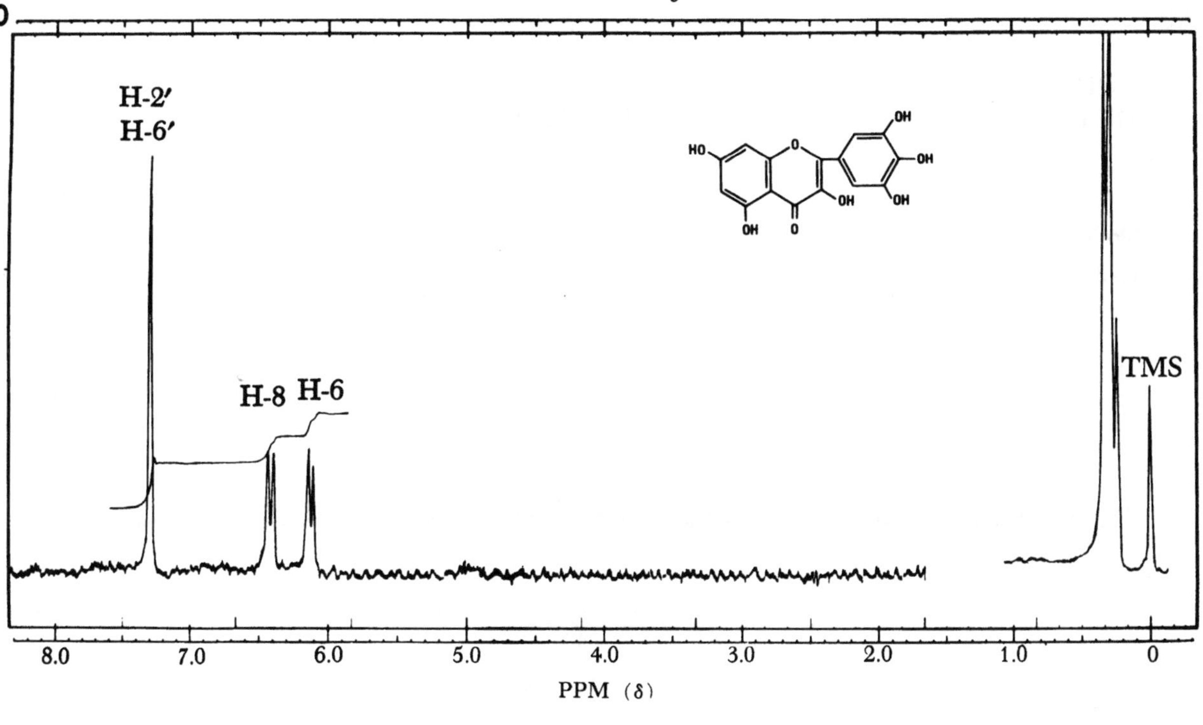

NMR of TMS Ether of 3′,5,5′-Trihydroxy-3,4′,6,7-tetramethoxy-flavone in CCl_4

61

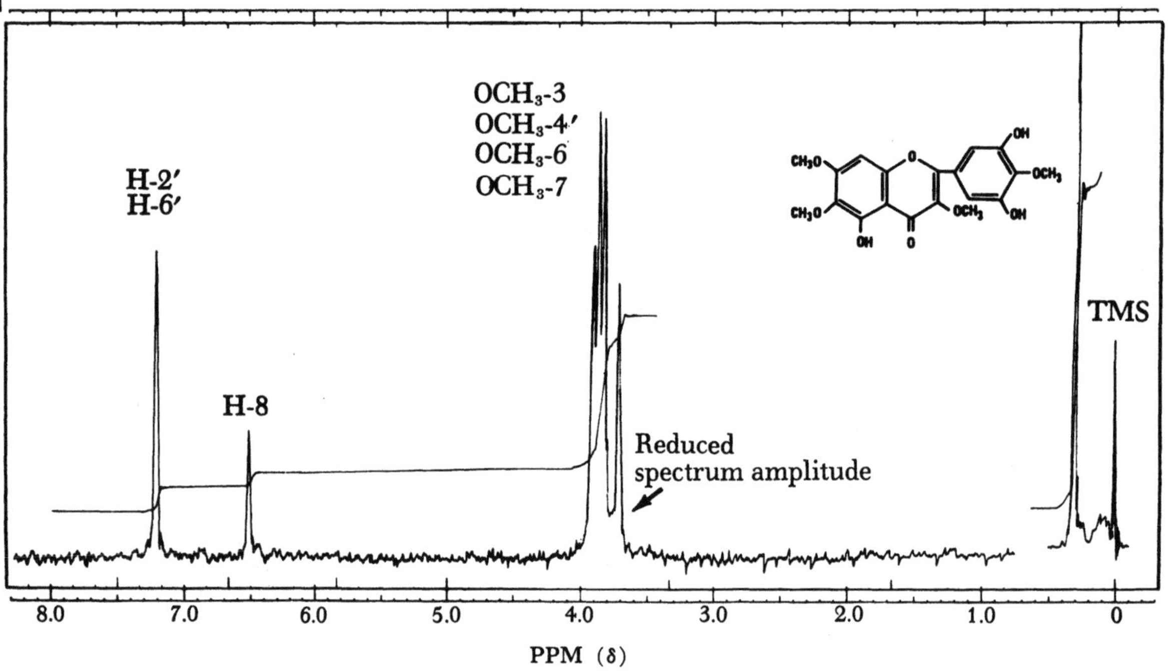

NMR of 8-Carbomethoxy-5-hydroxy-6-methylisoflavone in $CDCl_3$

62

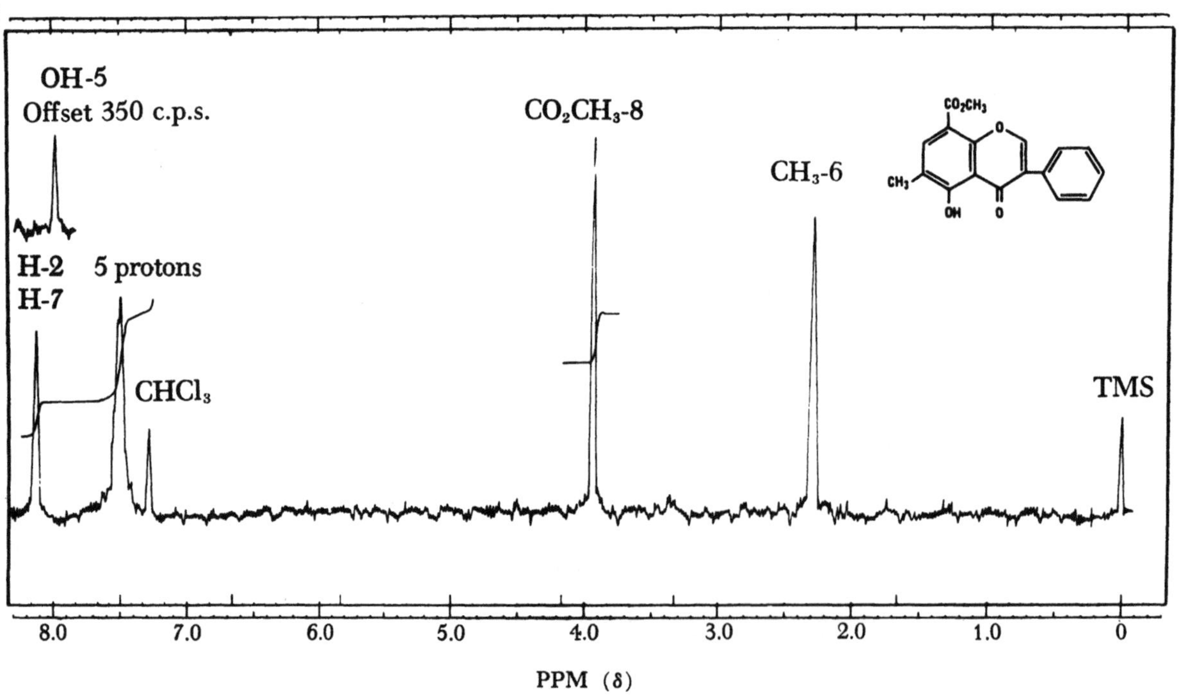

NMR of 8-Carbomethoxy-5- hydroxy-6-methylisoflavone in DMSO (deuterated)

63

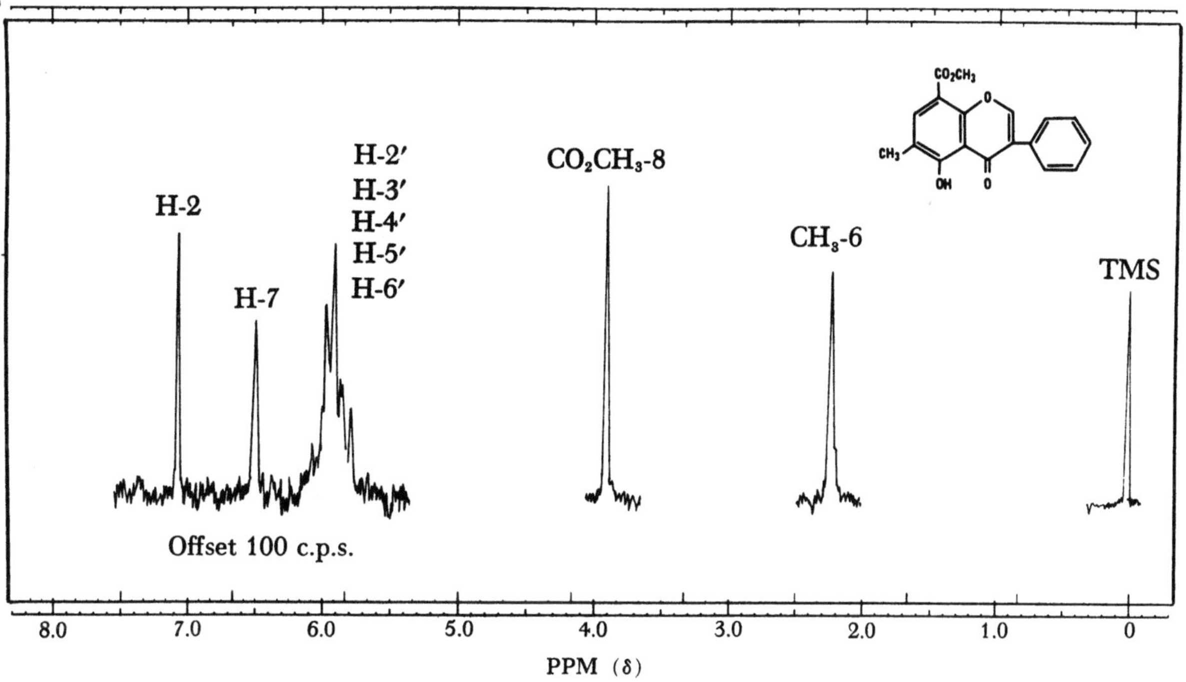

NMR of 5-Methoxy-8-methylisoflavone in $CDCl_3$

64

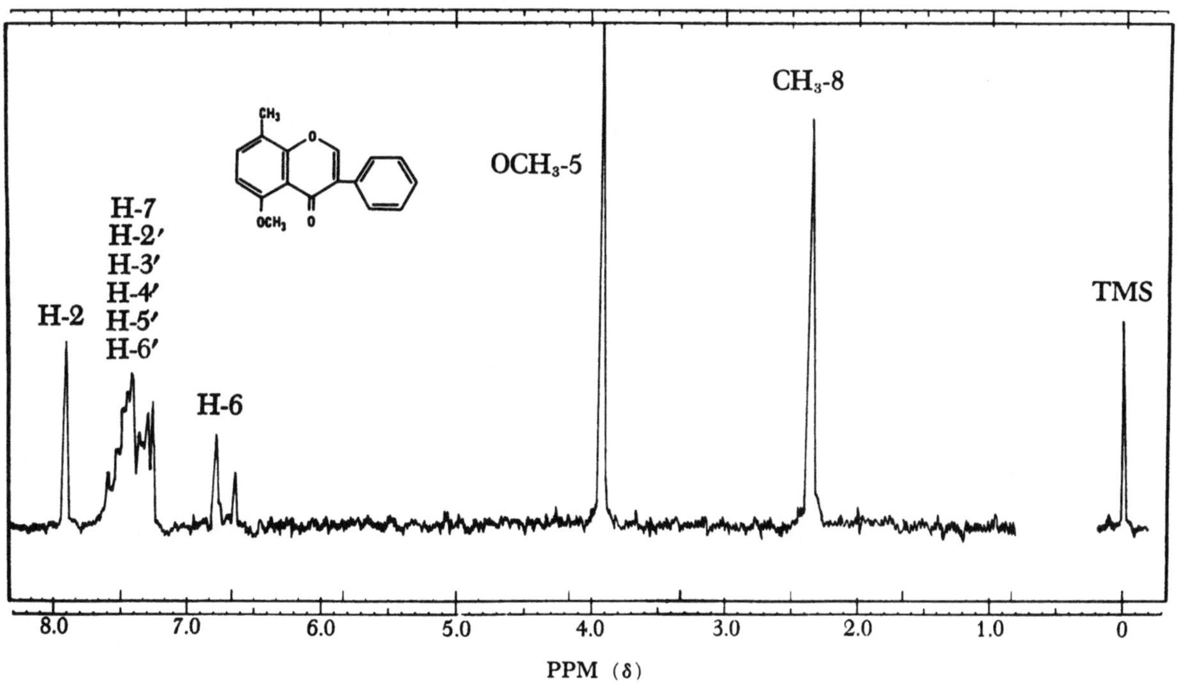

NMR of 7-Acetyloxy-6-carbomethoxyisoflavone in $CDCl_3$

65

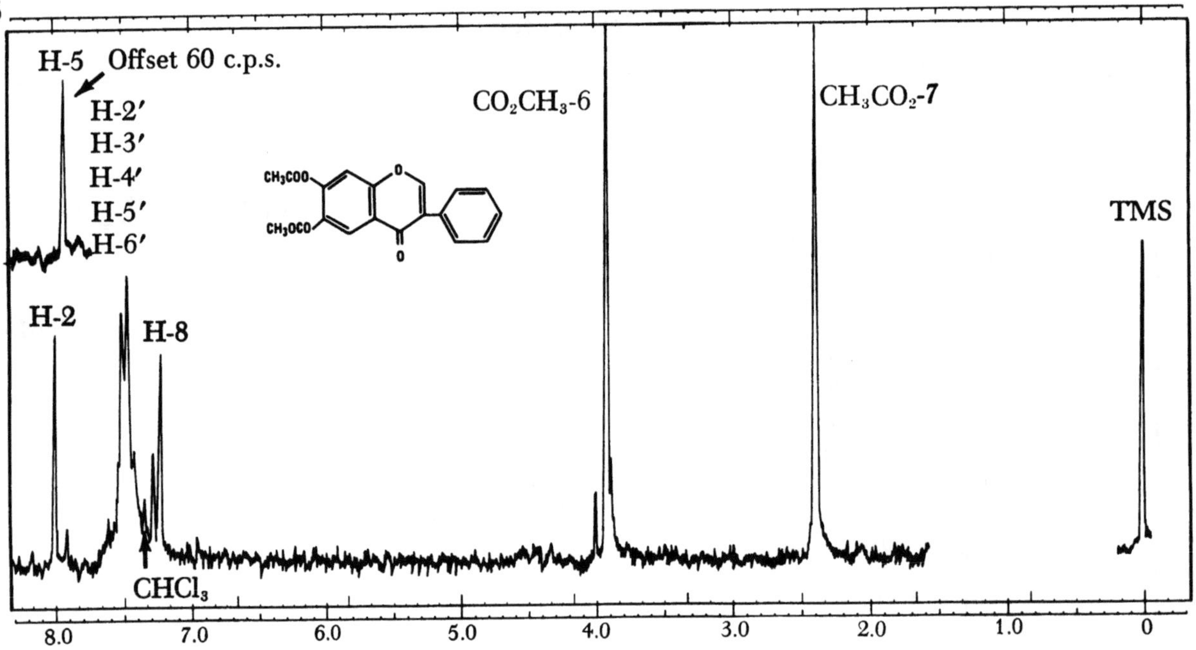

NMR of TMS Ether of Formononetin 7-O-Glucoside in CCl_4

66

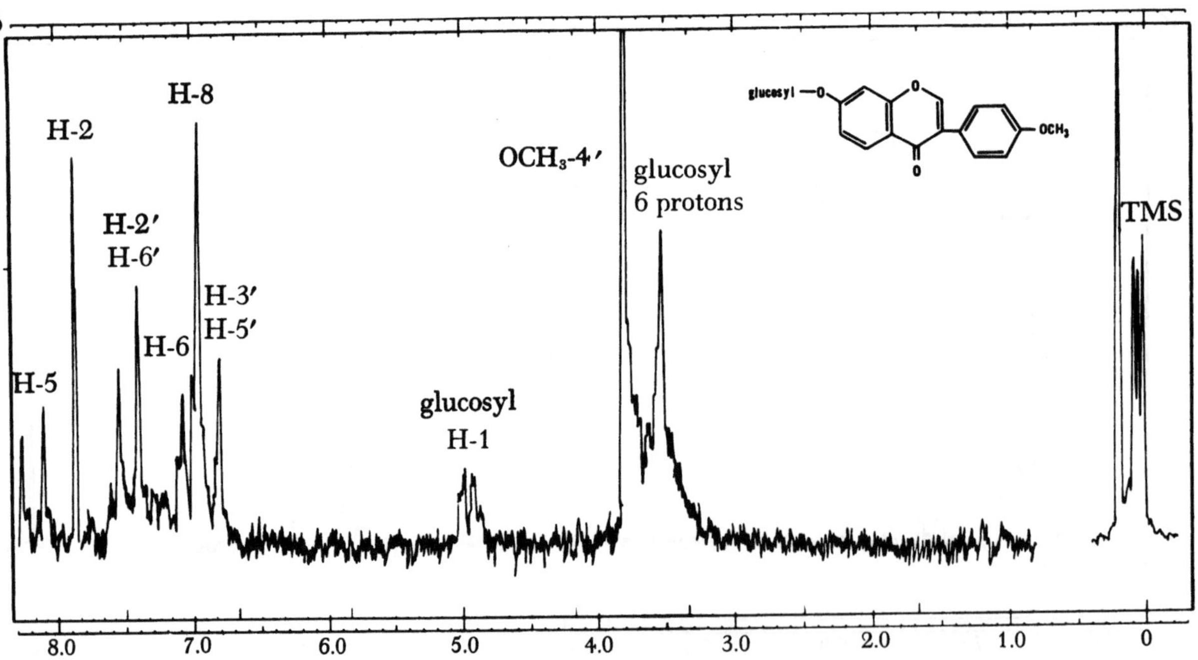

NMR of 5,7-Dimethoxy-8-methylisoflavone in $CDCl_3$

67

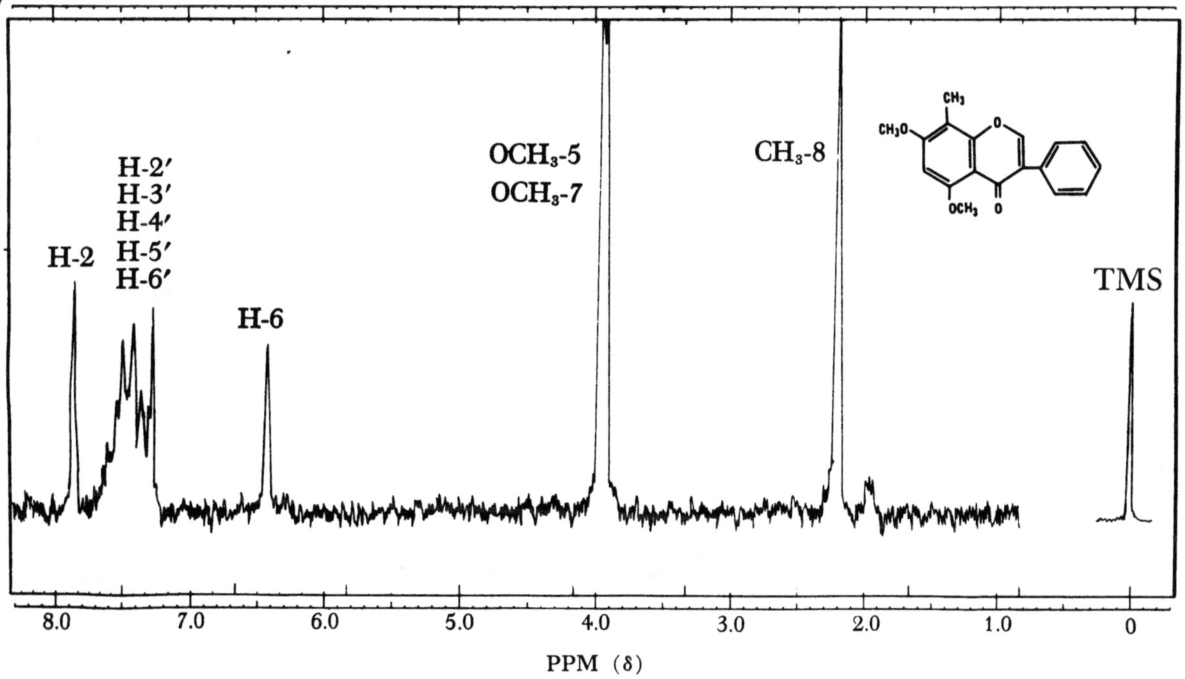

NMR of TMS Ether of Sphaerobioside in CCl_4

68

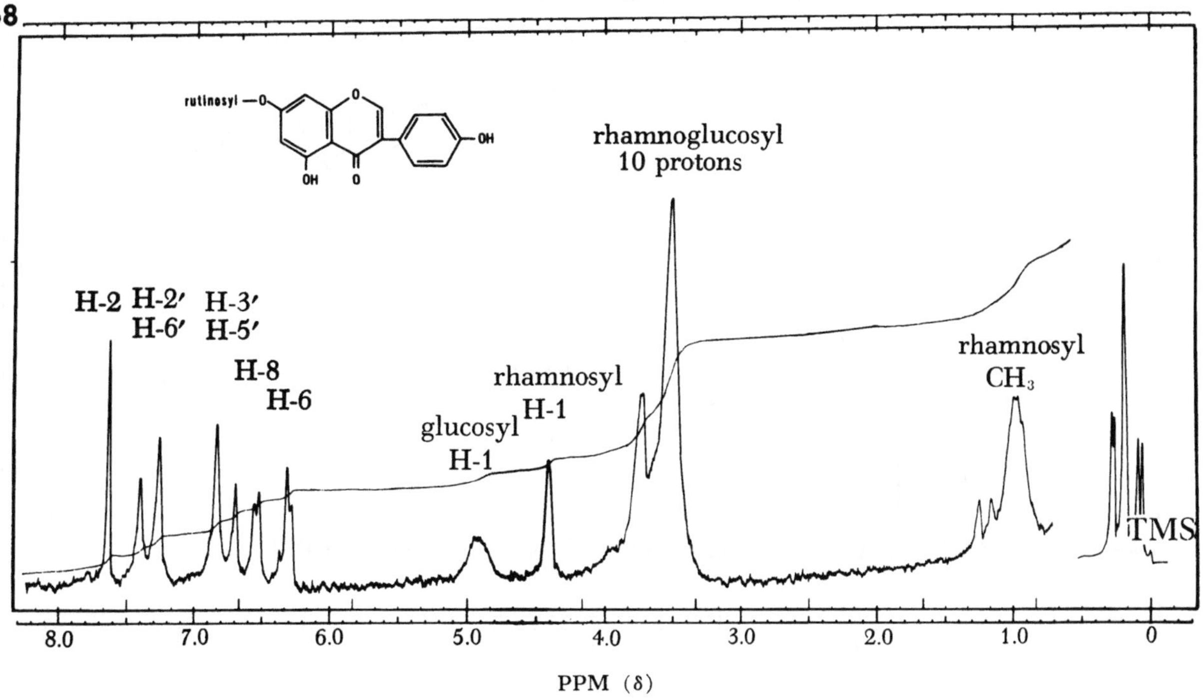

NMR of Sphaerobioside Acetate in $CDCl_3$

69

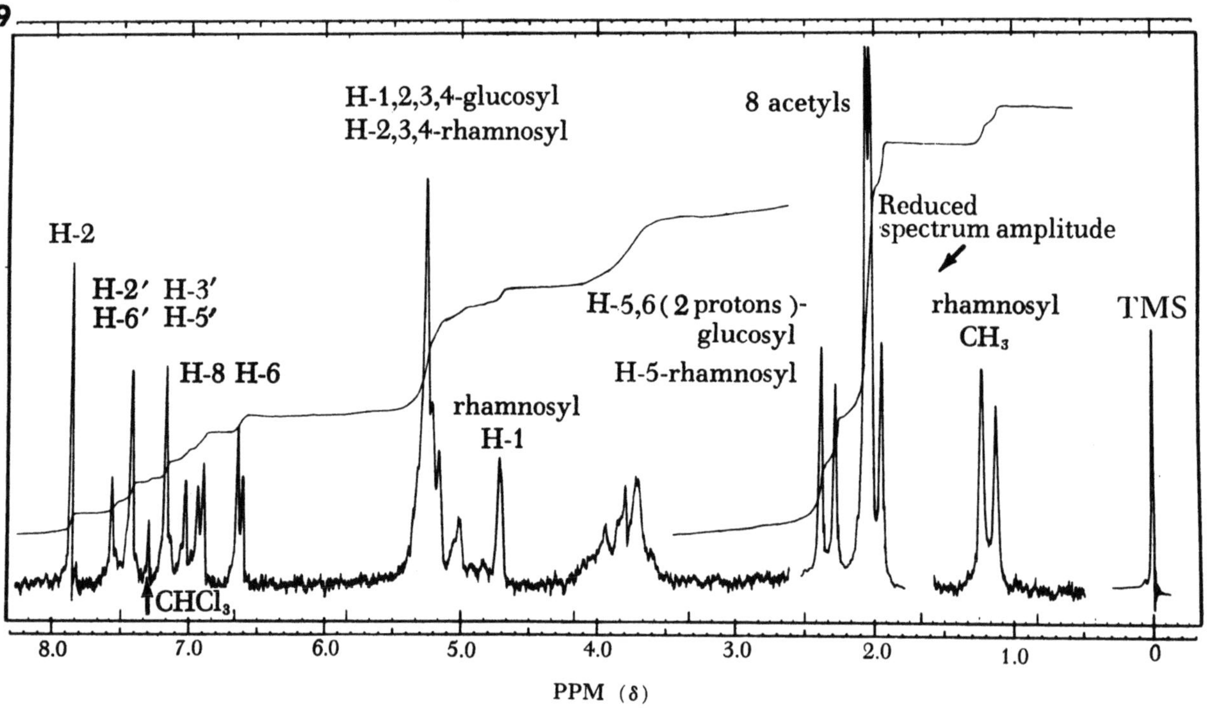

NMR of TMS Ether of Biochanin A in CCl_4

70

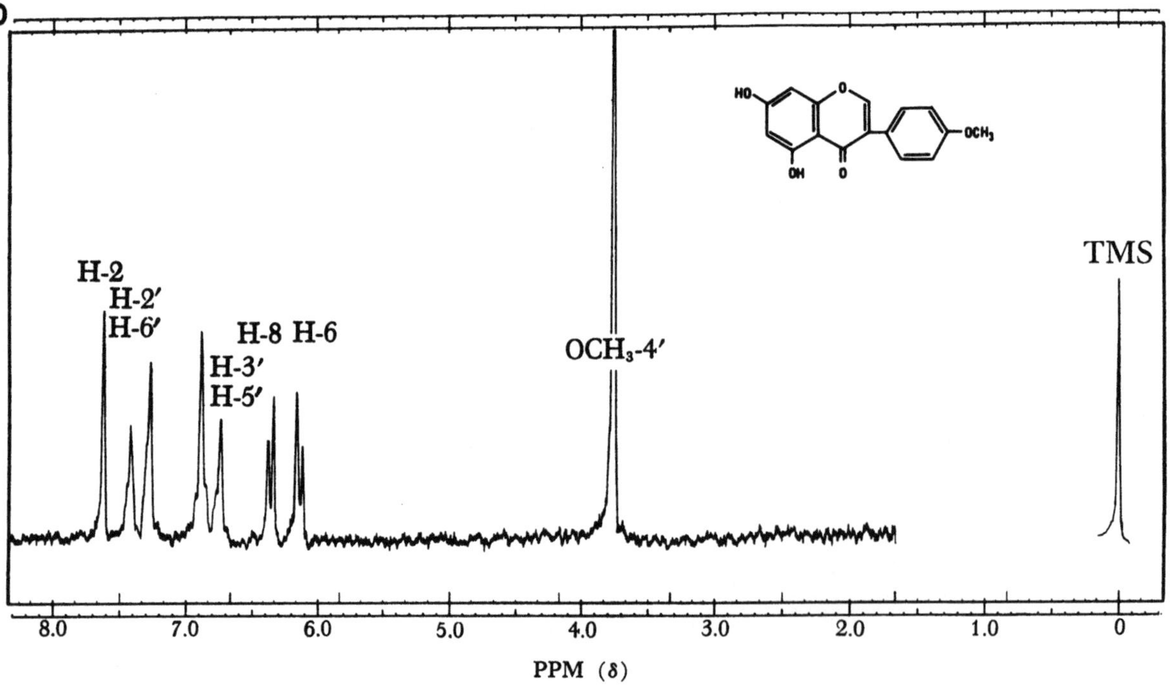

NMR of TMS Ether of Genistein 5-Methyl Ether in CCl_4

71

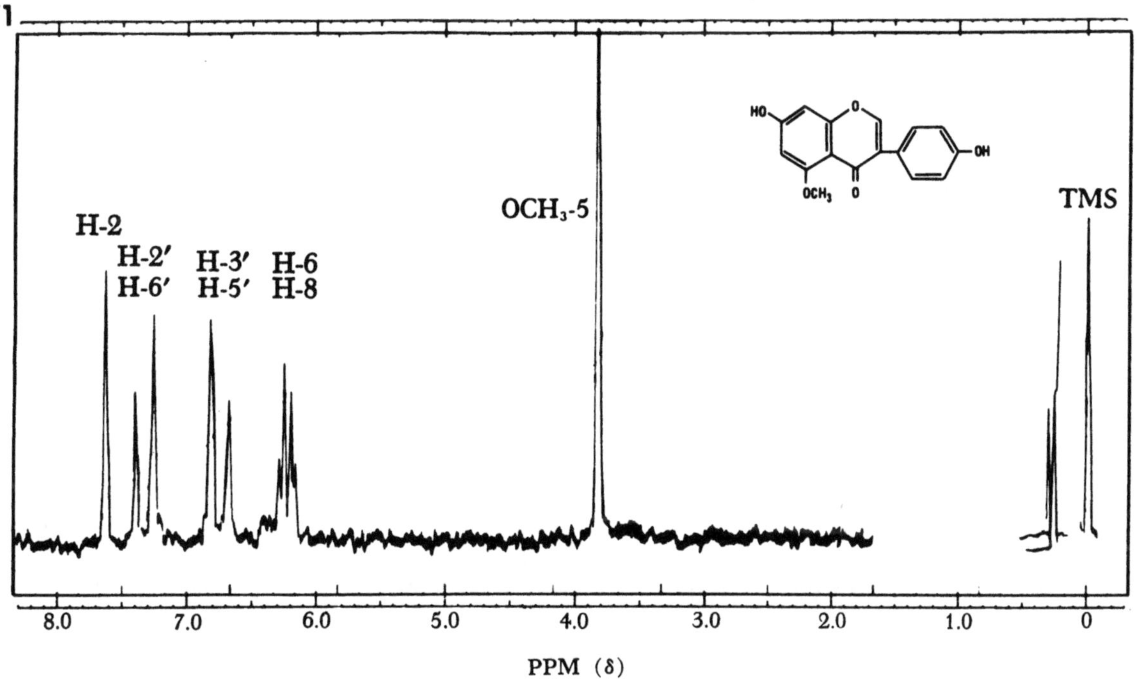

NMR of TMS Ether of Genistein 4′,5-Dimethyl Ether in CCl_4

72

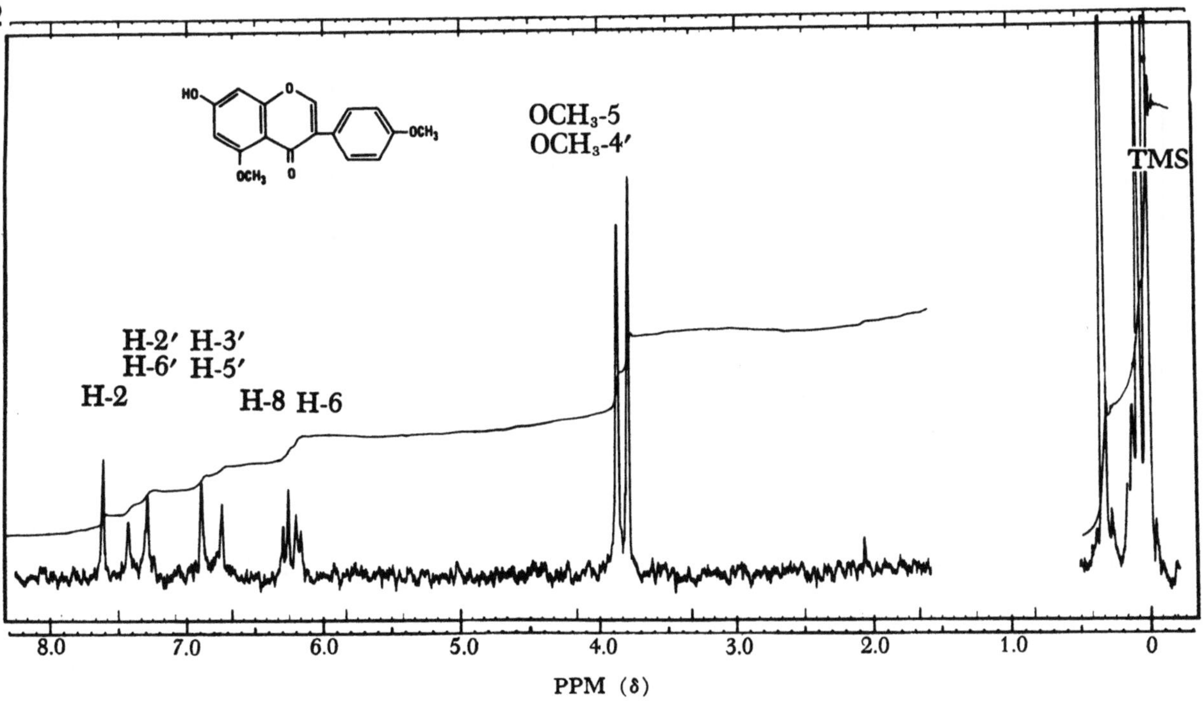

NMR of TMS Ether of 5,7-Dihydroxy-4′-methoxy-8-methylisoflavone in CCl_4

73

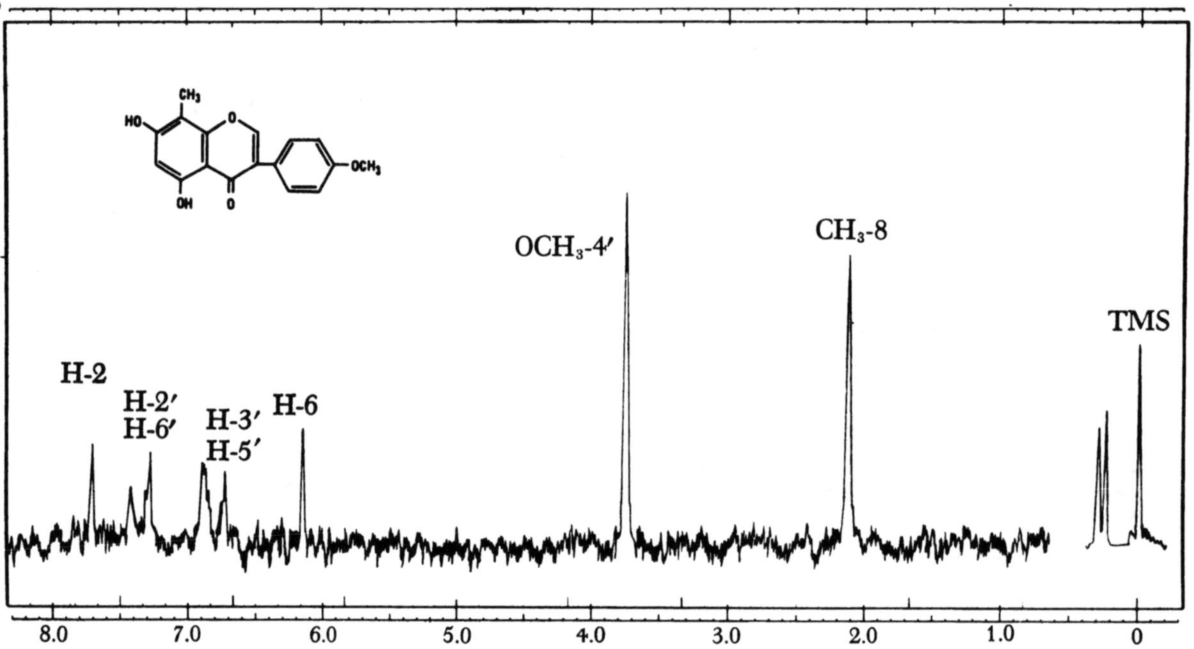

NMR of 4′,5,7-Trimethoxy-6-methylisoflavone in $CDCl_3$

74

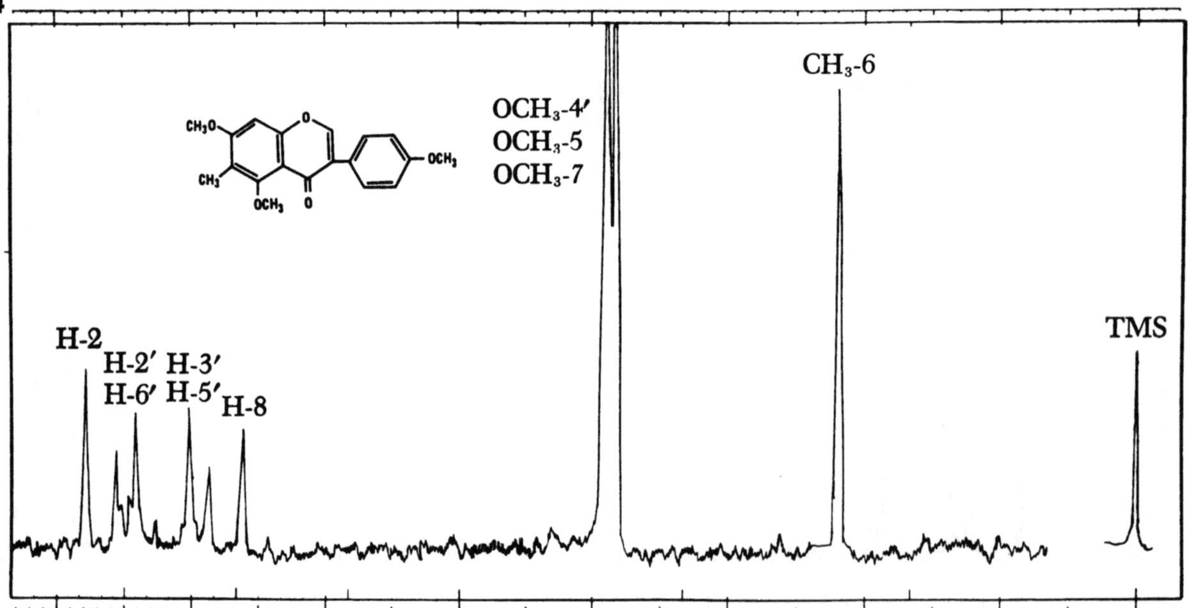

NMR of 4′,5,7-Trimethoxy-8-methylisoflavone in $CDCl_3$

75

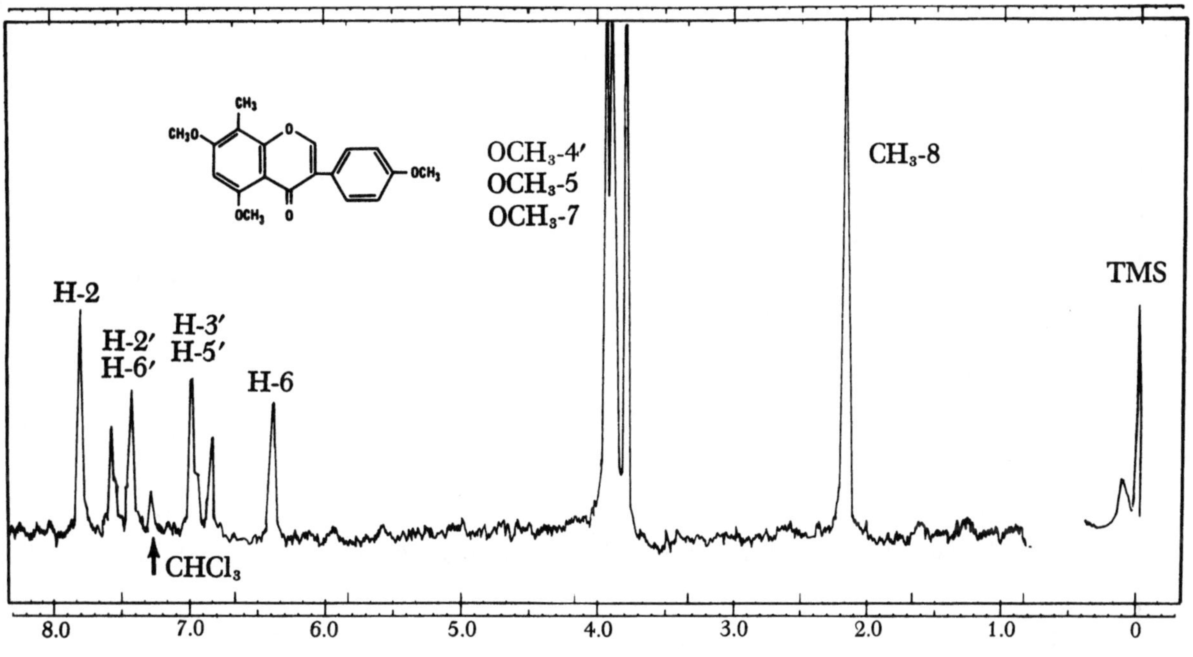

NMR of TMS Ether of 2-Carbethoxy-5,7-dihydroxy-4′-methoxy-8-methylisoflavone in CCl_4

76

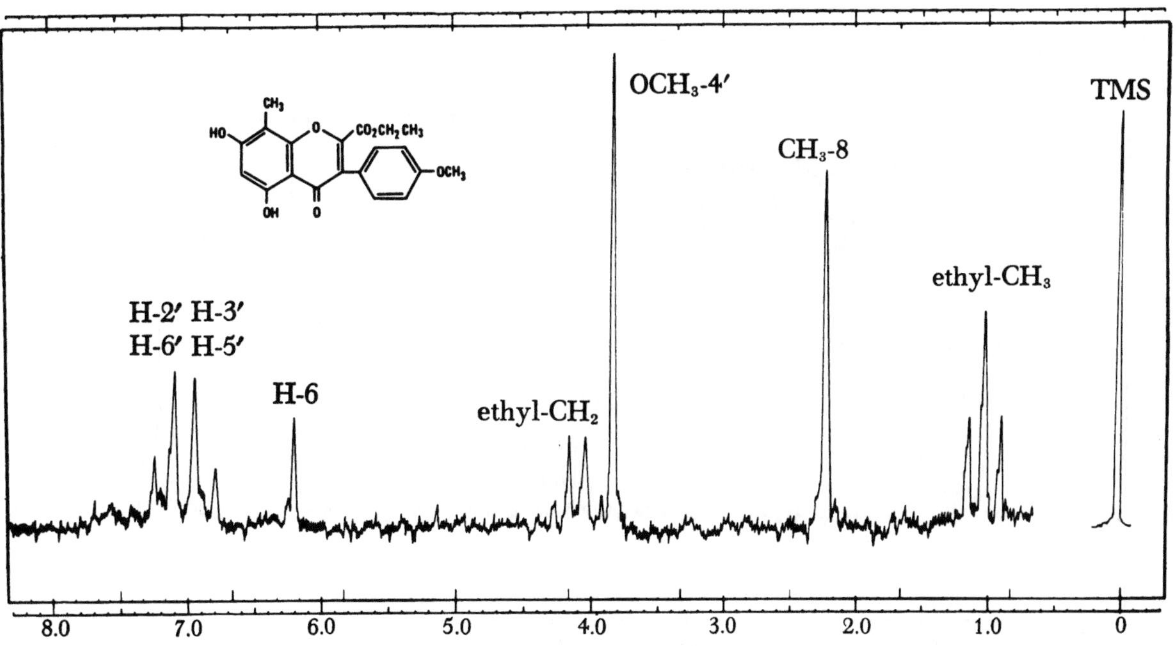

NMR of 2-Carbethoxy-5,7-dihydroxy-4′-methoxy-8-methyliso-
flavone in $CDCl_3$

77

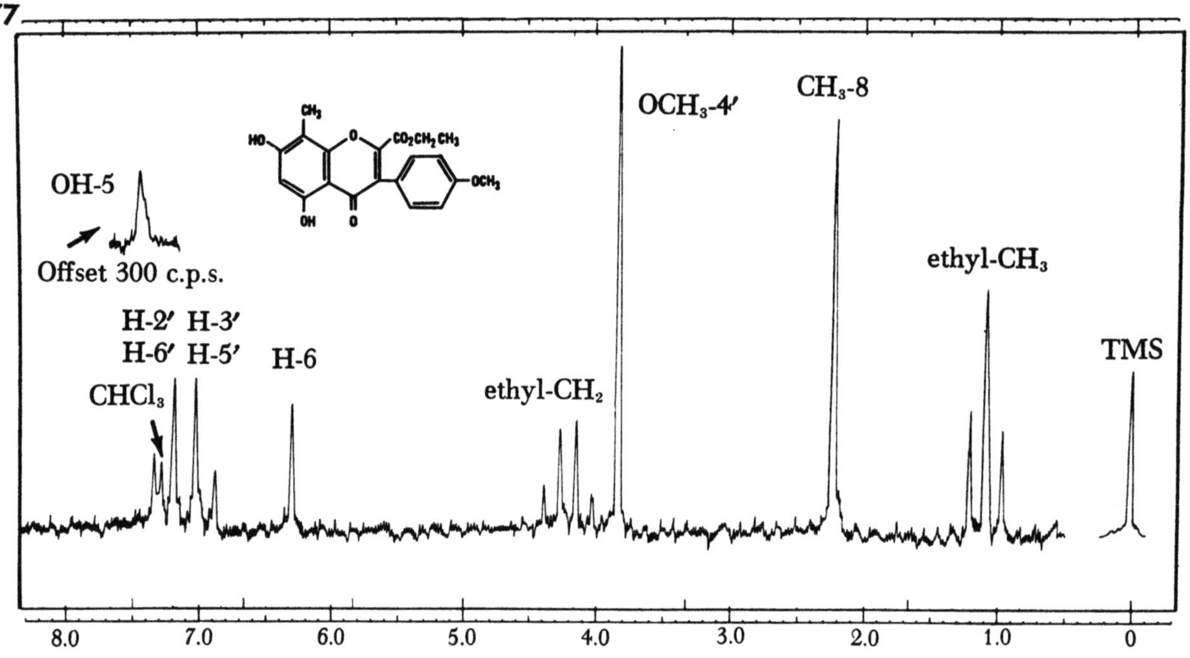

NMR of TMS Ether of 2-Carbethoxy-5,7-dihydroxy-4′-methoxy-
6-methylisoflavone in CCl_4

78

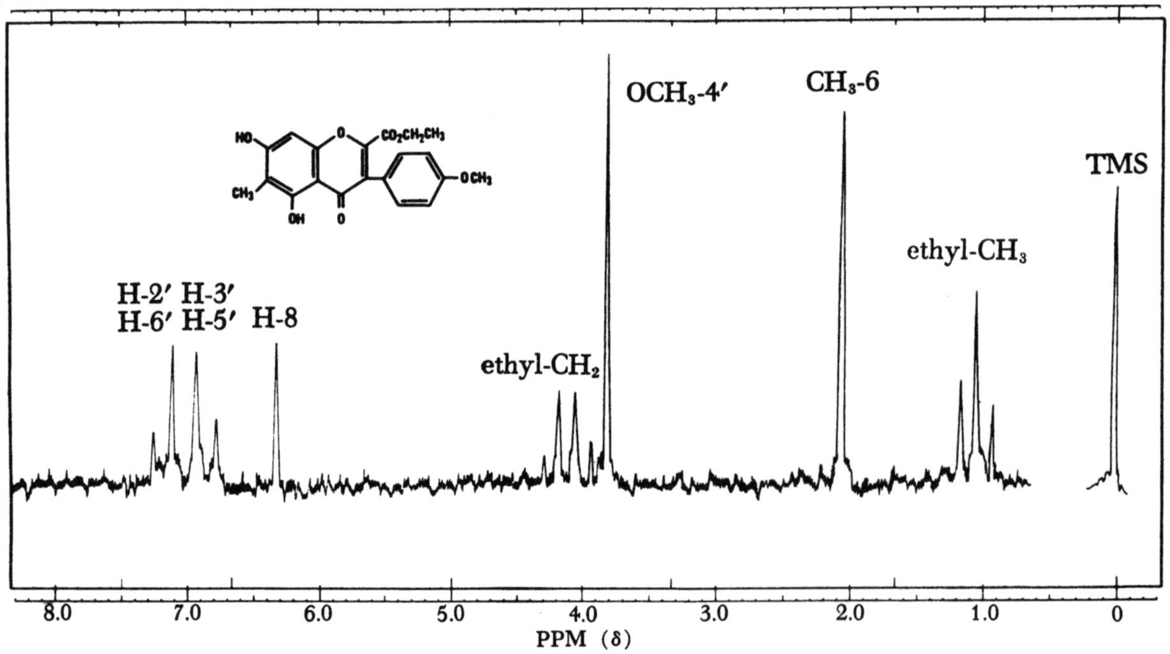

NMR of 2-Carbethoxy-5,7-dihydroxy-4′-methoxy-6-methylisoflavone in $CDCl_3$

79

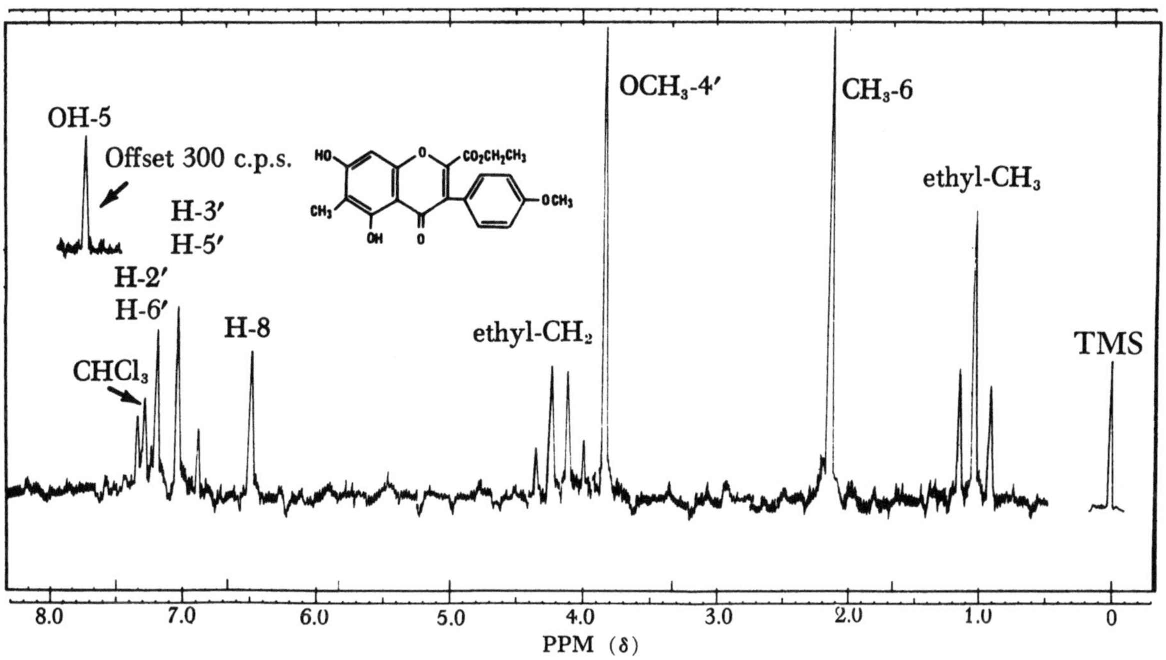

NMR of TMS Ether of Texasin 7-O-Glucoside in CCl_4

80

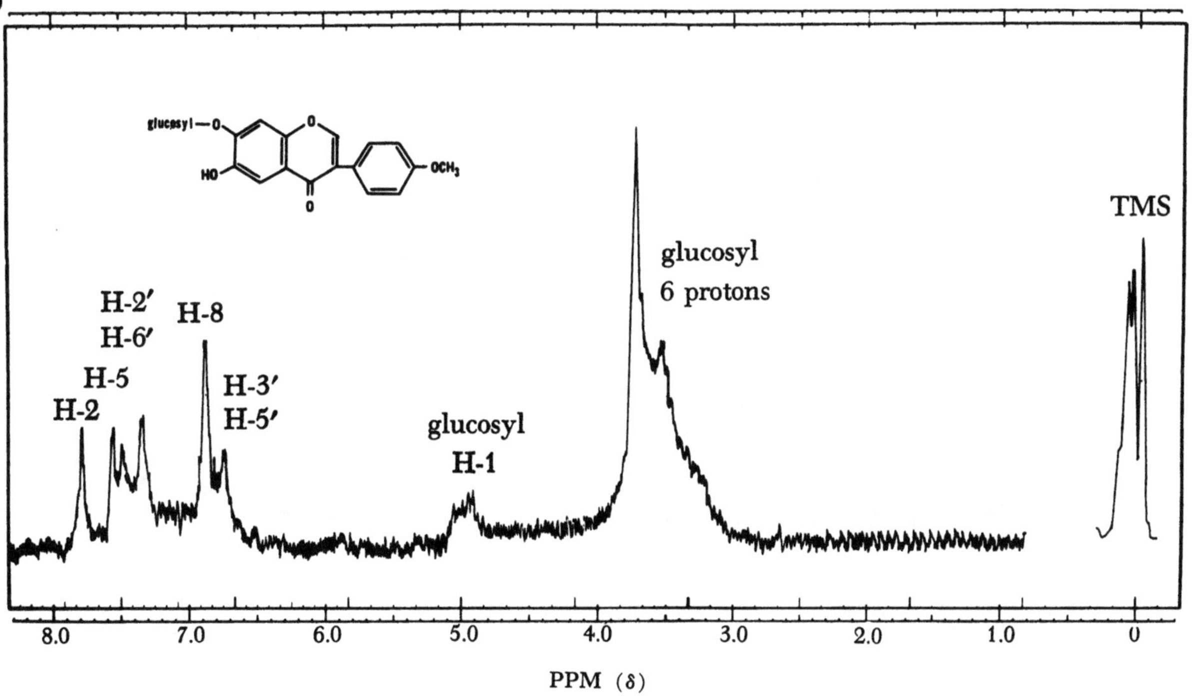

NMR of Afrormosin in $CDCl_3$

81

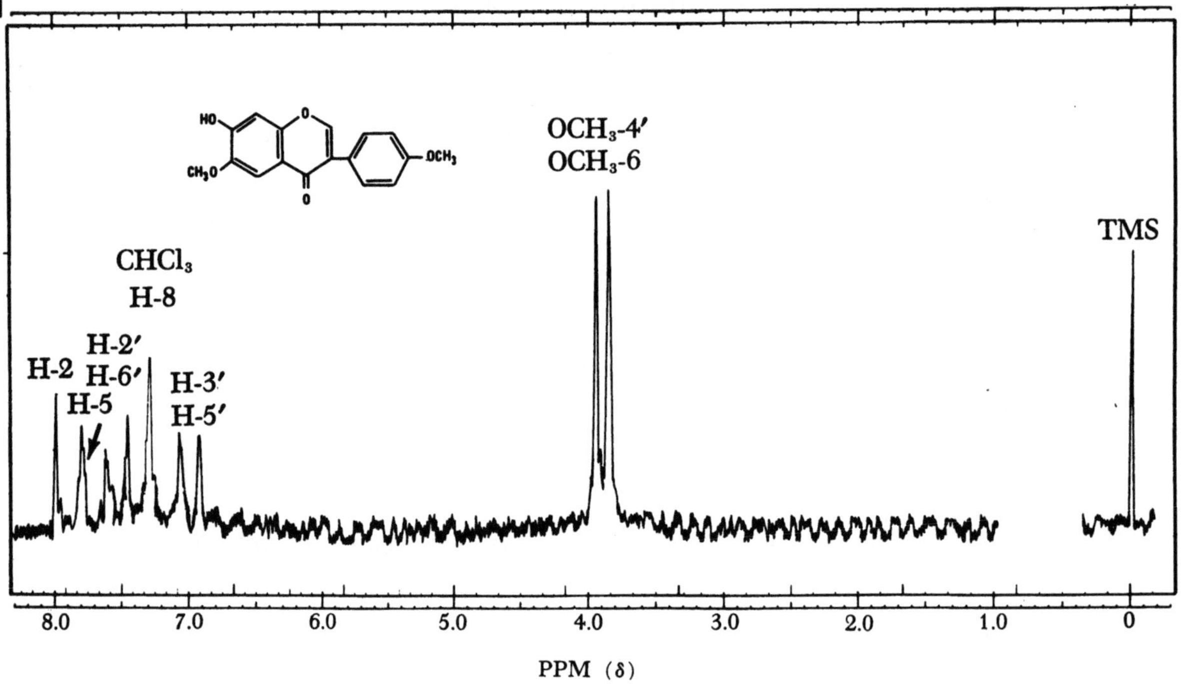

NMR of TMS Ether of Pseudobaptisin in CCl_4

82

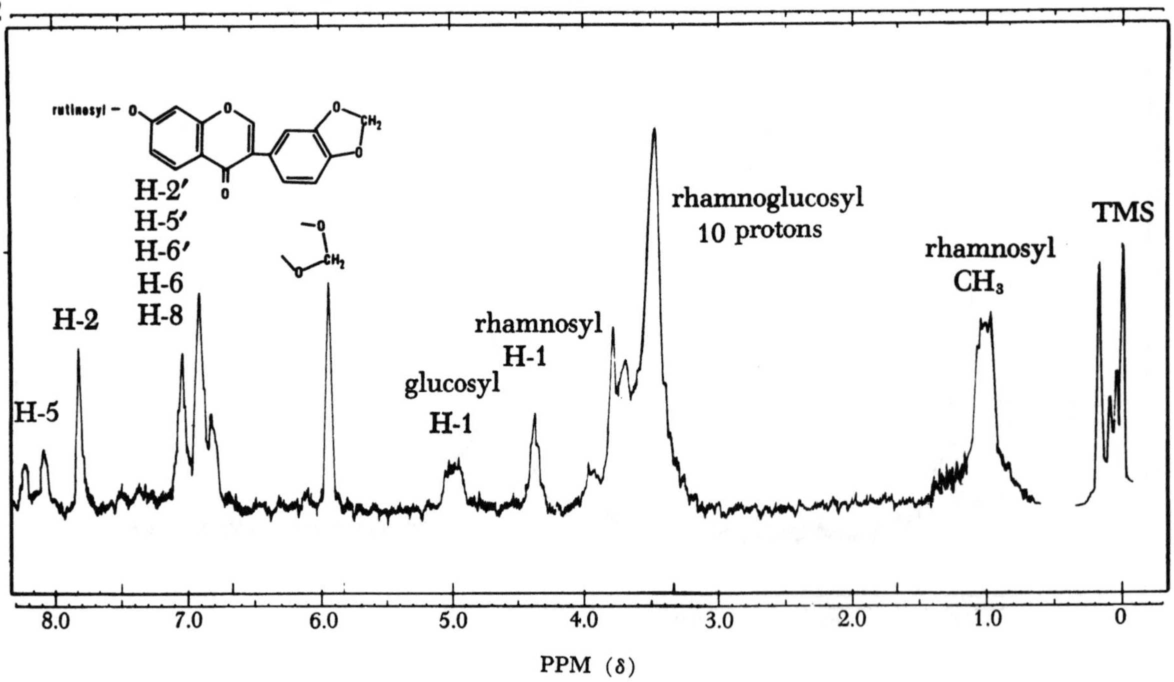

NMR of Pseudobaptisin in DMSO (Deuterated)

83

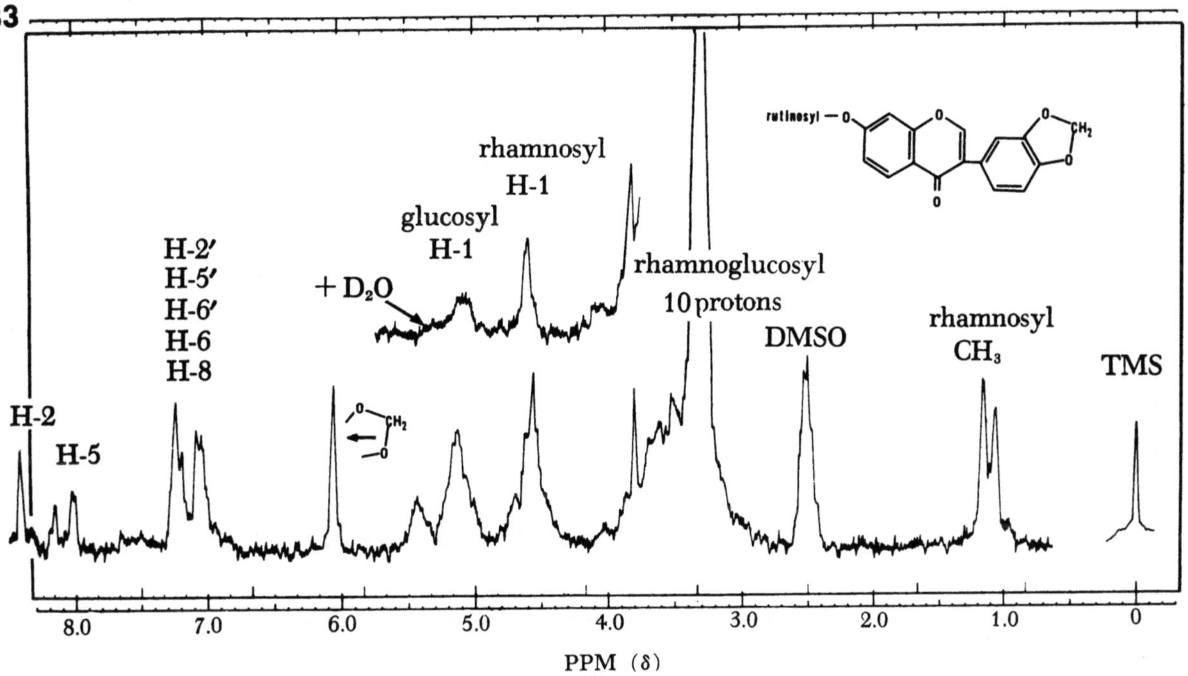

NMR of 5-Hydroxy-7,8-dimethoxyisoflavone in $CDCl_3$

84

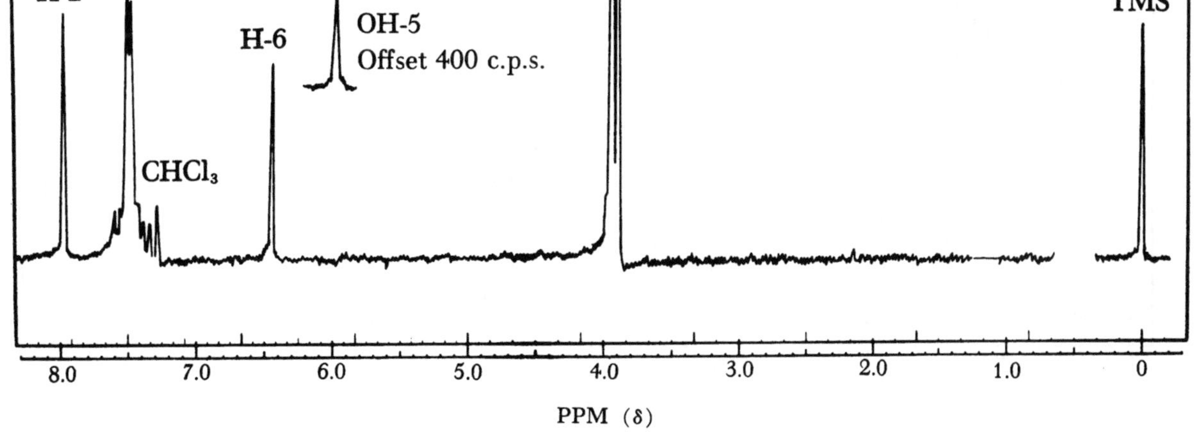

NMR of TMS Ether of Orobol in CCl_4

85

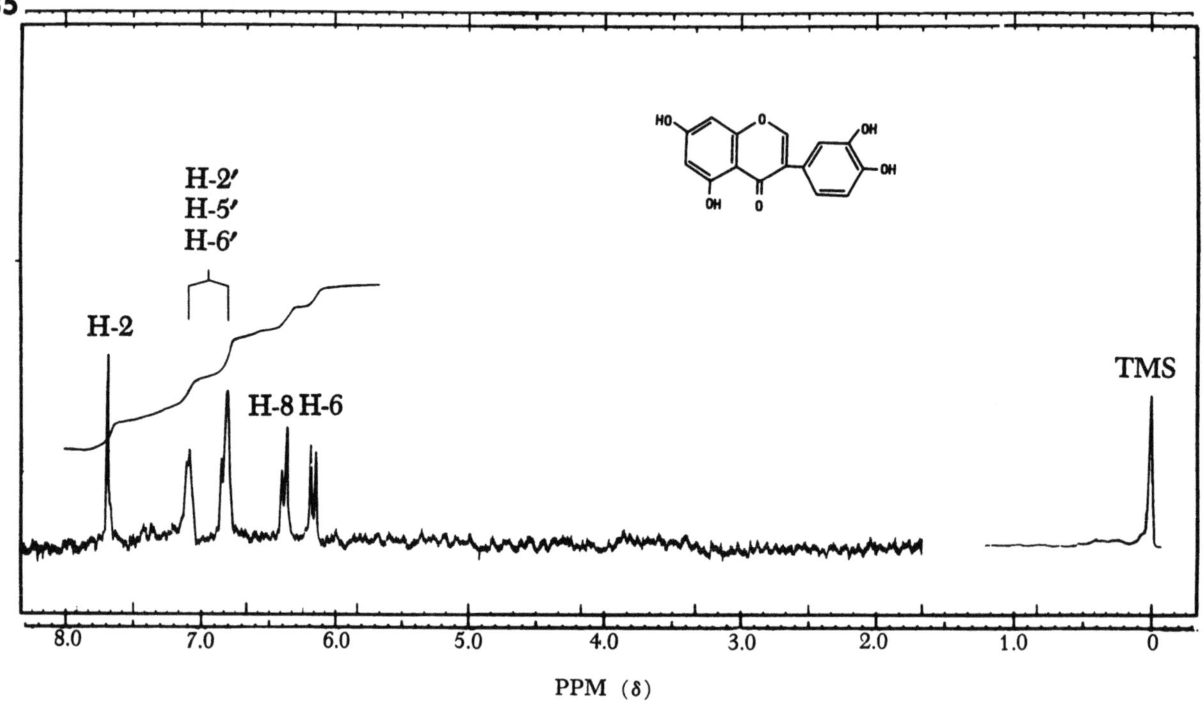

NMR of TMS Ether of Orobol 7-O-Glucoside in CCl_4

86

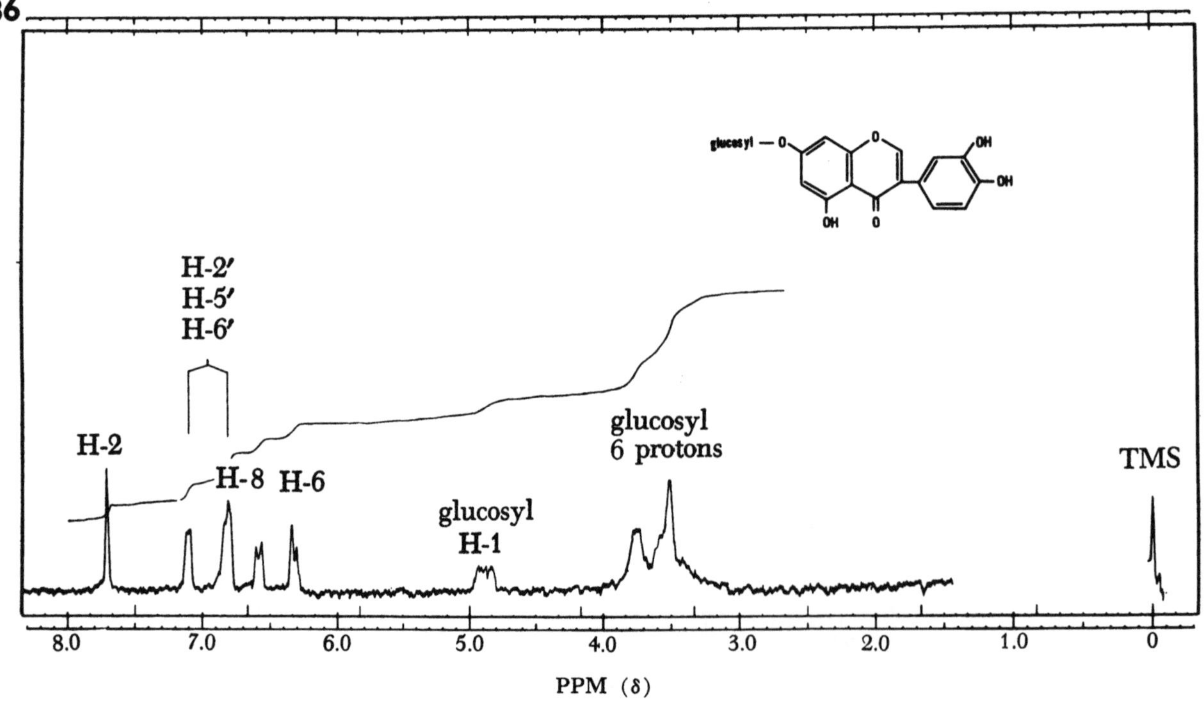

NMR of TMS Ether of Irisolidone in CCl₄

87

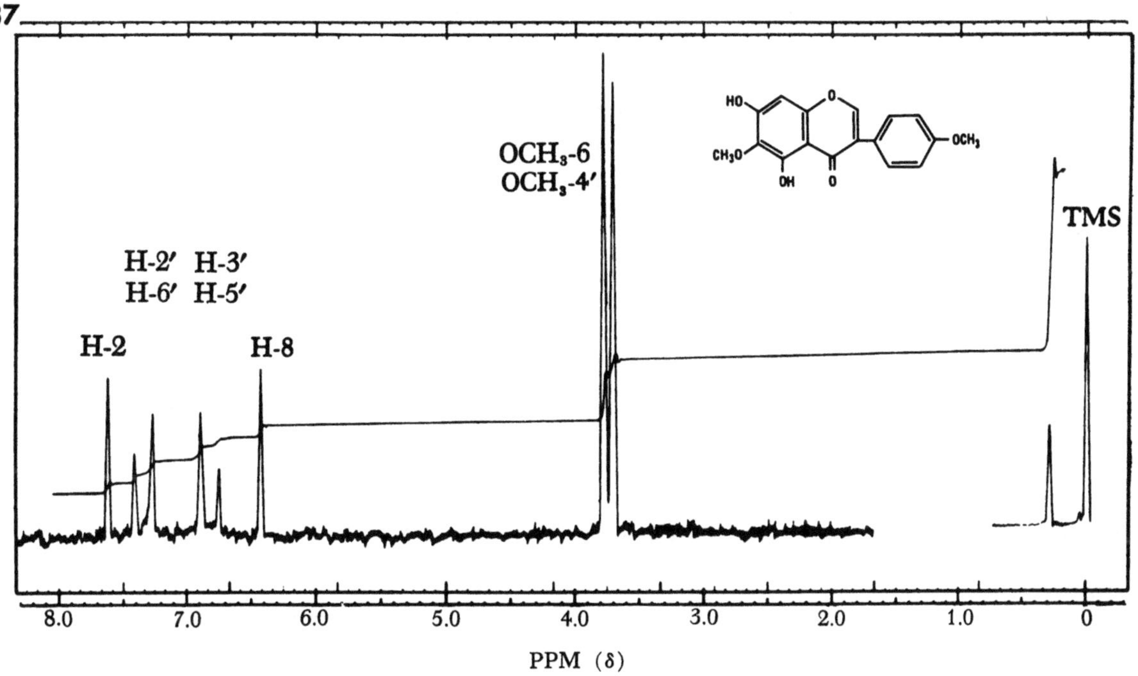

NMR of TMS Ether of Tectoridin in CCl₄

88

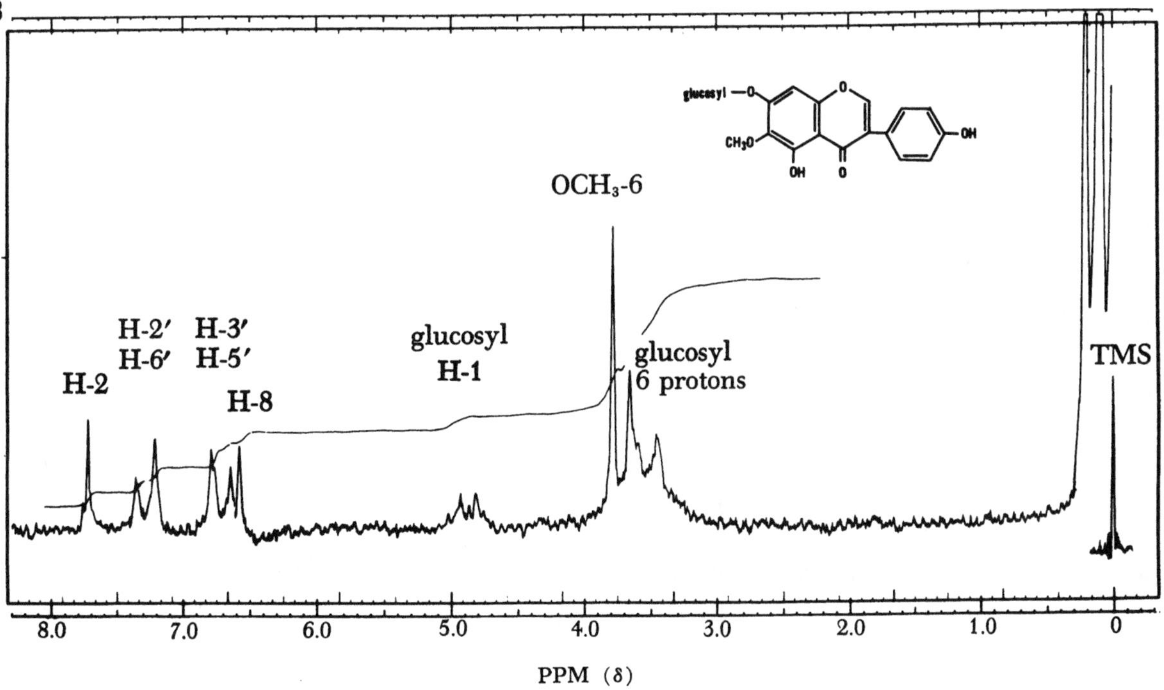

NMR of TMS Ether of Irigenin in CCl₄

89

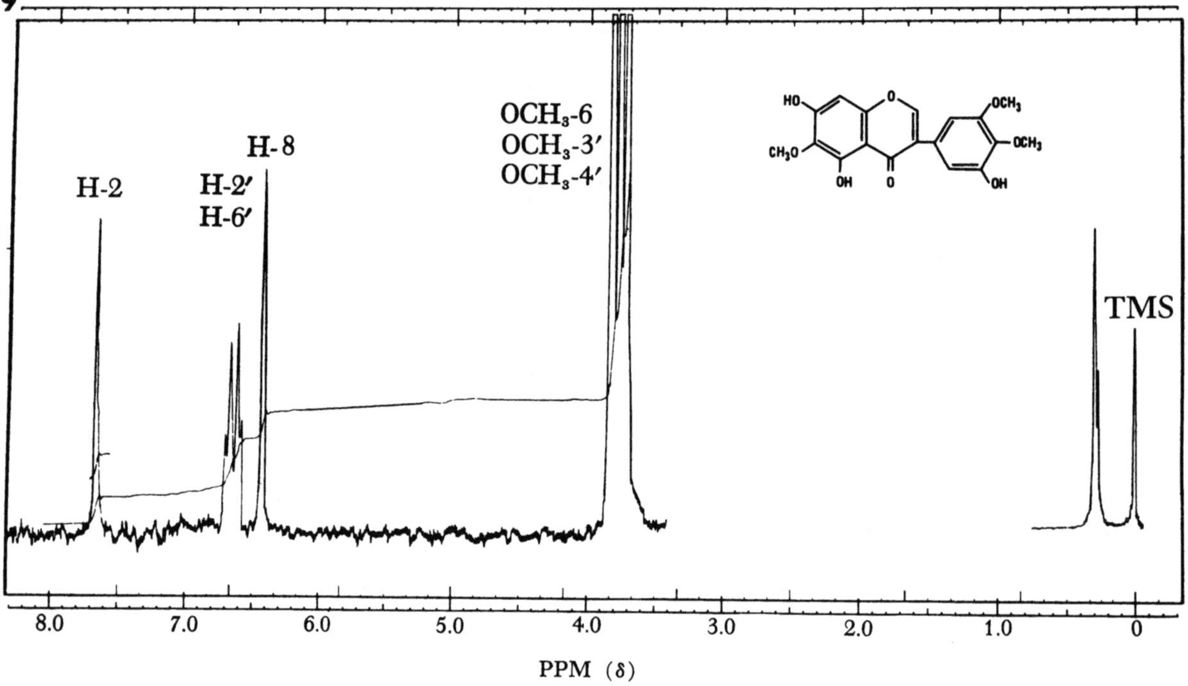

NMR of TMS Ether of Iridin in CCl₄

90

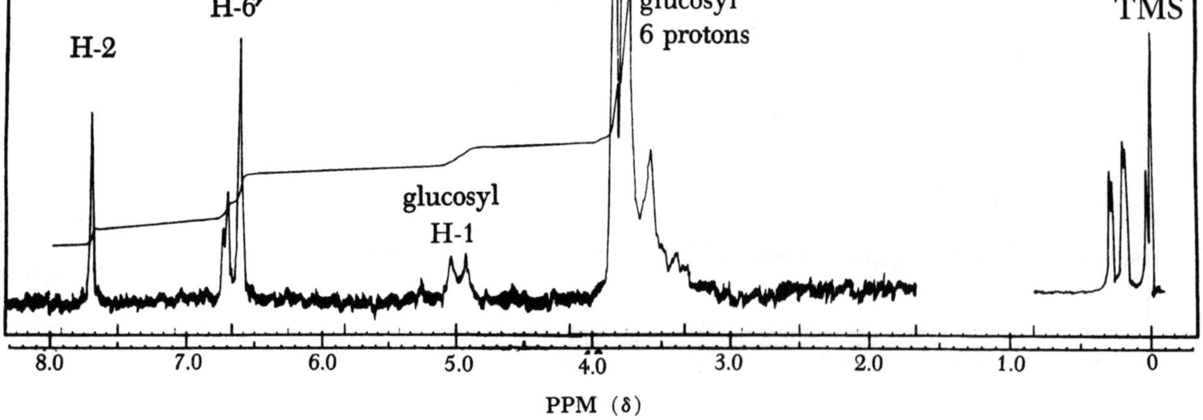

NMR of TMS Ether of Lecontin in CCl_4

91

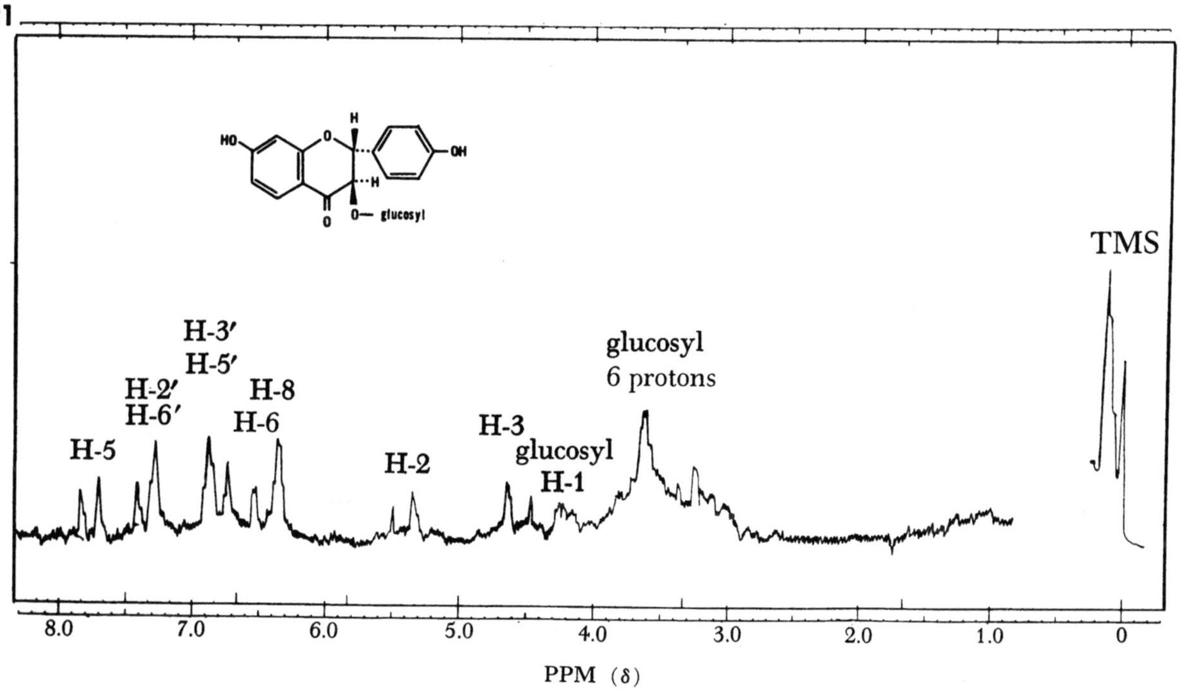

NMR of TMS Ether of Dihydrokaempferol in CCl_4

92

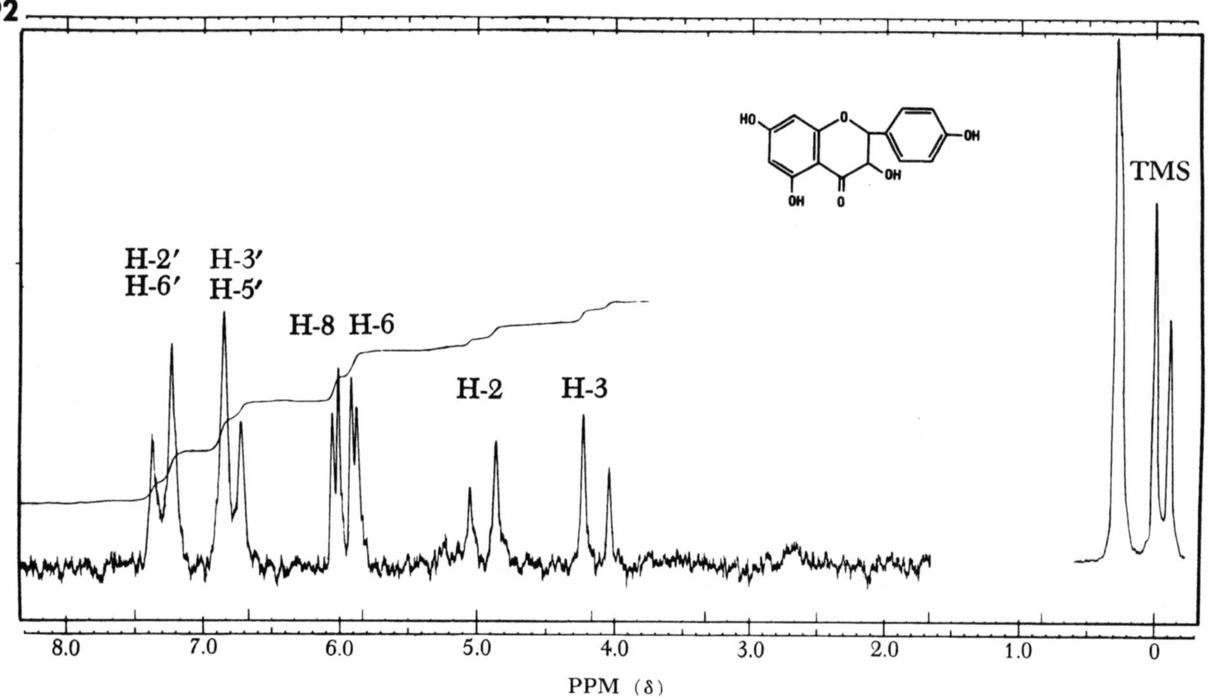

NMR of TMS Ether of Engeletin in CCl_4

93

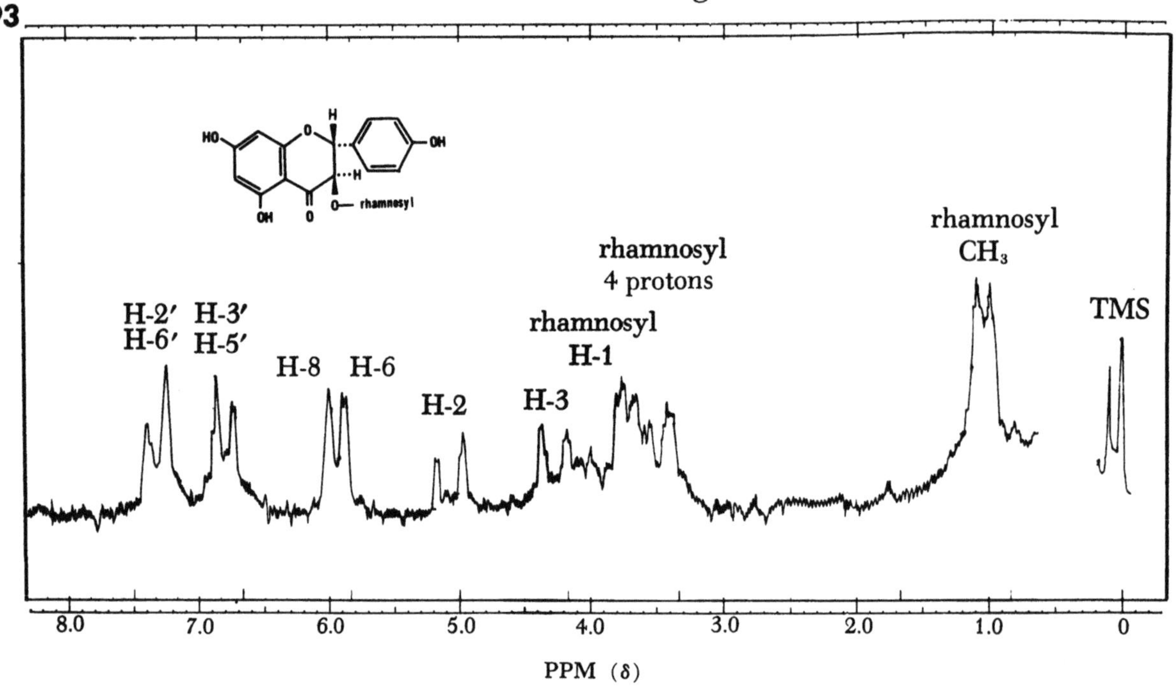

NMR of TMS Ether of Dihydrofisetin in CCl_4

94

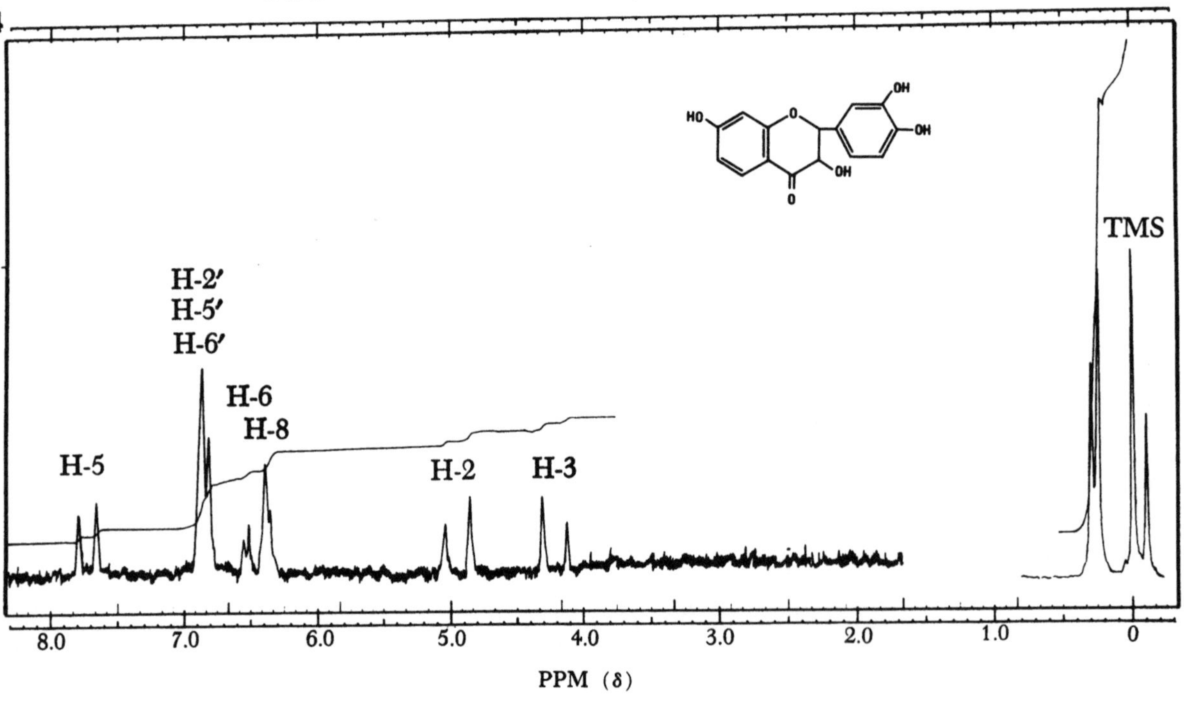

NMR of TMS Ether of Dihydroquercetin in CCl_4

95

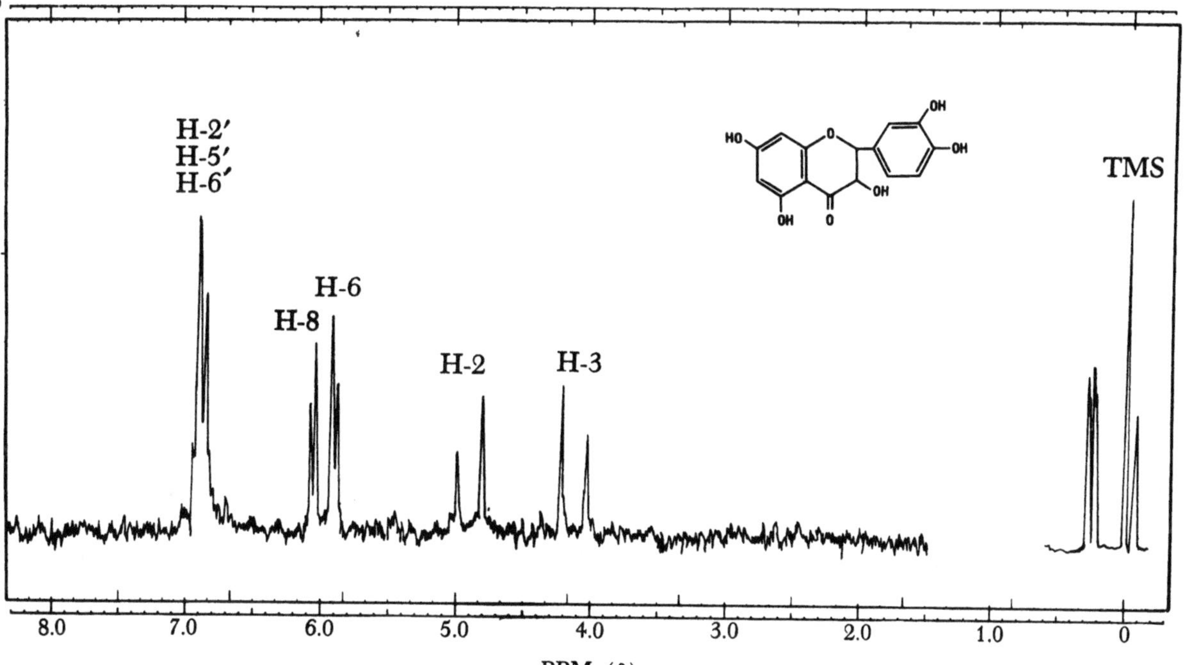

NMR of TMS Ether of Astilbin in CCl_4

96

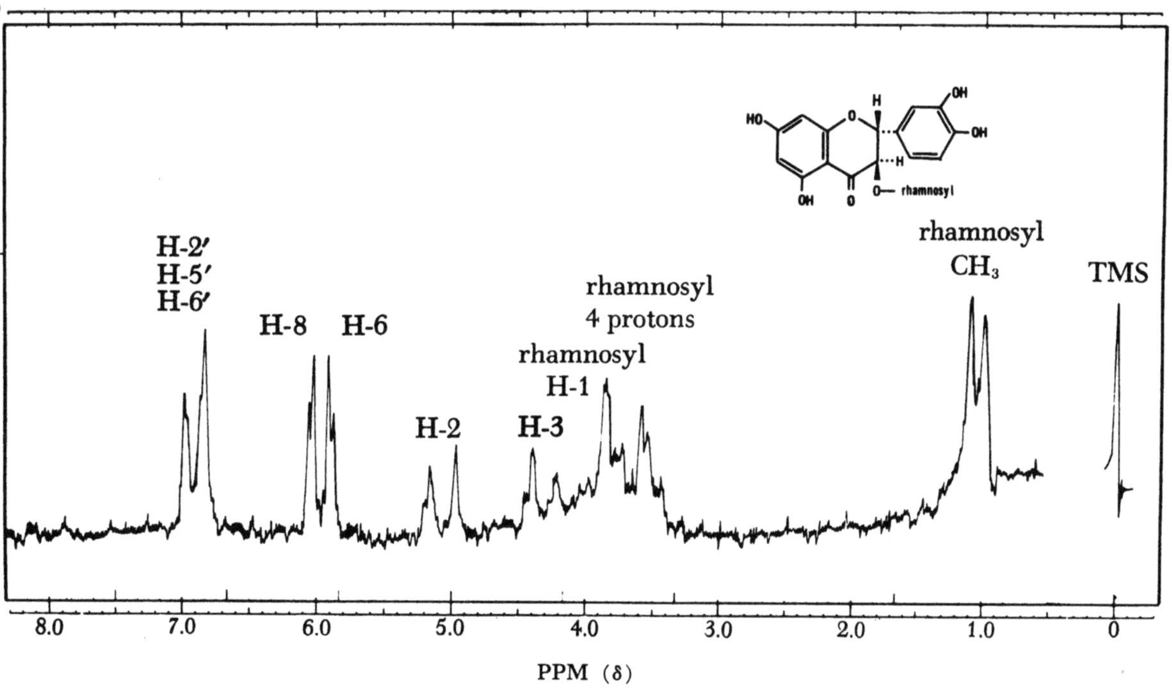

NMR of TMS Ether of Dihydrorobinetin in CCl_4

97

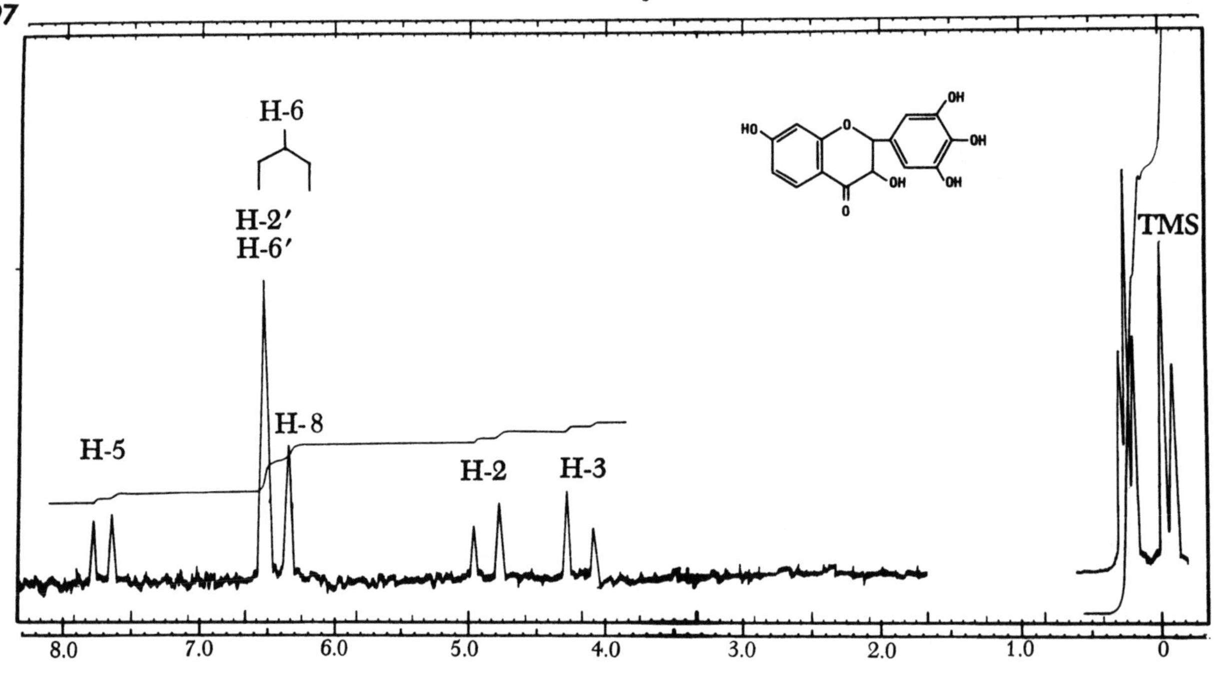

NMR of TMS Ether of 6-Hydroxyflavanone in CCl_4

98

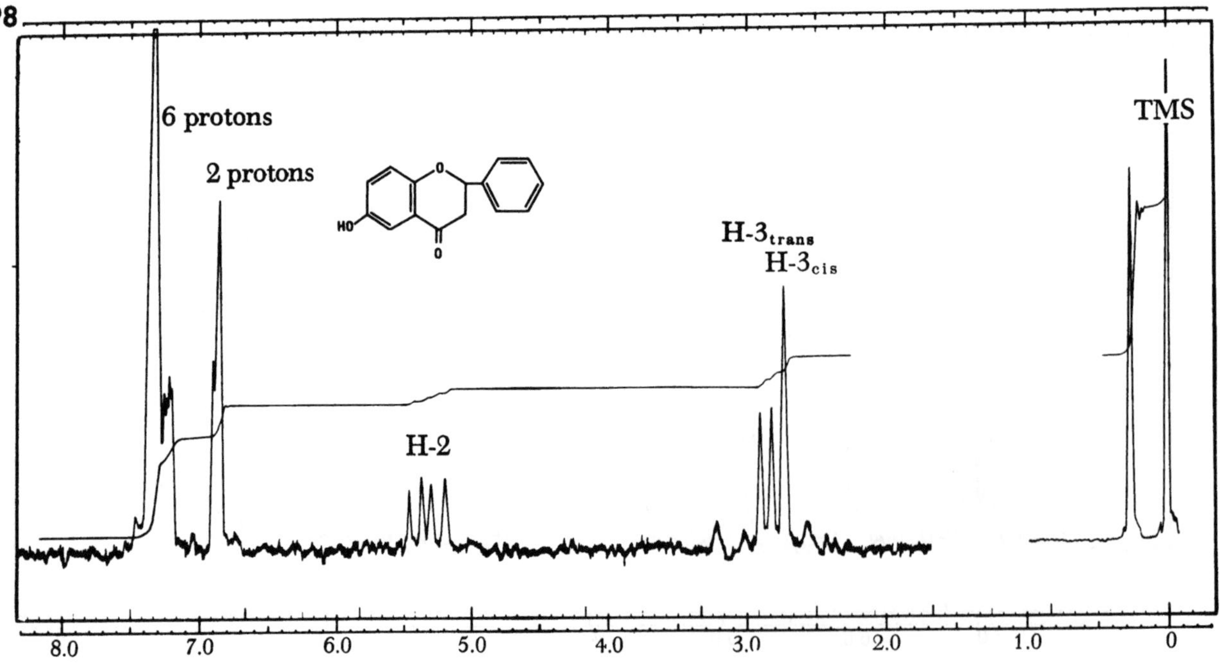

NMR of TMS Ether of Liquiritigenin in CCl_4

99

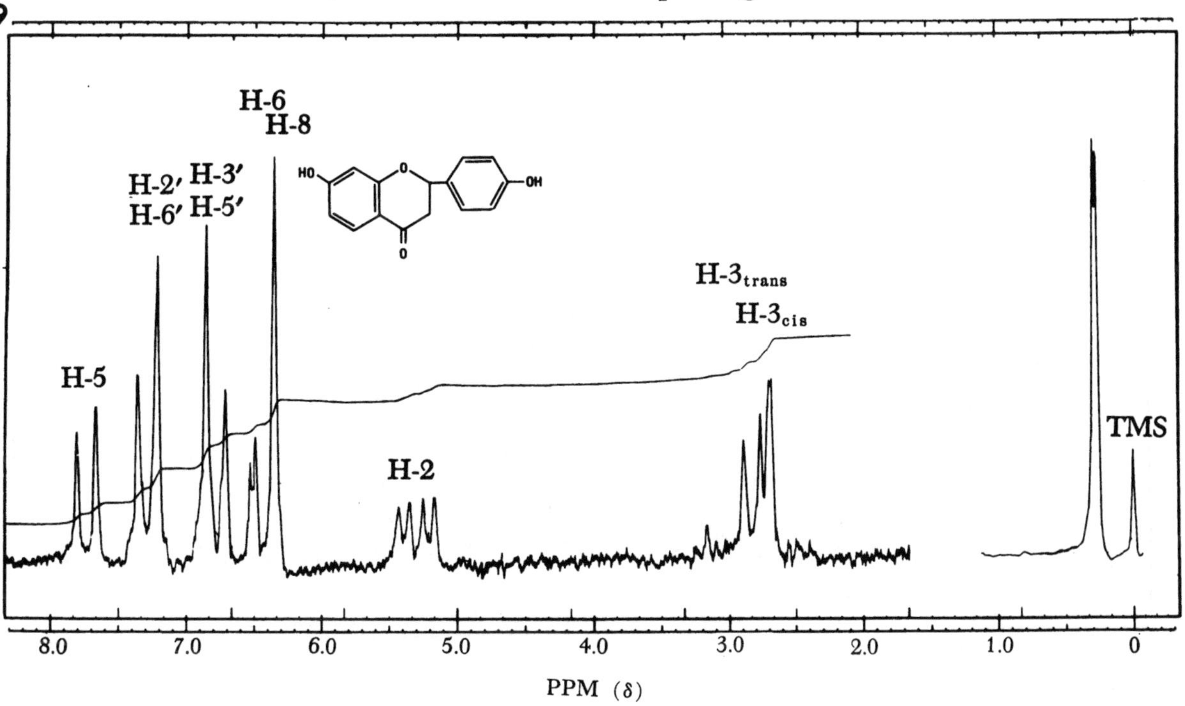

NMR of TMS Ether of Naringenin in CCl_4

100

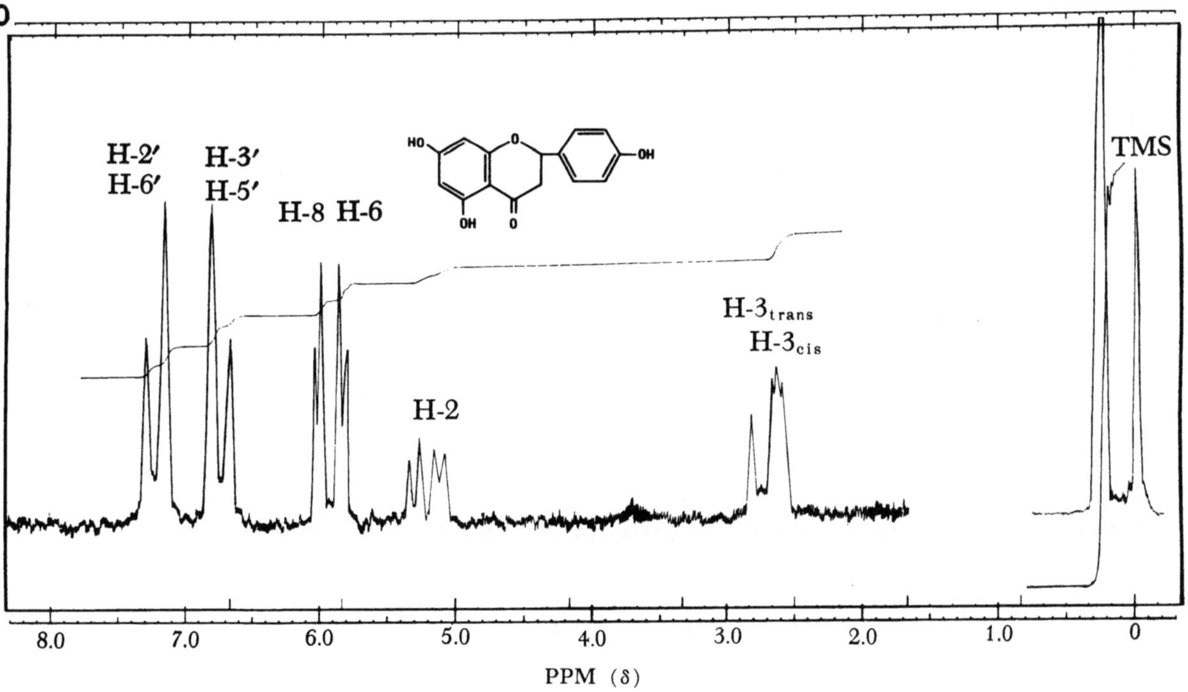

NMR of TMS Ether of Naringin in CCl_4

101

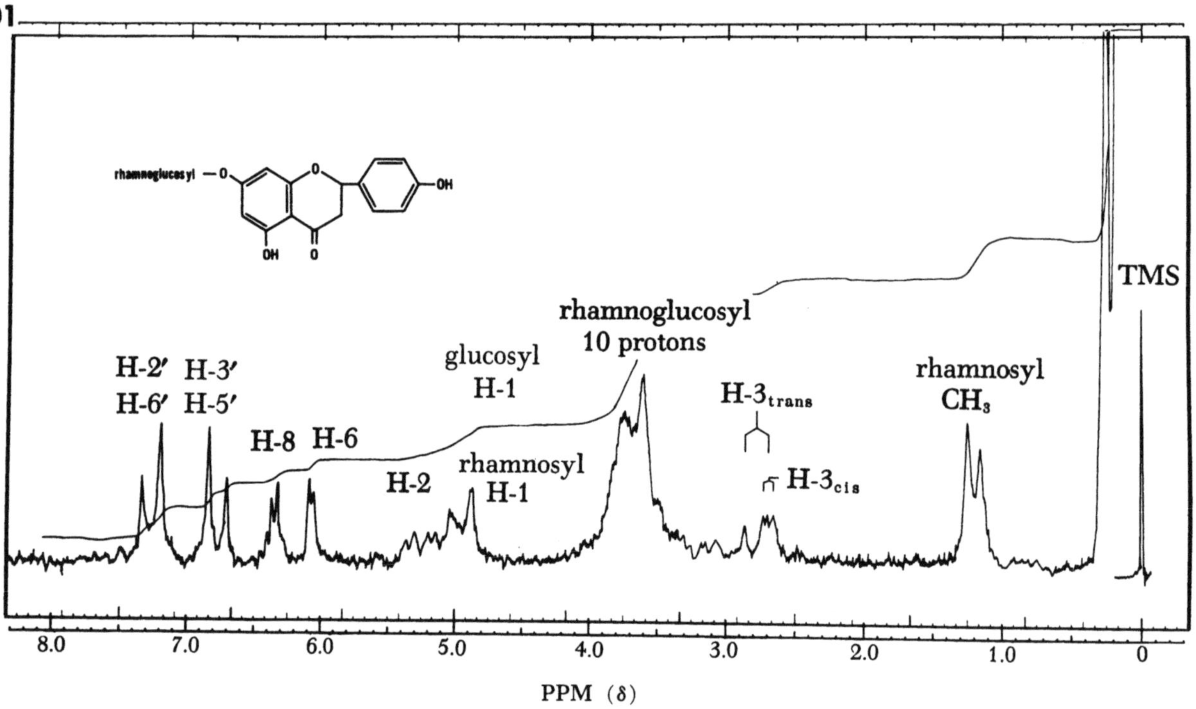

NMR of TMS Ether of Sakuranetin in CCl_4

102

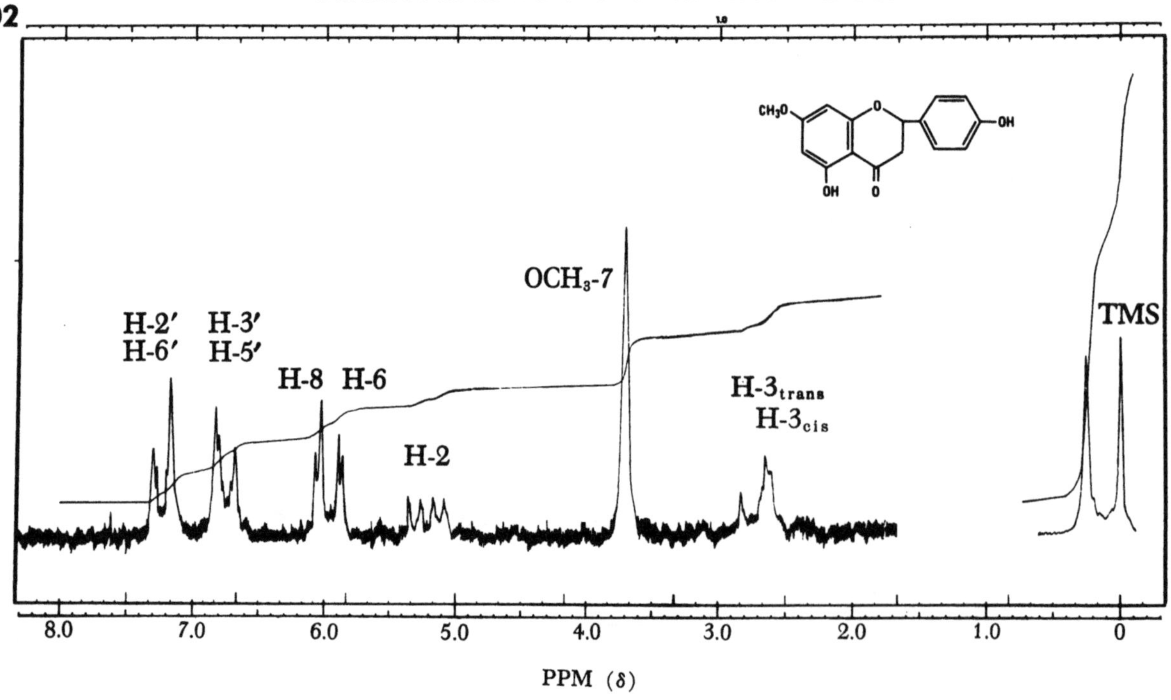

NMR of TMS Ether of Sakuranin in CCl_4

103

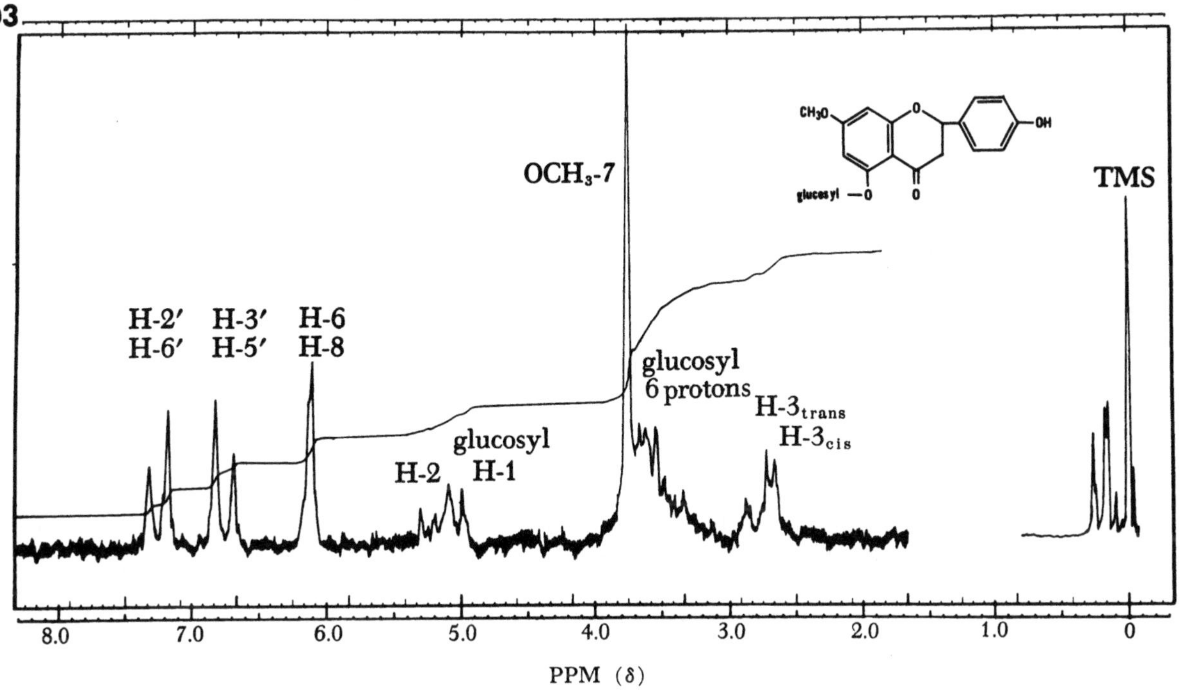

NMR of Naringenin 4′,7-Dimethyl Ether in $CDCl_3$

104

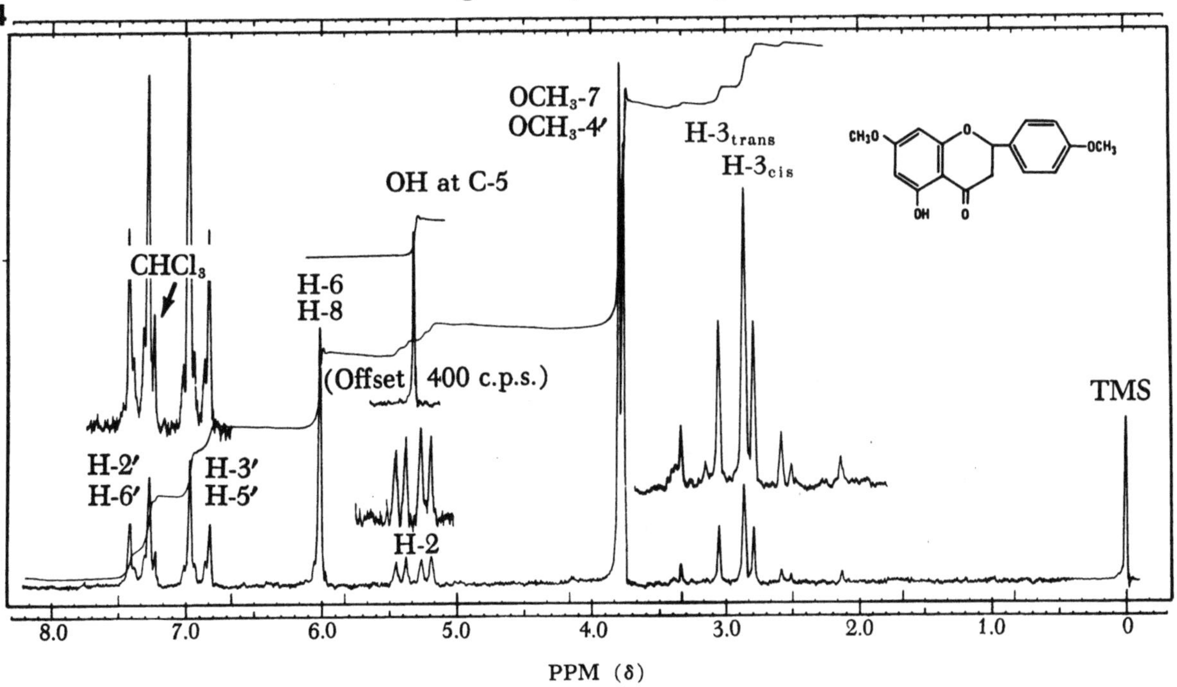

NMR of TMS Ether of Homoeriodictyol in CCl_4

105

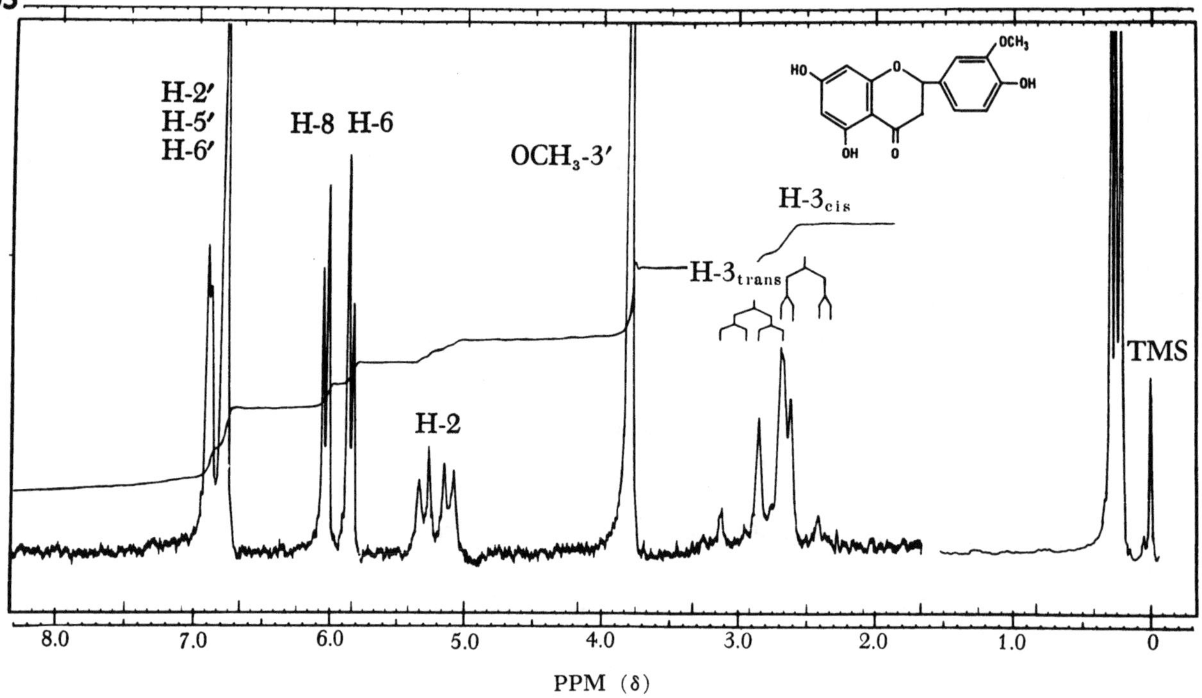

NMR of TMS Ether of Hesperetin in CCl_4

106

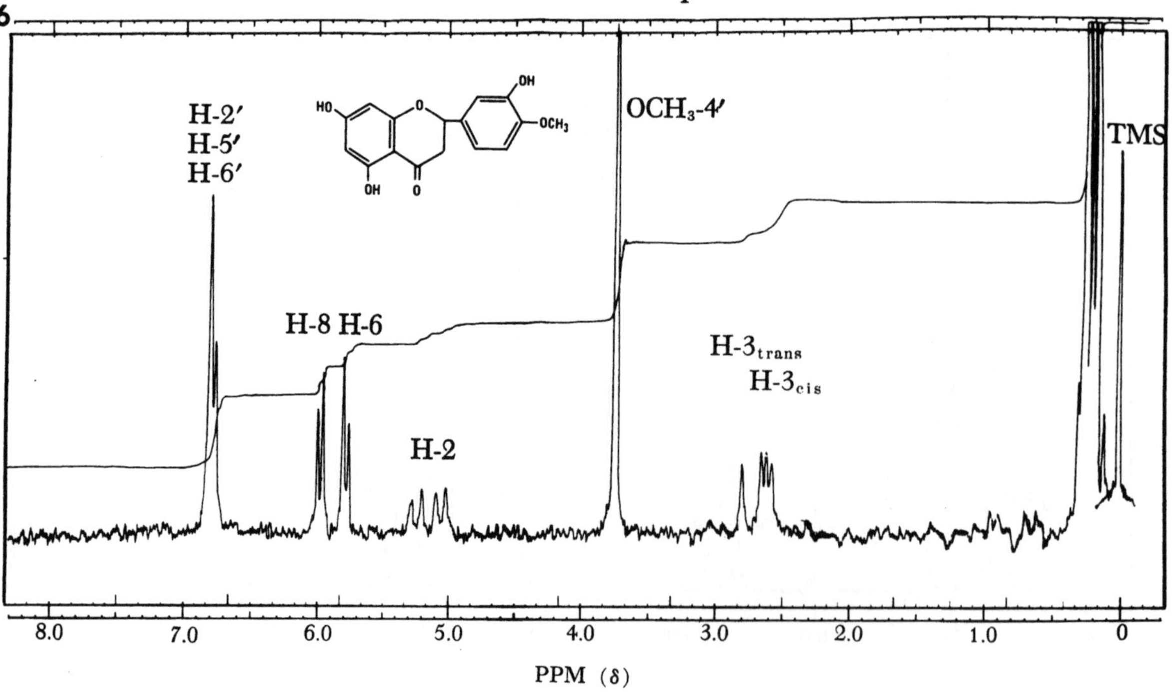

NMR of TMS Ether of Hesperidin in CCl_4

107

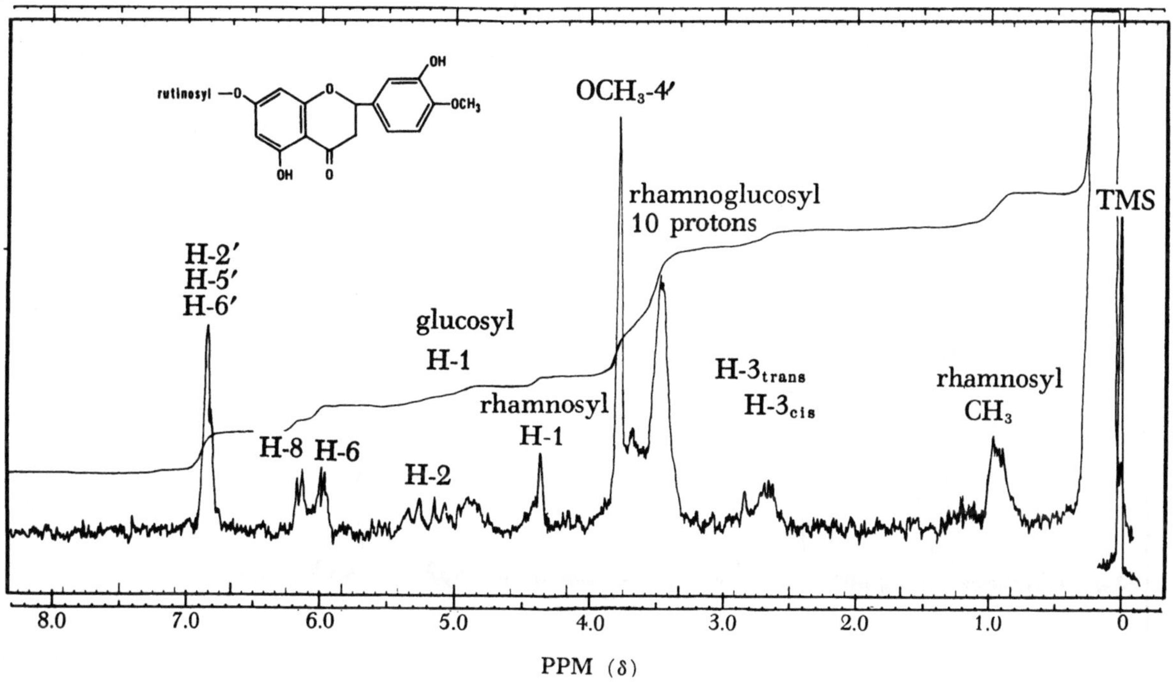

NMR of TMS Ether of Neohesperidin in CCl_4

08

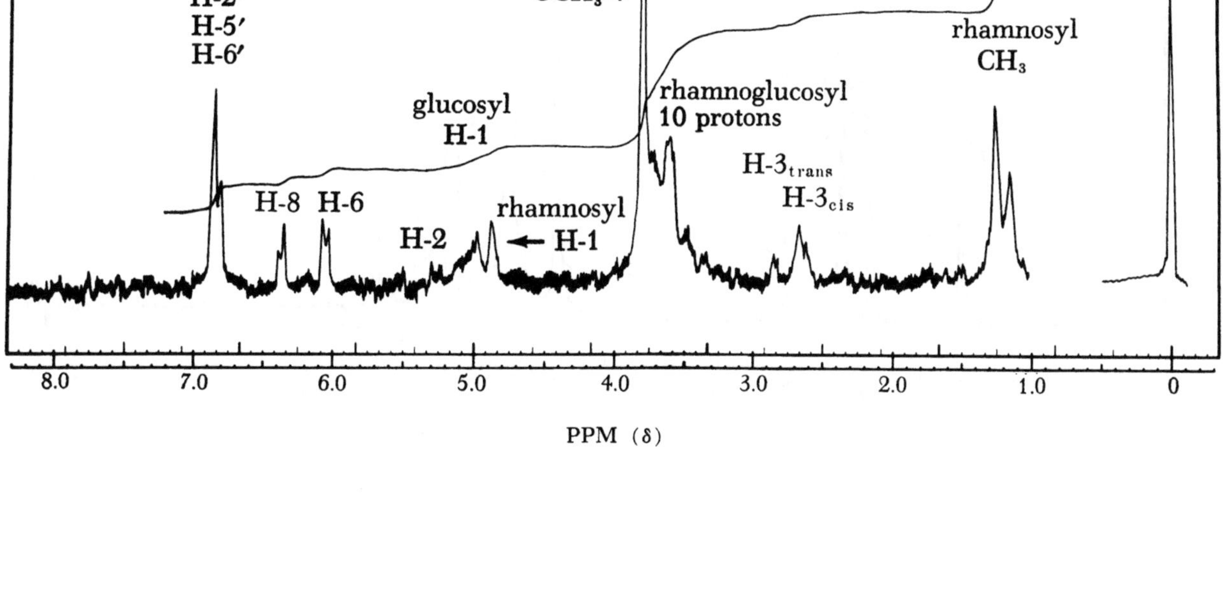

NMR of TMS Ether of 5,7-Dihydroxy-3′,4′-dimethoxyflavanone in CCl_4

109

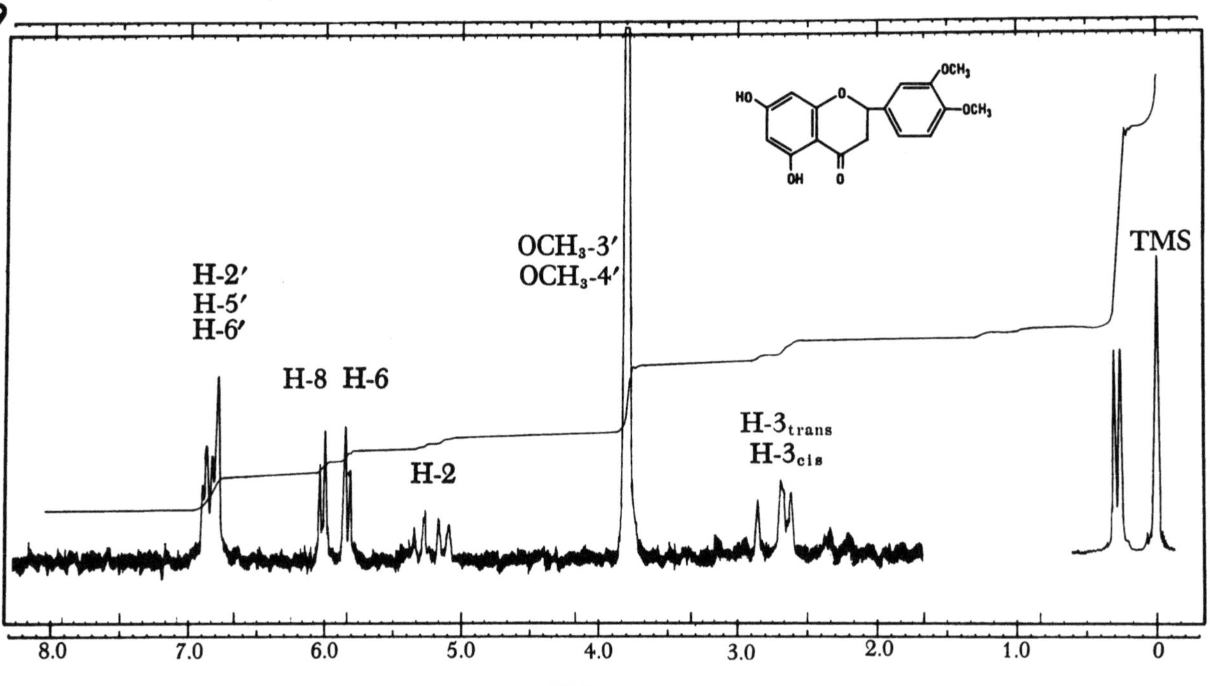

NMR of Hesperetin Quinone in DMSO (Deuterated)

110

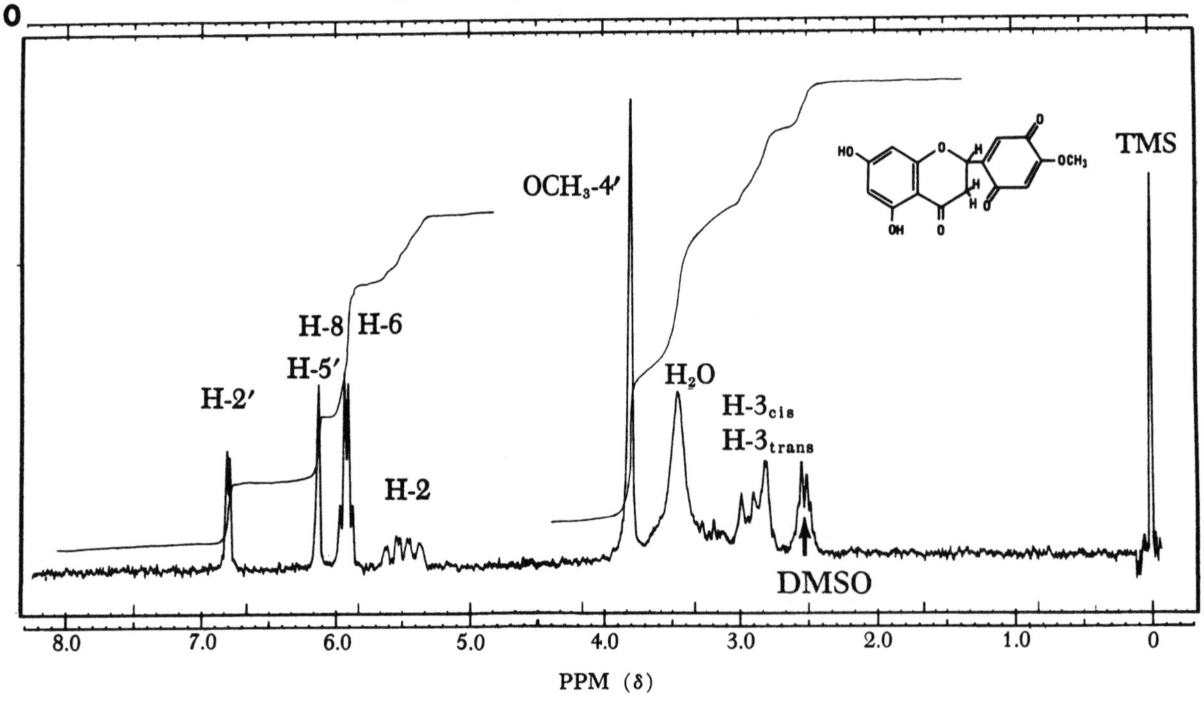

NMR of 4,4′-Dimethoxychalcone in $CDCl_3$

111

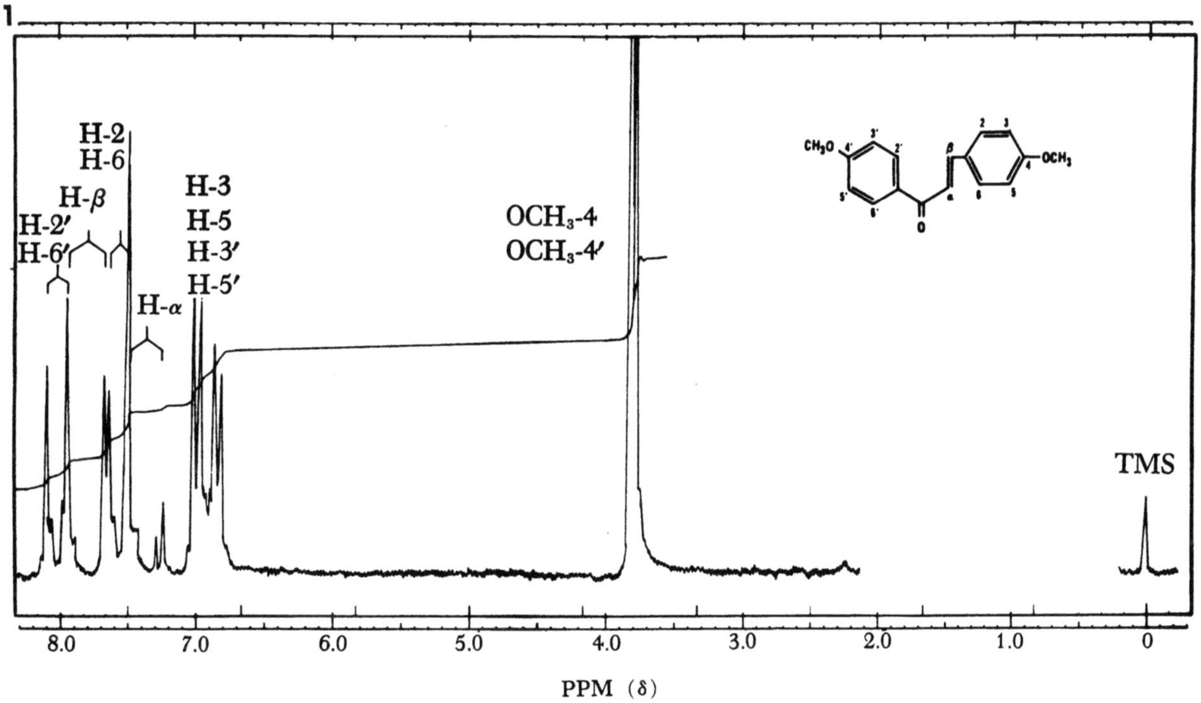

NMR of TMS Ether of 2′,4-Dihydroxy-4′-methoxychalcone in CCl_4

12

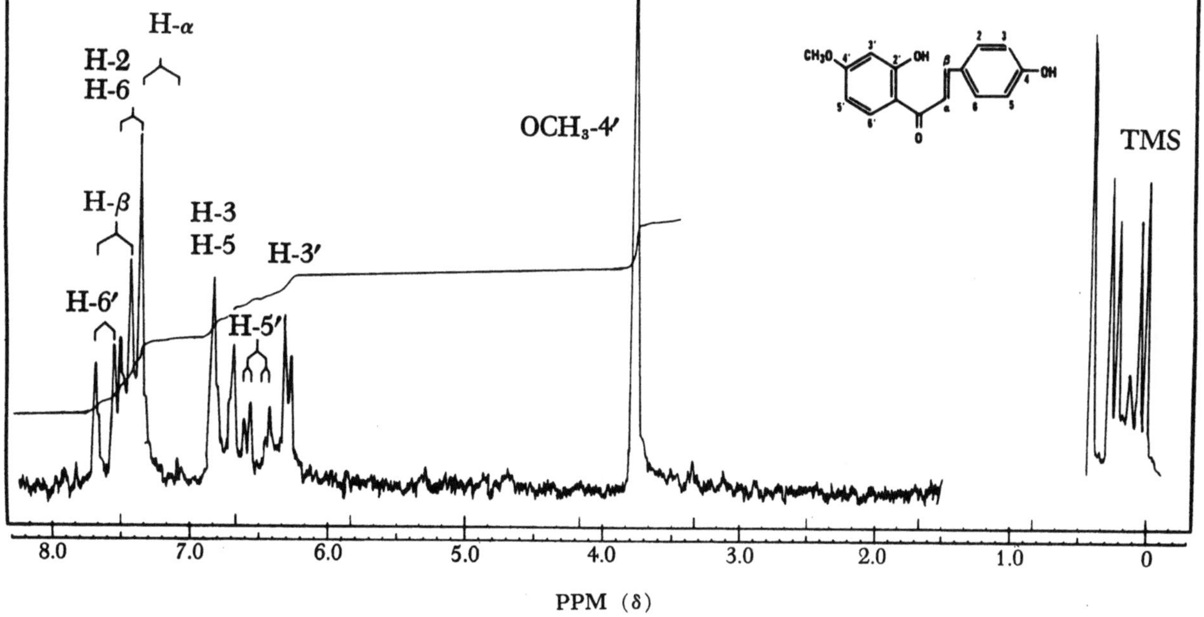

NMR of 2′-Hydroxy-4,4′-dimethoxychalcone in $CDCl_3$

113

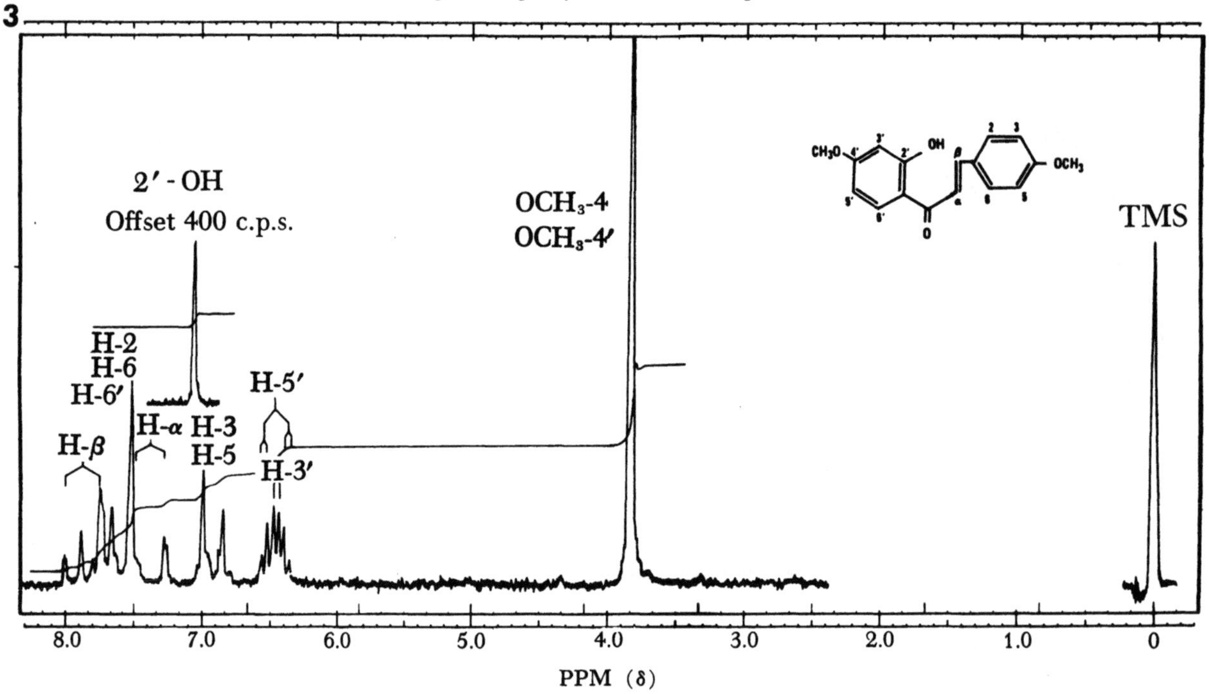

NMR of TMS Ether of 2′,3,4,4′-Tetrahydroxychalcone in CCl_4

114

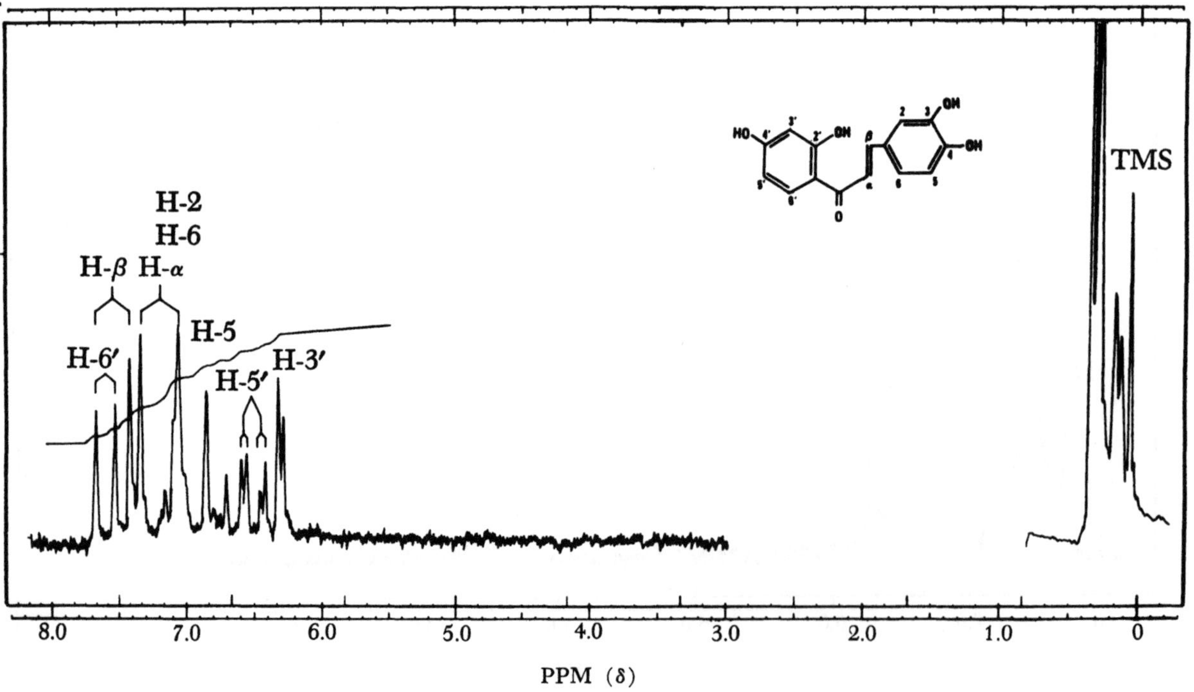

NMR of TMS Ether of 2′,4′-Dihydroxy-3,4-dimethoxychalcone in CCl_4

115

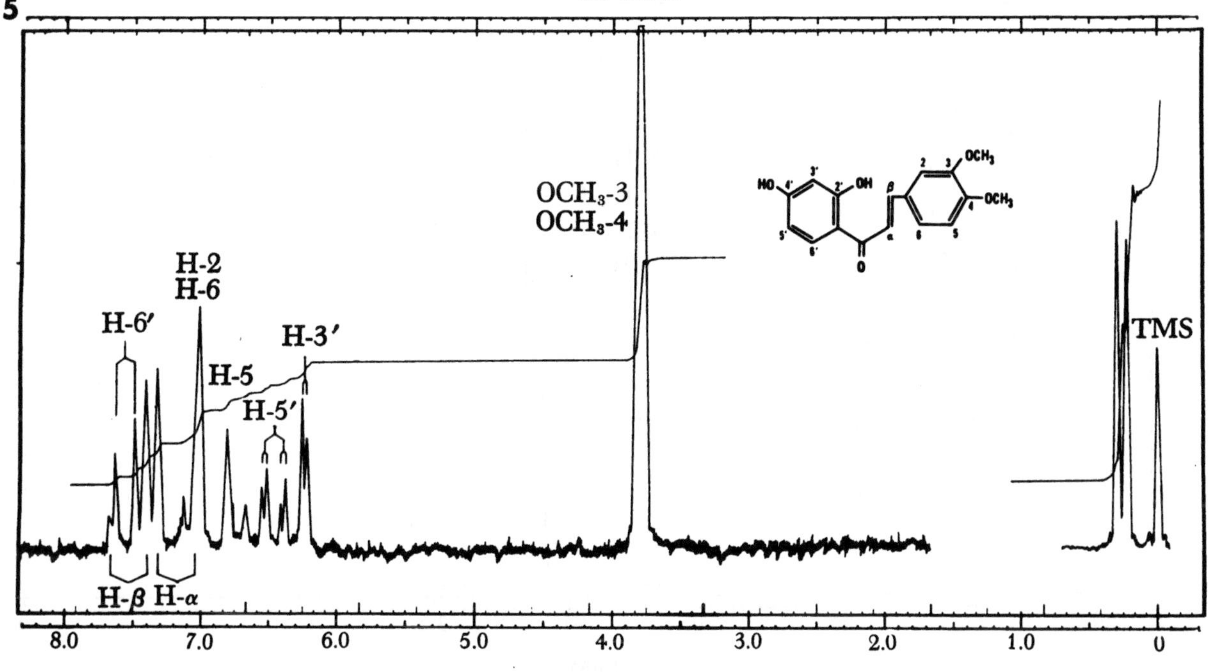

NMR of 2′-Hydroxy-4,4′,6′-trimethoxychalcone in $CDCl_3$

116

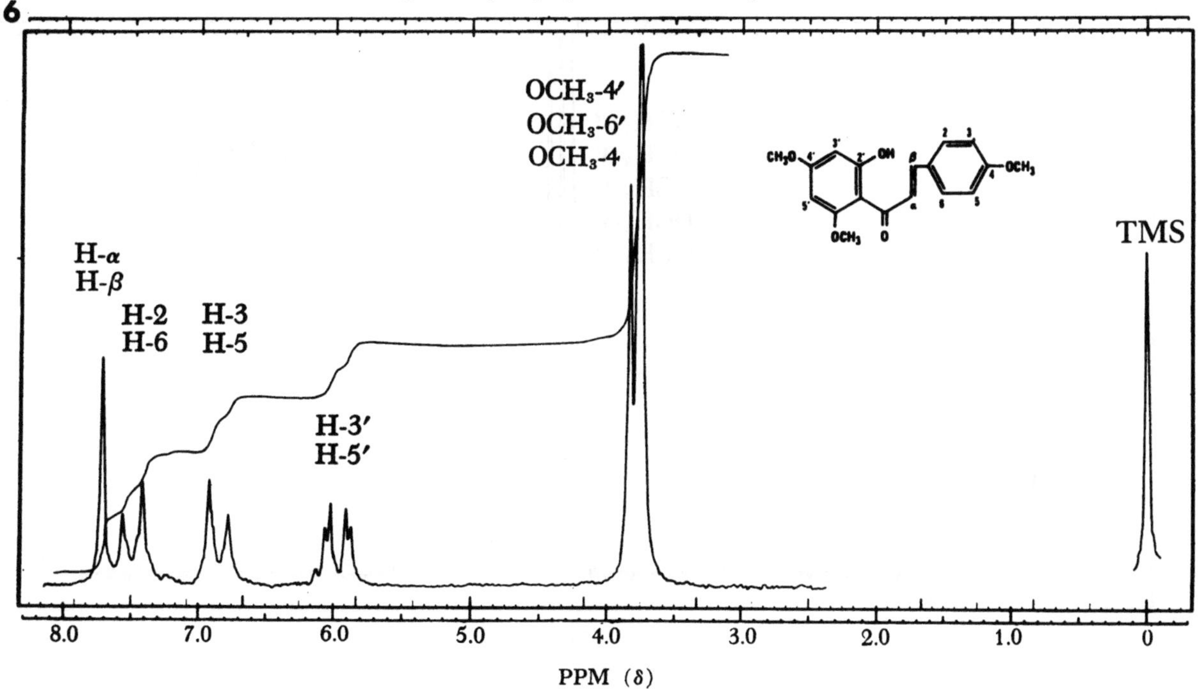

NMR of 2′,4,4′,6′-Tetramethoxychalcone in $CDCl_3$

117

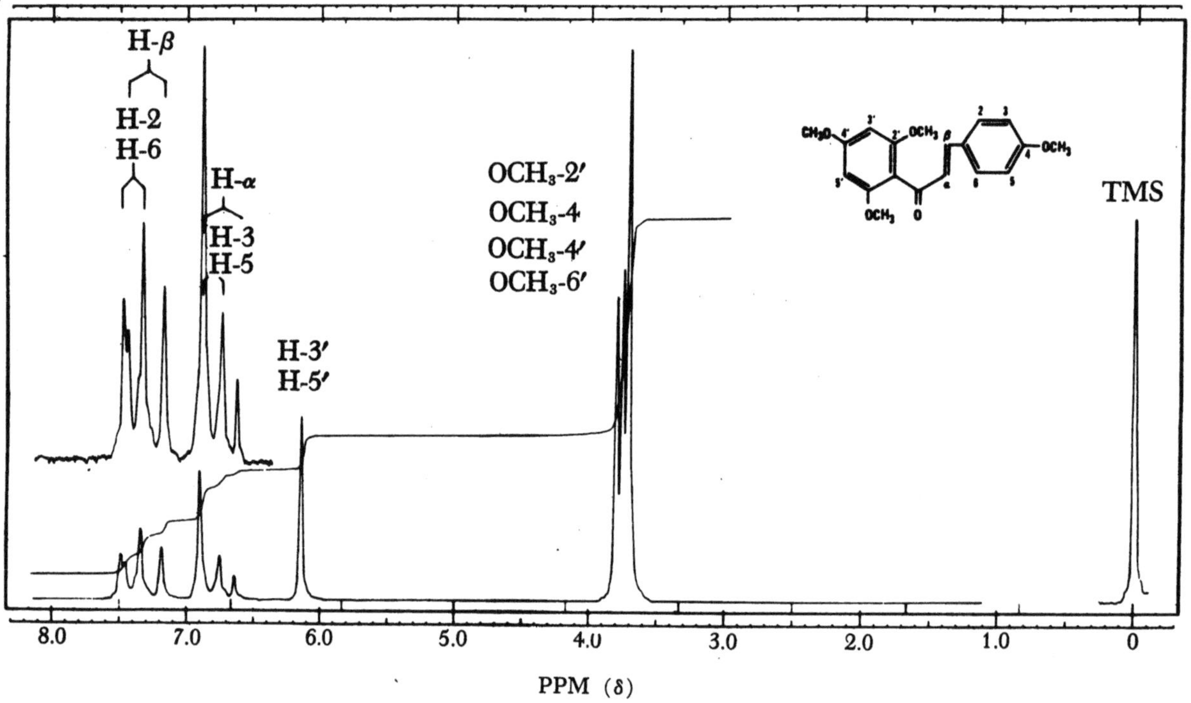

NMR of 2′,3′,4′,5′,6′-Pentamethoxychalcone in $CDCl_3$

118

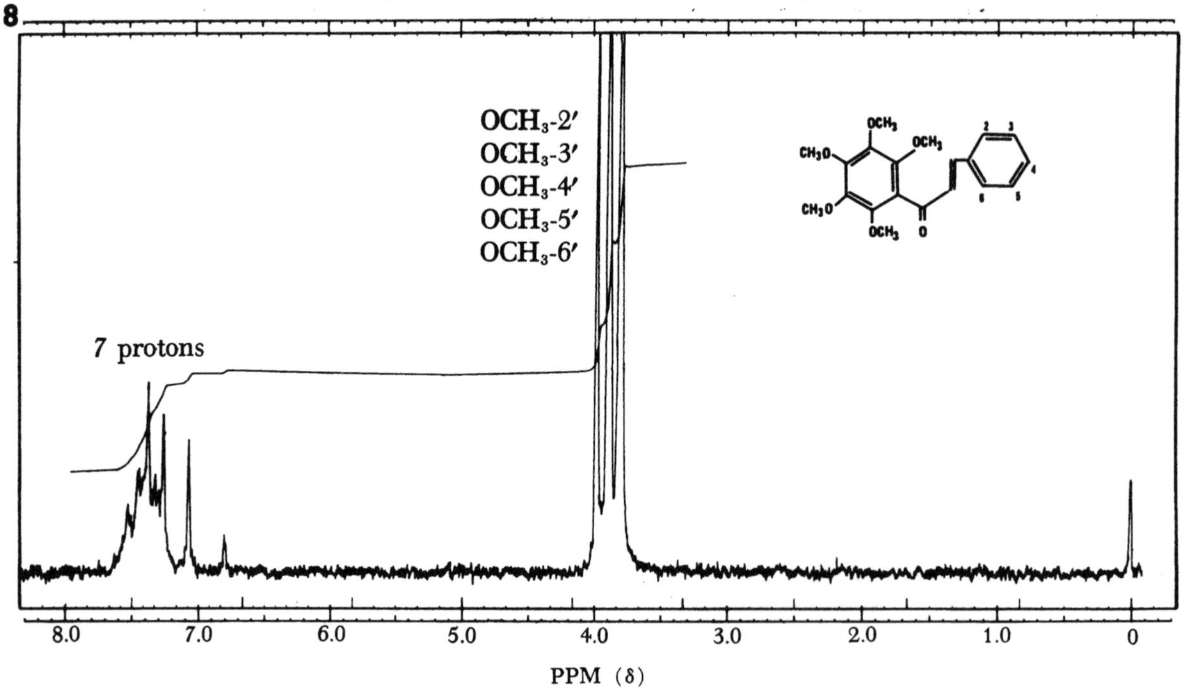

NMR of 2′,3,4,4′,6′-Pentamethoxychalcone in $CDCl_3$

119

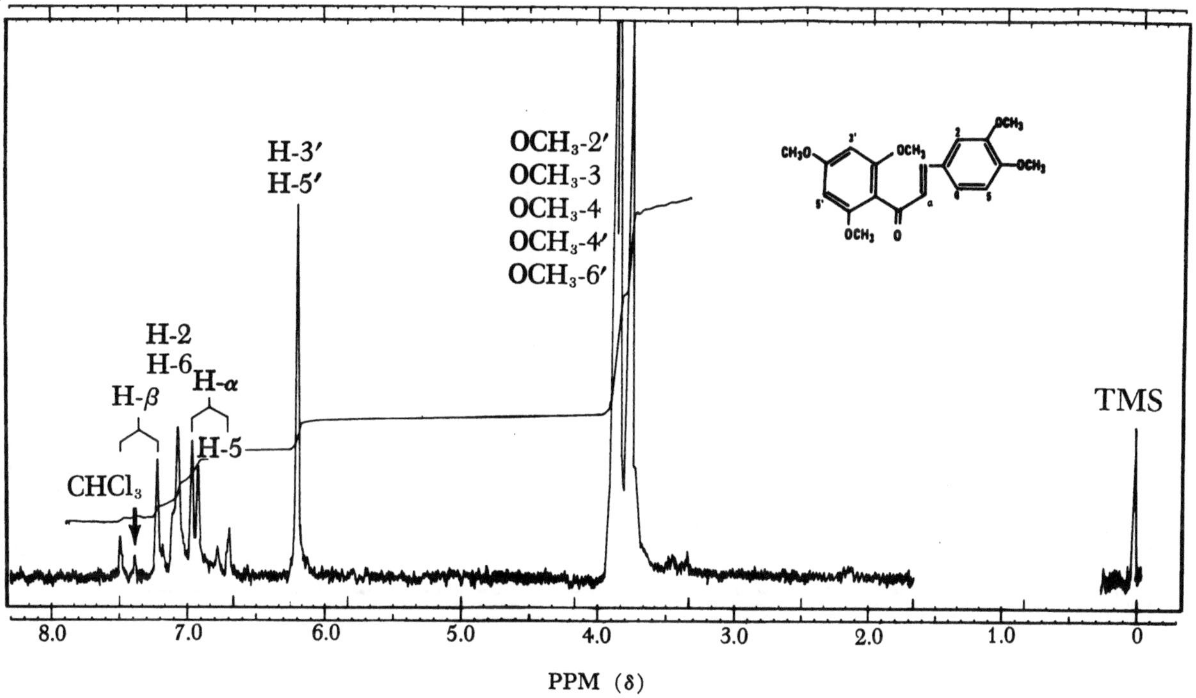

NMR of Chalcone Epoxide in CCl_4

120

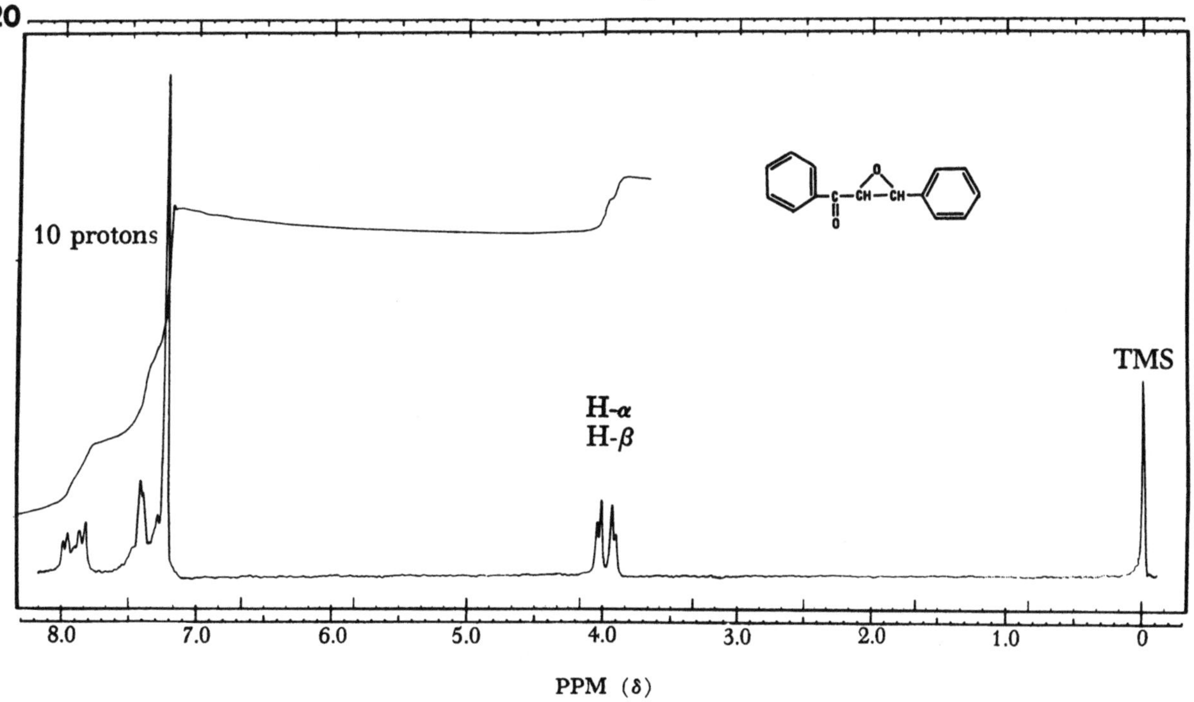

NMR of 2′,4,4′,6-Tetramethoxychalcone Epoxide in $CDCl_3$

121

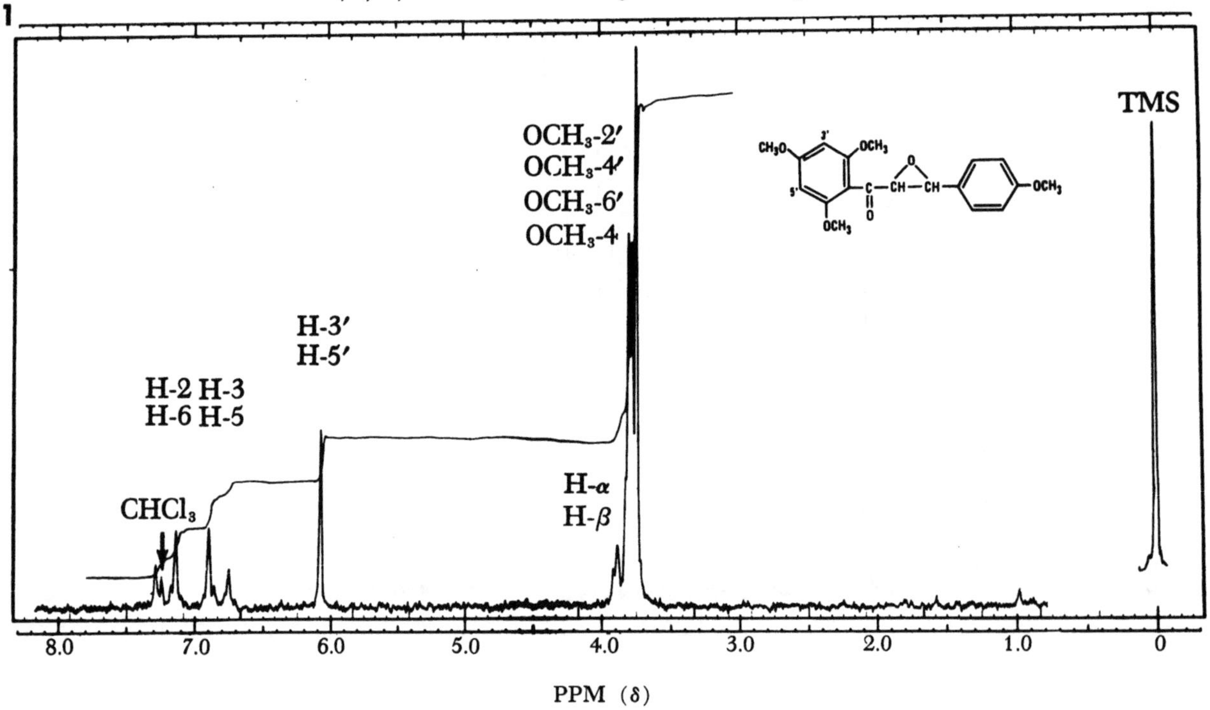

NMR of TMS Ether of 4′,7-Dihydroxyaurone in CCl_4

122

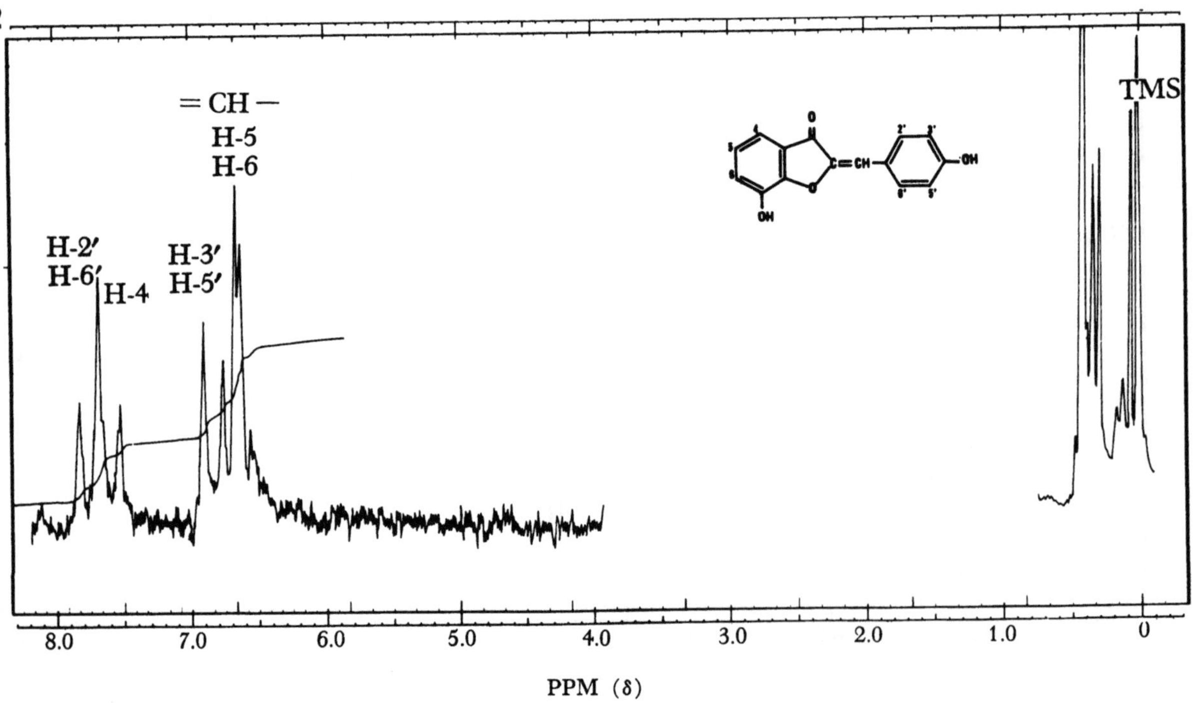

NMR of TMS Ether of 6-Hydroxy-4′-methoxyaurone in CCl_4

123

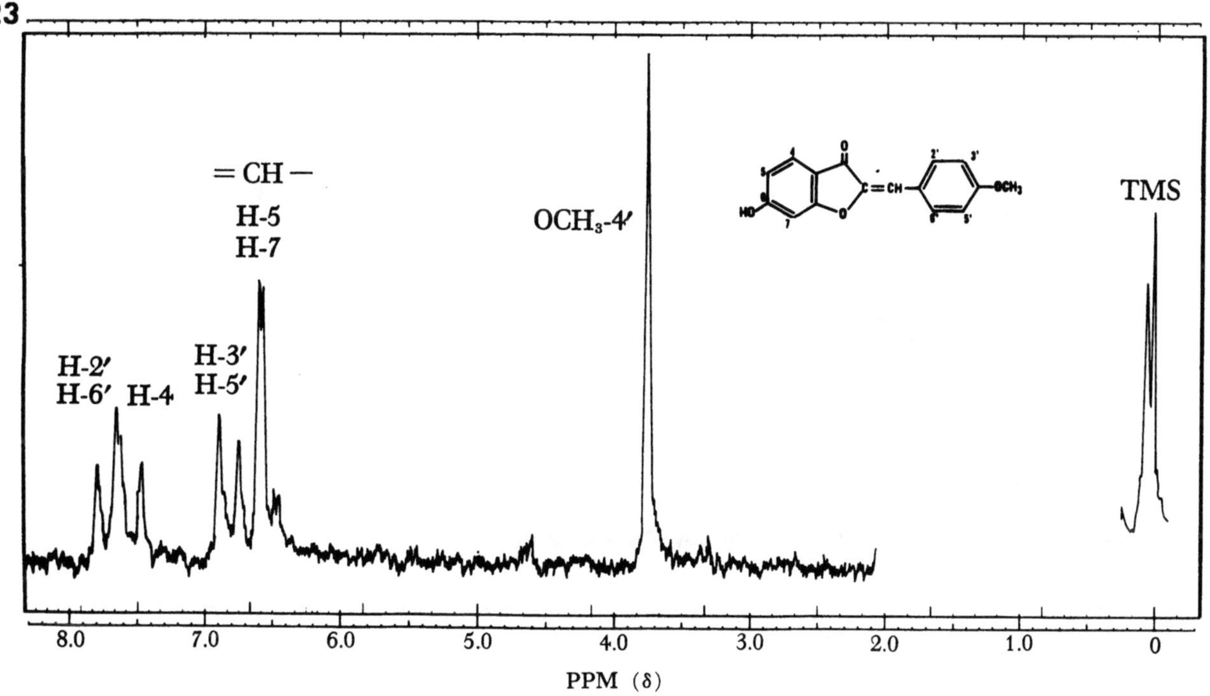

NMR of 4,4′,6-Trimethoxyaurone in $CDCl_3$

124

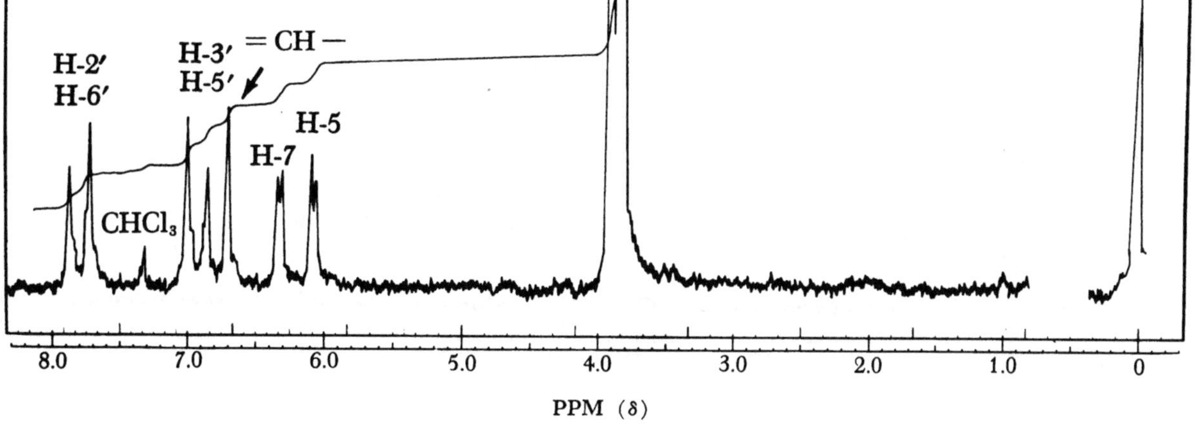

NMR of TMS Ether of 3′,4′,6,7-Tetrahydroxyaurone in CCl_4

125

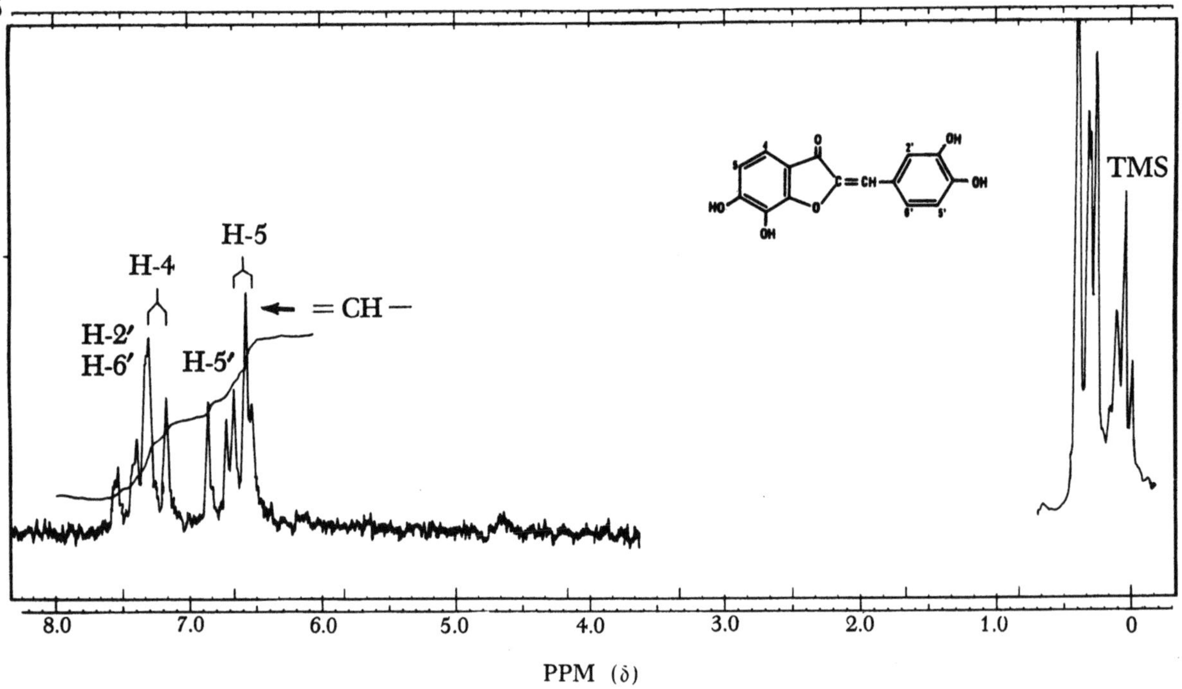

NMR of 3′,4′,6,7-Tetramethoxyaurone in $CDCl_3$

126

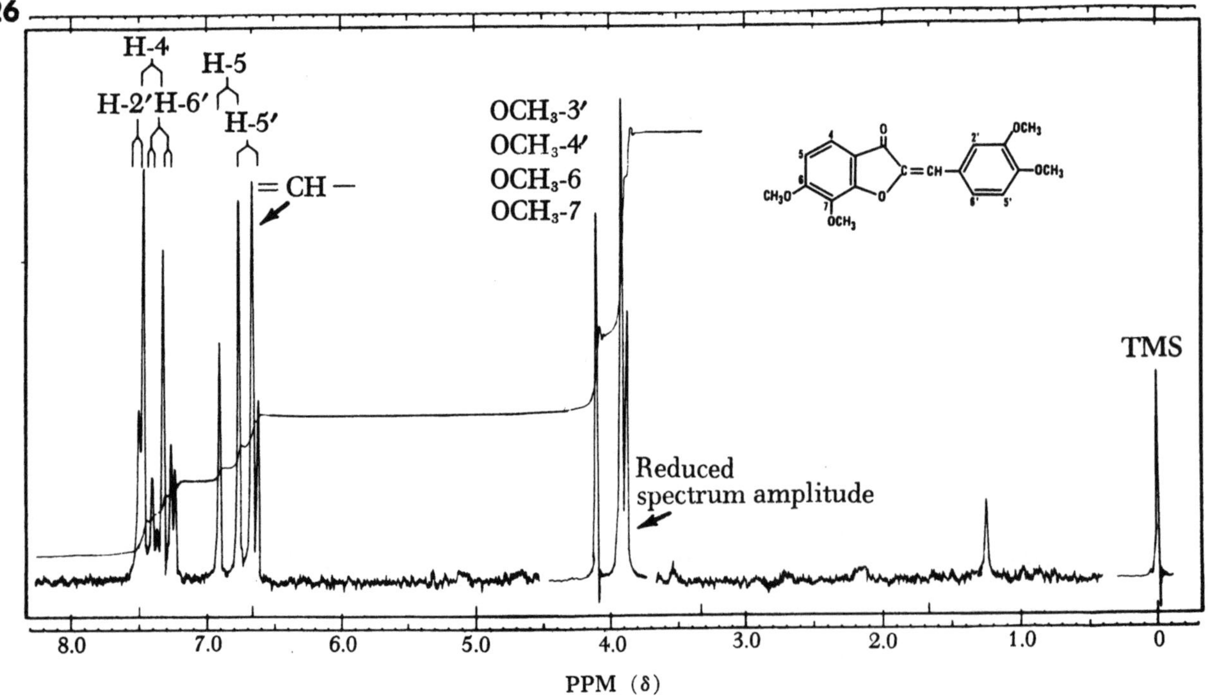

127 NMR of 4′-Benzyloxy-4,6,7-Trimethoxyaurone in $CDCl_3$

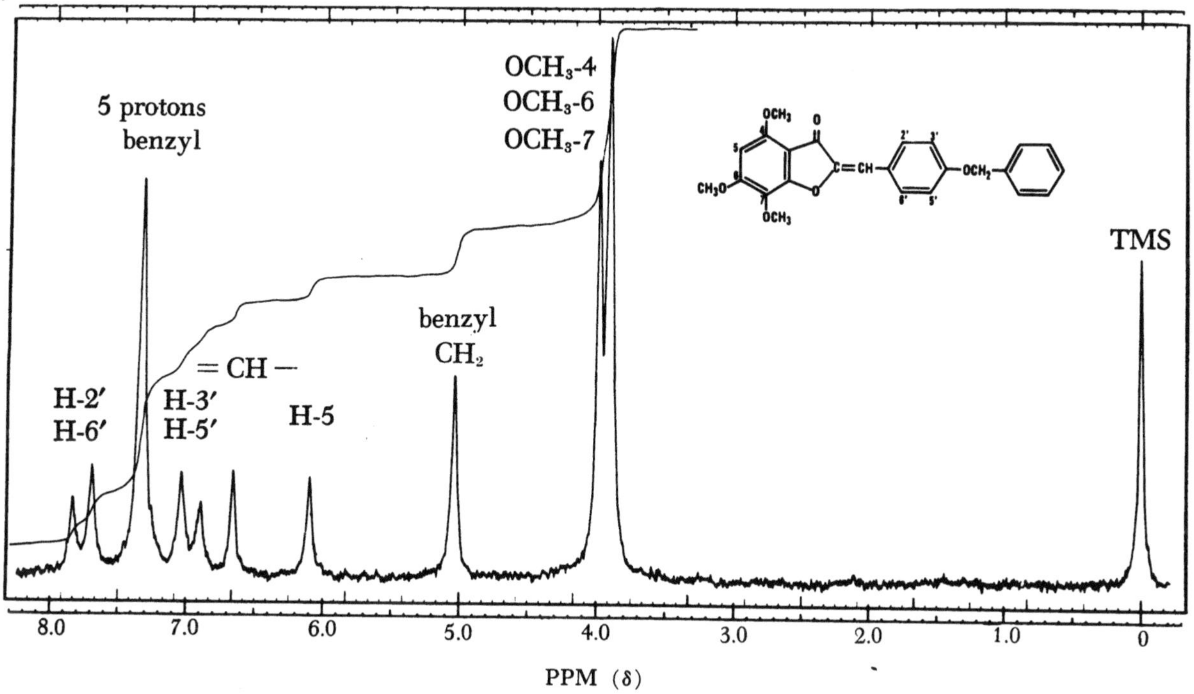

128 NMR of TMS Ether of Amentoflavone in CCl_4

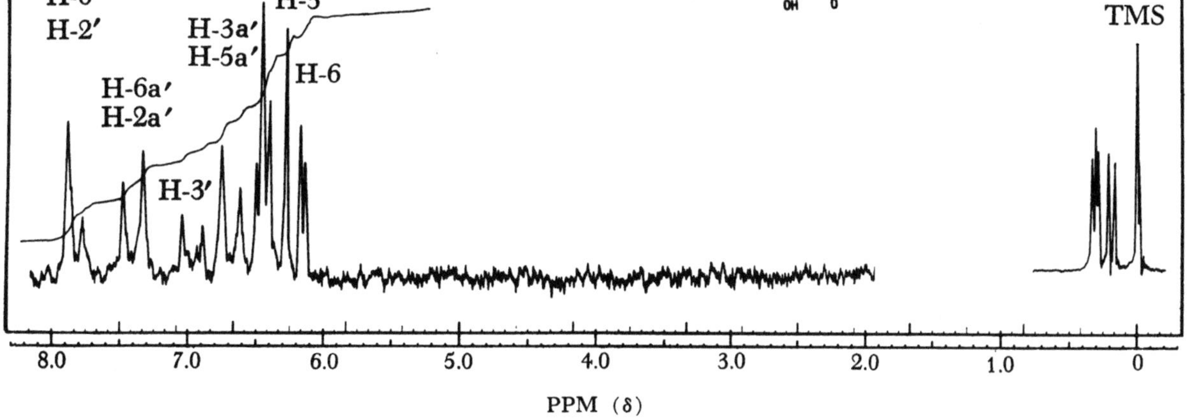

Subject Index

Absolute configuration 28, 30, 267
Acacetin 43, 48, 55, 57
—, UV spectra of 90
Acacetin 7-O-glucoside 43, 55, 58
— —, UV spectra of 91
Acetylated aglycone 31, 45, 271
Acetylation of TMS ethers 255, 259
7-Acetyloxy-6-carbomethoxyisoflavone 277
—, NMR spectrum of 312
Acid hydrolysis 24
Afrormosin 14, 166, 169, 173, 278
—, NMR spectrum of 320
—, structure of 20
—, UV spectra of 195
Aglycone, degradation 28
—, identification 27
—, methylation and acetylation 31, 45
— (see also UV and NMR)
Aluminum chloride for UV spectroscopy 35
— — — —, chalcones and aurones 229
— — — —, 5-deoxy-7-hydroxyflavones 53
— — — —, flavones and flavonols 50—56
— — — —, isoflavones, flavanones, and dihydroflavonols 171
Amentoflavone 43, 46, 49, 55, 58, 279
—, NMR spectrum of 343
—, UV spectra of 110
Ammonia, paper chromatographic spot detection 12, 13, 21
Amurensin 40
p-Anisidine hydrochloride, spray reagent 27
Anthochlor pigments 227
Anthocyanase 25
Anthocyanins 52
Apigenin 6, 10, 43, 46, 48, 53, 55, 57, 275
—, NMR spectrum of 280
—, structure of 8
—, UV spectra of 81
— 7-O-glucoside 12, 43, 46, 55, 57, 275
— —, NMR spectrum of 282
— —, structure of 8
— —, UV spectra of 82
— 7-O-neohesperidoside 43, 46, 55, 57, 275
— —, NMR spectrum of 281
— —, NMR spectrum of sugar 270
— —, structure of 269
— —, UV spectra of 83
— 7-O-neohesperidoside acetate 275
— — —, NMR spectrum of 281
— 7-O-rhamnoglucoside 4
— —, structure of 8
Arabinose identification 32
A-Ring oxygenation, effect on UV spectra 44, 45, 46
A-Ring protons (NMR), C-6 and C-8 261—265
— — —, C-5 264
Artemetin 43, 56, 60, 277
—, NMR spectrum of 308
—, UV spectra of 155
Astilbin 40, 170, 172, 174, 278
—, NMR spectrum of 327
—, UV spectra of 255
Aurone 40
Aurones, NMR spectra index 274
—, — — interpretation 254—272
—, numbering system 13, 227
—, paper chromatography 13
—, thin-layer chromatography 22
—, UV spectra index 230
—, UV spectra interpretation 227, 230

Baicalein (5,6,7-trihydroxyflavone) 43, 45, 57
— —, UV spectra of 77
Baicalin (5,6,7-trihydroxyflavone 7-O-glucuronide) 43, 55, 57
— — —, UV spectra of 78
Baker-Venkataraman transformation 28
Bands I and II 41
— — —, chalcones and aurones 227
— — —, flavones and flavonols 42
— — —, isoflavones, flavanones and dihydroflavonols 166
Baptigenin 169, 173
—, UV spectra of 199
Baptisia australis 20
— *lecontei* flavonoids 4, 7—10, 16—18, 30
Bathochromic shifts (see UV and Detection)
Bayin (see 5-Deoxyvitexin)
4-Benzyloxy-2,5-dihydroxy-3,6-dimethoxyacetophenone 28
4-Benzyloxy-3,6-dimethoxy-2,5-di(5-benzyloxy-3,4-dimethoxy benzoyloxy) acetophenone 29
Biochanin A 14, 169, 171, 173, 277
—, NMR spectrum of 314
—, structure of 21
—, UV spectra of 190
Boric acid for UV spectroscopy 35
B-Ring oxygenation, effect on UV spectrum 44, 166, 227
B-Ring protons (NMR), C-2′ and C-6′ 265—267
— — —, C-3′ and C-5′ 265—267

Calycosin (3′,7-dihydroxy-4′-methoxyisoflavone) 6, 14
— —, structure of 9
— 7-O-glucoside 7
— —, structure of 9

Calycosin 7-O-rhamnoglucoside 7
— —, structure of 9
Carbon tetrachloride (NMR) 255, 267
2-Carboxy-5,7-dihydroxyisoflavone 169, 171, 173
—, UV spectra of 178
2-Carbethoxy-5,7-dihydroxy-4′-methoxy-6-methylisoflavone 278
—, NMR spectrum of (CCl_4) 318
—, NMR spectrum of ($CDCl_3$) 319
2-Carbethoxy-5,7-dihydroxy-4′-methoxy-8-methylisoflavone 277
—, NMR spectrum of (CCl_4) 317
—, NMR spectrum of ($CDCl_3$) 318
2-Carboxy-6,7-dihydroxy-4′-methoxyisoflavone 169, 173
—, UV spectra of 194
8-Carbomethoxy-5-hydroxy-6-methylisoflavone 277
—, NMR spectrum of ($CDCl_3$) 310
—, NMR spectrum of (DMSO-d_6) 311
C. A. T. (see time-averaging computer)
Celite 17, 18
Cellulose, column chromatography 17
—, thin-layer chromatography 21, 22
Centaurein 42, 56, 59, 277
—, NMR spectrum of 307
—, UV spectra of 150
Chalcones, NMR spectra index 274
—, — — interpretation 254—272
—, numbering system 13, 227
—, paper chromatography 13
—, thin-layer chromatography 22
—, UV spectra index 230
—, — — interpretation 227—230
Chalcone 40
— α,β-epoxide 279
— —, NMR spectrum of 339
Charcoal 16
Chelation 50
Chemical shifts, α and β-protons (chalcones) 267
— —, acetylated C-glycosides 271
— —, benzylic protons (aurones) 267
— —, C-methyl protons 272
— —, C-2 protons 267
— —, C-2′ and C-6′ protons 265—267
— —, C-3 protons 262, 267
— —, C-3′ and C-5′ protons 265—267
— —, C-5 protons 264
— —, C-6 and C-8 protons 261—265
— —, downfield on acetylation 31, 45
— —, hydroxyl protons 254
— —, methoxyl and acetoxyl protons 271
— —, solvent induced 267, 272
— —, sugar protons 268
— —, table of 260
— —, TMS ether protons 272
Chlorophyll, removal 19
Chromatography cabinet 4, 5
Chrysin 43, 48, 55, 57
—, UV spectra of 68
Chrysoeriol 43, 46, 49, 55, 58
—, UV spectra of 101
C-Glycosides, identification of sugar 31
— in *Lemna minor* 12
— isomerization 11, 24
C-Glycosides, NMR 267, 269, 271
—, paper chromatography 11, 12
Circular dichroism 28, 30
C-Methyl protons (NMR) 272
Column chromatography
— —, aglycones 19
— —, general 17
Complexes — $AlCl_3$/flavones and flavonols 51
Cotton effect 30
Coumarins 18
Coupling constants (see proton of interest under chemical shift)
C-Ring protons (NMR) 267
— — —, α- and β-protons (chalcones) 267
— — —, benzylic protons (aurones) 267
— — —, C-2, C-3 262, 267
Cytisoside 263, 275
—, NMR spectrum of 284

Daidzein 6, 9, 166, 169, 173
—, structure of 21
—, UV spectra of 179
— 7-O-glucoside (Daidzin) 7, 173
— — —, structure of 9
— — —, UV spectra of 180
— 7-O-rhamnoglucoside 7
— —, structure of 9
Degradation technique 28
Demethoxysudachitin 21, 275
—, NMR spectrum of 290
—, structure of 19
Demethylation, selective 29
5-Deoxy-flavonoids, NMR spectra 264—265
—, relative *Rf* values 10
—, UV spectra 44, 53, 56, 166, 169
5-Deoxyvitexin (Bayin) 43, 46, 48, 53, 57
— —, UV spectra of 72
Desalting 32
Detection (UV) of, 3,4′-dihydroxyl system 47, 167
— — 6,7-dioxygenated isoflavones 166
— — ortho-dihydroxyl groups 50, 52, 169—171, 228
— — 3,3′,4′-trihydroxyl system 47, 50
— — 5,6,7; 5,7,8; and 3′,4′,5′-trihydroxyl systems 47, 50, 168—170, 228, 229
— — of free hydroxyl groups, 3-hydroxyl 45, 48, 52—55
— — — — —, 4-hydroxyl 228
— — — — —, 5-hydroxyl 52—55, 169—172
— — — — —, 6-hydroxyl 51, 52, 169, 170, 228
— — — — —, 7-hydroxyl 48, 51, 169
— — — — —, 8-hydroxyl 51, 52, 169
— — — — —, 2′-hydroxyl 229
— — — — —, 3′-hydroxyl 50, 169
— — — — —, 4′-hydroxyl 45, 48—50, 169, 228
Determination of *Rf* values 9
De-trimethylsilylation 258, 259
Deuteriochloroform solvent (NMR) 254, 267
Deuteriodimethyl sulfoxide solvent (NMR) 254—267
Diaxial protons 267, 268

Diazomethane 30, 259
5′,7-Dibenzyloxy-3′,4′,5,6,8-pentamethoxyflavone 28, 29
5′-7-Dibenzyloxy-6-(5-benzyloxy-3,4-dimethoxybenzoyloxy)-3′,4′,5,8-tetramethoxyflavone 29
Diequatorial protons 269
4′,7-Di-O-ethylvitexin 40
Diglycosides, column chromatography 17, 20
—, distinction from tri- and mono- 27
—, hydrolysis 24
—, NMR 269, 271
—, paper chromatography 7, 10, 11
—, thin-layer chromatography 20—22
Dihydrofisetin 170, 174, 278
—, NMR spectrum of 326
—, UV spectra of 220
Dihydroflavonols, absolute configuration 28, 30
—, column chromatography 17—20
—, 3-O-glucosides 30
—, NMR spectra index 274
—, — — interpretation 254—272
—, paper chromatography 13
—, thin-layer chromatography 20—22
—, UV spectra index 173
—, — — interpretation 165—173
Dihydrokaempferol 40, 172, 174, 278
—, NMR spectrum of 325
—, UV spectra of 222
Dihydroquercetin 278
—, NMR spectrum of 327
Dihydrorobinetin 170, 174, 278
—, NMR spectrum 328
—, UV spectra of 226
3′,4′-Dihydroxyaurone 40, 229, 230
—, UV spectra of 243
4′,7-Dihydroxyaurone 279
—, NMR spectrum of 340
5,7-Dihydroxyaurone 227, 230
—, UV spectra of 244
6,7-Dihydroxyaurone 227, 229, 230
—, UV spectra of 245
2,2′-Dihydroxychalcone 229, 230
—, UV spectra of 235
2′,4-Dihydroxychalcone 229, 230
—, UV spectra of 236
3,4-Dihydroxychalcone 229, 230
—, UV spectra of 234
(+)-4′,7-Dihydroxy-dihydroflavonol 7
—, structure of 8
4′,7-Dihydroxyflavonol 6
—, structure of 8
5,7-Dihydroxyflavanone (see also pinocembrin) 40, 45
2′,3-Dihydroxyflavone 40
3′,4′-Dihydroxyflavone 40, 43, 46, 50, 57
—, UV spectra of 66
4′,7-Dihydroxyflavone 6, 43, 46, 48, 57
—, structure of 8
—, UV spectra of 70
4′,7-Dihydroxyflavone 7-O-glucoside 6
—, structure of 8
4′,7-Dihydroxyflavone 7-O-rhamnoglucoside 7, 43, 46, 57
4′,7-Dihydroxyflavone 7-O-rhamnoglucoside,
—, structure of 8
—, UV spectra of 71
2′,4′-Dihydroxy-3,4-dimethoxychalcone 279
—, NMR spectrum of 337
5,7-Dihydroxy-3′,4′-dimethoxyflavanone 278
—, NMR spectrum of 334
5,7-Dihydroxy-3′,4′-dimethoxyflavone 43, 49, 55, 58, 275
—, NMR spectrum of 287
—, UV spectra of 104
5,7-Dihydroxyisoflavone 166, 169, 171, 173
—, UV spectra of 176
2′,4-Dihydroxy-4′-methoxychalcone 279
—, NMR spectrum of 335
5,7-Dihydroxy-2′-methoxyflavone 43, 48, 55, 58, 275
—, NMR spectrum of 284
—, UV spectra of 92
3′,7-Dihydroxy-4′-methoxyisoflavone (see Calycosin)
5,7-Dihydroxy-4′-methoxyisoflavone (see Biochanin A)
6,7-Dihydroxy-4′-methoxyisoflavone (see Texasin)
4′,7-Dihydroxy-5-methoxyisoflavone 40
5,7-Dihydroxy-4′-methoxy-8-methylisoflavone 277
—, NMR spectrum of 316
5,7-Dihydroxy-3′,4′,5′-trimethoxyflavone 43, 49, 55, 58
—, UV spectra of 106
3′,5-Dihydroxy-3,4′,7,8-tetramethoxyflavone 258
4,4′-Dimethoxychalcone 279
—, NMR spectrum of 335
3′,4′-Dimethoxyflavone 43, 57
—, UV spectra of 67
3,4-Di-O-methylgallic acid 28
5,7-Dimethoxyisoflavone 173
—, UV spectra of 176
5,7-Dimethoxy-8-methylisoflavone 277
—, NMR spectrum of 313
Dimethyl sulfate 30, 259
Diosmetin 40, 43, 49, 55, 58, 275
—, NMR spectrum of 287
—, UV spectra of 103
— 7-O-rutinoside (see Diosmin)
— triacetate, structure of 45
Diosmin 270, 271, 275
—, NMR spectrum of 288
—, structure of 269
— acetate 275
— —, NMR spectrum of 288
Distinguishing, 6-C- and 8-C-glycosides 11, 12, 263
—, 6-C- and 8-C-methyl groups 272
—, neohesperidosides and rutinosides 269
—, C-2 protons 269
—, C-3, C-6 and C-8 protons 262
Downfield shifts, on acetylation 31, 45
— — on solvent change 267, 272
Durethan Bk 40 F 18

Egger's solvent 17, 22
Engeletin 170, 172, 174, 278

Engeletin, NMR spectrum of 326
—, UV spectra of 223
Enzymatic hydrolysis 25
Eriodictyol 170, 172, 174
—, UV spectra of 217
Extinction coefficient 38, 39
— table 40
Extraction (solvent) 17, 19

Ferric chloride oxidations 31
Fisetin 6, 40, 42, 47, 49, 51, 56, 59
—, structure of 8
—, UV spectra of 123
— 3-O-glucoside 43, 46, 51, 59
—, UV spectra of 124
— 7-O-rhamnoglucoside 6
—, structure of 8
Flavanone 40, 43, 45
Flavanones, column chromatography 17—20
—, NMR spectra index 274
—, — — interpretation 254, 272
—, paper chromatography 13—15
—, thin-layer chromatography 20—22
—, UV spectra index 173
—, — — interpretation 165—173
Flavone 40, 45, 57
—, UV spectra of 62
Flavones, column chromatography 17—20
—, NMR spectra index 274
—, — — interpretation 254—272
—, numbering system 13, 261
—, paper chromatography 13—15
—, synthesis 28
—, thin-layer chromatography 20—22
—, UV spectra index 57
—, — — interpretation 40, 41—57
Flavonoid aglycones, column chromatography 19, 20
— —, numbering system 13, 261
— —, paper chromatography 13—15
— glycosides (see Glycosides)
Flavonoids from *Baptisia australis* 20
— — — *lecontei* 4—7, 17, 30
— — *Hymenoxys scaposa* 9, 19, 29, 262
— — *Lemna minor* 12
Flavonols, column chromatography 17—20
—, NMR spectra index 274
—, — — interpretation 254—272
—, paper chromatography 13—15
—, thin-layer chromatography 20—22
—, UV spectra index 57
—, — — interpretation 40—57
Flow rate (polyamide columns) 18
Formononetin 14, 21, 169, 173
—, structure of 20
—, UV spectra of 181
— 7-O-glucoside 173, 277
— —, NMR spectrum of 312
— —, UV spectra of 182
— — tetraacetate 173
— — —, UV spectra of 183
Free hydroxyl group (see Detection)
(+)-Fustin 7
— structure of 8
— 3-O-glucoside 7, 30, 170, 174
(+)-Fustin 3-O-glucoside, structure of 8
— —, UV spectra of 221

Galactosides, NMR spectrum of 269
Galangin 40, 42, 44, 46, 49, 56, 58, 254
—, UV spectra of 115
— 3-methyl ether 43, 49, 55, 58
—, UV spectra of 116
Garbanzol 170, 174
—, UV spectra of 219
Gas chromatography, sugars 26
Genistein 169, 171, 173
—, structure of 21
—, UV spectra of 184
— 4′,5-dimethyl ether 277
— — —, NMR spectrum of 315
— 5-methyl ether 169, 173, 277
— — —, NMR spectrum of 315
— — —, UV spectra of 188
— 7-O-rhamnoglucoside, structure of 167
Genistin 171, 173
—, UV spectra of 185
Glucose, identification 27
—, NMR spectrum of 268
β-Glucosidase 25
β-Glucuronidase 25
Glucuronides 24
Glucosides, acid hydrolysis 24, 30
—, enzymatic hydrolysis 25
—, location of sugar 27, 30
—, NMR spectra of 268
—, sugar analysis 26—32
— (see also mono-, di-, tri-, and C-glucosides)
Gossypetin 40, 42, 47, 49, 51, 56, 59
—, UV spectra of 159
Gossypetin hexamethyl ether 42, 60
— — —, UV spectra of 162
Gossypin 42, 47, 49, 51, 56, 60
—, UV spectra of 160
Gossypitrin 42, 47, 51, 56, 60
—, UV spectra of 161

Hepta-O-methylvitexin 271
Herbacetin 8-methyl ether 42, 47, 49, 56, 59, 276
— — —, NMR spectrum of 303
— — —, UV spectra of 125
Hesperetin 26, 40, 278
—, NMR spectrum of 332
—, quinone 279
— —, NMR spectrum of 334
Hesperidin 23, 172, 174, 255, 278
—, NMR spectrum of 333
—, structure of 26
—, UV spectra of 218
— octa-trimethylsilyl ether, preparation of 255
3,4′,5,6,7,8-Hexamethoxyflavone 277
—, NMR spectrum of 309
Hexamethyldisilizane 26, 255—259
Homoeriodictyol 40, 278
—, NMR spectrum of 332
Hydrochloric acid for UV spectroscopy 35
Hydrolysis of glycosides 24, 30
— —, enzymatic 25
— —, partial 24, 25

Hydrolysis of glycosides, relative rates 24
— —, side reactions 24
— of TMS ether (see Trimethylsilyl ether derivatives)
— — — (see De-Trimethylsilylation)
3′-Hydroxyaurone 40
4′-Hydroxyaurone 40, 175, 230
—, UV spectra of 242
2-Hydroxychalcone 175, 230
—, UV spectra of 231
2′-Hydroxychalcone 40
4′-Hydroxychalcone 175, 230
—, UV spectra of 232
2′-Hydroxy-4,4′-dimethoxychalcone 279
—, NMR spectrum of 336
3-Hydroxy-3′,4′-dimethoxyflavone 42, 46, 56, 58
—, UV spectra of 118
7-Hydroxy-3′,4′-dimethoxyflavone 43, 57
—, UV spectra of 76
5-Hydroxy-7,8-di-methoxyisoflavone 278
—, NMR spectrum of 321
7-Hydroxy-4′,6-dimethoxyisoflavone (see Afrormosin)
6-Hydroxyflavanone 278
—, NMR spectrum of 328
3-Hydroxyflavone 43, 56, 58
—, UV spectra of 112
5-Hydroxyflavone 40, 43, 45, 55, 57
—, UV spectra of 63
7-Hydroxyflavone 40, 43, 45, 48, 57
—, UV spectra of 64
6-Hydroxygenistein 14, 129, 168—171, 173
—, UV spectra of 200
6-Hydroxygenistein 7-O-rhamnoglucoside 23, 30, 31
7-Hydroxyisoflavone 169, 173
—, UV spectra of 175
Hydroxyl groups, detection (see Detection)
— protons, chemical shifts 254
6-Hydroxy-4′-methoxyaurone 230, 279
—, NMR spectrum 341
—, UV spectra of 246
2′-Hydroxy-4′-methoxychalcone 229, 230
—, UV spectra of 233
3-Hydroxy-4′-methoxyflavone 42, 46, 56, 58
—, UV spectra of 113
7-Hydroxy-4′-methoxyisoflavone (see Formononetin)
7-Hydroxy-4′-methoxyflavone (pratol) 43, 48, 57
—, UV spectra of 73
7-Hydroxy-3′,4′-methylenedioxyisoflavone (see Pseudobaptigenin)
7-Hydroxypentamethoxyquercetagetin 30
3′-Hydroxy-4,4′,6-trimethoxyaurone 230
—, UV spectra of 250
6-Hydroxy-3′,4,4′-trimethoxyaurone 230
—, UV spectra of 249
2′-Hydroxy-4,4′,6′-trimethoxychalcone 279
—, NMR spectrum of 337
Hymenoxin 21, 43, 49, 55, 258, 275
—, NMR spectrum of 291
—, structure of 19
—, UV spectra of 109
Hymenoxys scaposa 4, 9, 16, 19, 29, 262
Hyperin 276
—, NMR spectrum of 295

Identification of aglycones 28
— of sugars 26
— (see also UV and NMR)
Index of NMR spectra 274
Indexes of UV spectra, aurones and chalcones 230
— — —, flavones and flavonols 57
— — —, isoflavones, flavanones and dihydroflavonols 173
Internal reference (NMR) (see Tetramethylsilane)
Iridin (3′,5,7-trihydroxy-4′,5′,6-trimethoxy isoflavone 7-O-glucoside) 37, 171, 174, 278
— NMR spectrum of 324
— UV spectra of 210
Irigenin 169, 171, 174, 278
—, NMR spectrum of 324
—, UV spectra of 209
Irisolidone 31, 169, 171, 173, 278
—, NMR spectrum of 323
—, UV spectra of 203
Isoflavone 40
Isoflavones, column chromatography 17—20
—, NMR spectra index 272
—, — — interpretation 254—272
—, numbering system 13
—, paper chromatography 13—15
—, thin-layer chromatography 20—22
—, UV spectra index 173
—, — — interpretation 165—172
Isolation of flavonoids
— — by column chromatography 17
— — by paper chromatography 13
— — by thin-layer chromatography 20
Isoorientin 31, 32, 42, 46, 48, 50, 55, 56, 58
—, structure of 12
—, UV spectra of 98
Isoorientin 3′-methyl ether, structure of 12
Isoorientin 3′,4′,5,7-tetra-O-methyl ether 31
Isoorientin 4′-O-glucoside (lutonarin), structure of 12
Isoprenoid side chains 25
Isoquercitrin 40
Isorhamnetin 42, 47, 49, 56, 59, 276
—, NMR spectrum of 298
—, UV spectra of 138
— 3-O-galactoside 42, 47, 49, 55, 59, 276
— —, NMR spectrum of 299
— —, UV spectra of 139
— 3-O-rutinoside 42, 47, 49, 55, 59
— —, UV spectra of 140
Isovitexin (Saponaretin) 43, 46, 48, 55, 57, 275
— —, NMR spectrum of 283
— —, structure of 12, 257
— —, UV spectra of 84
— glucoside, structure of 12
— hepta-trimethylsilyl ether, preparation of 257

Jaceidin 42, 47, 49, 56, 59
—, UV spectra of 148

Jacein 42, 47, 56, 59, 277
—, NMR spectrum of 307
—, UV spectra of 149

Kaempferol 42, 44, 46, 47, 49, 56, 58, 275
—, NMR spectrum of 292
—, UV spectra of 119
— 7-O-neohesperidoside 42, 46, 47, 56, 59, 275
— —, NMR spectrum of 293
— —, UV spectra of 120
— — acetate 276
— — —, NMR spectrum of 293
— 3-robinoside 7-O-rhamnoside (Robinin), 42, 46, 58, 59, 275
— — — —, NMR spectrum of 292
— — — —, UV spectra of 121
Kaempferol 4′-methyl ether 42, 49, 56, 59
—, UV spectra of 122

Lanceolarin 171, 173
—, UV spectra of 191
Lecontin 7, 30, 278
—, NMR spectrum of 325
—, structure of 8
Lemna minor 11, 12
Leptosidin 229, 230
—, UV spectra of 248
Liquiritigenin 6, 170, 174, 278
—, NMR spectrum of 329
—, UV spectra of 212
Location of sugar linkage 27
Lucenin-1 12, 42, 46, 49, 50, 55, 58,
—, UV spectra of 100
Luteolin 6, 12, 40, 42, 46, 48, 50, 55, 58, 262, 275
—, NMR spectrum of 263, 286
—, structure of 8
—, UV spectra of 95
— 7-O-glucoside 6, 12, 25, 43, 46, 50, 55, 58, 275
— —, NMR spectrum of 285
— —, structure of 8
— —, UV spectra of 96
— 7-O-rhamnoglucoside 23
— —, structure of 24
— 7-O-rutinoside 6, 12, 43, 46, 50, 55, 58, 275
— —, NMR spectrum of 285
— —, structure of 8
— —, UV spectra of 97
Lutonarin (isoorientin 4′-O-glucoside) 12

Magnesol for column chromatography 17
Maritimetin (see 3′,4′,6,7,-Tetrahydroxyaurone)
Methoxylated flavonoids, column chromatography 19
3-Methoxyflavone 40
4′-Methoxyflavone 43, 57
—, UV spectra of 65
Methoxyl protons (NMR) 271
5-Methoxy-8-methylisoflavone 277
—, NMR spectrum of 311
Methylation of hydroxyl groups 30
— — —, relative rates 31
— — — when trimethylsilylated 255, 259

Methyl protons (NMR) 271
Microcell, use in NMR analysis 256
Molecular extinction coefficient (see Extinction)
Morin 40, 42, 47, 49, 56, 59, 276
—, NMR spectrum of 302
—, UV spectra of 143
Myricetin 40, 42, 44, 47, 49, 51, 56, 60, 277
—, NMR spectrum of 309
—, UV spectra of 163

Naringenin 40, 170, 172, 174, 278
—, NMR spectrum of 329
—, UV spectra of 215
Naringenin 4′,7-dimethyl ether 278
— — —, NMR spectrum of 331
Naringin 40, 278
—, NMR spectrum of 330
Neohesperidin 278
—, NMR spectrum of 333
Neohesperidosides, distinction from rutinosides 269
— (see also Diglycosides)
Nevadensin 43, 49, 55, 58
—, UV spectra of 107
Nona-O-methylxylosylvitexin 271
Norwogonin (5,7,8-trihydroxyflavone) 42, 45, 48, 55, 57
Nuclear magnetic resonance spectroscopy 254—273
— — — —, A-ring protons 261
— — — —, B-ring protons 265
— — — —, chemical shifts (see chemical shifts)
— — — —, C-methyl protons 272
— — — —, C-ring protons 267
— — — —, effect of acetylation 31, 271
— — — —, index of spectra 274
— — — —, methoxyl and acetoxyl protons 271
— — — —, solvent shifts 267, 272
— — — —, sugar protons 268
— — — —, table of chemical shifts 260
— — — —, TMS ether protons 272
Numbering systems (see compound type e.g. Flavone etc.)
Nylon, column chromatography 18

Optical density 39
— rotatory dispersion 28, 30
Orientin 31, 32, 43, 46, 48, 50, 55, 58
—, structure of 12
—, UV spectra of 99
— 3′,4′,5,7-tetra-O-methyl ether 31
Orobol 6, 14, 169, 171, 173, 278
—, NMR spectrum of 322
—, structure of 8
—, UV spectra of 204
— 7-O-glucoside 171, 173, 278
— —, NMR spectrum of 222
— —, structure of 167
— —, UV spectra of 205
— 7-O-rhamnoglucoside 7, 171, 174
— —, structure of 8
— —, UV spectra of 206
Ortho-dihydroxyl groups (see Detection)
Oxidation of C-glycosides 31
Oxygenation pattern

Oxygenation pattern, effect on NMR spectrum 261—268
— —, — on UV spectrum 44, 166, 227

Paper chromatographic analysis of *Baptisia lecontei* flavonoids 4—7, 17
— — — *Hymenoxys scaposa* flavonoids 9
— — — *Lemna minor* flavonoids 12
— chromatography 3, 11
— —, aglycones 14
— —, general accessories 4
— —, preparative 13, 14
— —, solvents 4
— —, spot color and flavonoid structure 12
— —, sugars 27, 32
— —, UV spectrum from 36
Patuletin 42, 47, 49, 51, 56, 59
—, UV spectra of 151
— 3-O-glucoside 42, 47, 49, 51, 56, 59, 276
— —, NMR spectrum of 306
— —, UV spectra of 152
— 3-O-rhamnoglucoside
— —, structure of 259
— 3-O-rutinoside 42, 47, 49, 51, 56, 59, 277
— —, NMR spectrum of 306
— —, UV spectra of 153
Patulitrin 42, 47, 51, 56, 59, 264, 276
—, methylation of 30
—, NMR spectrum of (CCl_4) 305
—, — — of (DMSO-d_6) 305
—, structure of 30, 262
—, UV spectra of 154
Penduletin 40, 43, 47, 55, 59, 276
—, NMR spectrum of 304
—, UV spectra of 145
Pendulin 40, 43, 56, 59, 276
—, NMR spectrum of 304
—, UV spectra of 146
2′,3,4,4′,5′-Pentahydroxychalcone 40
2′,3,4,4′,6′-Pentamethoxychalcone 279
—, NMR spectrum of 339
2′,3′,4′,5′,6′-Pentamethoxychalcone 279
—, NMR spectrum of 338
3,5,6,7,8-Pentamethoxyflavone 43, 59, 276
—, NMR spectrum of 302
—, UV spectra of 147
Phenol (as eluent) 16
Phosphors 21
Pinocembrin 40, 170, 172, 174
—, UV spectra of 211
Polyamide, column chromatography 17—19, 26
—, poly-caprolactam 18
—, poly-vinylpyrrolidone 17, 22, 26
—, preparation from pellets 18, 21
—, TLC 17, 31
Polypenco 66D, 17, 26
Poly-vinylpyrrolidone (Polyclar AT) 17, 22, 26
Pomiferin 171, 174
—, UV spectra of 208
Pratensein 169, 171, 174
—, UV spectra of 207
Pratol (see 7-Hydroxy-4′-methoxyflavone)
Pridham's method of sugar analysis 27
Prunetin 171, 173
—, UV spectra of 189
Prunin 40
Pseudobaptigenin 6, 14, 169, 173
—, structure of 9
—, UV spectra of 197
Pseudobaptisin 7, 173, 259, 278
—, NMR spectrum of (CCl_4) 320
—, — — of (DMSO-d_6) 278
—, structure of 9
—, UV spectra of 198
Pseudobaptisin hexatrimethylsilyl ether 259
Purification by paper chromatography 13, 14, 27, 36
— by sublimation 15
— using charcoal 16

Qualitative sugar analysis 27
Quantitative sugar analysis 27
Quercetagetin hexamethyl ether 43, 59, 277
— — —, NMR spectrum of 308
— — —, UV spectra of 158
Quercetagetin 3,3′,4′,5,6-pentamethyl ether 43, 49, 59
—, UV spectra of 157
Quercetin 42, 44, 47, 49, 51, 54, 56, 59, 276
—, NMR spectrum of 294
—, structure of 20
—, UV spectra of 126
— 3-L-arabinoside 40
— 3,7-O-diglucoside 42, 47, 51, 55, 59, 276
— —, NMR spectrum of 295
— —, UV spectra of 131
— 3-O-galactoside 42, 46, 49, 51, 55, 59
— —, UV spectra of 125
— 3-O-glucoside 7-O-rhamnoside 42, 47, 51, 55, 59, 269, 276
— — —, NMR spectrum of 297
— — —, UV spectra of 132
— 3-O-glucoside 7-O-rutinoside 42, 47, 51, 55, 59, 276
— — —, NMR spectrum of 297
— — —, UV spectra of 133
— 3-methyl ether 42, 47, 49, 51, 55, 59
— — —, UV spectra of 134
— 3-methyl ether 4′-O-glucoside 7-O-diglucoside 42, 55, 59
— — — — —, UV spectra of 135
— 3′,4′,5,6,7-pentamethyl ether 42, 56, 59
— — —, UV spectra of 156
— 3-O-rhamnoglucoside (see also Rutin) 25, 42, 47, 49, 51, 55, 59, 269, 276
— — —, NMR spectrum of 296
— — —, UV spectra of 130
— 7-O-rhamnoside 42, 47, 51, 56, 59
— —, UV spectra of 127
— 3,3′,7-trimethyl ether 40
Quercitrin 42, 47, 49, 51, 55, 59, 276
—, NMR spectrum of 294
—, UV spectra of 129

Rate of methylation 31
Reagents for UV spectroscopy 35
Rf values (see UV spectra), determination 9
— —, effect of flavonoid structure on 10
— —, isoflavones 14
— —, structural information from 10

Rhamnetin 42, 47, 51, 56, 59, 276
—, NMR spectrum of 298
—, UV spectra of 137
Rhamnoglucosides (see Diglycosides, Neohesperidosides and Rutinosides)
Rhamnose, identification 27
—, NMR 269
Rhamnosylvitexin 43, 46, 48, 55, 57
—, UV spectra of 87
Robinetin 42, 47, 49, 51, 56, 59, 276
—, NMR spectrum of 303
—, UV spectra of 144
Robinin (see Kaempferol 3-O-robinoside 7-O-rhamnoside)
Rutin 23, 25, 42, 47, 49, 51, 55, 59, 276
—, NMR spectrum of 296
—, structure of 10, 20, 27
—, UV spectra of 130
— acetate 276
— —, NMR spectrum of 296
Rutinosides (see Diglycosides; also Distinguishing rutinosides from neohesperidosides)

Sakuranetin 278
—, NMR spectrum of 330
Sakuranin 174, 278
—, NMR spectrum of 331
—, structure of 168
—, UV spectra of 216
Saponaretin (see Isovitexin)
Saponarin 39, 43, 46, 55, 57, 275
—, NMR spectrum of 283
—, UV spectra of 85
Scaposin 21, 23, 275
—, degradation 28
—, NMR spectrum of 291
—, synthesis 28, 29
Sciadopitysin 43, 55, 58
—, UV spectra of 111
Scoparin 43, 46, 49, 55, 58
—, UV spectra of 102
Scopoletin 7
—, structure of 9
Scopoletin 7-O-glucoside 7
— —, structure of 9
Separation, flavonoids from sugars 16, 24
—, glycosides from aglycones 21
—, rutin from quercetin 20
Sephadex (for column chromatography) 20
Silica Gel (column chromatography) 17
Sodium acetate-boric acid for UV spectroscopy, chalcones and aurones 228
— — — — —, flavones, flavonols 50
— — — — —, isoflavones, flavanones, dihydroflavonols 170
— acetate for UV spectroscopy 35, 39
— — — —, chalcones and aurones 228
— — — —, flavones, flavonols 47—50
— — — —, isoflavones flavanones, dihydroflavonols 169
— methoxide for UV spectroscopy 35
— — — —, chalcones and aurones 228
— — — —, flavones, flavonols 45
— — — —, isoflavones, flavanones, dihydroflavonols 167
Solvents, column chromatography 17—20
—, extraction 17, 19
—, NMR spectroscopy 254, 255
—, paper chromatography 4
Solvent shifts (NMR) 267, 272
Sophoricoside 169, 171, 173
—, UV spectra of 187
Sphaerobioside 7, 171, 177, 277
—, NMR spectrum of 313
—, structure of 8
—, UV spectra of 186
Sphaerobioside acetate 277
— —, NMR spectrum of 314
Spot color (paper chromatography) 12, 13
Starch (for column chromatography) 17
Structural variations in flavonoids, effect on *Rf* value 10
Sublimation 15
Substitution (see oxygenation)
Sugars, gas chromatography 26
—, identification 26, 27, 31, 32
—, in C-glycosides 31
—, location in glycoside 30
—, NMR 268
—, paper chromatography 27
—, trimethylsilylation 27
Synthesis of scaposin 28

Talc for TLC 21
Tamarixetin 276
—, NMR spectrum of 299
Tamarixetin 7-O-neohesperidoside 59, 276
— —, NMR spectrum of 300
— —, UV spectra of 141
Tamarixetin 7-O-neohesperidoside acetate 42, 56, 276
— — —, NMR spectrum of 300
— 7-O-rutinoside 42, 56, 59, 276
— —, NMR spectrum of 301
— —, UV spectra of 142
Taxifolin 170, 172, 174
—, UV spectra of 223
Tectochrysin 43, 55, 57, 275
—, NMR spectrum of 280
—, UV spectra of 69
Tectoridin 171, 173, 278
—, NMR spectrum of 323
—, UV spectra of 202
Tectorigenin 14, 169, 171, 173
—, UV spectra of 201
3′,4′,6,7-Tetrahydroxyaurone (maritimetin) 228, 230, 279
— —, NMR spectrum of 342
— —, UV spectra of 247
2′,3,4,4′-Tetrahydroxychalcone 229, 230, 279
—, NMR spectrum of 336
—, UV spectra of 241
2′,3,5,7-Tetrahydroxyflavone 40
3,3′,4′,7-Tetrahydroxyflavone (see also Fisetin) 40
3′,4′,5,7-Tetrahydroxyisoflavone (see Orobol)
4′,5,6,7-Tetrahydroxyisoflavone (see 6-Hydroxygenistein)
3′,4′,6,7-Tetramethoxyaurone 279
—, NMR spectrum of 342
2′,4,4′,6′-Tetramethoxychalcone 279

2′,4,4′,6′-Tetramethoxychalcone, NMR spectrum of 338
2′,4,4′,6-Tetramethoxychalcone α,β-epoxide 279
— —, NMR spectrum of 340
3,4′,7,8-Tetramethoxy-3′,5-di-trimethylsilyloxy-flavone, de-trimethylsilylation of 258
2′,5,6,7-Tetramethoxyisoflavone 40
Tetramethylsilane, internal standard (NMR) 255—260
Texasin 14, 166, 169, 176
—, structure of 20
—, UV spectra of 192
Texasin 7-O-glucoside 173, 278
— —, NMR spectrum of 319
— —, UV spectra of 193
Thin-layer chromatography, cellulose 21, 22
— —, polyamide 17, 20—22, 31
— —, silica gel 19, 20
— —, talc 21
Time-averaging computer 256
Tricin 42, 46, 49, 55, 58
—, UV spectra of 105
Triglycosides, distinction from di- and mono-, 27
3′,4,4′-Trihydroxyaurone 40
4′,6,7-Trihydroxyaurone 40
2,2′,4-Trihydroxychalcone 229, 230
—, UV spectra of 239
2′,3,4-Trihydroxychalcone 229, 230
—, UV spectra of 238
2′,3′,4′-Trihydroxychalcone 229, 230
—, UV spectra of 237
2′,4,4′-Trihydroxychalcone 229, 230
—, UV spectra of 240
5,6,7-Trihydroxyflavanone 142
—, UV spectra of 213
4′,5,7-Trihydroxyflavanone 40, 166
— (see also Naringenin)
5,6,7-Trihydroxyflavanone 7-O-glucuronide 172, 174
— —, UV spectra of 214
3′,4′,7-Trihydroxyflavonol 3-O-glucoside 7
— —, structure of 8
3,3′,4′-Trihydroxyflavone 42, 47, 51, 56, 58
—, UV spectra of 117
3,4′,7-Trihydroxyflavone 42, 46, 47, 49, 56, 58
—, UV spectra of 114
3′,4′,7-Trihydroxyflavone 6, 43, 46, 50, 57
—, structure of 8
—, UV spectra of 74
3,5,7-Trihydroxyflavone (see Galangin)
5,6,7-Trihydroxyflavone (see also Baicalein) 48, 55, 170, 171
5,7,8-Trihydroxyflavone (see also Norwogonin) 42, 48, 55, 57
— —, UV spectra of 79
3′,4′,7-Trihydroxyflavone 7-O-glucoside 6
— —, structure of 8
5,6,7-Trihydroxyflavone 7-O-glucuronide (see also Baicalin) 43, 55, 57
— — —, UV spectra of 78
5,7,8-Trihydroxyflavone 7-O-glucuronide 43, 55, 57
— —, UV spectra of 80
3′,4′,7-Trihydroxyflavone 7-O-rhamnoglucoside 7, 46, 50, 57
3′,4′,7-Trihydroxyflavone 7-O-rhamnoglucoside, structure of 8
— —, UV spectra of 75
3′,4′,7-Trihydroxyisoflavone 14, 169, 173
—, UV spectra of 196
4′,5,7-Trihydroxyisoflavone (see Genistein)
Trihydroxyl systems (see Detection)
3′,5,5′-Trihydroxy-3,4′,6,7-tetramethoxyflavone 43, 56, 60, 277
—, NMR spectrum of 310
—, UV spectra of 164
5,5′,7-Trihydroxy-3′,4′,6,8-tetramethoxyflavone (see Scaposin)
3′,5,7-Trihydroxy-4′,5′,6-trimethoxyisoflavone 7-O-glucoside (see Iridin)
4,4′,6-Trimethoxyaurone 279
—, NMR spectrum of 341
4,6,7-Trimethoxy-4′-benzyloxyaurone 279
—, NMR spectrum of 343
4′,5,7-Trimethoxyflavone 40
4′,5,6-Trimethoxy-7-hydroxyisoflavone 31
4′,5,7-Trimethoxy-6-methylisoflavone 277
—, NMR spectrum of 316
4′,5,7-Trimethoxy-8-methylisoflavone 277
—, NMR spectrum of 317
2′,5,7-Trimethoxy-8-methylisoflavone 40
Trimethylchlorosilane 26, 255—260
3,3′,7-Tri-O-methylquercetin 40
Trimethylsilylation, reagents for 26, 255
—, sugars 26
Trimethylsilyl ether derivatives
— — —, hydrolysis 255, 258, 259
— — — of hindered hydroxyl groups 257
— — —, methylation and acetylation of 259
— — —, preparation of 255—259
— — —, table of chemical shifts 260

UV fluorescent phosphors 21
— — spots 12, 13
— light detection of spots on paper chromatograms 13
Ultraviolet spectroscopy 33—250
— —, chalcones and aurones 227—230
— —, detection of free hydroxyl groups (see Detection)
— —, effect of acetylation 45
— —, — of methylation, glycosylation 45, 166, 228
— —, — of oxygenation pattern 44, 166, 227
— —, — of time delay 37
— —, flavones and flavonols 41—57
— —, isoflavones, flavanones, dihydroflavonols 165—173
— —, indexes of UV spectra 57, 173, 230
— —, procedures 35
— —, reagents 35
UV viewing lamp 4

Vicenin-1, structure of 12
Violanthin 43, 46, 48, 55, 57
—, UV spectra of 89
Vitexin 23, 43, 46, 48, 55, 57, 263, 270, 271, 275
—, NMR spectrum of 282
—, structure of 12

Vitexin, UV spectra of 86
—, 2″-O-xylosyl derivative (see Xylosylvitexin)

Water, effect on NMR spectrum 254
Wessely-Moser rearrangement 11, 24
Whatman 3 MM chromatographic paper 3

Xanthomicrol 43, 46, 55, 58, 275
—, NMR spectrum of 290
—, UV spectra of 108

2″-O-Xylosylvitexin 4, 23, 25, 43, 46, 48, 55
—, nona-O-methyl derivative 271
—, structure of 270

Zapotin 43, 58, 275
—, NMR spectrum of 289
—, UV spectra of 94
Zapotinin 43, 55, 58, 275
—, NMR spectrum of 289
—, UV spectra of 93

Errata

The U.V. Spectra of the Flavones, Flavonols, Isoflavones, Flavanones, Dihydroflavonols, Chalcones, and Aurones in Sections V-6, VI-6, and VII-6 have, been listed in the subject index not under Spectrum numbers, but under page numbers, which were not printed in the Spectra pages. To facilitate finding the Spectra, they are listed below with the Spectrum number printed in boldface, after the page number given in the subject Index.

Page No.		No. of U.V. spectra	Page No.		No. of U.V. spectra	Page No.		No. of U.V. spectra	Page No.		No. of U.V. spectra	Page No.		No. of U.V. spectra
62	=	**1**	97	=	**36**	132	=	**71**	177	=	**106**	212	=	**141**
63	=	**2**	98	=	**37**	133	=	**72**	178	=	**107**	213	=	**142**
64	=	**3**	99	=	**38**	134	=	**73**	179	=	**108**	214	=	**143**
65	=	**4**	100	=	**39**	135	=	**74**	180	=	**109**	215	=	**144**
66	=	**5**	101	=	**40**	136	=	**75**	181	=	**110**	216	=	**145**
67	=	**6**	102	=	**41**	137	=	**76**	182	=	**111**	217	=	**146**
68	=	**7**	103	=	**42**	138	=	**77**	183	=	**112**	218	=	**147**
69	=	**8**	104	=	**43**	139	=	**78**	184	=	**113**	219	=	**148**
70	=	**9**	105	=	**44**	140	=	**79**	185	=	**114**	220	=	**149**
71	=	**10**	106	=	**45**	141	=	**80**	186	=	**115**	221	=	**150**
72	=	**11**	107	=	**46**	142	=	**81**	187	=	**116**	222	=	**151**
73	=	**12**	108	=	**47**	143	=	**82**	188	=	**117**	223	=	**152**
74	=	**13**	109	=	**48**	144	=	**83**	189	=	**118**	224	=	**153**
75	=	**14**	110	=	**49**	145	=	**84**	190	=	**119**	225	=	**154**
76	=	**15**	111	=	**50**	146	=	**85**	191	=	**120**	226	=	**155**
77	=	**16**	112	=	**51**	147	=	**86**	192	=	**121**	231	=	**156**
78	=	**17**	113	=	**52**	148	=	**87**	193	=	**122**	232	=	**157**
79	=	**18**	114	=	**53**	149	=	**88**	194	=	**123**	233	=	**158**
80	=	**19**	115	=	**54**	150	=	**89**	195	=	**124**	234	=	**159**
81	=	**20**	116	=	**55**	151	=	**90**	196	=	**125**	235	=	**160**
82	=	**21**	117	=	**56**	152	=	**91**	197	=	**126**	236	=	**161**
83	=	**22**	118	=	**57**	153	=	**92**	198	=	**127**	237	=	**162**
84	=	**23**	119	=	**58**	154	=	**93**	199	=	**128**	238	=	**163**
85	=	**24**	120	=	**59**	155	=	**94**	200	=	**129**	239	=	**164**
86	=	**25**	121	=	**60**	156	=	**95**	201	=	**130**	240	=	**165**
87	=	**26**	122	=	**61**	157	=	**96**	202	=	**131**	241	=	**166**
88	=	**27**	123	=	**62**	158	=	**97**	203	=	**132**	242	=	**167**
89	=	**28**	124	=	**63**	159	=	**98**	204	=	**133**	243	=	**168**
90	=	**29**	125	=	**64**	160	=	**99**	205	=	**134**	244	=	**169**
91	=	**30**	126	=	**65**	161	=	**100**	206	=	**135**	245	=	**170**
92	=	**31**	127	=	**66**	162	=	**101**	207	=	**136**	246	=	**171**
93	=	**32**	128	=	**67**	163	=	**102**	208	=	**137**	247	=	**172**
94	=	**33**	129	=	**68**	164	=	**103**	209	=	**138**	248	=	**173**
95	=	**34**	130	=	**69**	175	=	**104**	210	=	**139**	249	=	**174**
96	=	**35**	131	=	**70**	176	=	**105**	211	=	**140**	250	=	**175**

MIX
Papier aus verantwortungsvollen Quellen
Paper from responsible sources
FSC® C105338

If you have any concerns about our products,
you can contact us on
ProductSafety@springernature.com

In case Publisher is established outside the EU,
the EU authorized representative is:
Springer Nature Customer Service Center GmbH
Europaplatz 3, 69115 Heidelberg, Germany

Printed by Libri Plureos GmbH
in Hamburg, Germany